UNIFIED INTEGRATION

This is a volume in
PURE AND APPLIED MATHEMATICS

A Series of Monographs and Textbooks

A list of recent titles in this series appears at the end of this volume.

UNIFIED INTEGRATION

E. J. McShane
Department of Mathematics
University of Virginia
Charlottesville, Virginia

1983

ACADEMIC PRESS, INC.
(Harcourt Brace Jovanovich, Publishers)
Orlando San Diego San Francisco New York London
Toronto Montreal Sydney Tokyo São Paulo

ACADEMIC PRESS, INC.
Orlando, Florida 32887

United Kingdom Edition published by
ACADEMIC PRESS, INC. (LONDON) LTD.
24/28 Oval Road, London NW1 7DX

Library of Congress Cataloging in Publication Data

McShane, E. J. (Edward James), Date
Unified integration.

(Pure and applied mathematics ;)
Includes index.
1. Integrals. I. Title. II. Series: Pure and applied mathematics (Academic Press) ; .
QA3.P8 [QA312] 510s 82-16266
ISBN 0-12-486260-8 [515.4]

AMS (MOS) 1980 Subject Classifications: 28-01, 28-A20, 28-A25 and 60-A20

PRINTED IN THE UNITED STATES OF AMERICA

83 84 85 86 9 8 7 6 5 4 3 2 1

Contents

Preface

Year by year the importance of integration processes, both for pure mathematics and for its applications, has steadily increased. Almost every undergraduate student of mathematics or physics or engineering studies enough calculus to meet Riemann integrals in one dimension and in higher dimensions, "improper" integrals, line integrals, and surface integrals—roughly, all the nineteenth-century types of integral. But the twentieth century brought advances in integration that were indispensable in analysis and that later proved beautifully adapted to probability theory and to such applications as quantum theory, communication theory, and the optimal control of systems with random noises. For all of these applications and for others too, one should be versed in the ideas associated with the theory of the Lebesgue integral.

This is ominous news for the undergraduate student whose chief interest is in a science or in engineering, and even for the student of mathematics whose preference is for some field other than analysis. For them it is discouraging to find that for further advances, they have to abandon the Riemann integral to which they had devoted so much time and to start all over with a new concept of integration. Most students simply lack the time to add so much to their studies.

A logically sound but pedagogisally unacceptable alternative is to discard the Riemann integral and teach Lebesgue integration from the beginning of calculus. But this ignores the experimental evidence that the several customary ways of introducing the Riemann integral all appeal to students as more natural and easily comprehended than any of the customary ways of introducing the Lebesgue integral.

A way out of this apparent impasse was opened in 1957, when J. Kurzweil published for an integral of a function of one variable a definition that closely resembled Riemann's definition, yet was more general; in fact, Kurzweil's integral is more general than Lebesgue's. This integral, called the "Riemann-complete" integral, was studied in depth by R. Henstock. Another slight modification of the definition produces an integral (we call it the "gauge-integral") exactly equivalent to the Lebesgue integral. As far as generality is concerned, this integral will serve the needs of mathematicians and of scientists and engineers who are working at the very frontiers of their fields. There remains the crucial question: Can we take advantage of the close resemblance to Riemann integration to produce a unified theory that can be taught to students who have no intention of becoming research mathematicians, with just about the same level of difficulty as is encountered in ordinary courses, and that can also go from beginning calculus to the graduate level without ever abandoning earlier work and starting again (as usually now happens when Lebesgue integration is met)?

This book is an attempt to show that this is indeed possible.

In Chapter I the integral is defined, for real-valued functions of one real variable, in a way that should be easily accessible even to beginners. This definition is never abandoned; it is extended stage by stage to apply to functions of more general types. Nothing is first studied and then discarded, as the Riemann integral is customarily discarded in favor of the Lebesgue integral. Moreover, each extension consists of a rewording of the first definition that is clearly needed in order to make it applicable in the new situation, so that the treatment of integration is unified. Thus, for example, no special definition need be given for the integrals of unbounded functions and integrals over unbounded intervals; the integral at the beginning of Chapter I already applies to these.

Chapter V is to some extent a departure from the program of defining and applying various forms of the gauge integral. Line integrals are in fact defined by a procedure like that in Chapter I, but surface integrals posed a problem. In many texts the level of mathematical rigor reaches its nadir in the treatment of surface area and surface integrals. Here area is defined and computed with the help of a convergence lemma proved in Section 8. But no new integral is introduced; the treatment of area is merely an application of the theorems on integration proved in preceding chapters. A currently widespread treatment of surface integrals, along with generalizations of the theorems of Gauss, Green, and Stokes, is based on the use of exterior differential forms. But in this treatment differentiation is central. The integral is a type of Newton's integral, and a degree of smoothness is required that is unsuitable for some applications and is quite unlike that in other chapters. In the second-last of the many versions of this text the concept of cross-product was extended to apply to several factors, each a vector or covector in r-space, and integrals over higher-dimensional analogues of surfaces in higher-dimensional spaces were defined by a limit passage from integrals over finite unions of linear images of simplices. For these integrals, a generalized form of Stokes's theorem was proved, under smoothness hypotheses resembling those of Theorem V-9-3. So in the second-last version I showed (at least to my own satisfaction) that the gauge integral can furnish us with a theory of surface integrals that applies to a satisfactorily large class of analogues of surfaces in spaces of arbitrarily high dimensionality. But this surface integral was no more than an application of the integral in preceding chapters, and the requisite preparatory development of determinant theory and geometry of complexes was exceedingly long. So in this the final version all discussion of surface integrals has simply been omitted.

My most optimistic hope is that over the course of sufficiently many years the unified treatment of integration will become widely accepted, and that writers of textbooks for various courses in undergraduate mathematics will adopt and adapt the appropriate parts of this book, using its proofs or replacing them by such improvements as will be thought of. For this hope to be realized, it is necessary that this book be quite inclusive, covering large parts of the subject matter that is found in the contemporary assortment of courses on advanced calculus, applied mathematics, introductory analysis, probability theory, etc., and reaching from the beginning of calculus to the level of graduate courses in mathematics. This means that anyone who uses this book must pick and choose, just as is the case with so many of today's textbooks. Picking and choosing is the privilege, or rather the annoying responsibility, of the reader. The author does not have this privilege. The proofs must all be exhibited in detail. There is no possible recourse to proofs "to be found in other books."

In the meanwhile, while waiting and hoping for general acceptance of unified integration, this book (perhaps supplemented by a textbook on devices for solving differential equations) could serve as a textbook for a course replacing "advanced calculus," which is itself a blanket name for an assortment of courses. The quick objection, that it is all integration and no differentiation, is not a valid one. True, differentiation plays a role secondary to that of integration, but this merely resognizes a present-day trend in analysis. Many mathematical concepts and procedures have been found to gain in generality and convenience when expressed in terms of integration rather than of differentiation. For example, the density of a mass distribution can be defined by a type of differentiation; but it proves better to think of the density as that function whose integral over every region is equal to the mass contained in that region. For such a use of this book as textbook Chapter I would be merely a quick review of elementary calculus, and Chapter VII would be too advanced to use at all. The last four sections of Chapter II would be omitted, and the last three of Chapter III. Chapter IV from Section 9 on would be omitted unless probability theory is to be included; Sections 5–8 would be at the instructor's discretion. My own preference would be to omit Chapter V, to avoid squeezing out material that I prefer. I would include as much of Chapter VI as time permits.

Between the times of completion of the manuscript of this book and of its being set in type two related books have appeared. Both are concerned primarily with what we have called Henstock's "Riemann-complete" integral, rather than with the integral of this gauge-integral of this book. The first of these is *Introduction à l'ànalyse* (Université de Louvain, CABAY, 1979), by Professor Jean Mawhin, of the Université Catholique de Louvain. In the preface he states "For integration, we have adopted, and adapted to the level of this course, a recent approach due to Kurzweil, Henstock and McShane which by technically minor but conceptually important modifications of the classical definition of Reimann leads to the integrals of Perron and Lebesgue. This presentation, of which this work contains perhaps the first systematic treatment at an introductory level, allows in our opinion a more natural progression, without modification of definition, from elementary integral calculus to the advanced and always difficult aspects of the Lebesgue integral." Professor Mawhin's book is very readable and can be recommended to any student for whom the French language has no terrors.

The second of the books referred to is "The Generalized Riemann Integral," by

Robert M. McLeod of Kenyon College (No. 20 of the Carus Mathematical Monographs, Mathematical Association of America, 1980). The author investigates the Riemann-complete integral, which (slightly inconveniently for me) he calls the "gauge-integral."

These two books are especially welcome to me because they present extensive accounts of the Riemann-complete integral, which I have barely mentioned. My preference is for the gauge-integral as in this book, because it is equivalent to that workhorse of contemporary analysis, the Lebesgue integral; the Riemann-complete integral, being equivalent to the Perron integral, is certainly not devoid of interest, but could not be included without the possibility of confusing beginning students and the certainty of lengthening an already long book.

All of the more important assertions, whether theorems, lemmas, corollaries, or definitions, are numbered consecutively. Thus the eighth named assertion in Section 10 of Chapter III is the theorem that the family of measurable sets is a σ-algebra. It is therefore named Theorem 10-8. Elsewhere in Chapter III it is referred to as Theorem 10-8; in other chapters it is referred to as Theorem III-10-8.

Acknowledgments

Several times I have used parts of earlier versions of this book in teaching undergraduate classes, and I owe thanks to many of the students in those classes for indicating places where improvement was needed. One of those classes was taught by Professor H. N. Ward and me jointly. Professor Ward not only suggested and produced improvements; he contributed Section 8 of Chapter II, at which early stage he showed by example that several useful evaluations, previously carried out by use of the Riemann integral, can be performed with saving of labor by use of the Lebesgue integral.

Professor R. B. Kirchner of Carleton College used the earlier chapters for a class taught at Carleton College, and favored me with comments. Professor Washek Pfeffer, of the University of California at Davis, performed the tour de force of basing a successful undergraduate course on a brief note "A unified theory of integration" in the *American Mathematical Monthly* (Vol. 80, 1973), inventing the necessary details. I have enjoyed conversation with him, and was sorely tempted to follow his suggestion that the theory of the Riemann-complete integral be included; an estimate of the resulting thickness of this book dissuaded me. Professor W. L. Duren, Jr., read the earlier chapters and made numerous helpful suggestions, especially about style. (Incidentally, he suggested the name "gauge-integral.") Finally, I thank Miss Alida Ward for her highly cooperative assistance.

0

Introduction

1. Functions, Intervals, and Limits

In discussing integration, we shall use many concepts that are familiar to everyone who has had a beginning course in college mathematics. However, since different texts use different notation it is desirable to list the definitions and symbols that we shall use.

The empty set will be denoted by $\varnothing$. If A and B are sets, the **union** of A and B (consisting of all things that belong to A, or to B, or to both of them) is denoted by $A \cup B$. The **difference** $A \setminus B$ is the set of all things that belong to A but do not belong to B. The **intersection** of A and B, consisting of all things that belong to A and also belong to B, is denoted by $A \cap B$. Two sets are **disjoint** if their intersection is the empty set $\varnothing$; that is, if there is nothing that belongs to both of them. A collection $\mathscr{K}$ of sets is **pairwise disjoint** if, whenever A and B are different sets that belong to the collection $\mathscr{K}$, $A \cap B = \varnothing$.

There are various ways of wording the definition of **function**, but they all have this in common: if f is a function, there is a set D, called the **domain** of the function, such that for each element x of D the function f assigns to x a uniquely determined **functional value** that corresponds to x and is denoted by $f(x)$. Usually we shall denote such a function by the symbol $x \mapsto f(x)$ (x in D) – omitting the note (x in D) if there is no danger of confusion. Sometimes we shall denote the function by the letter f alone and occasionally by other symbols that will be explained when they are used. But we shall never use the symbol $f(x)$ to mean the function f, or $x \mapsto f(x)$. The symbol $f(x)$ always denotes a specific quantity, namely the functional value that corresponds to the particular element x of D.

If f is a function defined on a domain D, g a function defined on a domain D', and for each x in D the value $f(x)$ is in D', then the **composite** of g and f assigns to each x in D the functional value $g(f(x))$. This composite function $x \mapsto g(f(x))$ is sometimes denoted by $g \circ f$, sometimes by $g(f(\cdot))$. But there are a few familiar exceptions to this notational custom. If g is the function $y \mapsto |y|$ (y real) and f is a function $x \mapsto f(x)$ (x in D) whose values are real numbers, the composite function

$x \mapsto |f(x)|$ is usually designated by $|f|$. Likewise, with the same kind of f, if g is the function $y \mapsto -y$, the composite $x \mapsto -f(x)$ is customarily written $-f$. Similarly, if f and g are both real-valued functions with the same domain D, their composite with the "plus" function, $x \mapsto f(x) + g(x)$, is usually denoted by $f + g$, and the composite with the "times" function, $x \mapsto f(x) \cdot g(x)$, is usually denoted by $f \cdot g$.

The set of all real numbers will be denoted by R. For discussing limits it is convenient to introduce two other elements, not themselves real numbers, denoted by ∞ and $-\infty$. These have the order relations

$$-\infty < x < \infty \qquad (\text{all } x \text{ in } R).$$

The set obtained by adjoining these two new elements to R is called the **extended real number system**; we shall denote it by $\bar{R}$. The statement "x is finite" means "x is in R," that is, x is in $\bar{R}$ but $x \neq \infty$ and $x \neq -\infty$. In the extended real number system we adopt some computation rules for ∞ and $-\infty$, all but one of which are just what anyone would expect. These are as follows:

$\infty + x = x + \infty = \infty$ unless $x = -\infty$.
$(-\infty) + x = x + (-\infty) = -\infty$ unless $x = \infty$.
If $c > 0$, then $c \cdot \infty = \infty \cdot c = \infty$ and $c \cdot (-\infty) = (-\infty) \cdot c = -\infty$.
If $c < 0$, then $c \cdot \infty = \infty \cdot c = -\infty$ and $c \cdot (-\infty) = (-\infty) \cdot c = \infty$.
$0 \cdot \infty = \infty \cdot 0 = 0$.

The last is perhaps unexpected, but it is convenient in integration theory.

Some people use the word **interval** in a broader sense, others in a narrower. We shall use it in the broader sense. If a and b are any members of $\bar{R}$, the sets

$$(a, b) = \{x \text{ in } R : a < x < b\},$$

$$(a, b] = \{x \text{ in } R : a < x \leqq b\},$$

$$[a, b) = \{x \text{ in } R : a \leqq x < b\},$$

$$[a, b] = \{x \text{ in } R : a \leqq x \leqq b\}$$

are all intervals in $\bar{R}$. If a and b are both in R (that is, are finite), they are all **bounded** intervals. (Those who prefer the narrower sense of the word "interval" would not apply it unless a and b were finite.) Note that the parenthesis at either end reminds us to use the sign $<$ at that end, the square bracket reminds us to use the sign $\leqq$. The first of these four intervals is called **open**, the second is **left-open**, the third is **right-open**, and the fourth is **closed.** In particular, if $a > b$, all four of $(a, b), [a, b), (a, b], [a, b]$ are empty; there is no x in R that satisfies the conditions $a \leqq x$ and $x \leqq b$ or the similar conditions with $<$ in place of either or both of the signs $\leqq$. Accordingly, we regard the empty set $\varnothing$ as an open, left-open, right-open, and closed interval in R.

Another case that occurs often is that in which $a = -\infty$, or $b = \infty$, or both. Since $x > -\infty$ for every x in R, the set $(-\infty, b]$ in R is the same as the set of all x

with $x \leqq b$; the other condition, $x > -\infty$, is automatically satisfied. Likewise the half-line consisting of all x greater than a is the same as the set $(a, \infty]$; the other condition, $x \leqq \infty$ (or even $x < \infty$), is always satisfied.

Similarly, the set of all x in $\bar{R}$ such that $a < x$ and $x \leqq b$ is a left-open interval in $\bar{R}$, and likewise for the other kinds of intervals. However, the open intervals in $\bar{R}$ will be referred to most often, and with these there is a special convention. All intervals of the form (a, b) with a and b in $\bar{R}$ (that is, all sets of x satisfying $a < x$ and $x < b$) are open intervals in $\bar{R}$; but besides these, the intervals $[-\infty, b)$, $(a, \infty]$, and $[-\infty, \infty]$ are all included among the open intervals. They are, respectively, the sets $\{x \text{ in } \bar{R} : x < b\}$, $\{x \text{ in } \bar{R} : x > a\}$, and $\bar{R}$ itself. (For those who know some topology, $[-\infty, b)$ is an interval that is an open set in the space $\bar{R}$, so it makes sense to call it an open interval.)

If x is any point in $\bar{R}$, we define a **neighborhood** of x to be an open interval in $\bar{R}$ that contains x. A similar definition applies if we replace $\bar{R}$ by R; but as it happens, we shall not use the concept of neighborhood in R.

Thus, among the neighborhoods in $\bar{R}$ of 5 we have $(4, 5.001)$, $(-8, \infty)$, $(-8, \infty]$, and $[-\infty, \infty]$, this last being $\bar{R}$. The neighborhoods of ∞ consist of the intervals $(a, \infty]$ for all $a < \infty$ in $\bar{R}$, and also of $\bar{R}$ itself.

When the sets R and $\bar{R}$ are equipped with a topology, that is, when neighborhoods are defined, the combination of the set and the topology is called a **one-dimensional space**. The set $\bar{R}$ with the neighborhoods just defined is denoted by $\bar{R}^1$, and R with its neighborhoods is R^1.

As an example of a use of the concept of neighborhood, let us consider an extended-real-valued function f defined on a set D in $\bar{R}$. Let x_0 be a point of $\bar{R}$ such that every neighborhood of x_0 contains at least one point of D different from x_0. Let y be in $\bar{R}$. The statement

$$\lim_{x \to x_0} f(x) = y$$

means that for each neighborhood V of y there is a neighborhood U of x_0 such that for all x in $D \cap U$ different from x_0, $f(x)$ is in V. (The reader should pause at this point to verify that this definition agrees with whatever formulation of the concept of limit he or she is accustomed to.)

In particular, a **sequence** is defined to be a function on the set Z of all positive integers. By custom, the functional value corresponding to integer n is denoted by some such symbol as a_n. The above definition takes the form: a sequence $a_1, a_2, \ldots$ of numbers in $\bar{R}$ has a limit y in $\bar{R}$ if for each neighborhood V of y there is a number $\bar{n}$ in R such that a_n is in V whenever $n > \bar{n}$.

There is a useful relationship between the concept of a limit, as defined by means of neighborhoods, and the limit of a sequence.

THEOREM 1-1 *Let f be extended-real-valued on a set D in $\bar{R}$, and let x_0 be a point of $\bar{R}$ each neighborhood of which contains a point of D different from x_0.*

Let y be a point of $\bar{R}$. Then in order that

$$\lim_{x \to x_0} f(x) = y$$

it is necessary and sufficient that whenever $x_1, x_2, x_3, \ldots$ is a sequence of points of D all different from x_0 and with x_0 as limit, the sequence $f(x_1), f(x_2), \ldots$ have limit y.

Suppose first that $\lim_{x \to x_0} f(x) = y$. Let V be any neighborhood of y. There is a neighborhood U of x_0 such that if x is in $D \cap U$ and $x \neq x_0$, $f(x)$ is in V. If $x_1, x_2, \ldots$ are different from x_0 and tend to x_0, x_n is in U for all large n, so $f(x_n)$ is in V. Then by definition $f(x_n)$ has limit y.

The foregoing part of the proof used no properties at all of the neighborhoods. The converse will depend on one property.

> For each x_0 in $\bar{R}$ there is a sequence of neighborhoods $U_1, U_2, U_3, \ldots$ such that $U_1 \supset U_2 \supset U_3 \supset \cdots$, and for each neighborhood U of x_0 there is an n for which $U_n \subset U$.

If x_0 is in R, we choose $U_n = (x_0 - 1/n,\ x_0 + 1/n)$. If $x_0 = \infty$, we choose $U_n = (n, \infty]$. If $x_0 = -\infty$, we choose $U_n = [-\infty, -n)$.

Suppose, then, that it is false that $\lim_{x \to x_0} f(x) = y$. Then there exists a neighborhood V of y to which no neighborhood U corresponds as in the definition of limit; that is, every neighborhood U of x_0 contains a point $x \neq x_0$ of D with $f(x_0)$ not in V. In particular, for each n there is an $x_n \neq x_0$ in $D \cap U_n$ with $f(x_n)$ not in V. If U is any neighborhood of x_0, for all large n, U contains U_n and therefore contains x_n. So x_n tends to x_0, and $x_n \neq x_0$, but it is false that $f(x_n)$ tends to y. This completes the proof.

If A is any nonempty interval in R or $\bar{R}$, by the **interior** of A we mean the largest open interval in $\bar{R}^1$ that is contained in A and by the **closure** of A we mean the smallest closed interval in $\bar{R}^1$ that contains A. We denote the interior of A by A^{o} and the closure of A by $\bar{A}$ or by A^-. If $a < b$ and A is any one of the intervals (a, b), $[a, b)$, $(a, b]$, $[a, b]$, then A^- is $[a, b]$. If A is any one of these four intervals and a and b are finite, then A^{o} is (a, b). But the situation is different if $a = -\infty$ or if $b = \infty$. For if $b < \infty$, the interval $[-\infty, b)$ is already open and is the largest open interval contained in itself, so $[-\infty, b)^{\mathrm{o}} = [-\infty, b)$. Likewise if a is finite, $(a, \infty]^{\mathrm{o}} = (a, \infty]$, and $[-\infty, \infty]^{\mathrm{o}} = (\bar{R}^1)^{\mathrm{o}} = \bar{R}^1$.

When A is empty, we define $A^{\mathrm{o}} = A^- = \varnothing$.

If a and b are points of $\bar{R}$ with $a < b$, all four intervals (a, b), $(a, b]$, $[a, b)$, $[a, b]$ have length $b - a$. This may be ∞. If A is any one of the four intervals, since $b - a$ is the measure of the length of A, we denote it by $m_{\mathrm{L}} A$. If A consists of a single point or is empty, we define $m_{\mathrm{L}} A = 0$.

LEMMA 1-2 *Let $A_1, \ldots, A_k$ be pairwise disjoint intervals, and let $B_1, \ldots, B_n$ be intervals (not necessarily pairwise disjoint) such that $B_1 \cup \cdots \cup B_n \supset$*

$A_1 \cup \cdots \cup A_k$. *Then*

$$m_L A_1 + \cdots + m_L A_k \leqq m_L B_1 + \cdots + m_L B_n.$$

Let $c_1, \ldots, c_m$ be all the end-points of all the intervals $A_1, \ldots, A_k, B_1, \ldots, B_n$, arranged in increasing order. Define $C_j = (c_j, c_{j+1}]$ $(j = 1, \ldots, m-1)$. These are pairwise disjoint and nondegenerate. If A_i is any one of the intervals $A_1, \ldots, A_k$, the lower and upper ends of A_i are numbers c_p, c_q, and the intervals C_h with interior $C_h^\circ \subset A_i$ are $C_p, \ldots, C_{q-1}$. Then

$$\begin{aligned} m_L A_i &= c_q - c_p \\ &= (c_{p+1} - c_p) + (c_{p+2} - c_{p+1}) + \cdots + (c_q - c_{q-1}) \\ &= m_L C_p + m_L C_{p+1} + \cdots + m_L C_{q-1} \\ &= \sum \{m_L C_h : C_h^\circ \subset A_i\}. \end{aligned}$$

By a similar proof, if B_j is any one of the intervals $B_1, \ldots, B_n$, then

$$m_L B_j = \sum \{m_L C_h : C_h^\circ \subset B_j\}.$$

By addition, these equations yield

$$\text{(A)} \qquad \sum_{i=1}^{k} m_L A_i = \sum_{i=1}^{k} \sum \{m_L C_h : C_h^\circ \subset A_i\},$$

$$\text{(B)} \qquad \sum_{j=1}^{n} m_L B_j = \sum_{j=1}^{n} \sum \{m_L C_h : C_h^\circ \subset B_j\}.$$

If C_h is any one of the $C_1, \ldots, C_{m-1}$ with interior contained in some A_i, each interior point of C_h is in some B_j. The lower end-point of this B_j is one of the numbers $c_1, \ldots, c_m$ that cannot be greater than c_h, and the upper end-point is one that cannot be less than c_{h+1}, so (c_h, c_{h+1}) is contained in B_j and $m_L C_h$ is among the numbers in the right member of (B). Because the A_i are pairwise disjoint, no interval C_h can occur twice in the right member of (A), but it then has to occur at least once (possibly several times) in the right member of (B). So the right member of (B) is at least as great as the right member of (A). This completes the proof.

COROLLARY 1-3 *If $A_1, \ldots, A_k$ are pairwise disjoint intervals whose union is an interval A, then*

$$\text{(C)} \qquad m_L A = m_L A_1 + \cdots + m_L A_k.$$

By Lemma 1-2, with $n = 1$ and $B_1 = A$, the left member of (C) is at least as great as the right member. By applying Lemma 1-2 with the set $\{A_1, \ldots, A_k\}$ replaced by $\{A\}$ and the set $\{B_1, \ldots, B_n\}$ replaced by $\{A_1, \ldots, A_k\}$, we find that the right member of (C) is at least as great as the left member. So they are equal.

The property of m_L specified in Corollary 1-3 is called **additivity**, or, more specifically, **finite additivity**.

LEMMA 1-4 *Let A be any interval in $\bar{R}$. Then*

(i) *if c is any number such that $c > m_L A$, there is a left-open interval G with $m_L G < c$ whose interior contains the closure A^- of A;*

(ii) *if c is any number such that $c < m_L A$, there is a bounded left-open interval F with $m_L F > c$ whose closure F^- is contained in the interior A^{o} of A.*

If $c > m_L A$, $m_L A$ must be finite. Choose any positive number ε such that $2\varepsilon < c - m_L A$. If a and b are the end-points of A, the interval $G = (a - \varepsilon, b + \varepsilon]$ has the required properties, and (i) is proved.

For (ii), if $m_L A = 0$, we choose F to be the empty set; then $F^- = \varnothing$, which is contained in A^{o}. Otherwise $m_L A > 0$, and we can and do choose a point e interior to A. If $m_L A = \infty$, either the upper end-point of A is ∞, in which case we choose $F = (e, e + c + 1]$, or else the lower end-point is $-\infty$, in which case we choose F to be $(e - c - 1, e]$. In either case, F^- is contained in A^{o} and $m_L F = c + 1$. If $m_L A$ is finite, let the end-points of A be a and b, and let $\varepsilon = (m_L A - c)/3$, which is positive. We choose F to be $(a + \varepsilon, b - \varepsilon]$. This has closure $[a + \varepsilon, b - \varepsilon]$ contained in the interior of A, and $m_L F = (b - a) - 2\varepsilon > c$.

2. Bounds

If a and b both belong to $\bar{R}$, then $a = b$ or $a < b$ or $b < a$. We use the symbol $a \vee b$ to denote the larger of a and b; formally, $a \vee b = a$ if $a > b$ or if $a = b$, and $a \vee b = b$ if $b > a$. Likewise $a \wedge b$ is the smaller of a and b. As just stated, if a and b are in $\bar{R}$, then both $a \vee b$ and $a \wedge b$ have meaning. By an easy mathematical induction we can prove that if $a_1, \ldots, a_n$ all belong to $\bar{R}$, there is a least member of the set $\{a_1, \ldots, a_n\}$, denoted by $a_1 \wedge a_2 \wedge \cdots \wedge a_n$, or by $\wedge\{a_1, \ldots, a_n\}$, or by $\min\{a_1, \ldots, a_n\}$, that is a member of the set (say, is equal to a_k) and has the property that $a_j \geqq a_k$ for all a_j in the set. Likewise there is a greatest member in the set.

As usual, we define $|x|$ to be x if $x \geqq 0$ and to be $-x$ if $x < 0$. This is the same as saying that $|x| = x \vee (-x)$. When f and g are functions on the same domain D, we have already defined $f + g$ to be the function whose value at each x in D is $f(x) + g(x)$. In the same way, we define $f \vee g$ to be the function whose value at x is $f(x) \vee g(x)$, $f \wedge g$ to be the function whose value at x is $f(x) \wedge g(x)$, and $|f|$ to be the function whose value at x is $|f(x)|$.

When B is an infinite subset of $\bar{R}$, it is not always true that B has a greatest member or a least member. For example, if B is the set of all positive numbers in $\bar{R}$, B has no smallest member; if x is any member (and therefore is positive), $x/2$ is positive and therefore is a member of B smaller than x. But there are serviceable substitutes for the ideas of "smallest member" and "greatest member." A number a in $\bar{R}$ is a **lower bound** for B if $a \leqq x$ for every x that belongs to B. It is a fundamental property of $\bar{R}$ (which we shall not attempt to prove) that every nonempty set B in $\bar{R}$ has a greatest lower bound. That is, if $B \neq \varnothing$ and $B \subset \bar{R}$, there is a number a^* in $\bar{R}$ with the two properties: (i) a^* is a lower bound for B; (ii) for every number a that is a lower bound for B, $a^* \geqq a$. (Property (ii) makes a^* the greatest number that is a lower bound for B.)

For example, if B is the set of all positive numbers in $\bar{R}$, 0 is a lower bound for B (since $0 \leqq x$ for all positive x), and if a is any lower bound for B, it must be $\leqq 0$ (otherwise it would be positive, and $a/2$ would be a member of B less than a, contradicting the assumption that a is a lower bound for B).

The greatest lower bound of B is also known as the **infimum** of B and is denoted by g.l.b. B or by $\inf B$. We shall prefer the name "infimum" and the symbol $\inf B$.

From the greatest lower bound property it is easy to deduce (by change of signs) that every nonempty set B in $\bar{R}$ has a **least upper bound**, which is a number b^* with the following properties: (i) b^* is an upper bound for B; that is, if x is in B, then $x \leq b^*$, (ii) if b is any upper bound for B, then $b^* \leqq b$. The least upper bound of B is also called the **supremum** of B and is denoted by l.u.b. B or by $\sup B$. We shall prefer the name "supremum" and the symbol $\sup B$.

The fact that every nonempty set contained in $\bar{R}$ has a supremum and an infimum is of fundamental importance in the study of limits. For example, if we knew only the rational numbers, it would be easy to show that the set B of all *known* numbers (that is, all rational numbers) whose square is less than 2 has an upper bound (for example, 5), but it can also be shown that it has no (rational) least upper bound. Let b be any rational number. It can be written as a fraction $b = p/q$ in which p and q are integers and $q > 0$. If b is an upper bound for B, we must have $p/q = b \geqq 1$, since 1 is in B. For b to be the supremum of B it would have to satisfy the condition $p \geqq q \geqq 1$ and one of the conditions $p^2 = 2q^2$, $p^2 < 2q^2$, $p^2 > 2q^2$. We shall show that in none of these cases can b be the supremum of B.

Case 1. $p^2 = 2q^2$. We factor the positive integers p and q into prime factors. Of the prime factors of p, a number $N(p)$ are equal to 2; of the factors of q, $N(q)$ are equal to 2. Then p^2 has $2N(p)$ factors 2 and $2q^2$ has $1 + N(q) + N(q)$ factors 2. Since $2N(p)$ is even and $1 + 2N(q)$ is odd, it is not possible that $p^2 = 2q^2$. So this case never occurs.

Case 2. $p^2 < 2q^2$. In this case,

$$\left[p + \frac{1}{4p}\right]^2 \Big/ q^2 - 2 = \left[\frac{1}{2} + \frac{1}{16p^2} - (2q^2 - p^2)\right] \Big/ q^2 .$$

Since $2q^2 - p^2$ is a positive integer it is at least 1; and since $1/16p^2 < 1/2$ the quantity in parentheses on the right is negative. So $[p + 1/(4p)]/q$ is a rational number greater than p/q whose square is less than 2, and it must belong to B. This contradicts the assumption that b is an upper bound for B.

Case 3. $p^2 > 2q^2$. Then

$$p^2 - 2\left(q + \frac{1}{8q}\right)^2 = (p^2 - 2q^2) - \frac{1}{2} - \frac{1}{32q^2}.$$

Since $p^2 - 2q^2$ is a positive integer, it is at least 1; and since $1/32q^2 < 1/2$, the right side of this equation is positive. This implies

$$\left[p\Big/\left(q + \frac{1}{8q}\right)\right]^2 > 2.$$

If x is rational and

$$x > p\Big/\left[\left(q + \frac{1}{8q}\right)\right],$$

then

$$x^2 > p^2\Big/\left(q + \frac{1}{8q}\right)^2 > 2,$$

and x is not in B. So $p/[q + 1/(8q)]$ is a rational upper bound for B and is less than b, and in this case also b is not the supremum of B.

To us, who are (presumably) familiar with the real numbers, the fact that the set of rationals with square less than 2 has no *rational* least upper bound is a matter of rather mild interest. To the Greek philosophers of about 25 centuries ago it was gravely upsetting. In the language of the geometry that they were so brilliantly developing, the side and diagonal of a square were incommensurable; there was no unit of length so small that side and diagonal were both whole-number multiples of that unit. Instead of collapsing, they developed a theory of ratios that enabled them to handle such "incommensurables." Today this is done with the help of real numbers. However, in fairness it should be mentioned that certain mathematical philosophers still regard contemporary treatments of real numbers with deep distrust, feeling that some of the assertions that sound neat and precise in formulas or sentences do not in fact convey any idea within the capacity of the human mind.

The least-upper-bound property of $\bar{R}$ enters later proofs by way of two of its consequences, which we now prove.

THEOREM 2-1 *If $a_1, a_2, a_3, \ldots$ is an ascending sequence of numbers in $\bar{R}$ (that is, $a_1 \leqq a_2 \leqq a_3 \leqq \cdots$), then a_n approaches a limit as n increases, and this limit is* $\sup\{a_1, a_2, a_3, \ldots\}$.

Denote the supremum of the set of a_n by b^*. If all a_n are $-\infty$, so is b^*, and the conclusion is trivial. Otherwise $b^* > -\infty$. Let U be any neighborhood of b^*. Since $b^* \neq -\infty$, U contains a number b' less than b^*. Since b^* is the least upper bound and $b' < b^*$, b' is not an upper bound, so there is an integer n' such that $a_{n'} > b'$. Then for all $n > n'$ we have

$$b' < a_{n'} \leqq a_n \leqq b^*.$$

Since, therefore, all a_n with $n > n'$ are in the interval $[b', b^*]$, both ends of which are in U, they are all in U. By definition,

$$\lim_{n \to \infty} a_n = b^*.$$

THEOREM 2-2 *Let $\mathscr{K}$ be a collection of closed intervals in $\bar{R}$ such that if $[a_1, b_1]$ and $[a_2, b_2]$ are any two members of $\mathscr{K}$, there is a point that belongs to both of them. Then there is a point a^* in $\bar{R}$ that belongs to all the intervals in the collection $\mathscr{K}$.*

Let A be the set of all lower end-points of intervals that belong to $\mathscr{K}$; that is, a is in A if and only if there is a number b for which $[a, b]$ is in the collection $\mathscr{K}$. If a is any member of A and b_1 is the upper end of any interval of collection $\mathscr{K}$, then $a \leqq b_1$, for then there is a number b such that $[a, b]$ belongs to $\mathscr{K}$ and there is a number a_1 such that $[a_1, b_1]$ belongs to $\mathscr{K}$. So, by hypothesis there is a number x that belongs to both $[a, b]$ and $[a_1, b_1]$. Since x is in $[a, b]$, it must be true that $a \leqq x$; since x belongs to $[a_1, b_1]$, it must be true that $x \leqq b_1$. These two statements together imply that $a \leqq b_1$.

Now let a^* be the least upper bound of the set A, and let $[a_1, b_1]$ be any interval that belongs to collection $\mathscr{K}$. As we have just shown, $b_1 \geqq a$ for every a in set A, so b_1 is an upper bound for A and must be at least as large as the least upper bound a^* of A. Therefore $a^* \leqq b_1$. On the other hand, a_1 is in A, and a^* is an upper bound for A, so $a_1 \leqq a^*$. These two inequalities together imply $a_1 \leqq a^* \leqq b_1$, so a^* belongs to $[a_1, b_1]$, which is an arbitrary interval of collection $\mathscr{K}$.

From the least-upper-bound property of $\bar{R}$ we can prove the following theorem, which implies the intermediate-value property of continuous functions.

THEOREM 2-3 *Let f be real-valued and continuous on an interval B in $\bar{R}$, and let x_1 and x_2 be points of B with $x_2 > x_1$. If $f(x_1) < f(x_2)$ and c is a number such that $f(x_1) < c < f(x_2)$, there exists a number $\bar{x}$ in (x_1, x_2) such that $f(\bar{x}) = c$ and*

$$\text{(A)} \qquad f(x) \geqq c \qquad (\bar{x} < x \leqq x_2).$$

Let E be the set of numbers x_3 in $[x_1, x_2]$ such that $f(x) \geqq c$ for all x in $(x_3, x_2]$. Then E is not empty; it contains x_2. Let $\bar{x}$ be its infimum. If x is in $(\bar{x}, x_2]$, it is greater than the infimum $\bar{x}$ of E, so there is a point x_3 in E such that $x_3 < x$. Then x is in $(x_3, x_2]$, so by definition of E, $f(x) \geqq c$. Therefore conclusion (A) is valid. It remains to show that $f(\bar{x}) = c$.

If $f(\bar{x})$ were less than c, $[-\infty, c)$ would be a neighborhood of $f(\bar{x})$, and by the continuity of f there would exist a neighborhood U of $\bar{x}$ such that for all x in $U \cap [x_1, x_2]$, $f(x)$ is in $[-\infty, c)$ and is less than c. In particular, $\bar{x}$ could not be x_2, and $U \cap [x_1, x_2]$ would contain a point x greater than $\bar{x}$. This x is in U, so $f(x) < c$, which contradicts the fact that (A) has been established. So we cannot have $f(\bar{x}) < c$.

If $f(\bar{x})$ were greater than c, $(c, \infty]$ would be a neighborhood of $f(\bar{x})$, and there would exist a neighborhood U of $\bar{x}$ such that $f(x) > c$ for all x in $U \cap [x_1, x_2]$. In particular, $\bar{x}$ could not be x_1, so there would exist an x_3 in $U \cap [x_1, x_2]$ such that $x_3 < \bar{x}$. For every x in $(x_3, x_2]$, either x is in $(x_3, \bar{x}]$ and is in U, whence $f(x) > c$; or x is in $(\bar{x}, x_2]$ and $f(x) \geqq c$ because (A) is satisfied. In any case $f(x) \geqq c$, so x_3 belongs to E. But this contradicts the definition of $\bar{x}$ as the infimum of E, so we cannot have $f(\bar{x}) > c$. This and the preceding paragraph show that $f(\bar{x}) = c$, which completes the proof.

The following simple exercises are designed merely to allow the reader to verify that he understands the definitions.

EXERCISE 2-1 Prove that the intersection of two left-open intervals in $\bar{R}^1$ is a left-open interval in $\bar{R}^1$.

EXERCISE 2-2 Prove that if A_1 and A_2 are neighborhoods in $\bar{R}^1$ of a point x_0 of $\bar{R}^1$, their intersection $A_1 \cap A_2$ is a neighborhood of x_0.

EXERCISE 2-3 Prove that if x_0 and x_1 are two different points of $\bar{R}^1$, there exist a neighborhood of x_0 and a neighborhood of x_1 with no points in common.

EXERCISE 2-4 Show that in $\bar{R}^1$, $\varnothing$ and $\bar{R}^1$ are intervals that are both open and closed. Are there any other intervals in $\bar{R}^1$ that are both open and closed?

EXERCISE 2-5 Show that if A is a neighborhood of a point in $\bar{R}^1$, $A^{\text{o}} = A$.

EXERCISE 2-6 Show that if A is a closed interval in $\bar{R}^1$, $A^{-} = A$.

I

Elementary Properties of the Integral in One-Dimensional Space

1. A Heuristic Approach to the Definition of the Integral

For many centuries there has been interest in the problems of finding the area of a plane figure and the volume of a solid figure. These interested Greek mathematicians as purely mathematical problems more than 2000 years ago, and they also arise in highly practical applications of mathematics.

In ancient times the concept of area was felt to have an intrinsic meaning. There was no need to define the area of a plane figure; the figure *had* an area, in some sense not clearly described, and the problem was to find some way of computing that area. We shall begin by taking this ancient point of view and shall try to use our primitive feelings about area to lead us to a way of calculating the area under some curves of an especially simple type. When we finally state precisely what we mean by the area and formulate a way of computing it, we shall find that the same ideas apply to a much larger class of curves than we considered to start with, and to many problems other than finding areas.

Suppose that f is a real-valued function defined and nonnegative on the extended real number system $\bar{R}$. Then f determines a point-set that could be called "the region in the upper half-plane below the graph of f"; it is the set

(A) $$\{(x, y) \text{ in plane} : x \text{ in } R,\ 0 \leqq y < f(x)\}.$$

But this is too wordy, and with little danger of misunderstanding we can call it simply "the region under f" and its area "the area under f."

From the properties of the area assumed true by the ancient geometers the following statement can be proved.

(B) Let f and g be defined and nonnegative on $\bar{R}$, and let $[c, d]$ be a bounded interval. Assume that f and g are identical outside $[c, d]$ and that their difference is not greater than a number h in $[c, d]$. Let J_f and J_g be the areas under f and g, respectively. Then

$$|J_f - J_g| \leqq h(d - c).$$

We shall not stop to prove this; in this section, we are trying not to prove any theorems but to invent a process for computing areas. However, if any reader finds (B) unclear or hard to believe, he will find in Exercise 1-1 a sketch of a proof of (B) that would have been considered complete a few generations ago.

Suppose now that f is a function defined, nonnegative, and continuous on $\bar{R}$ that is identically 0 outside a bounded interval $[a, b]$. Let J denote the area under f. We set ourselves first the task of finding a method of calculating arbitrarily close approximations to J. We name a number ε that will be the largest error we will tolerate in the approximation. This ε may be 10 acres or a millionth of a square millimeter, but it is a positive number, and once specified it remains fixed.

Because f is continuous on $\bar{R}$ and the number

$$\varepsilon' = \varepsilon/(b - a + 3)$$

is positive, for each $\bar{x}$ in $[a, b]$ there is an open interval $(u(\bar{x}), v(\bar{x}))$ containing $\bar{x}$ such that for every point x in the interval

(C) $$|f(x) - f(\bar{x})| < \varepsilon'.$$

The intersection of $(u(\bar{x}), v(\bar{x}))$ with $(a - 1, b + 1)$ is also an open interval that contains $\bar{x}$; that is, it is a neighborhood of $\bar{x}$. We give it the name $\gamma(\bar{x})$, (read "gamma of $\bar{x}$"); then for every x in $\gamma(\bar{x})$, x is in $(a - 1, b + 1)$ and (C) is valid. For $\bar{x} < a$ we define $\gamma(\bar{x}) = [-\infty, a)$, and for $\bar{x} > b$ we define $\gamma(\bar{x}) = (b, \infty]$. Since f vanishes outside $[a, b]$, it is clear that for $\bar{x}$ not in $[a, b]$, f is identically 0 on $\gamma(\bar{x})$, and again (C) holds for x in $\gamma(\bar{x})$.

There are infinitely many ways in which R can be subdivided into a finite collection of intervals $\{A_1, \ldots, A_k\}$. Such a set of intervals is often called a "partition" of R; however, we shall not have much need of this name. We shall try to find a function constant on each interval of a partition that is near f. This is obviously not always possible; the partition has to be fine enough to let us find such a function. The test for fineness that we shall use is based on the neighborhoods $\gamma(\bar{x})$ defined in the preceding paragraph. A partition is fine enough to use if it satisfies this condition:

(D) To each interval A_i in the collection $\{A_1, \ldots, A_k\}$ there corresponds a point $\bar{x}_i$ in $\bar{R}$ such that the neighborhood $\gamma(\bar{x}_i)$ contains the closure A_i^- of A_i. (This is the closed interval with the same end-points as A_i.)

If $\{A_1, \ldots, A_k\}$ is a partition of R that satisfies (D) with points $\bar{x}_1, \ldots, \bar{x}_k$, we define a function g by setting

(E) $$g(x) = f(\bar{x}_i) \qquad (x \text{ in } A_i;\ i = 1, \ldots, k).$$

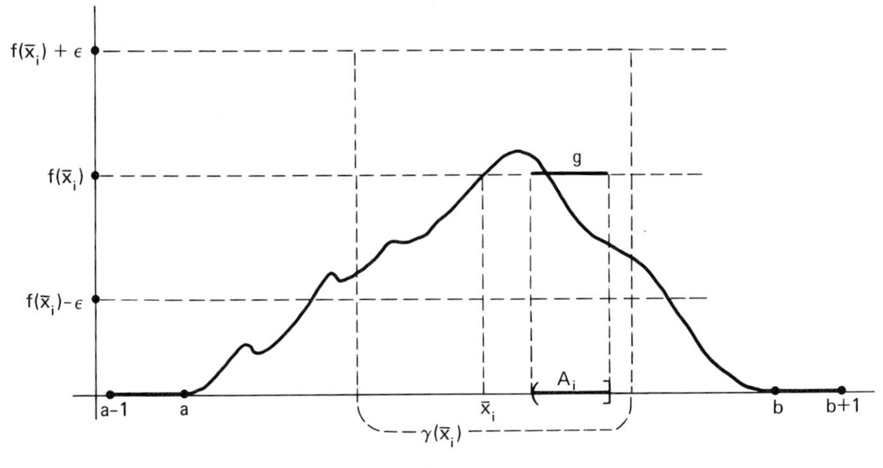

Fig. I-1

If x is any point in R, it belongs to some A_i, which is contained in $\gamma(\bar{x}_i)$ (Fig. I-1). Therefore, by (C),

(E) $$|g(x) - f(x)| = |f(\bar{x}_i) - f(x)| < \varepsilon'.$$

If, in particular, x is not in $[a - 1, b + 1]$, $\bar{x}_i$ cannot belong to $[a, b]$; for if $\bar{x}_i$ were in $[a, b]$, $\gamma(\bar{x}_i)$ would be contained in $(a - 1, b + 1)$ and could not contain the point x of A_i^-. So $\bar{x}_i$ is outside $[a, b]$, and $f(\bar{x}_i)$ must be 0. Therefore,

(G) if x is in $R\backslash[a - 1, b + 1]$,

$$g(x) - f(\bar{x}_i) = f(x) = 0.$$

Now g coincides with f outside of $[a - 1, b + 1]$ and differs from f by less than ε' in $[a - 1, b + 1]$, so by (B) the area J_g under g satisfies

(H) $$|J_g - J| \leqq \varepsilon'([b + 1] - [a - 1]).$$

Since $\varepsilon' = \varepsilon/(b - a + 3)$, this implies

(I) $$|J_g - J| < \varepsilon.$$

The region under g can be subdivided into the k parts

$$\begin{aligned} G_i &= \{(x, y) \text{ in plane} : x \text{ in } A_i,\ 0 \leqq y < g(x)\} \\ &= \{(x, y) \text{ in plane} : x \text{ in } A_i,\ 0 \leqq y < f(\bar{x}_i)\}. \end{aligned}$$

If $f(\bar{x}_i) \neq 0$, $\bar{x}_i$ must belong to $[a, b]$ and A_i is contained in $\gamma(\bar{x}_i)$, which is contained in the bounded interval $[a - 1, b + 1]$. Thus, in this case the region G_i is a rectangle whose base is the interval A_i and whose altitude is $f(\bar{x}_i)$. So,

(J) $$\text{area } G_i = f(\bar{x}_i)\, m_L A_i.$$

If $f(\bar{x}_i) = 0$, there are no points (x, y) with x in A_i and $0 \leqq y < f(\bar{x}_i) = 0$, so G_i is empty and has area 0. Therefore, in this case too, (J) is valid. We add for $i = 1, \ldots, k$; the sum of the areas of the G_i is the area J_g under g, so by (J),

(K) $$J_g = f(\bar{x}_1)\, m_L A_1 + \cdots + f(\bar{x}_k)\, m_L A_k.$$

We have now finished our first task. We have found a computable expression – the right member of (K)) – which by (I) differs by less than the tolerable amount ε from the area J. Let us stop to summarize what we have done so far.

(L) Let f be a function defined, continuous and nonnegative on $\bar{R}$ that vanishes outside a bounded interval, and let ε be a positive number. Then it is possible to assign a neighborhood $\gamma(\bar{x})$ to each point $\bar{x}$ in $\bar{R}$ in such a way that whenever $A_1, \ldots, A_k$ are finitely many pairwise disjoint intervals whose union is R, and $\bar{x}_1, \ldots, \bar{x}_k$ are points in $\bar{R}$ such that for each i the closure $\overline{A_i}$ of A_i is contained in $\gamma(\bar{x}_i)$, the sum

$$f(\bar{x}_1)\, m_L A_1 + \cdots + f(\bar{x}_k)\, m_L A_k$$

differs from the area under f by less than ε.

The procedure in (L) will be used over and over again, and we sorely need more convenient expressions for the things described in (L). To begin with, it makes no difference whether the A_i are open or closed, or left- or right-open; all that we meet in (L) are the closure of A_i and the length of A_i, which are the same whether or not the end-points of A_i belong to it. So, to forestall tedious and unimportant discussion, we make the agreement once and for all that the A_i are left-open; each A_i contains its right end-point unless that end-point is ∞, and no A_i contains its left end-point. Next, the A_i constitute a way of subdividing, or partitioning, R into subintervals, and such a set of A_i is often called a "partition" of R. However, we do not have much use for such partitions. In the foregoing discussion each A_i was matched up with, or allotted to, a point $\bar{x}_i$ of $\bar{R}$. The useful objects were the pairs $(\bar{x}_i, A_i)$.

DEFINITION 1-1 *An **allotted partition** is a finite set of pairs* $\{(\bar{x}_1, A_1), \ldots, (\bar{x}_k, A_k)\}$ *in which the* A_i *are pairwise disjoint left-open intervals in* R *and the* $\bar{x}_1$ *are points of* $\bar{R}$. *If* B *is the union of the* A_i, *the allotted partition is an **allotted partition of*** B. *The* A_i *are called the **intervals of the allotted partition**; the* $\bar{x}_i$ *are its **evaluation points**.*

As a aid to memory, we shall reserve the capital script letter $\mathscr{P}$, with or without affixes (such as $\mathscr{P}_3$ or $\mathscr{P}'$) for allotted partitions.

The sum that occurs in (L) is named the *partition-sum* corresponding to the allotted partition and the function. In general, we use the following definition.

DEFINITION 1-2 *Let $\mathscr{P}$ be an allotted partition*

$$\mathscr{P} = \{(\bar{x}_1, A_1), \ldots, (\bar{x}_k A_k)\},$$

and let f be an extended-real-valued function on $\bar{R}$. Then the ***partition-sum*** *corresponding to $\mathscr{P}$ and f is the sum*

$$S(\mathscr{P}; f) = f(\bar{x}_1)\, m_L A_1 + \cdots + f(\bar{x}_k)\, m_L A_k,$$

provided that this sum exists.

In (L) we assigned to each point $\bar{x}$ in $\bar{R}$ a neighborhood $\gamma(\bar{x})$ of $\bar{x}$. This is just another way of saying that we defined a function γ on $\bar{R}$ whose functional values, instead of being numbers, are neighborhoods in $\bar{R}$. This function was used in (D) as a measuring tool to gauge whether or not the allotted partition was fine enough to give a good estimate of J. We shall therefore call it a *gauge*.

DEFINITION 1-3 *A* ***gauge*** *on a set B in $\bar{R}$ is a function $\bar{x} \mapsto \gamma(\bar{x})$ (x in B) such that for each $\bar{x}$ in B, $\gamma(\bar{x})$ is a neighborhood of $\bar{x}$.*

An allotted partition $\mathscr{P}$ that passes the fineness test (D) with a gauge γ will be said to be *γ-fine.*

DEFINITION 1-4 *Let γ be a gauge on $\bar{R}$, and let*

$$\mathscr{P} = \{(\bar{x}_1, A_1), \ldots, (\bar{x}_k, A_k)\}$$

be an allotted partition of a set $A_1 \cup \cdots \cup A_k$ contained in R. The allotted partition $\mathscr{P}$ is ***γ-fine*** *if for each i in $\{1, \ldots, k\}$ the closure A_i^- of A_i is contained in $\gamma(\bar{x}_i)$.*

We can and shall shorten the expression "γ-fine allotted partition" to simply "γ-fine partition," since only allotted partitions can be γ-fine.

With the help of these definitions we can express (L) less wordily, as follows:

(M) Let f be a function defined, continuous, and nonnegative on $\bar{R}$ that vanishes outside a bounded interval, and let ε be a positive number. Then there exists a gauge γ on $\bar{R}$ such that for each γ-fine partition $\mathscr{P}$ of R,

$$|S(\mathscr{P}; f) - \text{area under } f| < \varepsilon.$$

Given any one positive ε, the procedure in (M) does not determine the area J exactly; it merely gives us an interval of length 2ε in which the number J must lie, since when we calculate $S(\mathscr{P}; f)$, we know that J is between $S(\mathscr{P}; f) - \varepsilon$ and $S(\mathscr{P}; f) + \varepsilon$. But when we have the possibility of carrying out approximation (M) for every positive ε, the complete battery of tests, one for each ε, is enough to determine J exactly. For, suppose that someone claims that some different number J' is the area. We choose $\varepsilon = |J - J'|/2$, and we find an interval of length less than 2ε that contains the value of the area. Then it contains J, and since the distance from J to J' is 2ε, it cannot also contain the false claimant J'. Thus, each false estimate for the area is ruled out by some test or other, and only the correct value J passes all the tests. So we have established this rule for computing the area:

(N) Let f be a function defined, continuous, and nonnegative on $\bar{R}$ that vanishes outside a bounded interval, and let J be the area under f. Then J is the only number with the property that for each positive ε there is a gauge γ on $\bar{R}$ such that for every γ-fine partition $\mathscr{P}$ of R,

$$|S(\mathscr{P}; f) - J| < \varepsilon.$$

The last part of (N) strongly resembles the definition of the limit of a sequence, or of the limit of a function. In fact, we shall introduce the following terminology.

DEFINITION 1-5 *Let S be a function defined for each allotted partition in a class K of allotted partitions, and let L be a number. The statement that L is the* ***gauge-limit*** *of S means that for each positive ε there is a gauge γ on $\bar{R}$ such that for every γ-fine partition $\mathscr{P}$ that belongs to K,*

$$|S(\mathscr{P}) - L| < \varepsilon.$$

Now the description (N) of the procedure for calculating the area J simplifies to the following form:

(O) Let f be defined, nonnegative, and continuous on $\bar{R}$, and let it vanish outside a bounded interval. The area J under f is the gauge-limit of $S(\mathscr{P}; f)$.

EXERCISE 1-1 Using the customary assumptions of elementary plane geometry, prove (B). *Suggestion*: Define

$$F' = \{(x, y) \text{ in plane} : x \text{ in } [c, d],\ 0 \leqq y < f(x)\},$$

$$G' = \{(x, y) \text{ in plane} : x \text{ in } [c, d],\ 0 \leqq y < g(x)\}.$$

Since f and g coincide outside $[c, d]$,

$$J_f - J_g = \text{area } F' - \text{area } G'.$$

The region

$$\{(x, y) \text{ in plane} : x \text{ in } [c, d],\ 0 \leqq y < f(x) + h\}$$

contains G' and is the union of the disjoint sets

$$\{(x, y) \text{ in plane} : x \text{ in } [c, d],\ h \leqq y < f(x) + h\},$$

$$\{(x, y) \text{ in plane} : x \text{ in } [c, d],\ 0 \leqq y < h\}.$$

The first of these is congruent to F' and the second is a rectangle with base $[c, d]$ and altitude h. Thus,

$$\text{area } G' \leqq \text{area } F' + h(d - c).$$

Similarly,

$$\text{area } F' \leqq \text{area } G' + h(d - c).$$

2. Definition of the Integral

In the preceding section we considered functions f that are nonnegative and continuous on $\bar{R}$ and vanish outside some bounded interval. But this was merely to make it highly plausible that the gauge-limit of the partition-sums exists and is equal to the area. If f is any nonnegative real-valued function on $\bar{R}$ with $f(\infty) = f(-\infty) = 0$, it is easy to see that the partition-sum $S(\mathscr{P}; f)$ will be finite whenever $\mathscr{P}$ is γ-fine and γ is a gauge such that $\gamma(\bar{x})$ is bounded whenever $f(\bar{x}) \neq 0$. So it is possible for the gauge-limit of $S(\mathscr{P}; f)$ to exist. We cannot say that its value is the same as the intuitive value of the area under f because f can be so complicated that our intuitions fail to give any idea of the area. But we can take care of this in a simple way by *defining* the area under f as the gauge-limit of $S(\mathscr{P}; f)$.

Often we are given a point set B on the x-axis, and we want to know the area of the region that is under f and is also over the set B on the x-axis. This is the set

(A) $$\{(x, y) \text{ in plane}: x \text{ in } B,\ 0 \leqq y < f(x)\}.$$

This is really nothing new, in spite of appearances. We define a new function f_B by setting

(B) $$f_B(x) = \begin{cases} f(x) & (x \text{ in } B) \\ 0 & (x \text{ in } \bar{R} \setminus B). \end{cases}$$

The region under f_B is then the set

$$\{(x, y) \text{ in plane}: x \text{ in } R,\ 0 \leqq y < f_B(x)\}.$$

But this is the same as the set (A). For if (x, y) is in (A), then x is in R and $0 \leqq y < f(x) = f_B(x)$, so (x, y) is in the region under f_B. Conversely, if (x, y) is in the region under f_B, then $f_B(x) > 0$. So x has to be in B and $0 \leqq y \leqq f_B(x) = f(x)$, which implies that (x, y) is in A. Therefore the area of the region under f and over B is the same as the total area under f_B, which we have already defined to be the gauge-limit of $S(\mathscr{P}; f_B)$.

However, the process of finding the gauge-limit of partition-sums has applications in which f is not nonnegative. For example, let $f(x)$ be the rate at which water is flowing into a tank at time x, expressed in gallons per second; this is a negative number if water is flowing out at time x. If f is a step-function, with constant value g_i on each of the intervals $A_1, \ldots, A_k$ of time, it is easy to see that the total influx is $g_1 m_L A_1 + \cdots + g_k m_L A_k$. By an argument like that in Section 1, if f is continuous and vanishes outside a bounded interval, the total influx is the gauge-limit of $S(\mathcal{P}; f)$. If B is a set of times (for instance, the time-interval $B = [a, b]$), the influx during times B is the same as the total influx would be if the rate of flow were changed to 0 at all times outside B and were left unchanged (equal to $f(x)$) at times x in B. So the total influx during the set B of times is the gauge-limit of $S(\mathcal{P}; f_B)$.

It is clear that the gauge-limit of partition-sums is a mathematical entity of varied applications and deserves a name of its own, not one (such as "area" or "influx") that comes from some particular application. We shall call it the "integral."

DEFINITION 2-1 *Let B be a set contained in R, and let f be a real-valued function defined on a domain that contains B. Then f is* ***integrable over*** *B if the gauge-limit of $S(\mathcal{P}; f_B)$ exists, where f_B is the function such that $f_B(x) = f(x)$ for x in B and $f_B(x) = 0$ for x in $\bar{R} \backslash B$. If $S(\mathcal{P}; f_B)$ has gauge-limit J, J is called the* ***integral*** *of f over B and is denoted by*

$$\int_B f(x)\, m_L(dx).$$

Integrals such as this, defined as gauge-limits of partition-sums, are the chief objects of study in this book. If f is a function as in Definition 2-1, when we speak of the integral of f we shall mean the integral defined in Definition 2-1. But sometimes we need to compare this integral with other integrals, such as the integral defined in elementary calculus. In such cases we shall identify the integral defined in Definition 2-1 by calling it the **gauge-integral.**

Because this definition is fundamentally important to us, we shall write it out again in greater detail.

DEFINITION 2-2 (Definition 2-1 expanded) *Let B be a subset of R and f a function with values in R that is defined on a set D such that $B \subset D \subset \bar{R}$. Define f_B to be the function on $\bar{R}$ such that $f_B(x) = f(x)$ if x is in B and $f_B(x) = 0$ on the rest of $\bar{R}$. Then the statement that*

$$J = \int_B f(x)\, m_L(dx)$$

means that for each positive number ε there exists a gauge γ on $\bar{R}$ such that whenever $\mathcal{P}$ is a γ-fine partition of R, the partition-sum $S(\mathcal{P}; f_B)$ defined in Definition 1-2 exists and

$$|S(\mathcal{P}; f_B) - J| < \varepsilon.$$

When B is an interval with end-points a and b, a notation for the integral more like that in elementary calculus is

$$\int_a^b f(x)\, m_L(dx).$$

But the end-points a and b do not uniquely specify the interval B; it could be open, closed, left-open, or right-open. Soon we shall see that this makes no difference in the value of the integral; it has the same value over (a, b) as over $(a, b]$, etc. But until we have proved this we avoid the appearance of ambiguity by defining, whenever a and b ($> a$) are points of $\bar{R}$,

$$\int_a^b f(x)\, m_L(dx) = \int_B f(x)\, m_L(dx),$$

where B is the *left-open* interval in R with end-points a and b.

The requirement that $S(\mathscr{P}; f_B)$ shall exist for all γ-fine partitions $\mathscr{P}$ of R causes no trouble, even though $m_L A_i$ can be ∞ for some intervals A_i of $\mathscr{P}$. All we have to do is to use only gauges γ such that $\gamma(\bar{x})$ is bounded for all $\bar{x}$ such that $f(\bar{x}) \neq 0$. For then in the product $f_B(\bar{x}_i)\, m_L A_i$, either the first factor is a finite nonzero number and the second factor is finite because A_i is contained in the bounded neighborhood $\gamma(\bar{x}_i)$, or else the first factor is 0 and the product is 0 no matter what $m_L A_i$ is.

In using this definition, it is important to have clearly in mind how sparse the assumptions are. It would be easy to read into our definitions some concepts that are not there, or to carry them over from some previous study of calculus. We shall, therefore, point out some requirements that are *not* in our theory. For the benefit of any reader who may be familiar with the Riemann integral, these remarks will serve to mark the differences between the two kinds of integral.

First, the function f need not be bounded. Each individual value $f(x)$ is finite, but there need not be any finite N such that $|f(x)| \leqq N$ for all x in B. Second, B need not be a bounded set nor be an interval. The A_i in the partitions need not be numbered from left to right. They could be, but often it is more convenient to number them in some other way. They are not even assumed to be nonempty. If some A_i in an allotted partition $\mathscr{P}$ of R is empty, it contributes the amount $f(\bar{x}_i) m_L A_i = 0$ to the partition-sum $S(\mathscr{P}; f)$, and if we discard it from $\mathscr{P}$, the remaining pairs still form an allotted partition of R. If $\mathscr{P}$ contains some pairs with empty intervals, we could thus always "clean up" $\mathscr{P}$ by discarding them; but often it is more convenient not to bother to do this. The $\bar{x}_i$ are not assumed to be all different. In fact, it will often be convenient to use partitions $\mathscr{P}$ in which several pairs $(\bar{x}_i, A_i)$, $(\bar{x}_j, A_j), \ldots$ have $\bar{x}_i = \bar{x}_j = \cdots$ but have different A_i. But more important than any of these remarks are the next two.

Whenever γ is a gauge, for each x in $\bar{R}$ the neighborhood $\gamma(x)$ has positive length. But we do not require that there be any positive number δ such that all of these neighborhoods have length of at least δ. Suppose that we had been more demanding in our definition of gauge and had required that for every gauge there exist a positive δ such that for all finite $\bar{x}$ (or for all $\bar{x}$ in a bounded interval over which we wish to integrate f) the neighborhood $\gamma(\bar{x})$ contain the interval $(\bar{x} - \delta, \bar{x} + \delta)$. Then we would have obtained in Definition 2-1 an integral with some good properties, but we would not be able to prove the extremely important Theorem II-4-2. (Side-remark to experts: The integral thus defined would have been equivalent to the Riemann integral.)

We have not assumed that, in the pairs $(\bar{x}_i, A_i)$, the evaluation-point $\bar{x}_i$ should be contained in the closure A_i^- of the interval A_i. In the preceding section, this assumption was not needed at all. The importance of the gauge was to ensure that for partitions that are γ-fine, each interval A_i^- should be contained in the neighborhood $\gamma(\bar{x}_i)$, on which f differed little from $f(\bar{x}_i)$. Whether $\bar{x}_i$ were in A_i^- had nothing to do with this. If we had been more demanding in our definition of allotted partition and had required that for each pair the evaluation-point should lie in the closure of the interval, we would have obtained an integral with many interesting properties, but we would not be able to prove the useful theorems and corollaries found in Sections 2 and 3 of Chapter II. (Side-remark to experts: The integral thus obtained would have been equivalent to the Denjoy integral, whereas the gauge-integral is equivalent to the Lebesgue integral.)

The question has been asked, "Why define γ for all points of $\bar{R}$ when it is used only at evaluation-points of the allotted partitions $\mathscr{P}$?" We need to have γ defined at all points of $\bar{R}$ because whatever allotted partition $\mathscr{P}$ we encounter, we must be able to decide whether or not it is γ-fine. For this, we must know $\gamma(\bar{x})$ at all evaluation-points $\bar{x}$ of $\mathscr{P}$; and every point $\bar{x}$ in $\bar{R}$ is an evaluation-point in infinitely many allotted partitions. (Exercise 2-1 will illustrate another good reason.)

We have been using the letter x to denote the "independent variable" in the function f, but this is a matter of no importance. There is, for instance, a function f defined on the interval [2, 6] whose value at each number in [2, 6] is the square of that number. This function is equally well denoted by $f: x \mapsto x^2$ $(2 \leqq x \leqq 6)$, or by $f: u \mapsto u^2$ $(2 \leqq u \leqq 6)$, or by $f: A \mapsto A^2$ $(2 \leqq A \leqq 6)$. Similarly, if $\mathscr{P}$ is an allotted partition $\{(\bar{x}_i, A_i), \ldots, (\bar{x}_k, A_k)\}$ of R, it could equally well be denoted by $\{(u_1, B_1), \ldots, (u_k, B_k)\}$, where $u_i = \bar{x}_i$ and $B_i = A_i$. The integral defined in Definition 2-1 depends on the function f but not on the letter that we choose to use to name values of the "independent variable." Consequently, instead of the symbols introduced in Definition 2-1, we could just as well have denoted J by either of the symbols

$$\int_B f(u)\, m_L(du), \qquad \int_B f(v)\, m_L(dv).$$

The only symbols that we should avoid using in place of x are those that already have a fixed meaning, either by universal agreement or by having already been assigned a fixed meaning in the discussion in which the integral occurs. Thus, the integral of f over R should not be denoted by

$$\int_R f(2)\, m_L(d2).$$

Also, if B is an interval $(a, x]$, the integral of f over B can be denoted by

$$\int_a^x f(u)\, m_L(du),$$

but it should not be denoted by either of the symbols

$$\int_a^x f(a)\, m_L(da), \qquad \int_a^x f(x)\, m_L(dx)$$

because a and x have been assigned fixed meanings, namely, the lower and upper ends of B. This is worth emphasizing because, in spite of the fact that the last of these symbols is unacceptable, it is widely used.

EXERCISE 2-1 Show that if $\bar{x} \mapsto \gamma(\bar{x})$ is a function that assigns a neighborhood $\gamma(\bar{x})$ to every point $\bar{x}$ of $\bar{R}$ *except one*, there may not exist any γ-fine partition of R. *Suggestion*: If $\gamma(\bar{x}) = [-\infty, 0)$ for $\bar{x} < 0$ and $\gamma(\bar{x}) = (0, \infty]$ for $\bar{x} > 0$, there exists no γ-fine pair $(\bar{x}, A)$ with 0 in A.

EXERCISE 2-2 If $f(x) = 0$ for $x \neq 7$ and $f(7) = 10$, then

$$\int_R f(x)\, m_L(dx) = 0.$$

Suggestion: For $\varepsilon > 0$, take $\gamma(7) = (7 - \varepsilon/22,\ 7 + \varepsilon/22)$, and for $\bar{x} \neq 7$ take $\gamma(\bar{x}) = R$. In any γ-fine partition, the pairs $(\bar{x}_i, A_i)$ with $f(\bar{x}_i) \neq 0$ have $A_i \subset \gamma(7)$.

EXERCISE 2-3 Prove in detail that if a particle has velocity $v(x)$ at each time x in a time-interval $[a, b]$ and v is a continuous function on $[a, b]$ with $v(a) = v(b) = 0$, the distance moved during the time-interval $[a, b]$ is

$$\int_a^b v(x)\, m_L(dx).$$

EXERCISE 2-4 (i) Show that if $\gamma(\bar{x})$ is a neighborhood of $\bar{x}$, the set $-\gamma(\bar{x}) = \{y \text{ in } \bar{R}: -y \text{ in } \gamma(\bar{x})\}$ is a neighborhood of $-\bar{x}$.

(ii) Show that if γ is a gauge on $\bar{R}$, and for each $\bar{x}$ in $\bar{R}$ we define

$$\gamma'(\bar{x}) = -\gamma(-\bar{x}),$$

γ' is a gauge on $\bar{R}$.

(iii) Show that

$$\mathscr{P} = \{(\bar{x}_1, (a_1, b_1]), \ldots, (\bar{x}_k, (a_k, b_k])\}$$

is a γ-fine partition of R if and only if

$$\mathscr{P}' = \{(-\bar{x}_1, (-b_1, -a_1]), \ldots, (-\bar{x}_k, (-b_k, -a_k])\}$$

is a γ'-fine partition of R.

(iv) Show that if $f'(x) = f(-x)$ for all x in R, and $\mathscr{P}$ and $\mathscr{P}'$ are as in (iii), then

$$S(\mathscr{P}'; f') = S(\mathscr{P}; f).$$

(v) Show that, with the same notation, f' is integrable over R if and only if f is integrable over R, and in that case

$$\int_R f'(x)\, m_L(dx) = \int_R f(x)\, m_L(dx).$$

EXERCISE 2-5 Show that from part (iv) of Exercise 2-4 it follows that if f is an odd function on R, meaning that $f(-x) = -f(x)$ for all x in R, and if f is integrable over R, then its integral over R is 0.

3. Examples

It is possible to use Definition 2-1 directly, without any theoretical preliminaries, to evaluate a great number of particular examples of integrals, such as

$$\text{(A)} \qquad \int_a^b x^n\, m_L(dx) = \frac{b^{n+1} - a^{n+1}}{n+1},$$

where n is any positive integer. But the details are tedious, and there are two more convenient paths, each with its advantages, open to us. First, most readers of this book will have enough acquaintance with calculus to have encountered some definition, usually one in which step-functions above and below f are used to give overestimates and underestimates, of the integral of f. Soon we shall discuss this method of defining the integral, and in Theorem 7-2 we shall show that

> whenever f has an integral $\int f(x)\, dx$ in that elementary sense, it also has a gauge-integral $\int f(x)\, m_L(dx)$, and the two integrals have the same value.

This means that nothing of the elementary theory of integration has to be abandoned. The reader will almost surely have met a large number of simple examples of integration problems in which the integral has the elementary meaning, and every one of these is also an example for the gauge-integral. This makes it superfluous to exhibit a long list of gauge-integrals of elementary functions. For example, the reader already knows that (A) holds for the elementary integral, so it holds for the gauge-integral also.

Second, there are two (perhaps not empty) subsets of readers who might prefer to avoid any reference to the elementary integral. One is the presumably small subset of readers who have never encountered calculus before; the other is the subset of readers whose mathematical tastes are so well developed that they prefer this theory to be self-contained, even to the extent of relegating differential calculus to a secondary role. For them, the least strenuous course is to wait until we have proved some theorems about the integral that will make the evaluations easier. For example, after we have proved the theorems in Section 7 we can establish (A) with less labor than by direct application of Definition 2-1 (see Exercises 7-3, 7-4, and 7-5). After we prove the formula for integration by parts in Section 11 we shall be able to establish (A) with hardly any effort; but this proof uses differential calculus in proving the formula for integration by parts. Theorem II-9-4 is a strong form of the theorem on integration by parts that does not even mention derivatives; with Exercise 11-2 it gives an easy proof of (A) that is contained entirely in integration theory.

Likewise, the equation

$$\text{(B)} \qquad \int_a^b \cos x \, m_L(dx) = \sin b - \sin a$$

can be proved laboriously from the definition of the gauge-integral. But we know that its analog holds for the elementary integral, and after we prove in Section 7 that the elementary integral is a special case of the gauge-integral, we shall know that (B) holds for the gauge-integral also. However, the proof of (B) in elementary calculus uses a knowledge of the properties of the trigonometric functions. In Section 6 of Chapter II we shall present a derivation of (B) that not only does not presuppose knowledge of formulas of elementary calculus but does not even require knowledge of trigonometry. In that section theorems about the integral are used to let us define the trigonometric functions and to prove the fundamental theorems about them, including (B).

For these reasons we shall be content with working out two examples. The first is very simple, but besides exemplifying the use of the definition of the gauge-integral, it provides us with an evaluation of an integral that we shall often use. The second example will show that the gauge-integral exists for functions that cannot be integrated by the integral of elementary calculus. In it we integrate a rather queer function that is a favorite toy of mathematicians.

Before presenting the examples, we define a kind of function that is quite convenient to use. If we apply Definition 2-2 when f is identically 1 and B is a subset of $\bar{R}$, we obtain the function 1_B, which has value 1 on B and 0 elsewhere. This function is called the **indicator** of the set B, for the obvious reason that its value at x indicates whether or not x belongs to B. (It is also called the **characteristic function** of B, but we shall not use that name for it.) Formally,

DEFINITION 3-1 *Let B be a subset of $\bar{R}$. Then the **indicator** of B is the function 1_B defined by*

$$1_B(x) = \begin{cases} 1 & (x \text{ in } B); \\ 0 & (x \text{ in } \bar{R} \backslash B). \end{cases}$$

If A is a bounded closed interval $[a, b]$, the figure under the graph of 1_A is a rectangle with base $m_L A$ and altitude 1. Unless our somewhat pictorial discussion in Section 1 has led us astray, the integral of 1_A must be equal to the area of this rectangle, which is $1 \cdot m_L A$. We now show that this is in fact true.

LEMMA 3-2 *Let A be a bounded interval in R. Then 1_A is integrable over R, and*

(C) $$\int_R 1_A(x)\, m_L(dx) = m_L A.$$

Let ε be positive. To prove the lemma we must exhibit a gauge γ on $\bar{R}$ such that for every γ-fine partition $\mathscr{P}$ of R,

(D) $$|S(\mathscr{P}; 1_A) - m_L A| < \varepsilon.$$

Here, as usual, ingenuity is called for in selecting the gauge, and industry is called for in showing that (D) is satisfied. By Lemma 1-4 in the Introduction, we can and do select intervals F, G such that $A \subset G^{\text{O}}$ and $F^- \subset A$ and

(E) $$m_L G < m_L A + \varepsilon \qquad \text{and} \qquad m_L F > m_L A - \varepsilon.$$

Denote G^{O} by (a', b') and F^- by $[a'', b'']$. If x is any point of A, it is in the open interval G^{O}, and we choose G^{O} for $\gamma(x)$. If x is any point of $\bar{R} \backslash A$, it is not in the closed interval F^-. We therefore can and do select for $\gamma(x)$ a neighborhood of x that is disjoint from F^-. For example, if $x < a''$, we can select $\gamma(x) = [-\infty, a'')$, and if $x > b''$ we can select $\gamma(x) = (b'', \infty]$. Now for every x in $\bar{R}$, $\gamma(x)$ is a neighborhood of x, so γ is a gauge on $\bar{R}$. Also, if x is in A, then $\gamma(x) \subset G^{\text{O}}$, and if x is not in A, then $\gamma(x)$ is disjoint from F^-. It remains to show that γ does what we want it to do.

Let $\mathscr{P} = \{(\bar{x}_1, A_1), \ldots, (\bar{x}_k, A_k)\}$ be a γ-fine partition of R. There will be a certain number of pairs, say h of them, for which $\bar{x}_i$ is in A. By renumbering the pairs if necessary, we can bring it about that $\bar{x}_i$ is in A if $i = 1, \ldots, h$ and is not in A if $i = h + 1, \ldots, k$. Then $1_A(\bar{x}_i) = 1$ if $i = 1, \ldots, h$ and $1_A(\bar{x}_i) = 0$ if $i = h + 1, \ldots, k$, so

(F) $$\begin{aligned} S(\mathscr{P}; 1_A) &= 1_A(\bar{x}_1)\, m_L A_1 + \cdots + 1_A(\bar{x}_h)\, m_L A_h \\ &\quad + 1_A(\bar{x}_{h+1})\, m_L A_{h+1} + \cdots + 1_A(\bar{x}_k)\, m_L A_k \\ &= m_L A_1 + \cdots + m_L A_h. \end{aligned}$$

Since $\mathscr{P}$ is γ-fine, for each i, A_i^- is contained in $\gamma(\bar{x}_i)$. If $i \leq h$, $\bar{x}_i$ is contained in A, and $\gamma(\bar{x}_i)$ is G^{O} by definition of γ. So $A_1, \ldots, A_h$ are pairwise disjoint intervals all contained in G, and by Lemma 1-2 in the Introduction,

$$m_L A_1 + \cdots + m_L A_h \leqq m_L G.$$

By (E) and (F), this implies

$$S(\mathscr{P}; 1_A) < m_L A + \varepsilon. \tag{G}$$

If x is any point of F, it is in one of the intervals $A_1, \ldots, A_k$, since the union of these intervals is all of $\bar{R}$. Suppose it is in A_i. Then x is in A_i, and because $\mathscr{P}$ is γ-fine, A_i^- is contained in $\gamma(\bar{x}_i)$, so $\gamma(\bar{x}_i)$ contains the point x of F. But if i were greater than h, $\gamma(\bar{x}_i)$ would not contain any point of F, so we must have $i \leqq h$. Therefore the intervals $A_1, \ldots, A_h$ have a union that contains all of F, and by Lemma 1-2 in the Introduction,

$$m_L A_1 + \cdots + m_L A_h \geqq m_L F. \tag{H}$$

By (E) and (F) this implies

$$S(\mathscr{P}; 1_A) > m_L A - \varepsilon. \tag{I}$$

Inequalities (G) and (I) imply that (D) is satisfied, and by Definition 2-1 the lemma is proved.

At this point it is tempting to show some examples of functions too complicated to be integrable in the sense of Riemann, but integrable in our sense. These examples, however, will appear in a natural way as we continue our study of integration. Moreover, the point of the integral we are using is not that it can successfully integrate some especially weird integrands. Rather, its virtue is that we can prove powerful theorems about it that allow us to perform mathematical manipulations not possible with the Riemann integral. In fact, if our purpose were to strain after generality we could easily increase the power of the integral by redefining the statement that $\mathscr{P}$ is γ-fine to mean that for each i, $A_i^- \subset \gamma(\bar{x}_i)$ and $\bar{x}_i$ is in the closure $\bar{A}_i$. But this change would give us an integral that lacks some of the desirable properties of our integral, so we give up the extra generality in favor of the greater convenience.

Nevertheless, we shall exhibit one more example, partly because it shows a different trick for defining the gauge and partly because it is one that comes quickly to the mind of any experienced analyst.

EXAMPLE 3-3 *Let f be the indicator of the set of rational numbers in* $(0, 1)$ *so that* $f(x) = 1$ *if x is rational and* $0 < x < 1$ *and* $f(x) = 0$ *for all other x in* $\bar{R}$. *Then*

$$\int_R f(x)\, m_L(dx) = 0.$$

Before we start the proof of this assertion, we point out that in every nonempty open subinterval of $(0, 1)$ there are rational numbers and irrational numbers. Thus, f is discontinuous at all points of $[0, 1]$, and its Riemann integral cannot exist.

Every rational number in $(0, 1)$ can be uniquely expressed "in lowest terms" as a fraction p/q in which p and q are positive integers with no common divisor. Let ε be positive. If x is a rational number in $(0, 1)$, expressed in lowest terms as p/q, we define

$$\gamma(x) = \left(x - \frac{\varepsilon}{2^{p+q+1}}, x + \frac{\varepsilon}{2^{p+q+1}}\right);$$

if x is any other number in $\bar{R}$, we define $\gamma(x) = \bar{R}$. Then γ is a gauge. Let $\mathscr{P} = \{(\bar{x}_1, A_1), \ldots, (\bar{x}_k, A_k)\}$ be a γ-fine partition of R. We shall prove

$$\text{(J)} \qquad |S(\mathscr{P}; f) - 0| < \varepsilon.$$

There is no loss of generality in assuming that the numbering has been chosen so that $\bar{x}_i$ is rational and in $(0, 1)$ for $i = 1, \ldots, h$ but not for $i = h + 1, \ldots, k$. Then $f(\bar{x}_i) = 1$ for $i = 1, \ldots, h$, and $f(\bar{x}_i) = 0$ for $i = h + 1, \ldots, k$, and therefore

$$\text{(K)} \qquad S(\mathscr{P}; f) = \sum_{i=1}^{h} m_{\mathrm{L}} A_i.$$

If x is a rational number in $(0, 1)$ expressed in lowest terms as p/q, the length of $\gamma(x)$ is $\varepsilon/2^{p+q}$. If we add the lengths of all these $\gamma(x)$, we get a number smaller than the sum of $\varepsilon/2^{p+q}$ for all positive integers p and q. But if we sum first over q and then over p, using the formula for the sum of a geometric series, we find that the sum of the lengths of all the $\gamma(x)$ for x rational and in $(0, 1)$ is less than ε. Each A_i is contained in some one of these $\gamma(x)$, so the set of pairwise disjoint intervals $A_1, \ldots, A_h$ is contained in the union of finitely many of the $\gamma(\bar{x}_i)$—say, in $\gamma(\bar{x}_1), \ldots, \gamma(\bar{x}_N)$—with $\bar{x}_i$ rational and in $(0, 1)$. By Lemma 1-2 in the Introduction,

$$m_{\mathrm{L}} A_1 + \cdots + m_{\mathrm{L}} A_h \leqq m_{\mathrm{L}} \gamma(\bar{x}_1) + \cdots + m_{\mathrm{L}} \gamma(\bar{x}_N).$$

The right member of this inequality is less than ε, so by (K), (J) is satisfied, and the proof is complete.

4. Existence of γ-Fine Partitions

In Section 1 we remarked that, whatever testing procedure we were proposing for obtaining assurance that a partition-sum should be close to the area sought, it should be possible to find some admissible set of pairs that passed the test. We have been quietly ignoring this point, and we must return to it. Since we have adopted Definition 2-1, we have to show that for every gauge γ there exists at least one γ-fine partition of R. To many beginners this will seem obvious, as it would have seemed to mathematicians of the early nineteenth century. It really is not so very obvious, as Exercise 4-1 will indicate. In any case, it is the duty of the author of a book such as this to show that all statements in it can be deduced logically from the assumed properties of the real number system. There are some users of mathematics who feel that existence theorems are mathematical hairsplitting, unworthy of the attention of anyone who wants to get down to the real uses of the subject. We shall attempt here to convince such people that existence theorems can be important. In the case of the existence of γ-fine partitions, the existence theorem is so important that if there were even one single gauge γ^* for which no γ^*-fine partition of R existed, the whole theory based on Definition 2-1 would be nonsense, as we now show.

A little reflection should convince the reader that Definition 2-2 can be rephrased thus:

Let B be a subset of $\bar{R}$ and f a function defined and real valued on a set D such that $B \subset D \subset \bar{R}$. Let f_B coincide with f on B and be 0 on the rest of $\bar{R}$, and let J be real. The statement

$$J = \int_B f(x)\, m_L(dx)$$

means that for each positive number ε there exists a gauge γ on $\bar{R}$ such that for every partition $\mathscr{P}$ of R the two statements

(i) $\mathscr{P}$ is γ-fine,
(ii) $|S(\mathscr{P}; f_B) - J| \geqq \varepsilon$

are not both true.

Now let B be any subset of $\bar{R}$, f any function on B, and J any number. If there exists a gauge γ^* such that there is no γ^*-fine partition of R, the statement

$$\int_B f(x)\, m_L(dx) = J$$

is valid. For whatever ε is, we can choose $\gamma = \gamma^*$, and for every allotted partition $\mathscr{P}$ of R the two statements (i) and (ii) cannot both be true, because (i) is false. Thus, if a gauge γ^* exists for which there are no γ^*-fine partitions of R, every function is integrable over every set and has every number for its integral. An integration theory in which

$$\int_0^\infty 7\, m_L(dx) = -37$$

is a valid statement is a theory fit only for consignment to the wastebasket.

In several proofs we shall need a simple remark that we display as a lemma.

LEMMA 4-1 *If $C_1, \ldots, C_h$ are pairwise disjoint sets in $\bar{R}$, and $\mathscr{P}_j$ is an allotted partition of C_j $(j = 1, \ldots, h)$, then the union $\mathscr{P}$ of the $\mathscr{P}_j$ (which is the set of all pairs $(\bar{x}, A)$ that belong to at least one $\mathscr{P}_j$) is an allotted partition of $C_1 \cup \cdots \cup C_h$.*

Each member of $\mathscr{P}$ belongs to some $\mathscr{P}_j$, so it is a pair $(\bar{x}, A)$ with $\bar{x}$ in $\bar{R}$ and A a left-open interval in $\bar{R}$. Every point x in an interval of the set $\mathscr{P}$ is in an interval of some allotted partition $\mathscr{P}_j$, so it is contained in the set C_j, of which $\mathscr{P}_j$ is an allotted partition, and is therefore in $C_1 \cup \cdots \cup C_h$. Conversely, if x is in the union of the C_j, it is in some C_j, and is hence in the interval A of some pair $(\bar{x}, A)$ that belongs to $\mathscr{P}_j$. So the union of the intervals in the set $\mathscr{P}$ of pairs is $C_1 \cup \cdots \cup C_h$. If A_1 and A_2 are intervals that belong to different pairs $(\bar{x}_1, A_1)$, $(\bar{x}_2, A_2)$ of the set $\mathscr{P}$, either $(\bar{x}_1, A_1)$ belongs to an allotted partition $\mathscr{P}_j$ of one set C_j and $(\bar{x}_2, A_2)$ belongs to the allotted partition $\mathscr{P}_k$ of a different set C_k – in which case A_1 and A_2 are disjoint because they are contained in the respective disjoint sets C_j, C_k – or else $(\bar{x}_1, A_1)$ and $(\bar{x}_2, A_2)$ belong to the same allotted partition $\mathscr{P}_j$, and then they are disjoint because the intervals of the allotted partition $\mathscr{P}_j$ are pairwise disjoint. In any case, A_1 and A_2 are disjoint left-open intervals, and $\mathscr{P}$ is an allotted partition of $C_1 \cup \cdots \cup C_h$.

We shall now prove a statement stronger than the mere existence of γ-fine partitions. Later we shall have a need for the stronger conclusion.

★THEOREM 4-2 *Let γ be a gauge on $\bar{R}$ and let B be a left-open interval in R. Then there exists a γ-fine partition $\mathscr{P}$ of B such that for each pair $(\bar{x}, A)$ in $\mathscr{P}$, $\bar{x}$ is in the closure A^-.*

For each positive integer n we define $\mathscr{Q}_1[n]$ to be the set of intervals $\{Q(n,0), \ldots, Q(n, 2 \cdot 4^n + 1)\}$, where

$$\text{(A)} \qquad Q(n,0) = (-\infty, -2^n],$$

$$Q(n,j) = (-2^n + (j-1)2^{-n}, -2^n + j2^{-n}] \qquad (j = 1, \ldots, 2 \cdot 4^n),$$

$$Q(n, 2 \cdot 4^n + 1) = (2^n, \infty).$$

These are pairwise disjoint left-open intervals in R, and their union is R. Each interval in $\mathscr{Q}_1[n+1]$ is obtained by subdividing an interval that belongs to $\mathscr{Q}_1[n]$, so if A' is an interval that belongs to $\mathscr{Q}_1[n']$ and A'' is an interval that belongs to $\mathscr{Q}_1[n'']$, and $n'' \geqq n'$, then either A' and A'' are disjoint or $A'' \subset A'$.

For convenience in this proof, we define a *special partition* of an interval B' to be a γ-fine partition $\mathscr{P}$ such that for each pair $(\bar{x}, A)$ in $\mathscr{P}$, either A is empty or it is the intersection of B' with an interval that belongs to one of the sets $\mathscr{Q}_1[1]$, $\mathscr{Q}_1[2], \ldots$, and $\bar{x}$ is in the closure A^-. We shall prove a statement that is stronger than the conclusion of the theorem, namely,

there exists a special partition of B.

We shall prove this by assuming that B has no special partition and finding that this leads us to a contradiction. Suppose that B has no special partition. If each nonempty intersection $B \cap Q(1,j)$ $(j = 0, \ldots, 2 \cdot 4 + 1)$ had a special partition, by Lemma 4-1 the union of these special partitions would be an allotted partition $\mathscr{P}$ of B. If $(\bar{x}, A)$ is a pair that belongs to $\mathscr{P}$ with $A \neq \emptyset$, it belongs to a special partition of some interval in the set $B \cap Q(1,0), \ldots, B \cap Q(1, 2 \cdot 4 + 1)$, say, to $B \cap Q(1,j')$. Then $\bar{x}$ is in A^-, A^- is contained in $\gamma(\bar{x})$, and A is the intersection of $B \cap Q(1,j')$ with an interval $Q(n,j'')$ that belongs to one of the sets $\mathscr{Q}_1[n]$. But then this $Q(n,j'')$ must be contained in $Q(1,j')$, so

$$A = [B \cap Q(1,j')] \cap Q(n,j'') = B \cap Q(n,j'').$$

Now $\mathscr{P}$ satisfies all the requirements in the definition of special partition, so B has a special partition, contrary to assumption. So it must be true that at least one of the intersections $B \cap Q(1,0), \ldots, B \cap Q(1, 2 \cdot 4 + 1)$ has no special partition. We choose one such intersection and give it the name B_1.

If each nonempty intersection $B_1 \cap Q(2,j)$ $(j = 0, 1, \ldots, 2 \cdot 4^2 + 1)$ had a special partition, by the same argument as in the preceding paragraph B_1 would have a special partition, which is false. So there must be at least one intersection $B_1 \cap Q(2,j)$ that is nonempty and has no special partition. We choose one such intersection and name it B_2. Next we consider the intersections $B_2 \cap Q(3,j)$ $(j = 0, \ldots, 2 \cdot 4^3 + 1)$ and repeat the argument; and we continue this process indefinitely. We thus obtain a sequence of left-open intervals $B_0, B_1, B_2, B_3, \ldots$, where B_0 is merely another name for B, such that each B_{i+1} is the intersection of the preceding B_i with one of the intervals in $\mathscr{Q}_1[i]$, and each B_i fails to have any special partition.

Each closure B_i^- is a nonempty closed interval in $\bar{R}$, and if $j \geqq i \geqq 0$, $B_j^- \subset B_i^-$. By Theorem 2-2 in the Introduction, there is a point y of $\bar{R}$ that is contained in all the closures B_i^-. Let B_i^- be denoted by $[a_i, b_i]$. Then $a_i \leqq y \leqq b_i$ for all i. We distinguish three cases.

Case 1 $y = -\infty$. In this case $a_i = -\infty$ for all i, so the intersection $B_{i-1} \cap Q(i,j)$ that is named B_i must be $B_i \cap Q(i,0)$. But then B_i is contained in $[-\infty, -2^i]$, and since -2^i $(i = 1, 2, 3, \ldots)$ tends to $y = -\infty$, it must be in the neighborhood $\gamma(y)$ for all large i. Therefore, in case 1, $B_i^- \subset \gamma(y)$ for all large i.

Case 2 $y = \infty$. By essentially the same proof as in case 1 we see that $B_i^- \subset \gamma(y)$ for all large i.

Case 3 y is finite. Then for all large i, $-2^i < y < 2^i$, so the $Q(i,j)$ whose intersection with B_{i-1} is B_i cannot be either $Q(i,0)$ or $Q(i, 2 \cdot 4^j + 1)$; it has to be one of the intervals in between, all of which have length 2^{-i}. Therefore $b_i - a_i \leqq 2^{-i}$, and y (which is between them) cannot differ from a_i or from b_i by more than they differ from each other. So both a_i and b_i tend to y as i increases. Therefore, for all large i, a_i and b_i are both in $\gamma(y)$, and the interval B_i^- is contained in $\gamma(y)$.

Now we fix an i such that B_i^- is contained in $\gamma(y)$, as we have just shown to be possible in all cases, and we define $\mathscr{P}$ to consist of the single pair $\{(y, B_i)\}$. This is an allotted partition of B_i, and it is γ-fine, because B_i^- is contained in $\gamma(y)$. Since $B_i = B_{i-1} \cap Q(i,j)$ for some j, it is also true that $B_i = B_i \cap Q(i,j)$. So $\mathscr{P}$ is a special partition of B_i. But the construction B_i has no special partition. This contradiction proves the theorem.

Before we prove that the integral is unique, we shall explain a notational convention that we are going to use. It is an obvious simplification to write mA instead of $m_{\mathrm{L}}A$, and we shall do this, but only under circumstances that we now specify. The following pages contain many theorems in whose proofs no properties of m_{L} are used except the following very simple ones.

(i) For every left-open interval A, $m_{\mathrm{L}}A$ is defined and $\geqq 0$.

(ii) Whenever $A_1, \ldots, A_k$ are pairwise disjoint left-open intervals whose union is a left-open interval A,

$$m_{\mathrm{L}}A = m_{\mathrm{L}}A_1 + \cdots + m_{\mathrm{L}}A_k.$$

(See Corollary 1-3 in the Introduction.)

In such theorems, and only in such theorems, we shall use the abbreviation mA for $m_{\mathrm{L}}A$. These theorems will be indicated by a star (★), as in Theorem 4.2. (In contrast, we do not use m for m_{L} in, e.g., Lemma 3.2 because of its reference to Lemma 1-4 in the Introduction, which is based on the specific formula for $m_{\mathrm{L}}A$.)

This convention has the appearance of emphasizing the predilection, common among mathematicians, for deducing conclusions from the barest minimum of hypotheses. In reading Chapters I and II the reader, if he so chooses, may regard the convention in this way, but it has a quite practical use in Section III-7 and in later sections.

The integral is defined as a gauge-limit, and it will have many properties of the other kinds of limit that are familiar to us. As a guide to thinking, let us look at the traditional proof that if a sequence of numbers $a_1, a_2, a_3, \ldots$ has a (finite) limit A, and a sequence $b_1, b_2, b_3, \ldots$ has a (finite) limit B, then the sequence $a_1 + b_1$, $a_2 + b_2$, $a_3 + b_3, \ldots$ has limit $A + B$. Let ε be any positive number. Since a_i tends to A, there is an integer n' such that if $i > n'$, $|a_i - A| < \varepsilon/2$. Since b_i tends to B, there exists an integer n'' such that if $i > n''$, $|b_i - B| < \varepsilon/2$. *There exists an integer n such that if $i > n$, then $i > n'$ and $i > n''$*; for example, we can choose $n = n' \vee n''$. Now, for all i greater than n we have $|a_i - A| < \varepsilon/2$ because $i > n'$, and we have $|b_i - B| < \varepsilon/2$ because $i > n''$. So,

$$\begin{aligned} |(a_i + b_i) - (A + B)| &= |(a_i - A) + (b_i - B)| \\ &\leqq |a_i - A| + |b_i - B| \\ &< \varepsilon/2 + \varepsilon/2 = \varepsilon, \end{aligned}$$

and by definition $a_i + b_i$ has limit $A + B$.

In this proof the italicized words play a crucial role. If we were using partitions instead of integers and partition-sums in place of the a_i and b_i, the corresponding idea would be: *there is a gauge γ such that if $\mathscr{P}$ is γ-fine, it is γ'-fine and it is γ''-fine.* The next theorem and definition tell us how to construct such a γ, given γ' and γ''.

★THEOREM and DEFINITION 4-3 *Let $\gamma_1, \gamma_2, \ldots, \gamma_k$ be gauges on $\bar{R}$. Then the function $x \mapsto \gamma_1(x) \cap \gamma_2(x) \cap \cdots \cap \gamma_k(x)$ is a gauge on $\bar{R}$. We denote it by $\gamma_1 \cap \gamma_2 \cap \cdots \cap \gamma_k$. Every partition that is γ-fine is also γ_i-fine for $i = 1, \ldots, k$.*

If x is any point in $\bar{R}$, $\gamma_1(x), \gamma_2(x), \ldots, \gamma_k(x)$ are finitely many open intervals that contain x. Then their intersection is also an open interval that contains x. This proves that the function $x \mapsto \gamma_1(x) \cap \gamma_2(x) \cap \cdots \cap \gamma_k(x)$ is a gauge on $\bar{R}$. If $\mathscr{P} = \{(\bar{x}_1, A_1), \ldots, (\bar{x}_n, A_n)\}$ is a $\gamma_1 \cap \gamma_2 \cap \cdots \cap \gamma_k$-fine partition and i is any one of the numbers $1, \ldots, k$, for each pair $(\bar{x}_j, A_j)$ in $\mathscr{P}$ the closure A_j^- is contained in $\gamma_1(\bar{x}_j) \cap \cdots \cap \gamma_k(\bar{x}_j)$, which is contained in $\gamma_i(\bar{x}_j)$. So $\mathscr{P}$ is γ_i-fine.

It is now easy to prove that the integral is unique.

★THEOREM 4-4 *Let B be a subset of R, and let f be a real-valued function integrable over B. There cannot be two different numbers J', J'' such that*

$$\text{(B)} \qquad \int_B f(x)\, m_L(dx) = J' \qquad \text{and} \qquad \int_B f(x)\, m_L(dx) = J''.$$

Suppose that J' and J'' are two different numbers that satisfy both equations in (B). Define $\varepsilon = |J' - J''|/2$. This is positive, so by Definition 2-1 there is a gauge γ' on $\bar{R}$ such that if $\mathscr{P}$ is any γ'-fine partition of R,

$$\text{(C)} \qquad |S(\mathscr{P}; f_B) - J'| < \varepsilon.$$

Likewise there is a gauge γ'' on $\bar{R}$ such that if $\mathscr{P}$ is any γ''-fine partition of R,

$$\text{(D)} \qquad |S(\mathscr{P}; f_B) - J''| < \varepsilon.$$

By Theorem 4-2, there is a $\gamma' \cap \gamma''$-fine partition $\mathscr{P}$ of R. By Theorem 4-3, $\mathscr{P}$ is both γ'-fine and γ''-fine, so both (C) and (D) are valid. Then

$$\begin{aligned} |J' - J''| &= |[S(\mathscr{P}; f_B) - J''] + [J' - S(\mathscr{P}; f_B)]| \\ &\leqq |S(\mathscr{P}; f_B) - J''| + |S(\mathscr{P}; f_B) - J'| < 2\varepsilon. \end{aligned}$$

This contradicts the definition of ε, and the proof is complete.

The purpose of the next exercise is to convince the reader (if any doubt remains) that some seemingly obvious statements are really false. In probability theory there has been some use for allotted partitions such that for each pair $(\bar{x}_i, A_i)$ in the partition, $A_i \subset [\bar{x}_i, \infty]$.

EXERCISE 4-1 Define $\gamma(x) = [-\infty, 1)$ if $x < 1$ and $\gamma(x) = \bar{R}$ if $x \geqq 1$. Show that there does not exist any partition $\mathscr{P}$ of $(0, 1]$ such that for each pair $(\bar{x}_i, A_i)$ in

$\mathscr{P}$, $A_i \subset [\bar{x}_i, \infty]$. *Suggestion*: The point 1 would have to lie in an interval A_i with $\bar{x}_i < 1$. Then $(\bar{x}_i, A_i)$ could not be γ-fine.

EXERCISE 4-2 There is a very simple proof that if there exists a positive δ such that for every x in R, $\gamma(x) \supset (x - \delta, x + \delta)$ – or even if $\gamma(x) \supset [x, x + \delta)$ – there exists a γ-fine partition $\mathscr{P}$ for each left-open interval in R. Construct such a proof. Why is this result inadequate for our theory?

5. Elementary Computational Formulas

In the rest of this chapter we shall prove for the gauge-integral the analogs of the theorems proved (or at least asserted) in elementary calculus for the integral used there. Since our Definition 2-1 of the integral has a strong resemblance to the definition of the Riemann integral, it is not surprising that the proofs strongly resemble the familiar proofs of the corresponding theorems about the Riemann integral.

The basic computational property of the integral is its linearity, by which the integral of the sum of two integrable functions is the sum of their integrals and the integral of a constant multiple of an integrable function is the same multiple of its integral. To keep the notation from becoming complicated, we split the proof into two lemmas.

★LEMMA 5-1 *If f and g are integrable over a set B contained in R, their sum is integrable over B, and*

(A) $$\int_B [f(x) + g(x)]\, m(dx) = \int_B f(x)\, m(dx) + \int_B g(x)\, m(dx).$$

To simplify notation we define

(B) $$J_1 = \int_B f(x)\, m(dx), \qquad J_2 = \int_B g(x)\, m(dx).$$

Let ε be positive. By (B) there exists a gauge γ_1 on $\bar{R}$ such that for every γ_1-fine partition $\mathscr{P}$ of R

(C) $$|S(\mathscr{P}; f_B) - J_1| < \varepsilon/2.$$

Likewise there exists a gauge γ_2 on $\bar{R}$ such that for every γ_2-fine partition $\mathscr{P}$ of R

(D) $$|S(\mathscr{P}; g_B) - J_2| < \varepsilon/2.$$

Now let $\mathscr{P} = \{(\bar{x}_1, A_1), \ldots, (\bar{x}_k, A_k)\}$ be any $\gamma_1 \cap \gamma_2$-fine partition of R. By Theorem 4-2, $\mathscr{P}$ is both γ_1-fine and γ_2-fine, so both (C) and (D) are satisfied. But it is clear that $[f+g]_B = f_B + g_B$, so by rearranging terms,

$$\text{(E)}\quad S(\mathscr{P};[f+g]_B) = [f_B(\bar{x}_1) + g_B(\bar{x}_1)]mA_1 + \cdots + [f_B(\bar{x}_k) + g_B(\bar{x}_k)]mA_k$$
$$= [f_B(\bar{x}_1)mA_1 + \cdots + f_B(\bar{x}_k)mA_k)]$$
$$+ [g_B(\bar{x}_1)mA_1 + \cdots + g_B(\bar{x}_k)mA_k]$$
$$= S(\mathscr{P}; f_B) + S(\mathscr{P}; g_B).$$

From (E), (C), and (D) we obtain

$$|S(\mathscr{P};[f+g]_B) - [J_1 + J_2]| = |[S(\mathscr{P}; f_B) - J_1] + [S(\mathscr{P}; g_B) - J_2]|$$
$$\leqq |S(\mathscr{P}; f_B) - J_1| + |S(\mathscr{P}; g_B) - J_2)| < \varepsilon.$$

That is, the conditions of Definition 2-1 are satisfied if we take J to be the right member of (A) and γ to be $\gamma_1 \cap \gamma_2$, and the lemma is proved.

★LEMMA 5-2 *Let f be real-valued and integrable over a set $B \subset R$, and let c be a real number. Then cf is integrable over B, and*

$$\text{(F)}\qquad \int_B [cf(x)]\, m(dx) = c\int_B f(x)\, m(dx).$$

Let ε be positive. Then $\varepsilon/(|c| + 1)$ is positive, and by definition there exists a gauge γ such that for every γ-fine partition $\mathscr{P}$ of R,

$$\text{(G)}\qquad \left| S(\mathscr{P}; f_B) - \int_B f(x)\, m(dx)\right| < \frac{\varepsilon}{|c| + 1}.$$

If $\mathscr{P} = \{(\bar{x}_1, A_1), \ldots, (\bar{x}_k, A_k)\}$ is any γ-fine partition of R,

$$S(\mathscr{P}; (cf)_B) = cf_B(\bar{x}_1)mA_1 + \cdots + cf_B(\bar{x}_k)mA_k$$
$$= c[f_B(\bar{x}_1)mA_1 + \cdots + f_B(\bar{x}_k)mA_k] = cS(\mathscr{P}; f_B).$$

From this and (G),

$$\left| S(\mathscr{P}; (cf)_B) - c\int_B f(x)\, m(dx)\right| = \left| c\left\{ S(\mathscr{P}; f_B) - \int_B f(x)\, m(dx)\right\}\right|$$
$$= |c|\left| S(\mathscr{P}; f_B) - \int_B f(x)\, m_L(dx)\right|$$
$$\leqq |c|\frac{\varepsilon}{|c| + 1} < \varepsilon.$$

The conditions of Definition 2-1 are satisfied with cf in place of f and the right member of (F) in place of J, and the lemma is proved.

★THEOREM 5-3 *Let $f_1, \ldots, f_n$ be real-valued functions integrable over a set $B \subset R$, and let $c_1, \ldots, c_n$ be real numbers. Then $c_1 f_1 + \cdots + c_n f_n$ is integrable over B, and*

$$\text{(H)} \qquad \int_B [c_1 f_1(x) + \cdots + c_n f_n(x)]\, m(dx) = c_1 \int_B f_1(x)\, m(dx) + \cdots + c_n \int_B f_n(x)\, m(dx).$$

For $i = 1, \ldots, n$, define

$$\text{(I)} \qquad g_i = c_i f_i.$$

By Lemma 5-2, g_i is integrable over B, and

$$\text{(J)} \qquad \int_B g_i(x)\, m(dx) = c_i \int_B f_i(x)\, m(dx).$$

Then (H) can be written as

$$\text{(K)} \qquad \int_B [g_1(x) + \cdots + g_n(x)]\, m(dx) = \int_B g_1(x)\, m(dx) + \cdots + \int_B g_n(x)\, m(dx).$$

We prove this by induction. For $n = 1$, (K) holds. Assume that it holds for $n = h$. Then when there are $h + 1$ summands, we can write

$$g_1 + \cdots + g_{h+1} = [g_1 + \cdots + g_h] + g_{h+1}.$$

By the induction hypothesis, the function in brackets is integrable, and by hypothesis g_{h+1} is integrable, so by Lemma 5-1, the sum of the quantity in brackets and g_{h+1} is integrable; that is, $g_1 + \cdots + g_{h+1}$ is integrable. Moreover, by Lemma 5-1,

$$\begin{aligned}
\int_B [g_1(x) + \cdots + g_{h+1}(x)]\, m(dx) &= \int_B \{[g_1(x) + \cdots + g_h(x)] + g_{h+1}(x)\}\, m(dx) \\
&= \int_B [g_1(x) + \cdots + g_h(x)]\, m(dx) + \int_B g_{h+1}(x)\, m(dx) \\
&= \int_B g_1(x)\, m(dx) + \cdots + \int_B g_{h+1}(x)\, m(dx).
\end{aligned}$$

By induction, the conclusion holds for all positive integers n.

We shall often find it convenient to use step-functions in later proofs.

★DEFINITION 5-4 *Let B be a subinterval of R (it may be R itself). A function f is a* ***step-function*** *on B if it is defined and real-valued on B, and there exist finitely many pairwise disjoint subintervals $B_1, \ldots, B_h$ of B such that f is constant on each B_i and is 0 on $B \backslash [B_1 \cup \cdots \cup B_h]$.*

The next statement is a trivial calculation, which we dignify with the name of lemma just so we can refer to it.

★LEMMA 5-5 *Let f be a step-function on R that has the respective constant values $c_1, \ldots, c_h$ on the pairwise disjoint intervals $B_1, \ldots, B_h$ and is 0 outside the union of the B_j. Then for all x in R*

$$f(x) = c_1 1_{B_1}(x) + \cdots + c_h 1_{B_h}(x).$$

If x is not in any of the B_j, $f(x)$ is 0 by hypothesis and all terms in the right member are 0. If x is in the union of the B_j, it is in just one of them, say B_k. Then the left member has value c_k since f is constantly c_k on B_k; and in the right member all terms are 0 except the single term $c_k 1_{B_k}(x)$, which is equal to c_k.

THEOREM 5-6 *Let f be a step-function on $\bar{R}$ that has values $c_1, \ldots, c_h$ on the respective pairwise disjoint bounded intervals $B_1, \ldots, B_h$ and is 0 outside the union of the B_j. Then f is integrable over R, and*

$$\int_R f(x)\, m_L(dx) = c_1 m_L B_1 + \cdots + c_h m_L B_h.$$

We write f in the form shown in Lemma 5-5. Each indicator of an interval B_i is integrable, and its integral is $m_L B_i$, by Lemma 3-2. By Theorem 5-3, f is integrable, and its integral has the stated value.

Next we state a theorem that is in reality nothing more than an offer of a choice of notation.

★THEOREM 5-7 *Let f be defined on a subset B of R. Then f is integrable over B if and only if f_B is integrable over R, and in that case*

$$\int_R f_B(x)\, m(dx) = \int_B f(x)\, m(dx).$$

The equation

$$\text{(L)} \qquad [f_B]_R(x) = f_B(x)$$

holds for all x in R by definition of the left member; it holds for x in $\bar{R} \backslash R$ because for such x both members of (L) are 0. So (L) holds for all x in $\bar{R}$. Therefore if

$$\mathscr{P} = \{(\bar{x}_1, A_1), \ldots, (\bar{x}_k, A_k)\}$$

is any allotted partition of R,

$$\begin{aligned} S(\mathscr{P}; [f_B]_R) &= [f_B]_R(\bar{x}_1)mA_1 + \cdots + [f_B]_R(\bar{x}_k)mA_k \\ &= f_B(\bar{x}_1)mA_1 + \cdots + f_B(\bar{x}_k)mA_k = S(\mathscr{P}; f_B). \end{aligned}$$

It follows at once that if either of the sums $S(\mathscr{P}; [f_B]_R)$, $S(\mathscr{P}; f_B)$ has a gauge-limit, the other has the same gauge-limit, which is the conclusion of the theorem.

Theorem 5-6 has two useful corollaries.

COROLLARY 5-8 *If f is defined on a subset B of R and is 0 at all but finitely many points of B, it is integrable over B, and*

$$\int_B f(x)\, m_L(dx) = 0.$$

By Theorem 5-7, the conclusion of this corollary is equivalent to

$$\int_R f_B(x)\, m_L(dx) = 0.$$

Let $a_1, \ldots, a_h$ be the points of B at which $f(x) \neq 0$; these are also the points at which $f_B(x) \neq 0$. Each one-point set consisting of a_i alone is a closed interval $[a_i, a_i]$ of length 0. So f_B is a step-function on R, and by Theorem 5-6 its integral is

$$f_B(a_1)m_L[a_1, a_1] + \cdots + f_B(a_h)m_L[a_h, a_h] = 0.$$

COROLLARY 5-9 *If f and g are defined on a subset B of R, and f is integrable over B, and $g(x) = f(x)$ for all but finitely many points of B, then g is also integrable over B, and*

$$\int_B g(x)\, m_L(dx) = \int_B f(x)\, m_L(dx).$$

The difference $g - f$ is 0 except at finitely many points of B, so by Corollary 5-8, it is integrable over B and its integral is 0. By Theorem 5-3, the sum $f + [g - f]$, which is g, is integrable over B, and

$$\int_B g(x)\, m_L(dx) = \int_B f(x)\, m_L(dx) + 0.$$

6. Order Properties of the Integral

It is a simple but important fact that integration preserves order. That is, if f and g are two functions defined on and integrable over a set B and if $f \leqq g$—which means $f(x) \leqq g(x)$ for all x in B—then the integral of f is not greater than the integral of g. We shall now prove this and some related statements.

★THEOREM 6-1 *If f is nonnegative on and integrable over a set B, then*

$$\int_B f(x)\, m(dx) \geqq 0.$$

Suppose that this is false; the integral of f over B is a negative number $-\varepsilon$. There exists a gauge γ on $\bar{R}$ such that for every γ-fine partition $\mathscr{P}$ of R,

$$\text{(A)} \qquad \left| S(\mathscr{P}; f_B) - \int_B f(x)\, m(dx) \right| < \varepsilon.$$

By Theorem 4-2, there is a γ-fine partition

$$\mathscr{P} = \{(x_1, A_1), \ldots, (x_k, A_k)\}$$

of R. Since $f_B(x) = f(x) \geqq 0$ whenever x is in B, and $f_B(x) = 0$ when x is not in B,

$$S(\mathscr{P}; f_B) = f_B(x_1) mA_1 + \cdots + f_B(x_k) mA_k \geqq 0.$$

But since the integral of f is $-\varepsilon$, (A) now asserts that the sum of ε and a nonnegative number $S(\mathscr{P}; f_B)$ is less than ε, which is impossible. The theorem is proved.

★COROLLARY 6-2 *If f and g are both integrable over a set B, and $f(x) \leqq g(x)$ for all x in B, then*

$$\int_B f(x)\, m(dx) \leqq \int_B g(x)\, m(dx).$$

The difference $g - f$ is nonnegative on B. By Theorem 5-3 it is integrable over B, and by Theorem 6-1 its integral is nonnegative. Again by Theorem 5-3

$$\int_B g(x)\, m(dx) - \int_B f(x)\, m(dx) = \int_B [g(x) - f(x)]\, m(dx) \geqq 0.$$

The conclusion follows at once.

The next corollary is rather trivial, but it is still convenient.

★COROLLARY 6-3 *Let f be defined on a domain that contains a set A. Let f be nonnegative on A and integrable over A. Let B be a subset of A over which f is* *integrable. Then*

(B) $$\int_B f(x)\,m(dx) \leqq \int_A f(x)\,m(dx).$$

Note that this corollary merely formalizes the obvious conjecture that the area under the graph of f and over all of A is at least as great as the area under the graph and over part of A.

By Theorem 5-7, (B) is equivalent to

(C) $$\int_R f_B(x)\,m(dx) \leqq \int_R f_A(x)\,m(dx).$$

Consider the inequality

(D) $$f_B(x) \leqq f_A(x).$$

If x is not in A, it is not in B, and both members of (D) are 0. If x is in B, it is also in A, and both members of (D) are equal to $f(x)$. If x is in $A \backslash B$, the left member of (D) is 0 and the right member is $f(x)$, which is nonnegative by hypothesis. So (D) is valid for all x in $\bar{R}$. By Corollary 6-2, this implies that (C) is satisfied.

The next corollary is often useful in obtaining a crude but easy estimate of an integral in which the integrand is a product of two factors, one but not the other easily integrated.

★COROLLARY 6-4 (First Theorem of the Mean for Integrals) *Let f and φ be functions on a set B in R such that f and $f\varphi$ are integrable over B. Assume*

(i) *f is bounded; there are numbers m, M such that*

$$m \leqq f(x) \leqq M \qquad \textit{for all} \quad x \qquad \textit{in} \quad B,$$

(ii) *φ is nonnegative on B.*

Then

(E) $$m \int_B \varphi(x)\,m(dx) \leqq \int_B f(x)\varphi(x)\,m(dx) \leqq M \int_B \varphi(x)\,m(dx).$$

In particular, if B is a bounded closed interval and f is continuous on B, there exists a number x^* in B such that

(F) $$\int_B f(x)\varphi(x)\,m(dx) = f(x^*) \int_B \varphi(x)\,m(dx).$$

By hypotheses (i) and (ii),

$$m\varphi(x) \leqq f(x)\varphi(x) \leqq M\varphi(x)$$

for all x in B. By Theorem 6-1, inequality (E) holds. If f is continuous on a bounded closed interval B, we choose the infimum of f on B for m and the supremum of f on B for M. If the integral of φ over B is 0, by (E) so is the integral of $f\varphi$, so (F) holds no matter what point of B we choose for x^*. Otherwise the integral of φ is positive. By (E), the ratio

$$\text{(G)} \qquad \left[\int_B f(x)\varphi(x)\, m(dx)\right] \Big/ \left[\int_B \varphi(x)\, m(dx)\right]$$

is a number in the interval $[m, M]$. These are the least and greatest values of f on B, so by the intermediate value property of continuous functions there is a number x^* in B such that

$$\left[\int_B f(x)\varphi(x)\, m(dx)\right] \Big/ \left[\int_B \varphi(x)\, m(dx)\right] = f(x^*).$$

This implies that (F) is satisfied

EXERCISE 6-1 If f is continuous on a bounded closed interval $[a, b]$ [and is integrable from a to b], there exists a number x^* in $[a, b]$ such that

$$\int_a^b f(x)\, m_L(dx) = f(x^*)(b - a).$$

The words in brackets are superfluous, since every function continuous on $[a, b]$ is integrable over $[a, b]$. But since we have not yet proved this, we insert the superfluous words. To prove the statement of the exercise, we use Corollary 6-4.

EXERCISE 6-2 Let f be nondecreasing and integrable on $[1, 5]$. For each positive integer n, divide $[1, 5]$ into n equal parts; define S to be the step-function that on each of these subintervals is constantly equal to the supremum of f on the closure of that subinterval; and define s to be the step-function that on each subinterval is constantly equal to the infimum of f on that closure. Show that

$$\int_1^5 s(x)\, m_L(dx) \leqq \int_1^5 f(x)\, m_L(dx) \leqq \int_1^5 S(x)\, m_L(dx)$$

and that

$$\int_1^5 S(x)\, m_L(dx) - \int_1^5 s(x)\, m_L(dx) = 4\frac{f(5) - f(1)}{n}.$$

EXERCISE 6-3 As in Exercise 6-2 with $n = 8$, find lower and upper bounds for

$$\int_1^5 x^{-1}\, m_L(dx).$$

EXERCISE 6-4 As in Exercise 6-3, with $n = 8$, find lower and upper bounds for

$$\int_c^{5c} x^{-1}\, m_L(dx),$$

where c is a positive number.

EXERCISE 6-5 It has been known for about 2200 years that the area under the graph of the function $f: x \mapsto 1 - x^2$ and above $B = [-1, 1]$ is $\frac{4}{3}$. By dividing B into eight parts $A_1, \ldots, A_8$ of length $\frac{1}{4}$, defining S and s as in Exercise 6-2, and assuming that f is integrable over B, prove by the theory in this section that

$$1.06125 = \int_{-1}^{1} s(x)\, m_L(dx) \leqq \int_{-1}^{1} f(x)\, m_L(dx) \leqq \int_{-1}^{1} S(x)\, m_L(dx) = 1.56125.$$

EXERCISE 6-6 With the help of tables of natural logarithms and trigonometric functions of angles in radians, show that

$$0.9888 \int_0^{\pi/2} \sin x\, m_L(dx) \leqq \int_0^{\pi/2} (\sin x)\left[\cos \log\left(1 + \frac{x}{10}\right)\right] m_L(dx)$$
$$\leqq \int_0^{\pi/2} \sin x\, m_L(dx).$$

(We shall soon see that

$$\int_0^{\pi/2} \sin x\, m_L(dx) = 1.)$$

EXERCISE 6-7 Let x be time and $f(x)$ be the coordinate at time x of a particle moving along R^1. Suppose that the actual velocity of the particle at time x is

$$Df(x) = \begin{cases} 2.0001 & (0 \leqq x \leqq 4), \\ -2 & (4 < x \leqq 8), \end{cases}$$

but that because of experimental error in the velocity it is measured as

$$Df(x) = \begin{cases} 2 & (0 \leqq x \leqq 4), \\ -2 & (4 < x \leqq 8). \end{cases}$$

What is the resulting percentage of error in the estimate of the net distance traveled, $f(8)$? What is the percentage of error in the estimate of the total distance traveled, $|f(4) - f(0)| + |f(8) - f(4)|$?

7. Comparison with the Riemann Integral

Any reader of this book has presumably already had some acquaintance with calculus and has encountered something that in elementary treatments is usually called "the integral," without qualifying adjective. This is the integral defined in 1854 by B. Riemann. Different texts state the definition in other ways. An often-used version, published independently in 1875 by Giulio Ascoli in Italy, Gaston Darboux in France, H. J. S. Smith in England, and Karl J. Thomae in Germany, is the following.

First, if s is a step-function that has the constant values $c_1, \ldots, c_k$ on pairwise disjoint bounded intervals $A_1, \ldots, A_k$, we *define* its integral to be

$$\text{(A)} \qquad \int_R s(x)\,dx = c_1 m_L A_1 + \cdots + c_k m_L A_k.$$

This is the same number as our integral

$$\text{(B)} \qquad \int_R s(x)\,m_L(dx),$$

but now it is a definition, whereas in our theory the integral of s is defined by Definition 2-1 and the value of the integral is proved equal to this sum in Theorem 5-6. We preserve the distinction in concept (which involves no distinction in numerical value) by the difference in the symbol for the integral. Since the integral in (A) has the same value as the integral (B), we know by Corollary 6-2 that if $s_1 \leqq s_2$, then

$$\text{(C)} \qquad \int_R s_1(x)\,dx \leqq \int_R s_2(x)\,dx.$$

But we do not really have to appeal to Corollary 6-2, because (C) is easy to prove directly from the definition (A). Now let f be a function defined and bounded on a bounded closed interval $B = [a, b]$. If there is just one number J such that whenever s_1 and s_2 are step-functions on B such that $s_1 \leqq f \leqq s_2$, the inequalities

$$\int_a^b s_1(x)\,dx \leqq J \leqq \int_a^b s_2(x)\,dx$$

are satisfied, that number J is called the **Riemann integral** of f over B.

We shall now prove that if f is Riemann integrable from a to b, it is gauge-integrable over $[a, b]$, and the two integrals are equal. When this is proved, we will know that all the computations that we carried through for the Riemann integral are correct for the gauge-integrals of the same (Riemann-integrable) functions. We do not have to abandon anything about integrals that we learned earlier; the new integral takes in more territory, but it never stands in contradiction to the old (Riemann) integral. In preparation, we establish a lemma slightly more general than is needed for the proof; it has other uses too.

★LEMMA 7-1 *Let f be defined and real-valued on a subset D of R and let B be contained in D. If there exists a number J such that for every positive number ε there* *exist functions g, h that are gauge-integrable over B such that $g \leqq f \leqq h$ and*

$$\int_B g(x)\,m(dx) > J - \varepsilon, \qquad \int_B h(x)\,m(dx) < J + \varepsilon,$$

then f is gauge-integrable over B, and

$$\int_B f(x)\,m(dx) = J.$$

Let ε be positive. We can and do choose functions g, h integrable over B, satisfying $g \leqq f \leqq h$ on B, and

$$\text{(D)} \qquad \int_B g(x)\,m(dx) > J - \frac{\varepsilon}{2}, \qquad \int_B h(x)\,m(dx) < J + \frac{\varepsilon}{2}.$$

Since g is integrable over B, there exists a gauge γ_1 on $\bar{R}$ such that for every γ_1-fine partition $\mathscr{P}$ of R,

$$\text{(E)} \qquad \left| S(\mathscr{P}; g_B) - \int_B g(x)\,m(dx) \right| < \frac{\varepsilon}{2}.$$

Likewise there exists a gauge γ_2 on $\bar{R}$ such that if $\mathscr{P}$ is any γ_2-fine partition of R,

$$\text{(F)} \qquad \left| S(\mathscr{P}; h_B) - \int_B h(x)\,m(dx) \right| < \frac{\varepsilon}{2}.$$

Now define $\gamma = \gamma_1 \cap \gamma_2$ and let

$$\mathscr{P} = \{(x_1, A_1), \ldots, (x_k, A_k)\}$$

be any γ-fine partition of R. Since $\mathscr{P}$ is both γ_1-fine and γ_2-fine, both (E) and (F) are satisfied. For $j = 1, \ldots, k$ the inequalities

$$\text{(G)} \qquad g_B(x_j) \leqq f_B(x_j) \leqq h_B(x_j)$$

are satisfied; for if x_j is in B, (G) is the same as $g(x) \leqq f(x) \leqq h(x)$, which is true by hypothesis, and if x is not in B, all three numbers in (G) are 0. We multiply each member of (G) by the nonnegative number mA_j and sum over $j = 1, \ldots, k$. The result is

$$\text{(H)} \qquad S(\mathscr{P}; g_B) \leqq S(\mathscr{P}; f_B) \leqq S(\mathscr{P}; h_B).$$

But since (D), (E), and (F) hold, we have

$$J - \varepsilon < \int_B g(x)\,dx - \frac{\varepsilon}{2} < S(\mathscr{P}; g_B)$$

and

$$J + \varepsilon > \int_B h(x)\,m(dx) + \frac{\varepsilon}{2} > S(\mathscr{P}; h_B).$$

These inequalities, with (H), show that

$$J - \varepsilon < S(\mathscr{P}; f_B) < J + \varepsilon;$$

and since this holds for every γ-fine partition $\mathscr{P}$ of R, by definition J is the integral of f over B.

THEOREM 7-2 *Let $B = [a, b]$ be a bounded closed interval, and let f be Riemann integrable from a to b. Then f is gauge integrable from a to b, and the two integrals are equal.*

Let J be the Riemann integral of f from a to b, and let ε be positive. There exists a step-function g on B such that $g \leqq f$ and

$$\int_a^b g(x)\,dx > J - \varepsilon.$$

For otherwise, for every pair of step-functions g, h on B such that $g \leqq f \leqq h$, we would have

$$\int_a^b g(x)\,dx \leqq J - \varepsilon < J \leqq \int_a^b h(x)\,dx,$$

and J would not be the only number that separates the integrals of all step-functions $g \leqq f$ from the integrals of all step-functions $h \geqq f$. But this would contradict the hypothesis that J is the Riemann integral of f. Likewise there exists a step-function h such that $h \geqq f$ and

$$\int_a^b h(x)\,dx < J + \varepsilon.$$

Since the step-functions g and h are gauge integrable by Theorem 5-6, the hypotheses of Lemma 7-1 are satisfied. So the gauge-integral of f over B exists and has the value J, which is the value of the Riemann integral of f from a to b.

Lemma 7-1 has a simple but useful corollary.

★COROLLARY 7-3 *Let f be real-valued on a set B contained in R. Let there exist functions g, h gauge integrable over B such that $g \leqq f \leqq h$ on B and*

$$\int_B g(x)\,m(dx) = \int_B h(x)\,m(dx).$$

Then f is gauge integrable over B, and its integral is equal to the integrals of g and of h.

Define J to be the integral of g over B. For every positive ε the functions g and h satisfy the specifications in the hypotheses of Lemma 7-1, so by that lemma, f is integrable and its integral is J.

We have been simplifying notation by writing mA in place of $m_L A$, but only in those theorems, etc., that will remain valid even after replacement of the length-function $m_L(a, b] = b - a$ by certain other functions, as will happen in the next chapter. Now we can use a simplified notation even for those examples in which we are required to use length-measure. We shall define

$$\text{(I)} \qquad \int_B f(x)\,dx = \int_B f(x)\,m_L(dx)$$

whenever the integral in the right member exists, even though it is not a Riemann integral. Up to now we have reserved the symbol in the left member of (I) for Riemann integrals. But by Theorem 7-2, whenever the Riemann integral of f over an interval B exists, the integral in the right member of (I) also exists and has the same value. So if we use the abbreviation (I), no confusion should result.

EXERCISE 7-1 Show that if f is real-valued and nondecreasing on a bounded interval $[a, b]$, it is Riemann integrable (and therefore gauge integrable) over $[a, b]$. *Suggestion*: Use the construction in Exercise 6-2, subdividing $[a, b]$ into 2^n parts of equal length. The corresponding intervals

$$\left[\int_a^b s(x)\,m_L(dx),\ \int_a^b S(x)\,m_L(dx)\right]$$

have lengths that approach 0, and there is a point J in all of them by Theorem 2-2 in the Introduction.

EXERCISE 7-2 Show that if k and c are positive, the integrals

$$\int_1^k x^{-1}\,m_L(dx), \qquad \int_c^{kc} x^{-1}\,m_L(dx)$$

are equal. *Suggestion*: See Exercises 6-3 and 6-4.

EXERCISE 7-3 Let $f(x) = x$ on $[a, b]$. Subdivide $[a, b]$ into n parts of equal length by points $x_0 = a, x_1, \ldots, x_n = b$, and define s and S as in Exercise 6-2. Show that

$$\int_a^b s(x)\,dx = x_0(x_1 - x_0) + x_1(x_2 - x_1) + \cdots + x_{n-1}(x_n - x_{n-1}),$$

$$\int_a^b S(x)\,dx = x_1(x_1 - x_0) + x_2(x_2 - x_1) + \cdots + x_n(x_n - x_{n-1}).$$

As n increases, by Exercise 7-1 these both tend to the integral of x over $[a, b]$. So, therefore, does their average. Show that this average is $(b^2 - a^2)/2$ for all n, whence

$$\int_a^b x\,dx = \frac{b^2 - a^2}{2}.$$

EXERCISE 7-4 Let $f(x) = x^2$ on $[0, b]$, $b > 0$. Subdivide $[0, b]$ into n parts of equal length by points $x_0 = 0, x_1, \ldots, x_n = b$, and define s and S as in Exercise 6-2. Show that

$$\int_0^b s(x)\,dx = x_0^2(x_1 - x_0) + x_1^2(x_2 - x_1) + \cdots + x_{n-1}^2(x_n - x_{n-1}),$$

$$\int_0^b S(x)\,dx = x_1^2(x_1 - x_0) + x_2^2(x_2 - x_1) + \cdots + x_n^2(x_n - x_{n-1}).$$

The right members tend to the integral of x^2 over $[0, b]$, and so therefore does the number

$$x_0 x_1(x_1 - x_0) + x_1 x_2(x_2 - x_1) + \cdots + x_{n-1}x_n(x_n - x_{n-1}),$$

which is between them. The average of the three must also tend to the integral. Deduce from this that

$$\int_0^b x^2\,dx = \frac{b^3}{3}.$$

EXERCISE 7-5 Extend the method used in the preceding exercise to deduce that

$$\int_0^b x^n\,dx = \frac{b^{n+1}}{n+1} \qquad (n = 1, 2, 3, \ldots).$$

8. Additivity of the Integral

If an interval A is subdivided into finitely many pairwise disjoint subintervals, the length of A is the sum of the lengths of the subintervals. We shall now show that a similar statement is true for the integrals of a function f over intervals; the integral over A is the sum of the integrals over the subintervals. But it is just as easy to prove an even stronger result, which we now state.

★THEOREM 8-1 *Let $E_1, \ldots, E_h$ be pairwise disjoint sets in R, and let E be their union. If f is a function defined on E and integrable over each E_j, f is integrable over*

E, and

$$\int_E f(x)\,m(dx) = \sum_{j=1}^{h} \int_{E_j} f(x)\,m(dx).$$

The equation

$$\text{(A)} \qquad f_E(x) = f_{E_1}(x) + \cdots + f_{E_h}(x)$$

holds for all x, the terms being defined as in Definition 2-2. For if x is not in E, it is not in any E_j, and all terms in both members of A are 0. If x is in E, there is just one j for which x is in E_j. Then the left member of (A) is $f(x)$. In the right member, the jth term is $f(x)$; all the others are 0.

By hypothesis, each f_{E_j} is integrable over R. By (A) and Theorems 5-7 and 5-3, f_E is integrable, and the equation in the conclusion of the theorem is valid.

It is customary to extend the meaning of the integral "from a to b" in the following way.

DEFINITION 8-2 *Let f be defined and integrable over an interval $[a, b]$, where $a \leqq b$. Then we define*

$$\int_a^a f(x)\,dx = 0, \qquad \int_b^a f(x)\,dx = -\int_a^b f(x)\,dx.$$

In the next theorem there is an unnecessarily strong hypothesis. We shall assume that f is integrable over every subinterval of an interval $[a, b]$. In the next chapter we shall show that all we need is to assume that f is integrable over $[a, b]$ itself; then it will automatically be integrable over every subinterval of $[a, b]$. However, we have not yet proved this, so we shall include the extra hypothesis. But we shall enclose it in brackets as a reminder that it will soon be proved superfluous.

THEOREM 8-3 *Let f be integrable over [every subinterval of] an interval $[a, b]$. Let u, v, w be points of $[a, b]$, not necessarily distinct and not necessarily in any particular order. Then*

$$\text{(B)} \qquad \int_u^w f(x)\,dx = \int_u^v f(x)\,dx + \int_v^w f(x)\,dx.$$

If $u = v$, the first term on the right is 0, and the other term is the same as the left member, so (B) holds. Similarly (B) holds if $v = w$. If $u = w$, the left member is 0, and (B) holds by Definition 8-2. There remains the principal case, in which u, v, and w are all different.

If $u < v < w$, the three integrals in (B) are the integrals of f over the three sets

$$E = (u, w], \qquad E_1 = (u, v], \qquad E_2 = (v, w].$$

The sets E_1 and E_2 are disjoint; f is integrable over each of them; and E is their union, so (B) holds by Theorem 8-1.

Let p, q, r be the three numbers u, v, w rearranged in increasing order. We have just seen that

$$\int_p^q f(x)\,dx + \int_q^r f(x)\,dx = \int_p^r f(x)\,dx,$$

which, by transposing one term and using Definition 8-2, can be rewritten in the form

$$\text{(C)} \qquad \int_p^q f(x)\,dx + \int_q^r f(x)\,dx + \int_r^p f(x)\,dx = 0.$$

By a similar rearrangement, the equation (A) to be proved takes the form

$$\text{(D)} \qquad \int_u^v f(x)\,dx + \int_v^w f(x)\,dx + \int_w^u f(x)\,dx = 0.$$

Six cases are possible. In the first three, the three numbers u, v, w are respectively equal to p, q, r, or to q, r, p, or to r, p, q. In all three cases, the three integrals in (D) are the same as the three integrals in (C). This is known to be true; so (D) is true. In the remaining three cases, the three numbers u, v, w are, respectively, equal to p, r, q, or to r, q, p, or to q, p, r. In all three cases, each integral in (D) is the negative of one of the three integrals in (C), so again (D) is true. The proof is complete.

EXERCISE 8-1 Show that

$$\int_0^y 1\,dx = y$$

for all y in R.

EXERCISE 8-2 Assuming that if $a \leqq b$, then

$$\int_a^b x^n\,dx = \frac{b^{n+1} - a^{n+1}}{n+1},$$

show that

$$\int_0^y x^n\,dx = \frac{y^{n+1}}{n+1}$$

for all y in R.

9. The Fundamental Theorem of the Calculus, First Part

Ideas closely related to integration and differentiation go back far in history, even to the ancient Greeks. But the beginning of the modern flowering of calculus and of mathematical analysis was the realization by Newton and Leibniz that integration and differentiation are inverse operations. Differentiation and integration thus became parts of one unified theory. The inverse relationship can be thought of as having two parts. First, under suitable hypotheses on f, if we define F on an interval $[a, b]$ by setting

$$F(x) = F(a) + \int_a^x f(u)\,du,$$

then the derivative of F is f. Second, if f is the derivative of a function F, then

$$\int_a^b f(x)\,dx = F(b) - F(a).$$

This second part enables us to solve a multitude of problems of the form "find the integral of f from a to b." Because the combination of these two parts was of great importance in the development of calculus, the two together are called the "fundamental theorem of the calculus." The theorem itself has several different forms, with varying degrees of strength of assumptions about f and F. In this section we shall prove a version of the first part in which the hypotheses are quite strong (and the proof correspondingly easy) but which is still of sufficient power to cover the usual applications in ordinary calculus. In it we have a superfluous hypothesis: the function f will be assumed to be integrable over every subinterval of an interval B. Later we shall show that all we need to assume is that f is integrable over B; it will then automatically be integrable over every subinterval of B. But since we have not yet proved this, we include the unnecessary words, enclosing them in brackets as a reminder that later they will be proved superfluous.

Suppose, then, that f is integrable over [every subinterval of] an interval $[a, b]$. If c is any point of $[a, b]$, the integral of f from c to u exists for every u in $[a, b]$. Its value is determined by the upper limit u, so it is a function of u. We give the name **indefinite integral of** f to this function and also to every function that differs from it only by a constant. This statement we formalize in a definition.

DEFINITION 9-1 *Let f be integrable over [every subinterval of] the interval $[a, b]$, and let F be defined on $[a, b]$. If any one of these three statements is true, all are true:*

(i) *There is a number c in* $[a, b]$ *and there is a constant k such that for all x in* $[a, b]$

$$F(x) = k + \int_c^x f(u)\, du.$$

(ii) *For every number c in* $[a, b]$, *the function*

$$x \mapsto F(x) - \int_c^x f(u)\, du$$

is constant on $[a, b]$.

(iii) *For every* x' *and* x'' *in* $[a, b]$,

$$F(x'') - F(x') = \int_{x'}^{x''} f(u)\, du.$$

In that case, we say that F is an ***indefinite integral of*** f *on* $[a, b]$.

This is easy. If the function in (ii) has a constant value k, the equation in (i) is satisfied. If (i) holds and x' and x'' are in $[a, b]$, then by (i) and Theorem 8-3,

$$F(x'') - F(x') = \left(k + \int_c^{x''} f(u)\, du\right) - \left(k + \int_c^{x'} f(u)\, du\right)$$
$$= \int_{x'}^{x''} f(u)\, du,$$

so (iii) is satisfied. If (iii) is satisfied, we choose any c in $[a, b]$ and write the equation in (iii) with c in place of x' and x in place of x''. Since $F(c)$ is a constant, (i) holds with $k = F(c)$.

We shall now prove the simple form of the first part of the fundamental theorem that states that if f is continuous and F is an indefinite integral of f, then f is the derivative of F.

THEOREM 9-2 *Let f be integrable over [every subinterval of] an interval* $[a, b]$, *and let F be an indefinite integral of f. Then at each point x of* $[a, b]$ *at which f is continuous, F has a derivative, and* $DF(x) = f(x)$. *In particular, if f is continuous on* $[a, b]$, *F is differentiable at every point of* $[a, b]$, *and its derivative is identically equal to f.*

We split the proof into two parts, each of which has other uses.

(A) Under the hypotheses of the theorem, if f is continuous at a point $\bar{x}$ of $[a, b]$, to each positive ε there corresponds a neighborhood $\gamma(\bar{x})$ such that if x' and x'' $(> x')$ are two points of $[a, b]$ both in $\gamma(\bar{x})$,

(B) $$\left| \frac{F(x'') - F(x')}{x'' - x'} - f(\bar{x}) \right| < \varepsilon.$$

Since f is continuous at $\bar{x}$, there is a neighborhood $\gamma(\bar{x})$ such that for each point x of $[a,b]$ in $\gamma(\bar{x})$

(C) $$f(\bar{x}) - \varepsilon/2 < f(x) < f(\bar{x}) + \varepsilon/2.$$

Let x' and x'' be two points of $[a,b]$ in $\gamma(\bar{x})$, with $x'' > x'$. Then (C) holds for all x in $[x', x'']$, so by Corollary 6-2

$$\int_{x'}^{x''}\left[f(\bar{x}) - \frac{\varepsilon}{2}\right]dx \leqq \int_{x'}^{x''} f(x)\,dx \leqq \int_{x'}^{x''}\left[f(\bar{x}) + \frac{\varepsilon}{2}\right]dx.$$

By Theorem 5-3 and Definition 9-1, this implies

$$[f(\bar{x}) - \varepsilon/2](x'' - x') \leqq F(x'') - F(x') \leqq [f(\bar{x}) + \varepsilon/2](x'' - x').$$

We divide both members by the positive number $x'' - x'$, and (B) follows.

(D) Suppose that F is a function on $[a,b]$, $\bar{x}$ is a point of $[a,b]$, and l is a number, and to each positive ε there corresponds a neighborhood $\gamma(\bar{x})$ of $\bar{x}$ such that whenever x' and x'' are two different points of $[a,b]$ with $x' \leqq \bar{x} \leqq x''$, it is true that

(E) $$\left|\frac{F(x'') - F(x')}{x'' - x'} - l\right| < \varepsilon.$$

Then F has a derivative at $\bar{x}$, and $DF(\bar{x}) = l$.

With the ε and $\gamma(\bar{x})$ of statement (D), let x be any point of $[a,b]$ different from $\bar{x}$ and in $\gamma(\bar{x})$. If $x > \bar{x}$, we choose $x' = \bar{x}$, $x'' = x$, and the conditions on x' and x'' are satisfied. By hypothesis, (E) holds, so

(F) $$\left|\frac{F(x) - F(\bar{x})}{x - \bar{x}} - l\right| < \varepsilon.$$

If $x < \bar{x}$, we choose $x' = x$, $x'' = \bar{x}$ and by the same argument obtain (F) with x and $\bar{x}$ interchanged. But this interchange leaves the value of the quotient in (F) unaltered, so (F) holds for all $x \neq \bar{x}$ in $[a,b] \cap \gamma(\bar{x})$. By definition, $DF(\bar{x}) = l$.

Now if f is continuous at the point $\bar{x}$ of $[a,b]$, we first apply (A) and then apply (D) with $l = f(\bar{x})$. We find that $DF(\bar{x})$ exists and is equal to $f(\bar{x})$. The remaining conclusion of the theorem obviously follows from this, by applying it at each point of $[a,b]$ separately.

Beginners applying this theorem often make the mistake of forgetting the hypothesis that $\bar{x}$ is a point of continuity and using it even at discontinuities of f. This can lead to incorrect results. For example, let f be defined by

$$f(x) = -1 \quad (x < 0); \qquad f(x) = 1 \quad (x > 0); \qquad f(0) = 0.$$

This is called the **signum function** and is denoted by sgn. If $[a,b]$ is any interval containing 0 in its interior, we easily compute that for every number u in $[a, b]$

$$\int_0^u f(x)\,dx = |u|.$$

So the function $u \mapsto F(u) = |u|$ is an indefinite integral of $x \mapsto \operatorname{sgn} x$, but at 0 it does not have a derivative.

As a second example, let f be a function that is 0 at all but finitely many points of $[a, b]$. Then the indefinite integral F of f satisfies

$$F(u) = \int_a^u f(x)\,dx = 0$$

for all u in $[a, b]$. This has derivative equal to 0 at all points of $[a, b]$, and it is different from $f(x)$ at the points at which $f(x) \neq 0$.

If f is continuous, by Theorem 9-2, F has a property stronger than continuity; it is everywhere differentiable in $[a, b]$. We can easily show that if f is integrable and bounded, its indefinite integral F still has a property stronger than mere continuity, even if f is not continuous. This property we now define.

★DEFINITION 9-3 *Let F be a function defined and real-valued on a set E of real numbers. Then F is* ***Lipschitzian*** *(or* ***Lipschitz continuous****, or* ***satisfies a Lipschitz condition****) on E if there is a number L such that whenever x' and x'' are in E,*

$$|F(x'') - F(x')| \leqq L|x'' - x'|.$$

If F is Lipschitzian on E, it is continuous on E. For let ε be positive. For each real $\bar{x}$ let $\gamma(\bar{x})$ be the interval

$$(\bar{x} - \varepsilon/(L + 1),\ \bar{x} + \varepsilon/(L + 1)).$$

If $\bar{x}$ is in E and x is a point of E in $\gamma(\bar{x})$, then

$$|x - \bar{x}| < \varepsilon/(L + 1)$$

and

$$|F(x) - F(\bar{x})| \leqq L|x - \bar{x}| < \varepsilon.$$

The converse is false; a function continuous on E need not be Lipschitzian on E. For example, if $F(x) = x^2$ on $\mathscr{R}$, F is continuous. But given any number L, it fails to work in Definition 9-3; for if $x' = |L|$ and $x'' = |L| + 1$,

$$|F(x'') - F(x')| = (|L| + 1)^2 - L^2 = 2|L| + 1 > L|x'' - x'|.$$

As another example, if F is defined by

$$F(x) = x^{1/2} \qquad (0 \leqq x \leqq 3),$$

F is continuous. But for any positive L we can choose $x' = 0$, x'' a positive number less than 3 and less than L^{-2}. Then

$$|F(x'') - F(x')| = (x'')^{1/2} - 0 = [x'' - x']/(x'')^{1/2} > L|x'' - x'|.$$

We now show that any indefinite integral F of a bounded function f is Lipschitzian.

THEOREM 9-4 *Let f be a bounded function integrable over [every subinterval of] $[a,b]$, and let F be an indefinite integral of f. Then F is Lipschitzian on $[a,b]$.*

By hypothesis there exists a number C such that

$$-C \leqq f(x) \leqq C \qquad \text{for all} \quad x \quad \text{in} \quad [a,b].$$

Let x' and x'' be any two points of $[a,b]$. Without loss of generality we may assume that $x' < x''$. By Corollary 6-2

$$-C(x'' - x') \leqq \int_{x'}^{x''} f(x)\,dx \leqq C(x'' - x').$$

By Definition 9-1, this is the same as

$$-C(x'' - x') \leqq F(x'') - F(x') \leqq C(x'' - x'),$$

whence

$$|F(x'') - F(x')| \leqq C|x'' - x'|,$$

and the proof is complete.

EXERCISE 9-1 Prove the following two statements:

(i) If F_1 and F_2 are both indefinite integrals of the same function f on an interval $[a,b]$, the difference $F_2 - F_1$ is constant on $[a,b]$.
(ii) If F_1 and F_2 are both differentiable at every point of $[a,b]$ and their derivatives are everywhere equal, then $F_2 - F_1$ is constant on $[a,b]$.

Show that neither of these statements implies the other.

EXERCISE 9-2 Use Lemma 3-2 and the fundamental theorem 9-2 to calculate that if $f(x) = x$ for all x, then $Df = 1$. Similarly, use Exercises 7-3, 7-4, and 7-5 to prove that

$$Dx^2 = 2x, \qquad Dx^3 = 3x^2, \qquad Dx^n = (n-1)x^{n-1} \qquad (n = 2, 3, 4, \ldots).$$

(The point of this triviality is to move in the direction of deducing formulas of differential calculus from facts about integrals.)

EXERCISE 9-3 Let f be integrable over [every subinterval of] an interval $[a,b]$, and let c be in $[a,b]$. Prove that at every point $\bar{x}$ of $[a,b]$ at which f is continuous the derivative of the function

$$x \mapsto \int_x^c f(t)\,dt$$

is $-f(\bar{x})$.

EXERCISE 9-4 Let f be integrable over [every subinterval of] $[a, b]$ and let $x \mapsto u(x)$ be a function on an interval $[a^*, b^*]$ that takes values in $[a, b]$. Let $\bar{x}$ be a point of $[a^*, b^*]$ such that u has a derivative at $\bar{x}$ and f is continuous at $u(\bar{x})$. Prove that at $\bar{x}$

$$D \int_{a^*}^{u(x)} f(t)\,dt = f(u(\bar{x}))Du(\bar{x}).$$

Suggestion: Use the fundamental theorem and the chain rule for differentiation.

EXERCISE 9-5 Prove that if c and x are positive,

$$\int_c^{cx} u^{-1}\,du = \int_1^x u^{-1}\,du.$$

Suggestion: Calculate the derivatives of both members, using Exercise 9-4. Apply Exercise 9-1 (ii), and evaluate the constant by setting $x = 1$. The result of this exercise has been proved by other methods in Exercise 7-2.

10. The Fundamental Theorem of the Calculus, Second Part

We now turn to the second part of the fundamental theorem, already mentioned at the beginning of Section 9. It is convenient to introduce an expression that is probably already familiar to most readers.

DEFINITION 10-1 *Let f and F be real-valued functions defined on an interval B. The statement that F is a* ***primitive*** *or* ***antiderivative*** *of f on B means that for each x in B, $F'(x)$ exists and is equal to f.*

The second part of the fundamental theorem states that under suitable conditions, if F is a primitive of f on $B = [a, b]$, the integral of f over B is $F(b) - F(a)$. We prove two forms of this theorem. In the first we make the strong hypothesis that f is continuous. This makes the proof easy, and the theorem is still strong enough to allow us to compute the integrals usually encountered in applications.

THEOREM 10-2 *Let f be continuous on a bounded closed interval $[a, b]$ [and integrable over every subinterval of $[a, b]$], and let F be an antiderivative of f on $[a, b]$. Then*

$$\int_a^b f(x)\,dx = F(b) - F(a).$$

The hypothesis that f is integrable over every subinterval of $[a, b]$ is necessarily true because f is continuous on $[a, b]$. Since we have not yet proved this fact, we include the extra hypothesis, enclosing it in brackets to indicate that later it will be shown to be unnecessary.

Let us observe that every function g that has derivative 0 at each point of $[a, b]$ satisfies $g(b) = g(a)$. For by the theorem of the mean there is a point $\bar{x}$ in (a, b) such that

$$g(b) - g(a) = g'(\bar{x})(b - a),$$

and since $g'(\bar{x}) = 0$, the right member of this equation is 0.

If we define G on $[a, b]$ by

$$G(x) = \int_a^x f(u)\,du,$$

by Theorem 9-2, $G'(x) = f(x)$ for all x in $[a, b]$. By hypothesis, $F'(x) = f(x)$ for all x in $[a, b]$, so $G - F$ has derivative 0 at each point of $[a, b]$. By the preceding paragraph,

$$G(b) - F(b) = G(a) - F(a).$$

But $G(b)$ is the integral of f from a to b, and $G(a) = 0$, so this equation is the conclusion of the theorem.

Next we prove a stronger form of the second part of the fundamental theorem in which f does not have to be continuous or even bounded. Since the proof is necessarily a bit more difficult, beginners are advised to skip over this theorem.

THEOREM 10-3 *Let f and F be defined on a bounded closed interval $[a, b]$ in R, and let F be an antiderivative of f on $[a, b]$. If f is integrable over $(a, b]$, then*

(A) $$\int_a^b f(x)\,dx = F(b) - F(a).$$

Let ε be a positive number. For typographical simplicity we define

(B) $$B = (a, b], \qquad \varepsilon' = \frac{\varepsilon}{2b - 2a + 1}, \qquad J = \int_a^b f(x)\,dx.$$

By definition of DF, and since $DF = f$, for each $\bar{x}$ in $[a, b]$ there exists a neighborhood $\gamma_1(\bar{x})$ such that if x is a point of $\gamma_1(\bar{x}) \cap [a,b]$ different from $\bar{x}$,

(C) $$f(\bar{x}) - \varepsilon' < \frac{F(x) - F(\bar{x})}{x - \bar{x}} < f(\bar{x}) + \varepsilon'.$$

We can now show that

(D) if u and v are points of $\gamma_1(\bar{x}) \cap [a,b]$ such that $u \leqq \bar{x} \leqq v$, then

$$[f(\bar{x}) - \varepsilon'](v - u) \leqq F(v) - F(u) \leqq [f(\bar{x}) + \varepsilon'](v - u).$$

For if $u < \bar{x}$, (C) holds with u in place of x, and by multiplying all members by the positive number $\bar{x} - u$ we deduce

(E) $$[f(\bar{x}) - \varepsilon'](\bar{x} - u) \leqq F(\bar{x}) - F(u) \leqq [f(\bar{x}) + \varepsilon'](\bar{x} - u).$$

If $u = \bar{x}$, (E) is still obviously true. Likewise, if $v > \bar{x}$, (C) holds with v in place of x, and by multiplying all members by the positive number $v - \bar{x}$ we deduce

(F) $$[f(\bar{x}) - \varepsilon'](v - \bar{x}) \leqq F(v) - F(\bar{x}) \leqq [f(\bar{x}) + \varepsilon'](v - \bar{x}).$$

This too is evidently still valid if $v = \bar{x}$. So (E) and (F) hold in all cases, and by adding them member by member we obtain the conclusion of (D).

We extend the definition of γ_1 to all of $\bar{R}$ by setting $\gamma_1(\bar{x}) = \bar{R}$ for all $\bar{x}$ in $\bar{R} \setminus [a,b]$.

By (B), J is the integral of f from a to b, which is by definition the integral of f over B; and by Theorem 5-7, this is the same as the integral of f_B over R. So by definition of the integral, there exists a gauge γ_2 on $\bar{R}$ such that for every γ_2-fine partition $\mathscr{P}$ of R

(G) $$J - \varepsilon/2 < S(\mathscr{P}; f_B) < J + \varepsilon/2.$$

We now define γ to be $\gamma_1 \cap \gamma_2$.

By Theorem 4-2, there exists a γ-fine partition of $(a,b]$, denoted by $\{(\bar{x}_1, A_1), \ldots, (\bar{x}_k, A_k)\}$, such that for each i, $\bar{x}_i$ is in the closure A_i^-. Without loss of generality we may assume that the A_i are numbered in order from left to right. Then the end-points of the A_i are numbers $c_0 = a, c_1, \ldots, c_k = b$ in increasing order; $A_i = (c_{i-1}, c_i]$; and $c_{i-1} \leqq \bar{x}_i \leqq c_i$. Also, by Theorem 4-2, there exists a γ-fine partition of $(-\infty, a]$, which we denote by $\{(\bar{x}_{k+1}, A_{k+1}), \ldots, (\bar{x}_m, A_m)\}$, and there exists a γ-fine partition of (b, ∞), which we denote by $\{(\bar{x}_{m+1}, A_{m+1}), \ldots, (\bar{x}_n, A_n)\}$, such that each point $\bar{x}_j$ is contained in the closure of the corresponding A_j. We define $\mathscr{P}$ to be the set of all pairs $\{(\bar{x}_i, A_1), \ldots, (\bar{x}_n, A_n)\}$. By Lemma 4-1, this is a γ-fine partition of R. Then

(H) $$S(\mathscr{P}; f_B) = \sum_{i=1}^{n} f_B(\bar{x}_i) m_L A_i.$$

We shall prove the conclusion first under the extra hypothesis

(I) $$f(a) = f(b) = 0.$$

In this case, f_B vanishes identically on each interval A_i^- with $i = k + 1, \ldots, m$, since such intervals are contained in $[-\infty, a]$. Likewise, f_B vanishes identically on each closure $A_{m+1}^-, \ldots, A_n^-$. So all terms in the right member of (H) that have $i > k$ have value 0. For $i = 1, \ldots, k$, $\bar{x}_i$ is in A_i^-, which is contained in $[a, b]$. On $[a, b]$ we have $f(x) = f_B(x)$; for if x is in $[a, b]$, either it is in $(a, b]$, on which $f_B(x) = f(x)$ by definition, or else $x = a$, at which $f(a) = 0$ by (I) and $f_B(x) = 0$ by definition. Also,

$$m_L A_i = c_i - c_{i-1}.$$

So (H) implies

$$S(\mathscr{P}; f_B) = \sum_{i=1}^{k} f(\bar{x}_i)(c_i - c_{i-1}). \tag{J}$$

Since $\mathscr{P}$ is γ-fine, it is γ_1-fine, and by (D), for $i = 1, \ldots, k$

$$[f(\bar{x}_i) - \varepsilon'](c_i - c_{i-1}) \leqq F(c_i) - F(c_{i-1}) \leqq [f(\bar{x}_i) + \varepsilon'](c_i - c_{i-1}).$$

We add these inequalities member by member for $i = 1, \ldots, k$ and recall that $c_0 = a$ and $c_k = b$; the result is, by (J),

$$S(\mathscr{P}; f_B) - \varepsilon'(b - a) \leqq F(b) - F(a) \leqq S(\mathscr{P}; f_B) + \varepsilon'(b - a). \tag{K}$$

Since $\mathscr{P}$ is γ_2-fine, (G) holds. Since, by (B), $\varepsilon'(b - a) < \varepsilon/2$, (G) and (K) imply

$$J - \varepsilon < F(b) - F(a) < J + \varepsilon.$$

So the absolute difference between the two real numbers J and $F(b) - F(a)$ is less than the arbitrary positive number ε, and they are equal. The theorem is proved under the extra hypothesis (I).

If (I) is not satisfied, we define

$$c = [bf(a) - af(b)]/(b - a), \qquad m = [f(b) - f(a)]/(b - a),$$

and on R we define functions g and G by

$$g(x) = f(x) - c - mx, \qquad G(x) = F(x) - cx - mx^2/2.$$

Then $g(a) = g(b) = 0$, and for all real numbers x we have

$$DG(x) = DF(x) - c - mx = g(x).$$

So, by the part of the proof already completed,

$$\int_a^b g(x)\,dx = G(b) - G(a). \tag{L}$$

By Theorem 10-2, or else by Lemma 3-2 and Exercise 7-3,

$$\int_a^b (c + mx)\,dx = \left(cb + \frac{mb^2}{2}\right) - \left(ca - \frac{ma^2}{2}\right). \tag{M}$$

If we add (M) member by member to (L), we obtain (A), and the proof of the theorem is complete.

Theorem 10-3 has a simple but useful corollary.

Corollary 10-4 *Let f be defined on a bounded interval $[a, b]$ and be integrable over $(a, b]$. Let F be an antiderivative for f on $[a, b]$. Then for every u and v in $[a, b]$*

$$\int_u^v f(x)\,dx = F(v) - F(u).$$

By Theorem 10-3,

$$F(v) = F(a) + \int_a^v f(x)\,dx, \qquad F(u) = F(a) + \int_a^u f(x)\,dx.$$

If we substitute this in the conclusion of the corollary, it takes the form of the conclusion of Theorem 8-3.

Whenever a function F is defined on a set that includes points u and v, it is customary to use the notation

$$F(x)\Big|_u^v = F(v) - F(u).$$

With this notation, the conclusion of Corollary 10-4 takes the familiar form

$$\int_u^v f(x)\,dx = F(x)\Big|_u^v.$$

There is no point in listing examples in which Theorem 10-2 or Theorem 10-3 is applied. Every elementary textbook provides numerous examples and problems based on finding the antiderivative of a given function. Also, there are several published tables of integrals that are designed to allow us to find the antiderivatives of a great number of specific functions. The reader is undoubtedly well practiced in finding antiderivatives.

11. Substitution and Integration by Parts

The theorems on integration by substitution and integration by parts that we shall prove here are far from being the most general possible. However, the restricted cases that we consider are general enough to apply to many problems, and the proofs of the theorems are easy.

Theorem 11-1 *Let f be defined and continuous on an interval $[c, d]$, and let g be defined and continuously differentiable on an interval $[a, b]$ and take values that* *lie in* $[c, d]$. *Then*

$$\int_{g(a)}^{g(b)} f(x)\,dx = \int_a^b f(g(y))Dg(y)\,dy.$$

In the proof we assume, as will be shown in Corollary II-1-3, that if a function is continuous on an interval $[a, b]$, it is integrable from a to b. For each u in $[c, d]$ we define

$$F(u) = \int_{g(a)}^{u} f(x)\,dx.$$

Then by the fundamental theorem 9-2, we have $DF(u) = f(u)$ for all u in $[c, d]$, and so for all y in $[a, b]$

$$\frac{dF(g(y))}{dy} = DF(g(y))Dg(y) = f(g(y))Dg(y).$$

By Corollary 10-4

$$\int_a^b f(g(y))Dg(y)\,dy = \int_a^b \frac{dF(g(y))}{dy}\,dy$$

$$= F(g(b)) - F(g(a)) = \int_{g(a)}^{g(b)} f(x)\,dx - 0.$$

THEOREM 11-2 *Let* f, g, F, *and* G *be functions defined on an interval* $[a, b]$ *and such that for all* x *in* $[a, b]$,

$$DF(x) = f(x) \qquad \text{and} \qquad DG(x) = g(x).$$

Let the products fG *and* Fg *be integrable from* a *to* b. *Then*

$$\int_a^b f(x)G(x)\,dx = F(x)G(x)\Big|_a^b - \int_a^b F(x)g(x)\,dx.$$

For each x in $[a, b]$ we have

$$D[F(x)G(x)] = DF(x)G(x) + F(x)DG(x) = f(x)G(x) + F(x)g(x).$$

Both functions in the right member are integrable from a to b by hypothesis, so the left member is also integrable. By the fundamental theorem 10-2,

$$F(b)G(b) - F(a)G(a) = \int_a^b [f(x)G(x) + F(x)g(x)]\,dx,$$

from which the conclusion follows at once.

Theorem 11-2 contains as a special case the situation in which f and g are continuous on $[a, b]$. For then F and G are continuous, being differentiable, so the products fG and Fg are continuous. In Theorem II-9-4 we shall show that the conclusion of Theorem 11-2 still holds under much weaker hypotheses. All that we need to assume is that f and g are integrable over $[a, b]$ and that F and G are their indefinite integrals; no assumption is made about the derivatives of F and G.

We first use Theorem 11-1 to obtain easily a result already obtained with rather more effort in Exercises 7-2 and 9-5. Let c and u be positive, and for each x let $g(x) = cx$. By Theorem 11-1, with $f(x) = x^{-1}$, $a = 1$, and $b = u$,

$$\text{(A)} \qquad \int_c^{cu} x^{-1}\,dx = \int_1^u (cy)^{-1} \cdot c\,dy = \int_1^u y^{-1}\,dy.$$

As is customary, we define the (natural) logarithm of x to be

$$\text{(B)} \qquad \log x = \int_1^x t^{-1}\,dt \qquad (x > 0).$$

We can easily establish the fundamental properties of the logarithm.

THEOREM 11-3 *The function* $x \mapsto \log x$ $(x > 0)$ *defined by* (B) *has the properties*

(i) *if* $0 < x_1 < x_2$, *then* $\log x_1 < \log x_2$;
(ii) *if* x_1 *and* x_2 *are positive,*

$$\log(x_1 x_2) = \log x_1 + \log x_2;$$

(iii) $D \log x = 1/x$.

For (i),

$$\begin{aligned} \log x_2 - \log x_1 &= \int_1^{x_2} t^{-1}\,dt - \int_1^{x_1} t^{-1}\,dt \\ &= \int_{x_1}^{x_2} t^{-1}\,dt \geq \int_{x_1}^{x_2} x_2^{-1}\,dt = \frac{x_2 - x_1}{x_2} > 0. \end{aligned}$$

For (ii) we have by (A)

$$\begin{aligned} \log x_1 + \log x_2 &= \int_1^{x_1} t^{-1}\,dt + \int_1^{x_2} t^{-1}\,dt \\ &= \int_1^{x_1} t^{-1}\,dt + \int_{x_1}^{x_1 x_2} t^{-1}\,dt \\ &= \int_1^{x_1 x_2} t^{-1}\,dt = \log(x_1 x_2). \end{aligned}$$

Conclusion (iii) follows from (B) by the fundamental theorem.

Conclusion (ii) easily generalizes by induction to any finite number of positive factors. That is,

$$\log(x_1 x_2 \cdots x_n) = \log x_1 + \log x_2 + \cdots + \log x_n.$$

In particular,

$$\log 2^n = n \log 2,$$

and since $\log 2 > 0$, this has no finite upper bound. Since by conclusion (ii)

$$\log 2^n + \log(1/2^n) = \log 1 = 0,$$

we have $\log 2^{-n} = -n \log 2$, and therefore $\log x$ has no finite lower bound.

Theorems 11-1 and 11-2 are traditional workhorses in finding antiderivatives of functions. Elementary textbooks of calculus usually contain several pages of exercises in each of which some substitution or some integration by parts reduces an integral to a manageable form. We shall not give such exercises but shall state two exercises of a different type.

EXERCISE 11-1 If f is continuous and is integrable over R,

$$\int_R f(-x)\,dx = \int_R f(x)\,dx.$$

(Compare this with Exercise 2-4, in which a stronger result is established, but with more work.)

EXERCISE 11-2 Prove that for every nonnegative integer n,

$$\int_0^x t^n\,dt = \frac{x^{n+1}}{n+1}.$$

This is true for $n = 0$ by Lemma 3-2 and Definition 9-1. Proceed by induction, writing $f(t) = t^{n-1}$, $g(t) = t$.

12. Estimates of Integrals

When we wish to find the value of an integral

$$\int_a^b f(x)\,dx,$$

sometimes we can find an antiderivative of f, and then Theorem 10-3 lets us find the value of the integral. But we are not always so fortunate. The function f may be one that was determined by some experimental procedure, and then we have no analytic expression for f but merely a table of values. In this case we cannot hope to find an antiderivative. Or f may be given by some formula, but no known function (or at least no function known to us) can be found whose derivative is f. In this case the best we can do is to find approximations to the value of the integral by some computational procedure. There is no procedure that will work conveniently and accurately for all integrable functions, however complicated. Finding the integral of a highly discontinuous function (such as the indicator of the rational numbers) calls for ingenuity rather than for the routine application of a computational formula. But when f is of a fairly simple nature – for example, when f has several continuous derivatives – several good techniques for approximation are available.

Of course, the very definition of the integral gives us a way of approximating its value; to estimate the integral of f over an interval B we choose some sufficiently fine allotted partition $\mathscr{P}$ and compute $S(\mathscr{P}; f_B)$. But for serviceable approximating we must know ways of choosing the partitions that will yield close estimates, and we need to know an upper bound for the error that we make in using the approximation technique. We shall discuss three procedures, one simple but historically interesting, the other two more practical.

In the early nineteenth century mathematicians were vague about the concepts of limit, integral, etc. Augustin-Louis Cauchy did much to dispel this vagueness. In particular, his definition of the integral can be expressed in our terminology thus. If f is defined on an interval B, consider all those allotted partitions of B in which, in each pair (x, A), x is the left end-point of A. For each such partition form the partition-sum $S(\mathscr{P}; f_B)$. If this tends to a limit J as the length of the longest interval of the partition tends to 0, J is the integral (the "Cauchy integral") of f over B. It is obvious that our gauge-integral is a direct descendant of the Cauchy integral, which was widely used until in 1857 Riemann published a better one.

Accordingly, in the first of our approximation techniques we use the "Cauchy formula," denoted by Cf. If f is real-valued on an interval $B = [a, b]$, we first subdivide B into n subintervals each of length

(A) $$h = (b - a)/n$$

by points

(B) $$x_1 = a, \qquad x_2 = a + h, \ldots, \qquad x_{n+1} = b.$$

Then the "Cauchy partition" with these division points is

$$\mathscr{P} = \{(x_1, (x_1, x_2]), \ldots, (x_n, (x_n, x_{n+1}])\}.$$

We call the corresponding partition-sum the "Cauchy formula" for estimating the integral of f over B and denote it by $\mathrm{C}_B\mathrm{f}$.

(C) $$\mathrm{C}_B\mathrm{f} = [f(x_1) + \cdots + f(x_n)]h.$$

We have already applied this to monotonic functions, in Exercise 6-2. Suppose that f is nondecreasing and real-valued on B. Then by Lemma 3-2 or Theorem

5-6 and Corollary 6-2 we obtain for $i = 1, \ldots, n$

$$\text{(D)} \qquad f(x_i)h \leqq \int_{x_i}^{x_{i+1}} f(x)\,dx \leqq f(x_{i+1})h.$$

Adding for $i = 1, \ldots, n$ yields

$$\begin{aligned} \text{(E)} \qquad [f(x_1) + \cdots + f(x_n)]h &\leqq \int_a^b f(x)\,dx \\ &\leqq [f(x_2) + \cdots + f(x_{n+1})]h \\ &= [f(x_1) + \cdots + f(x_n)]h \\ &\quad + [f(x_{n+1}) - f(x_1)]h. \end{aligned}$$

This shows that the error in using the Cauchy formula as an approximation to the integral is not more than $[f(b) - f(a)]h$, which not only clearly goes to 0 as h decreases but also gives us a safe upper bound for our error. This can easily be extended to functions that have finitely many intervals of rising or of falling.

One obvious way to improve the estimate is to use intervals half as long. So, to the points x_i in (B) we adjoin other points

$$\text{(F)} \qquad y_i = (x_i + x_{i+1})/2.$$

The points x_i and y_i together subdivide B into subintervals of length $h/2$. If with each subinterval we associate its left end-point as evaluation-point, as in the Cauchy formula, then for nondecreasing f, the error is nonnegative. If we associate the right end-point as evaluation-point, by (E) the error is nonpositive. We can expect to reduce the error by using the left end as evaluation-point half the time and the right end the other half. So, we form an allotted partition $\mathscr{P}$ in which the intervals are

$$\text{(G)} \qquad (x_1, y_1], (y_1, x_2], (x_2, y_2], (y_2, x_3], \ldots, (y_n, x_{n+1}]$$

and the corresponding evaluation-points are

$$\text{(H)} \qquad x_1, x_2, x_2, x_3, x_3, \ldots, x_{n-1}, x_{n-1}, x_n, x_n, x_{n+1}.$$

The partition-sum corresponding to this allotted partition we denote by $\mathrm{T}_B\mathrm{f}$; then, since each of the subintervals has length $h/2$,

$$\text{(I)} \qquad \mathrm{T}_B\mathrm{f} = [f(x_1) + 2f(x_2) + 2f(x_3) + \cdots + 2f(x_n) + f(x_{n+1})](h/2).$$

This is called the "trapezoidal formula," Tf, for a reason that will soon become clear.

By defining the intervals of the partition by (G), but with a different choice of the y_i, we can obtain an approximation formula that for smooth f is much more accurate than the trapezoidal formula. We require that n be even. Instead of

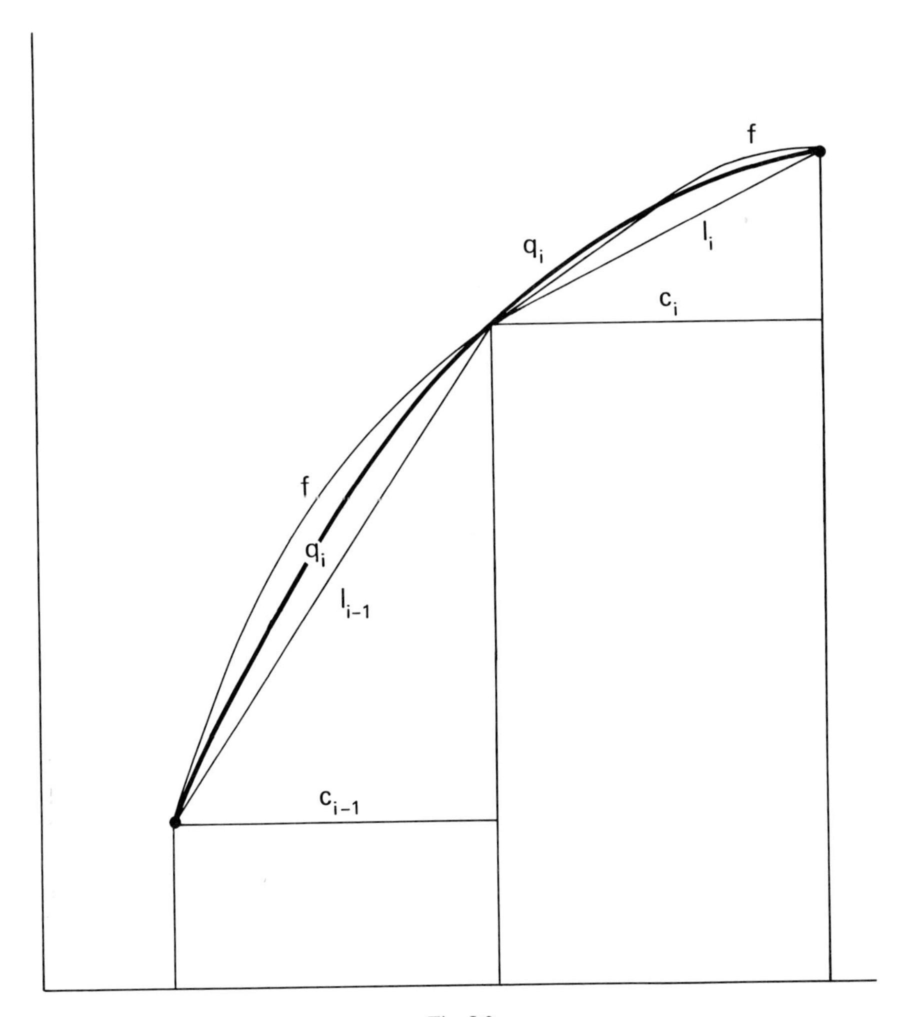

Fig. I-2

using (F), we define

(J) $$y_i = \begin{cases} (2x_i + x_{i+1})/3 & (i = 1, 3, 5, \ldots, n-1) \\ (x_i + 2x_{i+1})/3 & (i = 2, 4, 6, \ldots, n). \end{cases}$$

We again define the intervals of the partition by (G) and the evaluation-points by (H). Each interval in (G) is paired with an evaluation-point x_j, and its length is $h/3$ if j is odd and is $2h/3$ if j is even. We denote the partition-sum corresponding to this partition by S_Bf; then

(K) $$\text{S}_B\text{f} = [f(x_1) + 4f(x_2) + 2f(x_3) + 4f(x_4) + \cdots + 2f(x_{n-1}) + 4f(x_n) + f(x_{n+1})](h/3).$$

This is called the "Simpson formula" Sf for approximating the integral (Fig. I-2).

The choice of division-points, especially in Simpson's formula, seems quite arbitrary; but there is a simple geometrical explanation, and this explanation will also make it clear that we may expect the trapezoidal formula to be more accurate than Cauchy's formula and Simpson's formula to be more accurate than either. If we construct, as in Fig. I-2, the step-function that on each interval $(x_i, x_{i+1}]$ has the constant value $c_i = f(x_i)$, the integral of that step-function over each interval $(x_i, x_{i+1}]$ is $f(x_i)h$ and the sum of all of them is the Cauchy approximation $C_B f$. If l_i is the function linear on $(x_i, x_{i+1}]$ that coincides with f at x_i and at x_{i+1}, then

$$l_i(x) = [f(x_i)(x_{i+1} - x) + f(x_{i+1})(x - x_i)]2h^{-1}$$

and its integral from x_i to x_{i+1} is easily calculated to be

$$[f(x_i) + f(x_{i+1})](h/2).$$

If we add these for $i = 1, \ldots, n$, we obtain the trapezoidal formula. If f is positive, the integral from x_i to x_{i+1} is the area of the region over the interval $(x_i, x_{i+1}]$ and under the line that is the graph of l_i, and this figure is a trapezoid. Hence the name "trapezoidal rule." Evidently, on $(x_i, x_{i+1}]$ the graph of l_i may be expected to approximate the graph of f much more closely than the step-function.

By substituting x_{i-1}, x_i, and x_{i+1} for x in the quadratic

$$\begin{aligned} q_i(x) = f(x_i) &+ (x - x_i)[f(x_{i+1}) - f(x_{i-1})]/2h \\ &+ (x - x_i)^2[f(x_{i-1}) - 2f(x_i) + f(x_{i+1})]/2h^2 \end{aligned}$$

we verify that at those points the quadratic coincides with f. By integrating the quadratic from x_{i-1} to x_{i+1}, we obtain

$$\int_{x_{i-1}}^{x_{i+1}} q_i(x)\,dx = [f(x_{i-1}) + 4f(x_i) + f(x_{i+1})]\frac{h}{3}.$$

If we add these for all even i, we obtain Simpson's formula. Therefore $S_B f$ is the sum of the integrals over the intervals $(x_1, x_3], (x_3, x_5], \ldots, (x_{n-1}, x_{n+1}]$ of the quadratics which, on the respective intervals, coincide with f at both ends and the midpoint. Such a quadratic can be expected to be a much closer approximation to a smooth function f than the piecewise linear approximations that lead to the trapezoidal rule.

As a numerical test of the accuracy of the Cauchy, trapezoidal, and Simpson formulas, we shall apply them to two examples.

EXAMPLES 12-1

(i) $$\int_1^2 \frac{1}{x}\,dx = \log 2 = 0.6931471806 \cdots;$$

(ii) $$\int_0^1 (1 + x^2)^{-1}\,dx = \frac{\pi}{4} = 0.785398163397448 \cdots.$$

These are to be found in any elementary calculus text, as well as in this book: (i) is the definition of log 2, and in equation (BB) of Section II-6 we shall see that $x \mapsto (1 + x^2)^{-1}$ has $x \mapsto \arctan x$ as antiderivative.

If we choose $h = 0.25$, we find for (i)

$$f(1) = 1,$$
$$f(1.25) = 1/1.25 = 0.800000000000 \cdots,$$
$$f(1.5) = 1/1.5 = 0.666666666666 \cdots,$$
$$f(1.75) = 1/1.75 = 0.57142857142857 \cdots,$$
$$f(2) = 1/2 = 0.5000000000 \cdots.$$

The Cauchy formula (C) yields the estimate

$$(1 + 0.8000 \cdots + 0.666 \cdots + 0.571428571428 \cdots)(0.25) = 0.7595237 \cdots;$$

the error is $0.0663 \cdots$. The trapezoidal formula (I) yields

$$\begin{aligned} \mathrm{T}_B f &= (1 + 1.6000 \cdots + 1.333 \cdots + 1.14285714 \cdots + 0.5000 \cdots)(0.125) \\ &= 0.6970238; \end{aligned}$$

the error is 0.003876. Clearly, the trapezoidal formula is much better than the Cauchy formula.

If we apply the trapezoidal formula to (ii), we obtain

$$\mathrm{T}_B f = 0.7827941 \cdots,$$

the error being -0.0026040.

If we apply Simpson's formula to (i) with $h = 0.25$, we obtain

$$\begin{aligned} \int_1^2 \frac{1}{x}\,dx &\approx [1 + 4(0.8) + 2(0.66666666 \cdots) + 4(0.5714285714 \cdots) + 0.5]\frac{0.25}{3} \\ &= 0.6932539 \cdots. \end{aligned}$$

The error is 0.0001068. If we apply it to (ii) with $h = 0.25$, we obtain

$$\int_0^1 (1 + x^2)^{-1}\,dx \approx 0.7853921.$$

The error is only 0.0000060.

We can thus obtain an approximation to π that is correct to the fifth decimal place. With an electronic calculator, the calculation takes about 1 minute; even with no such aid it requires only a few minutes. Had we subdivided [0, 1] into 10 parts, the only error left with an 8-place calculator would be the error of rounding to 8 places; the calculation takes about $2\frac{1}{2}$ minutes. But this quick and easy computation involves: advanced technology in the form of the calculator; the great labor-saving device of Arabic numerals; and advanced pure mathematics in the form of theorems about integrals. We could dispense with the calculator, at the expense of performing tedious computations by hand. Dispensing with Arabic numerals would be worse; let the reader try to work out even the simplest of the calculations in the preceding paragraphs using Roman numerals! Dispensing with the calculus would be disastrous. Although the number π was defined long before calculus was thought of, estimating it to within, say, a hundredth of 1 percent using no mathematics but the theorems of elementary geometry would be extremely tedious, even if we allowed ourselves the use of Arabic numerals and an electronic calculator. This reflection should help us to appreciate the accomplishment of Archimedes, who obtained an estimate of π with error less than 1/4970 although he had no mechanical aids to computation, a system of writing numbers even more cumbersome than Roman numerals, and no calculus to draw on.

Although the numerical examples indicate that the trapezoidal method is rather accurate and the Simpson method even more so, it is important to have some estimate of the degree of accuracy we can count on in using them. The errors in the trapezoidal approximation $\mathrm{T}_B\mathrm{f}$ and the Simpson approximation $\mathrm{S}_B\mathrm{f}$ are specified in the following theorem.

THEOREM 12-2 *Let f be real-valued on an interval $B = [a, b]$, and let $[a, b]$ be subdivided into k subintervals all of length $h = (b - a)/k$. Then*

(i) *if f has continuous derivatives of orders up to 4 on $[a, b]$,*

$$\mathrm{T}_B\mathrm{f} - \int_a^b f(x)\,dx = \frac{[f'(b) - f'(a)]h^2}{12} + \varepsilon_{\mathrm{T}}(h),$$

where

$$|\varepsilon_{\mathrm{T}}(h)| \leqq h^4[(b - a)\sup\{|f^{iv}(x)| : a \leqq x \leqq b\}/150];$$

(ii) *if k is even and f has continuous derivatives of orders up to 6 on $[a, b]$,*

$$\mathrm{S}_B\mathrm{f} - \int_a^b f(x)\,dx = \frac{[f'''(b) - f'''(a)]h^4}{180} + \varepsilon_{\mathrm{S}}(h),$$

where

$$|\varepsilon_{\mathrm{S}}(h)| \leqq h^6[(b - a)\sup\{|f^{vi}(x)| : a \leqq x \leqq b\}/630].$$

We shall prove only the more important part (ii); the easier part (i) is left as an exercise. As before, we subdivide $[a, b]$ into an even number of subintervals by points

$$x_1 = a, \qquad x_2 = a + h, \ldots, \qquad x_{n+1} = b,$$

where

$$x_{i+1} - x_i = h = (b - a)/n$$

for all i. We consider a particular x_i in (a, b), and for convenience we denote it by

c. Then by the theorem of the mean, for each x in $[x_{i-1}, x_{i+1}]$ there is an $\bar{x}(x)$ between x and c such that

$$\text{(L)} \qquad f(x) = f(c) + (x-c)f'(c) + \cdots + (x-c)^6 f^{vi}(\bar{x}(x))/6!.$$

Let M be the supremum of $|f^{vi}(x)|$ on $[a, b]$. Then the number

$$\text{(M)} \qquad C_1 = [f^{vi}(\bar{x}(x_{i-1})) + f^{vi}(\bar{x}(x_{i+1}))]/2$$

is the average of two numbers in $[-M, M]$, so

$$|C_1| \leqq M,$$

and by applying (L) to x_{i-1} and to x_{i+1}, we obtain

$$\text{(N)} \qquad [f(x_{i-1}) + 4f(x_i) + f(x_{i+1})](h/3)$$
$$= 2hf(c) + h^3 f''(c)/3 + 2h^5 f^{iv}(c)/3 \cdot 4! + 2h^7 C_1/3 \cdot 6!.$$

The last term in (L) has absolute value at most $(x-c)^6 M/6!$, and it is integrable because all the other terms in (L) are integrable. We integrate both members of (L); there exists a number C_2 in $[-M, M]$ such that

$$\text{(O)} \qquad \int_{x_{i-1}}^{x_{i+1}} f(x)\,dx = 2hf(c) + \frac{2h^3 f''(c)}{3!} + \frac{2h^5 f^{iv}(c)}{5!} + \frac{C_2 \cdot 2h^7}{7!}.$$

By the theorem of the mean, for each x in $[x_{i-1}, x_{i+1}]$ there is an $\tilde{x}(x)$ between x and c such that

$$\text{(P)} \quad f'''(x) = f'''(c) + (x-c)f^{iv}(c) + (x-c)^2 f^{v}(c)2! + (x-c)^3 f^{vi}(\tilde{x}(x))/3!.$$

We apply this to x_{i+1} and to x_{i-1} and subtract; if we define

$$C_3 = [f^{vi}(\tilde{x}(x_{i-1})) + f^{vi}(\tilde{x}(x_{i+1}))]/2,$$

we obtain

$$\text{(Q)} \qquad f'''(x_{i+1}) - f'''(x_{i-1}) = 2hf^{iv}(c) + 2C_3 h^3/3!,$$

and

$$\text{(R)} \qquad |C_3| \leqq M.$$

From (N), (O), and (Q) we obtain

$$\text{(S)} \qquad [f(x_{i-1}) + 4f(x_i) + f(x_{i+1})]\frac{h}{3}$$
$$- \int_{x_{i-1}}^{x_{i+1}} f(x)\,dx - [f'''(x_{i+1}) - f'''(x_{i-1})]\frac{h^4}{180}$$
$$= h^7\left(\frac{2C_1}{3 \cdot 6!} - \frac{2C_2}{7!} - \frac{2C_3}{6 \cdot 180}\right).$$

The absolute value of the right member is, at most, $2h(h^6)M/630$. If we add the members of (S) for $i = 2, 4, 6, \ldots, n$ and recall that $nh = b - a$, we obtain conclusion (ii) of the theorem.

This theorem can be used to obtain an estimate of the error in approximating the integral of f by the Simpson formula Sf, even when f'''' is difficult or impossible to compute. Suppose that $[a, b]$ is subdivided into k subintervals, where k is a multiple of 4. We shall denote by $\mathrm{Sf}_{[h]}$ the estimate of the integral formed by using intervals of length $h = (b - a)/k$, and by $\int f$ the value of the integral of f from a to b. Then the error is $\mathrm{Sf}_{[h]} - \int f$. Now let us compute another estimate $\mathrm{Sf}_{[2h]}$ by using only the division-points $x_1, x_3, x_5, \ldots, x_{k+1}$. This is easy because we already have calculated the functional values at these points. The error in this estimate is $\mathrm{Sf}_{[2h]} - \int f$. But if $f''''(b) - f''''(a) \neq 0$ and h is small, the term $\varepsilon_S(h)$ in the expression for the error in Theorem 12-2 will be much smaller than $[f''''(b) - f''''(a)]h^4/180$, and we will have

$$\mathrm{Sf}_{[h]} - \int f \approx \frac{[f''''(b) - f''''(a)]h^4}{180} = \frac{[f''''(b) - f''''(a)](2h)^4/180}{16}$$

$$\approx \frac{\mathrm{Sf}_{[2h]} - \int f}{16} = \frac{(\mathrm{Sf}_{2h} - \mathrm{Sf}_h) + (\mathrm{Sf}_h - \int f)}{16},$$

whence

$$\mathrm{Sf}_{[h]} - \int f \approx \frac{\mathrm{Sf}_{[2h]} - \mathrm{Sf}_{[h]}}{15}.$$

That is, the difference between the better estimate $\mathrm{Sf}_{[h]}$ and the integral of f is roughly $\frac{1}{15}$ of the difference between the two estimates, $\mathrm{Sf}_{[2h]} - \mathrm{Sf}_{[h]}$.

For example, let us estimate the integral of $1/x$ from 1 to 2 with $h = \frac{1}{8}$. We have the accompanying tabulation of functional values. From these we compute

$$\mathrm{Sf}_{[2h]} = 0.6932539, \qquad \mathrm{Sf}_{[h]} = 0.6931544,$$

whence $[\mathrm{Sf}_{[2h]} - \mathrm{Sf}_{[h]}]/15 = 0.0000066$. This is a usably accurate estimate for the actual error, which is 0.0000073.

x	$1/x$	x	$1/x$
1.0	1.0000000 ···	1.625	0.61538461 ···
1.125	0.88888888 ···	1.75	0.57142857 ···
1.25	0.80000000 ···	1.875	0.53333333 ···
1.375	0.72727272 ···	2.0	0.50000000 ···
1.5	0.66666666 ···		

If it happens that $f''''(b) - f''''(a) = 0$, this reasoning will not apply, because then the ε_S terms will not be smaller than the others. On the other hand, in this case the estimate using the smaller h will be even closer to the correct result than a constant times h^4 for small h, so the procedure we have just considered will be conservative and will give us an overestimate of the error in the computation using the smaller value of h.

EXERCISE 12-1 In the example in the preceding two paragraphs, compute $f'''(2) - f'''(1)$ accurately and verify that the estimate we found is not far from the correct value.

EXERCISE 12-2 For Example 12-1 (ii) compute $f'''(1) - f'''(0)$. Use this to explain the startlingly accurate estimate for $\pi/4$ that we obtained by using Simpson's method.

EXERCISE 12-3 By subdividing $[0, 1]$ into 10 equal parts and using Simpson's formula, find $\arctan x$ for $x = 0, 0.2, 0.4, 0.6, 0.8, 1$. Check these values by means of a table of $\tan x$ for x in radians.

EXERCISE 12-4 Define the "midpoint formula" to be

$$\mathrm{M}_B\mathrm{f} = [f(x_1^*) + \cdots + f(x_k^*)]h,$$

where

$$x_i^* = [x_i + x_{i+1}]/2.$$

Apply this to Example 12-1(i) with $h = 0.25$; show that it gives an approximation having only about half as much error as that in the trapezoidal method.

EXERCISE 12-5 Complete the proof of Theorem 12-2.

EXERCISE 12-6 Show that

$$\mathrm{Mf} - \int_a^b f(x)\,dx = -\frac{[f'(b) - f'(a)]h^2}{24} + \varepsilon_{\mathrm{M}}(h),$$

where $|\varepsilon_{\mathrm{M}}(h)|$ is, at most, a constant multiple of h^4.

EXERCISE 12-7 Let $h = (b - a)/k$, with k even. Let $x_1 = a$, $x_3 = a + 2h$, $x_5 = a + 4h, \ldots, x_{k+1} = b$ be the division-points. With these points show that the weighted average $[2\mathrm{Mf} + \mathrm{Tf}]/3$ differs from the integral by, at most, a constant times h^4 and thus is much more accurate than the trapezoidal or the midpoint formula. Why does the author not claim this as a discovery of a new and accurate approximation method?

Integration in One-Dimensional Space: Further Development

1. A Condition for Integrability

To apply Definition I-2-2 of the integral to any specific function f, we must first know what the number J is. Then for each positive ε we must exhibit a gauge γ on $\bar{R}$ and show that with this γ, every γ-fine partition $\mathscr{P}$ of R gives us a partition-sum that satisfies $|S(\mathscr{P}; f_B) - J| < \varepsilon$. This is what we did in the examples in Section I-3. But even when we know the value of J, this is an inconvenient procedure. Also, in many cases of considerable importance we need to be able, without knowing J in advance, to show that the integral exists; the number J can then be determined by means of the partition-sums or in some other way, and it is useful to us just because it *is* the value of the integral.

A long-known test for convergence of a sequence is named for Augustin-Louis Cauchy. Roughly stated, the numbers $a_1, a_2, \ldots$ in a sequence get arbitrarily close to some fixed number (which is their limit) if and only if they ultimately stay arbitrarily close to each other. We shall not use this fact but shall prove a similar statement for partition-sums.

★THEOREM 1-1 *Let f be defined on a subset B of R. The integral*

$$\int_B f(x)\, m(dx)$$

exists if and only if the following condition is satisfied:

(A) *To each positive ε there corresponds a gauge γ on $\bar{R}$ such that whenever $\mathscr{P}'$ and $\mathscr{P}''$ are γ-fine partitions of R, $S(\mathscr{P}'; f_B)$ and $S(\mathscr{P}''; f_B)$ are finite and*

$$|S(\mathscr{P}'; f_B) - S(\mathscr{P}''; f_B)| < \varepsilon.$$

Suppose first that the integral exists; let J denote its value. Then by Definition I-2-2, for each positive ε there exists a gauge γ on $\bar{R}$ such that whenever $\mathscr{P}$ is a γ-fine partition of R,

$$\text{(B)} \qquad |S(\mathscr{P}; f_B) - J| < \varepsilon/2.$$

Let $\mathscr{P}'$ and $\mathscr{P}''$ be two γ-fine partitions of R. Then (B) holds for both of them. This implies, first, that $S(\mathscr{P}'; f_B)$ and $S(\mathscr{P}''; f_B)$ are finite and, second, that

$$|S(\mathscr{P}'; f_B) - S(\mathscr{P}''; f_B)| \leqq |S(\mathscr{P}'; f_B) - J| + |J - S(\mathscr{P}''; f_B)| < \varepsilon.$$

Therefore, condition (A) is satisfied.

Conversely, suppose that (A) is satisfied. For each positive ε we can and do select a gauge γ_ε on $\bar{R}$ such that whenever $\mathscr{P}'$ and $\mathscr{P}''$ are γ_ε-fine partitions of R, $S(\mathscr{P}'; f_B)$, and $S(\mathscr{P}''; f_B)$ are finite and

$$\text{(C)} \qquad |S(\mathscr{P}'; f_B) - S(\mathscr{P}''; f_B)| < \varepsilon/3;$$

and we can and do select a γ_ε-fine partition $\mathscr{P}[\varepsilon]$ of R. Let $C[\varepsilon]$ be the closed interval

$$\text{(D)} \qquad C[\varepsilon] = [S(\mathscr{P}[\varepsilon]; f_B) - \varepsilon/3, S(\mathscr{P}[\varepsilon]; f_B) + \varepsilon/3].$$

By (C), if $\mathscr{P}$ is any γ_ε-fine partition of R, $S(\mathscr{P}; f_B)$ is in $C[\varepsilon]$. If ε' and ε'' are positive numbers, the intersection $\gamma = \gamma_{\varepsilon'} \cap \gamma_{\varepsilon''}$ is a gauge on $\bar{R}$. By Theorem I-4-2, there is a γ-fine partition $\mathscr{P}$ of R. Since $\mathscr{P}$ is both $\gamma_{\varepsilon'}$-fine and $\gamma_{\varepsilon''}$-fine, $S(\mathscr{P}; f_B)$ is in both $C[\varepsilon']$ and $C[\varepsilon'']$. So by Theorem 2-2 in the Introduction, we know that there is a number J contained in all the intervals $C[\varepsilon]$ ($\varepsilon > 0$). If $\varepsilon > 0$ and $\mathscr{P}$ is any γ_ε-fine partition of R, both J and $S(\mathscr{P}; f_B)$ are in $C[\varepsilon]$, whose length is $2\varepsilon/3$. Therefore,

$$|S(\mathscr{P}; f_B) - J| < \varepsilon,$$

which shows that f is integrable over B and that its integral is J. The proof is complete.

From this we deduce an improvement of Lemma I-7-1.

★Corollary 1-2 *Let f be defined on a subset B of R. If for each positive number ε there exist functions g and h integrable over B and such that*

$$\text{(E)} \qquad g(x) \leqq f(x) \leqq h(x) \qquad \textit{for all } x \quad \textit{in} \quad B$$

and

$$\int_B h(x)\, m(dx) < \int_B g(x)\, m(dx) + \varepsilon,$$

then f is integrable over B.

Let ε be positive. We can and do choose a pair of functions g, h integrable over B that satisfy (E) and also satisfy

$$\text{(F)} \qquad \int_B h(x)\,m(dx) < \int_B g(x)\,m(dx) + \frac{\varepsilon}{3}.$$

Then

$$\text{(G)} \qquad g_B(x) \leqq f_B(x) \leqq h_B(x) \qquad (x \text{ in } \bar{R});$$

for if x is in B, this is the same as (E), and otherwise all three numbers in (G) are 0. There exists a gauge γ' on $\bar{R}$ such that for every γ'-fine partition $\mathscr{P}$ of R,

$$\text{(H)} \qquad \int_B g(x)\,m(dx) - \frac{\varepsilon}{3} < S(\mathscr{P}; g_B) < \int_B g(x)\,m(dx) + \frac{\varepsilon}{3},$$

and there exists a gauge γ'' on $\bar{R}$ such that for every γ''-fine partition $\mathscr{P}$ of R

$$\text{(I)} \qquad \int_B h(x)\,m(dx) - \frac{\varepsilon}{3} < S(\mathscr{P}; h_B) < \int_B h(x)\,m(dx) + \frac{\varepsilon}{3}.$$

Let $\gamma = \gamma' \cap \gamma''$, and let $\mathscr{P}$ be a γ-fine partition of R. Then (H) and (I) are satisfied, and from this and (G)

$$\text{(J)} \qquad \int_B g(x)\,m(dx) - \frac{\varepsilon}{3} < S(\mathscr{P}; g_B) \leqq S(\mathscr{P}; f_B)$$

$$\leqq S(\mathscr{P}; h_B) < \int_B h(x)\,m(dx) + \frac{\varepsilon}{3}.$$

Therefore the partition-sum $S(\mathscr{P}; f_B)$ lies in the interval

$$\left(\int_B g(x)\,m(dx) - \frac{\varepsilon}{3}, \int_B h(x)\,m(dx) + \frac{\varepsilon}{3}\right).$$

By (F) this interval has length less than ε; and since $S(\mathscr{P}'; f_B)$ and $S(\mathscr{P}''; f_B)$ are both in it whenever $\mathscr{P}'$ and $\mathscr{P}''$ are γ-fine partitions of R, the Cauchy condition (A) in Theorem 1-1 is satisfied, and the integral exists.

★COROLLARY 1-3 *Let f be continuous on a bounded closed interval $B = [a, b]$ in R. Then f is integrable over $(a, b]$.*

Let ε be positive; define

$$\varepsilon' = \varepsilon/2[mB + 1].$$

To each $\bar{x}$ in B there corresponds a neighborhood $\gamma(\bar{x})$ of $\bar{x}$ such that for all x in $\gamma(\bar{x}) \cap B$

$$\text{(K)} \qquad |f(x) - f(\bar{x})| < \varepsilon'.$$

For each $\bar{x}$ in $\bar{R} \setminus B$ we choose for $\gamma(\bar{x})$ a neighborhood of $\bar{x}$ disjoint from B. (For example, we could choose $\gamma(\bar{x}) = [-\infty, a)$ for $\bar{x} < a$ and $\gamma(\bar{x}) = (b, \infty]$ for $\bar{x} > b$.) Let

$$\mathscr{P} = \{(\bar{x}_1, A_1), \ldots, (\bar{x}_k, A_k)\}$$

be a γ-fine partition of $(a, b]$; such partitions exist, by Theorem I-4-2. Let s and S be the step-functions on $(a, b]$ that on each A_i have the respective constant values

$$\text{(L)} \qquad s(x) = f(\bar{x}_i) - \varepsilon', \qquad S(x) = f(\bar{x}_i) + \varepsilon'.$$

Each x in $(a, b]$ is in some A_i, and this A_i is contained in $\gamma(\bar{x}_i)$, so by (K),

$$\text{(M)} \qquad f(x) < f(\bar{x}_i) + \varepsilon' = S(x).$$

Likewise, for all x in $(a, b]$

$$\text{(N)} \qquad f(x) > s(x).$$

By definition (L),

$$\text{(O)} \quad \int_{(a,b]} S(x)\, m(dx) - \int_{(a,b]} s(x)\, m(dx) = \int_{(a,b]} 2\varepsilon'\, m(dx) = 2\varepsilon' m(a, b] < \varepsilon.$$

Inequalities (M), (N), and (O) show that the hypotheses of Corollary 1-2 are satisfied with s and S in place of g and h, and by that corollary the integral of f over $(a, b]$ exists.

EXERCISE 1-1 The Cauchy condition for convergence of $a_1, a_2, a_3, \ldots$ referred to above is that a_n has a limit in R if and only if for each positive ε there is an n_ε such that $|a_n - a_m| < \varepsilon$ whenever $n > n_\varepsilon$ and $m > n_\varepsilon$. Prove this. *Suggestion*: For each neighborhood $\gamma(\infty)$, let $m(\gamma)$ be the infimum and $M(\gamma)$ the supremum of values a_n with n in $\gamma(\infty)$. The proof is similar to that of Theorem 1-1.

EXERCISE 1-2 Prove that if f is continuous on a bounded closed interval $[a, b]$, it is Riemann integrable over the interval. *Suggestion*: Use the proof of Corollary 1-3.

2. Absolute Integrability

In proving two of the main theorems of this chapter, we shall need to make some elementary calculations. To avoid interrupting the chain of thought in those main proofs, we shall perform the easy calculations in advance and collect them in a lemma.

★LEMMA 2-1 *Let γ be a gauge on $\bar{R}$ and let f be a real-valued function on R. For $j = 1, \ldots, h$, let B_j be a subset of R and*

$$\mathscr{P}_j = \{(x_{j,1}, A_{j,1}), \ldots, (x_{j,k(j)}, A_{j,k(j)})\}$$

be an allotted partition of B_j. *Then*

(i) *if* $B_1 \subset B_2$, *the set of pairs*

$$\mathscr{P}' = \{(x_{1,i}, A_{1,i} \cap A_{2,j}) : i = 1, \ldots, k(1); j = 1, \ldots, k(2)\}$$

is an allotted partition of B_1, *and*

$$S(\mathscr{P}'; f) = S(\mathscr{P}_1; f), \tag{A}$$

and if $\mathscr{P}_1$ *is* γ-*fine, so is* $\mathscr{P}'$;

(ii) *if the* B_j *are pairwise disjoint, the set of pairs*

$$\mathscr{P}'' = \mathscr{P}_1 \cup \cdots \cup \mathscr{P}_h = \{(x_{i,j}, A_{i,j}) : i = 1, \ldots, h; j = 1, \ldots, k(i)\}$$

is an allotted partition of $B_1 \cup \cdots \cup B_h$, *and*

$$S(\mathscr{P}''; f) = S(\mathscr{P}_1; f) + \cdots + S(\mathscr{P}_h; f),$$

and if all the $\mathscr{P}_1$ *are* γ-*fine, so is* $\mathscr{P}''$.

For (i), we observe that the intersections $A_{1,i} \cap A_{2,j}$ are left-open intervals. Each point x of B_1 also belongs to B_2, so it is in just one interval $A_{1,i}$ of the allotted partition $\mathscr{P}_1$ of B_1 and in just one interval $A_{2,j}$ of the allotted partition $\mathscr{P}_2$ of B_2. So the intersections $A_{1,i} \cap A_{2,j}$ are pairwise disjoint left-open intervals whose union is B_1, and therefore $\mathscr{P}'$ is an allotted partition of B_1. Each x in $A_{1,i}$ also belongs to just one interval $A_{2,j}$, so $A_{1,i}$ is the union of the pairwise disjoint intervals $A_{1,i} \cap A_{2,1}, \ldots, A_{1,i} \cap A_{2,k(2)}$. By Corollary 1-3 in the Introduction,

$$mA_{1,i} = m(A_{1,i} \cap A_{2,1}) + \cdots + m(A_{1,i} \cap A_{2,k(2)}).$$

If we multiply both members by $f(x_{1,i})$ and sum for $i = 1, \ldots, k(1)$, we obtain

$$\sum_{i=1}^{k(1)} f(x_{1,i}) mA_{1,i} = \sum_{i=1}^{k(1)} \sum_{j=1}^{k(2)} f(x_{1,i}) m(A_{1,i} \cap A_{2,j}).$$

This is the same as equation (A). For each i and j, if $\mathscr{P}_1$ is γ-fine, we have

$$\gamma(x_{1,i}) \supset (A_{1,i})^- \supset (A_{1,i} \cap A_{2,j})^-,$$

so $\mathscr{P}'$ is also γ-fine. This completes the proof of (i).

For (ii), we observe that each point x of $B_1 \cup \cdots \cup B_h$ is in just one set B_j and is therefore in just one interval $A_{j,i}$ of the allotted partition $\mathscr{P}_j$ and not in any interval of any other allotted partition $\mathscr{P}_{j'}$. So, the intervals $A_{j,i}$ ($j = 1, \ldots, h$; $i = 1, \ldots, k(j)$) are pairwise disjoint left-open intervals whose union is $B_1 \cup \cdots \cup B_h$, and the set $\mathscr{P}''$ of pairs is an allotted partition of $B_1 \cup \cdots \cup B_h$. If each $\mathscr{P}_j$ is γ-fine, each pair $(x_{j,i}, A_{j,i})$ in $\mathscr{P}''$ is γ-fine because it belongs to the γ-fine partition $\mathscr{P}_j$, and therefore $\mathscr{P}''$ is γ-fine. Obviously,

$$S(\mathscr{P}''; f) = \sum_{j=1}^{h} \sum_{i=1}^{k(j)} f(x_{j,i}) mA_{j,i} = \sum_{j=1}^{h} S(\mathscr{P}_j; f).$$

The proof of (ii) is complete.

By virtue of Lemma 2-1 we can rephrase Theorem 1-1 thus. A function f on a set B in R is integrable over R if and only if to each positive ε there corresponds a gauge γ on $\bar{R}$ such that whenever $\mathscr{P}'$ and $\mathscr{P}''$ are γ-fine partitions of R (using the notation of Lemma 2-1), both $S(\mathscr{P}'; f_B)$ and $S(\mathscr{P}''; f_B)$ are finite and

$$\text{(B)} \qquad \left| \sum_{i=1}^{k} \sum_{j=1}^{h} [f_B(x_i') - f_B(x_j'')] m(A_i' \cap A_j'') \right| < \varepsilon.$$

But from inequality (B) alone we cannot deduce any bound on the sum

$$\text{(C)} \qquad \sum_{i=1}^{k} \sum_{j=1}^{h} |f(x_i') - f(x_j'')| m(A_i' \cap A_j'').$$

Clearly, the sum (C) is at least as great as the left member of (B). In general, $|a_1| + \cdots + |a_k|$ is at least as great as $|a_1 + \cdots + a_k|$, but it can be far greater. It is therefore surprising that we can nevertheless obtain a bound on the sums (C). This is not only surprising; it is useful, because on it depend the proofs of the extremely useful theorems of the next section.

Since the smallness of sums (C) is important to us, we give it a name.

★DEFINITION 2-2 *Let f be defined and real-valued on a set B contained in R. The statement that f is* ***absolutely integrable over*** *B means that to each positive ε there corresponds a gauge γ on $\bar{R}$ such that whenever*

$$\mathscr{P}' = \{(x_1', A_1'), \ldots, (x_k', A_k')\}$$

and

$$\mathscr{P}'' = \{(x_1'', A_1''), \ldots, (x_h'', A_h'')\}$$

are γ-fine partitions of R, it is true that

$$\text{(D)} \qquad \sum_{i=1}^{k} \sum_{j=1}^{h} |f_B(x_i') - f_B(x_j'')| m(A_i' \cap A_j'') < \varepsilon.$$

The principal theorem of this section, which will be used repeatedly in later parts of the book, states that a function is integrable if and only if it is absolutely integrable. In preparation for the proof of this theorem we shall establish a lemma that has several uses.

★LEMMA 2-3 *Let f be integrable over a subset B of R. Let ε be positive, and let γ be a gauge on $\bar{R}$ such that whenever $\mathscr{P}$ is a γ-fine partition of R,*

$$\left| S(\mathscr{P}; f_B) - \int_B f(x)\, m(dx) \right| < \varepsilon.$$

Let $\mathscr{P}'$ and $\mathscr{P}''$ be γ-fine partitions of the same set C in R. Then

$$|S(\mathscr{P}'; f_B) - S(\mathscr{P}''; f_B)| < 2\varepsilon.$$

If C were equal to R, this would be a triviality. The whole point of the lemma is that C is any subset of R that can have a γ-fine allotted partition, which means that C is any finite union of left-open intervals.

Since C is the union of the left-open intervals of the γ-fine partition $\mathscr{P}'$, its complement $R \setminus C$ consists of finitely many left-open intervals $B_1, \ldots, B_n$. For each one B_j of these we choose a γ-fine partition $\mathscr{P}_j$, as is possible by Theorem I-4-2. By Lemma 2-1, $\mathscr{P}' \cup \mathscr{P}_1 \cup \cdots \cup \mathscr{P}_n$ is a γ-fine partition of $C \cup B_1 \cup \cdots \cup B_n$, which is R; so by hypothesis

$$\text{(E)} \qquad \left| S(\mathscr{P}' \cup \mathscr{P}_1 \cup \cdots \cup \mathscr{P}_n; f_B) - \int_B f(x)\, m(dx) \right| < \varepsilon.$$

Likewise,

$$\text{(F)} \qquad \left| S(\mathscr{P}'' \cup \mathscr{P}_1 \cup \cdots \cup \mathscr{P}_n) - \int_B f(x)\, m(dx) \right| < \varepsilon,$$

so

$$\text{(G)} \qquad |S(\mathscr{P}' \cup \mathscr{P}_1 \cup \cdots \cup \mathscr{P}_n) - S(\mathscr{P}'' \cup \mathscr{P}_1 \cup \cdots \cup \mathscr{P}_n)| < 2\varepsilon.$$

By Lemma 2-1,

$$\text{(H)} \qquad S(\mathscr{P}' \cup \mathscr{P}_1 \cup \cdots \cup \mathscr{P}_n) = S(\mathscr{P}'; f_B) + \sum_{j=1}^{n} S(\mathscr{P}_j; f_B),$$

$$\text{(I)} \qquad S(\mathscr{P}'' \cup \mathscr{P}_1 \cup \cdots \cup \mathscr{P}_n) = S(\mathscr{P}''; f_B) + \sum_{j=1}^{n} S(\mathscr{P}_j; f_B).$$

If we substitute (H) and (I) in (G), we obtain the conclusion of the lemma.

★THEOREM 2-4 *Let f be defined and real-valued on a subset B of R. Then f is integrable over B if and only if it is absolutely integrable over B. If f is integrable over B, $\varepsilon > 0$, and γ is a gauge on $\bar{R}$ such that*

$$\left| S(\mathscr{P}; f_B) - \int_B f(x)\, m(dx) \right| < \frac{\varepsilon}{4}$$

whenever $\mathscr{P}$ is a γ-fine partition of R, then for every pair of γ-fine partitions

$$\text{(J)} \qquad \mathscr{P}' = \{(x_1', A_1'), \ldots, (x_k', A_k')\}, \qquad \mathscr{P}'' = \{(x_1'', A_1''), \ldots, (x_h'', A_h'')\}$$

of R it is true that

$$\text{(K)} \qquad \sum_{i=1}^{k} \sum_{j=1}^{h} |f_B(x_i') - f_B(x_j'')| m(A_i' \cap A_j'') < \varepsilon.$$

Suppose first that f is absolutely integrable over B. Let ε be positive; then there is a gauge γ' on $\bar{R}$ such that for every pair $\mathscr{P}'$, $\mathscr{P}''$ of γ'-fine partitions of

R, using notation (J), (K) is valid. There is also a gauge γ^* on $\bar{R}$ such that for every γ^*-fine partition $\mathscr{P}$ of R, $S(\mathscr{P}; f_B)$ is finite. (See the second paragraph after Definition I-2-2.) Now let $\gamma = \gamma' \cap \gamma^*$, and let $\mathscr{P}'$ and $\mathscr{P}''$ be γ-fine partitions of R, with the same notation (J). Because they are γ^*-fine, $S(\mathscr{P}'; f_B)$ and $S(\mathscr{P}''; f_B)$ are finite; and because they are γ'-fine, (K) is satisfied. Therefore,

$$\left| \sum_{i=1}^{k} \sum_{j=1}^{h} [f_B(x_i') - f_B(x_j'')] m(A_i' \cap A_j'') \right|$$
$$\leqq \sum_{i=1}^{k} \sum_{j=1}^{h} |f_B(x_i') - f_B(x_j'')| m(A_i' \cap A_j'') < \varepsilon.$$

As we saw in the paragraph containing (B), this is the same as the necessary and sufficient condition in Theorem 1-1 for the existence of the integral.

For the converse, suppose that f is integrable over B. Let ε be positive. There exists a gauge γ on $\bar{R}$ such that for every γ-fine partition $\mathscr{P}$ of R

$$\text{(L)} \qquad \left| S(\mathscr{P}; f_B) - \int_B f(x)\,dx \right| < \frac{\varepsilon}{4}.$$

Let $\mathscr{P}'$ and $\mathscr{P}''$ (with the same notation as before) be two γ-fine partitions of R. By Lemma 2-1, the two sets of pairs

$$\mathscr{P}' = \{(x_i', A_i' \cap A_j''): i = 1, \ldots, k, j = 1, \ldots, h\}$$

and

$$\mathscr{P}'' = \{(x_j'', A_i' \cap A_j''): i = 1, \ldots, k, j = 1, \ldots, h\}$$

are both γ-fine partitions of R. We divide the set of all pairs

$$(i, j) \qquad (i = 1, \ldots, k, j = 1, \ldots, h)$$

into two subsets. The subset I consists of all those (i, j) for which $f_B(x_i') \geqq f_B(x_j'')$; the subset II consists of the others, for which $f_B(x_i') < f_B(x_j'')$. Let C be the union of all $A_i' \cap A_j''$ for which (i, j) is in I. Then the two sets of pairs

$$\{(x_i', A_i' \cap A_j''): (i, j) \text{ in I}\},$$
$$\{(x_j'', A_i' \cap A_j''): (i, j) \text{ in I}\}$$

are both γ-fine partitions of C. By Lemma 2-3 and (L),

$$\left| \sum_{(i,j) \text{ in I}} [f_B(x_i') - f_B(x_j'')] m(A_i' \cap A_j'') \right| < \frac{\varepsilon}{2}.$$

But by definition of class I all the numbers in square brackets are nonnegative, so this can be written in the form

$$\text{(M)} \qquad \sum_{(i,j) \text{ in I}} |f_B(x_i') - f_B(x_j'')| m(A_i' \cap A_j'') < \frac{\varepsilon}{2}.$$

Likewise, the two sets of pairs

$$\{(x'_i, A'_i \cap A''_j) : (i, j) \text{ in II}\},$$

$$\{(x''_j, A'_i \cap A''_j) : (i, j) \text{ in II}\}$$

are both γ-fine partitions of $R \setminus C$, and by Lemma 2-3,

$$\left| \sum_{(i,j) \text{ in II}} [f_B(x''_j) - f_B(x'_i)] m(A'_i \cap A''_j) \right| < \frac{\varepsilon}{2}.$$

Again, the numbers in brackets are nonnegative, so we can put the absolute value bars around the square brackets. This yields

$$\text{(N)} \qquad \sum_{(i,j) \text{ in II}} |f_B(x'_i) - f_B(x''_j)| m(A'_i \cap A''_j) < \frac{\varepsilon}{2}.$$

But classes I and II together constitute all the pairs (i, j), so by adding (M) and (N) we obtain (K). This completes the proof.

EXERCISE 2-1 Let $B = [-1, 1]$, $f(x) = 1 - x^2$ on R. Take $\varepsilon = 0.1$ and find a gauge γ that serves in Definition 2-2.

EXERCISE 2-2 Prove from Definition 2-2 that if f is absolutely integrable over B, so are $|f|$, f^+, and f^-.

EXERCISE 2-3 Let A be a bounded interval with end-points a and b. Show directly (without using Theorem 2-4) that 1_A is absolutely integrable over R. *Suggestion*: Use $\gamma(\bar{x}) = [-\infty, a)$ if $\bar{x} < a$; $\gamma(\bar{x}) = (a, b)$ if $a < \bar{x} < b$; $\gamma(\bar{x}) = (b, \infty]$ if $\bar{x} > b$; $\gamma(a)$ and $\gamma(b)$ very short intervals. In sum (D) only those terms in which just one of x'_i, x''_j is an end-point of A are not equal to 0.

EXERCISE 2-4 Let f be Lipschitzian on a bounded closed interval $[a, b]$. Prove that it is absolutely integrable over $(a, b]$.

3. Integration of Composite Functions

Frequently, new functions enter a discussion in the form of composites of other functions. Thus, from a given function f on a set B we can form such composites as f^2, which is $x \mapsto [f(x)]^2$, or $\sin \circ f$, which is $x \mapsto \sin(f(x))$. If we know that f is integrable over B, it can be useful to know that some composite $g \circ f$ is also integrable over B. The next theorem gives this conclusion in some important special cases. We recall that a function g on a set D in R is **Lipschitzian,** or **Lipschitz continuous**, if there exists a finite number L such that for all y' and y in D,

$$|g(y') - g(y)| \leqq L|y' - y|.$$

Any L with which this is satisfied is called a **Lipschitz constant** for g.

★Theorem 3-1 *Let f be real-valued on and integrable over a subset B of R. Let g be a Lipschitzian function on a set D of reals that contains 0 and all the functional values $\{f(x);\ x \text{ in } B\}$ and has $g(0) = 0$. Then the composite function $g \circ f$ is integrable over B.*

Let L be a Lipschitz constant for g. Since f is integrable over B, by Theorem 2-4 it is absolutely integrable over B. So, if ε is any positive number, there exists a gauge γ on $\bar{R}$ such that for any two γ-fine partitions $\mathscr{P}'$, $\mathscr{P}''$ of R (for which we use the usual notation, as in Lemma 2-1),

$$\text{(A)} \qquad \sum_{i=1}^{k} \sum_{j=1}^{h} |f_B(x_i') - f_B(x_j'')| m_L(A_i' \cap A_j'') < \frac{\varepsilon}{L}.$$

Since L is a Lipschitz constant for g on D,

$$\text{(B)} \qquad |g(f_B(x_i')) - g(f_B(x_j''))| \leqq L|f_B(x_i') - f_B(x_j'')|.$$

But if x_i' is not in B, $f_B(x_i') = 0$, and by hypothesis $g(f_B(x_i')) = 0$. Also, since x_i' is not in B, $[g \circ f]_B$ has the value 0 at x_i', and so in this case the equation

$$\text{(C)} \qquad g(f_B(x_i')) = [g \circ f]_B(x_i')$$

holds. If x_i' is in B, we have by definition $f_B(x_i') = f(x_i')$ and $[g \circ f]_B(x_i') = [g \circ f](x_i')$, and (C) is true by definition of the composite $g \circ f$. So (C) holds for all i. Likewise,

$$\text{(D)} \qquad g(f_B(x_j'')) = [g \circ f]_B(x_j'').$$

If we substitute from (C) and (D) in (B), multiply both members by the nonnegative number $m_L(A_i' \cap A_j'')$, and add for all i and j, we find by (A)

$$\sum_{i=1}^{k} \sum_{j=1}^{h} |[g \circ f]_B(x_i') - [g \circ f]_B(x_j'')| m(A_i' \cap A_j'')$$
$$\leqq L \sum_{i=1}^{k} \sum_{j=1}^{h} |f_B(x_i') - f_B(x_j'')| m(A_i' \cap A_j'') < \varepsilon.$$

So the composite $g \circ f$ is absolutely integrable, and by Theorem 2-4 it is integrable over B.

To apply this to special cases, it is convenient to have a simple test for verifying that a given g is Lipschitzian. The next lemma contains such a test.

★Lemma 3-2 *Let g be real-valued and continuous on an interval D in R, and let L be a real number. If the set of points x in D at which either $Dg(x)$ does not exist or*

$|Dg(x)| > L$ is a finite set, then g is Lipschitzian on D and L is a Lipschitz constant for g.

Let x' and x'' be two points of D; we choose the notation so that $x' < x''$. In the open interval (x', x'') there are finitely many points at which g lacks a derivative or $|Dg(\bar{x})| > L$; we call these $x_1, \ldots, x_n$, in increasing order. Also, we define x_0 to be x' and x_{n+1} to be x''. By the theorem of the mean, for $i = 0, \ldots, n$ there is a point $\bar{x}_i$ in (x_i, x_{i+1}) such that

$$g(x_{i+1}) - g(x_i) = Dg(\bar{x}_i)[x_{i+1} - x_i].$$

Since $|Dg(\bar{x}_i)| \leqq L$, this implies

$$-L(x_{i+1} - x_i) \leqq g(x_{i+1}) - g(x_i) \leqq L(x_{i+1} - x_i).$$

By adding these inequalities member by member for $i = 0, \ldots, n$, we obtain

$$-L(x_{n+1} - x_0) \leqq g(x_{n+1}) - g(x_0) \leqq L(x_{n+1} - x_0),$$

which is the same as

$$|g(x'') - g(x')| \leqq L|x'' - x'|.$$

COROLLARY 3-3 *If f is integrable over a set $B \subset R$ and $c \geqq 0$, then $c \wedge f$ and $(-c) \vee f$ are integrable over B.*

Define g to be the function $y \mapsto c \wedge y$. This is continuous on R, and it has value 0 at $y = 0$. For $y < c$ it coincides with $y \mapsto y$ and has derivative 1; for $y < c$ it coincides with c and has derivative 0. So g is Lipschitzian, and $g \circ f$ is integrable over B by Theorem 3-1. But $g \circ f = c \wedge f$. The proof that $(-c) \vee f$ is integrable is similar, using $g(y) = (-c) \vee y$ for all y in R.

It is customary to define

$$f^+ = f \vee 0, \qquad f^- = (-f) \vee 0. \tag{E}$$

It follows readily that for all x in the domain of f,

$$f(x) = f^+(x) - f^-(x), \qquad f(x) = f^+(x) + f^-(x). \tag{F}$$

For if $f(x) \geqq 0$, $f^-(x) = 0$ and $f^+(x) = |f(x)| = f(x)$, whereas if $f(x) < 0$, then $f^+(x) = 0$ and $f^-(x) = -f(x) = |f(x)|$. In either case Eqs. (F) are satisfied.

COROLLARY 3-4 *If f is integrable over a subset B of R, so are f^+, f^-, and f. Moreover,*

$$0 \leqq \int_B f^+(x)\, m(dx) \leqq \int_B |f(x)|\, m(dx),$$

$$0 \leqq \int_B f^-(x)\, m(dx) \leqq \int_B |f(x)|\, m(dx),$$

$$\left| \int_B f(x)\, m(dx) \right| \leqq \int_B |f(x)|\, m(dx).$$

Since f is integrable over B, so is $-f$. By Corollary 3-3, with $c = 0$, both $f \vee 0$ and $(-f) \vee 0$ are integrable over B; that is, f^+ and f^- are integrable. Since $|f| = f^+ + f^-$, it too is integrable over B. The inequalities

$$0 \leqq f^+ \leqq |f|, \qquad 0 \leqq f^- \leqq |f|, \qquad f \leqq |f|, \qquad -f \leqq |f|$$

are obvious. By integration (recalling Theorem I-6-2) we obtain the first two inequalities in the conclusion, and also

$$\int_B f(x)\,m(dx) \leqq \int_B |f(x)|\,m(dx), \qquad -\int_B f(x)\,m(dx) \leqq \int_B |f(x)|\,m(dx).$$

If the integral of f is nonnegative, the first of these inequalities is the last inequality in the conclusion; if the integral of f is negative, the second of these is the last inequality in the conclusion. In any case, the conclusion is valid.

★COROLLARY 3-5 *If $f_1, \ldots, f_n$ are functions integrable over a set $B \subset R$, $f_1 \vee \cdots \vee f_n$ and $f_1 \wedge \cdots \wedge f_n$ are integrable over B.*

Suppose first that $n = 2$. It is easy to verify that

$$f_1 \vee f_2 = [f_1 + f_2 + |f_1 - f_2|]/2;$$

for if $f_1(x) \geqq f_2(x)$, we have

$$f_1(x) \vee f_2(x) = f_1(x) \qquad \text{and} \qquad |f_1(x) - f_2(x)| = f_1(x) - f_2(x),$$

whereas if $f_1(x) < f_2(x)$, we have

$$f_1(x) \vee f_2(x) = f_2(x) \qquad \text{and} \qquad |f_1(x) - f_2(x)| = f_2(x) - f_1(x).$$

By Corollary 3-4, $|f_1 - f_2|$ is integrable over B, so $f_1 \vee f_2$ is also.

We proceed by induction. Suppose that the conclusion holds for $n = k$ and that $f_1, \ldots, f_{k+1}$ are integrable over B. By the induction hypothesis, $f_1 \vee \cdots \vee f_k$ is integrable over B. Since f_{k+1} is also integrable over B, by the first part of this proof $[f_1 \vee \cdots \vee f_k] \vee f_{k+1}$ is integrable over B. But this function is the same as $f_1 \vee \cdots \vee f_{k+1}$, so the conclusion of the theorem concerning maxima holds for $n = k + 1$. By induction, it holds for all n.

The part of the conclusion concerning $f_1 \wedge \cdots \wedge f_n$ could be proved similarly, but it is easier to obtain it from the fact that

$$f_1 \wedge \cdots \wedge f_n = -[(-f_1) \vee \cdots \vee (-f_n)].$$

★COROLLARY 3-6 *If f is bounded on a set $B \subset R$ and is integrable over B, and n is a positive integer, f^n is also integrable over B.*

By hypothesis, there is a number c such that the values of $f(x)$ for x in B all lie in the set $D = [-c, c]$. On D define $g(y) = y^n$. This has a derivative ny^{n-1} that is bounded on D, so g is Lipschitzian on D. By Theorem 3-1 the composite function $g \circ f = f^n$ is integrable over B.

★COROLLARY 3-7 *If f and g are defined and bounded on a set $B \subset R$ and are integrable over B, their product fg is integrable over B.*

The functions f, g, and $f + g$ are all bounded on B and integrable over B. By Corollary 3-6 their squares are integrable over B, and so therefore is the combination

$$[(f+g)^2 - f^2 - g^2]/2.$$

But this combination is fg.

From Corollary 3-7 we obtain a corollary, but an unsatisfactory one.

★COROLLARY 3-8 *If f is bounded on a set $B \subset R$ and is integrable over B, and A is a bounded interval contained in B, f is integrable over A.*

If x is in A, the left member of the equation

$$f_A(x) = f_B(x) \cdot 1_A(x) \tag{G}$$

is $f(x)$, and the two factors in the right member are $f(x)$ and 1, respectively. If x is not in A, both members of (G) are 0. So f_A is the product of the two bounded functions f_B and 1_A, both of which are integrable over R, and therefore f_A is integrable over R. This is equivalent to saying that f is integrable over A, by Theorem I-5-7.

This corollary is unsatisfactory, even though it applies to many examples, because of the assumptions that f is bounded on B and A is bounded. It would be possible to show that f is integrable over A even if the boundedness hypotheses are omitted. The proof, however, is somewhat tedious, and since the result will be a simple consequence of a theorem that we shall prove in Section 4, we shall not prove it here.

EXERCISE 3-1 Prove that if f is integrable over B and there is a positive ε such that $|f(x)| \geqq \varepsilon$ for all x in B, then $1/f$ is integrable over B. *Suggestion*: Define $g(y) = 1/y$ for $|y| \geqq \varepsilon$, $g(y) = y/\varepsilon^2$ for $|y| < \varepsilon$.

EXERCISE 3-2 Prove that if f is bounded on B and integrable over B, and $\alpha \geqq 1$, $|f|^\alpha$ is integrable over B.

EXERCISE 3-3 Prove that if f is integrable over B and n is an even integer, $[1 + f^n]^{1/n}$ is integrable over B.

EXERCISE 4-4 Let the hypotheses of Theorem 3-1 be satisfied, and let L be a Lipschitz constant for g on the set of values assumed by f_B. Show that

$$\int_B g(f(x))\, m(dx) \leqq L \int_B |f(x)|\, m(dx).$$

By computing the integrals verify that when $B = [-4, 4]$, $f(x) = x^n$ on B, and $g(y) = y^2$ on R, $L = 2^{2n+1}$ is a Lipschitz constant for g on the set of values assumed by f_B, and the above inequality is satisfied. Give an example to show that in that inequality the right member cannot be replaced by

$$L\left|\int_B f(x)\, m(dx)\right|.$$

4. The Monotone Convergence Theorem

One of the most frequently encountered methods of defining functions makes use of composition of previously defined functions; and another uses limits of sequences of known functions. In Theorem 3-1 we saw how to deduce the integrability of a composite function from properties of the functions of which it is composed. In this section we wish to obtain conditions that will enable us to know that the limit of a sequence of functions is itself an integrable function and to find the value of its integral.

Let $f_0, f_1, f_2, \ldots$ be a sequence of functions all defined on a set D. The statement that as n increases, f_n **converges** to f_0, or, more emphatically, that f_n **converges pointwise** to f_0 on D, means that for each point x in D the sequence of numbers $f_n(x)$ converges to the number $f_0(x)$. If the f_n $(n = 1, 2, 3, \ldots)$ are known functions and we can show that for each point x of D the sequence of numbers $f_n(x)$ converges to a limit – which we can call $f_0(x)$ – we have thereby defined a function f_0 on D. If D is an interval and each f_n is known to be integrable over the interval D, it is tempting to conclude that the integrals of the f_n would necessarily converge to the integral of the limit f_0. This is not necessarily the case. Integrating is evaluating a limit, and two processes of taking limits cannot always be interchanged without affecting the result.

We need nothing as complicated as integration to show that this is so. Consider, for example, the function defined for all positive x and y by

$$f(x, y) = x^2/(x^2 + y^2).$$

If for fixed positive x we let y tend to 0, we obtain

$$\lim_{y \to 0} f(x, y) = 1,$$

so

$$\lim_{x \to 0} \lim_{y \to 0} f(x, y) = 1.$$

But if for fixed positive y we let x tend to 0, we obtain

$$\lim_{x \to 0} f(x, y) = 0,$$

so

$$\lim_{y\to 0}\lim_{x\to 0} f(x,y) = 0.$$

Since this noninterchangeability of limits is an important fact, it is worthwhile to look at some examples involving integrals that show how many possibilities there are.

Let f_0 be any function on $[0,1]$ such that $f_0(0) = 0$, and let $c_1, c_2, \ldots$ be a sequence of real numbers. For each positive integer n we define a function f_n on $[0,1]$ as follows.

$$f_n(0) = f_0(0),$$

$$f_n(x) = c_n \qquad (0 < x < 1/n),$$

$$f_n(x) = f_0(x) \qquad (1/n \leqq x \leqq 1).$$

Then for each x in $[0,1]$ we have $\lim f_n(x) = f_0(x)$. For if $x = 0$, all the $f_n(x)$ are equal to $f_0(x)$, whereas if $x > 0$, for all integers $n > 1/x$ we have $f_n(x) = f_0(x)$.

In particular, consider $f_0(x)$ identically 0. This has integral 0. The integral of f_n from 0 to 1 is easily computed to be c_n/n. If all c_n are 1, this tends to 0, which is the integral of f_0. If $c_n = n$, it has the limit 1, which is different from the integral of f_0. If $c_n = (-1)^n n$, c_n/n is alternately 1 and -1 and has no limit. If $c_n = n^2$, c_n/n increases without bound. So it is possible for f_0 to be integrable while the integrals of the f_n tend to the integral of f_∞, or tend to some other limit, or stay bounded but approach no limit, or increase without bound.

For another special case we choose $f_0(x)$ to be $1/x$ for $0 < x \leqq 1$, while $f_0(0) = 0$. Then f_0 is integrable over each interval $[1/n, 1]$ $(n = 1, 2, 3, \ldots)$ but is not integrable over $[0,1]$. If we choose any sequence of numbers $e_1, e_2, e_3, \ldots,$ we can select values of the c_n such that the integral of f_n from 0 to 1 is e_n. Thus, the values of the integrals of the f_n can have any kind of behavior that we choose, whereas f_0 is not integrable, not even if all the f_n have integral 0.

These examples show that, along with pointwise convergence of the functions f_n to a limit function f_0, some other condition has to be satisfied if we are to be able to conclude that f_0 is integrable and that its integral is the limit of the integrals of the f_n. Historically, the first such condition of any generality was uniform convergence. If $f_0, f_1, f_2, \ldots$ are all defined and real-valued on a set D, we say that f_n **converges to** f_0 **uniformly on** D if for each positive ε there is an n_ε such that for all $n > n_\varepsilon$ the inequality

$$|f_n(x) - f_0(x)| < \varepsilon$$

holds for all x in D. It was not until the last half of the nineteenth century that even the best mathematicians realized the distinction between pointwise convergence and uniform convergence. We have already seen examples showing the difference. In the first special case at the beginning of this section, with $f_0 = 0$, whether we choose $c_n = 1$, $c_n = (-1)^n n$, or $c_n = n^2$, f_n fails to converge uniformly to f_0. For if we choose $\varepsilon = \frac{1}{2}$, no matter how large n is, there will always be a point (namely, $x = \frac{1}{2}n$) at which $|f_n(x) - f_0(x)| > \varepsilon$.

We state the next theorem because of its historic interest, and we prove it because it is so easy to prove.

★THEOREM 4-1 *Let B be a bounded interval in R. Let $f_1, f_2, f_3, \ldots$ be functions integrable over B and converging uniformly to a limit function f_0 on B. Then f_0 is integrable over B, and*

$$\int_B f_0(x)\, m(dx) = \lim_{n\to\infty} \int_B f_n(x)\, m(dx).$$

Let ε be positive; define $\varepsilon_1 = \varepsilon/(2mB + 1)$. By hypothesis, there is an n_ε such that if $n > n_\varepsilon$,

$$\text{(A)} \qquad |f_n(x) - f_0(x)| < \varepsilon_1 \qquad \text{for all} \quad x \quad \text{in} \quad B.$$

Define $g_n = f_n - \varepsilon_1$, $h_n = f_n + \varepsilon_1$ on B. Since f_n and the constant ε_1 are integrable over B, so are g_n and h_n. By (A), $g_n \leqq f_0 \leqq h_n$ on B, and

$$\int_B h_n(x)\, m(dx) - \int_B g_n(x)\, m(dx) = \int_B 2\varepsilon_1\, m(dx) = 2\varepsilon_1 mB < \varepsilon.$$

So by Corollary 1-2, f_0 is integrable over B. Also, by (A), $f_0 - \varepsilon_1 < f_n < f_0 + \varepsilon_1$ on B for all $n > n_\varepsilon$, so by integrating over B

$$\int_B f_0(x)\, m(dx) - \varepsilon_1 mB \leqq \int_B f_n(x)\, m(dx) \leqq \int_B f_0(x)\, m(dx) + \varepsilon_1 mB.$$

Since $\varepsilon_1 mB < \varepsilon$, this implies

$$\left| \int_B f_n(x)\, m(dx) - \int_B f_0(x)\, m(dx) \right| < \varepsilon,$$

completing the proof.

The proof applies without change if the integrals are understood to be Riemann integrals. In this form it is the standard convergence theorem of the late nineteenth century. This theorem is not a triviality; it served well in many applications. But as mathematics advanced, both in pure mathematics and in its applications there arose with increasing frequency situations in which nonuniformly convergent sequences had to be investigated. To replace the older theorem we shall establish two theorems (one in this section) of much greater power. Their proofs are admittedly more difficult than that of the theorem about uniformly convergent sequences, but the theorems are so much stronger that the additional effort of proving them is soon repaid in effort saved, even in cases where it is possible but difficult to prove that the convergence is uniform. Besides that, and of more importance, the new theorems can be used in situations in which the older theorem cannot be used at all.

The first of these theorems applies to sequences of functions such that for each x in their domain, $f_1(x) \leqq f_2(x) \leqq f_3(x) \leqq \cdots$. We call such a sequence an **ascending sequence.** A sequence such that, for each x in the domain, it is true that $f_1(x) \geqq f_2(x) \geqq f_3(x) \geqq \cdots$ is called a **descending sequence**. A sequence is a **monotone sequence** if it is either an ascending sequence or a descending sequence.

For ascending sequences, at each x the numbers $f_n(x)$ either tend to a finite limit as n increases or else increase without bound and have ∞ as limit. We shall consider here only the case in which as n increases the $f_n(x)$ have a finite limit at each x. If the functions f_n are ascending and are integrable over a set B, their integrals also form a nondecreasing sequence of numbers, by Corollary I-6-2, so they too either tend to a finite limit or else increase without bound and have ∞ as limit.

★THEOREM 4-2 (Monotone Convergence Theorem) *Let B be a subset of R. Let $f_1, f_2, f_3, \ldots$ be a monotone sequence of functions on B, all integrable over B. Assume that the limit, as n increases, of $f_n(x)$ (which necessarily exists, finite or infinite) is a finite number for each x in B. Then the function whose value at each point x of B is*

$$\lim_{n\to\infty} f_n(x)$$

is integrable over B if and only if the integrals of the f_n over B are a bounded set of numbers, and in that case

$$\int_B \lim_{n\to\infty} f_n(x)\, m(dx) = \lim_{n\to\infty} \int_B f_n(x)\, m(dx).$$

It is sufficient to consider the case $f_1(x) \leqq f_2(x) \leqq f_3(x) \leqq \cdots$; the other case follows from this by a mere change of sign.

For convenience, we define

$$f_0(x) = \lim_{n\to\infty} f_n(x)$$

wherever the limit exists. Then $f_n(x) \leqq f_0(x)$ for all positive integers n and all x in B. So, if f_0 is integrable over B, by Corollary I-6-2 we have for all n

$$\int_B f_1(x)\, m(dx) \leqq \int_B f_n(x)\, m(dx) \leqq \int_B f_0(x)\, m(dx) < \infty,$$

and the sequence of the integrals of the f_n is bounded.

Conversely, suppose that the integrals

$$J_n = \int_B f_n(x)\, m(dx) \qquad (n = 1, 2, 3, \ldots)$$

have a finite upper bound. Since the J_n form a nondecreasing sequence, they tend to a finite limit, which we call J:

$$J = \lim_{n\to\infty} J_n.$$

We must show that this implies that the integral of f_0 over B exists and has the value J. We prove this first under the extra hypothesis

(B) $$f_1(x) \geqq 0 \qquad (x \text{ in } B).$$

The existence and value of the integral of a function f over B are unaffected by the values of f outside of B, so we may assume that $f_0, f_1, f_2, \ldots$ are all defined on all of $\bar{R}$ and are identically 0 on $\bar{R} \setminus B$. For each such f_j we have $(f_j)_B = f_j$, so the integral of f_j over B, which by definition is the gauge-limit of the sums $S(\mathscr{P}; (f_j)_B)$, is the same as the gauge-limit of the sums $S(\mathscr{P}; f_j)$.

Let ε be positive. Since J_n tends to J, we can and do choose a positive integer N such that

(C) $$J_N > J - \varepsilon/2.$$

For each x in $\bar{R}$ and each positive integer n, consider the inequality

(D) $$f_n(x) \geqq [(2J + \varepsilon)/(2J + 2\varepsilon)]f_0(x).$$

If $f_0(x) > 0$, the right member is less than $f_0(x)$ and $f_n(x)$ tends to $f_0(x)$ as n increases, so (D) is valid for all large n. If $f_0(x) = 0$, from the relation

$$0 \leqq f_1(x) \leqq f_n(x) \leqq f_0(x) = 0,$$

we see that all $f_n(x)$ are 0, and (D) holds for all n. In any case, for each x in $\bar{R}$ we can and do choose an integer greater than N, which we call $n(x)$, with which (D) is valid:

(E) $$n(x) > N \qquad \text{and} \qquad f_{n(x)}(x) \geqq [(2J + \varepsilon)/(2J + 2\varepsilon)]f_0(x).$$

For each positive integer n, the integral of f_n over R exists and has value J_n. We can therefore find a gauge γ'_n on $\bar{R}$ such that for every γ'_n-fine partition $\mathscr{P}$ of R,

(F) $$|S(\mathscr{P}; f_n) - J_n| < \varepsilon/2^{n+3}.$$

We define

$$\gamma_n = \gamma'_1 \cap \gamma'_2 \cap \cdots \cap \gamma'_n \qquad (n = 1, 2, 3, \ldots).$$

Obviously, for all x in $\bar{R}$

(G) $$\gamma_1(x) \supset \gamma_2(x) \supset \gamma_3(x) \supset \cdots.$$

By Theorem I-4-3, each γ_n is a gauge on $\bar{R}$, and if $\mathscr{P}$ is any γ_n-fine partition it is also γ'_n-fine, so that (F) holds. That is,

(H) if $\mathscr{P}$ is a γ_n-fine partition of R,

$$|S(\mathscr{P}; f_n) - J_n| < \varepsilon/2^{n+3}.$$

Let x be any point of $\bar{R}$. Since $\gamma_1, \gamma_2, \gamma_3, \ldots$ are gauges on $\bar{R}$, each of the sets

$$\gamma_1(x),\ \gamma_2(x),\ \gamma_3(x), \ldots$$

is a neighborhood of x. Just one of these has a subscript h equal to the number $n(x)$ that we chose in (E). So this set $\gamma_{n(x)}(x)$ is a neighborhood of x, which we rename $\gamma(x)$. Then the function γ on $\bar{R}$ defined by

$$\gamma(x) = \gamma_{n(x)}(x) \qquad (x \text{ in } \bar{R})$$

is a gauge on $\bar{R}$. We shall show that it is the gauge we have been seeking; that is, for every γ-fine partition $\mathscr{P}$ of R it is true that

$$\text{(I)} \qquad J - \varepsilon < S(\mathscr{P}; f_0) < J + \varepsilon.$$

Let

$$\text{(J)} \qquad \mathscr{P} = \{(x_1, A_1), \ldots, (x_k, A_k)\}$$

be any γ-fine partition of R. By (G) we have for $i = 1, \ldots, h$

$$A_i^- \subset \gamma(x_i) = \gamma_{n(x_i)}(x_i) \subset \gamma_N(x_i),$$

and therefore $\mathscr{P}$ is γ_N-fine. So, by (H), we have

$$|S(\mathscr{P}; f_N) - J_N| < \varepsilon/2^{N+3}.$$

Since $f_N(x) \leqq f_0(x)$ for all x in R, this and (C) imply

$$S(\mathscr{P}; f_0) \geqq S(\mathscr{P}; f_N) > J_N - \varepsilon/2^{N+3} > J - \varepsilon,$$

and the first inequality in (I) is satisfied.

Let M be the largest number in the set $\{n(x_1), \ldots, n(x_k)\}$. We define, for each positive integer h,

(K) $I[h]$ is the set of all integers in the set $\{1, \ldots, k\}$ such that $n(x_i) = h$.

Then the $I[h]$ are pairwise disjoint, and every i in $\{1, \ldots, k\}$ belongs to exactly one of the sets $I[1], \ldots, I[M]$. We also define

$$\text{(L)} \qquad A[h] = \bigcup\{A_i : i \text{ in } I[h]\},$$

$$\text{(M)} \qquad \mathscr{P}[h] = \{(x_i, A_i) : i \text{ in } I[h]\}.$$

For each h in $\{1, \ldots, M\}$ and each pair (x_i, A_i) in $\mathscr{P}[h]$, $n(x_i) = h$, so

$$A_i^- \subset \gamma(x_i) = \gamma_{n(x_i)}(x_i) = \gamma_h(x_i).$$

Therefore,

(N) $\mathscr{P}[h]$ is a γ_h-fine partition of $A[h]$.

For each i in $\{1, \ldots, k\}$ we can and do choose a γ_M-fine partition $\mathscr{P}_i^*$ of A_i, and we define

$$\mathscr{P}^*[h] = \bigcup_{i \text{ in } I[h]} \mathscr{P}_i^*.$$

By Lemma 2-1, $\mathscr{P}^*[h]$ is a γ_M-fine partition of $A[h]$, so by (G), if (x, A) is a pair that belongs to $\mathscr{P}^*[h]$,

$$A^- \subset \gamma_M(x) \subset \gamma_h(x).$$

Therefore,

(O) $\quad \mathscr{P}^*[h]$ is a γ_h-fine partition of $A[h]$.

From (N), (O), and (H) we deduce by Lemma 2-3

$$\text{(P)} \qquad S(\mathscr{P}[h], f_h) < S(\mathscr{P}^*[h], f_h) + \varepsilon/2^{h+2}.$$

For each pair (x_i, A_i) in $\mathscr{P}[h]$ we have $n(x_i) = h$ by (M) and (K), so by (E),

$$f_h(x_i) \geqq [(2J + \varepsilon)/(2J + 2\varepsilon)]f_0(x_i).$$

Therefore, the left member of (P) is not increased if we replace f_h by $[(2J + \varepsilon)/(2J + 2\varepsilon)]\, f_0$. The f_j are ascending and $M \geqq h$, so $f_M \geqq f_h$, and the right member of (P) is not decreased if we replace f_h by f_M. Therefore,

$$\text{(Q)} \qquad [(2J + \varepsilon)/(2J + 2\varepsilon)]S(\mathscr{P}[h], f_0) < S(\mathscr{P}^*[h], f_M) + \varepsilon/2^{h+2}.$$

We add these inequalities member by member for $h = 1, \ldots, M$. The sum of the $S(\mathscr{P}[h], f_0)$ is the sum of the terms $f_0(x_i)mA_i$ for all i in $I[1] \cup I[2] \cup \cdots \cup I[M]$, which is the sum over all i in $\{1, \ldots, k\}$. This is $S(\mathscr{P}; f_0)$. The sum of the $S(\mathscr{P}^*[h]; f_M)$ is the sum of the terms $f_M(x_i^*)mA_i^*$ for all pairs (x_i^*, A_i^*) in the set

$$\mathscr{P}^* = \mathscr{P}^*[1] \cup \cdots \cup \mathscr{P}^*[M],$$

and by Lemma 2-1, $\mathscr{P}^*$ is a γ_M-fine partition of $A[1] \cup \cdots \cup A[M]$, which is R. So by Lemma 2-1, the sum of the right members of inequalities (Q) is

$$S(\mathscr{P}^*; f_M) + \varepsilon/2^3 + \varepsilon/2^4 + \cdots + \varepsilon/2^{M+3}.$$

Combining these statements yields, with (H),

$$\begin{aligned} S(\mathscr{P}; f_0) &< [(2J + 2\varepsilon)/(2J + \varepsilon)][S(\mathscr{P}^*; f_M) + \varepsilon/4] \\ &< [(2J + 2\varepsilon)/(2J + \varepsilon)][J_M + \varepsilon/2^{M+3} + \varepsilon/4] < J + \varepsilon. \end{aligned}$$

So the second inequality in (I) is valid, and the theorem is proved under the extra hypothesis (B).

If the f_n form a nondecreasing sequence with bounded integrals but (B) is not satisfied, we define $g_n(x) = f_n(x) - f_1(x)$ for all x in $\bar{R}$. These form a nondecreasing sequence with bounded integrals, and by Theorem I-5-3 they satisfy (B). So, by the part of the proof already completed, the limit of the g_n is integrable over B, and its integral is the limit of the integrals of the g_n. But the limit of the g_n is

$\lim f_n - f_1$, so $\lim f_n$ must also be integrable. Also,

$$\begin{aligned}\int_B \lim_{n\to\infty} f_n(x)\,m(dx) &= \int_B \lim_{n\to\infty} g_n(x)\,m(dx) + \int_B f_1(x)\,m(dx)\\ &= \lim_{n\to\infty}\left\{\int_B f_n(x)\,m(dx) - \int_B f_1(x)\,m(dx)\right\} + \int_B f_1(x)\,m(dx)\\ &= \lim_{n\to\infty}\int_B f_n(x)\,m(dx),\end{aligned}$$

and the proof is complete.

Theorem 4-2 suggests the following natural and convenient extension of the use of the integration symbol.

★DEFINITION 4-3 *Let B be a subset of R, and let f be a function defined and real-valued on B but not integrable over B. If there exists an ascending sequence $f_1, f_2, f_3, \ldots$ of functions integrable over B and converging to f at each point of B, we define*

$$\int_B f(x)\,m(dx) = \infty;$$

if there exists a descending sequence $f_1, f_2, f_3, \ldots$ of functions integrable over B and converging to f at each point of B, we define

$$\int_B f(x)\,m(dx) = -\infty.$$

With this definition, the monotone convergence theorem takes the following easily remembered form.

★COROLLARY 4-4 *Let f be real-valued on a set B contained in R, and let $f_1, f_2, f_3, \ldots$ be a monotone sequence of functions integrable over B and converging to f at each point of B. Then*

$$\lim_{n\to\infty}\int_B f_n(x)\,m(dx) = \int_B f(x)\,m(dx).$$

Lemma I-3-2 can be improved thus.

LEMMA 4-5 *Let A be an interval (bounded or unbounded) in R. Then 1_A has an integral over R, and*

$$\text{(R)} \qquad \int_R 1_A(x)\,m(dx) = mA.$$

If A is bounded, this follows from Lemma I-3-2. If A is unbounded, let f_n $(n = 1, 2, 3, \ldots)$ be the indicator of $A \cap (-n, n]$. Then by Lemma I-3-2, f_n is integrable, and its integral is the length of $A \cap (-n, n]$, which is easily seen to increase without bound as n increases. For x not in A, $f_n(x) = 0$ for all n; for x in A, $f_n(x)$ is nondecreasing as n increases and is 1 for all n large enough so that x is in $(-n, n]$. By Theorem 4-2, 1_A is not integrable, so by Definition 4-3 the left member of (R) has the same value ∞ as the right member.

Definition 4-3 brings with it a linguistic oddity. The statements "f is integrable over B" and "f has an integral over B" no longer have the same meaning, since f can have an integral over B with value ∞ or $-\infty$ without being integrable over B. It is the pair of statements "f is integrable over B" and "f has a *finite* integral over B" that are equivalent.

In the next example the convergence is in fact uniform, so Theorem 4-1 would have been adequate. Nevertheless, by using Theorem 4-2 we are saved the labor of proving the uniformity of the convergence.

EXAMPLE 4-6 As usual, for $u > 0$ we define

$$\log u = \int_1^u (1/x)\,dx.$$

Then

$$\lim_{n\to\infty} n(u^{1/n} - 1) = \log u.$$

Define f and f_n by

$$f(x) = 1/x, \qquad f_n(x) = x^{(1/n)-1} \qquad (x > 0).$$

Since $x^{1/n}$ tends to 1 as n increases, $f_n(x)$ tends to $f(x)$ for all positive x. If $1 \leqq x \leqq u$, we have

$$f_1(x) \geqq f_2(x) \geqq f_3(x) \geqq \cdots;$$

if $u \leqq x \leqq 1$, we have

$$f_1(x) \leqq f_2(x) \leqq f_3(x) \leqq \cdots.$$

In either case Theorem 4-2 applies, and

$$n(u^{1/n} - 1) = \int_1^u f_n(x)\,dx \to \int_1^u f(x)\,dx = \log u.$$

EXERCISE 4-1 If $g, f_0, f_1, f_2, \ldots$ are functions on a set B, the statement that f_n converges to f_0 **uniformly relative to** g means that for every positive ε there exists an n_ε such that if $n > n_\varepsilon$, $|f_n(x) - f_0(x)| < \varepsilon g(x)$ for all x in B. As a special case, uniform convergence relative to 1 is ordinary uniform convergence. Prove that if $g, f_0, f_1, f_2, \ldots$ are defined on a set B and $g, f_1, f_2, \ldots$ are integrable over B and f_n converges to f_0 uniformly relative to g, then f_0 is integrable over B, and

$$\int_B f_0(x)\,dx = \lim_{n\to\infty} \int_B f_n(x)\,dx.$$

EXERCISE 4-2 Let f be continuous and bounded on a bounded interval B. Prove that f is Riemann integrable over B. *Suggestion*: Let the functions in Exercise I-7-1, for given n, be called s_n and S_n. Then $S_n - s_n$ is nonnegative and descending, so by Theorem 4-2 its gauge-integral tends to 0.

EXERCISE 4-3 Prove the following statements.

(i) $\int_{-\infty}^{0} c\,dx = \infty$ if $c > 0$, $= -\infty$ if $c < 0$.

(ii) $\int_1^\infty x^{-1}\,dx = \infty$. *Suggestion*: By Theorem I-11-3, $\log 2 > 0$ and $\log 2^n = n \log 2$.

(iii) $\int_1^0 x^{-1}\,dx = -\infty$.

(iv) $\int_{-1}^{1} x^{-1}\,dx$ is meaningless.

(v) $\int_{-\infty}^{\infty} [x \vee 0]\,dx = \infty$.

(vi) $\int_{-\infty}^{5} x\,dx = -\infty$.

(vii) $\int_{-\infty}^{\infty} x\,dx$ is meaningless.

5. Integrals of Products

The rest of this book will contain many applications of the monotone convergence theorem – one of the most important and useful theorems of the whole theory. In this section and the next we shall present some simple and direct applications that show its power by greatly improving some previously proved theorems.

★LEMMA 5-1 *Let B be a subset of R, and let f and g be nonnegative functions on B that have integrals, finite or ∞, over B. Then fg has an integral over B; and if there exists a function M integrable over B such that*

(A) $$f(x)g(x) \leq M(x)$$

for all x in B, then fg is integrable over B.

Most of the labor of the proof goes into the routine proof of an unexciting preliminary statement.

(B) If f is nonnegative and has an integral over B, there exists an ascending sequence $f_1, f_2, f_3, \ldots$ of nonnegative bounded functions integrable over B such that $f_n(x)$ converges to $f(x)$ for every x in B.

If the integral of f over B is ∞, by Definition 4-3 there is an ascending sequence $f'_1, f'_2, f'_3, \ldots$ of functions integrable over B and tending everywhere in B to f. If f is integrable over B, we can choose all of the f'_n equal to f. For each positive integer n we define

$$f_n = n \wedge (f'_n \vee 0).$$

These are nonnegative, and they are bounded, since they never exceed n. They are integrable over B by Corollary 3-3. Obviously if u', u'', and v are real numbers and $u' \leqq u''$, then

$$u' \vee v \leqq u'' \vee v \qquad \text{and} \qquad u' \wedge v \leqq u'' \wedge v.$$

Hence, for each x in B

$$f_{n+1}(x) = (n+1) \wedge (f'_{n+1}(x) \vee 0) \geqq n \wedge (f'_{n+1}(x) \vee 0)$$
$$\geqq n \wedge (f'_n(x) \vee 0) = f_n(x),$$

so the f_n form an ascending sequence on B. For each x in B and for all n greater than $f(x)$, n is greater than $f'_n(x) \vee 0$, so

$$f_n(x) = f'_n(x) \vee 0.$$

The right member of this equation tends to $f(x) \vee 0$ as n increases, and $f(x) \vee 0 = f(x)$. So, $f_n(x)$ converges to $f(x)$, and statement (B) is proved.

Now let f and g be nonnegative and have integrals over B. By (B) we can and do choose ascending sequences $f_1, f_2, f_3, \ldots$ and $g_1, g_2, g_3, \ldots$ of nonnegative bounded integrable functions that converge to the respective limits f and g everywhere in B. Then the products

$$f_1g_1, f_2g_2, f_3g_3, \ldots$$

are nonnegative and ascending, and they are integrable over B by Corollary 3-7, and they converge to fg everywhere in B. By Corollary 4-4, fg has an integral (finite or ∞) over B. In particular, if there is an integrable function M on B such that $fg \leqq M$, then for all positive integers n we have

$$f_ng_n \leqq fg \leqq M.$$

So the integrals of the f_ng_n never exceed the integral of M, and by Theorem 4-2, fg is integrable.

We can now establish a great improvement on the unsatisfactory Corollary 3-8.

★COROLLARY 5-2 *Let f be integrable over a subset B of R, and let A be an interval (not necessarily bounded) in R. Then f is integrable over $A \cap B$.*

By Theorem I-5-7, f_B is integrable over R, so by Corollary 3-4, f_B^+ and f_B^- are also. By Lemma 4-5, 1_A has an integral over R, and obviously

$$f_B^+ 1_A \leqq f_B^+,$$

which is integrable. So by Lemma 5-1, $f_B^+ 1_A$ is integrable over R. The equation

$$f_B^+(x)1_A(x) = f_{A\cap B}^+(x)$$

holds for all x in R; for if x is in $A \cap B$, both members are equal to $f^+(x)$, and otherwise both members are 0. So $f_{A\cap B}^+$ is integrable over R, and by Theorem I-5-7, f^+ is integrable over $A \cap B$. Similarly, f^- is integrable over $A \cap B$, and therefore so is $f^+ - f^-$, which is f.

COROLLARY 5-3 *If a and $b > a$ are in $\bar{R}$ and f is a function integrable from a to b, then for every c and x in the closed interval $[a, b]$ the indefinite integral*

$$F(x) = \int_c^x f(u)\,du \tag{C}$$

exists.

By hypothesis, f is integrable over $(a, b]$. If $a \leqq c \leqq x \leqq b$, the integral

$$\int_{(c,x]} f(u)\,du$$

exists by Corollary 5-2, and this is the right member of (C). If $a \leqq x \leqq c \leqq b$, the integral

$$\int_{(x,c]} f(u)\,du$$

exists by Corollary 5-2, and this is the negative of the right member of (C).

The next theorem is much more powerful than Corollary 3-7.

★THEOREM 5-4 *If f is integrable over R and g is bounded and is integrable over every bounded interval contained in R (in particular, if g is bounded and integrable over R), then fg is integrable over R.*

We first prove a preliminary statement.

(D) If g is nonnegative and is integrable over every bounded interval contained in R, g has an integral (finite or ∞) over R.

For each positive integer n, let $g_n = g_{(-n,n]}$. For each x in R, let $n^*(x)$ be the least positive integer n such that x is contained in $(-n, n]$. Then

$$\begin{aligned} g_n(x) &= 0 \qquad (n < n^*(x)) \\ &= g(x) \qquad (n \geqq n^*(x)). \end{aligned}$$

Hence the g_n are ascending and converge to g. By hypothesis, g is integrable over

$(-n, n]$, so by Theorem I-5-7, g_n is integrable over R. Then either g is integrable over R, or by Definition 4-3 it has integral ∞ over R.

If f is integrable over R, so are f^+ and f^-. If g is bounded (say, $|g(x)| \leqq M$ for all x in R), g^+ and g^- have the same bound. If g is integrable over every bounded interval in R, so are g^+ and g^-, so by (D) these have integrals over R. This is true in particular if g is integrable over R. Since

$$f^+g^+ \leqq Mf^+,$$

and Mf^+ is integrable over R, by Lemma 5-1, f^+g^+ is integrable over R. Similarly f^+g^-, f^-g^+, and f^-g^- are integrable over R, and therefore so is the combination

$$f^+g^+ - f^-g^+ - f^+g^- + f^-g^- = (f^+ - f^-)(g^+ - g^-) = fg.$$

The proof is complete.

Example 5-5 Let f be integrable over R, and let k be any real number. By Theorem 5-4 the functions $x \mapsto f(x)\cos kx$ and $x \mapsto f(x)\sin kx$ are integrable over R.

These integrals are important in the theory of the Fourier transform, which we shall study in some detail in Chapter VI.

Example 5-6 Let f be defined on $[0, \infty)$. If for some real number a the product $x \mapsto f(x)\exp(-ax)$ is integrable over $[0, \infty)$, then for every $b > a$ the product $x \mapsto f(x)\exp(-bx)$ is integrable over $[0, \infty)$. (Use the familiar properties of the exponential function, or else postpone this example until after the next section.)

EXERCISE 5-1 Prove that if g is nonnegative on R and has an integral (finite or ∞) over every bounded interval in R, it has an integral over R. (The proof resembles that of (D).)

EXERCISE 5-2 Show that if f and g are integrable over every bounded interval in R, and F and G are their indefinite integrals, fG and Fg are integrable over every bounded interval.

EXERCISE 5-3 Prove that $x \mapsto (1 + x^2)^{-1}\cos x$ is integrable over R. (Assume that $x \to 1/(1 + x^2)$ is integrable over R; this will be proved in Exercise 7-2.)

EXERCISE 5-4 Prove that if f is continuous and $x \to x^2 f(x)$ is bounded on R, f is integrable over R.

6. Power Series

It is rather difficult to say in a universally satisfactory way just what we mean by a series. It would be formally correct to say that a series of numbers is a sequence in which the odd-numbered places are occupied by numbers and the even-numbered places by plus signs. Thus, a series of numbers is a sequence

$$c_1 + c_2 + c_3 + \cdots$$

in which the c_i are numbers. This differs from the sequence $c_1, c_2, c_3, \ldots$ in a subjective way. When we write the symbol for the series, we are in the first place specifying the sequence of numbers, and in the second place we are announcing that we are going to perform on the sequence of numbers some sort of operation that is a generalization of ordinary addition. The simplest of such operations, and the only one that we shall use, is this. First we form the sequence of partial sums

$$s_1 = c_1, \quad s_2 = c_1 + c_2, \quad \ldots, \quad s_n = c_1 + \cdots + c_n, \quad \ldots.$$

If this sequence has a limit L as n increases, we say that the series is **convergent**, and we define

$$\sum_{n=1}^{\infty} c_n = L.$$

The series is **absolutely convergent** if the series $|c_1| + |c_2| + |c_3| + \cdots$ is convergent.

A simple and useful type of absolutely convergent series is the geometric series

$$a + ar + ar^2 + ar^3 + \cdots,$$

where the ratio r has absolute value less than 1. For then by elementary algebra the sum of the absolute values of the first n terms is

$$\begin{aligned} s_n &= |a|(1 + |r| + |r|^2 + \cdots + |r|^{n-1}) \\ &= |a|(1 - |r|^n)/(1 - |r|) \\ &= |a|/(1 - |r|) - [|a|/(1 - |r|)]|r|^n, \end{aligned}$$

and the right member of this equation tends to $|a|/(1 - |r|)$ as n increases.

We first establish some elementary theorems about sums of series.

★LEMMA 6-1 *If $c_1 + c_2 + c_3 + \cdots$ is a convergent series of numbers, the numbers c_n are bounded and tend to 0 as n increases.*

Define

$$L = \sum_{n=1}^{\infty} c_n.$$

For each positive number ε there is an $n(\varepsilon)$ such that if $n \geqq n(\varepsilon)$, then

$$|s_n - L| < \varepsilon/2.$$

Then for $n > n(\varepsilon)$

$$|c_n| = |s_n - s_{n-1}| \leqq |s_n - L| + |L - s_{n-1}| < \varepsilon,$$

which proves that c_n tends to 0 as n increases. To show that the c_n are bounded, take $\varepsilon = 1$. Then

$$|c_n| < 1 \qquad (n > n(1)).$$

The largest of the numbers

$$|c_1|, \ldots, |c_{n(1)}|, 1$$

is then an upper bound for all the $|c_n|$.

★Lemma 6-2 *Let $c_1 + c_2 + c_3 + \cdots$ be a series of real numbers and j be a positive integer. Then the series $c_1 + c_2 + c_3 + \cdots$ converges if and only if the series $c_{j+1} + c_{j+2} + c_{j+3} + \cdots$ converges, and in that case*

$$\sum_{n=1}^{\infty} c_n = c_1 + \cdots + c_j + \sum_{n=j+1}^{\infty} c_n.$$

Whenever $k > j$,

$$s_k = c_1 + \cdots + c_j + [c_{j+1} + c_{j+2} + \cdots + c_k].$$

If the quantity in square brackets in the right member converges to a limit L as k increases, the left member converges to the limit $c_1 + \cdots + c_j + L$, and conversely. This establishes the lemma.

★Lemma 6-3 *If $a_1 + a_2 + a_3 + \cdots$ and $b_1 + b_2 + b_3 + \cdots$ are convergent series of real numbers and u and v are real numbers, the series $(ua_1 + vb_1) + (ua_2 + vb_2) + (ua_3 + vb_3) + \cdots$ is convergent, and*

$$\text{(A)} \qquad \sum_{n=1}^{\infty} (ua_n + vb_n) = u \sum_{n=1}^{\infty} a_n + v \sum_{n=1}^{\infty} b_n.$$

For each positive integer k,

$$(ua_1 + vb_1) + \cdots + (ua_n + vb_n) = u[a_1 + \cdots + a_n] + v[b_1 + \cdots + b_n].$$

As k increases, the two terms in the right member of this equation tend to the two terms in the right member of (A), which establishes the conclusion.

★Lemma 6-4 (Comparison Test) *If $a_1 + a_2 + a_3 + \cdots$ and $b_1 + b_2 + b_3 + \cdots$ are series of numbers, and the latter is convergent, and for each* *positive integer n it is true that*

$$|a_n| \leqq b_n,$$

then the series $a_1 + a_2 + a_3 + \cdots$ *is convergent, and*

$$\text{(B)} \qquad \left| \sum_{n=1}^{\infty} a_n \right| \leqq \sum_{n=1}^{\infty} b_n.$$

Define

$$B = \sum_{n=1}^{\infty} b_n.$$

If the a_n are nonnegative, for each positive integer k

$$s_k = a_1 + \cdots + a_k \leqq b_1 + \cdots + b_k \leqq B.$$

As k increases, the s_k ascend, so by Theorem 2-1 in the Introduction they approach a limit not greater than B, and the series converges.

Under the hypotheses of the lemma, the nonnegative numbers

$$a_n^+ = a_n \vee 0, \qquad a_n^- = (-a_n) \vee 0$$

do not exceed b_n, so by the preceding paragraph the series

$$a_1^+ + a_2^+ + a_3^+ + \cdots, \qquad a_1^- + a_2^- + a_3^- + \cdots$$

converge. By Lemma 6-3 so does

$$(a_1^+ - a_1^-) + (a_2^+ - a_2^-) + (a_3^+ - a_3^-) + \cdots,$$

which is $a_1 + a_2 + a_3 + \cdots$.

Since

$$\left| \sum_{n=1}^{k} a_n \right| \leqq \sum_{n=1}^{k} |a_n| \leqq \sum_{n=1}^{k} b_n \leqq B$$

for all positive integers k, by letting k increase we obtain (B).

★COROLLARY 6-5 *If* $a_1 + a_2 + a_3 + \cdots$ *converges absolutely, it converges.*

Apply Lemma 6-4 with $b_n = |a_n|$.

A series $a_1 + a_2 + a_3 + \cdots$ of functions defined and real-valued on a set B is said to **converge uniformly** to a sum L (L being a function on B) if to each positive ε there corresponds an integer $n(\varepsilon)$ such that for all $n > n(\varepsilon)$,

$$|[a_1(x) + \cdots + a_n(x)] - L(x)| < \varepsilon$$

for all x in B. There is a comparison test for uniform convergence that is practically the same as that in Lemma 6-4.

★Lemma 6-6 (Weierstrass Comparison Test) *Let*

$$(C) \qquad a_1 + a_2 + a_3 + \cdots$$

be a series of functions defined and real-valued on a set B, and let $b_1 + b_2 + b_3 + \cdots$ *be a convergent series of nonnegative numbers. If*

$$|a_n(x)| \leqq b_n$$

for all positive integers n and all x in B, the series (C) *converges absolutely and uniformly on B to a sum L which is a function* $x \mapsto L(x)$ *on B.*

For each fixed x in B the numbers $|a_1(x)|$, $|a_2(x)|$, $|a_3(x)|, \ldots$ satisfy the hypotheses of Lemma 6-4, so the series of numbers $a_1(x) + a_2(x) + a_3(x) + \cdots$ is absolutely convergent. By Corollary 6-5, it has a finite sum $L(x)$. Let ε be positive. Since the series $b_1 + b_2 + b_3 + \cdots$ converges, there is an integer $n(\varepsilon)$ such that if $n > n(\varepsilon)$,

$$\left| \sum_{j=1}^{\infty} b_j - [b_1 + \cdots + b_n] \right| < \varepsilon.$$

By Lemma 6-2,

$$\sum_{j=n+1}^{\infty} b_j < \varepsilon.$$

Again by Lemma 6-2, for all x in B

$$|L(x) - [a_1(x) + \cdots + a_n(x)]| = \left| \sum_{j=n+1}^{\infty} a_j(x) \right| \leqq \sum_{j=n+1}^{\infty} b_j < \varepsilon,$$

so the convergence is uniform.

From the monotone convergence theorem we deduce a corollary concerning the integration of series.

★Theorem 6-7 *Let* $c_1, c_2, c_3, \ldots$ *be functions integrable over a set B in R such that for each x in B the series* $c_1(x) + c_2(x) + c_3(x) + \cdots$ *is absolutely convergent. If the numbers*

$$\int_B [|c_1(x)| + \cdots + |c_k(x)|]\, m(dx)$$

have a finite upper bound for all k, the sum

$$f(x) = \sum_{n=1}^{\infty} c_n(x)$$

defines a function f integrable over B, and

$$\int_B f(x)\, m(dx) = \sum_{n=1}^{\infty} \int_B c_n(x)\, m(dx).$$

First, suppose that the c_n are nonnegative. For each positive integer k the sum

$$s_k = c_1 + \cdots + c_k$$

is integrable over B, and

$$\text{(D)} \qquad \int_B s_k(x)\, m(dx) = \sum_{n=1}^{k} \int_B c_n(x)\, m(dx).$$

As k increases, the functions s_k ascend and tend everywhere in B to f. Their integrals over B remain bounded because by hypothesis the right member of (D) has a finite upper bound. By the monotone convergence theorem (Theorem 4-2), the limit

$$\lim_{k \to \infty} s_k = \sum_{n=1}^{\infty} c_n$$

of the s_k is integrable, and its integral is

$$\text{(E)} \qquad \int_B \left[\sum_{n=1}^{\infty} c_n(x) \right] m(dx) = \lim_{k \to \infty} \sum_{n=1}^{k} \int_B c_n(x)\, m(dx)$$

$$= \sum_{n=1}^{\infty} \int_B c_n(x)\, m(dx).$$

So the conclusion holds if the c_n are nonnegative.

If the c_n satisfy the hypotheses of the theorem, so do the nonnegative functions c_n^+ and c_n^-. So, as we have just proved, the sum of the c_n^+ and the sum of the c_n^- are integrable, and

$$\text{(F)} \qquad \int_B \left[\sum_{n=1}^{\infty} c_n^+(x) \right] m(dx) = \sum_{n=1}^{\infty} \int_B c_n^+(x)\, m(dx),$$

$$\text{(G)} \qquad \int_B \left[\sum_{n=1}^{\infty} c_n^-(x) \right] m(dx) = \sum_{n=1}^{\infty} \int_B c_n^-(x)\, m(dx).$$

If we subtract the members of (G) from those of (F) and apply Lemma 6-3, we obtain

$$\int_B \left[\sum_{n=1}^{\infty} (c_n^+(x) - c_n^-(x)) \right] m(dx) = \sum_{n=1}^{\infty} \int_B [c_n^+(x) - c_n^-(x)]\, m(dx).$$

This is the conclusion of the theorem.

If in the series of functions

$$c_1(x) + c_2(x) + c_3(x) + \cdots \qquad (x \text{ in } R)$$

the functions c_n have the form

$$c_n(x) = a_{n-1} x^{n-1} \qquad (n = 1, 2, 3, \ldots),$$

the series is called a **power series**. Power series are especially well behaved. We shall establish a theorem about their convergence properties.

★THEOREM 6-8 *Let*

(H) $$a_0 + a_1x + a_2x^2 + a_3x^3 + \cdots$$

be a power series. There exists a number r in the interval $[0, \infty]$ *such that if* $x > r$, *the series* (H) *fails to converge, and if* $x < r$, *the series* (H) *and the two series*

(I) $$a_0x + (a_1/2)x^2 + (a_2/3)x^3 + \cdots + (a_n/[n+1])x^{n+1} + \cdots,$$

(J) $$a_1 + 2a_2x + 3a_3x^2 + \cdots + na_nx^{n-1} + \cdots$$

all converge. Moreover if b is any number such that $0 \leqq b < r$, *all three series* (H), (I), (J) *converge uniformly on the closed interval* $[-b, b]$.

The number r is called the **radius of convergence** of (H), and the interval $(-r, r)$ is its **interval of convergence**.

Series (H) converges for $x = 0$ and possibly for other x. We define r to be the supremum of the absolute values of numbers x such that (H) is convergent. Then, evidently, (H) diverges for all x with $|x| > r$. If $r = 0$, the statements about numbers x and b such that $|x| < r$ and $0 \leqq b < r$ are trivially true; there are no such numbers. Suppose, then, that $r > 0$; let b be a number such that $0 < b < r$. By definition of r, there is a number y with $|y| > b$ such that the series

$$a_0 + a_1y + a_2y^2 + \cdots$$

is convergent. By Lemma 6-1, there is a number M such that

(K) $$|a_ny^n| \leqq M$$

for all nonnegative integers n. Then for all x in $[-b, b]$ we have for all nonnegative integers n

(L) $$|a_nx^n| \leqq |a_nb^n| = |a_ny^n| \cdot |b^n/y^n| \leqq M|b^n/y^n|.$$

Since $|b/y| < 1$, the last expression is the term in place $n + 1$ of a geometric series with ratio between -1 and 1, and this series converges. By Lemma 6-6, series (H) converges uniformly on $[-b, b]$.

By (L), the inequality

$$|(a_n/[n+1])x^{n+1}| \leqq b|a_nb^n| \leqq (bM)|b^n/y^n|$$

holds for all x in $[-b, b]$ and all nonnegative integers n. So, again using the comparison test in Lemma 6-6, we find that series (I) converges uniformly on $[-b, b]$.

For series (J) we first choose a number c between b and $|y|$. For all nonnegative integers n and all x in $[-b, b]$,

(M) $$|na_nx^{n-1}| \leqq |na_nb^{n-1}| = |(n/b)a_ny^n(b/c)^n| \cdot |c/y|^n \leqq \{(n/b)M(b/c)^n\}(c/|y|)^n.$$

We denote the quantity in braces in (M) by C_n. Then,

$$C_{n+1}/C_n = (1 + 1/n)(b/c),$$

which is less than 1 if $n > b/(c - b)$. So, as n increases, C_n ascends as long as $n \leqq b/(c - b)$, and after that it descends with limit 0. It therefore has a finite greatest value, which we call M'. Then by (M),

$$na_n x^{n-1} \leqq M'(c/y)^n.$$

The right member is term $n + 1$ in a convergent geometric series, so by Lemma 6-6, series (J) converges uniformly on $[-b, b]$.

The series (I) is the term-by-term integral of series (H); that is, each term in series (K) is the integral from 0 to x of the corresponding term in series (H). Likewise (K) is the term-by-term derivative of series (H). We have shown that if r is the radius of convergence of series (H), (H) and its term-by-term integral and its term-by-term derivative all converge uniformly on every interval $[-b, b]$ with $0 \leqq b < r$. If (I) converged for some x with $|x| > r$, its term-by-term derivative would converge uniformly on an interval $[-b', b']$ with $r < b' < |x|$. But the term-by-term derivative of series (I) is series (H), which cannot converge at b'. So (I) converges for all x with $|x| < r$ but not for any x with $|x| > r$, and the radius of convergence of series (I) is the same as that of series (H). Similarly we prove that the radius of convergence of series (J) is the same as that of (H).

Power series can be differentiated and integrated inside the interval of convergence $(-r, r)$ as though they were polynomials, as the next theorem shows.

THEOREM 6-9 *Let r (> 0) be the radius of convergence of the power series* (H). *Then the sum $H(x)$ of series H is continuous on $(-r, r)$, and at each point x of $(-r, r)$ it has a derivative that is the sum of the series* (J), *and the integral of H from 0 to x is the sum of the series* (I).

For each x in $(-r, r)$ we denote the sums of the three series (H), (I), (J) by $H(x)$, $I(x)$, $J(x)$, respectively. The partial sum

$$s_n(x) = a_0 + a_1 x + a_2 x^2 + \cdots + a_{n-1} x^{n-1}$$

converges to $H(x)$ uniformly on every bounded interval $[-b, b]$ contained in $(-r, r)$, so by Theorem 4-1 we have for each x in $(-r, r)$

$$\text{(N)} \quad \int_0^x H(u)\,du = \lim_{n\to\infty} \int_0^x s_n(u)\,du$$

$$= \lim_{n\to\infty} \left[a_0 x + \left(\frac{a_1}{2}\right)x^2 + \left(\frac{a_2}{3}\right)x^3 + \cdots + \left(\frac{a_{n-1}}{n}\right)x^n \right] = I(x).$$

As a by-product of (L) in the proof of Theorem 6-8, $H(x)$ is bounded on $[-b, b]$

for all b in $(0, r)$, so by Theorem I-9-4, I is continuous (in fact, Lipschitzian) on $[-b, b]$. That is, the term-by-term integral of any power series over any closed bounded interval contained in its interval of convergence is a continuous function. But (H) is itself the term-by-term integral of (J), which has the same interval of convergence, so the sum H is continuous on the interval of convergence. By the fundamental theorem, the indefinite integral (I) of the continuous function H is differentiable at each point x in $(-r, r)$, and its derivative is $H(x)$. But this applies equally well to the power series (J) and its term-by-term integral (H), so the sum H of series (H) is differentiable at each x in $(-r, r)$, and its derivative is the sum $J(x)$ of series (J).

Since every power series can be differentiated term by term at each point in its interval of convergence, the term-by-term derivative being a convergent power series, we can apply this conclusion to the power series (J) that is the term-by-term derivative of (H). We find that J is differentiable at each point x of $(-r, r)$, and its derivative is the sum of the series arising by term-by-term differentiation of the series (J). We can keep this up indefinitely; the sum H of the power series (H) has derivatives of all orders, and they can all be obtained by successive term-by-term differentiations of series (H), just as though H were a polynomial.

This yields a formula for the coefficients in a power series.

Lemma 6-10 *Let f be a function defined on a set that contains a neighborhood of the origin* 0, *and let*

$$c_0 + c_1x + c_2x^2 + c_3x^3 + \cdots$$

be a power series whose sum is equal to $f(x)$ for all x in a neighborhood of 0. *Then*

$$c_k = D^kf(0)/k! \qquad (k = 0, 1, 2, 3, \ldots).$$

This series is called the **Taylor's series**, or the **MacLaurin series**, for f.

The equation

$$f(x) = c_0 + \sum_{n=1}^{\infty} c_nx^n$$

holds for all x in a neighborhood $(-\varepsilon, \varepsilon)$ of 0. Setting $x = 0$ yields

$$f(0) = c_0.$$

The derivative of f is given on $(-\varepsilon, \varepsilon)$ by

$$Df(x) = 0 + c_1 + \sum_{n=2}^{\infty} nc_nx^{n-1}.$$

Setting $x = 0$ yields

$$Df(0) = c_1.$$

Differentiating again,

$$D^2f(x) = 2c_2 + \sum_{n=3}^{\infty} n(n-1)c_n x^{n-2},$$

and setting $x = 0$ yields

$$D^2f(0) = 2!\,c_2.$$

The continuation by induction is obvious.

As a corollary, if two power series both converge and have the same sum on a neighborhood $(-\varepsilon, \varepsilon)$ of 0, they have the same coefficients, since the coefficients are expressed by means of the derivatives of the sum by Lemma 6-10.

We shall apply these theorems to three examples in each of which the power series is written in the form

(O) $\quad a_0 + (a_1/1!)x + (a_2/2!)x^2 + \cdots + (a_{n-1}/[n-1]!)x^{n-1} + \cdots.$

The next statement applies to all three examples.

(P) If the numbers $a_0, a_1, a_2, \ldots$ are bounded, the series (O) converges for all x; its radius of convergence is ∞.

Let M be an upper bound for $|a_n|$, let x be any real number, and let h be a positive integer greater than $2|x|$. Then for all positive integers j

$$\left|\frac{a_{h+j}}{[h+j]!}x^{h+j}\right| \leqq \left|\frac{a_{h+j}}{h!}x^h\right| \left|\frac{x}{h+1}\right| \left|\frac{x}{h+2}\right| \cdots \left|\frac{x}{h+j}\right| \leqq \frac{M|x^h|}{h!}\left(\frac{1}{2}\right)^j.$$

The last expression is the general term of a convergent geometric series, so by Lemma 6-4 the series

$$(a_{h+1}/[h+1]!)x^{h+1} + (a_{h+2}/[h+2]!)x^{h+2} + \cdots$$

is convergent. By Lemma 6-2 the series (O) is convergent, and statement (P) is established.

EXAMPLE 6-11 The exponential function is defined by

$$\exp x = 1 + (1/1!)x + (1/2!)x^2 + \cdots + (1/n!)x^n + \cdots.$$

By (P), this converges for every x. By Theorem 6-9, the exponential function is differentiable for every x, and

(Q) $\quad D\exp x = 0 + 1 + (1/2!)2x + \cdots + (1/n!)nx^{n-1} + \cdots = \exp x.$

By repetition, every derivative of $\exp x$ is $\exp x$.

By the chain rule, for all real numbers y and z the derivative of the function

$$x \mapsto \exp(y - x)\exp(x + z)$$

is

$$[D\exp(y - x)]\exp(x + z) + \exp(y - x)D\exp(x + z)$$
$$= [-\exp(y - x)]\exp(x + z) + [\exp(y - x)]\exp(x + z) = 0.$$

So $\exp(y - x)\exp(x + z)$ is a constant for x in R, and its value at $x = 0$ is the same as its value at $x = y$. Since $\exp(y - y) = 1$, this yields

$$\text{(R)} \qquad \exp(y + z) = (\exp y)(\exp z).$$

This holds for all real z and y. In particular, when $z = -y$,

$$1 = (\exp y)(\exp[-y]).$$

Since $\exp y$ is evidently positive when $y > 0$, this equation shows that it is also positive when $y < 0$.

The function $x \mapsto \exp x$ increases more rapidly than any polynomial, as the next lemma shows.

LEMMA 6-12 *If $P: x \mapsto P(x)$ is a polynomial and c is a positive number,*

$$\lim_{x \to \infty} P(x)/\exp cx = 0.$$

Let P be the polynomial

$$P(x) = a_0 + a_1 x + \cdots + a_k x^k \qquad (x \text{ real}).$$

By definition of $\exp c$, we have for all positive x

$$\exp cx > (1/[k + 1]!)(cx)^{k+1}.$$

Hence

$$\left|\frac{P(x)}{\exp cx}\right| \leqq \left|\frac{P(x)}{1/[k + 1]!}(cx)^{k+1}\right| \leqq \frac{[k + 1]!}{c^{k+1}}\left[\left|\frac{a_0}{x^{k+1}}\right| + \left|\frac{a_1}{x^k}\right| + \cdots + \left|\frac{a_k}{x}\right|\right].$$

The last expression evidently tends to 0 as x increases, which implies the conclusion of the lemma.

From Lemma 6-12 and (R) we deduce

$$\lim_{x \to -\infty} \exp x = \lim_{y \to \infty} [\exp 0/\exp y] = 0.$$

Let c be any positive number. By the preceding equation, there exists an x' such that $\exp x' < c$; and by Lemma 6-12, there exists an x'' such that $\exp x'' > c$. Since the exponential function is continuous, there exists a number x such that

$$\exp x = c.$$

Thus the function $x \mapsto \exp x$ (x in R) takes on all positive numbers, and no other numbers, as functional values.

By the chain rule,

$$D \log(\exp x) = (\exp x)^{-1} D \exp x = 1.$$

So, by the fundamental theorem, for all real x

$$\log(\exp x) = \log \exp 0 + \int_0^x 1 \, dx = x_0.$$

If c is positive, there is an x in R such that $\exp x = c$. By the preceding equation,

(S) $$\exp x = c \quad \text{if and only if} \quad \log c = x.$$

If $c > 0$ and b is a positive integer, we have by Theorem I-11-3,

$$\log c^b = b \log c,$$

whence by (S),

(T) $$c^b = \exp(b \log c).$$

If b is not a positive integer, we have not yet defined c^b. We now accept (T) as the *definition* of c^b for all positive c and all real b. It follows readily from (T) and (R) that if $c > 0$ and a and b are real,

$$c^{a+b} = \exp([a + b] \log c) = \exp(a \log c + b \log c)$$
$$= \exp(a \log c) \exp(b \log c) = c^a c^b.$$

Likewise,

$$(c^b)^a = \exp(a \log c^b) = \exp(a[b \log c])$$
$$= \exp([ab] \log c) = c^{ab}.$$

The number e is defined to be $\exp 1$; then

$$e = \exp 1 = 2.718281828459045 \ldots .$$

By (S),

$$\log e = 1,$$

and by (T), for all real b

$$e^b = \exp b.$$

This gives us an alternate and frequently used notation for the exponential function.

The reader should approach the next example as though he had never learned anything about trigonometry.

EXAMPLE 6-13 The functions $x \mapsto \cos x$ and $x \mapsto \sin x$ (x real) are defined as the sums of the power series

$$\cos x = 1 - x^2/2! + x^4/4! - \cdots + (-1)^n x^{2n}/(2n)! + \cdots,$$

$$\sin x = x - x^3/3! + x^5/5! - \cdots + (-1)^n x^{2n+1}/(2n+1)! + \cdots.$$

By (P), these series converge for all x, and by Theorem 6-9,

(U) $$D\cos x = -\sin x, \qquad D\sin x = \cos x.$$

From the definition it is obvious that

(V) $$\cos(-x) = \cos x, \qquad \sin(-x) = -\sin x,$$

and also

(W) $$\cos 0 = 1, \qquad \sin 0 = 0.$$

If a and y are real numbers, by the chain rule and (U),

$$\begin{aligned} D[\cos(a-x)\cos(y+x) - \sin(a-x)\sin(y+x)] \\ = \sin(a-x)\cos(y+x) + \cos(a-x)[-\sin(y+x)] \\ + \cos(a-x)\sin(y+x) - \sin(a-x)\cos(y+x) \\ = 0. \end{aligned}$$

So the function

$$x \mapsto \cos(a-x)\cos(y+x) - \sin(a-x)\sin(y+x)$$

has the same value at $x = a$ as it has at $x = 0$, whence by (V) and (W),

(X) $$\cos(y+a) = \cos a \cos y - \sin a \sin y.$$

For each real a this holds for all real y, so we can differentiate with respect to y and obtain (changing signs)

(Y) $$\sin(y+a) = \cos a \sin y + \sin a \cos y.$$

If we set $y = -a$ in (X), we obtain, with (W) and (V),

(Z) $$1 = \cos^2 a + \sin^2 a \qquad (a \text{ in } R).$$

By (U) and the theorem of the mean, there is a number $\bar{x}$ in $(0, 2)$ such that

$$\sin 2 = (2 - 0)\cos \bar{x}.$$

If $\cos x$ were greater than $2^{-1/2}$ for all x in $(0, 2)$, this would imply

$$\sin 2 > 2^{1/2},$$

which is incompatible with (Z). So there are numbers in $[0, 2]$ for which

$$\cos x \leqq 2^{-1/2}.$$

We choose the least such number and we call it $\pi/4$. (There is a least, because $x \mapsto \cos x$ is continuous.) Since $\cos x > 2^{-1/2}$ for $0 \leqq x < \pi/4$, and by the theorem of the mean there is an $\bar{x}$ in $(0, \pi/4)$ for which

$$\sin \pi/4 - \sin 0 = (\pi/4) \cos \bar{x},$$

we see that $\sin \pi/4 > 0$. By (Z), with the fact that $\cos \pi/4 = 2^{-1/2}$, we obtain

$$\sin \pi/4 = 2^{-1/2}.$$

By (X) and (Y), with $a = y = \pi/4$,

$$\cos \pi/2 = 0, \qquad \sin \pi/2 = 1,$$

and again by (X) and (Y),

$$\cos \pi = -1, \qquad \sin \pi = 0,$$

and also

$$\text{(AA)} \qquad \cos 2\pi = 1, \qquad \sin 2\pi = 0.$$

These last, with (X) and (Y), imply that for all y

$$\cos(y + 2\pi) = \cos y, \qquad \sin(y + 2\pi) = \sin y.$$

So the functions cos and sin have period 2π.

To connect these functions with the sine and cosine of elementary trigonometry, we shall assume in advance the formula for the length of a curve. For each real t we define $P(t)$ to be the point in the plane whose x-coordinate is $\cos t$ and whose y-coordinate is $\sin t$. By (Z), $P(t)$ is on the unit circle, with center $(0, 0)$ and radius 1. As t increases, we see by equations (U) that if $P(t)$ has a positive y-coordinate, its x-coordinate is decreasing, and if $P(t)$ has a positive x-coordinate, its y-coordinate is increasing. In common parlance, as t increases, the motion of $P(t)$ on the unit circle is "counterclockwise." If $a < b$, the length of the arc described by $P(t)$ as t increases from a to b is

$$\int_a^b [(D \cos t)^2 + (D \sin t)^2]^{1/2} \, dt,$$

by the formula (not to be proved until Theorem V-5-3) for arc length. The integrand is 1, by (U) and (Z), so the arc length is $b - a$. In particular, if $t > 0$, the point $P(t)$ is the point reached by starting at $(1, 0)$ and moving counterclockwise a distance t around the unit circle. In trigonometry we say that the ray $OP(t)$ makes an angle of t radians with the positive x-axis. If $t < 0$, we reach $P(t)$ by moving clockwise a distance $|t|$ around the unit circle, and in this case too the ray $OP(t)$ makes an angle of t radians with the positive x-axis. So $\cos t$ and $\sin t$ are, respectively, the x-coordinate and the y-coordinate of the point that has distance 1 from the origin and is on the ray that makes angle t radians with the positive x-axis, in agreement with the definition in elementary trigonometry. Also, by

(AA), if we move a distance 2π around the unit circumference we are back to the starting point (0, 1). So, 2π is the length of the circumference of the circle with radius 1, and our number π is the same as the number π of geometry.

The other trigonometric functions are defined by the formulas

$$\tan x = \frac{\sin x}{\cos x}, \qquad \cot x = \frac{\cos x}{\sin x}, \qquad \sec x = \frac{1}{\cos x}, \qquad \csc x = \frac{1}{\sin x}$$

at all points x for which the denominators are not 0. In particular, $\tan x$ is continuous for $-\pi/2 < x < \pi/2$, and its derivative is

$$D \tan x = \frac{[D \sin x] \cos x - [D \cos x] \sin x}{\cos^2 x} = \sec^2 x = 1 + \tan^2 x.$$

So, the function $x \mapsto \tan x$ is increasing on $(-\pi/2, \pi/2)$. It follows readily from the preceding calculations that

$$\lim_{x \to \pi/2} \tan x = \infty, \qquad \lim_{x \to -\pi/2} \tan x = -\infty.$$

Hence, on $(-\pi/2, \pi/2)$, $\tan x$ takes on all real numbers as functional values, and it therefore has an inverse function, which we name the arc tangent. By the standard theorem of calculus on the derivatives of inverse functions, at $y = \tan x$ we have

(BB) $$D \arctan y = 1/D \tan x = 1/(1 + \tan^2 x) = 1/(1 + y^2).$$

We have already used this equation in Section I-12.

EXERCISE 6-1 If P is any polynomial and $c > 0$, the functions

$$x \mapsto P(x) \exp(-|x|), \qquad x \mapsto P(x) \exp(-x^2)$$

are integrable over R. (See Exercise 5-4.)

Remark The class of functions $x \to P(x) \exp(-x^2)$ includes the well-known Hermite functions, studied in Chapter VI.

EXERCISE 6-2 Prove that $(\sin x)/x$ tends to 1 as x tends to 0.

EXERCISE 6-3 Prove that if $t \geqq 0$, $\sin t \leqq t$. [Apply the theorem of the mean to $t \mapsto t - \sin t$ and recall (Z).]

EXERCISE 6-4 Use Lemma 6-12 to prove that

$$\lim_{x \to 0} P(1/x) \exp(-1/x^2) = 0$$

for every polynomial P.

EXERCISE 6-5 Show that the function f defined by

$$f(x) = \begin{cases} \exp(-1/x^2) & (x \neq 0), \\ 0 & (x = 0) \end{cases}$$

has continuous derivatives of all orders on R, all having the value 0 at 0.

Remark This example shows that it is possible for a function to have infinitely many derivatives and to have a convergent MacLaurin series whose sum is nevertheless different from $f(x)$ for every nonzero number x.

EXERCISE 6-6 Show that if

$$c_n(x) = 2^{-n} \sin 2^n x \qquad (n = 1, 2, 3, \ldots; x \text{ in } R),$$

the series $c_1 + c_2 + c_3 + \cdots$ converges absolutely and uniformly on R, but it cannot be differentiated term by term.

EXERCISE 6-7 Let $g(x) = \tan x$ where $\cos x \neq 0$, and $g(x) = 0$ where $\cos x = 0$. Show that the series

$$g(x) + g(x/2) + g(x/2^2) + \cdots + g(x/2^n) + \cdots$$

converges uniformly on every bounded interval in R.

EXERCISE 6-8 Let $a_1 + a_2 + a_3 + \cdots$ and $b_1 + b_2 + b_3 + \cdots$ be convergent series of nonnegative numbers. For each positive integer k define

$$c_k = a_1 b_k + a_2 b_{k-1} + \cdots + a_k b_1.$$

Show that the series $c_1 + c_2 + c_3 + \cdots$ converges, and

$$\sum_{n=1}^{\infty} c_n = \left[\sum_{n=1}^{\infty} a_n\right]\left[\sum_{n=1}^{\infty} b_n\right].$$

Suggestion: If n^* is the largest integer $\leqq n/2$,

$$\left[\sum_{k=1}^{n^*} a_k\right]\left[\sum_{k=1}^{n^*} b_k\right] \leqq \sum_{k=1}^{n} c_k \leqq \left[\sum_{k=1}^{n} a_k\right]\left[\sum_{k=1}^{n} b_k\right],$$

and the first and the last of these converge.

EXERCISE 6-9 Extend the conclusion in Exercise 6-8 to absolutely convergent series.

EXERCISE 6-10 Let the power series

$$a_0 + a_1 x + a_2 x^2 + \cdots, \qquad b_0 + b_1 x + b_2 x^2 + \cdots$$

both converge on an interval $(-c, c)$. Show that the product of their sums is the sum of the power series obtained by multiplying the two given power series like polynomials.

EXERCISE 6-11 Find the first four terms of the MacLaurin expansion of the function $x \mapsto \exp(\sin x)$.

EXERCISE 6-12 Find the power series for the function ("Fresnel integral")

$$x \mapsto \int_0^x \sin u^2 \, du.$$

EXERCISE 6-13 Let f be a function on R that satisfies the equation

$$Df = cf,$$

where c is a real number. Prove that there is a real number k such that

$$f(x) = k \exp cx \qquad (x \text{ in } R).$$

Suggestion: Compute the derivative of $x \mapsto f(x)\exp(-cx)$.

EXERCISE 6-14 (i) Find the MacLaurin series for $x \mapsto \log(1 + x)$.
(ii) Use the relations

$$\log(1 + x) = \int_1^{1+x} u^{-1} \, du = \int_0^x (1 + u)^{-1} \, du$$

and

$$(1 + u)^{-1} = 1 - u + u^2 - u^3 + u^4 - \cdots$$

to find the power series for the function $x \mapsto \log(1 + x)$. What is its interval of convergence?

EXERCISE 6-15 Define the hyperbolic functions cosh and sinh by

$$\cosh x = [\exp x + \exp(-x)]/2, \qquad \sinh x = [\exp x - \exp(-x)]/2.$$

Find power series expansions for these functions. Prove

$$D \cosh x = \sinh x, \qquad D \sinh x = \cosh x.$$

Prove

$$\cosh^2 x - \sinh^2 x = 1 \qquad (x \text{ in } R).$$

EXERCISE 6-16 Find the sum of the power series

$$1 - x^2 + x^4 - x^6 + \cdots.$$

Use this and (BB) to find by integration a power series expansion for the arc tangent function.

EXERCISE 6-17 Use Exercise 6-14 to prove that for every real number x,

$$\lim_{n \to \infty} (1 + x/n)^n = \exp x.$$

EXERCISE 6-18 If a principal P is left at interest at rate j for one interest period, the amount resulting is $P(1 + j)$; if left for another period, with compound interest, the amount is $[P(1 + j)](1 + j)$, and so on. Interest "at rate j per annum compounded n times a year" is defined to be interest at rate j/n compounded at time-intervals of length $1/n$ years. Show that the amount at the end of 1 year, when interest is at rate j per annum compounded n times a year, tends to $P \exp j$ as n increases. This is called "continuous compounding" at rate j per annum. With $P = 1000$ and $j = 0.06$, compute the amount at the end of 1 year if interest is compounded once a year; if compounded 12 times a year (monthly); if compounded 365 times a year (daily); if compounded continuously.

EXERCISE 6-19 In elementary mathematics, when $c > 0$ and m and n are positive integers, one defines $c^0 = 1$, $c^{-n} = 1/c^n$, $c^{m/n} = (\sqrt[n]{c^m})$, where the last is that positive number whose nth power is c^m. Show that the c^b defined in (T) has all these properties.

7. "Improper" Integrals

With the Riemann integral, no unbounded function can be integrated. However, in some cases a function f defined on an interval $[a, b]$ and having a single infinite discontinuity can be handled by a special device. Suppose, to be specific, that f is defined on $[a, b]$ and that for every number u such that $a < u < b$ the Riemann integral of f from u to b exists. It may happen that this integral tends to a limit as u tends to a, even though f is unbounded on $[a, b]$. In this case it has been customary to call the limit the "improper integral" of f from a to b and to denote it by the usual symbol.

Likewise, the Riemann integral cannot be used directly to find the integral of a function over an unbounded interval. But if f is defined on $[a, \infty)$, it may happen that the integral of f over every bounded subinterval $[a, u)$ exists and that this integral tends to a limit as u increases. In this case, it has been customary to call the limit the "improper integral" of f from a to ∞.

The gauge-integral applies directly to unbounded functions and to unbounded sets. We have no need for any special definition to cover such cases. Nevertheless, it is often convenient to use the processes described in the preceding paragraph, not to define the integral (which is already defined), but to compute its value. This is possible, as the next theorem shows.

THEOREM 7-1 *Let $(a, b]$ be an interval in $\bar{R}$, and let $a_1, a_2, a_3, \ldots$ and $b_1, b_2, b_3, \ldots$ be two sequences of real numbers such that*

$$a_1 \geqq a_2 \geqq a_3 \geqq \cdots \quad \text{and} \quad a_1 < b_1 \leqq b_2 \leqq b_3 \leqq \cdots$$

and

$$\lim_{n\to\infty} a_n = a, \qquad \lim_{n\to\infty} b_n = b.$$

Let f be defined on $(a, b]$ and integrable over each interval $(a_n, b_n]$. Then f is integrable from a to b if and only if the numbers

$$\text{(A)} \qquad \int_{a_n}^{b_n} |f(x)|\, dx$$

are bounded, and in that case the limit

$$\text{(B)} \qquad \lim_{n\to\infty} \int_{a_n}^{b_n} f(x)\, dx$$

exists, and

$$\text{(C)} \qquad \int_a^b f(x)\, dx = \lim_{n\to\infty} \int_{a_n}^{b_n} f(x)\, dx.$$

Observe that we could choose all the a_n to be a or all the b_n to be b. Also, a can be $-\infty$ and b can be ∞.

We defined the integral of f from a to b to be the integral of f over $(a, b]$, which is the integral of $f_{(a,b]}$ over R. But changing the integrand at a single point leaves the integral unaffected, so the integral of f from a_n to b_n is the same as its integral over the open interval

$$B(n) = (a_n, b_n).$$

Similarly the integral of f from a to b is equal to its integral over the open interval $B = (a, b)$.

If f is integrable over (a, b), so is $|f|$, and the numbers (A) cannot exceed the integral of $|f|$ over B. To prove the main part of the theorem, we assume that the numbers (A) are bounded. We first add the supplementary hypothesis

$$\text{(D)} \qquad f(x) \geqq 0 \quad \text{for all} \quad x \quad \text{in} \quad B.$$

Let f_n be $f_{B(n)}$, which coincides with f on $B(n)$ and is 0 on $\bar{R} \setminus B(n)$. By hypothesis, the integral

$$\int_{a_n}^{b_n} f(x)\, dx = \int_{B(n)} f(x)\, dx = \int_R f_n(x)\, dx$$

exists for each positive integer n.

For each positive integer n the inequality

$$f_n(x) \leqq f_{n+1}(x)$$

holds for all x in R. For if x is in $B(n)$, it is also in $B(n+1)$, and both members are equal to $f(x)$; and if x is not in $B(n)$, the left member is 0 and the right member is nonnegative. So the sequence $f_1, f_2, f_3, \ldots$ is ascending. It has a limit, which we call f_0. For each x in B, x is in $B(n)$ for all large n, so for such n we have $f_n(x) = f(x)$. So the limit $f_0(x)$ is also equal to $f(x)$, which is equal to $f_B(x)$, because x is in B. If x is not in B, it is not in any $B(n)$, and $f_B(x)$ and all the $f_n(x)$ and their limit $f_0(x)$ have the same value 0. Therefore, f_B is identically equal to the limit f_0 of the f_n. By the monotone convergence theorem, this limit is integrable, and its integral is the limit of the integrals over R of the functions f_n. So the conclusions of the theorem are established under the supplementary hypothesis (D).

Assume now that the hypotheses of the theorem are satisfied, but not necessarily (D). Since f is integrable over each $B(n)$, so are f^+ and f^-. By the part of the proof already completed, f^+ and f^- are integrable from a to b, and therefore so is $f = f^+ - f^-$. Moreover, (C) holds for both f^+ and f^-, so

$$\begin{aligned}\int_a^b f(x)\,dx &= \int_a^b f^+(x)\,dx - \int_a^b f^-(x)\,dx \\ &= \lim_{n\to\infty}\left[\int_{a_n}^{b_n} f^+(x)\,dx - \int_{a_n}^{b_n} f^-(x)\,dx\right] \\ &= \lim_{n\to\infty}\int_{a_n}^{b_n} f(x)\,dx.\end{aligned}$$

The proof is complete.

In the discussions of "improper integrals" in most texts, integrals of the type just discussed (but based on the Riemann integral over subintervals of B) are called "absolutely convergent improper integrals." Where f is integrable over (a, u) whenever $a < u < b$, and the integral of f over (a, u) tends to a limit as u tends to b, but the integral of $|f|$ over (a, u) fails to be bounded, it is customary to call the limit the "conditionally convergent improper integral" of f from a to b. In this case we could do likewise and define the "conditionally convergent (gauge) integral" of f in the same way. But such an integral would not be a gauge-integral, and the theorems that we have proved and shall prove about the gauge-integral would not apply to it. We prefer not to stretch the meaning of the word "integral" to cover such limits. We shall meet functions f for which the limit

$$\lim_{u\to b}\int_a^u f(x)\,dx$$

exists, although the integral of $|f|$ is unbounded, and this limit is sometimes important. In such cases we shall use the limit; we merely refuse to call it any kind of integral.

Theorem 7-1 has a convenient corollary.

COROLLARY 7-2 *Let (a, b) and the intervals (a_n, b_n) be as in Theorem 7-1. Let f and g be defined on $(a, b]$ and satisfy the inequality $|f| \leqq g$ on $(a, b]$. If f is integrable over each interval (a_n, b_n) and g is integrable from a to b, then f is integrable from a to b.*

Since for each positive integer n

$$\int_{a_n}^{b_n} |f(x)|\, dx \leqq \int_{a_n}^{b_n} g(x)\, dx \leqq \int_a^b g(x)\, dx,$$

the hypotheses of Theorem 7-1 are satisfied, so f is integrable from a to b.

In Section 8 we shall provide some nontrivial examples of such "improper" integrals. Here we give a few exercises, mostly banal.

EXERCISE 7-1 The function $x \to x^{-r}$ $(x > 0)$ is integrable over each bounded interval $(0, b]$ if $r < 1$, and then

$$\int_0^b x^{-r}\, dx = \frac{b^{1-r}}{1-r}.$$

EXERCISE 7-2 The function $x \mapsto x^{-r}$ $(x > 0)$ is integrable over $[a, \infty)$ for every positive a if $r > 1$, and then

$$\int_a^\infty x^{-r}\, dx = \frac{a^{1-r}}{r-1}.$$

EXERCISE 7-3 Prove that

$$\int_{-\infty}^{\infty} (1 + x^2)^{-1}\, dx = \pi.$$

(See equations (AA) and (BB) in Section 6.)

EXERCISE 7-4 Show that the function $x \mapsto \exp(-cx^2)$ is integrable over R if $c > 0$. *Suggestion*:

$$\exp(-cx^2) = 1/\exp cx^2 < 1/(1 + cx^2).$$

Use Exercise 7-3.

EXERCISE 7-5 Prove that if $k^2 < 1$, the function

$$x \mapsto [(1 - x^2)(1 - k^2x^2)]^{-1/2}$$

is integrable from 0 to 1. *Suggestion*: Use the identity

$$1 - x^2 = (1 - x)(1 + x).$$

The integrand does not exceed a constant multiple of $(1 - x)^{-1/2}$.

EXERCISE 7-6 Prove that an unthinking application of the fundamental theorem to the integral

$$\int_{-1}^{+1} x^{-2}\, dx$$

would yield the ridiculous result that the integral has value 0. Show that the integral cannot exist.

EXERCISE 7-7 Find the integral from -3 to 3 of the function

$$x \to (9 - x^2)^{-1/2}.$$

EXERCISE 7-8 Find the values of p and q for which the integral

$$B(p, q) = \int_0^1 x^{p-1}(1 - x)^{q-1}\, dx$$

exists. (This is the "beta function.")

EXERCISE 7-9 Prove that $x \mapsto x/\sinh x$ is integrable over $R \setminus \{0\}$.

EXERCISE 7-10 Prove that if $0 < r < 1$, the integral

$$\int_0^1 (\log x)x^{-r}\, dx$$

exists.

8. Examples

The integrands in the examples we are about to exhibit are not pathological; they have discontinuities at only one or two points. All are within reach of the "improper Riemann integral" theory. The point is that by using the gauge-integral the results are obtained with a saving of labor that is slight in some cases, considerable in others.

We shall need an elementary calculation:

(A) If $p > 0$, then

$$\lim_{x \to 0+} x^p \log x = 0.$$

The function defined for positive u by $u \log u$ has derivative $1 \log u + u[1/u]$. This is negative for $u < 1/e$, where e is the number such that $\log e = 1$. So, if $0 < u < 1/e$,

$$0 > u \log u > (1/e)\log(1/e),$$

and $u \log u$ is bounded on $(0, 1/e)$. If we define $x = u^{2/p}$, then

$$x^p \log x = u^2 \log u^{2/p} = [2u/p][u \log u].$$

As x tends to 0, so does u, and the factor $[2u/p]$ tends to 0 whereas, as we have seen, the factor $u \log u$ remains bounded. So the product tends to 0.

EXAMPLE 8-1 If $p > 0$ and $b > 0$,

$$\int_0^b x^{p-1} \log x \, dx = [b^p/p^2][p \log b - 1].$$

We define

$$f(x) = x^{p-1}, \qquad F(x) = x^p/p, \qquad g(x) = 1/x, \qquad G(x) = \log x \qquad (x > 0).$$

If $c = 1 \wedge b$, then $x^{p-1} \log x < 0$ for $0 < x \leqq c$. Let $a_1, a_2, a_3, \ldots$ be a decreasing sequence of points in $(0, c)$ tending to 0. By integration by parts,

$$\int_{a_n}^c |x^{p-1} \log x| \, dx = -\int_{a_n}^c f(x)G(x) \, dx$$

$$= -\frac{c^p}{p} \log c + \frac{a_n^p}{p} \log a_n + \int_{a_n}^c \frac{x^p}{p} \frac{1}{x} \, dx.$$

As n increases, the last integral tends to a finite limit by Exercise 7-1, and the term before it tends to 0 by (A), so by Theorem 7-1, $x^{p-1} \log x$ is integrable from 0 to c. It is continuous from c to b, so it is integrable from 0 to b. To find the value of the integral, we again integrate by parts, this time from a_n to b, and obtain

$$\int_{a_n}^b x^{p-1} \log x \, dx = \frac{b^p}{p} \log b - \frac{a_n^p}{p} \log a_n - \int_{a_n}^b \frac{x^p}{p} \frac{1}{x} \, dx.$$

The second term in the right member tends to 0 by (A), and the last is the limit of $-b^p/p^2 + a_n^p/p^2$, which is $-b^p/p^2$. The proof is complete.

EXAMPLE 8-2 For all $p > -1$,

$$\int_0^1 \frac{-x^p}{1 - x} \log x \, dx = (p + 1)^{-2} + (p + 2)^{-2} + \cdots.$$

By the formula for the sum of a geometric progression,

$$(1 - x)^{-1} = \sum_{n=0}^{\infty} x^n$$

for $0 < x < 1$, and this formula continues to be correct for $x = 0$ if we understand $0^0 = 1$. For $0 < x < 1$ we define

$$f_n(x) = -x^{n+p}\log x, \qquad f(x) = -x^p(1-x)^{-1}\log x,$$

with $f(0) = f_n(0) = 0$. For $0 \leqq x < 1$ these are all nonnegative. If $0 < u < 1$, by the formula for integration by parts I-11-2,

$$\begin{aligned}\int_u^1 f_n(x)\,dx &= \int_u^1 [-\log x]x^{n+p}\,dx \\ &= -\frac{x^{n+p+1}\log x}{n+p+1}\bigg|_u^1 + \frac{1}{n+p+1}\int_u^1 x^{n+p+1}\frac{1}{x}\,dx \\ &= \frac{u^{n+p+1}\log u}{n+p+1} + \frac{1}{(n+p+1)^2}x^{n+p+1}\bigg|_u^1.\end{aligned}$$

By (A), as $u \to 0$ the first term in the right member tends to 0 and the second term obviously tends to $1/(n+p+1)^2$. So, by Theorem 7-1, f_n is integrable from 0 to 1, and

$$\int_0^1 f_n(x)\,dx = (n+p+1)^{-2}.$$

The series with terms $(n+p+1)^{-2}$ is well known to be convergent, so by Theorem 6-7 the sum of the series $f_0 + f_1 + \cdots$ is integrable, and its integral is the sum of the series with terms $(n+p+1)^{-2}$ for $n = 0, 1, 2, \ldots$. But

$$\sum_{n=0}^{\infty} f_n(x) = x^p \sum_{n=0}^{\infty} x^n[-\log x] = -x^p(1-x)^{-1}\log x,$$

which completes the proof.

It is interesting to notice that no matter how large p may be, the series $f_0 + f_1 + f_2 + \cdots$ does not converge to f uniformly on $[0, 1)$ or, in fact, on $(u, 1)$ for $0 < u < 1$. For the "tail" of the series is

$$\sum_{n \geqq N} [-x^{n+p}\log x] = (x-1)^{-1}x^{N+p}\log x,$$

and as x tends to 1 from below, this approaches 1, by de l'Hôpital's rule. So, no matter how large an N we choose, the sum of the first N terms differs from the limit by more than one-half somewhere in $(u, 1)$.

EXAMPLE 8-3 If p and q are positive,

$$\int_0^1 \frac{x^{p-1}}{1+x^q}\,dx = \frac{1}{p} - \frac{1}{p+q} + \frac{1}{p+2q} - \cdots.$$

For all x in $(0, 1)$ and all nonnegative integers n, define

$$f_n(x) = x^{p-1+2nq}(1 - x^q), \qquad f(x) = x^{p-1}(1 - x^q)^{-1}$$

and set $f_n(0) = f(0) = 0$. The f_n are nonnegative, and if $0 < u < 1$,

$$\int_u^1 f_n(x)\,dx = \int_u^1 x^{p-1+2nq}\,dx - \int_u^1 x^{p-1+2nq+q}\,dx$$

$$= \frac{x^{p+2nq}}{p + 2nq}\bigg|_u^1 - \frac{x^{p+2nq+q}}{p + 2nq + q}\bigg|_u^1.$$

As $u \to 0$, u^{p+2nq} and $u^{p+2nq+q}$ tend to 0, so by Theorem 7-1,

$$\int_0^1 f_n(x)\,dx = (p + 2nq)^{-1} - (p + 2nq + q)^{-1}.$$

By the formula for the sum of a geometric progression, for x in $[0, 1)$

$$\begin{aligned} f(x) &= x^{p-1}(1 + x^q)^{-1} \\ &= x^{p-1}(1 - x^q + x^{2q} - x^{3q} + \cdots) \\ &= x^{p-1}([1 - x^q] + x^{2q}[1 - x^q] + x^{4q}[1 - x^q] + \cdots) \\ &= x^{p-1}(1 - x^q)(1 + x^{2q} + x^{4q} + \cdots) \\ &= \sum_{n=0}^{\infty} f_n(x). \end{aligned}$$

Since f is integrable from 0 to 1, the series of the integrals of the f_n converges, and by Theorem 6-7, the integral of f is the sum of the integrals of the f_n. This is the statement to be proved.

EXERCISE 8-1 Use the equation proved in Example 8-3 to derive the formulas

$$\log 2 = 1 - \tfrac{1}{2} + \tfrac{1}{3} - \tfrac{1}{4} + \cdots, \qquad \pi/4 = 1 - \tfrac{1}{3} + \tfrac{1}{5} - \tfrac{1}{7} + \cdots.$$

EXAMPLE 8-4 If $r > 1$ and $k > 0$, the function $x \to (1 + kx^r)^{-1}$ is integrable from 0 to ∞.

It is integrable from 0 to 1, being continuous. For $n > 1$ it is integrable from 1 to n, being continuous, and

$$\int_1^n (1 + kx^r)^{-1}\,dx \leqq k^{-1}\int_1^n x^{-r}\,dx = \frac{1 - n^{1-r}}{k(r - 1)},$$

which is bounded for all n. So by Theorem 7-1, the integrand is integrable from 1 to ∞. By Theorem I-8-1, it is integrable from 0 to ∞.

EXAMPLE 8-5 If f is integrable from 0 to ∞ and $r > 1$, and $k > 0$, the integral

$$\int_0^\infty \frac{f(x)}{1 + kx^r}\,dx$$

exists, and it tends to 0 as $k \to \infty$.

Since $x \mapsto (1 + kx^r)^{-1}$ is bounded on $[0, \infty)$ and integrable over $[0, \infty)$, by Theorem 5-4 the integral exists. If k takes on any ascending sequence of values with limit ∞, $|f(x)|/(1 + kx^r)$ is a descending sequence with limit 0 for each x in $(0, \infty)$. By the monotone convergence theorem, the limit of the integral of $|f(x)|/(1 + kx^r)$ is 0, whence the conclusion follows.

EXAMPLE 8-6 If f is bounded on $(0, \infty)$ and is integrable over every subinterval of $(0, \infty)$, and $k > 0$ and $r > 1$, the integral

$$\int_0^\infty \frac{f(x)}{1 + kx^r}\,dx$$

exists, and it tends to 0 as $k \to \infty$.

The existence of the integral follows from Theorem 5-4 and Example 8-4; the value of the limit is found as in Example 8-5.

The rest of the examples in this section have to do with the gamma-function $x \mapsto \Gamma(x)$ $(x > 0)$.

DEFINITION 8-7 *If* $t > 0$,

$$\Gamma(t) = \int_0^\infty x^{t-1} e^{-x}\,dx.$$

To show that this integral exists for all positive t, we consider the intervals $(0, 1]$ and $(1, \infty)$ separately. On $(1, \infty)$ we write

$$x^{t-1} e^{-x} = [x^{-2}][x^{t+1}/e^x].$$

The first factor is integrable over $(1, \infty)$ by Exercise 7-2. If we fix on any integer $n > t + 1$, we have

$$e^x = 1 + x/1! + \cdots + x^n/n! + \cdots > x^n/n!,$$

so the second factor tends to 0 as $x \to \infty$, and it is bounded on $(0, \infty)$. Being continuous, it is integrable over every subinterval of $(0, \infty)$, so the product of the two factors is integrable over $(0, \infty)$ by Theorem 5-4. On $(0, 1]$ the factor x^{t-1} is integrable by Exercise 7-1 if $0 < t < 1$ and by continuity if $t \geqq 1$. Since $x \mapsto e^{-x}$ is bounded and continuous on $(0, 1]$, it is integrable, and by Theorem 5-4, $x \mapsto x^{t-1} e^{-x}$ is integrable over $(0, 1]$. Since we have already shown that it is integrable over $(1, \infty)$, it is integrable over $(0, \infty)$.

THEOREM 8-8 *If $t > 0$, $\Gamma(t+1) = t\Gamma(t)$.*

For each positive integer n, by integration by parts

$$\int_{1/n}^{n} x^t e^{-x}\,dx = -e^{-x}x^t\Big|_{1/n}^{n} - \int_{1/n}^{1} [tx^{t-1}][-e^{-x}]\,dx.$$

As $n \to \infty$, $-e^{-n}n^t$ tends to 0, since with m an integer $> t$,

$$0 < e^{-n}n^t < n^t/[n^m/m!].$$

Also, $e^{-1/n}(1/n)^t$ tends to 0. The other terms tend to $\Gamma(t+1)$ and $t\Gamma(t)$, respectively, by Theorem 7-1, completing the proof.

It is trivially easy to compute that $\Gamma(1) = 1$. From this and Theorem 8-8 we obtain by induction that

(B) For every positive integer n,

$$\Gamma(n) = (n-1)!.$$

EXAMPLE 8-9

$$\int_0^\infty e^{-x}\cos\sqrt{x}\,dx = 1 - \frac{1!}{2!} + \frac{2!}{4!} - \frac{3!}{6!} + \frac{4!}{8!} - \cdots.$$

We accept as known the expansion (see Example 6-13)

$$\cos y = 1 - \frac{y^2}{2!} + \frac{y^4}{4!} - \frac{y^6}{6!} + \cdots.$$

Then for all nonnegative x

$$e^{-x}\cos\sqrt{x} = e^{-x} - \frac{e^{-x}x}{2!} + \frac{e^{-x}x^2}{4!} - \cdots. \tag{C}$$

Define f_n ($n = 0, 1, 2, 3, \ldots$) by

$$f_n(x) = e^{-x}x^n/(2n!).$$

Then by Definition 8-7 and (B),

$$\left|\int_0^\infty f_n(x)\,dx\right| = \frac{\Gamma(n+1)}{(2n)!} = \frac{n!}{(2n)!}.$$

Since $n!/(2n)! \leqq 1/2^n$, this series converges. The partial sums of the series $f_1 + f_3 + f_5 + \cdots$ satisfy the hypotheses of the monotone convergence theorem, so the series sum $f_1 + f_3 + f_5 + \cdots$ is integrable, and its integral from 0 to ∞ is the sum of the integrals of the f_n for odd n. A similar statement holds for the sum $f_0 + f_2 + f_4 + \cdots$. Hence, the difference between the sum of the f_n with even n and the sum of the f_n with odd n, which by (C) is $e^{-x}\cos\sqrt{x}$, is integrable, and its integral from 0 to ∞ is the sum of the integrals of all the $(-1)^n f_n$. This completes the proof.

EXAMPLE 8-10 If t is positive and n is a positive integer,

$$\int_0^1 (1-x)^n x^{t-1}\,dx = \frac{n(n-1)(n-2)\cdots 2\cdot 1}{t(t+1)(t+2)\cdots(t+n)}.$$

The existence of the integral follows at once from Exercise 7-1 and Theorem 7-1, but the point is to find its value.

If we define f, g, F, and G by

$$\begin{aligned} f(x) &= x^{t-1} && (0 < t \leqq 1);\ f(0) = 0,\\ g(x) &= n(1-x)^{n-1} && (0 \leqq x \leqq 1),\\ F(x) &= x^t/t && (0 \leqq x \leqq 1),\\ G(x) &= -(1-x)^n && (0 \leqq x \leqq 1), \end{aligned}$$

the functions f and g are integrable, the former by Exercise 7-1 and the latter because it is continuous; F and G are continuous and bounded on $[0, 1]$; and $F'(x) = f(x)$ and $G'(x) = g(x)$ for $0 < x \leqq 1$. For $\varepsilon > 0$, by integration by parts,

$$\int_\varepsilon^1 (1-x)^n x^{t-1}\,dx = -\left.\frac{(1-x)^n x^t}{t}\right|_\varepsilon^1 + \frac{n}{t}\int_\varepsilon^1 (1-x)^{n-1} x^t\,dx.$$

By Exercise 7-1 and Theorem 7-1, letting $\varepsilon \to 0$ yields

$$\int_0^1 (1-x)^n x^{t-1}\,dx = \frac{n}{t}\int_0^1 (1-x)^{n-1} x^t\,dx.$$

But now we can apply this same formula to the integral in the right member, with $n-1$ in place of n and t in place of $t-1$, and continue the process until we obtain

$$\int_0^1 (1-x)^n x^t\,dx = \frac{n}{t}\,\frac{n-1}{t+1}\,\frac{n-2}{t+2}\cdots\frac{1}{t+(n+1)}\int_0^1 x^{t+n-1}\,dx.$$

The integral in the right member is easily computed, and from this last equation we obtain the statement that we wished to prove.

EXAMPLE 8-11 For all positive t,

$$\Gamma(t) = \lim_{n\to\infty}\left\{t^{-1} n^t \prod_{k=1}^{n}\left(1+\frac{t}{k}\right)^{-1}\right\}.$$

For each positive n we define

$$\Gamma_n(t) = \int_0^n \left(1-\frac{x}{n}\right)^n x^{t-1}\,dx.$$

For each number ε such that $0 < \varepsilon < n$ and each integer k such that $1 \leqq k \leqq n$ we obtain by integration by parts

$$\int_\varepsilon^n \left(1 - \frac{x}{n}\right)^k x^{t-1+n-k}\,dx = \left(1 - \frac{x}{n}\right)^k (t+n-k)^{-1} x^{t+n-k} \Bigg|_\varepsilon^n + \frac{k(t+n-k)}{n} \int_\varepsilon^n \left(1 - \frac{x}{n}\right)^{k-1} x^{t+n-k}\,dx.$$

Both integrals exist if we replace ε by 0, by Exercise 7-1 and Corollary 7-2, so by Theorem 7-1,

$$\int_0^n \left(1 - \frac{x}{n}\right)^k x^{t-1+n-k}\,dx = \frac{k}{n(t+n-k)} \int_0^n \left(1 - \frac{x}{n}\right)^{k-1} x^{t+n-k}\,dx.$$

By applying this repeatedly to the definition of Γ_n, we obtain

$$\Gamma_n(t) = \frac{n(n-1)(n-2)\cdots 1}{n^n t(t+1)(t+2)\cdots(t+n-1)} \int_0^n \left(1 - \frac{x}{n}\right)^0 x^{t+n-1}\,dx$$

$$= n^t \frac{n(n-1)(n-2)\cdots 1}{t(t+1)\cdots(t+n)} = n^t t^{-1} \prod_{k=1}^n \left(1 + \frac{t}{k}\right)^{-1}.$$

So the problem now is to prove that

(D) $$\lim_{n\to\infty} \Gamma_n(t) = \Gamma(t) \qquad (t > 0).$$

We define f_n on $[0, \infty)$ by

$$f_n(x) = \begin{cases} \left(1 - \dfrac{x}{n}\right)^n x^{t-1} & \text{if } 0 < x < n, \\ 0 & \text{for all other } x. \end{cases}$$

Then

(E) $$\Gamma_n(t) = \int_0^\infty f_n(x)\,dx.$$

We now prove that for all nonnegative x,

(F) $$f_1(x) \leqq f_2(x) \leqq f_3(x) \leqq \cdots \quad \text{and} \quad f_n(x) \to e^{-x}x^{t-1},$$

the right member being given the value 0 at $x = 0$. This is trivial at $x = 0$. If $x > 0$, $f_n(x)$ is 0 for $n \leqq x$, and for $n > x$ we have

$$\log\left(1 - \frac{x}{n}\right)^n = -n\left(\frac{x}{n} + \frac{x^2}{2n^2} + \frac{x^3}{3n^3} + \cdots\right)$$

$$= -\left(x + \frac{x^2}{2n} + \frac{x^3}{3n^2} + \frac{x^4}{4n^3} + \cdots\right).$$

This clearly increases as n increases, and its limit is $-x$. So its exponential $(1 - x/n)^n$ is nondecreasing and tends to e^{-x}, which implies our statement about the $f_n(x)$. By the monotone convergence theorem 4-2, we find that the limit of the f_n is integrable from 0 to ∞, and

$$\int_0^\infty e^{-x}x^{t-1}\,dx = \lim_{n\to\infty}\int_0^\infty f_n(x)\,dx.$$

The left member is $\Gamma(t)$, and by (E), the integral in the right member is $\Gamma_n(t)$, so (D) is established and the proof is complete.

9. Continuity of the Indefinite Integral; Integration by Parts and Substitution

In Theorem I-9-4 we showed that if f is bounded and integrable over $[a, b]$, its indefinite integral has a property stronger than continuity; it is Lipschitzian. Here we shall show that if f is integrable, even though unbounded, its indefinite integral still has a property stronger than continuity. This property is called *absolute continuity*.

DEFINITION 9-1 *Let f be defined and real-valued on an interval B in R. Then f is* ***absolutely continuous*** *on B if for each positive ε there is a positive number δ such that whenever $(x_1', x_1''], \ldots, (x_k', x_k'']$ are pairwise disjoint subintervals of B with total length less than δ, the inequality*

$$\sum_{j=1}^{k} |f(x_j'') - f(x_j')| < \varepsilon$$

is satisfied.

It is obvious that if f is absolutely continuous on $B\,(\subset R)$, it is continuous on B. For let ε be positive, and let δ be as in Definition 9-1. Let x be any point of B. Define U to be the open interval $(x - \delta, x + \delta)$. If $\bar{x}$ is any point of $B \cap U$, we choose $x_1' = x$ and $x_1'' = \bar{x}$ if $x < \bar{x}$, and we choose $x_1' = \bar{x}$ and $x_1'' = x$ if $x \geqq \bar{x}$. Then the sum in Definition 9-1 has just one term, and the inequality in Definition 9-1 is

$$|f(\bar{x}) - f(x)| = |f(x_1'') - f(x_1')| < \varepsilon.$$

This shows that f is continuous at x.

THEOREM 9-2 *Let f be integrable over an interval B (not necessarily bounded) contained in R, and let F be an indefinite integral of f. Then F is absolutely continuous on B.*

Since f is integrable over B, by Theorem 2-4, Corollary 3-3, and Lemma I-5-1, so are $|f|$ and $|f| \wedge n$ for every positive integer, and so is the function

$$\text{(A)} \qquad f_n = |f| - |f| \wedge n.$$

Since $|f| \wedge n \leqq |f|$, f_n is nonnegative, and for each x it is 0 whenever $n > |f(x)|$. So $f_n(x)$ tends to 0 for all x in B. Also, for each x in B and each positive integer n, either $|f(x)| \geqq n + 1$, in which case

$$|f(x)| \wedge n = n < n + 1 = |f(x)| \wedge (n + 1),$$

or else $|f(x)| < n + 1$, in which case

$$|f(x)| \wedge n \leqq |f(x)| = |f(x)| \wedge (n + 1).$$

In either case, this and (A) imply

$$f_n(x) \geqq f_{n+1}(x).$$

So the f_n satisfy the hypotheses of the monotone convergence theorem 4-2, and the integrals of the f_n converge to the integral of their limit 0. Therefore, if ε is any positive number, we can and do choose an n such that

$$\int_a^b f_n(x)\,dx < \frac{\varepsilon}{2}.$$

We define $\delta = \varepsilon/2n$.

Now let $(x_1', x_1''], \ldots, (x_k', x_k'']$ be pairwise disjoint subintervals of B with total length less than δ. Then by (A),

$$\begin{aligned}\text{(B)} \qquad \sum_{j=1}^{k} |F(x_j'') - F(x_j')| &= \sum_{j=1}^{k} \left| \int_{x_j'}^{x_j''} f(x)\,dx \right| \\ &\leqq \sum_{j=1}^{k} \int_{x_j'}^{x_j''} |f(x)|\,dx \\ &= \sum_{j=1}^{k} \int_{x_j'}^{x_j''} f_n(x)\,dx + \sum_{j=1}^{k} \int_{x_j'}^{x_j''} [|f(x)| \wedge n]\,dx.\end{aligned}$$

In the last expression, the first sum is the integral of f_n over the union of the intervals $(x_j', x_j'']$, which is contained in B. Since $f_n \geqq 0$, by Corollary I-6-3 this integral cannot be greater than the integral of f_n over all of B. In the second sum, the integrands are never greater than n, so the sum is not decreased if we replace $|f(x)| \wedge n$ by n. This gives us

$$\sum_{j=1}^{k} |F(x_j') - F(x_j'')| \leqq \int_B f_n(x)\,dx + \sum_{j=1}^{k} \int_{x_j'}^{x_j''} n\,dx.$$

The first term in the right member is less than $\varepsilon/2$ by choice of n. The second term is n times the sum of the lengths of the intervals $(x_j', x_j'']$, which is less than

$n\delta = n\varepsilon/2n$. So the sum in the left member is less than ε, and F is absolutely continuous on B.

It is worth mentioning, although it is too difficult to prove here, that the converse of Theorem 9-2 is also true; if a function f is absolutely continuous on an interval, there exists a function f such that F is an indefinite integral of f.

We have already seen that the indefinite integral F, being absolutely continuous on B, is continuous at each finite x in B. But if B is unbounded, either a is $-\infty$ or b is ∞, and the absolute continuity of F on B does not guarantee that it is continuous at the ends. However, this is easily proved true, as follows.

COROLLARY 9-3 *Let f be integrable over an interval B (not necessarily bounded) in R, a and b being the end-points of B. Then every indefinite integral of f is continuous on the closure $[a, b]$ of B.*

Let $\bar{x}$ be a point of $[a, b]$. If x is finite, F is already known to be continuous at $\bar{x}$. If $\bar{x} = \infty$, for each sequence of points $a < b_1 < b_2 < b_3 < \cdots$ tending to $\bar{x}$ we have by Theorem 7-1

$$\lim_{n\to\infty} F(b_n) = \lim_{n\to\infty} \left\{ F(a) + \int_a^{b_n} f(x)\,dx \right\} = F(a) + \int_a^{\infty} f(x)\,dx = F(\bar{x}).$$

This holds for all such sequences $b_1, b_2, b_3, \ldots$, so (as we saw in the Introduction), F is continuous at ∞. A similar proof applies if $\bar{x} = -\infty$.

We can now establish a strengthened form of the formula for integration by parts, in which the interval of integration can be unbounded and there is no mention of derivatives. Both features will prove useful in several places. In particular, the extension to unbounded intervals is needed in many applications.

THEOREM 9-4 *Let B be an interval (not necessarily bounded) in R, and let a and b be its end-points. Let f and g be functions integrable over B, and let F and G be indefinite integrals of f and g, respectively. Then*

$$\int_a^b f(x)G(x)\,dx = F(x)G(x)\Big|_a^b - \int_a^b F(x)g(x)\,dx.$$

Since f and g are integrable over B, so is $|f| + |g|$. Let H be defined on $[a, b]$ by

$$H(x) = \int_a^x [|f(u)| + |g(u)|]\,du.$$

Then $H(b)$ is finite, so if $\varepsilon > 0$, we can and do choose finitely many points $y_0 = 0 < y_1 < y_2 < \cdots < y_k = H(b)$ such that

$$\text{(C)} \qquad y_j - y_{j-1} < \varepsilon/(H(b) + 1) \qquad (j = 1, 2, \ldots, k).$$

By Corollary 9-3, H is continuous on $[a, b]$, so we can find points $x_0 = a < x_1 < x_2 < \cdots < x_k = b$ such that

$$\text{(D)} \qquad H(x_i) = y_i \qquad (i = 1, 2, \ldots, k).$$

We now define functions F^*, g_* on $[a, b]$ by setting

$$\text{(E)} \quad F^*(x) = F(x_i), \qquad G_*(x) = G(x_{i-1}) \qquad (x_{i-1} < x \leqq x_i,\ i = 1, 2, \ldots, k),$$

with $F^*(a) = F(x_1)$ and $G_*(a) = G(a)$. These are bounded, having values in the interval $[0, H(b)]$, and they are constant on each interval $(x_{i-1}, x_i]$. By Corollary 5-2, f and g are integrable over each interval $(x_{i-1}, x_i]$, so the same is true of fG_* and F^*g. Since F and G are indefinite integrals of f and g, respectively,

$$\begin{aligned}
\text{(F)} \qquad \int_{x_{i-1}}^{x_i} f(x)G_*(x)\,dx + \int_{x_{i-1}}^{x_i} F^*(x)g(x)\,dx
&= \int_{x_{i-1}}^{x_i} f(x)G(x_{i-1})\,dx + \int_{x_{i-1}}^{x_i} F(x_i)g(x)\,dx \\
&= [F(x_i) - F(x_{i-1})]G(x_{i-1}) + F(x_i)[G(x_i) - G(x_{i-1})] \\
&= F(x_i)G(x_i) - F(x_{i-1})G(x_{i-1}).
\end{aligned}$$

By Theorem I-8-1, fG_* and F^*g are integrable over the union $(a, b]$ of the intervals $(x_{i-1}, x_i]$ $(i = 1, 2, \ldots, k)$, and by adding equations (F) for $i = 1, 2, \ldots, k$,

$$\text{(G)} \qquad \int_a^b f(x)G_*(x)\,dx + \int_a^b F^*(x)g(x)\,dx = F(b)G(b) - F(a)G(a).$$

For every x in $(x_{i-1}, x_i]$ we have by Definition I-9-1 of indefinite integral,

$$\begin{aligned}
|F^*(x) - F(x)| = |F(x_i) - F(x)| &= \left| \int_x^{x_i} f(u)\,du \right| \\
\text{(H)} \qquad &\leqq \int_x^{x_i} |f(u)|\,du \leqq \int_{x_{i-1}}^{x_i} |f(u)|\,du \\
&\leqq H(x_i) - H(x_{i-1}) = y_i - y_{i-1} < \frac{\varepsilon}{H(b) + 1}.
\end{aligned}$$

Likewise,

$$\text{(I)} \qquad |G_*(x) - G(x)| < \varepsilon/(H(b) + 1).$$

From (G), (H), and (I) we deduce

$$\begin{aligned}
&\int_a^b f(x)G(x)\,dx + \int_a^b F(x)g(x)\,dx - F(b)G(b) - F(a)G(a)\\
&\quad = \left|\int_a^b f(x)[G(x) - G_*(x)]\,dx + \int_a^b [F(x) - F^*(x)]g(x)\,dx\right|\\
&\quad \leqq \int_a^b |f(x)||G(x) - G_*(x)|\,dx + \int_a^b |F(x) - F^*(x)||g(x)|\,dx\\
&\quad \leqq \frac{\varepsilon}{H(b)+1}\int_a^b \{|f(x)| + |g(x)|\}\,dx\\
&\quad = \frac{\varepsilon}{H(b)+1}[H(b)] < \varepsilon.
\end{aligned}$$

If we substitute in this from (G), we obtain

$$\left|\int_a^b f(x)G(x)\,dx + \int_a^b F(x)g(x)\,dx - F(b)G(b) + F(a)G(a)\right| < \varepsilon.$$

The first member in this string of inequalities and equations is a fixed positive number, and we have just proved that it is less than the arbitrary positive number ε, so it must be 0. This establishes our theorem.

Among the many uses of this theorem is a proof of an improved form of the theorem on integration by substitution, Theorem I-11-1. In proving this it is convenient to establish a lemma that also has uses in the next chapter.

LEMMA 9-5 *Let f be integrable over an interval B whose end-points are a and b, and let F be an indefinite integral of f. Then for every positive integer n*

$$\text{(J)} \qquad \int_a^b F(u)^n f(u)\,du = \frac{F(b)^{n+1} - F(a)^{n+1}}{n+1}.$$

If f were continuous, this would follow from the substitution theorem I-11-1, with $f(x) = x^n$ and $g(x) = F(x)$. The point of this lemma is that no continuity requirements are imposed on f, merely its integrability. The existence of the integral in (J) follows from Theorem 5-4, since F^n is bounded and continuous and therefore is integrable over every bounded interval.

The proof is by induction. If we understand $F(u)^0$ to mean 1 for all u, the conclusion is immediate for $n = 0$, since then (J) is part of the definition of indefinite integral, I-9-1. Suppose that (J) holds for $n = k$. If we define

$$g(x) = F(x)^k f(x), \qquad G(x) = \int_a^x g(u)\,du,$$

by Theorem 9-4 we have for all x in $[a, b]$

$$\text{(K)}\qquad \int_a^x F(u)^{k+1}f(u)\,du = \int_a^x F(u)g(u)\,du$$

$$= F(x)G(x) - F(a)G(a) - \int_a^x f(u)G(u)\,du.$$

But by the induction hypothesis, equation (J) holds for $n = k$, so

$$\text{(L)}\qquad G(x) = [F(x)^{k+1} - F(a)^{k+1}]/(k+1).$$

Since $G(a) = 0$, (K) and (L) imply

$$\text{(M)}\qquad \int_a^x F(u)^{k+1}f(u)\,du = \frac{F(x)^{k+2}}{k+1} - \frac{F(x)F(a)^{k+1}}{k+1}$$

$$- \frac{1}{k+1}\int_a^x F(u)^{k+1}f(u)\,du + F(a)^{k+1}\frac{F(x) - F(a)}{k+1}$$

$$= \frac{F(x)^{k+2} - F(a)^{k+2}}{k+1} - \frac{1}{k+1}\int_a^x F(u)^{k+1}f(u)\,du.$$

If we transpose the integral in the right member to the left member and multiply both members by $(k + 1)/(k + 2)$, we obtain (J), with $k + 1$ in place of k. So by induction (J) holds for all nonnegative integers n.

In the next theorem we shall assume that a function f is continuous on a bounded closed interval $[\alpha, \beta]$ and also has the property that for every positive number ε there is a polynomial p such that $|f(x) - p(x)| < \varepsilon$ for all x in $[\alpha, \beta]$. This last requirement is, in fact, unnecessary. In Theorem VI-8-3 we shall show that if f is continuous on $[\alpha, \beta]$, it necessarily can be approximated in this way by polynomials. Since we have not yet proved this statement, in the next theorem we shall include the unnecessary hypothesis, enclosing it in brackets as a reminder that it is really superfluous.

Theorem 9-6. *Let f be continuous on a bounded closed interval $[\alpha, \beta]$ in R [and have the property that for each positive ε there exists a polynomial p such that $|f(x) - p(x)| < \varepsilon$ for all x in $[\alpha, \beta]$]. Let $\dot{g}$ be a function integrable over an interval B (not necessarily bounded) with end-points a and b, and let g be an indefinite integral of $\dot{g}$ whose values lie in $[\alpha, \beta]$. Then*

$$\text{(N)}\qquad \int_a^b f(g(u))\dot{g}(u)\,du = \int_{g(a)}^{g(b)} f(x)\,dx.$$

If again we understand y^0 to mean 1 even if $y = 0$, for each nonnegative integer n we have, by Lemma 9-5,

$$\int_a^b g(u)^n\dot{g}(u)\,du = \frac{g(b)^{n+1} - g(a)^{n+1}}{n+1},$$

and by the fundamental theorem,

$$\int_{g(a)}^{g(b)} x^n\,dx = \frac{x^{n+1}}{n+1}\bigg|_{g(a)}^{g(b)} = \frac{g(b)^{n+1} - g(a)^{n+1}}{n+1}.$$

Therefore,

$$\text{(O)} \qquad \int_a^b g(u)^n \dot{g}(u)\,du = \int_{g(a)}^{g(b)} x^n\,dx.$$

If f is a polynomial

$$f(x) = \sum_{n=0}^{k} a_n x^n,$$

by multiplying both members of (O) by a_n and summing over $n = 1, 2, \ldots, k$ we obtain the conclusion (N) for the polynomial f.

Now let ε be any positive number. Since $\dot{g}$ is integrable from a to b, so is its absolute value. We define

$$\text{(P)} \qquad \varepsilon' = \frac{\varepsilon}{\int_a^b |\dot{g}(u)|\,du + \beta - \alpha + 1};$$

this is positive. We can and do choose a polynomial p such that for all x in $[\alpha, \beta]$

$$\text{(Q)} \qquad |f(x) - p(x)| < \varepsilon'.$$

We already know that if we replace f by p in (N), the difference of the two members of (N) is 0, so

$$\text{(R)} \qquad \left| \int_a^b f(g(u))\dot{g}(u)\,du - \int_{g(a)}^{g(b)} f(x)\,dx \right|$$

$$= \left| \int_a^b [f(g(u))\dot{g}(u) - p(g(u))\dot{g}(u)]\,du - \int_{g(a)}^{g(b)} [f(x) - p(x)]\,dx \right|$$

$$\leqq \int_a^b |f(g(u)) - p(g(u))| \cdot |\dot{g}(u)|\,du + \left| \int_{g(a)}^{g(b)} |f(x) - p(x)|\,dx \right|$$

$$\leqq \varepsilon' \int_a^b \dot{g}(u)\,du + \varepsilon' |g(b) - g(a)|.$$

Since $g(a)$ and $g(b)$ are in $[\alpha, \beta]$, $|g(b) - g(a)| \leqq \beta - \alpha$, and by (R) and (P), the left member of (R) is less than ε. But the left member of (R) is a fixed positive number, and by (R) it is less than an arbitrary positive number ε, so it must be 0. This completes the proof.

As another example of a use of Theorem 9-4, we prove a lemma that has some use in the theory of Fourier series.

LEMMA 9-7 *Let a be a real number. Let f be integrable over every bounded subinterval of $(a, \infty]$, and let the indefinite integral*

$$x \mapsto F(x) = \int_a^x f(u)\,du \qquad (a \leqq x < \infty)$$

be bounded, say, $|F(x)| \leqq M < \infty$ for $a \leqq x < \infty$. Let φ be nonpositive and integrable over $(a, \infty]$, and let Φ be the indefinite integral

$$x \mapsto \Phi(x) = -\int_x^\infty \varphi(u)\,du = \int_\infty^x \varphi(u)\,du.$$

Then $f \cdot \Phi$ is integrable over every interval $(a, x]$ $(a < x < \infty]$, and the limit

$$\lim_{x\to\infty} \int_a^x f(u)\Phi(u)\,du$$

exists, and

$$\left|\lim_{x\to\infty} \int_a^x f(u)\Phi(u)\,du\right| \leqq M\Phi(a).$$

By Theorem 9-4, f and F are integrable over every interval $[a, x]$ with $a < x < \infty$, and

$$\text{(S)} \qquad \int_a^x f(u)\Phi(u)\,du = F(x)\Phi(x) - F(a)\Phi(a) - \int_a^x F(u)\varphi(u)\,du.$$

The first term in the right member tends to 0 as x increases because $|F(x)| \leqq M$ and $\Phi(x)$ tends to 0. The second term is 0 because $F(a) = 0$. By hypothesis, φ is integrable over $[a, \infty)$, and F is bounded and continuous and therefore is integrable over every bounded subinterval of $[a, \infty)$. So $\varphi_{[a,\infty)}$ is integrable over R, and $F_{[a,\infty)}$ is integrable over every bounded interval and is bounded. By Theorem 5-4, their product is integrable over R; that is, $F\varphi$ is integrable over $[a, \infty)$. By Corollary 9-3,

$$\text{(T)} \qquad \lim_{x\to\infty} \int_a^x F(u)\varphi(u)\,du = \int_a^\infty F(u)\varphi(u)\,du.$$

By Corollary I-6-2, since $-\varphi \geqq 0$,

$$-M\int_a^\infty [-\varphi(u)]\,du \leqq \int_a^\infty F(u)[-\varphi(u)]\,du \leqq M\int_a^\infty [-\varphi(u)]\,du,$$

whence, by (S) and (T),

$$\begin{aligned}\left|\lim_{x\to\infty} \int_a^x f(u)\Phi(u)\,du\right| &= \left|\int_a^\infty F(u)\varphi(u)\,du\right| \\ &\leqq \int_a^\infty |F(u)|[-\varphi(u)]\,du \\ &\leqq M\int_a^\infty [-\varphi(u)]\,du = M\Phi(a).\end{aligned}$$

EXERCISE 9-1 Prove that

$$\lim_{x \to \infty} \int_0^x \frac{\sin u}{u}\, du$$

exists and is finite. *Suggestion*: The integrand is bounded on $(0, 1]$, and Lemma 9-7 applies to the integral over $[1, x)$.

EXERCISE 9-2 Prove the second theorem of the mean for integrals, which is the following:

Let $[a, b]$ be a bounded interval. Let f and φ be integrable over $[a, b]$, and let φ be either nonnegative or nonpositive on $[a, b]$. Let Φ be an indefinite integral of φ. Then there exists a number ξ in $[a, b]$ such that

$$\int_a^b f(u)\Phi(u)\, du = \Phi(a) \int_a^\xi f(u)\, du + \Phi(b) \int_\xi^b f(u)\, du.$$

Suggestion: Define

$$F(x) = \int_a^x f(u)\, du.$$

Apply Theorem 9-4 to the integral of $f\Phi$. By Corollary I-6-4, there is a number ξ in $[a, b]$ for which

$$\int_a^b F(u)\varphi(u)\, du = F(\xi) \int_a^b \varphi(u)\, du = F(\xi)[\Phi(b) - \Phi(a)].$$

EXERCISE 9-3 Prove that if $a < 0$ and b is not 0,

$$\int_0^\infty \exp(ax) \sin bx\, dx = -\frac{b}{a^2 + b^2},$$

$$\int_0^\infty \exp(ax) \cos bx\, dx = -\frac{a}{a^2 + b^2}.$$

Suggestion: Compute $D[\exp(ax) \sin bx]$ and $D[\exp(ax) \cos bx]$. From these find antiderivatives for $\exp(ax) \sin bx$ and $\exp(ax) \cos bx$.

EXERCISE 9-4 Let F be a function continuous on an interval $[a, b]$. Let f be a function integrable over $[a, b]$ such that for all points x of $[a, b]$ except those in a finite set E, $DF(x)$ exists and is equal to $f(x)$ and f is continuous at x. Prove that, for c and e in $[a, b]$,

$$\int_c^e f(x)\, dx = F(e) - F(c).$$

Suggestion: If $[c, e]$ has no point of E in its interior, for each positive integer n

$$\int_{c+1/n}^{e-1/n} f(x)\,dx = F\left(e - \frac{1}{n}\right) - F\left(c + \frac{1}{n}\right).$$

By Theorem 9-2 and the continuity of F, the conclusion holds for every $(c, e]$ with no point of E in its interior. Every $(c, e]$ is the union of such intervals.

EXERCISE 9-5 Use Exercise 9-4 to show that if a, b, c, and e are real, and $a \neq 0$, and neither c nor e is $-b/a$,

$$\int_c^e \log|ax + b|\,dx = \left(e + \frac{b}{a}\right)\log|ae + b| - \left(c + \frac{b}{a}\right)\log|ac + b| - e + c.$$

What happens if c or e is $-b/a$?

10. The Dominated Convergence Theorem

The theorems proved in the preceding sections are sufficient to allow us to generalize to gauge-integrals all the material concerning integration that is ordinarily found in texts on advanced calculus. In the remainder of this chapter we shall establish some further properties of the integral that should be known to anyone who wishes to be expert in integration theory. If the gauge-integral is being introduced in an undergraduate course in calculus or advanced calculus, however, it is likely that the time available will be too short to make it advisable to take up these more advanced topics. In this case the student should postpone reading the rest of this chapter, and will therefore have to pass over some parts of later chapters. Nevertheless, most of Chapters III, V, and VI will still be available. The reader who wishes to work through everything in this book should at some future time return to these last sections of Chapter II.

From the monotone convergence theorem we can deduce the following easy corollary, which we shall not dignify with a number because we shall not refer to it after this section:

(A) Let $b, f_0, f_1, f_2, f_3, \ldots$ be real-valued functions on a set B in R such that

(i) $b, f_1, f_2, f_3, \ldots$ are integrable over B;
(ii) $|f_n(x)| \leq b(x)$ for all x in B and all positive integers n;
(iii) $\lim_{n\to\infty} f_n(x) = f_0(x)$ for all x in B;
(iv) the sequence $f_1, f_2, f_3, \ldots$ is either ascending or descending.

Then f_0 is integrable over B, and

$$\lim_{n\to\infty} \int_B f_n(x)\,m(dx) = \int_B f_0(x)\,m(dx).$$

From hypothesis (ii) we see that $-b \leqq f_n \leqq b$ for all n, so by Corollary I-6-2,

$$-\int_B b(x)\,m(dx) \leqq \int_B f_n(x)\,m(dx) \leqq \int_B b(x)\,m(dx).$$

So the integrals of the f_n are bounded. Since the other hypotheses of Theorem 4-2 have been assumed satisfied, the conclusion of statement (A) follows by Theorem 4-2.

It is both surprising and useful that statement (A) continues to be valid if we simply throw out hypothesis (iv). The resulting theorem, called the "dominated convergence theorem," will be used many times in later pages. From the point of view of techniques of proof, it is a bit surprising that this great strengthening of (A) is proved by repeated use of (A) itself, which seems rather like lifting ourselves by our bootstraps.

THEOREM 10-1 (The Dominated Convergence Theorem) *Let $b, f_0, f_1, f_2, f_3, \ldots$ be real-valued functions on a set B in R such that*

(i) *$b, f_1, f_2, f_3, \ldots$ are integrable over B;*
(ii) *$|f_n(x)| \leqq b(x)$ for all x in B and all positive integers n;*
(iii) *$\lim_{n\to\infty} f_n(x) = f_0(x)$ for all x in B.*

Then f_0 is integrable over B, and

$$\lim_{n\to\infty} \int_B f_n(x)\,m(dx) = \int_B f_0(x)\,m(dx).$$

For each positive integer n and each m that is either ∞ or an integer greater than n, we define a function $g_{n,m}$ on B by setting

$$g_{n,m}(x) = \inf\{f_j(x) : n \leqq j < m\} \qquad (x \text{ in } B). \tag{B}$$

By hypothesis, for each x in B the numbers $f_j(x)$ are all in the closed interval $[-b(x), b(x)]$, so by (B),

$$-b(x) \leqq g_{n,m}(x) \leqq b(x). \tag{C}$$

If $q > p \geqq n$, $g_{n,q}(x)$ is the infimum of the numbers $\{f_j(x) : n \leqq j < q\}$. It is thus a lower bound for the subset $\{f_j(x) : n \leqq j < p\}$ and therefore cannot exceed the greatest lower bound of that subset, $g_{n,p}(x)$. So,

(D) if $n \leqq p < q$, then $g_{n,q}(x) \leqq g_{n,p}(x)$.

If n is a positive integer, x is in B, and ε is positive, by (B) there exists an integer k such that $k \geqq n$ and

$$f_k(x) < g_{n,\infty}(x) + \varepsilon.$$

Then for every integer $m > k$, by (D) and (B),

$$g_{n,\infty}(x) \leqq g_{n,m}(x) \leqq f_k(x) < g_{n,\infty}(x) + \varepsilon.$$

So for all x in B

(E)
$$\lim_{m\to\infty} g_{n,m}(x) = g_{n,\infty}(x).$$

For each positive integer m greater than n, the function $g_{n,m}$ is integrable over B by Corollary 3-5. This, together with (C), (D), and (E), shows that for each positive integer n, the hypotheses of statement (A) are satisfied with $f_0, f_1, f_2, \ldots$ replaced by $g_{n,\infty}, g_{n,n}, g_{n,n+1}, \ldots$. So, by a first use of (A),

(F) $g_{n,\infty}$ is integrable over B.

If $q > n$ and x is in B, $g_{n,\infty}(x)$ is the infimum of the set of numbers $\{f_j(x) : j \geqq n\}$. Then it is a lower bound for the subset $\{f_j(x) : j \geqq q\}$ and cannot exceed the greatest lower bound $g_{q,\infty}(x)$ of that subset. So if $q > n$, then $g_{q,\infty} \geqq g_{n,\infty}$, and

(G) the functions $g_{n,\infty}$ $(n = 1, 2, 3, \ldots)$ form an ascending sequence.

If x is in B and $\varepsilon > 0$, then since $f_n(x)$ converges to $f_0(x)$, there is an n' such that

(H) if $j > n'$, then $f_0(x) - \varepsilon/2 < f_j(x) < f_0(x) + \varepsilon/2$.

So, if $n > n'$, the inequality in (H) is satisfied for all $j \geqq n$, whence

$$\begin{aligned} f_0(x) - \varepsilon < f_0(x) - \varepsilon/2 &\leqq \inf\{f_j(x) : j \geqq n\} \\ &= g_{n,\infty}(x) \leqq f_n(x) < f_0(x) + \varepsilon/2. \end{aligned}$$

That is, for all x in B

(I)
$$\lim_{n\to\infty} g_{n,\infty}(x) = f_0(x).$$

By (C), (F), (G), and (I), the hypotheses of statement (A) are satisfied with $g_{n,\infty}$ in place of f_n. So, by a second use of (A),

(J) f_0 is integrable over B, and

$$\lim_{n\to\infty} \int_B g_{n,\infty}(x)\, m(dx) = \int_B f_0(x)\, m(dx).$$

We now define another collection of functions $h_{n,m}$ on B by setting

(K)
$$h_{n,m}(x) = \sup\{f_j(x) : n \leqq j < m\}.$$

We can discuss these as we did the $g_{n,m}$, with no changes except a few reversals of inequalities, and we find that they satisfy an equation analogous to (J), namely,

(L)
$$\lim_{n\to\infty} \int_B h_{n,\infty}(x)\, m(dx) = \int_B f_0(x)\, m(dx).$$

Now let ε be positive. By (J), there is an integer n' such that if $n > n'$,

$$\text{(M)} \qquad \int_B g_{n,\infty}(x)\,m(dx) > \int_B f_0(x)\,m(dx) - \varepsilon.$$

By (L), there is an n'' such that if $n > n''$,

$$\text{(N)} \qquad \int_B h_{n,\infty}(x)\,m(dx) < \int_B f_0(x)\,m(dx) + \varepsilon.$$

If $n > \max\{n', n''\}$, both (M) and (N) are satisfied, while by (B) and (K),

$$g_{n,\infty}(x) \leqq f_n(x) \leqq h_{n,\infty}(x) \qquad (x \text{ in } B),$$

so,

$$\text{(O)} \qquad \int_B g_{n,\infty}(x)\,m(dx) \leqq \int_B f_n(x)\,m(dx) \leqq \int_B h_{n,\infty}(x)\,m(dx).$$

Inequalities (M), (N), and (O) show that if $n > \max\{n', n''\}$,

$$\int_B f_0(x)\,m(dx) - \varepsilon < \int_B f_n(x)\,m(dx) < \int_B f_0(x)\,m(dx) + \varepsilon,$$

so that

$$\lim_{n\to\infty} \int_B f_n(x)\,m(dx) = \int_B f_0(x)\,m(dx).$$

This and (J) complete the proof.

Theorem 10-1 is in a sense unnecessary because anything that can be established with its help can also be established by means of the monotone convergence theorem. This is so because Theorem 10-1 was itself proved by using that theorem. However, the dominated convergence theorem can often be used conveniently where the monotone convergence theorem would be inconvenient. The main theorem of the next section is an example of such a situation.

It is interesting to compare the dominated convergence theorem with the uniform convergence theorem 4-1, which it somewhat resembles. Let B be a bounded interval in R, and let $f_1, f_2, f_3, \ldots$ be functions integrable over B and converging at each point of B to a limit function f. To use Theorem 4-1, we would need to know that for *every* positive ε there is an $n(\varepsilon)$ such that if $n \geqq n(\varepsilon)$,

$$\text{(P)} \qquad |f_n(x) - f(x)| < \varepsilon \qquad (x \text{ in } B).$$

To use Theorem 10-1, all we need to know is that *there is at least one* number ε such that (P) holds. For then we will have $|f_n(x) - f_m(x)| < 2\varepsilon$ whenever m and n are both at least equal to $n(\varepsilon)$, so that for $n \geqq n(\varepsilon)$

$$|f_n(x)| \leqq |f_{n(\varepsilon)}(x)| + 2\varepsilon \qquad (x \text{ in } B).$$

The right member of this inequality is integrable over B, so it serves as b in the hypotheses of Theorem 10-1. Obviously, even if we can show that (P) holds for every positive ε, it may be much easier to find *one* large ε with which (P) holds. So, the use of Theorem 10-1 can never call for more effort than the use of Theorem 4-1, and it may save much work.

EXERCISE 10-1 Let f be bounded and integrable over an interval $[-a, a]$ and let f have left and right limits $f(0-)$, $f(0+)$ at 0. Prove that

$$\lim_{h\to 0+} \int_{-a}^{a} \frac{h}{h^2 + x^2} f(x)\, dx = \pi \frac{f(0-) + f(0+)}{2}.$$

(Extend f, setting it $= 0$ outside $[-a, a]$. Substitute $x = hu$. Use Theorem 10-1.)

11. Differentiation under the Integral Sign

Suppose that f is a function $(x, \alpha) \mapsto f(x, \alpha)$ defined and real-valued for all x in a set B in R and all α in an interval (α', α''); that is, f is a function on $B \times (\alpha', \alpha'')$. If for each α in (α', α'') the function $x \mapsto f(x, \alpha)$ is integrable over B, then for each such α we can define $F(\alpha)$ to be the number

$$\text{(A)} \qquad F(\alpha) = \int_B f(x, \alpha)\, m(dx).$$

Let us also assume that for each x in B, f has a partial derivative with respect to α, which we choose to denote by f_α; that is, for each α in (α', α'') the limit

$$\text{(B)} \qquad f_\alpha(x, \alpha) = \lim_{\beta\to\alpha} \frac{f(x, \beta) - f(x, \alpha)}{\beta - \alpha}$$

exists. If we integrate this over B, the result is the combined effect of two limit processes, the first being the differentiation of f with respect to α and the second being the integration of f_α over B. The result of applying these two limit processes in reversed order (if this can be done at all) is, first, to compute F by (A) and then to compute the derivative of F. Often, in the course of some mathematical investigation, we find it desirable to effect this interchange. But we already know that interchanging the order of performing two limit processes can produce a different final result; this was discussed at the beginning of Section 4. The object of this section is to show that under suitable conditions the interchange is permissible; the derivative of F can be found by first computing the partial derivative of f with respect to α and then integrating that partial derivative.

★THEOREM 11-1 *Let B be a set in R and A an open interval in R. Let f be a real-valued function $(x, \alpha) \mapsto f(x, \alpha)$ defined for all x in B and all α in A such that for each* *such x and α the partial derivative*

$$f_\alpha(x, \alpha) = \frac{\partial f(x, \alpha)}{\partial \alpha}$$

exists, and such also that for each α in A the integral

$$F(\alpha) = \int_B f(x, \alpha)\, m(dx)$$

exists and is finite. If there exists a function b integrable over B such that for all x in B and all α in A

$$|f_\alpha(x, \alpha)| \leqq b(x),$$

then for each α in A, F has a derivative with respect to α, and

$$\text{(C)} \qquad DF(\alpha) = \int_B f_\alpha(x, \alpha)\, m(dx).$$

Let α be a point of A and let $\alpha_1, \alpha_2, \alpha_3, \ldots$ be a sequence of points of A, all different from α and tending to α as n increases. If for notational convenience we define

$$q_n(x) = [f(x, \alpha_n) - f(x, \alpha)]/[\alpha_n - \alpha],$$

by definition of F we have

$$\text{(D)} \qquad \frac{F(\alpha_n) - F(\alpha)}{\alpha_n - \alpha} = \int_B q_n(x)\, m(dx).$$

By the theorem of mean value, for each x in B there is a number $\alpha(x)$ between α and α_n such that

$$q_n(x) = f_\alpha(x, \alpha(x)).$$

This implies that

$$|q_n(x)| \leqq b(x) \qquad (x \text{ in } B).$$

Moreover, the definition of derivative implies that the limit of $q_n(x)$ is $f_\alpha(x, \alpha)$ for each x in B. So, by the dominated convergence theorem 10-1, we have

$$\text{(E)} \qquad \lim_{n \to \infty} \int_B q_n(x)\, m(dx) = \int_B f_\alpha(x, \alpha)\, m(dx).$$

Combining this with (D) yields

$$\text{(F)} \qquad \lim_{n \to \infty} \frac{F(\alpha_n) - F(\alpha)}{\alpha_n - \alpha} = \int_B f_\alpha(x, \alpha)\, m(dx).$$

Rewording what we have thus far shown, the function f_α is integrable over B, and for every sequence $\alpha_1, \alpha_2, \alpha_3, \ldots$ of points of A different from α and tending to α the limit of $[F(\alpha_n) - F(\alpha)]/[\alpha_n - \alpha]$ exists and is equal to the right member of (C). By Theorem 1-1 in the Introduction, this implies that the limit of $[F(\beta) - F(\alpha)]/[\beta - \alpha]$ as β tends to α exists and is equal to the right member of (C). That is to say, F has a derivative at α, and this derivative is the right member of (C), which is the conclusion of the theorem.

As an example showing the use of this theorem, we derive some formulas related to the normal distribution in probability theory. We shall need the next lemma.

Lemma 11-2

$$\int_{-\infty}^{\infty} \exp\left(-\frac{x^2}{2}\right) dx = \sqrt{2\pi}.$$

This is proved in a multitude of texts on probability theory, and we too shall prove it later, in equation (LL), Section IV-5. For the moment we accept it as true.

If t and σ are positive numbers, for each positive number k we make the substitution

$$x = (t^{1/2}/\sigma)y$$

in the integral

$$\int_{-k}^{k} \exp\left(-\frac{x^2}{2}\right) dx.$$

By Theorem I-11-1, we obtain

$$\text{(G)} \qquad \int_{-k}^{k} \exp\left(-\frac{x^2}{2}\right) dx = \frac{t^{1/2}}{\sigma} \int_{-\sigma k t^{-1/2}}^{\sigma k t^{-1/2}} \exp\left(-\frac{t y^2}{2\sigma^2}\right) dy.$$

We let k increase without bound; by Lemma 11-2 and Theorem 7-1, we obtain

$$\text{(H)} \qquad \int_{-\infty}^{\infty} \exp\left(-\frac{t y^2}{2\sigma^2}\right) dy = \left(\frac{2\pi}{t}\right)^{1/2} \sigma.$$

From this we can derive a whole infinite sequence of useful formulas, as follows.

Lemma 11-3 *Define $C_0 = 1$, $C_n = 1 \cdot 3 \cdot 5 \cdots (2n - 1)$ for $n = 1, 2, 3, \ldots$. Then for every positive number σ and every nonnegative integer n*

$$\text{(I)} \qquad \frac{1}{\sigma(2\pi)^{1/2}} \int_{-\infty}^{\infty} y^{2n} \exp\left(-\frac{y^2}{2\sigma^2}\right) dy = C_n \sigma^{2n}.$$

The identity

$$\text{(J)} \qquad \int_{-\infty}^{\infty} y^{2n} \exp\left(-\frac{ty^2}{2\sigma^2}\right) dy = (2\pi)^{1/2} t^{-n-1/2} \sigma^{2n+1} C_n \qquad \left(\frac{1}{2} < t < \frac{3}{2}\right)$$

is valid for $n = 0$, by (H). Suppose that it is valid for a nonnegative integer $k - 1$. Then for all t in $(\frac{1}{2}, \frac{3}{2})$

$$\text{(K)} \qquad \int_{-\infty}^{\infty} y^{2k-2} \exp\left(-\frac{ty^2}{2\sigma^2}\right) dy = (2\pi)^{1/2} t^{-k+1/2} \sigma^{2k-1} C_{k-1}.$$

Let us denote the integrand in the left member of (K) by $f(y, t)$. Then its partial derivative with respect to t is

$$\text{(L)} \qquad f_t(y, t) = (-y^{2k}/2\sigma^2) \exp(-ty^2/2\sigma^2).$$

For all t in $(\frac{1}{2}, \frac{3}{2})$ the right member of (L) has absolute value at most

$$(y^{2k}/2\sigma^2) \exp(-y^2/4\sigma^2),$$

which is integrable over R (see Exercise 6-1).

So by Theorem 11-1, we deduce from (K)

$$\text{(M)} \qquad \int_{-\infty}^{\infty} -\frac{y^{2k}}{2\sigma^2} \exp\left(-\frac{ty^2}{2\sigma^2}\right) dy = (2\pi)^{1/2}\left(-\frac{1}{2}\right)(2k-1) t^{-k-1/2} \sigma^{2k-1} C_{k-1}.$$

But by the definition of the C_n, we have

$$(2k - 1)C_{k-1} = C_k,$$

so (M) implies that (J) holds for $n = k$. By induction, (J) holds for all nonnegative integers n. If in (J) we set $t = 1$ and divide both members by $\sigma(2\pi)^{1/2}$, we obtain (I), and the proof is complete.

From the relation

$$\lim_{x \to 0} \frac{\sin x}{x} = \lim_{x \to 0} \frac{\sin x - \sin 0}{x - 0} = D \sin 0 = \cos 0 = 1$$

we see that if we adopt the convention that the function

$$x \mapsto (\sin x)/x$$

is assigned the value 1 at $x = 0$, it is continuous on R. It is therefore integrable over every bounded interval in R, but it is not integrable over R. Nevertheless, we shall prove

$$\text{(N)} \qquad \lim_{x \to \infty} \int_0^x \frac{\sin u}{u} du = \frac{\pi}{2}.$$

For each nonnegative number α the function

$$x \mapsto x^{-1} \exp(-\alpha x)$$

is nonincreasing and continuously differentiable on $(0, \infty)$, and it tends to 0 as x increases. By the second theorem of mean value (Exercise 9-2), if $0 < a < b$, there is a number ξ in $[a, b]$ such that

$$\text{(O)} \qquad \left|\int_a^b u^{-1} \exp(-\alpha u) \sin u\, du\right|$$

$$= \left|a^{-1} \exp(-\alpha a) \int_a^{\xi} \sin u\, du + b^{-1} \exp(-\alpha b) \int_{\xi}^b \sin u\, du\right|$$

$$\leqq 2a^{-1} \exp(-\alpha a) + 2b^{-1} \exp(-\alpha b).$$

If $0 \leqq a < b_1 < b_2 < \cdots$ is an ascending sequence with limit ∞, from (O) we obtain for positive integers n and $m > n$

$$\int_{b_n}^{b_m} u^{-1} \exp(-\alpha u) \sin u\, du \leqq 4b_n^{-1} \exp(-\alpha b_n).$$

So, the sequence of numbers

$$\int_a^{b_n} u^{-1} \exp(-\alpha u) \sin u\, du$$

converges, by the Cauchy test (Exercise 1-1), and so, by Theorem 1-1 in the Introduction, the limit

$$\text{(P)} \qquad F(a, \alpha) = \lim_{b \to \infty} \int_a^b u^{-1} \exp(-\alpha u) \sin u\, du$$

exists. Also, by (O),

$$\text{(Q)} \qquad F(a, \alpha) \leqq 2a^{-1} \exp(-\alpha a).$$

We now prove that for $a \geqq 0$

$$\text{(R)} \qquad \lim_{\alpha \to 0} F(a, \alpha) = F(a, 0).$$

Let ε be positive and choose $b > \max(a, 6/\varepsilon)$. Then for all nonnegative α and all x greater than b

$$\int_a^x u^{-1} \exp(-\alpha u) \sin u\, du = \int_a^b + \int_b^x u^{-1} \exp(-\alpha u) \sin u\, du.$$

Letting x increase and recalling (Q), we find

$$\left|F(a, \alpha) - \int_a^b u^{-1} \exp(-\alpha u) \sin u\, du\right| = |F(b, \alpha)| \leqq \frac{2}{b}.$$

If α and β are nonnegative, we deduce

$$\left|F(a, \alpha) - F(a, \beta) - \int_a^b u^{-1} [\exp(-\alpha u) - \exp(-\beta u)] \sin u\, du\right| \leqq \frac{4}{b}.$$

Since $u \mapsto (\sin u)/u$ is continuous, and therefore bounded, on $[a, b]$, we can choose β so close to α that the integral in this last inequality has absolute value less than $2/b$. Then

$$|F(a, \alpha) - F(a, \beta)| \leqq 6/b < \varepsilon.$$

That is, as α tends to β, $F(a, \alpha)$ tends to $F(a, \beta)$ and, in particular, (R) is satisfied.

For all positive α the function $u \mapsto [u^{-1} \sin u] \exp(-\alpha u)$ is integrable from any nonnegative b to ∞, for the first factor is bounded and continuous and therefore is integrable over every bounded interval, and the second factor is integrable from b to ∞. So by Theorem 7-1, Eq. (P) takes the form

$$\text{(S)} \qquad F(b, \alpha) = \int_b^\infty u^{-1} \sin u \exp(-\alpha u)\, du$$

for all positive α and all nonnegative b. If we denote the integrand by $f(u, \alpha)$, its partial derivative with respect to α is

$$f_\alpha(u, \alpha) = -\sin u \exp(-\alpha u).$$

Therefore, for each positive ε, the function $u \mapsto \exp(-\varepsilon u)$ is an upper bound for $f_\alpha(u, \alpha)$ whenever $\varepsilon < u < \infty$ and $\alpha > \varepsilon$, and it is integrable over $[0, \infty)$, so it serves as b in Theorem 11-1. By that theorem, for all α in (ε, ∞)

$$\text{(T)} \qquad DF(0, \alpha) = -\int_0^\infty \sin u \exp(-\alpha u)\, du.$$

But ε is an arbitrary positive number, so (T) holds for all positive α. The integral in (T) can be calculated by the methods of elementary calculus. All we need to do is to notice that an antiderivative of $u \mapsto -\sin u \exp(-\alpha u)$ is

$$u \mapsto [(\cos u + \alpha \sin u) \exp(-\alpha u)]/(1 + \alpha^2),$$

as can be verified by differentiation. Then by (T) we obtain

$$\text{(U)} \qquad \begin{aligned} DF(0, \alpha) &= \lim_{x \to \infty} [\exp(-\alpha x)(\cos x + \alpha \sin x)/(1 + \alpha^2)]\big|_0^x \\ &= -1/(1 + \alpha^2). \end{aligned}$$

This last is the derivative of $\operatorname{arc\,cot} \alpha$, so if $0 < \alpha < \beta$,

$$\text{(V)} \qquad F(0, \beta) - F(0, \alpha) = \operatorname{arc\,cot} \beta - \operatorname{arc\,cot} \alpha.$$

In this we let β increase without bound. Then $\operatorname{arc\,cot} \beta$ tends to 0. For all nonnegative u the function $u^{-1} \sin u \exp(-\beta u)$ tends to 0 as β increases, remaining at most equal to the integrable function $[\sup|u^{-1} \sin u|] \exp(-u)$ if $\beta > 1$. By the dominated convergence theorem, $F(0, \beta)$ tends to 0 as β increases, so (V) yields

$$F(0, \alpha) = \operatorname{arc\,cot} \alpha.$$

From this and (R),

$$F(0,0) = \pi/2,$$

which, with (P), implies (N).

Theorem 11-1 has a partial generalization whose easy proof we leave as an exercise.

THEOREM 11-4 *Let the hypotheses of Theorem* 11-1 *be satisfied with B an interval in R and with* $m = m_L$. *Let g and h be functions differentiable on A with values in B. Then the function*

$$\alpha \to \int_{g(\alpha)}^{h(\alpha)} f(x,\alpha)\,dx$$

has a derivative for each α *in A, and the value of the derivative is*

$$f(h(\alpha),\alpha)Dh(\alpha) - f(g(\alpha),\alpha)Dg(\alpha) + \int_{g(\alpha)}^{h(\alpha)} f_\alpha(x,\alpha)\,dx.$$

(The proof is merely an application of the chain rule.)

EXERCISE 11-1 For positive a and b, define

$$J(a,b) = \int_0^\infty [\exp(-ax) - \exp(-bx)]x^{-1}\,dx.$$

Show that this exists, and by differentiating the function

$$\alpha \mapsto J(\alpha, b) \qquad (a^* < \alpha < b^*),$$

where $0 < a^* < \min(a,b)$ and $b^* > \max(a,b)$, show that

$$J(a,b) = \log b - \log a.$$

[The derivative is $1/\alpha$, and $J(b,b) = 0$.]

EXERCISE 11-2 From any table of integrals we can find that on $(-\pi,\pi)$ an indefinite integral of $(a + \cos x)^{-1}$ is

$$2[a^2 - 1]^{-1/2} \arctan\{[(a-1)^{1/2}(a+1)^{-1/2}]\tan x/2\},$$

if $a > 1$. Show that

$$\int_0^\pi (a + \cos x)^{-2}\,dx = \pi a(a^2 - 1)^{-3/2}.$$

Suggestion: Integrate $(a + \cos x)^{-1}$ from 0 to y, where $0 < y < \pi$. Differentiate with respect to a. Let y tend to π; use Theorem 7-1.

EXERCISE 11-3 Let k be any nonzero real number. Let f be integrable over every bounded interval in R. Show that the function

$$x \mapsto y(x) = k^{-1} \int_0^x f(u) \sin k(x-u)\, du$$

is a solution of the differential equation

$$D^2 y + k^2 y = f(x).$$

EXERCISE 11-4 If b and α are positive,

$$\int_0^b (1+\alpha x)^{-1}\, dx = \alpha^{-1} \log(1+\alpha b).$$

Use this and Theorem 7-1 to show that

$$\int_0^b x(1+\alpha x)^{-2}\, dx = \alpha^{-2} \log(1+\alpha b) - b\alpha^{-1}(1+\alpha b)^{-1}.$$

Verify this by computing the last integral by elementary methods.

EXERCISE 11-5 This exercise is included to show that using the conclusion of a theorem without verifying that its hypotheses are satisfied can lead to errors. We shall give an argument that purports to evaluate the integral in Exercise 11-1, but the value obtained is incorrect. The reader should locate the error in the argument.

If $b > 0$,

$$\lim_{x \to 0} \frac{\exp(-bx) - 1}{x} = [D \exp(-bx)](0) = -b;$$

so if we understand the function $x \mapsto x^{-1}[\exp(-bx) - 1]$ to have value $-b$ at 0, it is continuous on $[0, \infty)$. If a and b are positive, by the substitution $x = (a/b)u$ we obtain

$$\begin{aligned}
\int_0^\infty x^{-1}[\exp(-bx) - 1]\, dx &= \int_0^\infty \frac{b}{au}[\exp(-au) - 1]\frac{a}{b}\, du \\
&= \int_0^\infty u^{-1}[\exp(-au) - 1]\, du \\
&= \int_0^\infty x^{-1}[\exp(-ax) - 1]\, dx.
\end{aligned}$$

Transposing, we deduce

$$\int_0^\infty x^{-1}[\exp(-bx) - \exp(-ax)]\, dx = 0.$$

But by Exercise 11-1, this is false.

12. Sets of Measure 0

The starting point of our theory of integration was an interval function m_L, defined for all intervals in $\bar{R}$. Using this, we defined an integral, and in particular we gave a meaning to the symbol

$$\text{(A)} \qquad \int_R 1_B(x)\, m_L(dx)$$

for a large class of sets B, including all intervals. Temporarily, we denote this integral by μB. If B happens to be an interval, we have shown in Lemma 4-5 that μB is the same as mB. If B happens to be the union of finitely many pairwise disjoint intervals, μB is the sum of their lengths. It therefore is worthy of being regarded as an expression for a generalization of the concept of length for a larger class of sets than merely the intervals. This also permits a simplification of notation. Since $\mu B = mB$ whenever B is an interval, so that mB has a meaning, there will be no ambiguity in meaning if we abandon the extra symbol μB and simply write mB to denote the value of the integral (A) whenever that integral exists. The sets for which this happens will be called *measurable sets*, or *m-measurable sets*, and the value of the integral in (A) will be called the *measure* (or the *m-measure*) of B, whether it is finite or infinite. This we state formally.

★Definition 12-1 *Let B be any set in $\bar{R}$. The statement that B **is measurable** (or **m-measurable**) means that 1_B has an integral (finite or ∞) over R, and the **measure** (or **m-measure**) of B is defined to be*

$$mB = \int_R 1_B(x)\, m(dx).$$

The statement that b has finite measure will have the obvious meaning that B is measurable and mB is finite.

Measurable sets are of great importance, and we shall study them later. But in this section we shall consider only the simple case of sets of measure 0. These sets are of importance in the theory of integration precisely because they are of no importance in computing integrals. Changing an integrand on a set of measure 0 has no effect on the integral, as we now show.

★Theorem 12-2 *Let f be defined and real-valued on R and have an integral (finite or infinite) over R, and let g be defined and real-valued on R and equal to f at all points of R except those of a set of measure 0. Then g also has an integral over R, and its integral over R is equal to the integral of f over R.*

Suppose first that the integral of f is finite. Define $h = g - f$, and for each positive integer n define $h_n = |h| \wedge n$. If E is the set of measure 0 on which

$f(x) \neq g(x)$, we have

$$\text{(B)} \qquad h_n(x) \leqq n 1_E(x) \qquad (x \text{ in } R);$$

for if x is in E, h_n is at most n, and the right member of (A) is n, whereas if x is not in E, $h(x) = 0$, so $h_n(x) = 0$. By hypothesis, the right member of (B) has integral 0, so by Corollary I-7-3, the left member has integral 0. For each x in R, $h_n(x) = |h(x)|$ for all n greater than $|h(x)|$, so the left member of (B) converges to $|h(x)|$ as n tends to ∞. The h_n are obviously an ascending sequence, so by the monotone convergence theorem, the integral of the limit $|h|$ is the limit of the integrals of the h_n, which is 0. Since $|h|$ has integral 0, $f + |h|$ and $f - |h|$ are integrable, and since $h = g - f$,

$$f - |h| \leqq g \leqq f + |h|.$$

The first and last functions in this inequality have the same integral, so by Corollary I-7-3, g also has the same integral, which is the integral of f. The proof is complete for the case of integrable f.

If the integral of f is ∞, by definition there exists an ascending sequence $f_1, f_2, f_3, \ldots$ of integrable functions tending everywhere to f and having integrals that tend to ∞. Let g be equal to f except on a set E with measure 0. For each n, define g_n by setting

$$g_n(x) = \begin{cases} f_n(x) & (x \text{ in } R \setminus E), \\ g(x) & (x \text{ in } E). \end{cases}$$

By the part of the proof already completed, for each n the function g_n is integrable over R and has the same integral as does f_n. For each x in $R \setminus E$, the sequence $g_1(x), g_2(x), g_3(x), \ldots$ is the same as $f_1(x), f_2(x), f_3(x), \ldots$, so it is a nondecreasing sequence and has the limit $f(x)$, which is equal to $g(x)$. For x in E the sequence $g_1(x), g_2(x), g_3(x), \ldots$ has all its terms equal to $g(x)$, so it is nondecreasing and has limit $g(x)$. Now the functions $g_1, g_2, g_3, \ldots$ form an ascending sequence of integrable functions with integrals tending to ∞, and they converge everywhere to g. Thus, by definition, g has integral ∞, which is the same as the integral of f. Similarly, if f has integral $-\infty$, so has g, and the proof is complete.

It is customary to say that a set E is **countably infinite**, or **denumerably infinite**, if there exists a sequence $a_1, a_2, a_3, \ldots$ in which each member of E occurs just once. The set E is **countable** if it is either finite or countably infinite. An obvious example of a countable set is the set of all positive integers. A less obvious one is the set of all positive rationals. To show that this set is countable, we write the sequence of fractions $\frac{1}{1}; \frac{1}{2}, \frac{2}{1}; \frac{1}{3}, \frac{2}{2}, \frac{3}{1}; \frac{1}{4}, \frac{2}{3}, \frac{3}{2}, \frac{4}{1}; \ldots$. The rule by which this sequence is written is too obvious to need detailed description. If from this sequence we reject all fractions in which the numerator and denominator have a common factor, we are left with a sequence that contains each positive rational just once.

Another useful example of a countable set is the set of all rationals. Let us first arrange all the positive rationals in a sequence $r_1, r_2, r_3, \ldots,$ as we have just shown we can do. Then the sequence $0, r_1, -r_1, r_2, -r_2, r_3, -r_3, \ldots$ contains all rationals.

The next theorem lists three useful devices for recognizing that a set has measure 0. Parts (ii) and (iii) are used with great frequency because they allow us to manipulate sets of measure 0 with ease.

★THEOREM 12-3 *Let E be a set in R.*

(i) *If for each positive ε there exists a countable collection of sets $E(1)$, $E(2)$, $E(3), \ldots,$ each of finite measure, such that E is contained in the union of the $E(i)$ and the sum of the measures of the $E(i)$ is less than ε, then $mE = 0$.*

(ii) *If E is contained in a set $E(1)$ with measure 0, then $mE = 0$.*

(iii) *If E is the union of countably many sets, each of measure 0, then $mE = 0$.*

In (i) we may as well assume that there is an infinite sequence of sets $E(1)$, $E(2)$, $E(3), \ldots$ with the properties specified, since if there is only a finite number n of them, we can define $E(j)$ to be the empty set $\varnothing$ for $i > n$ without disturbing the hypotheses. For each positive integer n we define

$$s_n = 1_{E(1)} \vee \cdots \vee 1_{E(n)}.$$

As n increases, s_n ascends, but it never rises above 1, so it approaches a finite-valued limit function, which we call h. If x is in E, it is in the union of the $E(i)$, so there is a k such that x is in $E(k)$. Then $1_{E(k)}(x) = 1$, so $s_n(x) = 1$ for all $n \geqq k$. This implies

$$h(x) = \lim_{n\to\infty} s_n(x) = 1 \geqq 1_E(x). \tag{C}$$

If x is not in E, the right member is 0, so this inequality is still valid.

Since

$$s_n \leqq 1_{E(1)} + \cdots + 1_{E(n)},$$

by integration over R we obtain

$$\int_R s_n(x)\, m(dx) \leqq mE(1) + \cdots + mE(n).$$

The right member of this inequality is less than ε, so by the monotone convergence theorem, h is integrable over R, and

$$\int_R h(x)\, m(dx) = \lim_{n\to\infty} \int_R s_n(x)\, m(dx) \leqq \lim_{n\to\infty} [mE(1) + \cdots + mE(n)] < \varepsilon.$$

By (C) and Lemma I-7-1, with $g = 0$ and $J = 0$, we see that the integral of 1_E over R exists and is equal to 0, so $mE = 0$.

If E is contained in the union of countably many sets $E(1), E(2), E(3), \ldots$, each of measure 0, then for every positive ε the conditions in (i) are satisfied with these $E(i)$, so by (i), $mE = 0$. In particular, if the countable set of sets $E(i)$ contains only $E(1)$, we obtain (ii), and if the set E is not merely contained in the union of the $E(i)$ but is the union of the $E(i)$, we obtain (iii).

It is customary to say that a statement about a point x is true "almost everywhere" if it is true for all points except those in a set of measure 0 (which may be empty). The words "almost everywhere" are often abbreviated to "a.e." Two functions f, g are called *equivalent* if they are defined on the same set and $f(x) = g(x)$ a.e. in that set. Thus, Theorem 12-2 states that if f has an integral and g is equivalent to f, then g also has an integral, and its integral is equal to that of f.

By Lemma I-3-2, every set consisting of a single point has m_L-measure 0. By Theorem 12-3, every countable set has m_L-measure 0. By Theorem 12-2, every function defined on a set B and equal to 0 except on a countable subset of B has integral 0 with respect to m_L over B. Thus, we again obtain Example I-3-3, this time as a special case of a general statement.

Theorem 12-2 has a partial converse.

★THEOREM 12-4 *If f is nonnegative on a set B, and its integral over B is* 0, *then $f(x) = 0$ a.e. in B.*

Let n be a positive integer. Since the integral of f is 0, so is the integral of nf. Define $f_n(x) = \min\{nf(x), 1\}$ for all x. By Corollary I-7-3, the integral of f_n is 0. We now prove that for all x,

$$f_1(x) \leqq f_2(x) \leqq f_3(x) \leqq \cdots \qquad \text{and} \qquad \lim_{n\to\infty} f_n(x) = \mathbf{1}_E(x),$$

where E is the set of all x at which $f(x) > 0$. The inequalities are obvious, since $nf(x)$ increases with n. If x is in E, $f(x) > 0$, and for all n greater than $1/f(x)$ we have $nf(x) > 1$ and $f_n(x) = 1$, which is $\mathbf{1}_E(x)$. If x is not in E, $f(x) = 0$, so $nf(x) = 0$ for all n, and $f_n(x) = 0$ for all n. Therefore, the limit of $f_n(x)$ is 0, which is $\mathbf{1}_E(x)$. By the monotone convergence theorem, $\mathbf{1}_E$ is integrable and its integral is the limit of the integrals of the f_n, which is 0. So E has measure 0.

Although we spoke of equivalence of functions when discussing real-valued functions, the idea is quite general. Two functions f, g on R, with values in any space whatever, are equivalent if $f(x) = g(x)$ for all x except those in a set of measure 0. The reader will have no trouble with the details of the proof that this kind of equivalence is indeed an equivalence relation. That is, for all f, f is equivalent to f; if f is equivalent to g, then g is equivalent to f; and if f is equivalent to g and g is equivalent to h, then f is equivalent to h. By Theorem 12-2, in every class of real-valued functions all equivalent to each other, either none is integrable, or else all are integrable and have the same integral. This permits an easy extension of the concept of integral to certain extended-real-valued functions; that is, to certain functions with values in $\bar{R}$. If f is an extended-real-valued function on R that is finite-valued except on a set N of measure 0, there are infinitely many real-valued functions g on R that are equivalent to f; for example, we could take $g(x)$ to be 0 on N and equal to $f(x)$ elsewhere. All such functions are equivalent to each other, so by Theorem 12-2, either none has an integral or else all have the same integral. In the former case we say that f is not integrable. In the latter case we extend the idea of integrability by saying that f is integrable, and we define its integral to be the common value of the integrals of all the real-valued functions equivalent to f. Clearly, Theorem 12-2 continues to be valid for these extended integrals too. More than that; all the theorems of Chapters I and II have strengthened forms. For example, Corollary I-6-2 can be improved thus:

★THEOREM 12-5 *Let B be any subset of R, and let f_1 and f_2 be extended-real-valued functions integrable over B such that $f_1(x) \leqq f_2(x)$ a.e. in B. Then*

$$\int_B f_1(x)\, m(dx) \leqq \int_B f_2(x)\, m(dx). \tag{D}$$

Let N_1 be the set of x in B on which $f_1(x) = \pm\infty$, N_2 the set on which $f_2(x) = \pm\infty$, and N_3 the set on which $f_1(x) > f_2(x)$. By hypothesis, these have measure 0, and therefore so has their union N. For $i = 1, 2$, define $g_i(x) = f_i(x)$ for all x in $B \setminus N$ and $g_i(x) = 0$ for all x in $\bar{R} \setminus (B \setminus N)$. Then g_1 and g_2 are finite-valued and are equivalent to $(f_1)_B$ and $(f_2)_B$, respectively, and $g_1(x) \leqq g_2(x)$ for all x in $\bar{R}$. By Corollary I-6-2,

$$\int_B g_1(x)\, m(dx) \leqq \int_B g_2(x)\, m(dx). \tag{E}$$

By Theorem 12-2, as generalized to extended-real-valued functions, the two members of (E) are equal to the corresponding members of (D), so (D) is valid.

However, even though this extension is clearly possible, it is not so clear that it is worth the trouble. This may be more believable if we show a strengthened form of the monotone convergence theorem in which one hypothesis is missing; this improved form will be quite useful.

★THEOREM 12-6 (Monotone Convergence Theorem) *Let $f_1, f_2, f_3, \ldots$ be a sequence of extended-real-valued functions all integrable over a subset B of R and such that either*

$$f_1(x) \leqq f_2(x) \leqq f_3(x) \leqq \cdots \qquad \text{a.e. in } B, \tag{F}$$

or else

$$f_1(x) \geqq f_2(x) \geqq f_3(x) \geqq \cdots \qquad \text{a.e. in } B. \tag{G}$$

Let f be a function on B such that

(H) $$\lim_{n\to\infty} f_n(x) = f(x) \qquad \text{a.e. in } B.$$

Then f is integrable over B if and only if the integrals of the f_n over B are a bounded set of numbers, and in that case

$$\int_B f(x)\,m(dx) = \lim_{n\to\infty}\int_B f_n(x)\,m(dx).$$

To be specific, we assume that (F) holds a.e. in B. Let N_0 be the set of x in B at which (F) fails to be true; for each n let N_n be the set of x in B at which $f_n(x)$ is ∞ or $-\infty$; and let N_L be the set of x in B at which the limit relation (H) fails to hold. These all have measure 0 by hypothesis, so their union N has measure 0. For each n we define $g_n(x)$ to be $f_n(x)$ if x is in $B \setminus N$ and to be 0 if x is in N, and we define $g(x)$ analogously. Then the g_n are real-valued, and $g_1(x) \leqq g_2(x) \leqq \cdots$ for all x in B, and $g_n(x)$ tends to $g(x)$ for all x in B. But we cannot yet apply Theorem 4-2 because we do not know that g is real-valued. It clearly cannot be $-\infty$ anywhere, but it can be $+\infty$ at some points.

It follows at once from Theorem 12-5 that if f is integrable, the f_n have bounded integrals. Suppose, then, that the integrals of the g_n are bounded. As in the proof of Theorem 4-2, it is enough to prove the conclusion when $g_n \geqq 0$; the general case follows from this by subtracting g_1 from each term in the sequence. Let M be an upper bound for the integrals of the g_n over B, and let E be the set of all x in B such that $g(x) = \infty$. We must show that $m_L E = 0$. For each positive integer j the functions g_n/j $(n = 1, 2, 3, \ldots)$ are integrable over B, and so are the functions $(g_n/j) \wedge 1$, by Corollary 3-3. Also,

(I) $$\int_B \left[\frac{g_n(x)}{j} \wedge 1\right] m(dx) \leqq \int_B \frac{g_n(x)}{j}\,m(dx) \leqq \frac{M}{j}.$$

For each fixed j the functions $(g_n/j) \wedge 1$ form an ascending sequence that converges to $(g/j) \wedge 1$. By (I), their integrals are bounded, so by Theorem 4-2 $(g/j) \wedge 1$ is integrable over B, and

(J) $$\int_B \left[\frac{g(x)}{j} \wedge 1\right] m(dx) = \lim_{n\to\infty}\int_B \left[\frac{g_n(x)}{j} \wedge 1\right] m(dx) \leqq \frac{M}{j}.$$

Since g is nonnegative, the functions $(g/j) \wedge 1$ $(j = 1, 2, 3, \ldots)$ form a descending sequence. By (J), their integrals tend to 0. If x is any point of $B \setminus E$, $g(x)$ is finite, so for all j greater than $g(x)$ we have $(g(x)/j) \wedge 1 = g(x)/j$, and the equation

(K) $$\lim_{j\to\infty} (g(x)/j) \wedge 1 = 1_E(x)$$

holds, the right member being 0. If x is in E, $g(x)/j = \infty$ for all positive integers j, so $(g(x)/j) \wedge 1 = 1$ for all j, and (K) is still valid. By Theorem 4-2,

1_E is integrable over B, and by (I),

$$\int_R 1_E(x)\,m(dx) = \int_B 1_E(x)\,m(dx) = \lim_{j\to\infty}\int_B \left[\frac{g(x)}{j} \wedge 1\right] m(dx) = 0,$$

and so E has measure 0.

We now define $h_n(x) = g_n(x)$ and $h(x) = g(x)$ for all x in $B \setminus E$ and $h_n(x) = h(x) = 0$ for all x in E. These functions satisfy all the hypotheses of Theorem 4-2, and the integrals of the h_n are the same as the integrals of the f_n, so h is integrable, and its integral is the limit of the integrals of the f_n. But h and f differ only on the set $N \cup E$, which has measure 0, so f is also integrable over B, its integral being equal to the integral of h. This completes the proof.

Theorem 12-6 allows the same extension of the idea of integral as in Definition 4-3; if f is extended-real-valued on B but is not integrable over B, and there exists an ascending sequence of functions integrable over B and converging everywhere to f, we say that f has an integral over B, and the value of the integral is ∞. A similar definition applies to functions with integral $-\infty$. If we use this extension, the monotone convergence theorem takes an especially simple and easily remembered form.

★THEOREM 12-7 *Let $f_1, f_2, f_3, \ldots$ be a sequence of functions integrable over a set B and ascending (or descending) on a set $B \setminus E$, where $mE = 0$. Let f be equal to the limit of the f_n at almost all points of B. Then f has an integral over B, and*

$$\int_B f(x)\,m(dx) = \lim_{n\to\infty}\int_B f_n(x)\,m(dx).$$

The extension of the dominated convergence theorem is easy, but it is worth stating because the theorem is so important.

★THEOREM 12-8 *Let $b, f_0, f_1, f_2, f_3, \ldots$ be extended-real-valued functions on a set B in R such that*

(i) *$b, f_1, f_2, f_3, \ldots$ are integrable over B;*
(ii) *for $n = 1, 2, 3, \ldots$, $|f_n(x)| \leqq b(x)$ a.e. in B;*
(iii) *$\lim_{n\to\infty} f_n(x) = f_0(x)$ a.e. in B.*

Then f_0 is integrable over B, and

$$\int_B f_0(x)\,m(dx) = \lim_{n\to\infty}\int_B f_n(x)\,m(dx).$$

Let N_0 be the set of measure 0 on which the equation in (iii) fails to hold, and for each positive integer n let N_n be the set of measure 0 on which the inequality in (ii) fails to hold. Define N to be the union of $N_0, N_1, N_2, \ldots$. This has measure 0. For all x in $B \setminus N$, define $b^*(x) = b(x)$ and $f_j^*(x) = f_j(x)$ ($j = 0, 1, 2, \ldots$); for all x in N, define $b^*(x)$ and all the $f_j^*(x)$ to be 0. Then $b^*, f_1^*, f_2^*, \ldots$ are integrable over B and have the same integrals as $b, f_1, f_2, \ldots$, respectively, since they differ from those functions only on a set of measure 0. Also, $f_n^*(x)$ tends to $f_0^*(x)$ for all x in B, and $|f_n^*(x)| \leqq b^*(x)$ for all positive integers n and all x in B. By Theorem 10-1, the limit f_0^* is integrable over B, and

$$\int_B f_0^*(x)\, m(dx) = \lim_{n\to\infty} \int_B f_n^*(x)\, m(dx) = \lim_{n\to\infty} \int_B f_n(x)\, m(dx).$$

Since f_0 differs from f_0^* only on a set of measure 0, it too is integrable over B, and it has the same integral as f^*. This completes the proof.

By use of the theorems in this section and Section 10, we can prove a generalization of the fundamental theorem I-10-2.

THEOREM 12-9 *Let F be a function Lipschitzian on a closed interval $[a, b]$ [and having a derivative $DF(x)$ at almost all points x of $[a, b]$]. Let f be a function on $[a, b]$ that coincides with DF at almost all points at which DF exists. Then f is integrable over $[a, b]$ with respect to m_L, and*

$$\int_a^b f(x)\, dx = F(b) - F(a).$$

The hypothesis in square brackets is superfluous. It is possible, but difficult, to show that if F is Lipschitzian, it necessarily has a derivative at almost all points of $[a, b]$. However, if we wish to apply Theorem 12-9 to the computation of an integral, this superfluous hypothesis does no harm, since we cannot use the theorem unless we know what $DF(x)$ is at almost all x.

For each positive integer n we define

(L) $$x_{n,j} = a + j(b - a)/n \qquad (j = 0, 1, \ldots, n),$$

and we define s_n to be the step-function on $(a, b]$ which for $j = 1, \ldots, n$ has the constant value

(M) $$s_n(x) = [F(x_{n,j}) - F(x_{n,j-1})]/(x_{n,j} - x_{n,j-1})$$

on the interval $(x_{n,j-1}, x_{n,j}]$. Then

(N) $$\int_a^b s_n(x)\, dx = \sum_{j=1}^{n} \int_{x_{n,j-1}}^{x_{n,j}} \frac{F(x_{n,j}) - F(x_{n,j-1})}{x_{n,j} - x_{n,j-1}}\, dx$$
$$= \sum_{j=1}^{n} [F(x_{n,j}) - F(x_{n,j-1})] = F(b) - F(a).$$

By hypothesis, there exists a number L such that for all positive integers n and for $j = 1, \ldots, n$

$$|F(x_{n,j}) - F(x_{n,j-1})| \leqq L|x_{n,j} - x_{n,j-1}|.$$

This and (M) imply that for all x in $(a, b]$ we have

$$|s_n(x)| \leqq L. \tag{O}$$

By hypothesis, $DF(x)$ exists for all x in $(a, b]$ except those in a set N_1 with $m_L N_1 = 0$. The set N_2 of all points $x_{n,j}$ $(n = 1, 2, 3, \ldots ; j = 0, 1, \ldots, n)$ is countable, so $m_L N_2 = 0$. Therefore their union $N = N_1 \cup N_2$ also has measure 0. If x is in $(a, b] \setminus N$, for each positive integer n there is a number $j(n)$ in the set $\{1, \ldots, n\}$ such that

$$x_{n,j(n)-1} < x < x_{n,j(n)}.$$

Then, by (M), we calculate

$$s_n(x) - DF(x) = \left\{\frac{x - x_{n,j(n)-1}}{x_{n,j(n)} - x_{n,j(n)-1}}\right\}\left[\frac{F(x) - F(x_{n,j(n)-1})}{x - x_{n,j(n)-1}} - DF(x)\right]$$
$$+ \left\{\frac{x_{n,j(n)} - x}{x_{n,j(n)} - x_{n,j(n)-1}}\right\}\left[\frac{F(x_{n,j(n)}) - F(x)}{x_{n,j(n)} - x} \quad DF(x)\right].$$

As n increases, the two expressions in brackets tend to 0, and the factors that multiply them remain between 0 and 1. Therefore, the left member of the equation tends to 0, and

$$\lim_{n \to \infty} s_n(x) = DF(x) \tag{P}$$

for all x in $(a, b]$ except those in the set N of measure 0. By (O), (P), (N) and the dominated convergence theorem, f is integrable from a to b, and

$$\int_a^b f(x)\,dx = \lim_{n \to \infty} \int_a^b s_n(x)\,dx = F(b) - F(a).$$

EXERCISE 12-1 Let $f_1, f_2, f_3, \ldots$ be functions nonnegative on a set B in R and integrable over B, such that the series of the integrals of the f_n is convergent (has a finite sum). Prove that for almost all x in B, the series

$$f_1(x) + f_2(x) + f_3(x) + \cdots$$

is convergent. Prove also that if for each n we define

$$m_n = \inf\{f_n(x) : x \text{ in } B\},$$

then the series $m_1 + m_2 + m_3 + \cdots$ is convergent.

EXERCISE 12-2 Define $f(x) = 1/x$ for $x \neq 0$, $f(0) = 0$. For each positive integer n, define $f_n = f \wedge n$. Show that on $(-1, 1]$ the f_n ascend and each one has an integral, but their limit does not have an integral. Show also that the functions $f_n 1_{(-1/n, 1]}$ ascend, and each has an integral, and their limit also has an integral, but the integral of the limit of the f_n is different from the limit of the integrals of the f_n.

EXERCISE 12-3 Exercise 12-2 shows that in Theorem 12-6 we cannot replace the words "all integrable over a subset B of R" by "all having integrals over a subset B of R." Show that if the f_n ascend, it is sufficient to replace "all integrable over R" by "all having integrals $\neq -\infty$ over R."

EXERCISE 12-4 Show that if $E \subset R$, and for each positive ε there is a sequence of intervals whose union contains E and the sum of whose lengths is less than ε, then $m_{\text{L}}E = 0$.

EXERCISE 12-5 Let $f_1, f_2, f_3, \ldots$ be functions integrable over a set $B \subset R$ such that the series $b_1 + b_2 + b_3 + \cdots$ converges, where

$$b_n = \int_B |f_n(x)|\, m(dx).$$

Prove that the series $f_1(x) + f_2(x) + f_3(x) + \cdots$ converges absolutely for all points x of B except those in a set N of measure 0, and the sum of the series is integrable over B, and

$$\int_{B \setminus N} \left[\sum_{n=1}^{\infty} f_n(x) \right] m(dx) = \sum_{n=1}^{\infty} \int_B f_n(x)\, m(dx).$$

Suggestion: Define $s_n(x) = f_1(x) + \cdots + f_n(x)$, $s_n^*(x) = |f_1(x)| + \cdots + |f_n(x)|$. Then s_n^* has a limit b on B which has a finite integral over B and therefore is finite except on a set N of measure 0, and $|s_n(x)| \leqq b(x)$ on $B \setminus N$.

EXERCISE 12-6 Let $f_0, f_1, f_2, \ldots$ all be integrable over a set B in R, and assume

$$\lim_{n \to \infty} \int_B |f_n(x) - f_0(x)|\, m(dx) = 0.$$

Prove that there is an integrable function b and a subsequence $f_{n(1)}, f_{n(2)}, f_{n(3)}, \ldots$ of the given sequence such that $f_{n(j)}$ converges to f_0 uniformly relative to b (cf. Exercise 4-1). *Suggestion*: Choose increasing integers $n(1), n(2), n(3), \ldots$ such that for each j, the integral of $|f_{n(j)} - f_0|$ is less than 4^{-j}. Define

$$b = \sum_{j=1}^{\infty} 2^j |f_{n(j)} - f_0|.$$

EXERCISE 12-7 Prove that if f is Riemann integrable from a to b, it is continuous at all points of $(a, b]$ except those in a set of m_{L}-measure 0. *Suggestion*: For positive ε let $E(\varepsilon)$ be the set of points x in $(a, b]$ such that in every neighborhood of x there is a point x' with $|f(x') - f(x)| > \varepsilon$. For each positive integer n choose step-functions $s \leqq f \leqq S$ with $\int [S - s]\, dx < \varepsilon/n$. Then

$$\varepsilon^{-1}[S(x) - s(x)] \geqq 1_{E(\varepsilon)}(x)$$

except perhaps at the discontinuities of S and s, which form a set of measure 0. This implies that the Riemann integral of $1_{E(\varepsilon)}$ is 0, so $E(\varepsilon)$ has measure 0. Take the union for $\varepsilon = 1, \frac{1}{2}, \frac{1}{3}, \ldots$.

EXERCISE 12-8 Prove that if f is bounded on a bounded closed interval B in R, and the set D of points of B at which f is discontinuous has $m_L D = 0$, f is Riemann integrable from a to b. *Suggestion*: Define s_n and S_n as in Exercise 4-2. Then $S_n - s_n$ descends and tends to 0 except on D.

13. Approximation by Step-Functions

In this section we shall prove a theorem that has several uses but would be worth proving even if it were never referred to again. There is a sense in which integrable functions can be very different from step-functions; a function can be integrable even though it is everywhere discontinuous. But as a numerical measure of the amount by which two functions f and g differ from each other, we can use the integral of the absolute value of their difference $|f - g|$. It is somewhat surprising that if f is integrable, for every positive ε there are step-functions that differ from f in this sense by less than ε. If the reader will carry out the simple exercise of proving that for every step-function s there is for each positive ε a continuous function c such that $|s - c|$ has an integral less than ε, he will see that it follows that every integrable f has continuous c such that the integral of $|f - c|$ is less than ε. Thus, in spite of the fact that our integration procedure allows us to integrate functions that seem very different from step-functions or continuous functions, we can approximate every integrable f by step-functions or continuous functions whose behavior, as far as integration is concerned, is nearly that of f. So, when we are casting about for a means of solving an unfamiliar problem or proving a new theorem, we can tentatively reason as though the integrable functions involved were actually step-functions or continuous functions. If we can handle the situation in this simple case, we have at least a reasonable hope that we can modify the methods a bit and deal with the general case.

★THEOREM 13-1 *Let f be integrable over R, and let ε be positive. Let γ be a gauge on $\bar{R}$ such that $\gamma(x)$ is a bounded interval whenever x is in R and such also that for every γ-fine partition $\mathscr{P}$ of R*

(A)
$$\left| S(\mathscr{P}; f_R) - \int_R f(x)\, m(dx) \right| < \varepsilon.$$

Let

$$\mathscr{P} = \{(x_1, A_1), \ldots, (x_k, A_k)\}$$

be a γ-fine partition of R. Let s be the function on R that on each A_j has the constant value $f_R(x_j)$. Then s is a step-function, and

$$\text{(B)} \qquad \int_R |f(x) - s(x)|\, m(dx) < 5\varepsilon.$$

For each j in $\{1,\ldots,k\}$, if A_j is unbounded, the neighborhood $\gamma(x_j)$ that contains A_j^- is unbounded, so x_j is not in R (that is, it is not finite), and on A_j the constant $f_R(x_j)$ of s is 0. So s is a step-function on R, and it is therefore integrable by Theorem I-5-6. Then $f - s$ is integrable over R by Theorem I-5-3, and $|f - s|$ is integrable over R by Corollary 3-4. Therefore there exists a gauge γ_1 on $\bar{R}$ such that for every γ_1-fine partition $\mathscr{P}'$ of R,

$$\text{(C)} \qquad \left| S(\mathscr{P}'; |f - s|_R) - \int_R |f(x) - s(x)|\, m(dx) \right| < \varepsilon.$$

For each x in $\bar{R}$ we define $\gamma^*(x)$ to be the intersection of $\gamma(x)$, $\gamma_1(x)$, and all of those neighborhoods $\gamma(x_1), \ldots, \gamma(x_k)$ that contain x. This is the intersection of finitely many neighborhoods of x, so it is itself a neighborhood of x, and therefore the function $x \mapsto \gamma^*(x)$ is a gauge on $\bar{R}$. For each interval A_j $(j = 1, \ldots, k)$ we choose a γ^*-fine partition $\mathscr{P}'_j$ of A_j such that for each pair $(\bar{x}, A)$ in the partition $\mathscr{P}'_j$, $\bar{x}$ is in the closure A^- of A; this is possible by Theorem I-4-2. We take the union $\mathscr{P}'$ of all the partitions $\mathscr{P}'_1, \ldots, \mathscr{P}'_k$. By Lemma 2-1, this union is an allotted partition

$$\mathscr{P}' = \{(x'_1, A'_1), \ldots, (x'_h, A'_h)\}$$

of the union of the A_j, which is R. Clearly, $\mathscr{P}'$ has the property that for each i, x'_i is contained in the closure of A'_i, and A'_i is contained in a single one of the intervals A_j.

For each i in $\{1, \ldots, h\}$ we define a number $j(i)$ in the set $\{1, \ldots, k\}$ as follows. If A'_i is bounded, x'_i is in the closure $(A'_i)^-$, which is bounded, so x'_i is in R. It therefore belongs to exactly one of the intervals A_j. We define $j(i)$ to be the integer j such that x'_i is in A_j. If A'_i is unbounded, there is just one integer j such that $A'_i \subset A_j$; this integer we name $j(i)$.

We now prove that

(D) for each pair $(x_{j(i)}, A'_i)$ $(i = 1, \ldots, h)$, the closure $(A'_i)^-$ is contained in $\gamma(x_{j(i)})$.

Suppose first that A'_i is bounded. By definition of $j(i)$, x'_i is contained in $A_{j(i)}$, which is contained in $\gamma(x_{j(i)})$. By definition, $\gamma^*(x'_i)$ is the intersection of several neighborhoods, one of which is $\gamma(x_{j(i)})$, so

$$\gamma^*(x'_i) \subset \gamma(x_{j(i)}).$$

Since $\mathscr{P}'$ is γ^*-fine, the closure $(A'_i)^-$ is contained in $\gamma^*(x'_i)$ and therefore is contained in $\gamma(x_{j(i)})$, and (D) is valid. Suppose, on the other hand, that A'_i is unbounded. Then by definition of $j(i)$, $A'_i \subset A_{j(i)}$. So $(A'_i)^-$ is contained in $A^-_{j(i)}$, which is contained in $\gamma(x_{j(i)})$ because $\mathscr{P}$ is γ-fine, and again (D) is valid.

From (D) it follows that the partition

$$\mathscr{P}'' = \{(x_{j(1)}, A'_1), \ldots, (x_{j(h)}, A'_h)\}$$

is a γ-fine partition of R. Since $\mathscr{P}'$ is also γ-fine, and (A) holds for all γ-fine partitions of R, by Theorem 2-4 we have

$$\text{(E)} \qquad \sum_{n=1}^{h} \sum_{i=1}^{h} |f_R(x'_n) - f_R(x_{j(i)})| m(A'_n \cap A'_i) < 4\varepsilon.$$

The intersection $A'_n \cap A'_i$ is empty unless $n = i$, so (E) implies

$$\text{(F)} \qquad \sum_{i=1}^{h} |f_R(x'_i) - f_R(x_{j(i)})| mA'_i < 4\varepsilon.$$

For each i such that A'_i is bounded, x'_i is in the bounded closure of A'_i and is therefore in R, so

$$s_R(x'_i) = s(x'_i).$$

Since in this case x'_i is in $A_{j(i)}$ and s has the constant value $f(x_{j(i)})$ on $A_{j(i)}$, this implies

$$\text{(G)} \qquad f_R(x_{j(i)}) = s_R(x'_i).$$

For each i such that A'_i is unbounded, by definition of $j(i)$ we have $A'_i \subset A_{j(i)}$. So, both A'_i and $A_{j(i)}$ are unbounded, and the neighborhoods $\gamma^*(x'_i)$, $\gamma(x_{j(i)})$ that contain them are unbounded. This cannot happen unless neither x'_i nor $x_{j(i)}$ is in R, so in this case both members of (G) are 0, and again (G) is satisfied. If we substitute this in (F) we obtain

$$\text{(H)} \qquad \sum_{i=1}^{h} |f_R(x'_i) - s_R(x'_i)| mA'_i < 4\varepsilon.$$

But the left member of (H) is $S(\mathscr{P}'; |f - s|_R)$, so (H) and (C) imply that (B) is satisfied. The proof is complete.

★Corollary 13-2 *Let f be integrable over R, let ε be positive, and let γ be a gauge on $\bar{R}$ such that $\gamma(x)$ is bounded whenever x is in R and such also that for every γ-fine partition $\mathscr{P}$ of R*

$$\left| S(\mathscr{P}; f_R) - \int_R f(x)\, m(dx) \right| < \varepsilon.$$

Let $\{(x_1, A_1), \ldots, (x_h, A_h)\}$ be a set of pairs such that the A_j are pairwise disjoint and for $j = 1, \ldots, h$, $A_j^- \subset \gamma(x_j)$. Then

$$\text{(I)} \qquad \sum_{j=1}^{h} \left| f_R(x_j) mA_j - \int_{A_j} f(x)\, m(dx) \right| < 5\varepsilon.$$

The complement $\bar{R} \setminus (A_1 \cup A_2 \cup \cdots \cup A_h)$ of the union of the A_j is the union of finitely many pairwise disjoint left-open intervals. We choose a γ-fine partition of each of these, and we form the union of these partitions and denote it by

$$\{(x_{h+1}, A_{h+1}), \ldots, (x_k, A_k)\}.$$

Then the set

$$\mathscr{P} = \{(x_1, A_1), \ldots, (x_k, A_k)\}$$

is a γ-fine partition of R, by Lemma 2-1.

We define s as in Theorem 13-1. By that theorem, with Corollary I-6-2, Theorem I-8-1, and Corollary 3-4,

$$\text{(J)} \qquad \begin{aligned} 5\varepsilon &> \int_R |f(x) - s(x)|\, m(dx) \\ &\geq \int_{A_1 \cup A_2 \cup \cdots \cup A_h} |f(x) - s(x)|\, m(dx) \\ &= \sum_{j=1}^{h} \left| \int_{A_j} |f(x) - s(x)|\, m(dx) \right| \\ &\geq \sum_{j=1}^{h} \left| \int_{A_j} [f(x) - s(x)]\, m(dx) \right| \\ &= \sum_{j=1}^{h} \left| \int_{A_j} f(x)\, m(dx) - \int_{A_j} s(x)\, m(dx) \right|. \end{aligned}$$

But on A_j the function s has the constant value $f_R(x_j)$, so

$$\int_{A_j} s(x)\, m(dx) = f_R(x_j) mA_j.$$

If we substitute this in (J), we obtain (I). The proof is complete.

We now show that an integrable function can also be approximated by the step-function which, on each interval of a sufficiently fine partition, is constantly equal to the average value of the function on that interval.

★COROLLARY 13-3 *Let f be integrable over R, and let ε be positive. Let γ be a gauge on $\bar{R}$ such that $\gamma(x)$ is bounded whenever x is in R and such also that for every γ-fine partition $\mathscr{P}$ of R*

$$\left| S(\mathscr{P}; f_R) - \int_R f(x)\, m(dx) \right| < \varepsilon.$$

Let $\{(x_1, A_1), \ldots, (x_k, A_k)\}$ be a γ-fine partition of R, and let g be the function which *on each bounded interval A_i with $mA_i > 0$ has the constant value*

$$g(x) = \left\{ \int_{A_i} f(x)\, m(dx) \right\} \Big/ mA_i,$$

and elsewhere is 0. *Then g is a step-function, and*

$$\int_R |f(x) - g(x)|\, m(dx) < 10\varepsilon.$$

It is obvious that g is a step-function. Let s be defined as in Theorem 13-1. By that theorem,

$$\text{(K)} \qquad \int_R |f(x) - s(x)|\, m(dx) < 5\varepsilon.$$

If A_i is bounded and $mA_i > 0$, both s and g are constant on A_i, and by their definitions, if $\bar{x}$ is any point of A_i,

$$\text{(L)} \qquad \begin{aligned} \int_{A_i} |g(x) - s(x)|\, m(dx) &= |g(\bar{x}) - s(\bar{x})| mA_i \\ &= |g(\bar{x}) mA_i - s(\bar{x}) mA_i| \\ &= \left| \int_{A_i} f(x)\, m(dx) - \int_{A_i} s(x)\, m(dx) \right|. \end{aligned}$$

Therefore, for such A_i

$$\text{(M)} \qquad \int_{A_i} |g(x) - s(x)|\, m(dx) \leqq \int_{A_i} |f(x) - s(x)|\, m(dx).$$

If $mA_i = 0$, both members of (M) are 0; and if A_i is unbounded, both g and s are identically 0 on A_i, so the left member of (M) is 0. Therefore, (M) holds for $i = 1, \ldots, k$. We add these k inequalities member by member. The A_i are pairwise disjoint, and their union is R, so we obtain

$$\text{(N)} \qquad \int_R |g(x) - s(x)|\, m(dx) \leqq \int_R |f(x) - s(x)|\, m(dx).$$

The inequality

$$\text{(O)} \quad \int_R |f(x) - g(x)|\, m(dx) \leqq \int_R |f(x) - s(x)|\, m(dx) + \int_R |s(x) - g(x)|\, m(dx),$$

with (N) and (K), yields the conclusion of the corollary.

We can now establish an improvement on Theorem 12-4, obtaining the same conclusion from weaker hypotheses.

★Theorem 13-4 *Let f be a function integrable over R such that for every bounded left-open interval A in R*

$$\text{(P)} \qquad \int_A f(x)\,m(dx) = 0.$$

Then $f(x) = 0$ almost everywhere in R.

Let ε' be any positive number; define

$$\varepsilon = \varepsilon'/10.$$

Since f is integrable, there exists a gauge γ_1 on $\bar{R}$ such that for every γ_1-fine partition $\mathscr{P}$ of R

$$\text{(Q)} \qquad \left| S(\mathscr{P}; f_R) - \int_R f(x)\,m(dx) \right| < \varepsilon.$$

Let γ_2 be any gauge on $\bar{R}$ such that $\gamma_2(x)$ is bounded whenever x is in R; define $\gamma = \gamma_1 \cap \gamma_2$. Then (Q) holds whenever $\mathscr{P}$ is γ-fine, so the hypotheses of Corollary 13-3 are satisfied. The function g of that corollary is 0 whenever either $mA_i = 0$ or A_i is unbounded, by definition of g; and when A_i is bounded and $mA_i > 0$, g is 0 on A_i because, by hypothesis, equation (P) holds. So g is identically 0, and by Corollary 13-3,

$$\text{(R)} \qquad \int_R |f(x)|\,m(dx) < 10\varepsilon = \varepsilon'.$$

The left member of (R) is a nonnegative number that is less than the arbitrary positive number ε', so it is 0. By Theorem 12-4, $f(x) = 0$ almost everywhere in R.

EXERCISE 13-1 Use Theorem 13-1 to obtain another proof of Theorem 9-4. *Suggestion*: For each n, choose step-functions f_n, g_n with

$$\int_a^b |f_n(x) - f(x)|\,dx < \frac{1}{n}, \qquad \int_a^b |g_n(x) - g(x)|\,dx < \frac{1}{n}$$

and let F_n, G_n be indefinite integrals with $F_n(a) = F(a)$, $G_n(a) = G(a)$.

EXERCISE 13-2 Show that Corollary 13-2 remains valid if in inequality (I) we replace $f_R(x_j)$ by the constant value on A_j of the function g of Corollary 13-3.

14. Differentiation of Indefinite Integrals

We know by Theorem I-9-2 that if a function F is an indefinite integral of a function f, then F has a derivative equal to $f(x)$ at each x at which f is continuous. But our integral operates on functions that may have no points of continuity at all. For such functions, the indefinite integral F will exist, but Theorem I-9-2 tells us nothing at all about its derivative. It is rather surprising that a theorem quite like Theorem I-9-2 holds for every integrable f, without any assumption of continuity. We cannot say that the indefinite integral F has derivative $f(x)$ at every point x, but we can come close to it; F has derivative $f(x)$ at all points x except those in a set of measure 0.

THEOREM 14-1 *Let f be integrable with respect to m_L over R, and let F be an indefinite integral of f. Then for almost all points x of R, F has a derivative, and $DF(x) = f(x)$.*

We shall in fact prove what seems to be a stronger statement but really is exactly equivalent to the conclusion. (Compare statement (A) in the proof of Theorem I-9-2.) For each interval A in R we define

$$F[A] = \int_A f(x)\,dx.$$

(The bracket distinguishes it from F.) If A has end-points a and b, by Definition I-9-1

$$F[A] = \int_a^b f(x)\,dx = F(b) - F(a).$$

We shall prove

(A) For almost all x in R, for each positive ε there exists a neighborhood U of x such that whenever A is a nondegenerate interval contained in U such that x is in A^-,

$$|F[A]/m_L A - f(x)| < \varepsilon.$$

This implies that F has derivative $f(x)$ at all points where this condition is satisfied. For let x be such a point; let ε be positive, and let U be as in (A). If x' is any point of U other than x, the closed interval A whose ends are x and x' is a nondegenerate interval in U whose closure contains x, so the inequality in (A) is satisfied. If $x' > x$, $F[A] = F(x') - F(x)$ and $m_L A = x' - x$, so

$$F[A]/m_L A = (F(x') - F(x))/(x' - x). \tag{B}$$

If $x' < x$, then $F[A] = F(x) - F(x')$ and $m_L A = x - x'$, so again (B) is satisfied. If we substitute this in the inequality in (A), it takes the form of the definition of the statement that the derivative of F is $f(x)$.

For each positive ε we define D_ε^+ to be the set of all points x in R for which there exists a neighborhood $\gamma_1(x)$ such that if A is any interval whose closure is contained in $\gamma_1(x)$ and contains x,

$$F[A]/m_L A \leqq f(x) + \varepsilon. \tag{C}$$

We denote by B_ε^+ the complement $R \setminus D_\varepsilon^+$ of D_ε^+. We shall first show that

(D) for each positive ε, $m_L B_\varepsilon^+ = 0$.

Let δ be any positive number. There exists a gauge γ such that if $\mathscr{P}$ is any γ-fine partition of R,

$$\left| S(\mathscr{P}; f_R) - \int_R f(x)\,dx \right| < \frac{\varepsilon\delta}{15}. \tag{E}$$

Let x be any point of B_ε^+. If, for every interval A whose closure is contained in $\gamma(x)$ and contains x, inequality (C) were satisfied, x would belong to D_ε^+, which it does not. So there is an interval A whose closure is contained in $\gamma(x)$ and contains x for which

$$F[A]/m_L A > f(x) + \varepsilon. \tag{F}$$

Let a and b be the end-points of A. We can and do choose an ascending sequence of rational numbers $a_1 < a_2 < a_3 < \cdots$ converging to a, and a descending sequence of rational numbers $b_1 > b_2 > b_3 > \cdots$ converging to b. Since a and b are in $\gamma(x)$, for all large n both a_n and b_n are in $\gamma(x)$; and since F is continuous by Corollary 9-3,

$$\lim_{n\to\infty} [(F(b_n) - F(a_n)]/(b_n - a_n) = [F(b) - F(a)]/(b - a).$$

By this and the definition of $F[A]$, corresponding to the point x of B_ε^+ we can and do choose an n so large that the interval

$$A(x) = (a_n, b_n]$$

has rational end-points, and its closure is contained in $\gamma(x)$, and

$$F[A(x)]/m_L A(x) > f(x) + \varepsilon. \tag{G}$$

Although there may be uncountably many points in B_ε^+, there can be only countably many intervals $A(x)$ (x in B_ε^+) because the $A(x)$ have rational end-points and there are only countably many intervals with rational end-points. So we can select countably many intervals $A(x_1)$, $A(x_2)$, $A(x_3), \ldots$ such that for every x in B_ε^+, $A(x)$ is one of the $A(x_i)$. Since the union of the $A(x)$ for all x in B_ε^+ contains B_ε^+, it is also true that

$$\bigcup_{i=1}^{\infty} A(x_i) \supset B_\varepsilon^+. \tag{H}$$

Let n be any positive integer. By interchanging their names, if necessary, the intervals $A(x_1), \ldots, A(x_n)$ can be rearranged so that they are of nonincreasing lengths. Suppose that this has already been done. We now subdivide the set $\{1, \ldots, n\}$ into a "selected" subset, which we call Sel, and a "rejected" subset, which we call Rej, according to the following rule. First, we assign 1 to Sel. Then, successively, if h is any number in the set $\{2, \ldots, n\}$ and all the numbers in

$\{1, \ldots, h-1\}$ have been assigned either to Sel or to Rej, either $A(x_h)$ is disjoint from all of those intervals $A(x_i)$ $(i = 1, \ldots, h-1)$ for which i belongs to Sel, or it has a point in common with one of them. In the former case we assign h to Sel; in the second case, we assign it to Rej. Clearly, the intervals $A(x_j)$ with j in Sel are pairwise disjoint. For each such j the closure $A(x_j)^-$ is contained in $\gamma(x_j)$, so by Corollary 13-2, with (B),

$$\text{(I)} \qquad \sum_{j \text{ in Sel}} \left| -f(x_j) m_{\mathrm{L}} A(x_j) + \int_{A(x_j)} f(x)\, dx \right| < 5 \frac{\varepsilon\delta}{15}.$$

Since x_j is in B_ε^+ and $A(x_j)$ is the interval associated with it, by (G),

$$\int_{A(x_j)} f(x)\, dx = F(A(x_j)) > f(x_j) m_{\mathrm{L}} A(x_j) + \varepsilon m_{\mathrm{L}} A(x_j).$$

So for each j in Sel the corresponding term in the left member of (I) is greater than $\varepsilon m_{\mathrm{L}} A(x_j)$, and by (I), we have

$$\text{(J)} \qquad \varepsilon \sum_{j \text{ in Sel}} m_{\mathrm{L}} A(x_j) < \frac{\varepsilon\delta}{3}.$$

For each interval B in R we denote by B^* the left-open interval with the same midpoint and three times the length. Thus,

$$\text{(K)} \qquad \text{if } B = (a, b],\ B^* = (2a - b, 2b - a].$$

We shall now prove

$$\text{(L)} \qquad \bigcup_{j \text{ in Sel}} A(x_j)^* \supset \bigcup_{i-1}^{n} A(x_i).$$

Since by the definition (K) of B^* it is evident that $A(x_j)^* \supset A(x_j)$,

$$\text{(M)} \qquad \bigcup_{i=1}^{n} A(x_i) = \left[\bigcup_{j \text{ in Sel}} A(x_j) \right] \cup \left[\bigcup_{k \text{ in Rej}} A(x_k) \right]$$
$$\subset \left[\bigcup_{j \text{ in Sel}} A(x_j)^* \right] \cup \left[\bigcup_{k \text{ in Rej}} A(x_k) \right].$$

Let x be any point that belongs to some $A(x_k)$ with k in Rej. There must be a number j in $\{1, \ldots, k-1\}$ that belongs to Sel for which $A(x_j)$ is not disjoint from $A(x_k)$; otherwise, by the selection rule, k would have been assigned to Sel. Let x^* be a point that belongs both to $A(x_k)$ and to $A(x_j)$. Denote $A(x_k)$ by $(a, b]$ and $A(x_j)$ by $(c, d]$. Since $k > j$ and the $A(x_i)$ have nondecreasing lengths,

$$d - c \geqq b - a.$$

Since x is in $(a, b]$ and x^* is in both $(a, b]$ and $(c, d]$,

$$x = x^* + (x - x^*) \leqq d + (b - a) \leqq d + (d - c),$$

and

$$x = x^* - (x^* - x) > c - (b - a) \geqq c - (d - c).$$

So x is in $(2c - d, 2d - c)$, which is $A(x_j)^*$. Thus, every point in the set in the second bracket in the right member of (M) is already in the set in the first bracket, and therefore (M) reduces to (L).

For convenience we define

$$U(n) = \bigcup_{i=1}^{n} A(x_i) \qquad (n = \infty, 1, 2, 3, \ldots),$$

and we denote the indicator of the interval $A(x_i)^*$ by f_i. Then (L) implies that for all x in R

$$1_{U(n)}(x) \leqq \sup\{f_j(x) : j \text{ in Sel}\} \leqq \sum_{j \text{ in Sel}} f_j(x).$$

The length of $A(x_j)^*$ is three times that of $A(x_j)$, so integrating this over R and recalling (J) yields

$$\text{(N)} \qquad \int_R 1_{U(n)}(x)\, dx \leqq \sum_{j \text{ in Sel}} m_L A(x_j)^* = 3 \sum_{j \text{ in Sel}} m_L A(x_j) < \delta.$$

As n increases, $1_{U(n)}$ ascends and converges everywhere in R to $1_{U(\infty)}$, so by (N) and the monotone convergence theorem,

$$\text{(O)} \qquad \int_R 1_{U(\infty)}(x)\, dx \leqq \delta,$$

while by (H),

$$\text{(P)} \qquad 0 \leqq 1_{B_\varepsilon^+} \leqq 1_{U(\infty)}.$$

Now, corresponding to the arbitrary positive number δ we have found an integrable function $1_{U(\infty)}$ with which (O) and (P) are satisfied. By Lemma I-7-1,

$$\int_R 1_{B_\varepsilon^+}(x)\, dx = 0,$$

so that

$$\text{(Q)} \qquad m_L B_\varepsilon^+ = 0.$$

Next, for each positive ε we define D_ε^- to be the set of all points x in R for which there exists a neighborhood $\gamma_2(x)$ such that if A is any interval whose closure is contained in $\gamma_2(x)$ and contains x,

$$\text{(R)} \qquad F[A]/m_L A \geqq f(x) - \varepsilon,$$

and we denote by B_ε^- the complement $R \setminus D_\varepsilon^-$ of D_ε^-. We could prove that B_ε^- has measure 0, just as we proved that $m_L B_\varepsilon^+ = 0$, but we do not even have to. Since

$-F$ is an indefinite integral of $-f$, and the B_ε^- just defined is the set B_ε^+ for the pair $-f$ and $-F$, the preceding proof implies that it has measure 0. Then the union $B_\varepsilon = B_\varepsilon^+ \cup B_\varepsilon^-$ has measure 0, and every point x not in B_ε is in both D_ε^+ and D_ε^-. So, for all nondegenerate intervals A whose closure is contained in $\gamma(x) = \gamma_1(x) \cap \gamma_2(x)$ and contains x, both (C) and (R) are satisfied, so that

$$\text{(S)} \qquad |F[A]/m_L A - f(x)| \leqq \varepsilon.$$

Now let B be the union

$$B = B_1 \cup B_{1/2} \cup B_{1/3} \cup B_{1/4} \cup \cdots.$$

This has measure 0, because each $B_{1/n}$ has measure 0. Let x be any point of R not in B, and let ε be any positive number. Choose an integer n such that $1/n < \varepsilon$. Since x is not in $B_{1/n}$, there is a neighborhood $\gamma(x)$ such that for all nondegenerate intervals A whose closures are contained in $\gamma(x)$ and contain x,

$$|F[A]/m_L A - f(x)| \leqq 1/n < \varepsilon.$$

This establishes statement (A) and completes the proof.

EXERCISE 14-1 Sometimes one is tempted to look for a subset E of a nondegenerate interval A such that for every subinterval B of A, $m_L(B \cap E) = (m_L B)/2$. Show that no such subset B can exist. Show that, in fact, there is no subset E of A such that for every subinterval B of A

$$0.01 m_L B \leqq m_L(B \cap E) \leqq 0.99 m_L B.$$

EXERCISE 14-2 What does Theorem 14-1 say when f is the indicator of the rational numbers?

15. Calculus of Variations

In practically every elementary calculus text it is stated that, under suitable hypotheses, if a curve is the graph of a function y that is the indefinite integral of some function $\dot{y}$, so that

$$\text{(A)} \qquad y(x) = y(x_1) + \int_{x_1}^{x} \dot{y}(u)\,du \qquad (x_1 \leqq x \leqq x_2),$$

and if y is smooth enough, the length of the curve is

$$\text{(B)} \qquad \int_{x_1}^{x_2} [1 + \dot{y}(x)^2]^{1/2}\,dx.$$

(In Theorem V-5-3 we shall establish this with no assumptions about $\dot{y}$ beyond mere integrability.) Likewise, if the graph of y lies in the upper half-plane, so that $y(x)$ is positive for all x in $[x_1, x_2]$, the area of the surface generated by revolving the curve about the x-axis is

$$\text{(C)} \qquad 2\pi \int_{x_1}^{x_2} y(x)[1 + \dot{y}(x)^2]^{1/2}\, dx.$$

The simplest type of problem studied in the branch of mathematics called the "calculus of variations" is that of finding the curve that, in a given subclass of the curves of type (A), gives the least value to some integral such as (B) or (C). Thus, the problem of minimizing (B) in the class of curves with given end-points is the (very easy) problem of finding the shortest curve with the given end-points. The problem of minimizing (C) is the (not so easy) problem of finding the surface of revolution bounded by the circles generated by rotating two given end-points about the x-axis which, among all such surfaces, has the least area. If the circles are not too far apart, this surface is the shape of the "soap film" bounded by the two circles.

To gain some useful generality, we shall consider problems in higher-dimensional spaces. The space of r dimensions is defined to be the set of all ordered r-tuples of real numbers. This is what we have already defined to be the Cartesian product of r spaces, each of which is the space R, so it is denoted by R^r. For the purposes of this section, we need to know hardly anything about it. We shall use the letters y and p to denote points of R^r, so that a point called y always has coordinates $y^{(1)}, \ldots, y^{(r)}$; that is,

$$y = (y^{(1)}, \ldots, y^{(r)}).$$

For the integrands in integrals of the type of (B) and (C) we shall always make this assumption.

(D) f is a function $(x, y, p) \mapsto f(x, y, p)$ defined for all (x, y) in a set G in R^{r+1} and all p in R^r, and f and all its partial derivatives of first and second order are continuous on the set $G \times R^r$ on which f is defined.

We shall assume that (x_1, y_1^*) and (x_2, y_2^*) are two points of G with $x_1 < x_2$. We shall also denote by $\mathscr{K}$ the class of all functions

$$x \mapsto y(x) \qquad (x_1 \leqq x \leqq x_2)$$

with values $y(x)$ in R^r such that $(x, y(x))$ is in G and

$$\text{(E)} \qquad y(x_1) = y_1^* \quad \text{and} \quad y(x_2) = y_2^*;$$

(F) for $i = 1, \ldots, r$, $y^{(i)}$ is the indefinite integral of some function $\dot{y}^{(i)}$, so that

$$y^{(i)}(x) = y^{(i)}(x_1) + \int_{x_1}^{x} \dot{y}^{(i)}(u)\, du;$$

(G) the integral

$$J[y] = \int_{x_1}^{x_2} f(x, y(x), \dot{y}(x))\, dx$$

exists.

The problem that we shall consider is that of finding a function in the class $\mathscr{K}$ that gives to the integral $J[y]$ its least value on the class $\mathscr{K}$. We shall not attempt to find conditions under which such a minimizing function can be guaranteed to exist. Instead, we shall assume that such a function exists and (under added hypotheses) find conditions that it has to satisfy so that we can distinguish it from the other functions in $\mathscr{K}$. This will usually enable us to find the minimizing function, *if* there is one of the type that we are allowing.

Suppose then that $x \mapsto y(x)$ $(x_1 \leqq x \leqq x_2)$ is a function that gives the integral J the least value that it has for any function in $\mathscr{K}$, and suppose in addition that y satisfies the hypotheses

(H) the functions $\dot{y}^{(i)}$ are bounded (say $|\dot{y}^{(i)}| \leqq M$),

(I) there is a positive ε such that every point (x, y) that satisfies the conditions

$$x_1 \leqq x \leqq x_2, \quad y^{(i)}(x) - \varepsilon \leqq y^{(i)} \leqq y^{(i)}(x) + \varepsilon \qquad (i = 1, \ldots, r)$$

is in G.

We shall make use of some functions called variations. A **variation** is a function $x \mapsto \eta(x)$ $(x_1 \leqq x \leqq x_2)$ such that, first, η vanishes at x_1 and at x_2:

(J) $$\eta^{(i)}(x_1) = \eta^{(i)}(x_2) = 0 \qquad (i = 1, \ldots, r),$$

and, second, for each i in $\{1, \ldots, r\}$ there is a bounded integrable function $\dot{\eta}^{(i)}$ such that

(K) $$\eta^{(i)}(x) = \int_{x_1}^{x} \dot{\eta}^{(i)}(u)\, du \qquad (x_1 \leqq x \leqq x_2).$$

(The importance of such functions in finding the minima of integrals $J[y]$ of type (G) accounts for the name "calculus of variations.") We can and do choose a positive number δ so small that for every number t in the interval $(-\delta, \delta)$ the functions $t\eta^{(i)}$ have absolute values less than ε and the functions $t\dot{\eta}^{(i)}$ have absolute values less than 1. Therefore, the functions

$$x \mapsto y(x) + t\eta(x)$$

will satisfy the equations

(L) $$y^{(i)}(x) + t\eta^{(i)}(x) = y^{(i)}(x_1) + t\eta^{(i)}(x_1) + \int_{x_1}^{x} [\dot{y}^{(i)}(u) + t\dot{\eta}^{(i)}(u)]\, du.$$

By (J), they have values y_1^* at x_1 and y_2^* at x_2. On the bounded closed set of points that satisfy the conditions in (I), the function f has continuous first-order partial derivatives, and therefore it is Lipschitzian in each variable separately. Theorem 3-1 does not apply to this function because there are too many variables; but the proof of Theorem 3-1 can be extended without trouble to include this case also. We omit details, since a more general result will be proved in Corollary VII-3-3. We thus find that the function

$$\text{(M)} \qquad x \mapsto f(x, y(x) + t\eta(x), \dot{y}(x) + t\dot{\eta}(x))$$

is integrable from x_1 to x_2 whenever t is in $(-\delta, \delta)$.

We have now shown that for all t in $(-\delta, \delta)$ the function $y + t\eta$ belongs to the class $\mathscr{K}$. When $t = 0$, it coincides with the minimizing function y, so the integral

$$J[y + t\eta] = \int_{x_1}^{x_2} f(x, y(x) + t\eta(x), \dot{y}(x) + t\dot{\eta}(x))\, dx$$

has its least value when $t = 0$. Therefore, if it has a derivative with respect to t, that derivative must be equal to 0 when $t = 0$. We shall now show that the derivative exists and find an expression for it.

For notational simplicity we shall denote the partial derivatives of f with respect to the 1st, 2nd, ..., $(2r + 1)$th variables by

$$f_x(x, y, p),\ f_{y^{(i)}}(x, y, p),\ f_{p^{(i)}}(x, y, p),$$

respectively. We shall also denote differentiation with respect to x by either d/dx or by a prime whenever it is convenient.

For each t in $(-\delta, \delta)$ and each x in $[x_1, x_2]$ we have, by the chain rule,

$$\text{(N)} \qquad \frac{\partial f(x, y(x) + t\eta(x), \dot{y}(x) + t\dot{\eta}(x))}{\partial t}$$

$$= \sum_{i=1}^{r} f_{y^{(i)}}(x, y(x) + t\eta(x), \dot{y}(x) + t\dot{\eta}(x))\eta^{(i)}(x)$$

$$+ \sum_{i=1}^{r} f_{p^{(i)}}(x, y(x) + t\eta(x), \dot{y}(x) + t\dot{\eta}(x))\dot{\eta}^{(i)}(x).$$

The points $(x, y(x) + t\eta(x), \dot{y}(x) + t\dot{\eta}(x))$ are in the bounded closed set for which the first $r + 1$ coordinates satisfy (I) and the last r do not exceed $M + 1$ in absolute value. The first factor in each term in the right member of (N) represents a function integrable from x_1 to x_2; the proof is the same as for the integrability of the function in (M). So each factor in the right member of (N) is a bounded integrable function, and therefore we can apply Theorem 11-1 to the integral $J[y + t\eta]$ and obtain

$$\text{(O)} \qquad \frac{dJ[y + t\eta]}{dt} = \int_{x_1}^{x_2} \frac{\partial f(x, y(x) + t\eta(x), \dot{y}(x) + t\dot{\eta}(x))}{\partial t}\, dx$$

$$= \int_{x_1}^{x_2} \left[\sum_{i=1}^{r} f_{y^{(1)}}(x, y(x) + t\eta(x), \dot{y}(x) + t\dot{\eta}(x))\eta^{(i)}(x) \right.$$

$$\left. + \sum_{i=1}^{r} f_{p^{(i)}}(x, y(x) + t\eta(x), \dot{y}(x) + t\dot{\eta}(x))\dot{\eta}^{(i)}(x) \right] dx.$$

As we have seen, this must be 0 at $t = 0$ because $J[y + t\eta]$ has its least value there. If we set $t = 0$ and perform an integration by parts on each of the r terms in the first sum, recalling (J), we obtain

$$\text{(P)} \quad \int_{x_1}^{x_2} \sum_{i=1}^{r} \left[-\int_{x_1}^{x} f_{y^{(i)}}(u, y(u), \dot{y}(u))\, du + f_{p^{(i)}}(x, y(x), \dot{y}(x)) \right] \dot{\eta}^{(i)}(x)\, dx = 0.$$

If $c_1, \ldots, c_r$ are any real numbers, from (J) and (K) we obtain

$$\text{(Q)} \qquad \int_{x_1}^{x_2} \left[\sum_{i=1}^{r} c_i \dot{\eta}^{(i)}(x) \right] dx = 0.$$

We subtract (Q) member by member from (P) and obtain the following result:

(R) When conditions (D) and (I) are satisfied, and y minimizes $J[y]$, and η is any variation, and $c_1, \ldots, c_r$ are any real numbers,

$$\int_{x_1}^{x_2} \sum_{i=1}^{r} \left[\int_{x_1}^{x} f_{y^{(i)}}(u, y(u), \dot{y}(u))\, du \right.$$
$$\left. + f_{p^{(i)}}(x, y(x), \dot{y}(x)) - c_i \right] \dot{\eta}^{(i)}(x)\, dx = 0.$$

Now we make the choice

$$\dot{\eta}^{(i)}(x) = -\int_{x_1}^{x} \left[f_{y^{(i)}}(u, y(u), \dot{y}(u))\, du + f_{p^{(i)}}(x, y(x), \dot{y}(x)) - c_i \right],$$

wherein

$$c_i = (x_2 - x_1)^{-1} \left\{ -\int_{x_1}^{x_2} \left(\int_{x_1}^{x} f_{y^{(i)}}(u, y(u), \dot{y}(u))\, du + f_{p^{(i)}}(x, y(x), \dot{y}(x)) \right) dx \right\}.$$

Each of these $\dot{\eta}^{(i)}$ is bounded, and their integrals from x_1 to x_2 all have value 0. Therefore, if we define $\eta^{(1)}, \ldots, \eta^{(r)}$ by (K), η is a variation. We can therefore substitute it in (R). The result is

$$\int_{x_1}^{x_2} \sum_{i=1}^{r} \left[f_{p^{(i)}}(x, y(x), \dot{y}(x)) - \int_{x_1}^{x} f_{y^{(i)}}(u(y(u), \dot{y}(u))\, du - c_i \right]^2 dx = 0.$$

The integrand is nonnegative, so by Theorem 12-4 it must be 0 at almost all points x in $[x_1, x_2]$. We have thus proved the following theorem.

THEOREM 15-1 *Let the function $(x, y, p) \mapsto f(x, y, p)$ have the continuity properties specified in* (D). *Let $x \mapsto y(x)$ be a function that belongs to the class $\mathscr{K}$, has the $\dot{y}^{(i)}(x)$ bounded on $[x_1, x_2]$, and satisfies* (I). *If y gives to the integral J its least value on the class $\mathscr{K}$, there are numbers $c_1, \ldots, c_r$ such that the equations*

$$\text{(S)} \qquad f_{p^{(i)}}(x, y(x), \dot{y}(x)) = \int_{x_1}^{x} f_{y^{(i)}}(u, y(u), \dot{y}(u))\, du + c_i$$

hold for almost all x in $[x_1, x_2]$.

When equations (S) are satisfied, for each variation η the value of $J[y + t\eta]$ is neither increasing nor decreasing as t goes through 0; at that point the value is momentarily stationary. Accordingly, we say that when (S) holds, the function y **makes J stationary,** or that y is a **stationary function** for J. This, of course, is not enough to ensure that y gives a greatest or least value to J. Even in one dimension, the function $x \mapsto x^3$ has a stationary value at $x = 0$, since its derivative there is 0, but it has neither maximum nor minimum there. However, by Theorem 15-1, we know that if y is minimizing, $\dot{y}$ is bounded, and the graph is strictly interior to G, y must make the integral stationary, and then we can determine what y is by solving (S).

If it happens that each $y^{(i)}$ has a continuous first derivative $y^{(i)\prime}$, by the fundamental theorem, (F) is satisfied with $\dot{y} = y'$. Then both members of (S) are continuous on $[x_1, x_2]$. Since they are equal at almost all x, they are by continuity equal at all x in $[x_1, x_2]$. So we can differentiate both members of (S) and obtain

$$\text{(T)} \qquad \frac{df_{p^{(i)}}(x, y(x), y'(x))}{dx} = f_{y^{(i)}}(x, y(x), y'(x)).$$

This is the "Euler–Lagrange" equation for the problem.

If the $y^{(i)}$ happen to be twice continuously differentiable, we can deduce from (T) another equation that is sometimes quite convenient. By the chain rule,

$$\frac{d}{dx}\left[f - \sum_{i=1}^{r} y^{(i)\prime}(x) f_{p^{(i)}}\right] = f_x + \sum_{i=1}^{r}\left\{f_{y^{(i)}} y^{(i)\prime}(x) + f_{p^{(i)}} y^{(i)\prime\prime}(x) - y^{(i)\prime\prime}(x) f_{p^{(i)}} - y^{(i)\prime}(x)\frac{d}{dx} f_{p^{(i)}}\right\},$$

where f and all its partial derivatives are understood to be evaluated at $(x, y(x), y'(x))$. If for the last term we substitute its value as given in (T), we obtain

$$\text{(U)} \quad \frac{d}{dx}\left[f(x, y(x), y'(x)) - \sum_{i=1}^{r} y^{(i)\prime}(x) f_{p^{(i)}}(x, y(x), y'(x))\right] = f_x(x, y(x), y'(x)).$$

As an example, we return to the problem of finding the curve with given end-points that generates the surface of revolution with least area. This requires us to minimize the expression (C), and clearly we minimize this by minimizing the integral in (C); the factor 2π can be omitted harmlessly. Now G is the half-plane consisting of all (x, y) with $y > 0$, and $r = 1$, and

$$f(x, y, p) = y[1 + p^2]^{1/2}.$$

Suppose that $x \mapsto y(x)$ has its graph in G and minimizes the integral, and that y has the form (A) with bounded $\dot{y}$. Since

$$f_p(x, y, p) = yp[1 + p^2]^{-1/2}, \qquad f_y(x, y, p) = [1 + p^2]^{1/2},$$

by Theorem 15-1 there is a constant c such that for almost all points x in $[x_1, x_2]$

$$\text{(V)} \qquad y(x)\dot{y}(x)[1 + \dot{y}(x)^2]^{-1/2} = \int_{x_1}^{x} [1 + \dot{y}(u)^2]^{1/2}\, du + c.$$

To simplify notation, we denote the right member of this equation by $F(x)$. If we multiply both members of (V) by $[1 + \dot{y}(x)^2]^{1/2}$, we obtain

$$y(x)\dot{y}(x) = [1 + \dot{y}(x)^2]^{1/2} F(x).$$

Since $y(x)$ is positive, this implies that $|F(x)| < y(x)$ and that $F(x)$ has the same sign as $\dot{y}(x)$. We square both members of the last equation; after a little manipulation, we find

$$\dot{y}(x)^2 = F(x)^2/[y(x)^2 - F(x)^2],$$

which, with the preceding statements, implies

$$\text{(W)} \qquad \dot{y}(x) = F(x)/[y(x)^2 - F(x)^2]^{1/2}.$$

This holds for almost all x, so the right member can be substituted for $\dot{y}(x)$ in (A); that is, we can and do suppose that (W) holds for all x in $[x_1, x_2]$. But now, by (W), the function $\dot{y}$ is continuous, so the integrand in the right member of (V) is continuous. By the fundamental theorem, the right member of (V) (which is $F(x)$) has a continuous derivative. Now, by (W), y' also has a continuous derivative. This allows us to apply equation (U). Since f is independent of x, the right member of (U) is 0, and (U) takes the form

$$\frac{d[y(x)(1 + y'(x)^2)^{-1/2}]}{dx} = 0.$$

Therefore there exists a constant b such that

$$y(x) = b[1 + y'(x)^2]^{1/2}.$$

It is easy to verify that the general solution of this equation is

$$\text{(X)} \qquad y(x) = b \cosh b^{-1}(x - a),$$

where cosh is the function defined in Exercise 6-15 and a is a constant. The graph of this function y is called a **catenary**. Given any points (x_1, y_1^*) and (x_2, y_2^*) with $x_2 > x_1$ and y_1^* and y_2^* positive, the constants b and a can be determined so that the curve will join the given end-points. If a solution exists that satisfies the conditions assumed in Theorem 15-1, this is it. If the catenary joining the end-points has any values of $y(x)$ that are nonpositive, we can be sure that no surface of revolution of the required type exists. This situation can be demonstrated by a simple experiment. If two wire circles of the same radius are held together and dipped in soap solution and then separated, when the separation is small the soap film will take the form of the surface obtained by rotating the graph of the catenary (X), but when the two circles are pulled further apart, the film will snap into two separate circles, one in each wire ring.

In mechanics, there is an important application of the concept of stationary function. Often, the positions of all the parts of a mechanical system are determined when a certain aggregate of numbers is known. These numbers are called the *coordinates* of the system. For example, if the system is a single particle in three-dimensional space, its position is determined by the three rectangular coordinates of the point it occupies. The position of a simple pendulum in a plane is determined by a single number – the angle that the cord makes with the vertical. It often happens that from some chosen initial position the system can be brought into any other position (or "state") by a continuous motion, and that the energy required to do this depends only on the coordinates of the final state. In this case the energy required is the "potential energy" of the system in the final state. If the coordinates that determine the state are $y^{(1)}, \ldots, y^{(r)}$, the potential energy will be a function of these coordinates, and we shall denote it by $V(y)$. If the system is in motion, the velocities of all its parts at time t will be determined by the time t, the coordinates $y^{(i)}$ at time t, and their rates of change

$$y^{(i)'}(t) = \frac{dy^{(i)}(t)}{dt}.$$

Therefore, there is a function $T(t, y, p)$ such that if the state of the system at time t is given by $y(t)$, its kinetic energy at time t will be $T(t, y(t), y'(t))$. A fundamental principle of mechanics (Hamilton's principle) asserts that the functions $y^{(i)}(t)$ that determine the state at time t are such as to make the integral

$$\int_{t_1}^{t_2} [T(t, y(y), y'(t)) - V(y(t))]\, dt$$

stationary for each pair of times t_1, t_2. This principle can be deduced from Newton's laws of motion and is so deduced in many books on mechanics. We shall not derive it but shall accept it as fundamental. We shall merely show by one very simple example that, at least for that example, Hamilton's principle is consistent with Newton's laws.

Let a particle of mass m be in the earth's gravity field. We choose rectangular axes with the $y^{(3)}$-axis vertically upward. The energy required to move the particle from the origin to the point $(y^{(1)}, y^{(2)}, y^{(3)})$ is $mgy^{(3)}$, where g is gravitational acceleration. Then

$$\text{(Y)} \qquad V(y) = mgy^{(3)}, \qquad T(t, y, p) = m[p^{(1)^2} + p^{(2)^2} + p^{(3)^2}]/2.$$

If we write equations (S), we see at once that the $y^{(i)}(t)$ are continuously differentiable, and by (T) we have

$$[my^{(1)'}(t)]' = 0, \qquad [my^{(2)'}(t)]' = 0, \qquad [my^{(3)'}(t)] = -mg.$$

The three right members are the three components of the force of gravity acting on the particle; the three left members are the three components of acceleration, each multiplied by m. So, the last three equations state that "mass times acceleration equals force," and for this system Hamilton's principle has given us back Newton's laws.

The great advantage of Hamilton's principle is that if we change from one coordinate system to another, changing the forms of the functions V and T accordingly, the integral of $T - V$ in the new coordinate system will be the same, for each function, as it was before. Therefore, what was a stationary curve in the original system will still be stationary in the new, and we are thus free to use any coordinate system that is convenient, rather than having to use rectangular systems.

The kinetic energy of a system is the sum of the kinetic energies of its particles, and in rectangular coordinates each particle has a kinetic energy that is given by an expression like the last one in (Y). If we change to other coordinates $y^{(1)}, \ldots, y^{(r)}$, the rates of change of the rectangular coordinates will be linear combinations of the $y^{(i)\prime}$, so the kinetic energy of the system will still be quadratic in the $y^{(i)\prime}$. That is, T will always have the form

$$\text{(Z)} \qquad T(t, y, p) = \sum_{i,j=1}^{r} a_{i,j}(t) p^{(i)} p^{(j)} + \sum_{i=1}^{r} b_i(t) p^{(i)}.$$

It can be left as an exercise to show that if the $a_{i,j}$ and b_i are continuous and the first sum in (Z) is positive unless all the $p^{(i)}$ are 0, the $y^{(i)}$ that make the integral of $T - V$ stationary will always be continuously differentiable and therefore will satisfy equations (T).

EXERCISE 15-1 Use Theorem 15-1 to show that if there is a shortest curve joining two points in R^r, it is a segment of a straight line.

EXERCISE 15-2 Let $r = 1$. Let $x \mapsto y(x)$ be part of a minimizing curve for an integral (G) that can be represented in the form $y \mapsto x(y)$. Change variables in the integral (G) so as to integrate with respect to y, and show that the equation (T) for the new integral is equation (U) for the original.

EXERCISE 15-3 Let y_0 be a "stationary point" for the potential energy V of a system. Show that by Hamilton's principle, the system can remain permanently in state y_0. (This is an "equilibrium state." It is stable if V has a minimum at y_0 but is unstable otherwise.)

EXERCISE 15-4 Let a simple pendulum consist of a particle of mass m suspended at the end of a weightless cord of length r and swinging in a plane. Let θ be the angle that the cord makes with the vertical. Show that

$$\frac{d^2\theta}{dt^2} = -\frac{g}{r}\sin\theta.$$

(The potential energy, measured from the lowest point of the arc of the pendulum, is $mg(1 - \cos\theta)$. The kinetic energy is $m/2$ times the square of the velocity $r(d\theta/dt)$.)

EXERCISE 15-5 A circular cylinder is made of material such that the density of the material is the same at any two points that have equal distances from the axis. Accept from physics the statement that there is a certain number I such that, if the cylinder is rotated with angular velocity θ' about any line parallel to the axis and lying in the surface of the cylinder, its kinetic energy will be $I\theta'^2/2$. Let r be the radius of the cylinder and m its mass. Show that if the cylinder rolls (without slipping) down a plane that makes angle A with the horizontal, its distance from the starting point will be a quadratic function of t. (If the reader is familiar with the computation of moments of inertia, he will be able to apply this to special cases such as the cylinder with all its mass on the surface, the cylinder with all its mass on the axis, and the homogeneous cylinder—obtaining the specific form of the quadratic function for each of them.)

EXERCISE 15-6 Let the kinetic energy of the system be given by equation (Z) with constant values for the $a_{i,j}$ and b_i. Show that, if the system changes in time in accord with Hamilton's principle, the sum of the kinetic and potential energies will remain constant:

$$T(t, y(t), y'(t)) + V(y(t)) = \text{const.}$$

Applications to Differential Equations and to Probability Theory

1. Ordinary Differential Equations

An ordinary differential equation is a relation between the values of a function $x \to y(x)$ on some interval in R, the values of x on that interval, and some derivatives of y. Such differential equations have been important since the first days of calculus, and their importance does not diminish. Newton invented calculus to be able to write and solve the differential equations that govern the motions of the planets. (Partial differential equations, involving functions of several independent variables and their partial derivatives, are also of great utility and theoretical importance, but we shall not discuss them.) The literature of differential equations is tremendous. All that we shall do is establish some of the basic theorems that underlie the great superstructure of special developments.

Let us start with the simplest case. This has one equation that involves one function and its first derivative and may be expressed in the form

$$\text{(A)} \qquad \frac{dy}{dx} = f(x, y(x)) \qquad (a \leqq x \leqq b).$$

We seek a function y that satisfies this equation and has at a given point x_0 of $[a, b]$ a given value c. If f is a function defined for all x in $[a, b]$ and all y, the right member of (A) is a real number for each x in $[a, b]$, and it may be possible to find a function y for which (A) is true at each x in $[a, b]$, or at least for each x in some nondegenerate subinterval that contains x_0. If f is continuous, solving (A) with the given initial condition $y(x_0) = c$ is equivalent to finding a function y on $[a, b]$ such that for each x in $[a, b]$

$$\text{(B)} \qquad y(x) = c + \int_{x_0}^{x} f(u, y(u))\, du.$$

For if (B) holds, clearly $y(x_0) = c$ by Definition I-8-2, and (A) is satisfied by the fundamental theorem I-9-2. Conversely, if (A) is satisfied, y has a derivative on $[a, b]$, so it is continuous on $[a, b]$. Therefore, $x \mapsto f(x, y(x))$ is continuous on $[a, b]$, and by the fundamental theorem I-10-2

$$y(x) - y(x_0) = \int_{x_0}^{x} f(u, y(u))\, du.$$

Since $y(x_0) = c$, this implies that (B) is satisfied.

However, even fairly simple applications present us with problems in which the f in equation (A) is not continuous. In this case there may not exist any y for which (A) holds at all points x in $[a, b]$. Suppose, for example, that $f(x, y) = 1$ when $x \geqq 0$ and $f(x, y) = -1$ when $x < 0$; we seek a solution of (A) with $y(0) = 0$. By the theorem of the mean, we would have for such a y:

$$y(x) = \begin{cases} 0 + (x - 0)(+1) & (x > 0), \\ 0 + (x - 0)(-1) & (x < 0). \end{cases}$$

That is, if y satisfied (A), we would have $y(x) = |x|$. But this only possible solution fails to satisfy (A) at $x = 0$, where it lacks a derivative. If we wish to continue to use (A), we shall have to allow some points (how many is not immediately clear) at which (A) fails to hold, and we shall have to add some sort of condition on the functions y to prevent really bad behavior, such as discontinuity at the points where the derivative is not required to exist. Clearly, form (A) of the differential equation can be used only in the presence of some safeguards, and we are not yet in a position to say just what those safeguards ought to be.

By contrast, if with the same f we use form (B) of the equation, we see readily that the function y whose value at x is $|x|$ satisfies (B) at all x in R. We need no precautions; the fact that (B) expresses y as an indefinite integral is sufficient to enforce enough good behavior on y to avoid the difficulties that we met when using (A).

The following is written for those readers who have read the last sections of Chapter II. The safeguards on y that have proved appropriate are that y shall be an indefinite integral; Eq. (A) is to hold at almost all points of $[a, b]$. If y satisfies (B), it is obviously an indefinite integral, and by Theorem II-14-1, y has at almost all points of $[a, b]$ a derivative that has value $f(x, y(x))$. Conversely, if y satisfies (A) almost everywhere in $[a, b]$, is an indefinite integral, and has value c at x_0, we shall show that it satisfies (B). For, let y be an indefinite integral of an integrable function z; then

$$\text{(C)} \qquad y(x) = k + \int_{x_0}^{x} z(u)\, du.$$

From this, by setting $x = x_0$, we see that $y(x_0) = k$. But by hypothesis, $y(x_0) = c$, so $k = c$. By Theorem II-14-1, except on a set N_1 with measure 0, we have

$dy/dx = z(x)$. By hypothesis, $dy/dx = f(x, y(x))$ except on a set N_2 of measure 0. So $z(x) = f(x, y(x))$ except on $N_1 \cup N_2$, which has measure 0. By Theorem II-12-2, the right member of (C) is unaltered if we replace $z(u)$ by $f(u, y(u))$ under the integral sign. This reduces (C) to (B). The condition that y be an indefinite integral implies that y is absolutely continuous, by Theorem II-9-2. The converse is true; but this is hard to prove, and we leave it to the last chapter of this book. We now have the safeguards – i.e., (A) shall hold almost everywhere and y shall be an indefinite integral – that guarantee that such solutions of (A) are also solutions of (B), and conversely. The two ways of studying the differential equation are logically equivalent. It does not follow that they are equally convenient.

We have seen that, in at least one simple case, form (B) of the differential equation has an advantage over form (A). In fact, there are four advantages. First, as we have seen, if we use (A) but allow dy/dx to fail to exist at some points (as we must if f has discontinuities), we have to put some limitations on the kind of functions that we will allow as solutions of (A), in order to exclude freaks; but if we use (B), such freaks are automatically excluded by equation (B) itself. Second, in proving theorems about solutions, integration theory, which applies to (B), furnishes us with more powerful mathematical tools than does differentiation theory, which applies to (A). Third, if we are obliged to compute an approximation to the solution by numerical methods, with or without the aid of a computer, there are efficient computational procedures for estimating integrals, so (B) is suitable for computational approximation of the solution. Fourth, on many applications (B) is a more natural way of modeling the system than is (A). For example, if v is velocity and a is acceleration, the statement

$$v(b) = v(0) + \int_0^b a(t)\,dt$$

can be thought of, pictorially though sloppily, as meaning that the change of velocity from time $t = 0$ to time $t = b$ is the "sum" of many tiny changes $a(t)\,dt$ in the velocity. For these reasons we shall base our discussions on form (B) of the equation, and we shall call this a "differential equation," although it is really the equivalent form (A) that deserves the name.

It is worthwhile to look at two simple examples of solutions of equations of form (B); first, because these two examples are important and useful, and second, because the process of solving them will exemplify the approximation procedure that will be used in establishing the existence theorem in the next section.

If $[a, b]$ is an interval in R and p is a real-valued function on $[a, b]$, the differential equation

$$\text{(D)} \qquad \frac{dy}{dx} = p(x)y(x)$$

with initial condition $y(a) = c$ can easily be solved provided that p is continuous. For then we define

(E) $$P(x) = \int_a^x p(u)\,du \qquad (a \leqq x \leqq b).$$

By the fundamental theorem I-9-2, P has derivative $p(x)$ at x, so by the chain rule, the function y defined by

(F) $$y(x) = c \exp P(x)$$

has derivative

(G) $$\frac{dy}{dx} = c[\exp P(x)][p(x)].$$

Since $P(a) = 0$ by (E), $y(a) = c$ by (F); this and (G) show that the function $y: x \mapsto c \exp P(x)$ is a solution of (D) with the given initial condition. If p is not continuous but is merely integrable, this elementary procedure cannot be used. Nevertheless, we shall show that the same expression is still a solution of the integrated form of the differential equation. That is;

EXAMPLE 1-1 If p is integrable over $[a, b]$, a solution of the equation

(H) $$y(x) = c + \int_a^x p(u)y(u)\,du \qquad (a \leqq x \leqq b)$$

is given by $y(x) = c \exp P(x)$, where

$$P(x) = \int_a^x p(u)\,du.$$

To prove this, we define a sequence of approximations to the solution of (H) by choosing an arbitrary (continuous) initial approximation and then using the right member of (H) to define successively better approximations. For the first approximation we choose the constant function defined by $y_1(x) = c$. We feed this into the right member of (H) to obtain a second approximation y_2, defined by

$$y_2(x) = c + \int_a^x p(u)y_1(u)\,du,$$

and continue the process; that is, having obtained the kth approximation y_k, we define the next approximation y_{k+1} by

(I) $$y_{k+1}(x) = c + \int_a^x p(u)y_k(u)\,du.$$

It is not difficult to calculate these functions explicitly. In fact, we can show by induction that for $k = 1, 2, 3, \ldots,$

$$\text{(J)} \qquad y_k(x) = [1 + P(x) + \cdots + P(x)^{k-1}/(k-1)!]c.$$

This holds for $k = 2$ by definition. If it holds for $k = n$, then by Lemma II-9-5,

$$\text{(K)} \qquad y_{n+1}(x) = c + \int_a^x p(u) \frac{1 + P(u) + \cdots + P(u)^{n-1}}{(n-1)!} c\, du$$

$$= \left[1 + P(x) + \frac{P(x)^2}{2!} + \cdots + \frac{P(x)^n}{n!}\right] c,$$

so equation (J) holds for $k = n + 1$ also. By induction, (J) holds for all integers greater than 1.

By Example II-6-11,

$$\text{(L)} \qquad \lim_{k \to \infty} y_k(x) = c \exp P(x)$$

for every x in R. Moreover, since P is continuous on $[a, b]$ its absolute value has a finite maximum value M on $[a, b]$. Then for each x in $[a, b]$ and each integer k greater than 1 we have

$$|y_k(x)| \leqq \left[\frac{1 + M + M^2}{2!} + \cdots + \frac{M^{k-1}}{(k-1)!}\right]|c|$$

$$\leqq \left[\sum_{j=0}^{\infty} \frac{M^j}{j!}\right]|c| = [\exp M]|c|.$$

Then for all x in $[a, b]$ and all integers k greater than 1

$$|y_k(x)p(x)| \leqq [|c| \exp M]|p(x)|,$$

and the right member defines a function integrable over $[a, b]$. By (K), (L), and the dominated convergence theorem,

$$\lim_{k \to \infty} \int_a^x p(u)y_k(u)\, du = \int_a^x \left[\lim_{k \to \infty} p(u)y_k(u)\right] du = c \int_a^x p(u) \exp P(u)\, du.$$

By this and (L), (I) yields the conclusion

$$c \exp P(x) = c + \int_a^x p(u)[c \exp P(u)]\, du$$

for all x in $[a, b]$. So the function defined by (F) is a solution of the equation (H). We could show that it is the only solution, but we shall not take the time to do this, because it will be a consequence of a theorem in the next section.

EXAMPLE 1-2 Let p and q be real-valued functions integrable over the interval $[a, b]$, and let c be a number. Then the equation

$$\text{(M)} \qquad y(x) = c + \int_a^x [p(u)y(u) + q(u)]\, du$$

has the solution

$$\text{(N)}\qquad y(x) = \left\{ c + \int_a^x q(u)\exp[-P(u)]\,du \right\} \exp P(x),$$

where

$$P(x) = \int_a^x p(u)\,du.$$

If y is defined by (N),

$$\begin{aligned}\text{(O)}\quad c + \int_a^x [p(u)y(u) + q(u)]\,du &= c + \int_a^x p(u)c\exp P(u)\,du \\ &\quad + \int_a^x p(u)\left\{ \int_a^u q(v)\exp[-P(v)]\,dv \right\} \exp P(u)\,du + \int_a^x q(u)\,du.\end{aligned}$$

By Example 1-1, the sum of the first two terms in the right member of (O) is $c\exp P(x)$, from which it follows at once that $\exp P$ is an indefinite integral of $p\exp P$. In the third term in the right member of (O), the quantity in braces is an indefinite integral, so we can apply integration by parts (Theorem II-9-4) to that term, and (O) will reduce to

$$\begin{aligned} c + \int_a^x [p(u)y(u) + q(u)]\,du &= c\exp P(x) + \left[\exp P(u) \int_a^u q(v)\exp[-P(v)]\,dv \right]_a^x \\ &\quad - \int_a^x \exp P(u)q(u)\exp[-P(u)]\,du + \int_a^x q(u)\,du \\ &= y(u). \end{aligned}$$

So y satisfies Eq. (M).

The formula (N) for the solution of (M) is to be found in every elementary text on differential equations, but it is established only for continuous functions p and q. We have shown that it holds for all integrable p and q.

2. Existence Theorems for Solutions of Differential Equations

The equations that we shall consider are those of the form

$$\text{(A)}\qquad y^{(i)}(x) = c^{(i)} + \int_{x_0}^x f^{(i)}(u, y^{(1)}(u), \ldots, y^{(r)}(u))\,du$$

($i = 1, \ldots, r$), where the $f^{(i)}(x, y^{(1)}, \ldots, y^{(r)})$ are defined for all x in an interval $[a, b]$ and for all real $y^{(1)}, \ldots, y^{(r)}$, and the $c^{(i)}$ are real numbers, and x_0 is a point of $[a, b]$. This is the integral form of the equation

$$\text{(B)} \qquad \frac{dy^{(i)}}{dx} = f^{(i)}(x, y^{(1)}(x), \ldots, y^{(r)}(x)) \qquad (a \leqq x \leqq b)$$

with initial conditions $y^{(i)}(x_0) = c^{(i)}$. At first glance, it might seem more general to consider equations involving derivatives of higher order, such as, for example,

$$\text{(C)} \qquad \frac{d^n y}{dx^n} = f(x, y(x), \ldots, y^{(n-1)}(x)),$$

where $y^{(k)}$ denotes the kth derivative of y. But an equation of type (C) can easily be changed to a set of equations of type (B) by introducing new variables $y^{(1)}, \ldots, y^{(n)}$ that satisfy

$$\text{(D)} \qquad \frac{dy^{(1)}}{dx} = y^{(2)}, \qquad \frac{dy^{(2)}}{dx} = y^{(3)}, \ldots$$

$$\frac{dy^{(n-1)}}{dx} = y^{(n)}, \qquad \frac{dy^{(n)}}{dx} = f(x, y(x), \ldots, y^{(n)}(x)).$$

If $y^{(1)}, \ldots, y^{(n)}$ satisfy (D), $y = y^{(1)}$ satisfies (C), so solving (C) is reduced to solving (D), which is a special case of (B).

For brevity, we shall write the single letter y to denote the r-tuple $(y^{(1)}, \ldots, y^{(r)})$. The set of all ordered r-tuples of real numbers will be called the *r-dimensional space R^r*. With this notation, equation (A) condenses into

$$\text{(E)} \qquad y^{(i)}(x) = c^{(i)} + \int_{x_0}^{x} f^{(i)}(u, y(u))\, du.$$

Also, in the rest of this section we shall frequently need to sum over the numbers $1, \ldots, r$. Usually we denote such a sum by $\Sigma_{j=1}^{k}$. We shall shorten this to Σ_j. Sums with respect to j will always be over $j = 1, \ldots, r$.

To show that solutions of equations (A) exist and are uniquely determined, we have to make fairly strong continuity assumptions about the dependence of the functions $f^{(i)}$ on the variables $y^{(j)}$. We shall assume that if j is any one of the numbers $1, \ldots, r$, the function of $y^{(j)}$ alone obtained by fixing the other variables is a Lipschitzian function, with Lipschitz constant l. This Lipschitz constant l does not have to have the same value at all points x. (Remember, the x was treated as constant and only $y^{(j)}$ was allowed to vary.) It can be a function of x; it can even be an unbounded function $x \mapsto l(x)$; but it has to be integrable over $[a, b]$. By contrast, the assumptions about the behavior of the $f^{(i)}$ as functions of x alone, for fixed $y^{(1)}, \ldots, y^{(r)}$, are much weaker. All we have to assume is that for fixed $y^{(1)}, \ldots, y^{(r)}$ the functions $x \mapsto f^{(i)}(x, y^{(1)}, \ldots, y^{(r)})$ are integrable over $[a, b]$. With these hypotheses we shall prove the following theorem.

THEOREM 2-1 *Let the r functions* $(x, y) \mapsto f^{(i)}(x, y)$ *be defined and real-valued for all x in an interval* $[a, b]$ *in R and all y in* R^r. *Assume that there exists a function l integrable over* $[a, b]$ *such that for each i and j in* $\{1, \ldots, r\}$, *each set of numbers* $(x, y^{(1)}, \ldots, y^{(j-1)}, y^{(j+1)}, \ldots, y^{(r)})$ *with x in* $[a, b]$, *and each pair of real numbers u, v*

(F) $$|f^{(i)}(x, y^{(1)}, \ldots, y^{(j-1)}, u, y^{(j+1)}, \ldots, y^{(r)}) - f^{(i)}(x, y^{(1)}, \ldots, y^{(j-1)}, v, y^{(j+1)}, \ldots, y^{(r)})| \leqq l(x)|u - v|.$$

Assume also that for each set of real numbers $(y^{(1)}, \ldots, y^{(r)})$ *and each i in* $\{1, \ldots, r\}$ *the function*

$$x \mapsto f^{(i)}(x, y^{(1)}, \ldots, y^{(r)}) \qquad (a \leqq x \leqq b)$$

is integrable over $[a, b]$.

Then for each x_0 *in* $[a, b]$ *and each set of r real numbers* $(c^{(1)}, \ldots, c^{(r)})$ *there is a unique set of r functions*

$$x \mapsto y^{(i)}(x) \qquad (i = 1, \ldots, r;\ a \leqq x \leqq b)$$

such that

(G) $$y^{(i)}(x) = c^{(i)} + \int_{x_0}^{x} f^{(i)}(u, y(u))\, du.$$

The interesting part of the proof of this theorem begins with (J), below. It was devised by H. Picard and is named after him. But before we get to it, we have to establish some rather routine preliminaries. First,

(H) If the $f^{(i)}$ satisfy the hypotheses of Theorem 2-1, and u and v are points of R^r, then for all x in $[a, b]$ and all i in $\{1, \ldots, r\}$ it is true that

$$|f^{(i)}(x, u) - f^{(i)}(x, v)| \leqq l(x) \sum_j |u^{(j)} - v^{(j)}|.$$

We define $w_0 = v$ and $w_j = (u^{(1)}, \ldots, u^{(j)}, v^{(j+1)}, \ldots, v^{(r)})$ $(j = 1, \ldots, r)$. Then for each j

$$w_{j-1}^{(j)} = v^{(j)} \qquad \text{and} \qquad w_j^{(j)} = u^{(j)},$$

and for all $h \neq j$,

$$w_{j-1}^{(h)} = w_j^{(h)}.$$

So we can apply hypothesis (F) and obtain

$$|f^{(i)}(x, w_j) - f^{(i)}(x, w_{j-1})| \leqq l(x)|w_j^{(j)} - w_{j-1}^{(j)}| = l(x)|u^{(j)} - v^{(j)}|.$$

Therefore,

$$\begin{aligned} &|f^{(i)}(x, u) - f^{(i)}(x, v)| \\ &\quad = |[f^{(i)}(x, w_1) - f^{(i)}(x, w_0)] + \cdots + [f^{(i)}(x, w_r) - f^{(i)}(x, w_{r-1})]| \\ &\quad \leqq \sum_j |f^{(i)}(x, w_j) - f^{(i)}(x, w_{j-1})| \\ &\quad \leqq \sum_j l(x)|u^{(j)} - v^{(j)}|, \end{aligned}$$

and (H) is proved.

Next we shall prove

(I) if $f^{(i)}$ satisfies the hypotheses of Theorem 2-1 and $y^{(1)}, \ldots, y^{(r)}$ are continuous real-valued functions on $[a, b]$, then the function $x \mapsto f^{(i)}(x, y^{(1)}(x), \ldots, y^{(r)}(x))$ is integrable over $[a, b]$.

For each positive integer n we subdivide the interval $(a, b]$ into n congruent subintervals by points $x_h = a + h(b - a)/n$ $(h = 0, 1, \ldots, n)$, and we define $s_n^{(j)}$ to be the step-function of $(a, b]$ which on each interval $(x_{h-1}, x_h]$ has the constant value $y^{(j)}(x_h)$. For each h the function $x \mapsto f^{(i)}(x, y(x_h))$ is integrable over $[a, b]$ by hypothesis, so it is integrable over the subinterval $(x_{h-1}, x_h]$. This is the same as saying that $x \mapsto f^{(i)}(x, s_n(x))$ is integrable over $(x_{h-1}, x_h]$; and since this holds for $h = 1, \ldots, n$, that function is integrable over $(a, b]$. If x is in $(a, b]$, for each n the point x is in one of the n congruent intervals into which we have subdivided $(a, b]$; we denote the right end of this subinterval by $x_n^*(x)$. Then, by definition, $s_n(x) = y(x_n^*(x))$. Since

$$|x_n^*(x) - x| < (b - a)/n,$$

as n increases, $x_n^*(x)$ tends to x; and since each $y^{(j)}$ is continuous, $y(x_n^*(x))$ tends to $y(x)$. That is, $\lim_{n \to \infty} s_n(x) = y(x)$ for all x in $(a, b]$. Also, the continuous function

$$x \mapsto \sum_j |y^{(j)}(x)| \qquad (a \leqq x \leqq b)$$

has a finite upper bound M on $[a, b]$, so $\Sigma_j |s_n^{(j)}(x)|$ has the same upper bound, and by (H), with $u = s_n(x)$ and $v = 0$,

$$|f^{(i)}(x, s_n(x))| \leqq |f^{(i)}(x, 0)| + Ml(x).$$

Since the functions l and $x \mapsto f^{(i)}(x, 0)$ are integrable, by the dominated convergence theorem, the limit $x \mapsto f^{(i)}(x, y(x))$ of the integrable functions $x \mapsto f^{(i)}(x, s_n(x))$ is integrable, and (I) is proved.

Now we can begin the main part of the proof. We start with any set of r continuous real-valued functions $y_1^{(1)}, \ldots, y_1^{(r)}$, and we define $y_2, y_3, \ldots$ successively by the equations

$$\text{(J)} \qquad y_{n+1}^{(i)}(x) = c^{(i)} + \int_{x_0}^{x} f^{(i)}(u, y_n(u))\, du \qquad (n = 1, 2, 3, \ldots).$$

The statement "$y_n^{(1)}, \ldots, y_n^{(r)}$ exist and are continuous on $[a, b]$" is true for $n = 1$, since we chose the $y_1^{(i)}$ to be continuous. If it is true for $n = k$, by (I), the integral in (J) with $n = k$ exists, and it is a continuous function of x by Theorem II-9-2, so the statement is also true for $n = k + 1$. By induction, the statement is true for all positive integers n; all $y_n^{(i)}$ exist and are continuous on $[a, b]$.

In showing that the y_n converge to a solution of (G), the following computation is useful.

(K) Let the hypotheses of Theorem 2-1 be satisfied, and let L be defined by

$$L(x) = \int_{x_0}^{x} l(u)\,du \qquad (a \leqq x \leqq b).$$

If $y^{(1)}, \ldots, y^{(r)}, z^{(1)}, \ldots, z^{(r)}$ are functions continuous on $[a, b]$, and there exists a number C and a positive integer n such that

$$\text{(L)} \qquad \sum_j |y^{(j)}(x) - z^{(j)}(x)| \leqq C|L(x)|^{n-1} \qquad (a \leqq x \leqq b),$$

then

$$\text{(M)} \qquad \sum_j \left| \int_{x_0}^{x} [f^{(j)}(u, y(u)) - f^{(j)}(u, z(u))]\,du \right| \leqq \frac{Cr|L(x)|^n}{n} \qquad (a \leqq x \leqq b).$$

For $i = 1, \ldots, r$ we have by (H) and the hypothesis of (K),

$$\begin{aligned} |f^{(i)}(u, y(u)) - f^{(i)}(u, z(u))| &\leqq l(u) \sum_j |y^{(j)}(u) - z^{(j)}(u)| \\ &\leqq C|L(u)|^{n-1} l(u). \end{aligned}$$

Since L is nonnegative on $[x_0, x]$ if $x > x_0$ and is nonpositive on $[x, x_0]$ if $x < x_0$, this implies with the help of Lemma II-9-5 that

$$\begin{aligned} &\sum_j \left| \int_{x_0}^{x} [f^{(j)}(u, y(u)) - f^{(j)}(u, z(u))]\,du \right| \\ &\quad \leqq \sum_j \left| \int_{x_0}^{x} |f^{(j)}(u, y(u)) - f^{(j)}(u, z(u))|\,du \right| \\ &\quad \leqq Cr \left| \int_{x_0}^{x} |L(u)|^{n-1} l(u)\,du \right| = Cr \left| \int_{x_0}^{x} L(u)^{n-1} l(u)\,du \right| = Cr \left| \frac{L(x)^n}{n} \right|, \end{aligned}$$

which completes the proof.

The functions $y_1^{(j)}$ and $y_2^{(j)}$ $(j = 1, \ldots, r)$ are continuous on $[a, b]$ and therefore are bounded. So, there is a number M such that

$$\sum_j |y_2^{(j)}(x) - y_1^{(j)}(x)| \leqq M \qquad (a \leqq x \leqq b).$$

If we understand $L(x)^0$ to mean 1 even if $L(x) = 0$, this is hypothesis (L) of (K), with $n = 1$, $C = M$, $y = y_1$, and $z = y_2$. By definition,

$$\text{(N)} \qquad y_3^{(i)}(x) - y_2^{(i)}(x) = \int_{x_0}^{x} [f^{(i)}(u, y_2(u)) - f^{(i)}(u, y_1(u))]\,du,$$

so from (K) we obtain

$$\sum_j |y_3^{(j)}(x) - y_2^{(j)}(x)| \leqq \frac{Mr|L(x)|}{1}.$$

Now hypothesis (L) of (K) holds with $n = 2$, $C = Mr$, $y = y_2$, and $z = y_3$. Also, (N) remains valid if we replace y_3 by y_4, y_2 by y_3, and y_1 by y_2. So from (K) we obtain

$$\sum_j |y_4^{(j)}(x) - y_3^{(j)}(x)| \leqq \frac{Mr^2|L(x)|^2}{2!}.$$

We continue this process and obtain

$$\text{(O)} \qquad \sum_j |y_{n+1}^{(j)}(x) - y_n^{(j)}(x)| \leqq \frac{Mr^{n-1}|L(x)|^{n-1}}{(n-1)!}.$$

If we denote by L_1 the maximum value on $[a, b]$ of the continuous function $|L|$, we see by (O) that for each j and for $n > 1$, the nth term of the series

$$\text{(P)} \qquad y_1^{(j)}(x) + [y_2^{(j)}(x) - y_1^{(j)}(x)] + [y_3^{(j)}(x) - y_2^{(j)}(x)] + \cdots$$

has absolute value equal, at most, to the nth term of the series

$$\text{(Q)} \qquad 0 + M + MrL_1/1 + M(rL_1)^2/2! + M(rL_1)^3/3! + \cdots,$$

which is convergent. By the comparison test, (P) is absolutely convergent. Let $y^{(j)}(x)$ denote its sum. Then $y^{(j)}(x)$ is the limit as n increases of the sum of the first n terms of the series (P). But the sum of the first n terms of series (P) is $y_n^{(j)}(x)$. So,

$$\text{(R)} \qquad \lim_{n\to\infty} y_n^{(j)}(x) = y^{(j)}(x) \qquad (a \leqq x \leqq b;\ j = 1, \ldots, r).$$

Furthermore, by (O),

$$\text{(S)} \qquad \begin{aligned} \sum_{j=1}^{r} |y_n^{(j)}(x) - y_1^{(j)}(x)| &= \sum_{j=1}^{r} \left| \sum_{h=1}^{n-1} [y_{h+1}^{(j)}(x) - y_h^{(j)}(x)] \right| \\ &\leqq \sum_{h=1}^{n-1} \sum_{j=1}^{r} |y_{h+1}^{(j)}(x) - y_h^{(j)}(x)| \\ &\leqq \sum_{h=1}^{n-1} \frac{M(rL_1)^{h-1}}{(h-1)!} < M \exp(rL_1). \end{aligned}$$

From this and (H) we have for $i = 1, \ldots, r$ and $a \leqq x \leqq b$

$$|f^{(i)}(x, y_n(x)) - f^{(i)}(x, y_1(x))| \leqq l(x) M \exp(rL_1),$$

so for all positive integers n

$$\text{(T)} \qquad |f^{(i)}(x, y_n(x))| \leqq |f^{(i)}(x, y_1(x))| + l(x) M \exp(rL_1).$$

The function $x \mapsto f^{(i)}(x, y_1(x))$ is integrable by (I), and l is integrable by hypothesis, so the right member of (T) is the value at x of an integrable function. For each x in $[a, b]$

$$0 \leqq |f^{(i)}(x, y_n(x)) - f^{(i)}(x, y(x))| \leqq l(x) \sum_{j=1}^{r} |y_n^{(i)}(x) - y^{(i)}(x)|,$$

and $y_n^{(j)}(x)$ tends to $y^{(j)}(x)$ for each j and x, so

$$\lim_{n\to\infty} f^{(i)}(x, y_n(x)) = f^{(i)}(x, y(x)).$$

This and (T) allow us to apply the dominated convergence theorem and obtain

$$\lim_{n\to\infty} \int_{x_0}^{x} f^{(i)}(u, y_n(u))\, du = \int_{x_0}^{x} f^{(i)}(u, y(u))\, du.$$

If we apply this and (R) to (J), we obtain

$$y^{(i)}(x) = c^{(i)} + \int_{x_0}^{x} f^{(i)}(u, y(u))\, du \qquad (i = 1, \dots, r;\ a \leqq x \leqq b),$$

and so y is a solution of equations (G).

Suppose that z is also a solution of (G). Then by subtraction, if $1 \leqq i \leqq r$ and $a \leqq x \leqq b$,

$$\text{(U)} \qquad y^{(i)}(x) - z^{(i)}(x) = \int_{x_0}^{x} [f^{(i)}(u, y(u)) - f^{(i)}(u, z(u))]\, du.$$

If we let M be the maximum value on $[a, b]$ of the continuous function

$$\sum_{j=1}^{r} |y^{(j)} - z^{(j)}|,$$

inequality (L) of (K) holds with $n = 1$ and with $C = M$. By repeated applications of (K) we obtain for all positive integers n and all x in $[a, b]$

$$\text{(V)} \qquad \sum_{j=1}^{r} |y^{(j)}(x) - z^{(j)}(x)| \leqq \frac{M(r|L(x)|)^n}{n!}.$$

The right member of (V) is the $(n + 1)$th term of a convergent series—the series for $M \exp(r|L(x)|)$—so it tends to 0 as n increases. If the left member of (V) were positive, we could choose an n so large that the right member would be less than the left, which would be a contradiction. So, the left member of (V) is 0, and y and z are identical. This completes the proof of the theorem.

EXAMPLE 2-2 Let $a_{i,j}$ and b_i $(i, j = 1, \dots, r)$ be functions integrable over an interval $[a, b]$ in R; let $c^{(1)}, \dots, c^{(r)}$ be real numbers; and let x_0 be a point of $[a, b]$. Then the differential equations

$$\text{(W)} \qquad y^{(i)}(x) = c^{(i)} + \int_{x_0}^{x} \left[\sum_{j=1}^{r} a_{i,j}(u) y^{(j)}(u) + b_i(u) \right] du$$

have a unique solution on $[a, b]$. For if we define

$$f^{(i)}(x, y) = \sum_{j=1}^{r} a_{i,j}(x) y^{(j)} + b_i(x),$$

it is obvious that the functions $x \mapsto f^{(i)}(x, y)$ are integrable from a to b for each fixed y; and if we define

$$l(x) = \max\{|a_{i,j}(x)| : i, j = 1, \ldots, r\},$$

we see that l is integrable from a to b, and for each i and j in $1, \ldots, r$, each x in $[a, b]$, each y in R^r, and each pair of real numbers u and v

$$|f^{(i)}(x, y^{(1)}, \ldots, y^{(j-1)}, u, y^{(j+1)}, \ldots, y^{(r)}) \\ - f^{(i)}(x, y^{(1)}, \ldots, y^{(j-1)}, v, y^{(j+1)}, \ldots, y^{(r)})| \leqq l(x)|u - v|.$$

So all the hypotheses of Theorem 2-1 are satisfied.

In particular, this example shows that the solutions of the two linear differential equations exhibited in Section 1 are unique.

EXERCISE 2-1 It is advisable to use a calculator for this exercise. For the differential equation $Dy = y$, with $y(0) = 1$, calculate approximations to several of the Picard functions $y_0, y_1, y_2, \ldots$ on $[0, 0.4]$, taking $y_0 = 1$ and computing the integrals by the trapezoidal rule, and dividing the interval $[0, 0.4]$ into four parts of length 0.1.

EXERCISE 2-2 If we take $y_1 = 1$, the successive Picard approximations to the solution of

$$y(x) = 1 + \int_0^x y(u)\, du$$

are the partial sums of a certain power series. Calculate the coefficients of this series and identify it.

EXERCISE 2-3 The solution of the equation

$$D^2 y = -k^2 y \qquad (y(0) = c^{(1)},\ Dy(0) = kc^{(2)})$$

is the same as the function $y^{(1)}$ determined by the system

$$y^{(1)}(x) = c^{(1)} + k \int_0^x y^{(2)}(u)\, du,$$

$$y^{(2)}(x) = kc^{(2)} - k \int_0^x y^{(1)}(u)\, du.$$

Show that if we start with $y_1^{(1)}(x) = c^{(1)}$, $y_1^{(2)}(x) = kc^{(2)}$, the successive Picard approximations $y_1^{(1)}, y_2^{(1)}, y_3^{(1)}, \ldots$ are the partial sums of a power series. Identify the sum of the power series.

EXERCISE 2-4 Prove that, if to the hypotheses of Theorem 2-1 we add the assumption that the $f^{(i)}$ and all their first-order partial derivatives are continuous, the solutions $y^{(i)}$ have continuous first and second derivatives. Generalize this by proving that, if for some positive integer k the $f^{(i)}$ and all their partial derivatives of order $\leqq k$ are continuous, the solutions have continuous derivatives $Dy^{(i)}, \ldots, D^{k+1}y^{(i)}$. (Use the chain rule.)

EXERCISE 2-5 Verify that the differential equation

$$Dy = 4y^{3/4} \qquad (0 < x \leqq 1),$$

with the initial value $y(0) = 0$, has the two solutions

$$y(x) = 0, \qquad y(x) = x^4.$$

Why does this not contradict Theorem 2-1?

EXERCISE 2-6 We have shown that the function y defined by

$$y(x) = \exp(-1/x) \qquad (x > 0),$$

$$y(x) = 0 \qquad (x \leqq 0)$$

has continuous derivatives of all orders on R. It satisfies the differential equation

$$Dy(x) = a(x)y(x) \qquad (-1 \leqq x \leqq 1),$$

where $a(x) = x^{-2}$ if $x > 0$ and $a(x) = 0$ if $x \leqq 0$. The function $y = 0$ satisfies the same differential equation, and both functions have the same value 0 at $x = -1$. Why does this not contradict Theorem 2-1?

3. Effects of Change of Data

Sometimes we wish to approximate the solution of an equation

$$\text{(A)} \qquad y^{(i)}(x) = c^{(i)} + \int_{x_0}^{x} f^{(i)}(u, y(u))\, du \qquad (i = 1, \ldots, r)$$

by solving a different equation in which the $f^{(i)}$ are replaced by more manageable functions $g^{(i)}$. Or it may happen (in fact, it *always* happens!) in an application to some experimental situation that there is some inexactness in our knowledge of the $f^{(i)}$ and of the $c^{(i)}$. In either case, we have at hand the solutions of the substitute equations

$$\text{(B)} \qquad z^{(i)}(x) = c_1^{(i)} + \int_{x_0}^{x} g^{(i)}(u, z(u))\, du,$$

and we wish to know how far these can be from the desired solution of (A). In this section we shall establish such an estimate. First, however, we prove a lemma that has some interest in itself.

LEMMA 3-1 *Let $f^{(1)}, \ldots, f^{(r)}$ satisfy the hypotheses of Theorem 2-1; let y_1, y_2, and L be defined as in* (J) *and* (K) *of Section* 2, *and let y satisfy* (A). *If M is a number such that*

$$\text{(C)} \qquad \sum_{j=1}^{r} |y_2^{(j)}(x) - y_1^{(j)}(x)| \leqq M \qquad (a \leqq x \leqq b),$$

then for all x in $[a, b]$

$$\text{(D)} \qquad \sum_{j=1}^{r} |y^{(j)}(x) - y_1^{(j)}(x)| \leqq M \exp r|L(x)|$$

and

$$\text{(E)} \qquad \sum_{j=1}^{r} |y^{(j)}(x) - y_2^{(j)}(x)| \leqq M[\exp r|L(x)| - 1].$$

Turning back to the proof of Theorem 2-1, we recall that the sum of the series (P) in that proof was shown to be $y^{(j)}(x)$, so

$$|y^{(j)}(x) - y_1^{(j)}(x)| = \left| \sum_{n=1}^{\infty} [y_{n+1}^{(j)}(x) - y_n^{(j)}(x) \right|$$
$$\leqq \sum_{n=1}^{\infty} |y_{n+1}^{(j)}(x) - y_n^{(j)}(x)|.$$

Summing over $j = 1, \ldots, r$ and recalling that the terms in this series satisfy (O) in Section 2, we obtain

$$\sum_{j=1}^{r} |y^{(j)}(x) - y_1^{(j)}(x)| \leqq \sum_{n=1}^{\infty} \sum_{j=1}^{r} |y_{n+1}^{(j)}(x) - y_n^{(j)}(x)|$$
$$\leqq \sum_{n=1}^{\infty} \frac{M(r|L(x)|)^{n-1}}{(n-1)!}$$
$$= M \exp(r|L(x)|).$$

Likewise, the fact that the sum of the series (P) in Section 2 is $y^{(j)}(x)$ can be written in the form

$$y^{(j)}(x) = y_2^{(j)}(x) + [y_3^{(j)}(x) - y_2^{(j)}(x)] + [y_4^{(j)}(x) - y_3^{(j)}(x)] + \cdots$$

whence, transposing the first term in the right member to the left and summing over $j = 1, \ldots, r$, we find

$$\sum_{j=1}^{r} |y^{(j)}(x) - y_2^{(j)}(x)| \leq \sum_{n=2}^{\infty} \frac{M(r|L(x)|)^{n-1}}{(n-1)!} = M\{\exp r|L(x)| - 1\}.$$

The proof is complete.

With the help of this lemma we can estimate how far we are from the exact solution of the differential equation as soon as we have chosen y_1 and computed y_2. For if these functions satisfy (C), then the error in using y_2 as an approximation to the solution y will have to satisfy (E).

Now let us return to the estimation of the effect of changing the initial value from c to c_1 and replacing the functions $f^{(i)}$ by other functions $g^{(i)}$.

THEOREM 3-2 *Let the functions* $f^{(1)}, \ldots, f^{(r)}$ *satisfy the hypotheses of Theorem* 2-1, *and let* $g^{(1)}, \ldots, g^{(r)}$ *be real-valued functions defined on* $[a, b] \times R^r$. *Let* $c^{(i)}, c_1^{(i)}$ $(i = 1, \ldots, r)$ *be real numbers, and let* y *and* z *be solutions of equations* (A) *and* (B), *respectively. Let* M *be an upper bound for*

$$\sum_{i=1}^{r} \left| \int_{x_0}^{x} [f^{(i)}(u, z(u)) - g^{(i)}(u, z(u))] \, du \right|$$

on $[a, b]$. *Then for all* x *in* $[a, b]$

$$\sum_{j=1}^{r} |y^{(j)}(x) - z^{(j)}(x)| \leqq \left[M + \sum_{j=1}^{r} |c^{(j)} - c_1^{(j)}| \right] \exp(r|L(x)|).$$

We choose $y_1 = z$ and define $y_2, y_3, \ldots$ as in the proof of Theorem 2-1. Since the $z^{(i)}$ satisfy (B),

$$\begin{aligned} y_2^{(j)}(x) - y_1^{(j)}(x) &= c^{(j)} + \int_{x_0}^{x} f^{(j)}(u, y_1(u)) \, du - y_1^{(j)}(x) \\ &= c^{(j)} - c_1^{(j)} + \int_{x_0}^{x} [f^{(j)}(u, z(u)) - g^{(j)}(u, z(u)] \, du. \end{aligned}$$

So,

$$\begin{aligned} \sum_{j=1}^{r} |y_2^{(j)}(x) - y_1^{(j)}(x)| &\leqq \sum_{j=1}^{r} |c^{(j)} - c_1^{(j)}| \\ &\quad + \sum_{j=1}^{r} \left| \int_{x_0}^{x} [f^{(j)}(u, z(u)) - g^{(j)}(u, z(u))] \, du \right| \\ &\leqq \sum_{j=1}^{r} |c^{(j)} - c_1^{(j)}| + M. \end{aligned}$$

The conclusion now follows because of Lemma 3-1.

As an example of a use of Theorem 3-2, let us consider the motion of a simple pendulum consisting of a particle of mass m suspended by a weightless cord of length r. (If we wish to allow vibrations that reach more than a right angle from rest-position, we replace the cord by a weightless rigid rod.) The pendulum swings in a plane; we denote by θ the angle that the cord makes with the downward vertical direction. If the reader has worked through Section II-15, he will know from Exercise II-15-4 that

(F) $$\frac{d^2\theta}{dt^2} = -\frac{g \sin\theta}{r}.$$

If he has not, he can deduce (F) from the statement that the sum of the kinetic and potential energies of the particle remains constant (which was proved in Exercise II-15-6). The height of the particle above rest-position is $r - (r\cos\theta)$, so its potential energy is $mgr(1 - \cos\theta)$. Its velocity is $r\,d\theta/dt$, so its kinetic energy is $mgr[r\,d\theta/dt]^2/2$. Hence,

$$m\left[r\frac{d\theta}{dt}\right]^2\Big/2 + mgr - (mgr\cos\theta) = \text{const.}$$

By differentiating and dividing by mr, we obtain

$$r\left[\frac{d\theta}{dt}\right]\left[\frac{d^2\theta}{dt^2}\right] + g[\sin\theta]\left[\frac{d\theta}{dt}\right] = 0,$$

which implies (F).

Since this differential equation is not easy to solve, in elementary physics it is customary to replace it by the simpler equation

(G) $$\frac{d^2\theta}{dt^2} = -\frac{g}{r}\theta.$$

This is easily solved, and it is usually assumed without careful investigation that because the right members of the equations are nearly equal for small θ, the solutions must also be nearly equal and have nearly the same periods. But this means that we are comparing two solutions – one of (F), one of (G) – both near 0, so, at the least, we have to show that the difference between the solutions goes to 0 faster than the solutions do. We can do this with the help of Theorem 3-2.

We shall suppose that in both equations we seek the solution that starts with initial value θ_0 and initial velocity 0. Instead of solving the equations for θ, we shall solve them for the ratio $y^{(1)} = \theta/\theta_0$, writing $y^{(2)}$ for $Dy^{(1)}$. Then

$$Dy^{(2)} = D^2\left(\frac{\theta}{\theta_0}\right) = -\frac{g\sin(\theta_0 y^{(1)})}{r\theta_0},$$

and (F) takes the form

(H) $$y^{(i)}(t) = c^{(i)} + \int_0^t f^{(i)}(u, y(u))\,du,$$

where

$$f^{(1)}(t,y) = y^{(2)}, \qquad f^{(2)}(t,y) = -\frac{g\sin(\theta_0 y^{(1)})}{r\theta_0}, \qquad c^{(1)} = 1, \qquad c^{(2)} = 0.$$

Likewise, (G) takes the form

$$\text{(I)} \qquad y^{(i)}(t) = c^{(i)} + \int_0^t g^{(i)}(u, y(u))\, du \qquad (i = 1, 2),$$

where

$$g^{(i)}(t, y) = y^{(2)}, \qquad g^{(2)}(t, y) = -\frac{gy^{(1)}}{r},$$

with the same $c^{(i)}$. The $f^{(i)}$ and $g^{(i)}$ have partial derivatives that are 0 or 1, and the $f^{(2)}$ and $g^{(2)}$ have partial derivatives whose absolute values are at most g/r. So the hypotheses of Theorem 2-1 are satisfied with $l(t) = \max[1, g/r]$.

For notational simplicity we write C for $(g/r)^{1/2}$. Then (I) has solution

$$z^{(1)}(t) = \cos Ct, \qquad z^{(2)}(t) = -C \sin Ct.$$

Since $f^{(1)}(t, z(t)) = g^{(1)}(t, z(t))$, for the M in Theorem 3-2 we can take any upper bound of

$$\text{(J)} \qquad \left| \int_0^t C^2[\sin(\theta_0 \cos Cu) - \theta_0 \cos Cu]\, du \right|$$

on the interval $[0, b]$ on which we are solving the equations. By the theorem of mean value, for each real number v there is a number τ between 0 and v (inclusive) for which

$$\sin v = 0 + v + 0 + (-\cos \tau)v^3/3!.$$

Applying this with $v = \theta_0 \cos Cu$ yields

$$|C^2[\sin(\theta_0 \cos Cu) - \theta_0 \cos Cu]| \leqq C^2|\theta_0 \cos Cu|^3/3!,$$

and an upper bound for the integral (J) is

$$M = C^2 b|\theta_0^3/3!|.$$

Since

$$L(t) = \int_0^t l(u)\, du = t \max\left[1, \frac{g}{r}\right],$$

by Theorem 3-2

$$|y^{(1)}(t) - z^{(1)}(t)| \leqq (C^2 b\theta_0^3/3!)\left\{\exp\left(2t \max\left[1, \frac{g}{r}\right]\right)\right\}.$$

This shows that the period of the solution θ of (F), which is the same as the period of y, is arbitrarily close to the period $2\pi(r/g)^{1/2}$ of z when θ_0 is near 0. So the somewhat carefree reasoning of the elementary physics texts has led to the correct conclusion.

This example is a simple case of the convenient procedure of "linearization" of differential equations. The procedure takes many forms, but a common feature is that some complicated functions in the right member of the differential equation are replaced by linear functions of the $y^{(i)}$. Since this procedure can take many special forms, it is not easy to state a general theorem that will cover all of them. But Theorem 3-2 can be applied in at least a great number of cases to show that for solutions that stay near 0, the solutions of the exact equations and the solutions of the linearized substitutes do not differ much on a given bounded interval.

EXERCISE 3-1 Amend the equation for the simple pendulum by adding a small friction term,

$$\frac{d^2\theta}{dt^2} = -\frac{g}{r}\sin\theta - k\frac{d\theta}{dt}.$$

Show that the time of the first oscillation tends to that with zero friction as k tends to 0.

EXERCISE 3-2 Let k be nonnegative. Show that the solution $x \mapsto y(x, k)$ of the differential equation

$$\frac{dy}{dx} = y - ky^2$$

on $[0, \infty)$, with initial value $y(0) = 1$, can never be negative. (If it were, there would be a least x_2 with $y(x_2) = 0$ and a greatest x_1 in $[0, x_2)$ with $y(x_1) = \min\{1, 1/2k\}$, and y could not satisfy the theorem of the mean on $[x_1, x_2]$.) Show that for fixed X, $y(x, k)$ differs arbitrarily little from $\exp y$ on $[0, X]$ when k is near 0, but for all positive k, however small, $y(x, k)$ is bounded on $[0, \infty)$, whereas $\exp x$ is not.

EXERCISE 3-3 Use Theorem 3-2 to show that as θ_0 tends to π, the period of the solution of (F) increases without bound.

4. Computation of Approximate Solutions

One practical aspect of the study of differential equations consists of searching for ingenious devices by which, given the $f^{(i)}$, we can find expressions for the $y^{(j)}$ as combinations of "known" functions (meaning functions that have been named and tabulated). This is a chief activity in elementary courses in differential equations. But often, no such ingenious device can be found. In such cases we have to fall back on estimating the functions $y^{(i)}$ by some computational procedure. This means that in the interval $[a, b]$ we have chosen certain points, and at those points we are trying to find numbers that agree with the exact values of the $y^{(i)}$ to within an assigned tolerance. We shall suppose that we are to tabulate the estimates of the $y^{(i)}$ at points $x_1 = a, x_2, x_3, \ldots, x_{k+1} = b$. The value of the $y^{(i)}$ is assigned as a given initial value at some point x_j. We shall assume that this point is a. If it were b, we could change the independent variable from x to $-x$, so that the assigned value would be that at the left end-point of $[-b, -a]$; if it were given at an intermediate point, we could work both ways from that point. In all our approximation techniques we shall assume that we begin with the assigned value $y^{(i)}(a)$.

There are many different approximation procedures, with different degrees of complexity and of accuracy. Their study is part of the subject called *numerical analysis*. Some methods are designed for use with large computers. These we shall not discuss. Instead, we shall present some methods that are well adapted for use with the small and inexpensive electronic calculators that are in common use today. To save repetition, in all our discussions we shall assume that the functions $f^{(i)}$ are defined and have continuous partial derivatives of all orders up to the fifth on the set $[a, b] \times R^r$, and we shall subdivide $[a, b]$ into k subintervals all with the same length $h = (b - a)/k$ by points $x_1 = a$, $x_2 = a + h$, $x_3 = a + 2h, \ldots, x_{k+1} = b$.

If y is the solution of the differential equation

$$\text{(A)} \qquad y^{(i)}(x) = c^{(i)} + \int_a^x f^{(i)}(u, y(u))\, du,$$

then for $i = 1, \ldots, k$ we have

$$\text{(B)} \qquad y^{(i)}(x_{j+1}) = y^{(i)}(x_j) + \int_{x_j}^{x_{j+1}} f^{(i)}(u, y(u))\, du.$$

But we cannot compute the integral in the right member, because it involves the function f that we are trying to determine. We have to use some approximation procedure for the integral; and the choice of the approximation for the integral is what gives the character to the approximation method we are constructing.

In Section I-12 we named four formulas for approximating integrals. In the simplest, the Cauchy formula, the integral in (B) is estimated as the product of the length h of the interval of integration and the value of the integrand at the left end of that interval. If we use this approximation for the integral, we obtain a numerical solution method that one might reasonably expect to name the Cauchy method. Instead, we shall call it the Euler method, since Leonhardt Euler published it first, in 1768. It might seem mysterious that Euler published this procedure, which we deduce from the Cauchy formula, 21 years before Cauchy was born. There is really no mystery; Euler was led to the method by different considerations, based on geometric reasoning. Consider $r = 1$. Since the derivative of y at $x = a$ is $f(a, y(a))$, the tangent to the graph a of y has slope

$f(a, y(a))$ at a, and the tangent has equation $y = y(a) + (x - a)f(a, y(a))$. So, if h is small, the difference between the ordinate $y(a + h)$ at the point $x_2 = a + h$ and the ordinate $y(a) + hf(a, y(a))$ of the tangent will be much smaller than h. A repetition of the argument, interval by interval, leads to the Euler formula.

We shall use the symbol y_E to denote the approximation provided by Euler's method. Then, since we are using the Cauchy formula for the integral, we are led to the equation

(C) $$y_E^{(i)}(x_{j+1}) = y^{(i)}(x_j) + hf^{(i)}(x_j, y_E(x_j)).$$

We shall illustrate the use of this and the later methods by applying them to the differential equation

(D) $$y(x) = 1 + \int_0^x (u + y(u))\,du.$$

In all the calculations we shall carry many more decimal places than the accuracy of the method justifies so as to avoid hiding the error of the method under rounding errors. By Example 1-2, the solution of (D) is $y(x) = 2\exp x - x - 1$, so that

(E) $$y(1) = 3.436563656918090\cdots.$$

Then by (C), starting with $y(0) = 1$, we obtain the estimate in Table 4-1, using $h = 0.05$. The error in $y(1)$ is $-0.13\cdots$, showing that the method is not very accurate. This is not surprising; the Cauchy formula yields a poor approximation to the integral.

TABLE 4-1

x	y_E	$f(x, y_E)$
0.0	1	1
0.05	1.05	1.10
0.10	1.105	1.205
0.15	1.16525	1.31525
⋮	⋮	
1.0	3.3065943	

In Section I-12 we saw that the trapezoidal formula gave a much better approximation to the integral than the Cauchy formula did. If we use the trapezoidal formula for the integral, the method will be called the "trapezoidal method," and the approximation that it furnishes will be denoted by y_T. If we use the trapezoidal approximation to replace the integral in (B), we obtain

(F) $$y_T^{(i)}(x_{j+1}) = y_T^{(i)}(x_j) + [f^{(i)}(x_j, y_T(x_j)) + f^{(i)}(x_{j+1}, y_T(x_{j+1}))](h/2).$$

Here we have a difficulty. We cannot compute the right member of (F) until we know $y_T(x_{j+1})$, which is just what we are trying to determine. But if h is small enough for the method to be applicable, we can use (F) to determine $y_T(x_{j+1})$ to any desired degree of accuracy by successive approximations. We first make any sort of approximation to $f^{(i)}(x_{j+1}, y_T(x_{j+1}))$ and substitute it in the right member of (F). The resulting left member of (F) will then be a first approximation to $y_T(x_{j+1})$. With this we compute a second approximation to $f^{(i)}(x_{j+1}, y_T(x_{j+1}))$. Then, by (F), we compute a second approximation to $y_T(x_{j+1})$, and so on. If h is small enough, the successive approximations will converge to a number that satisfies (F). But in practice we are trying to approximate the solution of the differential equation to some selected number of decimal places, so we stop our successive approximations when we reach one that coincides with its predecessor to that many decimal places. The better our first approximation, the fewer steps we shall have to take in order to reach this degree of accuracy.

For the first step we have nothing available but the initial data, and there is no obvious guess for $f^{(i)}(x_2, y_T(x_2))$ that is better than $f^{(i)}(x_1, y_T(x_1))$. But after two lines of the computation are available, it will normally be better to assume that $f^{(i)}(x, y(x))$ increases as much from x_j to x_{j+1} as it increased from x_{j-1} to x_j. That is, we assume

$$f^{(i)}(x_{j+1}, y_T(x_{j+1})) - f^{(i)}(x_j, y_T(x_j)) \approx f^{(i)}(x_j, y_T(x_j)) - f^{(i)}(x_{j-1}, y_T(x_{j-1})),$$

whence

$$\text{(G)} \qquad f^{(i)}(x_{j+1}, y_T(x_{j+1})) = 2f^{(i)}(x_j, y_T(x_j)) - f^{(i)}(x_{j-1}, y_T(x_{j-1})).$$

We illustrate the use of the trapezoidal method by applying it to the same differential equation (D), this time with $h = 0.1$. In Table 4-2, when an estimate

TABLE 4-2

x	$y_T(x)$	$f(x, y_T(x))$
0.0	1	1
0.1		(1)
	(1.1)	(1.20)
	(1.11)	(1.21)
	(1.1105)	(1.2105)
	(1.110525)	(1.210525)
	(1.1105272)	(1.2105262)
	1.1105263	1.2105263
0.2		(1.421056)
	(1.2421052)	(1.4421052)
	(1.2431578)	(1.4431578)
	(1.2432104)	(1.4432104)
	(1.2432131)	(1.4432131)
	1.2423132	1.4432132

is replaced by a better one, instead of erasing it we enclose it in parentheses to show that is has been superseded. We show the calculations in detail for the first two lines. We shall not exhibit any further steps in the computation. The final result is

$$y_T(1) = 3.4410847.$$

The error is 0.00452. This is only about one-thirtieth as great as the error under the Euler method, even though for the trapezoidal method we used intervals twice as long as we did for the Euler method.

However, we saw in Section I-12 that the Simpson formula is more accurate than the trapezoidal. To use it, we replace equation (B) by

$$\text{(H)} \qquad y^{(i)}(x_{j+1}) = y^{(i)}(x_{j-1}) + \int_{x_{j-1}}^{x_{j+1}} f^{(i)}(u, y(u))\, du.$$

If we replace the integral in (I) by the approximation to it given by Simpson's formula, we obtain an approximation to y that we shall call the Simpson approximation and shall denote by y_S. Its successive values are given by the equation

$$\begin{aligned} \text{(I)} \qquad y_S^{(i)}(x_{j+1}) &= y_S^{(i)}(x_{j-1}) + [f^{(i)}(x_{j-1}, y_S(x_{j-1})) \\ &\quad + 4f^{(i)}(x_j, y_S(x_j)) + f^{(i)}(x_{j+1}, y_S(x_{j+1}))](h/3) \\ &= c^{(i)}(x_{j+1}) + f^{(i)}(x_{j+1}, y_S(x_{j+1}))(h/3), \end{aligned}$$

where in the last form of the right member we have denoted by $c^{(i)}(x_{j+1})$ the sum of all the terms in the right member that are independent of $y_S(x_{j+1})$; that is,

$$\text{(J)} \qquad c^{(i)}(x_{j+1}) = y_S^{(i)}(x_{j+1}) + [f^{(i)}(x_{j-1}, y_S(x_{j-1}) + 4f^{(i)}(x_j, y_S(x_j))](h/3).$$

This has the same difficulty that we met in the trapezoidal method; the quantity that we are trying to determine occurs in both members of the equation. As in the trapezoidal method, we overcome this by using successive approximations. We first compute $c^{(i)}(x_{j+1})$ by (J) and record it. We then make a first approximation to $f^{(i)}(x_{j+1}, y_S(x_{j+1}))$, for example, by using (G). With this we compute a first approximation to $y_S^{(i)}(x_{j+1})$ by (I). With this we compute a second approximation to $f^{(i)}(x_{j+1}, y_S(x_{j+1}))$, use (I) to get a second approximation to $y_S^{(i)}(x_{j+1})$, and so on.

However, the Simpson method poses another difficulty. To calculate $y_S(x_{j+1})$, we use data from the preceding two lines, and at the start we have only one line available. An obvious way out of this difficulty is to use the trapezoidal method to compute the second line and after that change to the Simpson method. But since the trapezoidal method is much less accurate than the Simpson method, this would introduce an unacceptably large error in the second line that would persist in all later lines. The simplest resolution of this difficulty is to start with an interval of length h' less than the length h that we plan to use in the Simpson method. If, for example, we choose the first interval to have length $h' = h/2$, we first compute $y^{(i)}(a + h/4)$ by the trapezoidal method. Having the lines for $x = a$ and $x = a + h/4$, we can use the Simpson method to estimate $y^{(i)}(a + h/2)$. Now, having the lines for $x = a$ and $x = a + h/2$, we can use the Simpson method to estimate $y^{(i)}(a + h)$, and from that point on we can use the Simpson method with the desired interval length h.

In our example (D) we shall use interval length $h = 0.2$, and we shall start by using the trapezoidal method with interval length $h' = h/4$ to estimate $y(0.05)$. The successive approximations are given in Table 4-3. We adopt this last value of

TABLE 4-3

x	$y_T(x)$	$f(x, y_T(x))$
0	1	1
0.05	(1.05)	(1.10)
	(1.0525)	(1.1025)
	(1.0525625)	(1.1025625)
	(1.0525640)	(1.1025640)
	1.0525641	1.1025641

$y_T(0.05)$ as the estimate for $y_S(0.05)$ and we use the Simpson method to estimate $y_S(0.10)$. The quantity defined in (J) we denote simply by $c(x_{j+1})$, and we compute that, with $h = 0.05$,

$$c(0.05) = 1 + [1 + 4(1.1025641)](0.05/3) = 1.0901709.$$

The successive estimates for $y_S(0.1)$ are given in Table 4-4. Next, with these

TABLE 4-4

x	$y_S(x)$	$f(x, y_S(x))$	$c(s)$
0	1	1	
0.05	1.0525641	1.1025641	
0.10		(1.2051282)	1.0901709
	(1.1102563)	(1.2102563)	
	(1.1103418)	(1.2103418)	
	1.1103432	1.2103432	

estimates for $y_S(0)$ and $y_S(0.1)$, we estimate $y_S(0.2)$ by Simpson's method. This time, with $h = 0.1$, we compute

$$c(0.1) = 1 + [1 + 4(1.2103432)](0.1/3) = 1.1947124.$$

The successive estimates are given in Table 4-5. We can now use the Simpson

TABLE 4-5

x	$y_S(x)$	$f(x, y_S(x))$	$c(s)$
0	1	1	
0.1	1.1103432	1.2103432	
0.2		(1.4206864)	1.1947124
	(1.2420686)	(1.4420686)	
	(1.2427813)	(1.4427813)	
	(1.2428051)	(1.4428051)	
	1.2428059	1.4428059	

method with $h = 0.2$. We omit the successive approximations and list the final results in Table 4-6.

TABLE 4-6

x	$y_S(x)$	$f(x, y_S(x))$	$c(x)$
0	1	1	
0.2	1.2428059	1.4428059	
0.4	1.5836588	1.9836588	1.4514149
0.6	2.0442520	2.6442520	1.8679686
0.8	2.6511105	3.4511105	2.4210365
1.0	3.4366052	4.4366052	3.1408316

The error in $y(1)$ is 0.00004.

It is possible, and in fact rather easy, to estimate the error in the estimate for $y_S(0.1)$ inherited from the use of the trapezoidal method in the first step. It can also be shown that the error in the trapezoidal method is not greater than a certain multiple of h^2, and the error in the Simpson method is not greater than a certain multiple of h^4. But since these statements are not referred to later in this book and their proofs are tedious, we omit them.

EXERCISE 4-1 The equation

$$\frac{dy}{dx} = \frac{5y}{x} \qquad (x > 0),$$

with $y(10) = 10^5$, has solution $y = x^5$. Apply the Simpson method to compute y at $10, 10.1, \ldots, 10.6$. Find the error.

EXERCISE 4-2 The solution of

$$y(x) = \int_0^x (1 + y(u)^2)\, du$$

for $0 \leqq x < \pi/2$ is $y(x) = \tan x$. Use the Simpson method to compute $\tan x$ for $x = 0.2$, 0.4, and 0.6 radians. Check with a table of $\tan x$ for x in radians.

EXERCISE 4-3 Given that $\tan 1.3 = 3.6021$ and $\tan 1.4 = 5.7979$, try to compute $\tan 1.5$ and $\tan 1.6$ by Simpson's method. What causes the trouble?

EXERCISE 4-4 Verify that if the interval-length is h, the error in computing $y^{(i)}(a + h)$ by the trapezoidal method is approximately

(L) $[f^{(i)}(a, y(a)) - 2f^{(i)}(a + h, y(a + h)) + f^{(i)}(a + 2h, y(a + 2h))](h/2)$.

Use this to show that in the example in the text, the error in the estimate of

$y_T(0.05)$ is about 0.000022. *Suggestion*: By the theorem of the mean, there is an $\bar{x}$ in $(a, a + 2h)$ such that the expression (L) is $D^2 f^{(i)}(\bar{x}, y(\bar{x}))h^3/12$. By Theorem I-12-2, there is an $\tilde{x}$ in $[a, a + h]$ such that the error in $y_T(a + h)$ is approximately $D^2 f^{(i)}(\tilde{x}, y(\tilde{x}))h^3/12$. If h is small, the two second derivatives are nearly equal.

EXERCISE 4-5 Let M be an upper bound for the absolute values of all the partial derivatives $\partial f^{(i)}/\partial y^{(j)}$. Given $y_T(x_j)$, choose a first approximation y_0 to $y_T(x_{j+1})$ and then, by the trapezoidal formula, compute successive approximations $y_1, y_2, \ldots$. Show that

$$\sum_{j=1}^{r} |y_{k+1}^{(j)} - y_k^{(j)}| \leqq \frac{Mrh}{2} \sum_{j=1}^{r} |y_k^{(j)} - y_{k-1}^{(j)}|.$$

Use this to prove that if h is chosen small enough so that $Mrh/2 \leqq c < 1$, the successive approximations will converge to their limit at least as fast as a geometric series with ratio c.

EXERCISE 4-6 State and prove an analog of Exercise 4-6 for Simpson's method.

5. Linear Differential Equations

A system of differential equations is *linear* if it has the form

$$\text{(A)} \qquad y^{(i)}(x) = y^{(i)}(x_0) + \int_{x_0}^{x} \left[\sum_{j=1}^{r} a_{i,j}(u) y^{(j)}(u) + b_i(u) \right] du,$$

where $a \leqq x \leqq b$. Here we have chosen to replace the symbol $c^{(i)}$ that we have used heretofore by the symbol $y^{(i)}(x_0)$, which is equal to $c^{(i)}$. We have already seen, at the end of Section 2, that such systems have unique solutions provided that the $a_{i,j}$ and the b_i are integrable over $[a, b]$. Now we shall discuss the solutions in greater detail.

The system (A) is *homogeneous* (linear) if the b_i are all 0, so that (A) has the form

$$\text{(B)} \qquad y^{(i)}(x) = y^{(i)}(0) + \int_{x_0}^{x} \left[\sum_{j=1}^{r} a_{i,j}(u) y^{(j)}(u) \right] du.$$

The system of equations (B) is the homogeneous system *corresponding* to the system (A). The solutions of homogeneous systems form a linear family, in the following sense.

LEMMA 5-1 *Let* $y_1 = (y_1^{(1)}, \ldots, y_1^{(r)})$ *and* $y_2 = (y_2^{(1)}, \ldots, y_2^{(r)})$ *be solutions of the homogeneous linear differential equations* (B), *and let* k_1 *and* k_2 *be real numbers. Then*

$$y = (k_1 y_1^{(1)} + k_2 y_2^{(1)}, \ldots, k_1 y_1^{(r)} + k_2 y_2^{(r)})$$

is also a solution of (B).

To prove this we need only write (B) with y_1 in place of y (which is permissible because y_1 is a solution) and multiply both members by k_1, then write (B) with y_2 in place of y and multiply both members by k_2, and then add the resulting equations member by member.

Let us recall from linear algebra that a set of vectors $y_1 = (y_1^{(1)}, \ldots, y_1^{(r)}), \ldots, y_n = (y_n^{(1)}, \ldots, y_n^{(r)})$ in R^r is *linearly dependent* if there exists a set of n numbers $k_1, \ldots, k_n$ not all 0 such that

$$\text{(C)} \quad k_1 y_1 + \cdots + k_n y_n = (k_1 y_1^{(1)} + \cdots + k_n y_n^{(1)}, \ldots, k_1 y_1^{(r)} + \cdots + k_n y_n^{(r)}) = (0, \ldots, 0).$$

The set $y_1, \ldots, y_n$ is *linearly independent* if it is not linearly dependent. A set of solutions of equations (B) always has the property that if they are linearly independent at any point x in $[a, b]$, they are linearly independent at all points x in $[a, b]$.

THEOREM 5-2 *Let* $a_{i,j}$ $(i, j = 1, \ldots, r)$ *be integrable over an interval* $[a, b]$, *and let* $y_1, \ldots, y_n$ *be solutions of the homogeneous linear differential equations* (B). *Then either the* $y_1(x), \ldots, y_n(x)$ *are linearly independent at all points* x *of* $[a, b]$, *or else they are linearly dependent at all points* x *of* $[a, b]$, *and there exists a linear combination* $k_1 y_1 + \cdots + k_n y_n$ $(k_1, \ldots, k_n$ *not all* 0) *that vanishes identically on* $[a, b]$.

Suppose that there exists a point x of $[a, b]$ such that the vectors $y_1(x), \ldots, y_n(x)$ are linearly dependent. We can use this x as the initial point x_0 and write the differential equations with the notation used in (B). Then by hypothesis, there exist numbers $k_1, \ldots, k_n$ not all 0 such that equations (C) are satisfied with $y_j = y_j(x_0)$. By Lemma 5-1, the linear combination $y = k_1 y_1 + \cdots + k_n y_n$ is also a solution of (B), and as we have just seen, its value at x_0 is $(0, \ldots, 0)$. But if the $y^{(i)}(x_0)$ are all 0, equations (B) have the obvious solution $y^{(i)}(x) = 0$ $(i = 1, \ldots, r; a \leq x \leq b)$; and by Theorem 2-1, this is the only solution. So the functions $y^{(i)} = k_1 y_1^{(i)} + \cdots + k_n y_n^{(i)}$ $(i = 1, \ldots, r)$ are all identically 0, and the vectors $y_1(x), \ldots, y_n(x)$ are linearly dependent at all points x in $[a, b]$.

We saw in Example 1-1 that the solution of the equation

$$y(x) = y(a) + \int_a^x p(u) y(u)\, du$$

is

$$y(x) = y(a)\exp \int_a^x p(u)\,du.$$

Likewise, the solution of

$$z(x) = z(a) - \int_a^x p(u)z(u)\,du$$

is

$$z(x) = z(a)\exp\left(-\int_a^x p(u)\,du\right).$$

The product of these functions is a constant; it is equal to $y(a)z(a)$ for all x in $[a, b]$. This can be generalized to r dimensions, as follows.

THEOREM 5-3 *Let $a_{i,j}$ $(i, j = 1, \ldots, r)$ be integrable over $[a, b]$. Let y satisfy*

(D) $$y^{(i)}(x) = y^{(i)}(x_0) + \int_{x_0}^x \left[\sum_{j=1}^r a_{i,j}(u)y^{(j)}(u)\right] du,$$

and let z satisfy

(E) $$z^{(i)}(x) = z^{(i)}(x_0) - \int_{x_0}^x \left[\sum_{j=1}^r a_{j,i}(u)z^{(j)}(u)\right] du.$$

Then the sum $y^{(1)}(x)z^{(1)}(x) + \cdots + y^{(r)}(x)z^{(r)}(x)$ is constant on $[a, b]$.

If in the integration-by-parts formula Theorem II-9-4 we replace a and b by x_0 and x, respectively, and transpose one term, we obtain the following statement.

(F) Let $\dot{f}$ and $\dot{g}$ be functions integrable over an interval $[a, b]$, and let f and g be indefinite integrals of $\dot{f}$ and $\dot{g}$, respectively. Then for x_0 and x in $[a, b]$

$$f(x)g(x) - f(x_0)g(x_0) = \int_{x_0}^x [\dot{f}(u)g(u) + f(u)\dot{g}(u)]\,du.$$

Equations (D) express $y^{(i)}$ as the indefinite integral of the quantity in brackets in (D), and equations (E) express $z^{(i)}$ as the indefinite integral of the negative of the quantity in brackets in (E). So by (F), we have for each i in $\{1, \ldots, r\}$

(G) $$y^{(i)}(x)z^{(i)}(x) - y^{(i)}(x_0)z^{(i)}(x_0)$$
$$= \int_{x_0}^x \left\{\left[\sum_{j=1}^r a_{i,j}(u)y^{(j)}(u)\right] z^{(i)}(u) + y^{(i)}(u)\left[-\sum_{j=1}^r a_{j,i}(u)z^{(j)}(u)\right]\right\} du.$$

If we add these equations member by member for $i = 1, \ldots, r$, the sum of the integrands in the right member is 0, so the left member of (G) is 0 for all x. The proof is complete.

The system of equations (E) is known as the system *adjoint* to the system (D). Note the interchange of the subscripts of $a_{i,j}$.

We shall use the standard notation

$$\text{(H)} \qquad \delta_i^j = \begin{cases} 1 & \text{if } i = j, \\ 0 & \text{if } i \neq j. \end{cases}$$

This is often called the "Kronecker δ."

If for each j and k in $\{1, \ldots, r\}$ we define y_j and z_k to be solutions of the respective equations

$$\text{(I)} \qquad y_j^{(i)}(x) = \delta_j^i + \int_a^x \left[\sum_{h=1}^r a_{i,h}(u) y_j^{(h)}(u) \right] du,$$

$$\text{(J)} \qquad z_k^{(i)}(x) = \delta_k^i - \int_a^x \left[\sum_{h=1}^r a_{h,j}(u) z_k^{(h)}(u) \right] du,$$

then for all x in $[a, b]$ we have, by Theorem 5-3,

$$\text{(K)} \qquad \sum_{i=1}^r z_k^{(i)}(x) y_j^{(i)}(x) = \sum_{i=1}^r \delta_k^i \delta_j^i = \delta_j^k.$$

In the notation of matrices, we can regard $z_k^{(i)}(x)$ as the element in row i and column k of a matrix Z of order r, and likewise we can regard $y_j^{(i)}(x)$ as the element in row i and column j of a matrix Y. The left member of (K) is then the ordinary matrix product ZY^{T} of Z and the transpose Y^{T} of Y. The right member of (K) is the element in row i and column k of the identity matrix I, so equation (K) states that $ZY^{\mathrm{T}} = I$. But from elementary linear algebra we know that if $ZY^{\mathrm{T}} = I$, then $Y^{\mathrm{T}}Z = I$. That is,

$$\text{(L)} \qquad \sum_{h=1}^r y_h^{(j)}(x) z_h^{(i)}(x) = \delta_j^i.$$

From this it follows that we do not need to solve Eqs. (I) and (J) separately. If we have found solutions of either one of them and arranged the solutions as columns of a matrix, the columns of the transpose of the inverse of that matrix are the solutions of the other equation.

Sometimes solutions are needed for several different equations of form (A), in all of which the $a_{i,j}$ are the same but the b_i are different. In such cases it has proved convenient to make use of the following theorem.

THEOREM 5-4 *Let $a_{i,j}$ and b_i be functions integrable over an interval $[a, b]$. Let $y_j^{(i)}$ and $z_k^{(i)}$ $(i, j, k = 1, \ldots, r)$ be solutions of equations* (I) *and* (J), *respectively. Then the solution of* (A) *(with $x_0 = a$) is*

$$\text{(M)} \qquad y^{(i)}(x) = \sum_{k=1}^r \left\{ y^{(k)}(a) + \sum_{h=1}^r \int_a^x b_h(u) z_k^{(h)}(u)\, du \right\} y_k^{(i)}(x).$$

For each i and k in $\{1, \ldots, r\}$, $y^{(i)}$ is an indefinite integral of the quantity in brackets in (A), and $z_k^{(i)}$ is an indefinite integral of the negative of the quantity in brackets in (J). By the formula for integration by parts, in the form (F), for each k in $\{1, \ldots, r\}$ and each x in $[a, b]$ we have

$$\begin{aligned}\sum_{h=1}^{r} y^{(h)}(x) z_k^{(h)}(x) = \sum_{h=1}^{r} y^{(h)}(a) z_k^{(h)}(a) \\ + \sum_{h=1}^{r} \int_a^x \Bigg\{ \Bigg[\sum_{i=1}^{r} a_{h,i}(u) y^{(i)}(u) + b_h(u) \Bigg] z_k^{(h)}(u) \\ + y^{(h)}(u) \Bigg[-\sum_{j=1}^{r} a_{j,h}(u) z_k^{(j)}(u) \Bigg] \Bigg\} du.\end{aligned}$$

Since $z_k^{(h)}(a) = \delta_k^h$ and the terms in the integrand that have factors $a_{j,h}$ cancel those with factors $a_{h,i}$, this reduces to

$$\sum_{h=1}^{r} y^{(h)}(x) z_k^{(h)}(x) = y^{(k)}(a) + \sum_{h=1}^{r} \int_a^x b_h(u) z_k^{(h)}(u)\, du.$$

We multiply both members by $y_k^{(i)}(x)$ and sum over $k = 1, \ldots, r$. Because of (L), the result simplifies to (M). The proof is complete.

The reason that this has been useful in some applications is that after Eqs. (I) have been solved, and Eqs. (J) have been solved by computing the transpose of the inverse matrix, and the results have been recorded, any equation of form (A) with the given $a_{i,j}$ can be solved by (M), which involves for each set of b_i merely the computation of the integral in (M), not the solution of a system of differential equations. Computing an integral, whether by theory or by numerical methods, is a much easier task than solving a differential equation.

EXERCISE 5-1 Show that if we have solved the differential equation in Example 1-1, by Theorem 5-4 we can obtain the solution of the equation in Example 1-2.

EXERCISE 5-2 Let (A) have the special form

(N)
$$y^{(1)}(x) = y^{(1)}(0) + \int_0^x [y^{(2)}(u) + b_1(u)]\, du,$$

$$y^{(2)}(x) = y^{(2)}(0) + \int_0^x [-y^{(1)}(u) + b_2(u)]\, du.$$

Write the corresponding homogeneous equations. Show that these have solutions

$$y_1^{(1)}(x) = \sin x, \qquad y_1^{(2)}(x) = \cos x,$$
$$y_2^{(1)}(x) = \cos x, \qquad y_2^{(2)}(x) = -\sin x.$$

Show that the transpose of the inverse of the matrix

$$\begin{pmatrix} \sin x & \cos x \\ \cos x & -\sin x \end{pmatrix}$$

has columns that are solutions of the adjoint equations. Write the general solution of equations (N).

Remark The general solution will involve integrals of functions of the form $b_i(u) \sin u$ and $b_i(u) \cos u$. Any table of integrals will present a great number of examples of integrals of such functions. So we have in this exercise a means of finding solutions for many particular cases, with an assortment of functions b_1 and b_2.

6. Differentiation of Solutions with Respect to Initial Values and Parameters

If the functions $f^{(i)}$ in the differential equation

$$\text{(A)} \qquad y^{(i)}(x) = c^{(i)} + \int_{x_0}^{x} f^{(i)}(u, y(u))\, du$$

satisfy the hypotheses of Theorem 2-1, we know that the equation has a unique solution. Obviously this solution depends on the choice of the initial values $c^{(i)}$, and if we wish to compare solutions corresponding to different initial values, we should indicate this dependence in the notation. So we shall change the symbols slightly, replacing the left member of (A) by $y^{(i)}(x, c)$, so that (A) takes the form

$$\text{(B)} \qquad y^{(i)}(x, c) = c^{(i)} + \int_{x_0}^{x} f^{(i)}(u, y(u, c))\, du.$$

We already know that each $y^{(i)}$ is a continuous function of the $c^{(i)}$, by Theorem 3-2. We shall now prove a stronger result; if the $f^{(i)}$ have continuous partial derivatives with respect to the $y^{(i)}$, the solutions $y^{(i)}(x, c)$ have partial derivatives with respect to the $c^{(j)}$, and these partial derivatives satisfy an equation that we shall specify. But in order to prove this it is desirable to use a slightly more convenient notation for partial derivatives than heretofore. We shall denote by $f^{(i)}_{y^{(j)}}$ that function which is the partial derivative of $f^{(i)}$ with respect to the $(j + 1)$th of its variables, which in the original notation was denoted by $y^{(j)}$. Thus, if

$$f(x, y^{(1)}, y^{(2)}) = x^3 \cos(y^{(1)} - y^{(2)}),$$

we have

$$f_{y^{(1)}}(u, v^{(1)}. v^{(2)}) = -u^3 \sin(v^{(1)} - v^{(2)}).$$

The notation for the partial derivatives is fixed as soon as we have stated the notation in which the function is first defined, and it does not change if we substitute different numbers for the independent variables.

Thus, if for some x in $[a, b]$, some i and j in $1, \ldots, r$, and some set of $r - 1$ numbers $y^{(1)}, \ldots, y^{(j-1)}, y^{(j+1)}, \ldots, y^{(r)}$ the function

$$u \mapsto f^{(i)}(x, y^{(1)}, \ldots, y^{(j-1)}, u, y^{(j+1)}, \ldots, y^{(r)})$$

has a derivative at a point v in R, this derivative is denoted by

$$f^{(i)}_{y^{(j)}}(x, y^{(1)}, \ldots, y^{(j-1)}, v, y^{(j+1)}, \ldots, y^{(r)}).$$

If $f^{(i)}$ satisfies the hypotheses of Theorem 2-1, by inequality (F) in that theorem we see that for all $u \neq v$

$$\begin{aligned}|[f^{(i)}&(x, y^{(1)}, \ldots, y^{(j-1)}, u, y^{(j+1)}, \ldots, y^{(r)}) \\ &- f^{(i)}(x, y^{(1)}, \ldots, y^{(j-1)}, v, y^{(j+1)}, \ldots, y^{(r)})]/(u - v)| \\ &\leqq l(x) \cdot |u - v|.\end{aligned}$$

If we let u tend to v, this yields

$$|f^{(i)}_{y^{(j)}}(x, y^{(1)}, \ldots, y^{(j-1)}, v, y^{(j+1)}, \ldots, y^{(r)})| \leqq l(x).$$

Consequently, if $f^{(i)}$ has a partial derivative with respect to $y^{(j)}$ at all points y in R^r, and $f^{(i)}$ satisfies the hypotheses of Theorem 2-1, the partial derivative satisfies the inequality

(C) $$|f^{(i)}_{y^{(j)}}(x, y)| \leqq l(x).$$

We can now prove our first differentiation theorem.

THEOREM 6-1 *Let the functions $f^{(1)}, \ldots, f^{(r)}$ satisfy the hypotheses of Theorem 2-1. Assume that for each x in $[a, b]$ and y in R^r, each $f^{(i)}$ has a partial derivative with respect to each $y^{(j)}$, and the functions*

$$y \mapsto f^{(i)}_{y^{(j)}}(x, y)$$

are continuous on R^r. Then for each k in $\{1, \ldots, r\}$ the functions $(x, c) \mapsto y^{(i)}(x, c)$ that satisfy (B) *have continuous partial derivatives with respect to $c^{(k)}$, and these partial derivatives coincide with the solutions $x \mapsto \eta^{(i)}(x)$ of the linear differential equations*

(D) $$\eta^{(i)}(x) = \delta^i_k + \int_{x_0}^{x} \left[\sum_{h=1}^{r} f^{(i)}_{y^{(h)}}(u, y(u, c)) \eta^{(h)}(u) \right] du.$$

As before, δ^i_k is the "Kronecker δ," with value 1 if $i = k$ and value 0 if $i \neq k$. Whenever $c = (c^{(1)}, \ldots, c^{(r)})$ is an r-vector and v is a real number, we denote by $c + v\delta_k$ the vector whose ith component is $c^{(i)} + v\delta^i_k$.

For real v we form the successive Picard estimates $y_1, y_2, y_3, \ldots$ of $y(x, c + v\delta_k)$ as in Section 2 by choosing

(E) $$y_1^{(i)}(x) = y^{(i)}(x, c) + v\eta^{(i)}(x);$$

then

(F) $$y_2^{(i)}(x) = [c^{(i)} + v\delta_k^i] + \int_{x_0}^{x} f^{(i)}(u, y(u, c) + v\eta(u))\, du.$$

By the mean-value theorem, for each i in $\{1, \ldots, r\}$ and each u in $[a, b]$ there is a point $\bar{y}_i(u, v)$ on the line segment joining $y(u, c)$ to $y(u, c) + v\eta(u)$ such that

$$\begin{aligned} &f^{(i)}(u, y(u, c) + v\eta(u)) \\ &\quad = f^{(i)}(u, y(u, c)) + \sum_{h=1}^{r} f_{y^{(h)}}^{(i)}(u, \bar{y}_i(u, v))v\eta^{(h)}(u). \end{aligned}$$

We substitute this in (F) and recall that $y(x, c)$ satisfies (B); the result is

$$y_2^{(i)}(x) = y^{(i)}(x, c) + v\delta_k^i + v \int_{x_0}^{x} \sum_{h=1}^{r} f_{y^{(h)}}^{(i)}(u, \bar{y}_i(u, v))\eta^{(h)}(u)\, du.$$

From this and (D),

(G) $$y_2^{(i)}(x) - y_1^{(i)}(x) = v \int_{x_0}^{x} \sum_{h=1}^{r} [f_{y^{(h)}}^{(i)}(u, \bar{y}_i(u, v)) - f_{y^{(h)}}^{(i)}(u, y(u, c))]\, du.$$

We define

(H) $$M(v) = \sum_{i=1}^{r} \int_{a}^{b} \sum_{h=1}^{r} |f_{y^{(h)}}^{(i)}(u, \bar{y}_i(u, v)) - f_{y^{(h)}}^{(i)}(u, y(u, c))|\, du.$$

Then by (G),

(I) $$\sum_{h=1}^{r} |y_2^{(h)}(x) - y_1^{(h)}(x)| \leqq vM(v) \qquad (a \leqq x \leqq b;\ v \text{ real}).$$

From this and Lemma 3-1, with (E) we find that for all x in $[a, b]$

(J) $$\sum_{j=1}^{r} |y^{(j)}(x, c + v\delta_k) - [y^{(j)}(x, c) + v\eta^{(j)}(x)]| \leqq vM(v) \exp r|L(x)|.$$

Let $v_1, v_2, v_3, \ldots$ be a sequence of nonzero numbers converging to 0. By (C), the integrand in (H) is at most $2r^2 l(u)$ for all v and all u in $[a, b]$, and l is integrable over $[a, b]$. As n increases, $y(u, c) + v_n\eta(u)$ converges to $y(u, c)$, and $y_i(u, v_n)$ is between them, so it too converges to $y(u, c)$. Since the partial derivatives of the $f^{(i)}$ are continuous, the integrand in (H) converges everywhere to 0. By the dominated convergence theorem, $M(v_n)$ converges to 0, so by (J),

$$\lim_{n \to \infty} \sum_{j=1}^{r} \left| \frac{y^{(j)}(x, c + v_n\delta_k) - y^{(j)}(x, c)}{v_n} - \eta^{(j)}(x) \right| = 0.$$

Hence, for $j = 1, \ldots, r$ and for all x in $[a, b]$

$$\lim_{v \to 0} \frac{y^{(j)}(x, c + v\delta_k) - y^{(j)}(x, c)}{v} = \eta^{(j)}(x).$$

That is, the partial derivative of $y^{(j)}(x, c)$ with respect to $c^{(k)}$ exists and is equal to $\eta^{(j)}(x)$. The proof is complete.

Suppose next that the $f^{(i)}$ are functions of x and y and also of another variable, a "parameter" w, that ranges over an interval (w_1, w_2). That is, each $f^{(i)}$ is a function

$$(x, y, w) \mapsto f^{(i)}(x, y, w) \qquad (a \leqq x \leqq b,\ y \text{ in } R^r,\ w_1 < w < w_2).$$

If for each w in (w_1, w_2) we use these functions for the $f^{(i)}$ in (A) and find that a unique solution exists, this solution will depend on w as well as on x, and so we shall denote it by $y(x, w)$. (There is no need to mention the $c^{(i)}$, since we are going to leave them unchanged in this discussion.) Then (A) takes the form

$$\text{(K)} \qquad y^{(i)}(x, w) = c^{(i)} + \int_{x_0}^{x} f^{(i)}(u, y(u, w), w)\, du.$$

If we knew that the $y^{(i)}(x, w)$ had a partial derivative with respect to w and that differentiation under the integral sign were permissible, we could differentiate both members of (K) with respect to w and obtain a differential equation that these partial derivatives would have to satisfy. But it is not evident that such a partial derivative exists. We shall now prove its existence; the proof needs little more than a change of notation in Theorem 6-1.

THEOREM 6-2 *Let the functions* $f^{(i)}$ $(i = 1, \ldots, r)$ *be defined on* $[a, b] \times R^r \times (w_1, w_2)$ *and satisfy the hypotheses of Theorem* 2-1 *for each fixed* w *in* (w_1, w_2). *Let all the* $f^{(i)}$ *have continuous partial derivatives with respect to* $y^{(1)}, \ldots, y^{(r)}, w$ *for all* (x, y, w) *in* $[a, b] \times R^r \times (w_1, w_2)$. *Then for each* w *in* (w_1, w_2) *the solutions* $y^{(i)}(x, w)$ *of equations* (K) *have partial derivatives with respect to* w, *and these partial derivatives coincide with the solutions* $\xi^{(1)}, \ldots, \xi^{(r)}$ *of the linear differential equations*

$$\text{(L)} \qquad \xi^{(i)}(x) = \int_{x_0}^{x} \left\{ \sum_{h=1}^{r} f^{(i)}_{y^{(h)}}(u, y(u, w), w)\xi^h(u) + f^{(i)}_w(u, y(u, w), w) \right\} du.$$

Let us define functions $g^{(j)}$ $(j = 1, \ldots, r + 1)$ on $[a, b] \times R^r \times (w_1, w_2)$ by the equations

$$g^{(i)}(x, z^{(1)}, \ldots, z^{(r+1)}) = f^{(i)}(x, z^{(1)}, \ldots, z^{(r)}, z^{(r+1)})$$

$(i = 1, \ldots, r)$,

$$g^{(r+1)}(x, z^{(1)}, \ldots, z^{(r+1)}) = 0.$$

We also use $c^{(r+1)}$ as another name for w. If $c^{(r+1)}$ is in (w_1, w_2) and the functions $z^{(j)}$, or $(x, c) \mapsto z^{(j)}(x, c)$ (x in $[a, b]$, $(c^{(1)}, \ldots, c^{(r)})$ in R^r), satisfy the equations

$$\text{(M)} \qquad z^{(j)}(x, c) = c^{(j)} + \int_{x_0}^{x} g^{(j)}(u, z(u, c))\, du \qquad (j = 1, \ldots, r + 1),$$

the last equation informs us that $z^{(r+1)}(x, c)$ is constantly equal to $c^{(r+1)}$, which is w, and the other r equations (M) inform us that $(z^{(1)}, \ldots, z^{(r)})$ is a solution of (K) with this value of w. Inconveniently, the hypotheses of Theorem 6-1 are not satisfied, because the $g^{(j)}$ are not defined on all of $[a, b] \times R^{r+1}$ but only on $[a, b] \times R^r \times (w_1, w_2)$. But a glance through the preceding proofs in this chapter will show that the only use we made of the hypothesis that the $f^{(i)}$ were defined on all of $[a, b] \times R^r$ was to ensure that the functional values $f^{(i)}(x, y(x))$ were defined for all functions y encountered in the proofs. In the case of equations (M), this is sure to happen whenever $c^{(r+1)}$ is a point of (w_1, w_2), for then the point $(z^{(1)}(x, c), \ldots, z^{(r)}(x, c))$ cannot get out of R^r, whereas $z^{(r+1)}(x, c)$ cannot get out of (w_1, w_2) because it is constantly equal to $c^{(r+1)}$. So we can apply Theorem 6-1 to Eqs. (M), and we find that at $(c^{(1)}, \ldots, c^{(r+1)})$ the solutions $z^{(j)}(x, c)$ have partial derivatives with respect to $c^{(r+1)}$ that coincide with the solutions ξ of the linear differential equation

$$\text{(N)} \qquad \xi^{(j)}(x) = \delta^j_{r+1} + \int_{x_0}^{x} \sum_{h=1}^{r+1} g^{(j)}_{z^{(h)}}(u, z(u, c))\xi^{(h)}(u)\, du$$

$(j = 1, \ldots, r + 1)$.

In the last of these equations the integrand is 0, so $\xi^{(r+1)}(x, c)$ is identically equal to 1. Likewise in the last of equations (M) the integrand is 0, so $z^{(r+1)}(x, c)$ is identically equal to $c^{(r+1)}$, which is w. So the first r of equations (N) reduce to equations (L) of the statement of the theorem, and the proof is complete.

Often we encounter differential equations, or systems of differential equations, that are not in the form (A) but can be transformed into that form. In such cases we could make the transformation, apply Theorem 6-1 or Theorem 6-2, and then transform back to the original form. But usually we can avoid this transformation back and forth by a simple technique which is easiest explained by means of an example.

Suppose that we are given a differential equation

$$\text{(O)} \qquad F(x, y, y', y'') = 0,$$

with initial conditions

$$\text{(P)} \qquad y(x_0) = c^{(1)}, y'(x_0) = c^{(2)}.$$

Assume that F is continuous and has continuous partial derivatives with respect to all variables, and that Eq. (O) can be solved for y'', yielding an equation

$$\text{(Q)} \qquad y''(x) = f^{(2)}(x, y(x), y'(x)).$$

As in the beginning of Section 2, we define $y^{(1)} = y$, $y^{(2)} = y'$ and find that the solution $y(x, c)$ of (O) and (P) is the same as the component $y^{(1)}(x, c)$ of the solution of

$$y^{(1)}(x, c) = c^{(1)} + \int_{x_0}^{x} y^{(2)}(u)\, du,$$

$$y^{(2)}(x, c) = c^{(2)} + \int_{x_0}^{x} f^{(2)}(u, y(u, c))\, du.$$

If $f^{(2)}$ is continuous and has continuous partial derivatives with respect to the $y^{(i)}$ and satisfies the hypotheses of Theorem 2-1, by Theorem 6-1 the solutions $y^{(i)}(x, c)$ will have continuous partial derivatives with respect to the $c^{(i)}$. That is, $y(x, c)$ and $y'(x, c)$ have continuous partial derivatives with respect to the $c^{(i)}$, and by (Q), so has $y''(x, c)$. If we write η_j for $\partial y(x, c)/\partial c^{(j)}$, by interchange of the order of differentiation, we find that η_j and its derivatives $\eta'(x)$ and $\eta''(x)$ with respect to x are all continuous. So by differentiating both members of (Q) with respect to $c^{(j)}$, we find that η_j must satisfy

$$F_x + F_y \eta_j + F_{y'} \eta_j' + F_{y''} \eta_j'' = 0, \tag{R}$$

where $F_{x'}$ etc., are evaluated at $(x, y(x, c), y'(x, c), y''(x, c))$.

A similar procedure can be applied to differential equations that involve a parameter.

As an example, let us return to the simple pendulum, governed by equation (F) of Section 3:

$$\frac{d^2\theta}{dt^2} = -\frac{g}{r} \sin\theta,$$

with initial conditions

$$\theta(0) = c^{(1)}, \qquad \theta'(0) = c^{(2)}. \tag{S}$$

We denote the solution of the differential equation with initial conditions (S) by $\theta(t, c^{(1)}, c^{(2)})$, so that

$$\frac{\partial^2\theta(t, c)}{\partial t^2} = -\frac{g}{r} \sin\theta(t, c). \tag{T}$$

Clearly, $\theta(t, 0, 0)$ is identically 0. We know by Theorem 6-1 that at $c = (0, 0)$ the partial derivative $\eta_2 = \partial\theta/\partial c^{(2)}$ exists, is continuous, and has continuous first and second derivatives with respect to t. By applying the chain rule to (T), we obtain

$$\frac{d^2\eta_2}{dt^2} = -\frac{g}{r} (\cos 0)\eta_2. \tag{U}$$

By (S),

$$\eta_2(0) = 0, \qquad \eta_2'(0) = 1,$$

so the solution of (U) with these initial values is

$$\eta_2(t) = (g/r)^{-1/2} \sin(g/r)^{1/2} t.$$

If ε is a small positive number, $\eta_2(t)$ is negative at the point $t_- = 2\pi(r/g)^{1/2} - \varepsilon$. So $\theta(t_-, 0, c^{(2)})$ has value 0 and a negative derivative with respect to $c^{(2)}$ at $(t_-, 0)$ and must be negative for all small positive $c^{(2)}$. Likewise, at $t_+ = 2\pi(r/g)^{1/2} + \varepsilon$, the value of $\theta(t_+, 0, c^{(2)})$ is positive for all small positive $c^{(2)}$. So, for all small positive $c^{(2)}$, $\theta(t, 0, c^{(2)})$ rises through 0 as t increases from t_- to t_+, and $\theta(t, 0, c^{(2)})$ has period that differs by less than ε from $2\pi(r/g)^{1/2}$.

Finally, it is worth remarking that we do not need to assume that the $f^{(i)}$ satisfy the hypotheses of Theorem 6-1 or of Theorem 6-2 at all points (x, y) with x in $[a, b]$ and y in R^r. It is enough to assume in Theorem 6-1 that the $f^{(i)}_{y^{(j)}}(x, y)$ are continuous in y and have absolute values at most equal to $l(x)$ at all points near those of the solution-curve $y(x, c)$; that is, for all (x, y) that satisfy

$$\text{(V)} \qquad a \leqq x \leqq b, \qquad |y^{(i)} - y^{(i)}(x, c)| < \delta \qquad (i = 1, \ldots, r),$$

where δ is some positive number. To prove this, we use an auxiliary function. For $u \geqq 0$ we define

$$\phi(u) = \begin{cases} u & (0 \leqq u \leqq \delta/2); \\ 3\delta/4 - (\delta - u)^2/\delta & (\delta/2 < u \leqq \delta); \\ 3\delta/4 & (\delta < u); \end{cases}$$

and for $u < 0$ we define $\phi(u) = -\phi(-u)$. Obviously, ϕ is continuous on each of the intervals $[0, \delta/2]$, $(\delta/2, \delta]$, and (δ, ∞), and the values join continuously at the ends of those intervals, so ϕ is continuous for $u \geqq 0$. Since $\phi(-u) = -\phi(u)$ and $\phi(0) = 0$, ϕ is continuous on all of R. Likewise, the derivatives of ϕ on the three subintervals, which have the respective values 1, $2(\delta - u)/\delta$, and 0, join continuously at the ends of the intervals, so ϕ' is continuous on all of R. Moreover, ϕ has values in $[-3\delta/4, 3\delta/4]$, and ϕ' has values in $[0, 1]$.

Now suppose that for each (x, y) satisfying (V) the functions $f^{(i)}$ have partial derivatives with respect to the $y^{(i)}$ that are continuous in y and have absolute values at most equal to $l(x)$. Define, for $i = 1, \ldots, r$, x in $[a, b]$, and y in R^r,

$$g^{(i)}(x, y) = f^{(i)}(x, y^{(1)}(x, c) + \phi(y^{(1)} - y^{(1)}(x, c)), \ldots, y^{(r)}(x, c) + \phi(y^{(r)} - y^{(r)}(x, c))).$$

The points at which the $f^{(i)}$ in the right member of this equation are evaluated are all in the set of points satisfying (V). By the chain rule, the partial derivatives of the $g^{(i)}$ with respect to the $y^{(j)}$ exist and are at most $l(x)$ in absolute value, so the $g^{(i)}$ satisfy the hypotheses of Theorem 6-1. Therefore, the partial derivatives of the solutions of the equations

$$\text{(W)} \qquad y^{(i)}(x, c) = c^{(i)} + \int_{x_0}^{x} g^{(i)}(u, y(u, c))\, du$$

exist and satisfy the differential equation deduced from Theorem 6-2. But by Theorem 3-2, these solutions will, for all c' near $(c^{(1)}, \ldots, c^{(r)})$, satisfy the inequalities

$$y^{(i)}(x, c') - y^{(i)}(x, c) < \delta/2 \qquad (i = 1, \ldots, r).$$

By the definitions of ϕ and of the $g^{(i)}$, for all such c' we have

$$g^{(i)}(x, y(x, c')) = f^{(i)}(x, y(x, c')).$$

So for such c' the solutions of (W) are also solutions of (B), and the partial derivatives of the solutions of (B) therefore exist and have the values specified in Theorem 6-1.

A like discussion applies to the extension of Theorem 6-2 to systems in which the $f^{(i)}(x, y, w)$ satisfy the hypotheses only for y that satisfy inequalities (V). The most significant applications of Theorem 6-2 are in situations requiring more background explanation than we can give here. We shall merely indicate one use that has been made of it.

The standard trajectory of a projectile is computed by solving a set of three second-order differential equations in which such quantities as air density, temperature, and velocity (wind) are assumed to be certain standard functions of altitude. If one of these, say density, is changed from the standard – say, from density $\rho(y)$ to density $\rho(y) + \Delta\rho(y)$ at altitude y – we wish to approximate the effect of the change on position $y(t)$ at time t. To do this we replace $\rho(y)$ by $\rho(y) + w\,\Delta\rho(y)$ in the equations of motion. By Theorem 6-2, the partial derivative of the position with respect to w satisfies (L), and the actual change in position (corresponding to $w = 1$) is close to 1 times that partial derivative. The function f_w in (L) will be a linear function of $\Delta\rho$. By Theorem 5-4, if we compute and record the solutions of Eqs. (M) and their adjoints with $\Delta\rho = 0$, we can compute the effect of any changes in density $\Delta\rho$ by computing a definite integral, as in (M) of Section 5. This has been used extensively in the past in preparing tables of corrections to firing tables and bombing tables to correct for wind, density, etc.

The following exercise is a trimmed-down version of the problem of finding how much to change the velocity of an interplanetary rocket at some intermediate point in order to bring it back into an orbit that meets the target.

EXERCISE 6-1 A body is moving in space so as to satisfy certain differential equations

$$\frac{dy}{dt} = f(t, y),$$

where $y = (y^{(1)}, y^{(2)}, y^{(3)})$. It is desired that at time T the position $y(T)$ should be a given point k. At an earlier time t_0 it is found that there is an error and that if the orbit is left unchanged, the position at time T will be $k + e$, where e is a triple of small numbers. Find how much to change $y(t_0)$ so as to correct the error.

Suggestion: Let η_j be the solution of

$$\eta_j^{(i)\prime}(t) = \sum_{h=1}^{r} f_{y^{(h)}}^{(i)}(t, y(t))\eta_j^{(h)}(t)$$

with value δ_j^i (the "Kronecker δ") at time T. If we change $y^{(i)}(t_0)$ by

$$-\sum_{j=1}^{3} e^{(j)}\eta_j^{(i)}(t_0),$$

the change in $y^{(i)}(T)$ will be nearly $-e^{(i)}$.

7. Definition of a More General Integral

In elementary courses on calculus the following problem is often presented. An amount of material is spread along the x-axis, the mass of the material in an interval A being denoted by mA. (This we shall suppose to be 0 if A lies outside a certain interval $[a, b]$.) We are to devise a reasonable mathematical procedure for finding the moment of inertia of this mass-distribution about the origin. We know that if the mass were concentrated at finitely many points $x_1, \ldots, x_k$, the mass at x_i being m_i, the moment of inertia would be $x_1^2 m_1 + \cdots + x_k^2 m_k$. So if we subdivide $[a, b]$ into pairwise disjoint subintervals $A_1, \ldots, A_k$ and for each i choose numbers c_i, C_i such that on A_k the values of x^2 are in $[c_i, C_i]$, it is reasonable to expect that the moment of inertia is between the two numbers

$$\sum_{i=1}^{k} c_i mA_i, \qquad \sum_{i=1}^{k} C_i mA_i.$$

By cutting $[a, b]$ into finer and finer intervals, these sums can be chosen arbitrarily close to each other. The one number that is between them, however fine the cutting up of $[a, b]$, can be taken to be the moment of inertia. But this procedure is practically the same as the Darboux method of defining the Riemann integral, with m in place of m_L, and we can reasonably write

$$\text{moment of inertia} = \int x^2\, dm.$$

Moreover, if the mass-distribution has at each point a density ρ in a sense familiar from elementary physics, and this ρ is sufficiently well behaved, we can use Riemann-integral techniques to reach a formula

$$\text{(A)} \qquad \int x^2\, dm = \int x^2 \rho(x)\, dx.$$

So far, because of the simplicity of the functions $x \mapsto x^2$ and $x \mapsto \rho(x)$, the Riemann integral is adequate. But the importance of probability theory has been steadily growing, and in that theory there is a somewhat similar situation for which the Riemann-type procedure is inadequate. In this theory we encounter mathematical models for experimental situations in which an experiment or trial is performed and the result of the experiment is a real number. For each interval A in R there is a nonnegative number $P(A)$ with the property that, within the demands of accuracy of the situation under study, we can assume that if a large number N of experiments is performed, the number of results that turn out to be in A is represented closely enough by $NP(A)$. When the result of the experiment is x, we record a number $f(x)$. What will be the average of the values of the recorded numbers $f(x)$ if a large number of experiments is performed?

Although this closely resembles the moment-of-inertia problem, it has the added difficulty that the needs of modern probability theory force us to consider functions f much more complicated than the easy $x \mapsto x^2$ of the moment-of-inertia problem. The Riemann-integral type of definition does not serve. Fortunately, a direct and simple extension of the gauge-integral has the power needed to prove all the theorems called for in probability theory.

At this point the author intrudes to make a personal remark. I do not believe that every teacher of every class in mathematics should require every student to work his way through every proof. But I believe that most students will feel greater confidence in the stated theorems if they know what the integrals mean and have proved at least a few of the simpler theorems, instead of being told that such procedures are possible and are presented in more advanced theories of integration that are too subtle for the present course. This does not relieve the author of the duty of exhibiting proofs intended to be comprehensible to an ordinary undergraduate, even though only a few will have the desire and the time to work through all the details.

For guidance in setting up the definition of the integral with respect to a mass-distribution, we proceed as in the beginning of Chapter I. We consider the example of the moment of inertia of a mass distributed along the x-axis. If we are to know how the material is distributed, it would seem reasonable to require that for each interval A on the axis we should know the mass mA of the material in A. But we can get along with slightly less than this. All we need to know is the mass mA in each *left-open* interval A in R. Then m is a function of left-open intervals in R. We shall suppose that the mass outside some bounded interval $[a, b]$ is 0. If $\varepsilon > 0$, for each point $\bar{x}$ of R we can find a neighborhood $\gamma(\bar{x})$ on which x^2 differs from $\bar{x}^2$ by less than ε. We define $\gamma(-\infty) = [-\infty, a)$ and $\gamma(\infty) = (b, \infty]$. Then γ is a gauge on $\bar{R}$. Let $\mathscr{P} = \{(\bar{x}_1, A_1), \ldots, (\bar{x}_k, A_k)\}$ be a γ-fine partition of R. If $\bar{x}_i$ is finite, the mass in A_i is mA_i and every point x in A_i^- has x^2 between $\bar{x}_i^2 - \varepsilon$ and $\bar{x}_i^2 + \varepsilon$, so we may accept it from physics that the moment of inertia about 0 of this part of the mass distribution is between $(\bar{x}_i^2 - \varepsilon)mA_i$ and $(\bar{x}_i^2 + \varepsilon)mA_i$. If $\bar{x}_i$ is $\pm\infty$, $mA_i = 0$, so the moment of inertia of this part of the mass-distribution is 0 and so is the term $\bar{x}_i^2 mA_i$. By adding, the moment of inertia of the total mass about 0 is between the numbers

$$\sum_{i=1}^{k} \bar{x}_i^2 mA_i - \varepsilon \sum_{i=1}^{k} mA_i, \qquad \sum_{i=1}^{k} \bar{x}_i^2 mA_i + \varepsilon \sum_{i=1}^{k} mA_i.$$

If we denote the moment of inertia about 0 by I_0 and the total mass by m, this implies that

$$\left| \sum_{i=1}^{k} \bar{x}_i^2 mA_i - I_0 \right| < \varepsilon m$$

for all γ-fine partitions of R. The sum in this inequality resembles what we called $S(\mathscr{P}; f)$ with $f(x) = x^2$, except that where that sum had $m_L A_i$, this one has mA_i. We shall denote it by a symbol that exhibits this difference:

$$S(\mathscr{P}; f; m) = \sum_{i=1}^{k} f(\bar{x}_i) mA_i.$$

Then the inequality above informs us that I_0 is the gauge-limit of the sums $S(\mathscr{P}; f; m)$, with $f(x) = x^2$. By analogy with our definition of the gauge-integral, it is natural to denote this gauge-limit by the symbol

$$\int_R x^2\, m(dx).$$

For another example, let $P(A)$ denote the probability that the outcome of a certain experiment is a number in the interval A. Suppose that when the outcome of the experiment is a number x, we record the value of a certain given function f at x. For simplicity we assume that f is continuous and that outside a bounded interval $[a, b]$ either f is 0 or the probability is 0. We perform a large number N of repetitions of the experiment and obtain outcomes $x_1, x_2, \ldots, x_N$. The recorded numbers are $f(x_1), f(x_2), \ldots, f(x_N)$. Since f is continuous, for each $\bar{x}$ in R we can find a neighborhood $\gamma(\bar{x})$ such that on $\gamma(\bar{x})$, $f(x)$ differs from $f(\bar{x})$ by less than ε. We take $\gamma(-\infty) = [-\infty, a)$ and $\gamma(\infty) = (b, \infty]$. Let $\mathscr{P} = \{(\bar{x}_1, A_1), \ldots, (\bar{x}_k, A_k)\}$ be a γ-fine partition of R. For each experiment we record not $f(x_i)$, but $f(\bar{x}_j)$, where A_j is the interval that contains x_i. This agrees with what happens in experimental situations, where we cannot know x_i with absolute accuracy but we lump together all the results that fall in the same short interval. Rounding to a chosen number of decimal places has the same effect. For each j the number of the experimental outcomes that fall in the interval A_j is approximately $NP(A_j)$, by the meaning of P. For each of these we recorded $f(\bar{x}_j)$, so the total of the numbers recorded for the experiments with outcome in A_j is approximately $f(\bar{x}_j)NP(A_j)$. The total of the N recorded numbers for all the experiments is the sum of these terms for $j = 1, \ldots, k$. As in the preceding paragraph, we define

$$S(\mathscr{P}; f; P) = \sum_{j=1}^{k} f(\bar{x}_j) P(A_j).$$

Then the average of the numbers recorded is $S(\mathscr{P}; f; P)$. The average of the values of f at the exact outcomes of the experiments is

$$\operatorname{Av} f(x) = [f(x_1) + \cdots + f(x_N)]/N.$$

For each i we recorded not $f(x_i)$, but a value $f(\bar{x}_1)$ of f that differed less than ε from $f(x_i)$. Hence, the average recorded value $S(\mathscr{P}; f; P)$ differs from Av $f(x)$ by less than ε:

$$|S(\mathscr{P}; f; P) - \text{Av}\, f(x)| < \varepsilon.$$

As N increases, if the theory of probability applies at all to our experimental procedure, the average Av $f(x)$ will tend to some number $E(f)$, or at least for large N it will be fairly sure to stay close enough to a number $E(f)$ so that the closeness of approximation is acceptable in the situation to which we are applying the theory. This $E(f)$ is called the *expectation* of f. The foregoing discussion now tells us that for each positive ε there is a gauge γ on $\bar{R}$ such that if $\mathscr{P}$ is γ-fine, $S(\mathscr{P}; f; P)$ differs from $E(f)$ by less than ε for all large N. But these two numbers do not depend on N, so we have shown that $E(f)$ is the gauge-limit of $S(\mathscr{P}; f; P)$. Again, by analogy with the definition of the gauge integral in Chapter I, we define this limit to be the gauge-integral of f with respect to P and we write

$$E(F) = \int_R f(x)\, P(dx).$$

These two examples are motivation enough for the following definition.

DEFINITION 7-1 *Let m be an extended-real-valued function of left-open intervals in R. Let f be a real-valued function defined on a set B contained in R. For each allotted partition*

$$\mathscr{P} = \{(\bar{x}_1, A_1), \ldots, (\bar{x}_k, A_k)\}$$

of R, define

$$S(\mathscr{P}; f_B; m) = \sum_{j=1}^{k} f_B(\bar{x}_j) m A_j,$$

provided that this sum exists. Then f has an ***integral*** *(more specifically, a* ***gauge-integral****) with respect to m over B if there exists a real number J such that the gauge-limit of $S(\mathscr{P}; f_B; m)$ is J. In detail, f has an integral with respect to m over B if there exists a number J such that for each positive ε there exists a gauge γ on $\bar{R}$ such that for every γ-fine partition $\mathscr{P}$ of R the partition-sum $S(\mathscr{P}; f_B; m)$ exists and differs from J by less than ε. Any such J is denoted by the symbol*

$$\int_B f(x)\, m(dx).$$

Since the integral defined in Definition 7-1 obviously includes the one defined in Definition I-2-1, every example of integral that we have met is an example of the gauge-integral of Definition 7-1. But this is not very interesting. In Section 12 we shall meet examples of a different type. But here we exhibit one example to show that the integral defined in Definition 7-1 includes cases that might not be expected.

EXAMPLE 7-2 For each left-open interval A in R, let mA be one-sixth of the number of points of the set $\{1, 2, \ldots, 6\}$ contained in A. Then every function f defined and real-valued on R is integrable over R, and

$$\int_R f(x)\, m(dx) = \frac{f(1) + \cdots + f(6)}{6}. \tag{B}$$

Anyone who has the slightest familiarity with probability theory will recognize the right member of (B) to be the expected value of f when the experiment consists of one roll of a fair die.

Let ε be positive. If x is one of the numbers $1, 2, \ldots, 6$, we define $\gamma(x)$ to be $(x - \frac{1}{2}, x + \frac{1}{2})$. If x is none of these six numbers, one of the open intervals $[-\infty, 1), (1, 2), (2, 3), \ldots, (5, 6), (6, \infty]$ contains x, and we choose that interval for $\gamma(x)$. Then $\gamma(1)$ contains 1, but $\gamma(x)$ does not contain 1 if $x \neq 1$; and likewise for 2, 3, 4, 5, 6.

Let $\mathscr{P} = \{(\bar{x}_1, A_1), \ldots, (\bar{x}_k, A_k)\}$ be any γ-fine partition of R. One of the A_i contains 1, since every number in R is in exactly one of the A_i. We may suppose the numbering of the pairs chosen so that A_1 is the interval that contains 1. Since $\mathscr{P}$ is γ-fine, A_1 is contained in $\gamma(\bar{x}_1)$, so the point 1 of A_1 is in $\gamma(\bar{x}_1)$. But $\gamma(x)$ does not contain 1 if $x \neq 1$, so $\bar{x}_1 = 1$. By repeating the discussion, we show that we can choose the numbering so that for $i = 1, 2, \ldots, 6$, i is in A_i and $\bar{x}_i = i$. Then for these i, A_i contains the single point i of the set $\{1, 2, \ldots, 6\}$, so by definition, $mA_i = \frac{1}{6}$. For $i = 7, \ldots, k$, A_i contains none of the points $1, 2, \ldots, 6$, so $mA_i = 0$. Therefore,

$$\begin{aligned} S(\mathscr{P}; f_R; m) &= f(1)mA_1 + \cdots + f(6)mA_6 + f(\bar{x}_7)mA_7 + \cdots + f(x_k)mA_k \\ &= [f(1) + f(2) + \cdots + f(6)]/6. \end{aligned}$$

The inequality

$$|S(\mathscr{P}; f_R; m) - [f(1) + f(2) + \cdots + f(6)]/6| < \varepsilon$$

holds because the left member is 0, and equation (B) is established.

EXERCISE 7-1 Let x_0 be a point of R. For each interval A in $\bar{R}$, define $mA = 1_A(x_0)$. Show that for every subset B of R and every function f defined and real-valued on B, f is integrable with respect to m over B, and its integral has the value $f_B(x_0)$.

8. Properties of the Integral

Since in Definition 7-1 we assumed nothing about m except that it is extended-real-valued for all left-open intervals in R, we can hardly expect to prove much about it without adding more hypotheses. Nevertheless, we can prove two important theorems without any further assumptions. The first of these is that the integral is unique; there cannot be any function f defined and real-valued on any set B in R such that two different numbers J', J'' both satisfy the requirements in Definition 1-7. The proof is the same as that of Theorem I-4-4 with no change except the replacement of m_L by m. The second is the linearity theorem; if $f_1, \ldots, f_k$ are integrable with respect to m over B, and $c_1, \ldots, c_k$ are real numbers, then $c_1 f_1 + \cdots + c_k f_k$ is integrable with respect to m over B, and

$$\int_B \left\{ \sum_{i=1}^{k} c_i f_i(x) \right\} m(dx) = \sum_{i=1}^{k} c_i \int_B f_i(x)\, m(dx).$$

The proofs of Lemmas I-5-1 and I-5-2 and of Theorem I-5-3 apply to this case with no change other than replacing m_L by m.

From now on we shall confine our attention to interval-functions m that are nonnegative and additive. These terms have the obvious meanings, which we now state.

DEFINITION 8-1 *Let m be a function defined on a set $\mathscr{A}$ of intervals in R, including the empty set. Then m is* ***nonnegative*** *if for every interval A in $\mathscr{A}$, $0 \leqq mA \leqq \infty$, and m is* ***additive*** *(more specifically,* ***finitely additive****) if whenever $A_1, \ldots, A_k$ are pairwise disjoint intervals in the set $\mathscr{A}$ whose union is an interval A in $\mathscr{A}$,*

$$mA = mA_1 + \cdots + mA_k.$$

Nonnegative additive interval-functions have a property like that described in Lemma 1-2 in the Introduction.

★LEMMA 8-2 *Let m be a nonnegative additive function on the set of all left-open intervals in R. Let $A_1, \ldots, A_k, B_1, \ldots, B_h$ be left-open intervals such that the A_i are pairwise disjoint and*

$$B_1 \cup \cdots \cup B_h \supset A_1 \cup \cdots \cup A_k.$$

Then

$$mA_1 + \cdots + mA_k \leqq mB_1 + \cdots + mB_h.$$

In particular, if the A_i are pairwise disjoint and are contained in an interval B_1, then $mA_1 + \cdots + mA_k \leqq mB_1$; and if A_1 is an interval contained in the union $B_1 \cup \cdots \cup B_h$, then $mB_1 + \cdots + mB_h \geqq mA_1$.

As in the proof of Lemma 1-2 in the Introduction, we denote by $c_1, \ldots, c_m$ the end-points of all the intervals A_i and B_j, named in increasing order, and we define

$$C_n = (c_n, c_{n+1}] \qquad (n = 1, \ldots, m - 1).$$

Each A_i is the union of all the C_n contained in it, and by additivity $mA_i = \Sigma\{mC_n : C_n \subset A_i\}$. Hence,

$$\text{(A)} \qquad \sum_{i=1}^{k} mA_i = \sum_{i=1}^{k} \sum \{mC_n : C_n \subset A_i\}.$$

Likewise,

$$\text{(B)} \qquad \sum_{j=1}^{h} mB_j = \sum_{j=1}^{h} \sum \{mC_n : C_n \subset B_j\}.$$

But each C_n that is contained in some A_i is in the union of the B_j, which contains A_i; and if an interior point of C_n is in some B_j, all of C_n is in that B_j, since no end-point of B_j can come between the end-points of C_n. So each term in the right member of (A) appears at least once in the right member of (B), and since all terms are nonnegative, the conclusion is established.

In Section I-4 we explained the notational convention that those theorems, corollaries, etc., whose proofs needed no properties of m_{L} except nonnegativeness and additivity were distinguished by a star (★) and by the use of the abbreviation m for m_{L}. It follows that all these theorems are valid, with unchanged proofs, for all of the nonnegative additive functions of left-open intervals that we are now considering. Henceforth, we shall make use of those theorems whenever convenient.

Even among the nonnegative additive interval-functions, however, there are some with undesirable properties, and we save trouble by excluding them from consideration. The interval-function that assigns $mA = \infty$ to every interval A in R is nonnegative, additive, and useless. If the empty set has measure $\neq 0$, no function except 0 is integrable, so the integration based on this m is worthless. However, our hypotheses will always imply that m is additive and that mA is defined and finite for some interval A. In this case, A, $\varnothing$, and $A \cup \varnothing$ are all intervals and m is defined on them, and A and $\varnothing$ have no points in common, so by additivity $m(A \cup \varnothing) = mA + m\varnothing$. This implies that $m\varnothing = 0$. In fact, we lose little generality and gain some convenience if we assume that mA is finite whenever A is a bounded left-open interval. This still leaves some undesirables. For example, for each left-open interval A in R let us define $mA = 1$ if the right end-point of A is ∞ and $mA = 0$ otherwise. This m is additive, nonnegative and not identically 0, but still for every function f real-valued on R we have

$$\text{(C)} \qquad \int_R f(x)\, m(dx) = 0.$$

To prove this we define $\gamma(x) = (x - 1, x + 1)$ if x is finite, and $\gamma(-\infty) = \gamma(\infty) = \bar{R}$. If $\mathscr{P} = \{(x_1, A_1), \ldots, (x_k, A_k)\}$ is any γ-fine partition of R, for each term $f_R(x_i)mA_i$ in $S(\mathscr{P}; f_R; m)$ either x_i is finite, so that A_i is contained in $(x_i - 1, x_i + 1)$ and $mA_i = 0$, or else x_i is $-\infty$ or ∞ and $f_R(x_i) = 0$. So $S(\mathscr{P}; f_R; m) = 0$, and (C) is satisfied. We prefer to exclude such undesirable interval-functions from consideration. We obtain a clue as to how to do this by imagining that m is the distribution of the mass of a collection of molecules spread along the x-axis, none of them at ∞ or at $-\infty$. Let A be a nonempty left-open interval in R with end-points a and b. We can choose points $a_1 > a_2 > a_3 > \cdots$ in A with a_n tending to a; and we choose $b_n = b$ if b is finite and $b_n = a_1 + n$ if $b = \infty$. The intervals $(a_n, b_n]$ are bounded, and their closures are contained in A. Every molecule of the distribution that is in A is also in $(a_n, b_n]$ for all large n, so it is reasonable to assume that $m(a_n, b_n]$ tends to mA. Likewise, if b is finite, we define $b_n = b + 1/n$ and $G_n = (a, b_n]$ for all n; if b is ∞ we define $G_n = (a, \infty) = A$ for all n. In either case G_n is a left-open interval whose interior contains A, and every molecule of the distribution that is outside A is also outside G_n for all large n. So it is reasonable to assume that mG_n tends to mA as n increases.

Intervals with this sort of limit property will be called *regular*. But we shall use this property only when the interval-functions are nonnegative and additive, which lets us simplify the definition somewhat.

★DEFINITION 8-3 *A function m is a* ***regular nonnegative additive interval-function*** *if*

(i) *it is defined and satisfies $0 \leqq mA \leqq \infty$ for all left-open intervals in R, and $mA < \infty$ for all bounded left-open intervals A;*

(ii) *it is finitely additive;*

(iii) *for each bounded left-open interval A in R and each number $c > mA$, there exists a left-open interval G with $A \subset G^{\mathrm{O}}$ and $mG < c$;*

(iv) *for each left-open interval A in R and each number $c < mA$, there exists a bounded left-open interval F with closure $F^{-} \subset A$ and $mF > c$.*

This property has a simple but useful consequence.

COROLLARY 8-4 *Let m be a regular nonnegative additive interval-function on R. Let A be a left-open interval in R. Then 1_A has an integral with respect to m over R, and*

$$\int_R 1_A(x)\, m(dx) = mA.$$

When A is bounded, the proof of Lemma I-3-2 applies with no change except to replace the words "Lemma 1-4 in the Introduction" with "Definition 8-3." When A is unbounded, for each positive integer n we define f_n to be the indicator of the interval $A \cap (-n, n]$. By the part of the proof already completed, f_n is integrable, and

$$\text{(D)} \qquad \int_R f_n(x)\, m(dx) = m(A \cap (-n, n]).$$

As n increases, the functions $f_1, f_2, f_3, \ldots$ form an ascending sequence tending to 1_A at every point of R. By Corollary II-4-4, the left member of (D) tends to the integral of 1_A over R. For every positive integer n, the right member of (D) is at most equal to mA, since $A \cap (-n, n] \subset A$. But if c is any number less than mA, there is a bounded left-open interval F with closure $F^- \subset A$ and $mF > c$. For all large n, $F^- \subset A \cap (-n, n]$, so $m(A \cap (-n, n]) > c$. Therefore, the right member of (D) tends to mA, and the proof is complete.

Theorem I-5-6, which states the formula for the integral of a step-function, does not generalize as stated because our interval-function is not necessarily defined and well-behaved for intervals that are not left-open. But Theorem I-5-6 can be replaced by an adequate substitute, as follows.

Theorem 8-5 *Let m be a regular nonnegative additive function of left-open intervals in R. Let f be a step-function on R that has values $c_1, \ldots, c_h$ on the respective pairwise disjoint left-open intervals $B_1, \ldots, B_h$ and is 0 outside the union of the B_j. Then f is integrable with respect to m over R, and*

$$\int_R f(x)\, m(dx) = c_1 mB_1 + \cdots + c_h mB_h.$$

This has the same simple proof as Theorem I-5-6, with the reference to Lemma I-3-2 replaced by a reference to Corollary 8-4.

EXERCISE 8-1 Let m be a regular nonnegative additive function of intervals in R, and let f be bounded and continuous on a bounded interval B in R. Prove that f is integrable with respect to m over B. *Suggestion*: See Exercise II-1-2.

EXERCISE 8-2 Let m be an interval-function defined for all left-open intervals in $\bar{R}$ and such that if A_1 and A_2 are disjoint left-open intervals in $\bar{R}$ whose union is an interval A, $mA = mA_1 + mA_2$. Prove that m is finitely additive. *Suggestion*: If $A_1, \ldots, A_n$ are pairwise disjoint left-open intervals whose union is an interval, we may suppose the A_i numbered from left to right. Then $A_1 \cup \cdots \cup A_{n-1}$ is a left-open interval. Use induction.

EXERCISE 8-3 Show that the function m defined in Exercise 7-1 is regular, nonnegative, and additive.

EXERCISE 8-4 Let c_1 and c_2 be real numbers, and let m_1 and m_2 be functions of intervals such that for every left-open interval A in $\bar{R}$ the numbers $m_1 A$ and

m_2A and the sum $c_1m_1A + c_2m_2A$ are defined and extended-real-valued. Show that if f is a real-valued function that is integrable with respect to m_1 and with respect to m_2 over a subset B of R, then f is integrable with respect to $m = c_1m_1 + c_2m_2$ over B, and

$$\int_B f(x)\,m(dx) = c_1 \int_B f(x)\,m_1(dx) + c_2 \int_B f(x)\,m_2(dx).$$

Use this and Exercise 7-1 to obtain Example 7-2 again.

9. Densities

In Section 7 we mentioned densities and also a formula (equation (A) of Section 7) involving them. Here we shall first state exactly what we mean by density and then prove that the generalization

$$\text{(A)} \qquad \int_R f(x)\,m(dx) = \int_R f(x)\rho(x)\,dx$$

of (A) of Section 7 holds under very weak assumptions. This equation is of considerable importance. Its left member is an integral of the form that we have been studying in this chapter, and we have seen that it has a theory almost as extensive as the integral with respect to m_L that we investigated in earlier chapters. But for it there is no fundamental theorem, and even computer techniques are not convenient. On the other hand, the right member of (A) is an integral for which the whole theory of Chapters I and II is available. If $f\rho$ has an antiderivative, we can use it to compute the right member of (A). Even if we have to resort to computer methods, the right member is the kind of integral for which standard computer programs are available.

If m is a nonnegative additive function of left-open intervals in R and we interpret mA to mean the amount of mass contained in interval A, it is in accord with the usage of elementary physics to say that the **mean density** in A of the mass-distribution is the ratio of the mass mA in A to the length m_LA of interval A. If $\bar{x}$ is any point of R, the **pointwise density** of m at $\bar{x}$ is the limit of the mean density in A as the closure A^- of A shrinks down on $\bar{x}$, provided that this limit exists. That is, the statement that $\rho(\bar{x})$ is the pointwise density of m at $\bar{x}$ means that for each positive ε there is a neighborhood $\gamma(\bar{x})$ of $\bar{x}$ such that whenever A is a nondegenerate interval such that $\bar{x}$ is in the closure A^- and $A \subset \gamma(\bar{x})$,

$$\text{(B)} \qquad |mA/m_LA - \rho(\bar{x})| < \varepsilon.$$

This closely resembles the definition of derivative. In fact, if m is a mass-distribution in a bounded interval $[a, b]$ and $(x', x'']$ is a subinterval of $[a, b]$, we have

$$\text{(C)} \qquad m(x', x''] = m(a, x''] - m(a, x'].$$

By statement (D) in the proof of Theorem I-9-2, if m has a pointwise density $\rho(\bar{x})$ at a point $\bar{x}$ of $[a, b]$, the function $x \mapsto m(a, x]$ has derivative $\rho(\bar{x})$ at $\bar{x}$. If m has a density ρ continuous on $[a, b]$, by the fundamental theorem I-10-2, with (C), the equation

$$\text{(D)} \qquad \int_A \rho(x)\, dx = mA$$

holds for every subinterval A of $[a, b]$. In fact, if m has a pointwise density $\rho(\bar{x})$ at each point $\bar{x}$ of $[a, b]$ and ρ is integrable over $[a, b]$, equation (D) holds by Theorem I-10-3. But we often wish to discuss mass-distributions that at some points fail to have a pointwise density. We could still prove that (D) holds if ρ is the pointwise density of m at almost all points of $[a, b]$ and m satisfies some other hypotheses. But all such proofs have the annoying feature of requiring assumptions about m and ρ that have nothing to do with the uses we wish to make of densities. We extricate ourselves from this difficulty just as we did in the case of differential equations; we use equation (D) to *define* density.

DEFINITION 9-1 *Let m be a function defined for all left-open intervals in R. Then a function ρ on R is a* **density** *for m if for every left-open interval A in R, ρ has an integral (finite or infinite) over A, and*

$$mA = \int_A \rho(x)\, dx.$$

As in the case of differential equations, this allows us to use the powerful theorems about integrals to draw conclusions about interval-functions and their densities. In fact, changing to the definition based on the integral is even more important here than it was in the study of differential equations. For, in physics we often meet mass-distributions in spaces of more than one dimension, and for such distributions we can use Definition 9-1 practically unchanged. Since the theory of the integral extends to higher dimensions with little change, the results we are about to obtain in this section will extend to spaces of higher dimensions. But the theories of differentiation and of pointwise densities are much harder to manage in such spaces.

We commented that it is difficult to start with the existence almost everywhere of a pointwise density and to conclude that (D) holds so that a density in the sense of Definition 9-1 exists. For those students who have worked through Section II-14, it is easy to prove the converse; if m has a density in the sense of Definition 9-1, it has a pointwise density almost everywhere. This is merely statement (A) in the proof of Theorem II-14-1, with changed notation.

The density defined in Definition 9-1 is not unique. If ρ is a density for m, then a function ρ_1 is a density for m if and only if it is almost everywhere equal to ρ. For if ρ_1 is almost everywhere equal to ρ, then for every left-open interval A in R the integral of ρ_1 over A is equal to the integral of ρ over A, by Theorem II-12-2, and this is equal to mA because ρ is a density. Conversely, if ρ and ρ_1 are both densities for m, then for every left-open interval A in R they have the same integral over A, so $\rho - \rho_1$ has integral 0 over every bounded left-open interval. By Theorem II-13-4, $\rho - \rho_1$ is almost everywhere equal to 0.

This ambiguity is of little importance. If an interval-function m has a density ρ, we can almost invariably use this ρ for all our needs, untroubled by the fact that m has other densities too.

Not all interval-functions have densities; the interval-function in Example 7-2 has no density. For our purposes, the most important interval-functions are the nonnegative ones, which correspond to distributions of masses and to probabilities. So nonnegative densities are especially important. It can be shown that if m is additive, nonnegative, and regular, then if it has any density, it has a nonnegative density. The proof is easy for anyone who has read Section II-14; without the developments in that section, it is quite difficult. We leave it as an exercise (Exercise 9-1). We can get along very well without it, especially since if it is possible to construct a density for a nonnegative interval-function, the construction will usually yield a nonnegative density at once. But the converse is of some interest.

THEOREM 9-2 *If m is an interval-function defined for all left-open intervals in R, and m has a nonnegative density ρ, then m is additive, nonnegative, and regular.*

From Definition 9-1, it is evident, by Theorem I-6-1, that m is nonnegative and, by Theorem I-8-1, that it is additive. To prove it regular, let A be any nonempty left-open interval in R with end-points a and b, and let c be any number less than mA. We choose points $a_1 > a_2 > a_3 > \cdots$ in A with limit a, and we choose $b_n = b$ if b is finite and $b_n = a_1 + n$ if $b = \infty$. Every point of A is in $(a_n, b_n]$ for all large n, so if we define

$$f_n(x) = \rho(x) \qquad (x \text{ in } (a_n, b_n]),$$
$$f_n(x) = 0 \qquad (x \text{ in } \bar{R} \setminus (a_n, b_n]),$$

we see that the f_n form an ascending sequence tending to ρ_A at all points of R. By the monotone convergence theorem, the integrals of the f_n tend to the integral over R of ρ_A. By Definition 9-1, this implies that $m(a_n, b_n]$ tends to mA. Therefore, we can choose an n for which $m(a_n, b_n] > c$; and the interval $(a_n, b_n]$ is a bounded left-open interval whose closure is contained in A. Likewise, if we define intervals G_n as in the discussion preceding Definition 8-3, we find that for every $c > mA$ there exists a left-open interval G_n such that $A \subset G_n^0$ and $mG_n < c$. So m is regular.

We can now prove the important computational formula (A) under two different sets of hypotheses. The second proof, in Theorem 9-4, has complete generality and is simply stated and easily remembered. But this proof uses the statements in the last sections of Chapter II, and the reader may have chosen to postpone the reading of those sections. For such readers we state and prove a theorem that has stronger hypotheses and yet is still general enough to cover many uses. Readers who have worked through the last sections of Chapter II should go directly to Theorem 9-4.

THEOREM 9-3 *Let m be an additive function of left-open intervals in R that has a nonnegative density ρ, and let f be a function on R that is the limit of a sequence of step-functions $s_1, s_2, s_3, \ldots$, each constant on left-open intervals. Then f is integrable with respect to m over R if and only if $f\rho$ is integrable with respect to m_{L} over R, and in that case*

$$\text{(E)} \qquad \int_R f(x)\, m(dx) = \int_R f(x)\rho(x)\, dx.$$

Suppose first that f is a step-function that has the nonzero values $c_1, \ldots, c_k$ on the respective bounded left-open intervals $A(1), \ldots, A(k)$. Then by Theorem 8-5 and Definition 9-1,

$$\begin{aligned}\int_R f(x)\, m(dx) &= \sum_{j=1}^{k} c_j m A(j) = \sum_{j=1}^{k} c_j \int_{A(j)} \rho(x)\, dx \\ &= \sum_{j=1}^{k} \int_R c_j 1_{A(j)}(x)\rho(x)\, dx = \int_R f(x)\rho(x)\, dx,\end{aligned}$$

so (E) holds for step-functions with left-open intervals of constancy.

Suppose next that f is nonnegative and satisfies the hypotheses of the theorem. For each positive integer k we define

$$\begin{aligned} f_k &= (f \wedge k) \cdot 1_{(-k,k]} = [(f \wedge k) \cdot 1_{(-k,k]}] \vee 0, \\ s_{n,k} &= [(s_n \wedge k) \cdot 1_{(-k,k]}] \vee 0. \end{aligned}$$

We shall prove

$$\text{(F)} \qquad \lim_{n \to \infty} s_{n,k}(x) = f_k(x) \qquad (x \text{ in } R).$$

If x is not in $(-k, k]$, this is trivial, since both members are 0. If x is in $(-k, k]$, then $f_k(x) = [f(x) \wedge k] \vee 0$ and $s_{n,k}(x) = [s(x) \wedge k] \vee 0$. By Lemma II-3-2,

$$|[f(x) \wedge k] \vee 0 - [s_n(x) \wedge k] \vee 0| \leqq |f(x) - s_n(x)|,$$

and the right member tends to 0 as n increases, so again (F) is valid.

By its definition, $s_{n,k}$ is a step-function with left-open intervals of constancy, and

$$\text{(G)} \qquad 0 \leqq s_{n,k} \leqq k 1_{(-k,k]},$$

and the last function is integrable with respect to m over R. By the dominated convergence theorem, with (F),

(H) $$\lim_{n\to\infty}\int_R s_{n,k}(x)\,m(dx) = \int_R f_k(x)\,m(dx).$$

Also, by (G),

$$0 \leqq s_{n,k}\rho \leqq k\rho 1_{(-k,k]},$$

and the last function is integrable with respect to m_L over R. Therefore,

(I) $$\lim_{n\to\infty}\int_R s_{n,k}(x)\rho(x)\,dx = \int_R f_k(x)\rho(x)\,dx.$$

Since (E) has been proved valid for step-functions with left-open intervals of constancy, the left members of (H) and (I) are equal, so the right members are equal, and for every k

(J) $$\int_R f_k(x)\,m(dx) = \int_R f_k(x)\rho(x)\,dx.$$

Both $f \wedge k$ and $1_{(-k,k]}$ are nonnegative and ascend as k increases, so the same is true of their product f_k. For each x in R, whenever k is greater than the greater of $|x|$ and $f(x)$, $f_k(x) = f(x)$. So the f_k are an ascending sequence with limit f everywhere in R. If either member of (E) is finite, both members of (J) have finite limits as k increases, and by the monotone convergence theorem, both integrals in (E) exist and they have the same value, so (E) is valid.

If f satisfies the hypotheses of the theorem and is integrable over R with respect to m, then f^+ and f^- are nonnegative and integrable with respect to m over R, and they both satisfy the hypotheses of the theorem. By the preceding part of the proof, both $f^+\rho$ and $f^-\rho$ are integrable with respect to m_L over R, and (E) holds for both of them. By subtraction, $f\rho = f^+\rho - f^-\rho$ is integrable with respect to m_L over R, and (E) is valid. Conversely, if $f\rho$ is integrable with respect to m_L over R, so are $(f\rho)^+$ and $(f\rho)^-$. But these are the same as $f^+\rho$ and $f^-\rho$, respectively. By the preceding part of the proof, f^+ and f^- are integrable with respect to m over R. Therefore, so is their difference f. This completes the proof.

We now proceed to the second proof of formula (A), under weaker hypotheses.

★THEOREM 9-4 *Let m be an extended-real-valued function of left-open intervals in R such that mA is finite whenever A is bounded. Let m have a nonnegative density ρ, and let f be defined and real-valued on R. Then f is integrable with respect to m over R if and only if $f\rho$ is integrable with respect to m_L over R, and in that case*

(K) $$\int_R f(x)\,m(dx) = \int_R f(x)\rho(x)\,dx.$$

We first prove this under the additional assumptions

(L) f is nonnegative;

(M) f has a finite upper bound $M - 1$ on R and vanishes outside a bounded interval B.

Let B be the interval $(a, b]$, and define C to be the interval $(a - 2, b + 2]$. Let ε be positive. Since the integral of ρ with respect to m_L over C is the finite number mC, ρ is integrable over C, and therefore ρ_C is integrable with respect to m_L over R. So there exists a gauge γ on $\bar{R}$ such that for every γ-fine partition $\mathscr{P}$ of R

$$\text{(N)} \qquad \left| S(\mathscr{P}; \rho_C; m_L) - \int_R \rho_C(x)\,dx \right| < \frac{\varepsilon}{5M}.$$

We may suppose that

$$\text{(O)} \qquad \gamma(x) \subset (x - 1, x + 1) \qquad (\text{all } x \text{ in } R);$$

otherwise we would merely have to replace $\gamma(x)$ by $\gamma(x) \cap (x - 1, x + 1)$. Let

$$\mathscr{P} = \{(x_1, A_1), \ldots, (x_k, A_k)\}$$

be any γ-fine partition of R. By Corollary II-13-2 and the definition of density,

$$\begin{aligned} \frac{\varepsilon}{M} &> \sum_{i=1}^{k} \left| \rho_C(x_i) m_L A_i - \int_{A_i} \rho_C(x)\,dx \right| \\ &= \sum_{i=1}^{k} |\rho_C(x_i) m_L A_i - m(A_i \cap C)|. \end{aligned}$$

We multiply the ith term in the last sum by $f_R(x_i)$, which is nonnegative and at most equal to $M - 1$, and we obtain

$$\text{(P)} \qquad \begin{aligned} \varepsilon &> \sum_{i=1}^{k} |f_R(x_i)\rho_C(x_i) m_L A_i - f_R(x_i) m(A_i \cap C)| \\ &\geqq \left| \sum_{i=1}^{k} f_R(x_i)\rho_C(x_i) m_L A_i - \sum_{i=1}^{k} f_R(x_i) m(A_i \cap C) \right|. \end{aligned}$$

For all i such that x_i is not in B, by hypothesis $f_R(x_i) = 0$, and therefore the equations

$$\text{(Q)} \qquad f_R(x_i)\rho_C(x_i) m_L A_i = f_R(x_i)\rho_R(x_i) m_L A_i,$$

$$\text{(R)} \qquad f_R(x_i) m(A_i \cap C) = f_R(x_i) m A_i$$

are satisfied. For all i such that x_i is in B, x_i is in C, and $\gamma(x_i)$ is contained in C, and A_i is contained in $\gamma(x_i)$. Thus, equations (Q) and (R) are satisfied in this case also. So (P) implies

$$\left| \sum_{i=1}^{k} f_R(x_i)\rho_R(x_i) m_L A_i - \sum_{i=1}^{k} f_R(x_i) m A_i \right| < \varepsilon,$$

or

$$|S(\mathscr{P}; (f\rho)_R; m_L) - S(\mathscr{P}; f_R; m)| < \varepsilon.$$

So if either of these two partition-sums has a limit, the other has the same limit, which is the conclusion of the theorem.

This is the mainspring of the proof. The rest consists merely of removing the extra hypotheses (L) and (M) by routine procedures.

Suppose first that the hypotheses of the theorem are satisfied, and also (L), but not necessarily (M). For each positive integer n we define

$$\text{(S)} \qquad f_n(x) = f(x) \wedge [n1_{(-n,n]}(x)].$$

If f is integrable with respect to m over R, we also know by Corollary 8-4 that $n1_{(-n,n]}$ is integrable with respect to m over R, so by (S) and Corollary II-3-5, f_n is integrable with respect to m over R. With f_n in place of f, the hypotheses of the theorem are satisfied, and also (L) and (M), so $f_n\rho$ is integrable with respect to m_L over R, and

$$\text{(T)} \qquad \int_R f_n(x)\rho(x)\,dx = \int_R f_n(x)\,m(dx) \qquad (n = 1, 2, 3, \ldots).$$

As n increases, f_n ascends, and by the monotone convergence theorem, the right member of (T) converges to the left member of (K). Again by the monotone convergence theorem, the limit $f\rho$ of the integrand in the left member of (T) is integrable with respect to m_L, and the limit of the left member of (T) is the right member of (K). So (K) is satisfied. If, on the other hand, $f\rho$ is integrable with respect to m_L over R, we also know that ρ is integrable with respect to m_L over $(-n, n]$, so $n1_{(-n,n]}\rho$ is integrable with respect to m_L over R, and by Corollary II-3-5, the function

$$f_n\rho = [f\rho] \wedge [n1_{(-n,n]}\rho]$$

is also m_L-integrable over R. By the part of the proof already completed, f_n is integrable with respect to m over R, and (T) is satisfied. The same use of the monotone convergence theorem shows that f is integrable with respect to m over R, so the theorem holds with the added hypothesis (L).

Finally, if f is integrable with respect to m, so are f^+ and f^-. These are nonnegative, so by the part of the proof already completed, $f^+\rho$ and $f^-\rho$ are integrable with respect to m_L over R, and (K) holds with f^+ in place of f and also with f^- in place of f. Then the difference $f \cdot \rho$ is integrable with respect to m_L, and by subtraction, (K) is satisfied. Conversely, if $f \cdot \rho$ is integrable with respect to m_L, so are $[f \cdot \rho]^+$ and $[f \cdot \rho]^-$. But since ρ is nonnegative,

$$[f \cdot \rho]^+ = f^+\rho, \qquad [f \cdot \rho]^- = f^-\rho.$$

Applying the part of the proof already completed to f^+ and f^-, we see that f^+ and f^- are integrable with respect to m. As before, (K) is satisfied.

EXERCISE 9-1 Let m be a nonnegative additive interval-function that has a density. Prove that m then has a nonnegative density. *Suggestion*: Let ρ be a density for m. For every positive integer n the function

$$x \mapsto m(-n, x] = \int_{-n}^{x} \rho(u)\, du$$

is nondecreasing, so wherever it has a derivative, that derivative is $\geqq 0$. By Theorem II-14-1, it has a derivative and the derivative is equal to $\rho(x)$ at all points of $(-n, \infty]$ except those in a set E_n with $m_L E_n = 0$. So $\rho(x)$ is nonnegative at all points of $(-n, \infty] \setminus E_n$, and therefore at all points of $R \setminus [\bigcup_{n=1}^{\infty} E_n]$. Then ρ^+ is a nonnegative density for m.

10. Measurable Functions and Measurable Sets

We introduce the subject matter of this section at this point for use in the applications to probability theory. The reader who chooses to postpone those applications can also postpone this section. Except in applications to probability theory, nothing in this section is referred to before Chapter VI, and even there only the first part of this section (through Theorem 10-5) is needed.

In order that a function f be integrable, it must not be too complicated and must not be too large. Often the proof that f is integrable is in two parts: in one it is shown to be not too large, and in the other it is shown to be not too complicated. The requirement that f be "not too large" is easily stated; if there is an integrable function g such that $|f| \leqq g$, f cannot be too large to be integrated. The requirement that it be "not too complicated" can be stated in several ways. One of these is suggested by the definition of measurable sets and by the dominated convergence theorem. A function is "not too complicated" if it is the limit of a sequence of integrable functions, even though it may not itself be integrable. We formalize this in a definition.

DEFINITION 10-1 *Let m be a nonnegative additive function of left-open intervals in $\bar{R}$, let B be a set in R, and let f be an extended-real-valued function on B. Then f is m-measurable on B if there exists a sequence $f_1, f_2, f_3, \ldots$ of extended-real-valued functions integrable with respect to m over B such that $f_n(x)$ converges to $f(x)$ for every x in B.*

It would make little difference if we were to consider only the case $B = R$. For f_n is integrable over B if and only if $f_{n,B}$ (equal to f_n on B and to 0 on the rest of $\bar{R}$) is integrable over R, and f_n converges to f on B if and only if $f_{n,B}$ converges to f_B on R.

If Definition 10-1 is adequate as a formalization of the idea of "not too complicated," every m-measurable function that is not too large should be integrable. Lemma 10-3 and Corollary 10-4 will show that this is indeed the case. But it is convenient to prepare for their proofs by establishing a simple but useful statement.

LEMMA 10-2 *Let $u_1, u_2, u_3, \ldots, v_1, v_2, v_3, \ldots, w_1, w_2, w_3, \ldots$ be sequences of numbers in $\bar{R}$ with the respective limits u, v, and w. Then*

(i) $\lim_{n \to \infty} u_n \vee v_n = u \vee v$;
(ii) $\lim_{n \to \infty} u_n \wedge v_n = u \wedge v$;
(iii) $\lim_{n \to \infty} (u_n \wedge v_n) \vee w_n = (u \wedge v) \vee w$.

For (i) and (ii) we may suppose that $u \geqq v$; otherwise we interchange u_n with v_n and u with v. If $u > v$, we choose a real number c between them. Then $(c, \infty]$ is a neighborhood of u and $[-\infty, c)$ is a neighborhood of v. Since u_n tends to u and v_n to v, for all large n it is true that u_n is in the neighborhood $(c, \infty]$ of u and v_n in the neighborhood $[-\infty, c)$ of v. So $v_n < u_n$, and $u_n \vee v_n = u_n$, and $u_n \wedge v_n = v_n$. Therefore, in this case

$$\lim_{n \to \infty} u_n \vee v_n = \lim_{n \to \infty} u_n = u = u \vee v,$$

$$\lim_{n \to \infty} u_n \wedge v_n = \lim_{n \to \infty} v_n = v = u \wedge v,$$

and (i) and (ii) are satisfied. If $u = v$, let $\gamma(u)$ be any neighborhood of u. Then both u_n and v_n tend to u, so for all large n they are both in $\gamma(u)$. Since $u_n \vee v_n$ is one of u_n, v_n, it is in $\gamma(u)$. Therefore $u_n \vee v_n$ tends to u, which is $u \vee v$, and (i) holds. Likewise (ii) holds in this case. Now (i) and (ii) are established in all cases. By (ii), $u_n \wedge v_n$ tends to $u \wedge v$, and by (i), $(u_n \wedge v_n) \vee w_n$ tends to $(u \wedge v) \vee w$, which establishes (iii).

LEMMA 10-3 *Let m be a regular nonnegative additive function of left-open intervals in $\bar{R}$. Let f be an extended-real-valued function on R. Then f is m-measurable on R if and only if for every pair of extended-real-valued functions g, h that are integrable over R and satisfy $g(x) \leqq h(x)$ for all x in R, the function $(f \wedge h) \vee g$ is integrable over R.*

Suppose that f is m-measurable on R. Then there exists a sequence $f_1, f_2, f_3, \ldots$ of functions integrable with respect to m over R and converging to f. For each n all three functions f_n, g, h are integrable with respect to m over R, so by Corollary II-3-5 the same is true of $(f_n \wedge h) \vee g$.

Since

$$g \leqq (f_n \wedge h) \vee g \leqq h,$$

we have

$$|(f_n \wedge h) \vee g| \leqq |g| \vee |h|,$$

and the last function is integrable. By Lemma 10-2, $(f_n \wedge h) \vee g$ converges everywhere to $(f \wedge h) \vee g$, so we can apply the dominated convergence theorem to conclude that $(f \wedge h) \vee g$ is integrable with respect to m over R.

Conversely, suppose that $(f \wedge h) \vee g$ is integrable with respect to m over R whenever g and h are integrable and $g \leqq h$. For each positive integer n we define

$$h_n = n1_{(-n,n]}.$$

This is integrable with respect to m over R, by Theorem 8-5, so the function

$$f_n = (f \wedge h_n) \vee (-h_n)$$

is integrable by hypothesis. At every point x in R, $h_n(x)$ tends to ∞ as n increases, so by Lemma 10-2

$$\lim_{n \to \infty} f_n(x) = (f(x) \wedge \infty) \vee (-\infty) = f(x).$$

Now f is everywhere the limit of the functions f_n, which are integrable with respect to m over R, so it is m-measurable.

COROLLARY 10-4 *Let m be a regular nonnegative additive function of left-open intervals in $\bar{R}$, and let f be an extended-real-valued function that is m-measurable on R. If there exists a function h integrable with respect to m over R such that*

$$|f(x)| \leqq h(x) \qquad (x \text{ in } R),$$

then f is integrable with respect to m over R.

If we define $g = -h$, both g and h are integrable with respect to m over R, and $g \leqq h$. By Lemma 10-3, $(f \wedge h) \vee g$ is integrable with respect to m over R. But $(f \wedge h) \vee g$ is identical with f.

Measurable functions can be manipulated even more freely than integrable functions, as the next theorem shows.

THEOREM 10-5 *Let m be a regular nonnegative additive function of left-open intervals in R, and let c be a number in $\bar{R}$. Then*

(i) *if f and g are extended-real-valued functions that are m-measurable on R, so are cf, f^+, f^-, $|f|$, $f \vee g$, $f \wedge g$, and fg; and so is $f + g$, if we interpret $f(x) + g(x)$ to mean 0 at all points x at which one of $f(x)$, $g(x)$ is ∞ and the other is $-\infty$;*

(ii) *if $f_1, f_2, f_3, \ldots$ is a sequence of m-measurable functions on R that converges everywhere in R to a limit f, then f is m-measurable on R.*

We can and do choose a sequence of finite real numbers $c_1, c_2, c_3, \ldots$ converging to c; and for each positive integer n we define h_n as before,

$$h_n = n1_{(-n,n]}. \tag{A}$$

This is a nonnegative step-function on R, and by Theorem 8-5 it is integrable over R. For each positive integer n we define

$$f_n = (f \wedge h_n) \vee (-h_n), \qquad g_n = (g \wedge h_n) \vee (-h_n).$$

These are bounded, and by Lemma 10-3, they are integrable with respect to m over R. As n increases, $h_n(x)$ tends to ∞ at each x in R, so by Lemma 10-2

(B) $$\lim_{n\to\infty} f_n(x) = (f(x) \wedge \infty) \vee (-\infty) = f(x),$$

and likewise,

(C) $$\lim_{n\to\infty} g_n(x) = g(x).$$

Hence, by Lemma 10-2, the functions

(D) $$c_n f_n, f_n^+, f_n^-, |f_n|, f_n \vee g_n, f_n \wedge g_n$$

converge to

(E) $$cf, f^+, f^-, |f|, f \vee g, f \wedge g,$$

respectively. Each of the functions in (D) is integrable, by Theorem I-5-3 and Corollaries II-3-4 and II-3-5, so by Definition 10-1 the functions in (E) are all m-measurable on R. At each point x in R at which neither $f(x)$ nor $g(x)$ is 0, by (B) and (C) we see that $f_n(x)g_n(x)$ tends to $f(x)g(x)$. If at x one of them, say $f(x)$, is 0, then $f_n(x) = 0$ for all n. Then $f_n(x)g_n(x)$ tends to 0, which is the value of $f(x)g(x)$. So $f_n g_n$ tends everywhere to fg, and fg is m-measurable. At each x at which $f(x)$ and $g(x)$ are not infinite of opposite signs, by (B) and (C) we see that $f_n(x) + g_n(x)$ tends to $f(x) + g(x)$. If one of them is ∞ and the other is $-\infty$, then one of $f_n(x)$, $g_n(x)$ is n and the other is $-n$. So

$$f_n(x) + g_n(x) = 0,$$

and this tends to 0, which we are accepting as the meaning of $f(x) + g(x)$ in this case. So $f_n + g_n$ tends everywhere to $f + g$, and therefore $f + g$ is m-measurable on R.

If $f_1, f_2, f_3, \ldots$ are all m-measurable and tend to f, by Lemma 10-3 the functions

$$g_n = (f_n \wedge h_n) \vee (-h_n)$$

are all integrable with respect to m over R. Since $h_n(x)$ tends everywhere to ∞, by Lemma 10-2,

$$\lim_{n\to\infty} g_n(x) = (f(x) \wedge \infty) \vee (-\infty) = f(x)$$

for all x in R. Therefore, f is m-measurable on R. The proof is complete.

Definition 10-1 strongly resembles the definition that f have an integral, finite or infinite. The next theorem shows that there is in fact a close relationship.

Theorem 10-6 *Let m be a regular nonnegative additive function of left-open intervals in R. Let B be a subset of R and f an extended-real-valued function on R. Then*

(i) *if f has an integral with respect to m over R, it is m-measurable on R;*
(ii) *f is m-measurable on R if and only if f^+ and f^- both have integrals over R;*
(iii) *B is m-measurable if and only if its indicator 1_B is an m-measurable function on R;*
(iv) *R is measurable.*

Comparison of Definitions II-4-3 and 10-1 makes (i) obvious. If f^+ and f^- have integrals, they are m-measurable by (i), so their difference $f = f^+ - f^-$ is m-measurable by Theorem 10-5. Conversely, if f is m-measurable, so are f^+ and f^- by Theorem 10-5. We again define h_n by (A). This is nonnegative and integrable, so by Lemma 10-3, the functions

$$g_n = f^+ \wedge h_n$$

are integrable with respect to m over R. As n increases, these ascend and tend everywhere to f^+, by Lemma 10-2, so by Definition II-4-3, f^+ has an integral with respect to m over R. Similarly, f^- has an integral, and (ii) is established.

If B is m-measurable, 1_B has an integral by Definition II-12-1, so 1_B is m-measurable by (i). If 1_B is m-measurable, $(1_B)^+$ (which is 1_B) has an integral over R by (ii). This establishes (iii).

By Corollary 8-4, for each positive integer n the function

$$f_n = 1_{(-n,n]}$$

is integrable with respect to m over R, so by (i) it is m-measurable on R. As n increases, $f_n(x)$ tends to $1_R(x)$ at all points x of R, so by Theorem 10-5, 1_R is an m-measurable function. By (ii), R is an m-measurable set.

The main reason that m-measurable sets are convenient to work with is that unions of countably many of them are again m-measurable. This property is important enough to be given a name.

Definition 10-7 *Let X be a nonempty set and $\mathscr{A}$ a family of subsets of X. Then $\mathscr{A}$ is a **σ-algebra** (of subsets of X) if it has the three properties*

(i) *the empty set $\varnothing$ belongs to $\mathscr{A}$;*
(ii) *whenever a set A belongs to $\mathscr{A}$, so does its complement $X \setminus A$;*
(iii) *whenever $A(1), A(2), A(3), \ldots$ is a sequence of sets all belonging to $\mathscr{A}$, the union of the $A(j)$ also belongs to $\mathscr{A}$.*

Theorem 10-8 *Let m be a regular nonnegative additive function of left-open intervals in $\bar{R}$. Then the family of all m-measurable subsets of R is a σ-algebra.*

Let $\mathscr{A}$ denote the family of m-measurable subsets of R. The indicator of the empty set $\varnothing$ is the function identically 0 on R, which is integrable with respect to m. So by Theorem 10-6 $\varnothing$ is m-measurable, and (i) in Definition 10-7 is satisfied. Let $A(1), A(2), A(3), \ldots$ be a sequence of sets belonging to $\mathscr{A}$. By Theorem 10-6, each indicator $1_{A(j)}$ is m-measurable, so for each positive integer n the function

$$1_{A(1)} \vee \cdots \vee 1_{A(n)}$$

is m-measurable by Theorem 10-5, and so is

$$\lim_{n \to \infty} [1_{A(1)} \vee \cdots \vee 1_{A(n)}].$$

This last is the indicator of the union of all the $A(j)$, so that union is m-measurable by Theorem 10-6. So (iii) in Definition 10-7 is satisfied. The set R is itself m-measurable by Theorem 10-6. If A is m-measurable, both 1_A and 1_R are m-measurable functions by Theorem 10-6, so $1_R - 1_A$ is m-measurable by Theorem 10-5. Since this is the indicator of $R \setminus A$, $R \setminus A$ is m-measurable by Theorem 10-6. So (ii) of Definition 10-7 is satisfied, and the proof is complete.

The function m of intervals that we started with had the property of finite additivity on the family of left-open intervals in $\bar{R}$. The measure m on the σ-algebra of m-measurable sets has a much stronger kind of additivity, which we now define.

DEFINITION 10-9 *If $\mathscr{A}$ is a collection of sets and F is an extended-real-valued function on $\mathscr{A}$, F is* ***countably additive*** *if, whenever $\mathscr{K}$ is a finite or denumerably infinite collection of pairwise disjoint sets belonging to $\mathscr{A}$ and the union U of the sets in $\mathscr{K}$ also belongs to $\mathscr{A}$, $F(U)$ is the sum of the $F(A)$ for all sets A in $\mathscr{K}$.*

THEOREM 10-10 *Let m be a nonnegative additive function of left-open intervals in $\bar{R}$. Then the measure m is a countably additive function on the family of m-measurable sets.*

Let K be a finite or denumerably infinite collection of pairwise disjoint m-measurable subsets of R. If K is finite, let the members of K be $A(1), \ldots, A(k)$, and let $U(k)$ be their union. Then

$$\text{(F)} \qquad 1_{U(k)} = 1_{A(1)} + \cdots + 1_{A(k)},$$

and the relation

$$\text{(G)} \qquad mU(k) = mA(1) + \cdots + mA(k)$$

follows by integration. If K is denumerably infinite, let its members be $A(1), A(2), A(3), \ldots$; let U be the union of all of them, and let $U(k)$ be the union $A(1) \cup \cdots \cup A(k)$. Again (F), and (G) hold. The set U is m-measurable by Theorem 10-8. If for any k the set $U(k)$ has infinite measure, the measure mU of the set U that contains $U(k)$ must also be infinite. If all $U(k)$ have finite measure, the indicators $1_{U(k)}$ are all integrable with respect to m over R, and they ascend as k increases and tend everywhere to 1_U. By the monotone convergence theorem II-12-7, 1_U has an integral over R, and

$$\int_R 1_U(x)\,m(dx) = \lim_{k\to\infty} \int_R 1_{U(k)}(x)\,m(dx) = \lim_{k\to\infty} mU(k)$$

$$= \lim_{k\to\infty} [mA(1) + \cdots + mA(k)] = \sum_{j=1}^{\infty} mA(j).$$

That is, in this case also the measure mU is the sum of the measures of the $A(j)$, and the proof is complete.

COROLLARY 10-11 *Let m be a regular nonnegative additive function of left-open intervals in R. Then:*

(i) *if $A(1), A(2), A(3), \ldots$ is a sequence of m-measurable sets such that $A(1) \subset A(2) \subset A(3) \subset \cdots$, then*

$$m \bigcup_{j=1}^{\infty} A(j) = \lim_{n\to\infty} mA(n);$$

(ii) *if $A(1), A(2), A(3), \ldots$ is a sequence of m-measurable sets such that $A(1) \supset A(2) \supset A(3) \supset \cdots$, and some $A(j)$ has finite measure, then*

$$m \bigcap_{j=1}^{\infty} A(j) = \lim_{n\to\infty} mA(n);$$

(iii) *if $A(1), A(2), \ldots$ is a finite or denumerably infinite collection of m-measurable sets, then*

$$m \bigcup_{j} A(j) \leqq \sum mA(j).$$

For (i), let U be the union of the $A(j)$. This is m-measurable by Theorem 10-8. If any $A(j)$ has $mA(j) = \infty$, then for the set U that contains $A(j)$ we must also have $mU = \infty$, and in this case (i) holds. If all $mA(j)$ are finite, we observe that the indicator $1_{A(j)}$ tends everywhere to the indicator 1_U. By the monotone convergence theorem II-12-7,

$$\int_R 1_U(x)\,m(dx) = \lim_{j\to\infty} \int_R 1_{A(j)}(x)\,m(dx) = \lim_{j\to\infty} mA(j).$$

The left member is mU, so the proof of (i) is complete.

Let the $A(j)$ be as in (ii). If some $A(j)$, say $A(j^*)$, has finite m-measure, we note that the sets $A(j)$ with $j > j^*$ must also have finite measure. Let V denote the intersection of all the $A(j)$. Since $1_{A(j)}$ tends everywhere to 1_V, by applying the monotone convergence theorem II-12-7 to $1_{A(j^*)}, 1_{A(j^*+1)}, 1_{A(j^*+2)}, \ldots$, we have

$$\int_R 1_V(x)\, m(dx) = \lim_{j\to\infty} \int_R 1_{A(j)}(x)\, m(dx) = \lim_{j\to\infty} mA(j),$$

and (ii) is established.

For (iii), if the set of $A(j)$ stops with $A(k)$, and $U(k)$ denotes the union $A(1) \cup \cdots \cup A(k)$,

$$1_{U(k)} = 1_{A(1)} \vee \cdots \vee 1_{A(k)} \leqq 1_{A(1)} + \cdots + 1_{A(k)}.$$

By integration, we obtain

$$\text{(H)} \qquad mU(k) \leqq mA(1) + \cdots + mA(k).$$

If the collection of $A(j)$ contains infinitely many sets, we observe that as k increases, the unions $U(k)$ expand and the union of all $U(k)$ is the same as the union of the $A(j)$. By (i),

$$m \bigcup_{j=1}^{\infty} U(j) = \lim_{n\to\infty} mU(n).$$

So by (H),

$$m \bigcup_{j=1}^{\infty} A(j) \leqq \lim_{n\to\infty} [mA(1) + \cdots + mA(n)] = \sum_{j=1}^{\infty} mA(j).$$

The proof is complete.

The next theorem states a relationship between m-measurable functions and m-measurable sets that is often used as the definition of m-measurability of functions. We shall need it in the applications to probability theory.

THEOREM 10-12 *Let m be a regular nonnegative additive function of left-open intervals in R. Let f be an extended-real-valued function that is m-measurable on R. Then for every number y in $\bar{R}$ the set $\{f \leqq y\}$ is m-measurable.*

Suppose, first, that y is in R. The set R is m-measurable, so its indicator (which is constantly 1) is m-measurable. By Theorem 10-5, so are the functions that are respectively constantly equal to y and constantly equal to $y + 1$, and by the same theorem, so is the function

$$g = y + 1 - (f \wedge [y + 1]) \vee y.$$

Since

$$(f \wedge [y + 1]) \vee y \leqq y + 1,$$

g is nonnegative. For each x such that $f(x) \leqq y$

$$f(x) \wedge [y + 1] = f(x) \qquad \text{and} \qquad (f(x) \wedge [y + 1]) \vee y = y,$$

so

(I) at all x such that $f(x) \leqq y$, $g(x) = 1$.

If $f(x) > y$, then

$$f(x) \wedge [y+1] > y \qquad \text{and} \qquad (f(x) \wedge [y+1]) \vee y > y,$$

so

(J) $0 \leqq g(x) < 1$ at all x such that $f(x) > y$.

By (I) and (J), at all x at which $f(x) \leqq y$, the powers $g(x)^n$ $(n = 1, 2, 3, \ldots)$ are all 1, and where $f(x) > y$, those powers converge to 0. That is,

$$\lim_{n \to \infty} g(x)^n = 1_{\{f \leqq y\}}.$$

The powers g^n are all m-measurable by Theorem 10-5, and so is their limit $1_{\{f \leqq y\}}$. By Theorem 10-6, if y is finite, the set $\{f \leqq y\}$ is m-measurable.

If $y = \infty$, the set $\{f \leqq y\}$ is R, which is m-measurable. If $y = -\infty$, the set $\{f \leqq y\}$ is the intersection of the sets $\{f \leqq -n\}$ for $n = 1, 2, 3, \ldots$. Since these sets have just been shown to be m-measurable, the complements $R \setminus \{f \leqq -n\}$ are m-measurable and the union of the complements is m-measurable. This union is the complement of $\{f \leqq -\infty\}$, so the set $\{f \leqq -\infty\}$ is m-measurable. The proof is complete.

In all the following exercises we understand that m is a regular nonnegative additive function of left-open intervals in R.

EXERCISE 10-1 Prove that a subset B of R is m-measurable if and only if $B \cap C$ has finite m-measure whenever C has finite m-measure.

EXERCISE 10-2 Prove that whenever C is an m-measurable subset of R, the m-measurable subsets of C form a σ-algebra.

EXERCISE 10-3 Show that conclusion (ii) of Corollary 10-11 is no longer valid if we omit the hypothesis that some $A(j)$ has finite measure. (Let m be m_{L} and $A(j) = \{x \text{ in } R : x > j\}$.)

EXERCISE 10-4 Let $W(1)$, $W(2)$, $W(3), \ldots$ be a sequence of sets of finite m-measure whose union contains R. Define

$$p(x) = \sum_{j=1}^{\infty} \frac{1_{W(j)}(x)}{2^j(1 + mW(j))}.$$

Prove that p is integrable with respect to m over R, and that $0 < p(x) \leqq 1$ for all x in R.

11. Applications to Probability Theory

We shall not attempt to discuss probability theory from the very beginning. Instead, we shall assume that the reader has at least some familiarity with the elements of probability theory, has worked through at least a few easy problems, and has used the expressions most commonly met in studying probability.

There are many situations in which an experiment or trial is repeatedly performed, and some coding scheme is used to record the outcome of the experiment by means of a single real number. Sometimes this number is of interest in its own right. For example, in rolling a die the number recorded might be the number that appears on the top of the die, which is of importance in games of chance. Or, the experiment might consist of choosing a particular acre in a wheat field and measuring the weight of the grain grown on it. In other cases the number recorded is merely an identifying number. For example, if a coin is tossed, it is usual to record a "1" if the coin falls "heads" and a "0" if the coin falls "tails." Or, if the experiment involves choosing a person from a population, the record might consist of a number that identifies the person chosen (such as his or her social security number). If the experiment has only countably many possible outcomes, we can number the possible outcomes with some or all of the integers and record for each trial the number that identifies the outcome.

This recording by a single real number can be carried far, but it is better not to carry it as far as possible. For example, if we roll two dice, die A and die B, it is natural to record the outcome as the pair of numbers (x, y) in which x is the number on die A and y is the number on die B. This can also be recorded as the single real number $6x + y$, so that the outcome "3 on die A, 2 on die B" is recorded as "20." The integers $7, 8, \ldots, 42$ serve to record all possible outcomes, but at the cost of losing the pictorial vividness of the representation by points (x, y) in the plane. When the experiment consists of measuring several quantities, each having a real-number value (for example, the components of a vector in three-dimensional space), the natural and visualizable representation is by a triple of real numbers (x, y, z). The mental picture is of a point in three-dimensional space. By an ingenious artifice, these triplets can each be represented by a single real number, but in doing so we lose all the pictorially graspable qualities of the three-dimensional representation. So, for such situations it will be profitable to wait until later, until we have advanced our theory of integration enough to be free of the restriction to one dimension. This we shall do in the next chapter.

Meanwhile, the problems in one-dimensional space are not at all trivial. In undergraduate courses in probability theory, the case of trials with countably many possible outcomes is often handled quite satisfactorily. But for distributions with a density, the teacher faces a difficulty. Ordinarily, the students entering the class have learned no integration theory beyond the Riemann theory, and the Riemann integral is inadequate for a satisfactory treatment of probability distributions with densities. We shall now show that the gauge-integral is adequate, and in fact it provides a unified theory that covers distributions with densities, distributions with countably many possibilities, and others that most students will never find out that they can handle.

As always in constructing mathematical models of experimental situations, we replace the measured quantities, which always have some possibility or even certainty of error, by precise mathematical quantities which are supposed to be acceptably close to them. This done, we carry out our reasoning on the inherently precise mathematical quantities that we have brought in as substitutes for the results of experiment. We shall do this here too; the "chance" that the result of an experiment will be in an interval A will be replaced by a precisely determined number $P(A)$, and we shall perform our mathematics with this precisely determined substitute $P(A)$ for the experimental quantity.

The fact that the $P(A)$ is to simulate the results of many trials puts certain requirements on it. Let $A_1, \ldots, A_k$ be left-open intervals, and let a large number N of trials be made, of which N_i give results that lie in interval A_i. Then N_i/N is close to $P(A_i)$, provided that N is large. This requires immediately that $P(A)$ be nonnegative. Also, if A_1 is empty, the number N_1 of results occurring in A_1 is 0, so $P(A_1) = 0$. If A_1 is all of R, the number of experiments with outcomes in A_1 is the same as the total number N of experiments, so $N_1/N = 1$ and $P(A_1) = 1$. If $A_1, \ldots, A_k$ are pairwise disjoint and their union is an interval A_0, the number N_0 of outcomes in A_0 is the sum of the numbers $N_1, \ldots, N_k$ of outcomes in the intervals $A_1, \ldots, A_k$, so $N_0/N = N_1/N + \cdots + N_k/N$. This indicates that we must have $P(A_0) = P(A_1) + \cdots + P(A_k)$, so P has to be additive. Finally, the discussion just before Definition 8-3 can be repeated here, using the outcomes of individual trials in place of the individual molecules in that discussion, with the result that we see that P should be assumed to be regular. We sum this up in the following definition.

★DEFINITION 11-1 *A **probability distribution** (in R) is a regular nonnegative additive function defined on the class of all left-open intervals in R and having the property that* $P(R) = 1$.

We have not specified that $P(\varnothing) = 0$, because that is a consequence of the additivity of P;

$$1 = P(R) = P(R \cup \varnothing) = P(R) + P(\varnothing) = 1 + P(\varnothing),$$

so $P(\varnothing) = 0$.

It often happens that the quantity having greatest interest is not the number that records the outcome of the experiment but some function of that number. If the experiment is the measuring of the amount of electric energy used by a person in a given month, the finance office of the electric company will be more interested in a function of that amount, namely, the amount that the person is to be charged for using that much electricity. When the recorded outcome of the experiment is merely an identifying number, it will surely be the case that some function of that number is what has more importance. The average of the social security numbers of the students in a given university is uninteresting. But from those numbers, with a file of recorded data, we can determine the average College Entrance Board Examination grade of the students in the university, which for some purposes can be an interesting number. In Section 1 we saw that if for each trial we record a function $f(x)$ of the outcome x of the trial, when the number of trials is large we may reasonably expect that the average of the recorded values of f will be closely approximated by

$$\text{(A)} \qquad Ef = \int_R f(x)\, P(dx),$$

if this integral exists and is finite. This suggests the following definition.

★DEFINITION 11-2 *Let P be a probability distribution in R. A function f defined and real-valued on R has* ***finite expectation*** *if it is integrable with respect to P over R. In that case the* ***expectation*** *Ef of f is defined to be the number defined by* (A).

Functions that are P-measurable play a large role in probability theory. In that theory such functions are called **random variables**. This name does not fit in well with modern mathematical usage, but it has been used so long and so often that we have no choice but to bow to custom and use it. For such functions we have already established many statements in the preceding section. In particular, by Theorem 10-6, every function with finite expectation is a random variable.

If B is a P-measurable subset of R, by Theorem 10-6 its indicator 1_B is a P-measurable function, and $0 \leqq 1_B \leqq 1$. Since both 0 and 1 are integrable with respect to P over R, by Lemma 10-3 1_B is integrable. Thus, for probability distributions a subset of R is P-measurable if and only if it has finite P-measure.

If B is a P-measurable subset of R, ordinary English usage allows us to speak of "the event that the outcome of the trial is a number in B." But in probability theory it is customary to depart from standard usage and transfer the name "event" from the occurrence to the set B. Thus, instead of speaking of "the event that the outcome is in B," we shall speak of "the event B."

★DEFINITION 11-3 *Let P be a probability distribution on R. A set B is an* ***event*** *if it is P-measurable.*

If we perform an experiment with probability distribution P, and B is an event, we record a "1" whenever the outcome x is in B and a "0" every time that it is not. That is, we record $1_B(x)$. The average value of this recorded number, taken for a large number N of trials, should be the mathematical model of the chance that the outcome is in B, which from that point of view could be thought of as the probability $P(B)$. But in accordance with our definition of $m(B)$ for measurable sets B, we have

$$P(B) = \int_R 1_B(x)\, P(dx) = E(1_B).$$

So the mathematical meaning we have given to $P(B)$ accords with what applications to chance events suggest should be the meaning of $P(B)$.

By Theorems 10-8 and 10-10, the family of events is a σ-algebra of subsets of R, and P is countably additive on that family. This is today the foundation of the mathematical theory of probability. By Theorem 10-12, if f is a random variable, the set $\{f \leqq y\}$ is an event whenever y is in $\bar{R}$.

The concepts of *moment* and of *absolute moment* are often useful in applications of probability theory.

★DEFINITION 11-4 *If f is a random variable and q is a positive integer, the qth* ***moment*** *of f is defined to be $E(f^q)$ if that expectation exists, and the qth* ***absolute moment*** *of f is defined to be $E(|f|^q)$. In particular, the qth* ***moment of the distribution*** *is defined to be the qth moment of the identity function $X: x \mapsto X(x) = x$, x in R, and the qth* ***absolute moment of the distribution*** *is the qth absolute moment of the identity function X.*

★COROLLARY 11-5 *Let f be a random variable, c a real number, and p and q integers such that $1 \leqq p \leqq q$. Then if f has a finite qth moment, $f + c$ has a finite pth moment.*

If n is a positive integer, f^n is a random variable, by Theorem 10-5. This is also true of f^0 if we understand $f(x)^0$ to mean 1 even at points x where $f(x) = 0$. If n is an integer and $0 \leqq n \leqq q$, the inequality

$$|f(x)|^n \leqq 1 + |f(x)|^q$$

holds for all x in R. For if $|f(x)| \leqq 1$, then $|f(x)|^n \leqq 1$, and if $|f(x)| > 1$, then $|f(x)|^n < |f(x)|^q$. By Corollary 10-4, f^n is integrable with respect to P over R for $n = 0, 1, \ldots, q$. By the binomial theorem,

$$[f + c]^p = \sum_{j=0}^{p} \binom{p}{j} f^{p-j} c^j.$$

Each term in the right member is integrable with respect to P over R, so $[f + c]^p$ is also integrable with respect to P over R for $1 \leqq p \leqq q$.

Two numbers useful in estimating the chance that f is far from its expectation are the *variance* and the *standard deviation*.

★DEFINITION 11-6 *If f is a random variable with finite expectation, its* ***variance*** *is defined to be $E([f - Ef]^2)$ and it is denoted by any one of the symbols* Var f

or σ^2 *or* σ_f^2. *The* ***standard deviation*** *of* f *is defined to be*

$$\sigma = \sigma_f = [E([f - Ef]^2)]^{1/2}.$$

In particular, if f is the identity function $X: x \mapsto x$ on R, the expectation of f is called the **mean** of the distribution, provided that it exists; and in that case the variance and the standard deviation of the identity function are called the **variance of the distribution** and the **standard deviation of the distribution**, respectively.

★COROLLARY 11-7 *If f is a random variable with finite second moment, f has finite expectation and finite variance, and*

$$\sigma_f^2 = E(f^2) - (Ef)^2.$$

By Corollary 11-5, if Ef^2 is finite, so is Ef. In the right member of the identity

$$\text{(B)} \qquad [f(x) - Ef]^2 = f(x)^2 - 2E(f)f(x) + (Ef)^2$$

all three terms are integrable with respect to P over R, so the left member has a finite integral, which by definition is the variance σ_f^2. Also, by integration in (B),

$$E([f - Ef]^2) = E(f^2) - 2E(f)E(f) + (Ef)^2,$$

which completes the proof.

If we know the variance of a random variable, we can make a rather crude but often useful estimate of the probability that the random variable differs from its expectation by as much as a given number c. This is the content of the next theorem, in which we prove **Chebyshev's inequality.**

★THEOREM 11-8 *If f is a random variable with finite second moment and c is a positive number, then*

$$P(|f| \geqq c) \leqq E(f^2)/c^2.$$

Let B be the set $\{|f| \geqq c\}$. This is the same as $\{-f^2 \leqq -c^2\}$, and since f^2 is a random variable, by Theorem 10-12, this is an event. For every x in R we have

$$f(x)^2 \geqq 1_B(x)c^2;$$

for if x is in B, the right member is c^2 and the left member is at least equal to c^2 by definition of B, and if x is not in B, the right member is 0. By integration,

$$E(f^2) \geqq E(1_B) \cdot c^2.$$

Since $E(1_B) = P(B) = P\{|f| \geqq c\}$, this completes the proof.

As an immediate consequence, if f has finite variance, then

$$\text{(C)} \qquad P(|f - Ef| \geqq c) \leqq \sigma_f^2/c^2.$$

For if we define $g = f - Ef$, $E(g^2) = \sigma_f^2$, and (C) is obtained by applying Chebyshev's inequality to g.

This applies to all distributions and therefore it is not nearly as good an estimate as can be made for some specific distributions. For example, without knowing anything more about the distribution than that it has a finite variance, by the above inequality we know that the probability that f is at least $3\sigma_f$ away from Ef is at most 1/9. But if we also know that the distribution is normal, we obtain the much better estimate that the probability that f is at least $3\sigma_f$ away from its expectation is about 0.0026, as can be found by use of a table of values of the normal distribution.

Probability distributions that have densities are of great importance both in theory and in applications. If a distribution P has a density, it is customary to denote the density by p rather than, as in Section 9, by ρ. Then from Theorem 9-3 we deduce the following.

★THEOREM 11-9 *Let P be a probability distribution on $\bar{R}$ that has a (non-negative) density p, and let f be defined and real-valued on R. Then f has finite expectation if and only if fp is integrable with respect to m_L over R, and in that case*

$$Ef = \int_R f(x)p(x)\,dx.$$

EXERCISE 11-1 Prove that if f is a random variable with finite variance, the following statements are true:

(i) $\operatorname{Var} X = \operatorname{Var} X^+ + \operatorname{Var} X^- + 2E(X^+)(EX^-)$;
(ii) $\operatorname{Var}|X| = \operatorname{Var} X^+ + \operatorname{Var} X^- - 2(EX^+)(EX^-)$;
(iii) $\operatorname{Var}|X| \leqq \operatorname{Var} X^+ + \operatorname{Var} X^- \leqq \operatorname{Var} X$;
(iv) $E(X^2) = E([X^+]^2) + E(X^-]^2)$.

EXERCISE 11-2 Let f have only the two values 5 and -5, each with probability 0.5. Show that with $c = 5$, equality holds in the Chebyshev inequality. (This shows that no estimate for $P\{f \geqq c\}$ better than the Chebyshev estimate can be correct for all random variables.)

12. Examples

At the end of Section 7 we presented an example that showed that the study of the probabilities associated with one roll of a fair die is a special case of our theory. We now show that a similar discussion applies to every experiment with only finitely many possible outcomes. Suppose that there are n possible outcomes. We label these with n different real numbers $x_1, \ldots, x_n$. These numbers may themselves be the numbers obtained by the experiment. In any case, we can name them "outcome 1, outcome 2, ..., outcome n" and thus label them with the numbers $1, \ldots, n$. However, as we mentioned in Section 7, this device may lose us some of the direct connection between outcomes and numbers representing them, so though it is always possible, it is not always the best thing to do. We shall denote the probability of outcome x_i by p_i. Then the p_i are nonnegative and their sum is 1. The probability that the outcome is in an interval A is the sum of those p_i for which x_i is in A, since for the outcome to be in A it must be one of those x_i for which x_i is in A. Then we can write

$$P(A) = \sum_{i=1}^{n} p_i 1_A(x_i).$$

It is obvious that this function of intervals is nonnegative and additive. We shall now prove a statement stronger than the statement that P is regular. The proof will make no use of the fact that the sum of the p_i is 1.

(A) If A is a left-open interval in R, there exists a bounded left-open interval F whose closure is contained in A and a left-open interval G whose interior contains A for which

$$P(F) = P(G) = P(A).$$

We may suppose that the x_i are numbered in increasing order. Let A be $\{x \text{ in } R : a < x \leqq b\}$, where a and b are in $\bar{R}$, and let $x_h, \ldots, x_k$ be the points x_i that are in A. Then x_h is not the left end-point of A, and we can find a point a' of A below x_h. Define $F = (a', x_k]$. Then F is bounded, and its closure is contained in A, and it contains x_i if $h \leqq i \leqq k$ and not otherwise, so $P(F) = P(A)$. If x_k is the greatest one x_n of the x_i, we take $b'' = \infty$; if $k < n$, we take b'' to be a number between b and x_{k+1}. We define G to be $(a, b'']$. Then G is left-open, and its interior contains A, and x_i belongs to G if $h \leqq i \leqq k$ and not otherwise, so $P(G) = P(A)$. Statement (A) is proved.

For this example we can show that every real-valued function f on R has a finite expectation, and

(B) $$Ef = f(x_1)p_1 + \cdots + f(x_n)p_n.$$

We omit the details because they are almost the same as those in the example at the end of Section 7, and besides that, (B) will be a special case of the next example.

In particular, if f is the indicator of a subset B of R, f is integrable with respect to P, and so B is an event, and

$$P(B) = E(1_B) = \sum \{p_i : x_i \text{ in } B\}.$$

Of course, if we are content to stop our study of probability with finite distributions, this whole chapter is a waste of time. We can use equation (A) to define the expectation of f, and we can carry through the whole study with nothing more than finite arithmetic operations. A similar remark applies to the next example too, in which the theory of infinite series would be enough without any need of integrals. But the theory presented in this chapter applies to finite distributions, to countable distributions, to distributions having densities, and to others that do not come into any of these categories, and so it presents a unified theory of random variables without any need of considering various special types one by one.

For our second example, we consider an experiment with countably many possible outcomes. These may be numbers obtained by whatever measurement constitutes the experiment. Or they could be merely labels; we could always label the outcomes with the positive integers, as we did in the finite case. We assume that outcome x_i has probability p_i. As in the finite case, the p_i are nonnegative and their sum is 1. For each interval A in R we define $P(A)$ to be the sum of the numbers p_i for all x_i (possibly infinitely many) for which x_i is in A. Thus, again,

$$P(A) = \sum_{i=1}^{\infty} 1_A(x_i)p_i. \tag{C}$$

This is obviously nonnegative. If $A(1), \ldots, A(k)$ are pairwise disjoint left-open intervals whose union is an interval A, then for each positive integer i we have

$$1_A(x_i) = 1_{A(1)}(x_i) + \cdots + 1_{A(k)}(x_i),$$

whence on multiplying by p_i and adding we find

$$P(A) = P(A(1)) + \cdots + P(A(k)),$$

proving that P is additive. Let A be any left-open interval in $\bar{R}$. To show that P is regular, we must show that for each positive number ε there exist left-open intervals F, G in R such that F is bounded, and its closure is contained in A, and A is contained in the interior of G, and

$$P(F) > P(A) - \varepsilon, \qquad P(G) < P(A) + \varepsilon. \tag{D}$$

We can and do choose an integer n such that

$$\sum_{j=n+1}^{\infty} p_j < \varepsilon. \tag{E}$$

Define, for each left-open interval A in $\bar{R}$,

$$mA = \sum_{j=1}^{n} 1_A(x_j)p_j; \tag{F}$$

that is, mA is the sum of p_j for all those j in $\{1, \ldots, n\}$ for which x_j is in A. Since the proof of (A) did not require that the sum of the p_i be 1, (A) applies to m, and by (A) there exist two left-open intervals F, G such that F is bounded, and

$F^{-} \subset A$ and $A \subset G^{\circ}$, and

(G) $$mF = mG = mA.$$

For every interval B

$$0 \leqq P(B) - mB = \sum_{j=n+1}^{\infty} 1_B(x_j)p_j \leqq \sum_{j=n+1}^{\infty} p_j < \varepsilon.$$

This and (G) imply that

$$P(G) < mG + \varepsilon = mA + \varepsilon \leqq P(A) + \varepsilon,$$
$$P(F) \geqq mF = mA > P(A) - \varepsilon.$$

So (D) is satisfied, and P is regular.

For each positive integer i the interval $(x_i - 1/n, x_i]$ $(i = 1, 2, 3, \ldots)$ contains x_i, so

$$P(x_i - 1/n, x_i] \geqq p_i.$$

The intersection of these intervals is the set $\{x_i\}$ consisting of the point x_i alone, so by Corollary 10-11,

(H) $$P(\{x_i\}) = \lim_{n \to \infty} P(x_i - 1/n, x_i] \geqq p_i.$$

Let X be the set of all x_i,

(I) $$X = \{x_1, x_2, x_3, \ldots\}.$$

By adding inequalities (H) member by member for all i and recalling the countable additivity of P (Theorem 10-10), we obtain

(J) $$P(X) \geqq \sum_{i=1}^{\infty} p_i = 1.$$

But $X \subset R$, and $P(R) = 1$, so $P(X) \leqq 1$. This and (J) imply

(K) $$P(X) = 1.$$

Therefore,

(L) $$P(R \setminus X) = 1 - 1 = 0.$$

Moreover, equality must hold in (H) for each positive integer i; otherwise the left member of (J) would be greater than 1, which it is not. So

(M) $$P\{x_i\} = p_i \qquad (i = 1, 2, 3, \ldots).$$

Every set B in R is the union of the set $B \cap (R \setminus X)$, which has P-measure 0 by (L), and the countable set of points x_i that are in B. By the countable additivity of P, with (M),

(N) $$P(B) = \sum \{p_j : x_j \text{ in } B\}.$$

We shall now prove

(O) If f is a function real-valued on R, f has finite expectation if and only if the series

(P) $$f(x_1)p_1 + f(x_2)p_2 + f(x_3)p_3 + \cdots$$

is absolutely convergent, and in that case

$$Ef = \sum_{j=1}^{\infty} f(x_j)p_j.$$

To simplify notation we let g_i denote the indicator of the set $\{x_i\}$ consisting of x_i alone, so that $g_i(x) = 1$ if $x = x_i$, and $g_i(x) = 0$ if $x \neq x_i$; and we define

$$X(n) = \{x_1, \ldots, x_n\}.$$

If f is real-valued on R, we have

(Q) $$f(x)g_i(x) = f(x_i)g_i(x) \qquad (x \text{ in } R);$$

for if $x = x_i$, this is obvious, and if $x \neq x_i$, both members are 0. The sum of $g_1, \ldots, g_n$ is the indicator of $X(n)$, so by (Q),

(R) $$\int_R f(x)1_{X(n)}(x)\,P(dx) = \sum_{i=1}^{n} \int_R f(x_i)g_i(x)\,P(dx)$$
$$= \sum_{i=1}^{n} f(x_i)P(\{x_i\}) = \sum_{i=1}^{n} f(x_i)p_i.$$

As n increases, the integrand in the left member of (R) tends everywhere to $f(x)1_X(x)$, and it is ascending if $f \geqq 0$. If the series (P) is absolutely convergent, we replace f by $|f|$ in (R). Then as n increases, the right member of (R) (with $|f|$ in place of f) remains bounded. By the monotone convergence theorem, $|f|1_X$ is integrable over R. Then, as n increases, the left member of (R) tends to the integral of $f \cdot 1_X$ over R, and the right member tends to the sum of series (P). Therefore, if series (P) is absolutely convergent, $f \cdot 1_X$ has finite expectation, and

(S) $$E(f \cdot 1_X) = \sum_{i=1}^{\infty} f(x_i)p_i.$$

Conversely, if $f \cdot 1_X$ has finite expectation, so has its absolute value $|f|1_X$. For every positive integer n, by (R),

$$\sum_{i=1}^{n} |f(x_i)p_i| = \int_R |f(x)|1_{X(n)}(x)\,P(dx) \leqq \int_R |f(x)|1_X(x)\,P(dx),$$

so the series (P) is absolutely convergent.

We have now shown that $f \cdot 1_X$ has finite expectation if and only if series (P) is absolutely convergent, and in that case $E(f \cdot 1_X)$ is the sum of the series (P).

But since $R \setminus X$ has P-measure 0, either both the functions f and $f \cdot 1_X$ have finite expectations or neither has, and if they have finite expectations, their expectations are equal. So statement (O) is established.

There is no point in exhibiting numerical examples of finite and countable distributions. All elementary textbooks of probability theory present many such examples.

Let us next look at some distributions that have densities. The first that we shall consider is the "uniform distribution in an interval (a, b)." For this, the probability measure $P(A)$ of an interval contained in (a, b) is proportional to its length; if A is disjoint from (a, b), $P(A) = 0$. For $A \subset (a, b)$ we must have $P(A) = cm_L A$, where c is the proportionality constant. In particular, if A is (a, b) itself, we have $P(a, b) = c(b - a)$. But $P(a, b) = P(R) - P((-\infty, a]) - P([b, \infty))$ $= 1 - 0 - 0$, so $c(b - a) = P(a, b) = 1$ and $c = (b - a)^{-1}$. This distribution has density

$$p = (b - a)^{-1} 1_{(a,b)}.$$

A function f on R has finite expectation if and only if fp is integrable over R by Theorem 11-9. That is, f has finite expectation if and only if it is integrable over (a, b), and in that case

$$\text{(T)} \qquad Ef = (b - a)^{-1} \int_a^b f(x)\, dx.$$

The first moment of the distribution is the expectation of the identity function X, defined by $X(x) = x$ for all x in R. The higher moments of the distribution are those of X, and likewise the variance and standard deviation of the distribution. For the first moment, or expectation, we set $f(x) = x$ in (T) and obtain

$$E(X) = (b - a)^{-1} \int_a^b x\, dx = \frac{b + a}{2}.$$

The second moment is

$$E(X^2) = (b - a)^{-1} \int_a^b x^2\, dx = \frac{1}{3(b - a)}[b^3 - a^3] = \frac{a^2 + ab + b^2}{3}.$$

By Corollary 11-7, the variance of X is

$$\sigma_X^2 = (a^2 + ab + b^2)/3 - [(b + a)/2]^2 = (b - a)^2/12,$$

so its standard deviation is $(b - a)/\sqrt{12}$.

Our next example is the normal distribution. The function $x \mapsto \exp(-x^2/2)$ is easily seen to be integrable over R (cf. Lemma II-11-2). As in Section II-11, we shall accept in advance the computation that the value of the integral is $(2\pi)^{1/2}$. This will be proved in Chapter IV; anyone who is too cautious to accept this in advance should replace $(2\pi)^{1/2}$ by C in the rest of this section and then replace C by $(2\pi)^{1/2}$ after reading the proof in Chapter IV. From this we have already deduced in Section II-11 that for each positive number σ

(U) $$\frac{1}{\sigma(2\pi)^{1/2}}\int_R \exp\left(-\frac{y^2}{2\sigma^2}\right)dy = 1.$$

From this, by substituting $z = y + m$, we obtain

(V) $$\frac{1}{\sigma(2\pi)^{1/2}}\int_R \exp\left(-\frac{[z-m]^2}{2\sigma^2}\right)dz = 1.$$

Consequently, the function p defined by

(W) $$p(z) = [1/\sigma(2\pi)^{1/2}]\exp(-[z-m]^2/2\sigma^2)$$

is a probability density; the corresponding distribution P is defined by setting $P(A)$ equal to the integral of p over A for all intervals A.

The expectation of this distribution is

$$\int_R z\,p(z)\,dz.$$

To compute this we make the substitution $y = z - m$ and obtain

$$\int_R zp(z)\,dz = \left[\frac{1}{\sigma(2\pi)^{1/2}}\right]\int_R y\exp\left(-\frac{y^2}{2\sigma^2}\right)dy + m\left[\frac{1}{\sigma(2\pi)^{1/2}}\right]\int_R \exp\left(-\frac{y^2}{2\sigma^2}\right)dy.$$

In the first integral in the right member, the integrand is an odd function of y, so its integral over every interval $(-b, b)$ is 0 (cf. Exercise I-11-1), and by letting b tend to ∞ we find that the first term in the right member is 0. By (U), the second term is m. So m is the expectation of the distribution P.

The variance of the distribution is the expectation of the function $z \mapsto [z-m]^2$, since m is the expectation. By the substitution $y = z - m$, we obtain

$$\text{variance} = \int_R [z-m]^2 p(z)\,dz = \left[\frac{1}{(2\pi)^{1/2}\sigma}\right]\int_R y^2\exp\left(-\frac{y^2}{2\sigma^2}\right)dy.$$

In Lemma II-11-3 we showed that the value of the integral in this equation is $\sigma^3(2\pi)^{1/2}$, so the variance of the distribution is σ^2. (Of course this explains the choice of the notation.)

When $m = 0$, the even moments of the distribution are given by the integrals

$$\left[\frac{1}{(2\pi)^{1/2}\sigma}\right]\int_R z^{2n}\exp\left(-\frac{z^2}{2\sigma^2}\right)dz.$$

These have been evaluated in Lemma II-11-3; the value of the 2nth moment is

$\sigma^{2n}[1 \cdot 3 \cdots (2n-1)]$. The odd moments are computed similarly, but with z^{2n} replaced by z^{2n+1}. For these, the integrands are all odd, so all the odd moments are 0.

EXERCISE 12-1 Let P be the uniform distribution on an interval (a, b); let n be a positive integer, and let P_n be the distribution that assigns probability $1/n$ to each of the points $a + (b-a)j/n$ $(j = 1, 2, \ldots, n)$. Compute the mean and the variance of P_n and show that as n increases, they tend to the mean and the variance of P, respectively. Prove that for every function f continuous on $[a, b]$

$$\lim_{n\to\infty} \int_R f(x)\, P_n(dx) = \int_R f(x)\, P(dx).$$

EXERCISE 12-2 Let P be a probability distribution that is bounded, so that there is a bounded interval $(a, b]$ with $P(a, b] = 1$. "Discretize" the distribution by choosing finitely many points $x_0 = a < x_1 < \cdots < x_n = b$ and defining P^* to be the distribution that assigns probability $P(x_{j-1}, x_j]$ to the point x_j $(j = 1, \ldots, n)$. Define the "mesh" of P^* to be the greatest of the numbers $x_j - x_{j-1}$, and let E^* be the expectation (integral with respect to P^*) that corresponds to P^*.

(i) Show that if f satisfies a Lipschitz condition of constant L on $[a, b]$,

$$|E(f) - E^*(f)| \leqq L \operatorname{mesh} P^*.$$

(ii) Show that for every function f continuous on $[a, b]$, if $P_1^*, P_2^*, P_3^*, \ldots$ is a sequence of discretizations with mesh P_n^* tending to 0, $E_n^*(f)$ tends to $E(f)$.

(iii) Show by means of an example that we cannot omit the word "continuous" from (ii).

EXERCISE 12-3 (Tail-End of Normal Distribution) Define

$$\varphi(x) = \frac{1}{\sqrt{2\pi}} \exp\left(-\frac{x^2}{2}\right), \qquad \Phi(x) = \int_{-\infty}^{x} \varphi(t)\, dt.$$

Prove the two inequalities (for $x > 0$)

(X) $$1 - \Phi(x) < \varphi(x)/x,$$

(Y) $$1 - \Phi(x) > [\varphi(x)/x][x^2/(1 + x^2)].$$

Suggestion: Since $D\varphi(x) = -x\varphi(x)$, integration by parts yields

(Z) $$1 - \Phi(x) = \frac{\varphi(x)}{x} - \int_x^\infty t^{-2}\varphi(t)\, dt.$$

The integrand is positive, and for t in $(x, \infty]$ it is less than $x^{-2}\varphi(t)$. Use these estimates to establish

$$0.001330 < 1 - \Phi(3) < 0.0014773.$$

Compare this with the estimate given by Chebyshev's inequality.

EXERCISE 12-4 (Better Estimate of Tail of Normal Distribution) Prove that for $x > 3^{1/4}$,

$$1 - \Phi(x) < [\varphi(x)/x](x^4 - x^2)/(x^4 - 3).$$

Show that $1 - \Phi(3) < 0.0013637$. (By tables, $1 - \Phi(3) = 0.00135$.) *Suggestion*: Perform another integration by parts in (Z).

Show that if we continue the estimates by integration by parts, we obtain a sequence of overestimates of $1 - \Phi(x)$ that have this peculiar behavior: first, each has an error that tends to 0 faster than the error in the preceding estimate as x increases; but, second, for each fixed positive x, the error first decreases and then, after several stages, starts increasing.

IV

Integration in Spaces of More Than One Dimension

1. Notation and Definitions

In applications, extension of the theory of the preceding chapters to spaces of more than one dimension is required. Most of these theorems can be extended with little effort. However, since the geometry of higher-dimensional spaces is intrinsically more complicated than that of the line, there are new possibilities to study in such spaces. We shall investigate some of these.

Much of the terminology of one-dimensional space can be adopted without change. The space R^r consists of all ordered r-tuples of points; in other words, R^r is the Cartesian product $R \times \cdots \times R$ (r factors). If we use any symbol to denote a point of R^r, the numbers in the r-tuple will be designated by superscripts $(1), \ldots, (r)$ attached to the same symbol. Thus, the point z of R^r is the r-tuple $(z^{(1)}, \ldots, z^{(r)})$. Similarly, $\bar{R}^r$ is the Cartesian product of r factors $\bar{R}$, and a point y of $\bar{R}^r$ will be designated by the r-tuple $(y^{(1)}, \ldots, y^{(r)})$ of points of $\bar{R}$.

An interval in $\bar{R}^r$ is a set that is the Cartesian product of r intervals in R. We shall use a notational convention like that for points: if the interval is denoted by A, it is the product of intervals $A^{(1)} \times \cdots \times A^{(r)}$ in $\bar{R}$. Intervals in R^r are similarly defined as Cartesian products of r intervals in R. If $A = A^{(1)} \times \cdots \times A^{(r)}$, the numbers $m_L A^{(1)}, \ldots, m_L A^{(r)}$ are called the **edge-lengths** of A. The interval A is **bounded** if its edge-lengths are finite; that is, if each $A^{(i)}$ is a bounded interval in R. It is a **cube** if all the edge-lengths are positive, finite, and equal. Interval A is **degenerate** if it is empty or if any one of the edge-lengths is 0.

Since each of the intervals in the Cartesian product $A^{(1)} \times \cdots \times A^{(r)}$ can be of any one of four kinds (open, closed, left-open, right-open), there are 4^r kinds of intervals in $\bar{R}^r$. But three of these 4^r kinds are especially useful to us. These are the **open** intervals, in which each $A^{(i)}$ is open; the **closed** intervals, in which each $A^{(i)}$ is closed; and the **left-open** intervals, in which each $A^{(i)}$ is left-open.

If $A = A^{(1)} \times \cdots \times A^{(r)}$ is an interval in $\bar{R}^r$, the **closure** of A is defined to be the smallest closed interval that contains A. This is easily seen to be the Cartesian product of the closures of the $A^{(i)}$. The **interior** A^{O} of A is defined to be the largest open interval contained in A. (For intervals in $\bar{R}$, this has already been discussed in Section 1 of the Introduction.) It is easy to see that

$$A^{\mathrm{O}} = [A^{(1)}]^{\mathrm{O}} \times \cdots \times [A^{(r)}]^{\mathrm{O}}.$$

So A is degenerate if and only if its interior is empty.

If x is a point of $\bar{R}^r$, we define a **neighborhood** of x (in $\bar{R}^r$) to be a Cartesian product $B^{(1)} \times \cdots \times B^{(r)}$ in which for $i = 1, \ldots, r$, $B^{(i)}$ is a neighborhood of $x^{(i)}$ in $\bar{R}$. As before, a **gauge** on $\bar{R}^r$ is a function γ defined on $\bar{R}^r$ such that, for each x in $\bar{R}^r$, $\gamma(x)$ is a neighborhood of x in $\bar{R}^r$.

As in the one-dimensional case, an **allotted partition** $\mathscr{P}$ of a set A in R^r is a finite set of pairs

$$\mathscr{P} = \{(\bar{x}_1, A_1), \ldots, (\bar{x}_k, A_k)\} \tag{A}$$

in which each $\bar{x}_i$ is a point of $\bar{R}^r$ and the A_i are pairwise disjoint left-open intervals whose union is A. If γ is a gauge on $\bar{R}^r$, a **γ-fine partition** is defined to be an allotted partition $\mathscr{P}$ – using the same notation (A) – such that for each i in $\{1, \ldots, k\}$, A_i^- is contained in the neighborhood $\gamma(x_i)$.

If A is an interval in R^r or in $\bar{R}^r$, the elementary measure $m_{\mathrm{L}}A$ of A is defined to be the product of the edge-lengths of A. Thus, if $A = A^{(1)} \times \cdots \times A^{(r)}$,

$$m_{\mathrm{L}}A = [m_{\mathrm{L}}A^{(1)}][m_{\mathrm{L}}A^{(2)}] \cdots [m_{\mathrm{L}}A^{(r)}]. \tag{B}$$

If $r = 2$, this is the same as the area of A; if $r = 3$, it is the volume of A. There is no word in common use for the elementary measure in spaces of more than three dimensions, so we shall use *elementary measure*, or sometimes just *measure*, for the product of edge-lengths in any number of dimensions. If A is a bounded interval, $m_{\mathrm{L}}A$ is finite, and if A is degenerate, $m_{\mathrm{L}}A = 0$. If A is unbounded, one of the factors $m_{\mathrm{L}}A^{(i)}$ is ∞. If A is degenerate, one of the other factors in the product of the edge-lengths is 0, so that $m_{\mathrm{L}}A = 0$; if A is nondegenerate, all the $m_{\mathrm{L}}A^{(i)}$ are positive and one of them is ∞, so $m_{\mathrm{L}}A = \infty$.

We use again the notational device introduced at the end of Section I-4. When a statement involves an integral with respect to m_{L} but its proof uses no properties of m_{L} except that it is a nonnegative additive function on the set of all left-open intervals in R^r, we shall abbreviate the symbol m_{L} to m, and we shall also prefix the titles of such statements with a star (★).

As before, if f is a function defined on a set B in $\bar{R}^r$, we define f_B to be the function on $\bar{R}^r$ defined by

$$\begin{aligned} f_B(x) &= f(x) \qquad (x \text{ in } B), \\ f_B(x) &= 0 \qquad (x \text{ in } \bar{R}^r \setminus B). \end{aligned} \tag{C}$$

If f is defined on $\bar{R}^r$ and

$$\mathscr{P} = \{(\bar{x}_1, A_1), \ldots, (\bar{x}_k, A_k)\}$$

is an allotted partition of some set in R^r, we define the **partition-sum** corresponding to $\mathscr{P}$ and f to be

$$\text{(D)} \qquad S(\mathscr{P}; f; m) = \sum_{i=1}^{k} f(\bar{x}_i) m A_i,$$

provided that this sum exists.

As in one dimension, we define the statement that the partition-sum $S(\mathscr{P}; f; m)$ has **gauge-limit** J to mean that to each positive number ε there corresponds a gauge γ on $\bar{R}^r$ such that for every γ-fine partition $\mathscr{P}$ of R^r, $S(\mathscr{P}; f; m)$ exists and

$$|S(\mathscr{P}; f; m) - J| < \varepsilon.$$

Now we can define the gauge-integral of a function f over a subset B of R^r just as we did in Definitions I-2-1 and III-7-1.

★DEFINITION 1-1 *Let m be a real-valued function of left-open intervals in R^r. Let B be a set contained in R^r and f a function defined and real-valued on a set D. Then f is* ***gauge-integrable over*** *B if B is contained in D and the gauge-limit of $S(\mathscr{P}; f_B; m)$ exists. If J is a number such that*

$$J = \textit{gauge-limit of} \quad S(\mathscr{P}; f_B; m),$$

we denote J by the symbol

$$\int_B f(x)\, m(dx).$$

2. Elementary Properties of Intervals and Measure

Many of the properties of the integral established in the preceding chapters carry over to integrals over sets in R^r. But to adapt their proofs we need to prove some rather obvious remarks about intervals and their (elementary) measures in $\bar{R}^r$.

Let A be a closed interval in $\bar{R}^r$. We can easily show that every point $\bar{x}$ of $\bar{R}^r \backslash A$ has a neighborhood disjoint from A. For let $A = [a^{(1)}, b^{(1)}] \times \cdots \times [a^{(r)}, b^{(r)}]$. If $\bar{x}$ is not in A, there is an i such that $\bar{x}^{(i)}$ is not in $[a^{(i)}, b^{(i)}]$. Merely to keep the notation simple, let us suppose this true for $i = 1$. Then either $\bar{x}^{(1)} < a^{(1)}$ or $\bar{x}^{(1)} > b^{(1)}$. In the former case we define $U = [-\infty, a^{(1)}) \times \bar{R} \times \cdots \times \bar{R}$, with

$r - 1$ factors $\bar{R}$; in the latter case we define $U = (b^{(1)}, \infty] \times \bar{R} \times \cdots \times \bar{R}$, with $r - 1$ factors $\bar{R}$. In either case, U is a neighborhood of $\bar{x}$ that contains no point of A.

In one dimension it was trivially easy to show that the elementary measure m_L is additive: if an interval A is the union of finitely many pairwise disjoint intervals $A_1, \ldots, A_k$, the length of A is the sum of the lengths of the A_i. In higher-dimensional spaces the result is still true, but the proof of that fact is tedious. Some people might think that this additivity is "intuitively evident" and that it is a waste of time to prove it. But even in the plane there are far more complicated dissections of an interval into subintervals than simple checkerboard patterns. Our definition of the elementary measure of A is that it is the product of the elementary measures (lengths) of the $A^{(i)}$, and the additivity of elementary measure in $\bar{R}^r$ should follow numerically from this definition, not from some "intuition" derived from physical feelings about the weights of the pieces of A. Besides that, who can honestly say that he has any clear-cut "intuitions" about 19-dimensional space? So, we shall prove the additivity of elementary measure.

LEMMA 2-1 *If $A_1, A_2, \ldots, A_k$ are pairwise disjoint intervals in R^r whose union is an interval A_0, then*

$$m_L A_0 = m_L A_1 + \cdots + m_L A_k.$$

If j is any one of the numbers $1, \ldots, r$ and c is a real number, it accords with custom to call the set $\{x \text{ in } \bar{R} : x^{(j)} = c\}$ a **hyperplane** (in $\bar{R}$). Temporarily, we extend the meaning of the word and call that set a hyperplane even if $c = \infty$ or $c = -\infty$. If A is an interval in R^r and c is in R, we say that A is cut into the pieces

$$A \cap \{x \text{ in } \bar{R}^r : x^{(j)} \leqq c\}, \qquad A \cap \{x \text{ in } \bar{R}^r : x^{(j)} > c\}$$

by the hyperplane $\{x \text{ in } \bar{R}^r : x^{(j)} = c\}$. These are disjoint intervals that may be empty or degenerate. If c is ∞ or $-\infty$, we define the two parts into which A is cut by $\{x \text{ in } \bar{R}^r : x^{(j)} = c\}$ to be A and the empty set $\varnothing$. We first prove the statement

(A) If an interval A in R^r is divided into the two parts A', A'' by a hyperplane $\{x \text{ in } \bar{R}^r : x^{(j)} = c\}$, then $m_L A = m_L A' + m_L A''$.

This is obvious if one of A', A'' is empty and the other is A. We must prove the equation if neither of A', A'' is empty. Let $A = A^{(1)} \times \cdots \times A^{(r)}$, and let the end-points of $A^{(i)}$ be $a^{(i)}$ and $b^{(i)}$. If we define

$$B = A^{(j)} \cap [-\infty, c], \qquad C = A^{(j)} \cap (c, \infty],$$

the two parts into which the hyperplane cuts A are

$$A^{(1)} \times \cdots \times A^{(j-1)} \times B \times A^{(j+1)} \times \cdots \times A^{(r)},$$

$$A^{(1)} \times \cdots \times A^{(j-1)} \times C \times A^{(j+1)} \times \cdots \times A^{(r)}.$$

We may suppose the notation chosen so that the first of these is called A' and the second A''. Then the elementary measure of A is

$$m_L A = (b^{(1)} - a^{(1)}) \cdots (b^{(r)} - a^{(r)}),$$

and $m_L A'$ differs from this only in having $(c - a^{(j)})$ in place of the jth factor $(b^{(j)} - a^{(j)})$, and $m_L A''$ differs from it only in having $(b^{(j)} - c)$ in place of the jth factor. It follows in all cases (even if some of the $b^{(i)} - a^{(i)}$ are 0 or some of them are ∞) that $m_L A' + m_L A'' = m_L A$, and (A) is proved.

For the intervals A_j we use the notation $A_j = A_j^{(1)} \times \cdots \times A_j^{(r)}$, with $a_j^{(i)}$ and $b_j^{(i)}$ the end-points of $A_j^{(i)}$. Let $\mathscr{H}$ be the set of hyperplanes

$$\{x \text{ in } \bar{R}^r : x^{(i)} = a_j^{(i)}\}, \qquad \{x \text{ in } \bar{R}^r : x^{(i)} = b_j^{(i)}\} \qquad (i = 1, \ldots, r; j = 1, \ldots, k).$$

There are at most $2rk$ hyperplanes in this set; let them be $H_1, \ldots, H_n$. For each j in $\{0, \ldots, k\}$ we first cut A_j into two parts, say A_j' and A_j'', by the hyperplane H_1. By (A), the total elementary measure of these two parts is the same as the elementary measure of A_j. Next we cut each of the two parts A_j', A_j'' into two parts by the hyperplane H_2. The total elementary measure of the two parts of A_j' is $m_L A_j'$, and the total elementary measure of the two parts of A_j'' is $m_L A_j''$, so the total elementary measure of the four parts of A_j is $m_L A_j$. We continue this process: having used $H_1, \ldots, H_m$ to cut A_j and its successive parts, we cut all the parts by H_{m+1} until we have used all the hyperplanes in $\mathscr{H}$. The total elementary measure of all the parts of A_j in the final dissection is the same as $m_L A_j$.

Evidently, the final set of parts of A_j arrived at after using all the hyperplanes in $\mathscr{H}$ does not depend on the order in which we use them. If $j > 0$, we can choose the numbering of the hyperplanes so that $H_1, \ldots, H_{2r}$ are the $2r$ hyperplanes

$$\{x \text{ in } \bar{R}^r : x^{(i)} = a_j^{(i)}\}, \qquad \{x \text{ in } \bar{R}^r : x^{(i)} = b_j^{(i)}\}, \qquad (i = 1, \ldots, r).$$

These cut A_0 into pairwise disjoint parts, one of which is A_j. Using the rest of the hyperplanes in $\mathscr{H}$ cuts these finer, but in the final result each of the intervals into which A_0 is cut either is contained in A_j or is disjoint from A_j.

Let $B_1, \ldots, B_q$ be all the parts of A_0 obtained after using all the hyperplanes in $\mathscr{H}$. Then

$$\text{(B)} \qquad m_L B_1 + \cdots + m_L B_q = m_L A_0.$$

But we can bracket these terms into subsets. Each B_m is contained in a single one of the intervals $A_1, \ldots, A_k$. Then by the above construction,

$$m_L A_1 = \sum \{m_L B_m : B_m \subset A_1\} \quad \cdots \quad m_L A_k = \sum \{m_L B_m : B_m \subset A_k\}.$$

If we add these equations member by member, the sum of the right members is the sum of $m_L B_m$ for all $B_m \subset A_0$, which by (B) is $m_L A_0$. This completes the proof.

The next theorem states another simple but essential property of the elementary measure m_L.

THEOREM 2-2 (i) *If B is an interval in R^r and c is a number such that $c < m_L B$, there exists a bounded left-open interval F such that the closure F^- is contained in B and $m_L F > c$.*

(ii) *If B is a bounded interval in R^r, and c is a number such that $c > m_L B$, there exists a left-open interval G such that B is contained in the interior G^0 of G and $m_L G < c$.*

For (i), if $m_L B = 0$, we choose F to be $\varnothing$. Otherwise, with B denoted by $B^{(1)} \times \cdots \times B^{(r)}$, we have $m_L B = (m_L B^{(1)}) \cdots (m_L B^{(r)})$. The product $c_1 \cdots c_r$ of r real numbers is a continuous function of the c_i, and its value when $c_i = m_L B^{(i)}$ $(i = 1, \ldots, r)$ is $m_L B$, so we can and do choose numbers $c_1, \ldots, c_r$ such that $c_i < m_L B^{(i)}$ $(i = 1, \ldots, r)$ and $c_1 \cdots c_r > c$. For each i, by Lemma 1-4 in the Introduction, there exists a bounded left-open interval $F^{(i)}$ with closure contained in $B^{(i)}$ and with $m_L F^{(i)} > c_i$. We define $F = F^{(1)} \times \cdots \times F^{(r)}$. Then the closure of F is in B, and

$$m_L F = (m_L F^{(1)}) \cdots (m_L F^{(r)}) > c_1 \cdots c_r > c.$$

For (ii), we again use the fact that the product $c_1 \cdots c_r$ is a continuous function and has value $m_L B < c$ when $c_i = m_L B^{(i)}$ $(i = 1, \ldots, r)$. We can therefore choose numbers $c_1, \ldots, c_r$ such that $c_i > m_L B^{(i)}$ for each i and $c_1 \cdots c_r < c$. For each i, by Lemma 1-4 in the Introduction, we can choose a left-open interval $G^{(i)}$ such that $B^{(i)}$ is contained in the interior of $G^{(i)}$ and $m_L G^{(i)} < c_i$. We define $G = G^{(1)} \times \cdots \times G^{(r)}$. Then the interior of G contains B, and

$$m_L G = (m_L G^{(1)}) \cdots (m_L G^{(r)}) < c_1 \cdots c_r < c.$$

This completes the proof.

3. Generalizations to Integration in R^r of Theorems in Preceding Chapters

Many of the theorems of Chapters I and II generalize to integrals over sets in R^r with little effort. In most cases the generalization will be found to consist merely of replacing R and $\bar{R}$ by R^r and $\bar{R}^r$, respectively. Because of this, we shall allow ourselves a convenient verbal inaccuracy. We shall use the words "by Theorem X-Y-Z" when we really mean "by the generalization to R^r of Theorem X-Y-Z."

The first statement that we shall generalize is Lemma I-3-2, on the value of the integral of the indicator of an interval. This lemma holds when R is replaced by R^r in its statement. Only two minor changes in the proof are required. The reference to Lemma 1-4 in the Introduction in justifying inequalities (E) is replaced by a reference to Theorem 2-2. When x is not in A, it has a neighborhood that contains no point of F^-; but now we justify this statement by referring to the second paragraph of Section 2 of this chapter. No other changes in proof are needed. The extension, Lemma II-4-5, of Lemma I-3-2 to unbounded intervals is also straightforward.

We shall prove a statement that is a little (and usefully) stronger than a direct generalization of Theorem I-4-2; the proof will be a slight modification of that of Theorem I-4-2. As there, for each positive integer n we define $\mathcal{Q}_1[n]$ to be the set of intervals $\{Q(n,0), \ldots, Q(n, 2 \cdot 4^n + 1)\}$, where

$$\text{(A)} \qquad Q(n,0) = (-\infty, -2^n],$$

$$Q(n,j) = (-2^n + (j-1)2^{-n}, -2^n + j2^{-n}] \qquad (j = 1, \ldots, 2 \cdot 4^n),$$

$$Q(n, 2 \cdot 4^n + 1) = (2^n, \infty).$$

For r and n both positive integers we define $\mathcal{Q}_r[n]$ to be the set of left-open intervals

$$C = C^{(1)} \times C^{(2)} \times \cdots \times C^{(r)}$$

in which each $C^{(j)}$ belongs to $\mathcal{Q}_1[n]$. We can now state our generalization of Theorem I-4-2.

THEOREM 3-1 *Let B be a left-open interval in R^r, and let γ be a gauge on $\bar{R}^n$. Then there exists a γ-fine partition*

$$\mathcal{P} = \{(\bar{x}_1, A_1), \ldots, (\bar{x}_k, A_k)\}$$

of B such that for each i in $\{1, \ldots, k\}$, either A_i is empty or it is the intersection $B \cap C$ of B with an interval C that belongs to one of the sets $\mathcal{Q}_r[1]$, $\mathcal{Q}_r[2]$, $\mathcal{Q}_r[3], \ldots$, and $\bar{x}_i$ is in A_i^-.

If B' is a left-open interval in R^r, a partition

$$\mathcal{P} = \{(\bar{x}_1, A_1), \ldots, (\bar{x}_k, A_k)\}$$

will be called a "special partition (of B')" if it is a γ-fine partition of B', and for each i such that $B' \cap A_i$ is not empty, $\bar{x}_i$ belongs to the closure A_i^- and A_i is the intersection of B' with an interval that belongs to one of the sets $\mathcal{Q}_r[n]$. The conclusion of the theorem is, then, that B has a special partition.

Suppose that B has no special partition. We shall find that this leads to a contradiction. If each nonempty intersection of B with an interval C in $\mathcal{Q}_r[1]$ had a special partition, we could show, just as in proving Theorem I-4-2, that the union of these special partitions would be a special partition of B, which we have assumed to have no special partition. So for some C in $\mathcal{Q}_r[1]$ the intersection $B \cap C$ is nonempty and has no special partition. We choose such an intersection and name it B_1. Next we consider all the intersections $B_1 \cap C$ in which C belongs to $\mathcal{Q}_r[2]$. By the same argument, one of these is nonempty and has no special partition. We choose such an intersection and name it B_2. Continuing the process yields a shrinking sequence of left-open intervals $B, B_1, B_2, B_3, \ldots$

such that each B_i is the intersection of B with an interval in the set $\mathcal{Q}_r[i]$. If we write

$$B_i = B_i^{(1)} \times \cdots \times B_i^{(r)},$$

as in the proof of Theorem I-4-2, for each j in $\{1, \ldots, r\}$ there is an extended real number $y^{(j)}$ that is contained in all the closures $[B^{(j)}]^-, [B_1^{(j)}]^-, [B_2^{(j)}]^-, \ldots$. Let y be the point $(y^{(1)}, \ldots, y^{(r)})$ of $\bar{R}^r$. The neighborhood $\gamma(y)$ is the Cartesian product

$$\gamma(y) = \gamma(y)^{(1)} \times \cdots \times \gamma(y)^{(r)}$$

in which each $\gamma(y)^j$ is a neighborhood of $y^{(j)}$. By the proof of Theorem I-4-2, there is a positive integer i such that for $j = 1, \ldots, r$

$$[B_i^{(j)}]^- \subset \gamma(y^{(j)}).$$

Then

$$B_i^- \subset \gamma(y),$$

and the single pair (y, B_i) is a special partition of B_i. But by the construction, B_i has no special partition. This contradiction establishes the theorem.

With this start, we can extend to integration in R^r all theorems, definitions, lemmas, and corollaries in Chapters I and II whose names and numbers are preceded by a star (★). In all cases, changes in proof are either unnecessary or trivial.

On the other hand, with few exceptions the theorems not introduced by a star are, for one reason or another, bound to one dimension and have no simple and direct extension to integration in higher-dimensional spaces. For this reason we abandon all theorems, lemmas, and corollaries in Sections 8, 9, 10, and 11 of Chapter I and in Sections 6, 7, and 9 of Chapter II except those introduced by a star. However, there are four theorems without a star whose statements and proofs extend to higher-dimensional spaces with changes that are, at most, trivial. These are Theorems I-5-6, Corollaries I-5-8 and I-5-9, and Theorem I-7-2. For convenience in reference we state their generalizations to integrals over sets in R^r.

THEOREM 3-2 *Let f be a step-function on R^r that has values $c_1, \ldots, c_h$ on the respective pairwise disjoint bounded intervals $B_1, \ldots, B_h$. Then f is integrable over R^r, and*

$$\int_{R^r} f(x)\, m_L(dx) = c_1 m_L B_1 + \cdots + c_h m_L B_h.$$

The proof is the same as that of Theorem I-5-6.

COROLLARY 3-3 *If f is defined on a subset B of R^r and is 0 at all but finitely many points of B, then f is integrable over B, and*

$$\int_B f(x)\, m_L(dx) = 0.$$

The proof is the same as that of Corollary I-5-8.

COROLLARY 3-4 *If f and g are defined on the same set B in R^r, and f is integrable over B, and $g(x) = f(x)$ for all but finitely many points x of B, then g is also integrable over B, and*

$$\int_B g(x)\, m_L(dx) = \int_B f(x)\, m_L(dx).$$

The proof is the same as that of Corollary I-5-9.

THEOREM 3-5 *If f is Riemann-integrable over a bounded interval B in R^r, it is also gauge-integrable over B, and the two integrals have the same value.*

The proof is the same as that of Theorem I-7-2.

Theorem 3-5 allows us to make the same notational simplification in R^r as we made in R. If the elementary measure is m_L, we replace the symbol

$$\int_B f(x)\, m_L(dx)$$

by either one of the simpler symbols

$$\int_B f(x)\, dx, \qquad \int_B f(x)\,(dx^{(1)} \cdots dx^{(r)}).$$

In the second of these the $dx^{(1)}$, etc., are held together in a parenthesis to remind us that in the definition of the m_L-measure of an interval with edge-lengths $m_L A^{(1)}, \ldots, m_L A^{(r)}$, these edge-lengths never occur separately; the m_L-measure of the interval is their product. This second notation is especially appropriate when the coordinates, instead of being distinguished by superscripts $(1), \ldots, (r)$, are distinguished by using different letters; for instance, points in R^3 might be labeled (x, y, z). When such notation is used for the points, it is reasonable to use such a symbol as

$$\int_B f(x, y, z)\,(dx\, dy\, dz)$$

for the integral. The symbol for the gauge-integral with respect to m_L now has the same appearance as that for the Riemann integral. But by Theorem 3-5, the two are equal whenever the Riemann integral exists, and the similarity should cause no confusion.

4. Iterated Integration

In the preceding section we extended many theorems from one dimension to higher dimensions, but the fundamental theorem was not among them. Thereby we lost an extremely useful device for computing integrals. To restore at least part of this convenience, we shall show that under reasonable hypotheses a "multiple integral" (that is, an integral over a set in a space R^r with $r > 1$) can be reduced to a succession of "simple" integrations (that is, integrations over sets in R). The Italian mathematician Guido Fubini was the first to prove a theorem of this type with power enough for the needs of modern mathematics. Nowadays it is customary to say that any theorem that asserts the possibility of reducing a multiple integral to an iteration of simple integrations is a *Fubini theorem*. In this chapter we shall prove two Fubini theorems. The one in this section applies to a fairly simple type of integrand, making it not too difficult to prove the theorem, and yet the theorem obtained is powerful enough to cover all the examples usually considered in advanced calculus or encountered in applications. In Section 7 we shall prove another theorem with weaker hypotheses (in fact, with no more hypotheses on f than mere integrability over some interval in R^r), but the proof requires the use of the later sections of Chapter II, which the reader may have chosen to pass over.

The simplest form of the Fubini theorem states that if f is a function on an interval $A = A^{(1)} \times A^{(2)}$ in R^2, the integral of f over A is equal to the integral over $A^{(1)}$ of the function defined on $A^{(1)}$ by the integral

$$\int_{A^{(2)}} f(x^{(1)}, x^{(2)})\, dx^{(2)},$$

provided that f satisfies suitable hypotheses. We shall extend this to higher-dimensional spaces. Let r, s, t be positive integers with $s + t = r$. Every point x in $\bar{R}^r$ can be written as $x = (u, v)$, where

$$u = (x^{(1)}, \dots, x^{(s)}), \qquad v = (x^{(s+1)}, \dots, x^{(r)})$$

are points of $\bar{R}^s$ and $\bar{R}^t$, respectively. Likewise, an interval $A = A^{(1)} \times A^{(2)} \times \cdots \times A^{(r)}$ in R^r can be written as

$$A = A(1) \times A(2),$$

where

$$A(1) = A^{(1)} \times \cdots \times A^{(s)}, \qquad A(2) = A^{(s+1)} \times \cdots \times A^{(r)}$$

are intervals in R^s and R^t, respectively. We shall write (u, v) for x and $A(1) \times A(2)$ for A whenever convenient; u, with or without affixes, will always be a point of $\bar{R}^s$ and v, with or without affixes, a point of $\bar{R}^t$.

Our first form of the Fubini theorem is the following.

THEOREM 4-1 *Let f be a bounded real-valued function on a bounded interval $A = A(1) \times A(2)$ in R^r such that there exists a sequence $s_1, s_2, s_3, \ldots$ of step-functions on A converging everywhere in A to f. Then for every u in $A(1)$ the function*

$$v \mapsto f(u, v) \qquad (v \text{ in } A(2))$$

is integrable over $A(2)$, and the intervals

$$\text{(A)} \qquad \int_{A(1)} \left\{ \int_{A(2)} f(u, v)\, dv \right\} du, \qquad \int_A f(x)\, dx$$

both exist and are equal.

Suppose first that f is the indicator of an interval $B = B(1) \times B(2)$ $(B(1) \subset R^s$, $B(2) \subset R^t)$ contained in A. Then for every point $x = (u, v)$ in $\bar{R}^r$

$$\text{(B)} \qquad 1_B(x) = 1_{B(1)}(u) 1_{B(2)}(v);$$

for the left member is 1 if and only if x is in B, which is true if and only if u is in $B(1)$ and v is in $B(2)$. For each fixed u in $\bar{R}^s$ the right member of (B) defines a step-function

$$v \mapsto 1_{B(1)}(u) 1_{B(2)}(v)$$

on R^t, so it is integrable over $A(2)$ and

$$\text{(C)} \qquad \int_{A(2)} 1_{B(1)}(u) 1_{B(2)}(v)\, dv = 1_{B(1)}(u) m_L B(2).$$

The function

$$u \mapsto 1_{B(1)}(u) m_L B(2)$$

is a step-function on R^s, so it is integrable over $A(1)$, and

$$\text{(D)} \qquad \int_{A(1)} 1_{B(1)} m_L B(2)\, du = m_L B(1) m_L B(2).$$

Since $m_L B(1)$ is the product of the first s edge-lengths of B and $m_L B(2)$ is the product of the last t edge-lengths of B, the right member of (D) is the product of all r edge-lengths of B, which is $m_L B$. But

$$\text{(E)} \qquad m_L B = \int_A 1_B(x)\, dx.$$

From (B), (C), (D), and (E) we deduce that the conclusion of the theorem is valid for this f.

Suppose next that f is a step-function on A, with values $c_1, \ldots, c_k$ on the pairwise disjoint subintervals $B_1, \ldots, B_k$ of A. For convenience we write f_j for the indicator of the interval B_j. Then

$$f = c_1 f_1 + \cdots + c_k f_k,$$

and by the preceding paragraph

$$\int_{A(1)} \left\{ \int_{A(2)} f(u, v)\, dv \right\} du = \sum_{j=1}^{k} c_j \int_{A(1)} \left\{ \int_{A(2)} f_j(u, v)\, dv \right\} du$$

$$= \sum_{j=1}^{k} c_j \int_A f_j(x)\, dx$$

$$= \int_A f(x)\, dx.$$

So the conclusion is valid whenever f is a step-function on A.

Suppose, finally, that the hypotheses of the theorem are satisfied. Then f_A is 0 outside A, and there is a real number M such that

(F) $$|f(x)| \leqq M \qquad (x \text{ in } A),$$

and there exists a sequence $s_1', s_2', s_3', \ldots$ of step-functions on A such that

(G) $$\lim_{N \to \infty} s_n'(x) = f(x) \qquad (x \text{ in } A).$$

For each positive integer n we define

(H) $$s_n(x) = [s_n'(x) \wedge M1_A(x)] \vee [-M1_A(x)] \qquad (x \text{ in } R^r).$$

This is a step-function, and its absolute value does not exceed the integrable function $M1_A$. The inequality

$$|f(x)| \leqq M1_A(x)$$

holds for all x in R^r, since the left member never exceeds M and is 0 where $1_A(x) = 0$. So

$$[f(x) \wedge M1_A(x)] \vee [-M1_A(x)] = [f(x)] \vee [-M1_A(x)] = f(x),$$

and by Lemma III-10-2 and (G),

(I) $$\lim_{n \to \infty} s_n(x) = \lim_{n \to \infty} [s_n'(x) \wedge M1_A(x)] \vee [-M1_A(x)]$$

$$= [f(x) \wedge M1_A(x)] \vee [-M1_A(x)] = f(x).$$

For each u in R^s, the functions

(J) $$v \mapsto s_n(u, v) \qquad (v \text{ in } R^t)$$

are step-functions, and they satisfy

(K) $$|s_n(u, v)| \leqq M1_{A(1)}(u)1_{A(2)}(v),$$

which for each fixed u is integrable over $A(2)$. By (I), the $s_n(u, v)$ converge for all v in R^t to the limit $f(u, v)$, so by the dominated convergence theorem the function $v \mapsto f(u, v)$ is integrable over R^t, and

$$\text{(L)} \qquad \int_{A(2)} f(u, v)\, dv = \lim_{n \to \infty} \int_{A(2)} s_n(u, v)\, dv.$$

By (K),

$$\left| \int_{A(2)} s_n(u, v)\, dv \right| \leqq M m_L A(2) 1_{A(1)}(u),$$

so

$$\text{(M)} \qquad \left| \lim_{n \to \infty} \int_{A(2)} s_n(u, v)\, dv \right| \leqq M m_L A(2) 1_{A(1)}(u).$$

The right member is integrable, as a function of u, over $A(1)$, so by (M) and (L) we can apply the dominated convergence theorem to obtain

$$\text{(N)} \qquad \int_{A(1)} \left\{ \int_{A(2)} f(u, v)\, dv \right\} du = \lim_{n \to \infty} \int_{A(1)} \left\{ \int_{A(2)} s_n(u, v)\, dv \right\} du.$$

Since the s_n do not exceed $M1_A$ in absolute value and, by (I), they converge everywhere in A to f, by the dominated convergence theorem f is integrable over A, and

$$\text{(O)} \qquad \int_A f(x)\, dx = \lim_{n \to \infty} \int_A s_n(x)\, dx.$$

But since the conclusion of the theorem holds for each step-function s_n, the right members of (N) and (O) are equal. So the left members are also equal, and the proof is complete.

It would be easy to show that the hypotheses of Theorem 4-1 are satisfied if f is continuous on the closure of A. But in many applications one is required to integrate a continuous function f over some bounded set G that can be specified as the set on which some other continuous function is positive (or nonnegative). For example, in the plane the interior of the triangle with vertices $(0, 0)$, $(2, 0)$, and $(0, 2)$ is the set on which $g > 0$, if we define

$$g(u, v) = u \wedge v \wedge (2 - u - v).$$

In three-space, the set consisting of the interior and surface of a sphere with center C and radius r is the set on which $g \geqq 0$ if we define for each x in three-dimensional space

$$g(x) = r - (\text{distance from } x \text{ to } C).$$

If we enclose G in an interval A, the integral sought is

$$\int_A f(x) 1_G(x)\, dx.$$

But although f is continuous, $f \cdot 1_G$ may be discontinuous at all boundary points of G. So the special case of Theorem 4-1 in which f is continuous is not adequate. However, another simple special case of Theorem 4-1 covers such applications.

COROLLARY 4-2 *Let A be a bounded interval, g a function continuous on the closure A^-, and*

$$G = \{x \text{ in } A : g(x) > 0\}.$$

Let f_1 and f_2 be functions continuous on A^-. Then the function f defined by

$$f(x) = f_1(x) \qquad (x \text{ in } A^-,\ g(x) > 0),$$

$$f(x) = f_2(x) \qquad (x \text{ in } A^-,\ g(x) \leqq 0)$$

satisfies the hypotheses of Theorem 4-1, *so that*

$$\int_{A(1)} \left\{ \int_{A(2)} f(u, v)\, dv \right\} du = \int_A f(x)\, dx.$$

Since the functions f_1 and f_2 are continuous on the bounded closed interval A^-, they are bounded; and since f is everywhere equal to one of them, it too is bounded. If

$$A^- = [a^{(1)}, b^{(1)}] \times \cdots \times [a^{(r)}, b^{(r)}],$$

we cut A by the r hyperplanes

$$\{x \text{ in } \bar{R} : x^{(i)} = \tfrac{1}{2}[a^{(i)} + b^{(i)}]\} \qquad (i = 1, \ldots, r)$$

and thus obtain 2^r subintervals $B_{1,1}, \ldots, B_{1,2^r}$ of A. This process is called "bisecting the edges of A." By bisecting the edges of each $B_{1,j}$ we obtain 2^{2r} intervals

$$B_{2,1}, \ldots, B_{2,2^{2r}},$$

and we continue this process. At the nth stage we have 2^{nr} intervals, each having its diagonals 2^{-n} times as long as the diagonal of A. In each $B_{n,j}^-$ we choose a point $x_{n,j}$ such that at $x_{n,j}$, $g(x)$ attains its least value on $B_{n,j}^-$. We define s_n to be the step-function such that

$$s_n(x) = f(x_{n,j}) \qquad \text{for all} \quad x \quad \text{in} \quad B_{n,j}.$$

It remains only to show that

$$\text{(P)} \qquad \lim_{n \to \infty} s_n(x) = f(x) \qquad (x \text{ in } A).$$

Let x be any point of A and for each positive integer n let $B_{n,j(n)}$ be that one of the intervals $B_{n,j}$ $(j = 1, \ldots, 2^{nr})$ that contains x. Then x and $x_{n,j(n)}$ both belong to $B^-_{n,j(n)}$, and their distance apart cannot be greater than the diagonal of $B^-_{n,j(n)}$, which is 2^{-n} times the diagonal of A. This implies

$$\text{(Q)} \qquad \lim_{n \to \infty} x_{n,j(n)} = x.$$

We consider two cases.

Case 1 $g(x) \leqq 0$. Since x is in $B^-_{n,j(n)}$, and on that set $g(x)$ has its least value at $x_{n,j(n)}$, this implies

$$g(x_{n,j(n)}) \leqq g(x) \leqq 0.$$

Therefore,

$$s_n(x) = f(x_{n,j(n)}) = f_2(x_{n,j(n)}),$$

$$f(x) = f_2(x).$$

Since f_2 is continuous, this and (Q) imply that (P) is satisfied.

Case 2 $g(x) > 0$. Define

$$\varepsilon = g(x)/2;$$

this is positive. Since g is continuous on A^-, there is a neighborhood $\gamma(x)$ such that for all x' in $A^- \cap \gamma(x)$, $g(x')$ differs from $g(x)$ by less than ε, whence

$$\text{(R)} \qquad g(x') > g(x) - \varepsilon = \varepsilon.$$

The interval $B_{n,j(n)}$ has edge-lengths that tend to 0 as n increases, so for all large n

$$B^-_{n,j(n)} \subset A^- \cap \gamma(x).$$

In particular, for all large n the point $x_{n,j(n)}$ of $B^-_{n,j(n)}$ is in $A^- \cap \gamma(x)$, so by (R)

$$g(x_{n,j(n)}) > \varepsilon.$$

This implies that

$$s_n(x) = f(x_{n,j(n)}) = f_1(x_{n,j(n)})$$

for all large n. But since $g(x) > 0$,

$$f(x) = f_1(x).$$

The last two equations, with (Q) and the continuity of f_1, imply that (P) is valid. So all the hypotheses of Theorem 4-1 are satisfied, and the proof is complete.

The hypotheses of Theorem 4-1 and Corollary 4-2 contain a restriction imposed merely to simplify notation, namely, that the first integration is with respect to $(x^{(s+1)}, \ldots, x^{(r)})$ and the second with respect to $(x^{(1)}, \ldots, x^{(s)})$. This restriction is unnecessary. We show this when $r = 2$, by a method that obviously applies in general. Let f satisfy the hypotheses of Theorem 4-1, and define

$$f'(v, u) = f(u, v) \qquad (u \text{ in } A(1),\ v \text{ in } A(2)).$$

Since f' satisfies the hypotheses of Theorem 4-1 on $A' = A(2) \times A(1)$, by that theorem

$$\int_{A'} f'(x)\, dx = \int_{A(2)} \left\{ \int_{A(1)} f'(v, u)\, du \right\} dv.$$

The left member is equal to the integral of f over A; this is obvious for step-functions, and it then holds for f by the dominated convergence theorem. So the integrals

$$\int_{A(2)} \left\{ \int_{A(1)} f(u, v)\, du \right\} dv, \qquad \int_A f(x)\, dx$$

both exist, and they are equal.

Often we need a slight generalization of Theorem 4-1 that applies to integrals of certain unbounded functions over unbounded intervals. The next corollary is such a generalization.

COROLLARY 4-3 *Let f be a real-valued function on an interval $A = A(1) \times A(2)$ in R^r. Assume that there exists a sequence $s_1, s_2, s_3, \ldots$ of step-functions on A that converges everywhere in A to f. Assume also that there exists a function g on A such that $|f| \leqq g$ on A, and for each u in $A(1)$ the function*

$$v \mapsto g(u, v) \qquad (v \text{ in } A(2))$$

is integrable over $A(2)$, and the iterated integral

$$\int_{A(1)} \left\{ \int_{A(2)} g(u, v)\, dv \right\} du$$

exists. Then for every u in $A(1)$ the function

$$v \mapsto f(u, v) \qquad (v \text{ in } A(2))$$

is integrable over $A(2)$, and the integrals

$$\int_{A(1)} \left\{ \int_{A(2)} f(u, v)\, dv \right\} du, \qquad \int_A f(x)\, dx$$

exist and are equal.

Consider first the case

$$f(x) \geqq 0 \qquad (x \text{ in } A).$$

For each positive integer n we define the intervals $W(n)$, $W(n, 1)$, and $W(n, 2)$ by

$$W(n) = \{x \text{ in } R^r : -n < x^{(i)} \leqq n,\ i = 1, \ldots, r\},$$
$$W(n, 1) = \{u \text{ in } R^s : -n < u^{(i)} \leqq n,\ i = 1, \ldots, s\},$$
$$W(n, 2) = \{v \text{ in } R^t : -n < v^{(j)} \leqq n,\ j = 1, \ldots, t\}.$$

Then the function f_n defined on R^r by

$$f_n(x) = f_A(x) \wedge (n1_{W(n)}(x)) \qquad (x \text{ in } R^r)$$

satisfies the hypotheses of Theorem 4-1 on the interval $A \cap W(n)$, so it can be integrated by iteration. Since it vanishes outside $W(n)$, by Theorem 4-1

$$\text{(S)} \qquad \int_{R^s} \left\{ \int_{R^t} f_n(u, v)\, dv \right\} du = \int_{A(1) \cap W(n,1)} \left\{ \int_{A(2) \cap W(n,2)} f_n(u, v)\, dv \right\} du$$
$$= \int_{A \cap W(n)} f_n(x)\, dx = \int_{R^r} f_n(x)\, dx.$$

As n increases, for each u in R^s the functions $v \mapsto f_n(u, v)$ ascend and tend to $f_A(u, v)$, never exceeding the function $v \mapsto g(u, v)$. This last is integrable over R^t, so by the monotone convergence theorem the integrals

$$\int_{R^t} f_n(u, v)\, dv$$

ascend and converge to

$$\int_{R^t} f_A(u, v)\, dv.$$

Since, for each u in R^s,

$$\int_{R^t} f_n(u, v)\, dv \leqq \int_{R^t} g(u, v)\, dv$$

and the last expression defines a function integrable over R^s, the integrals over R^s of the left members are bounded as n increases. By the monotone convergence theorem,

$$\lim_{n \to \infty} \int_{R^s} \left\{ \int_{R^t} f_n(u, v)\, dv \right\} du = \int_{R^s} \left\{ \int_{R^t} f_A(u, v)\, dv \right\} du.$$

By this and (S), the integrals of the f_n over R^r tend to a finite limit as n increases, and f_n tends everywhere to f_A. So, again by the monotone convergence theorem, the limit f_A is integrable over R^r, and

$$\lim_{n \to \infty} \int_{R^r} f_n(x)\, dx = \int_{R^r} f_A(x)\, dx.$$

The last two equations, with (S), show that

$$\int_{R^s}\left\{\int_{R^t} f_A(u,v)\,dv\right\}du = \int_{R^r} f_A(x)\,dx,$$

which is another way of writing the conclusion.

If f is not nonnegative but satisfies the hypotheses, both f^+ and f^- are nonnegative and satisfy the hypotheses. So, by the part of the conclusion already established,

$$\int_{A(1)}\left\{\int_{A(2)} f^+(u,v)\,dv\right\}du = \int_A f^+(x)\,dx,$$

$$\int_{A(1)}\left\{\int_{A(2)} f^-(u,v)\,dv\right\}du = \int_A f^-(x)\,dx,$$

all of which integrals exist. By subtraction, we obtain the conclusion of Corollary 4-3.

Many applications of Fubini's theorem (in the form of Corollary 4-2) are of the following type. A set G is defined by

$$G = \{(u,v) \text{ in } R^2 : a < u < b,\ L(u) < v = U(u)\},$$

where L and U are functions continuous on a bounded interval $[a,b]$ and $L \leqq U$. A function f is defined and continuous on G^-. We wish to compute

$$\int_G f(x)\,dx.$$

If A is any interval that contains G, this integral is the same as the integral of f_G over A. We define H on R^2 by

$$H(u,v) = (u-a) \wedge (b-v) \wedge (v-L(u)) \wedge (U(u)-v).$$

Then H is continuous on R^2, and

$$G = \{(u,v) \text{ in } R^2 : H(u,v) > 0\}.$$

Now by Corollary 4-2,

$$\int_G f(x)\,dx = \int_a^b\left\{\int_{L(u)}^{U(u)} f(u,v)\,dv\right\}du.$$

In such problems it is customary to denote the left member by

$$\int_G f(u,v)\,d(u,v) \qquad \text{or} \qquad \int_G f(u,v)\,(du\,dv).$$

In Definition II-12-1 we defined the measure of a set E to be the integral of its indicator, if that integral exists. Such measures can often be computed by iterated integration. Suppose that G is a bounded set and that A is a bounded interval that contains G, and that there is a function g continuous on A^- such that

$$G = \{x \text{ in } A : g(x) > 0\}.$$

(The last inequality could be replaced by $g(x) \geqq 0$.) Then the hypotheses of Corollary 4-2 are satisfied with $f_1 = 1$ and $f_2 = 0$, so by that corollary

$$\text{(T)} \qquad m_L G = \int_A 1_G(x)\,dx = \int_{A(1)} \left\{ \int_{A(2)} 1_G(u, v)\,dv \right\} du.$$

For each u_0 in R^s the set of points $\{(u_0, v) : (v \text{ in } R^t)\}$ is a t-dimensional subset of R^r; it is a line if $t = 1$, a plane if $t = 2$, etc. It intersects G in a subset, possibly empty, and the v-coordinates of the points in that intersection are what we call the *section* of G at u_0 and denote by $G[u_0]$. This is stated in the following definition.

★DEFINITION 4-4 *If G is a subset of R^r and u_0 is a point of R^s, the* **section** *of G at u_0 is the set*

$$G[u_0] = \{v \text{ in } R^t : (u_0, v) \text{ in } G\}.$$

Clearly, the indicator of the set $G[u_0]$ satisfies

$$1_{G[u_0]}(v) = 1_G(u_0, v) \qquad (v \text{ in } R^t).$$

The integral of the left member of this equation over $A(2)$ is $m_L G[u_0]$, so equation (T) can be written in the form

$$\text{(U)} \qquad m_L G = \int_{A(1)} m_L G[u]\,du.$$

As an example, let G be the interior of an ellipse, defined by

$$G = \{(u, v) \text{ in } R^2 : u^2/a^2 + v^2/b^2 < 1\}.$$

The section $G[u]$ is empty if $|u| \geqq a$, whereas if $|u| < a$, $G[u]$ is the open interval

$$G[u] = (-b[1 - u^2/a^2]^{1/2},\ b[1 - u^2/a^2]^{1/2}).$$

Hence,

$$m_L G[u] = 2b[1 - u^2/a^2]^{1/2},$$

and by (U),

$$m_L G = \int_{-a}^{a} 2b\left[1 - \frac{u^2}{a^2}\right]^{1/2} du.$$

In Section II-6 we saw that as y increases from $-\pi/2$ to $\pi/2$, the function $a \sin y$

increases from $-a$ to a. So we can integrate by substitution and obtain

$$m_L G = \int_{-\pi/2}^{\pi/2} 2b[1 - \sin^2 y]^{1/2}[a\cos y]\,dy$$

$$= 2ab \int_{-\pi/2}^{\pi/2} \cos^2 y\,dy$$

$$= ab \int_{-\pi/2}^{\pi/2} [1 + \cos 2y]\,dy$$

$$= ab\left[y + \frac{\sin 2y}{2}\right]_{-\pi/2}^{\pi/2}$$

$$= \pi ab.$$

In particular, if $a = b = 1$, the ellipse is a circle with radius 1, and its area is π. This shows that the π in Section II-6 is the same as the π of elementary geometry.

Integrals over sets in R^r can be reduced to a succession of r simple integrations by repeated use of Theorems 4-1 or Corollary 4-2. For example, let us compute the volume of the set G of points (x, y, z) in R^3 that lie above the plane $z = 7$ and below the paraboloid $z = 23 - x^2 - y^2$. This volume is the integral of 1_G over any interval in R^3 that contains G. Write u for (x, y). For each u in R^2, the section $G[u]$ is empty unless $x^2 + y^2 < 16$, and if that inequality holds, then

$$G[u] = (7, 23 - x^2 - y^2).$$

If we let C denote the circle

$$C = \{(x, y) \text{ in } R^2 : x^2 + y^2 < 16\},$$

by (U) we have

$$m_L G = \int_C m_L G[u]\,du = \int_C [16 - x^2 - y^2]\,du.$$

The set C is the same as the set of (x, y) in R^2 with $-4 < x < 4$ and

$$-[16 - x^2]^{1/2} < y < [16 - x^2]^{1/2},$$

so, as in the first of our examples,

$$m_L G = \int_{-4}^{4} \left\{ \int_{-[16-x^2]^{1/2}}^{[16-x^2]^{1/2}} (16 - x^2 - y^2)\,dy \right\} dx.$$

We leave it as an exercise in elementary calculus to verify that the last integral has the value 128π.

It is tempting to assume that if the iterated integral exists, the integral over the product-interval A (the "multiple integral") must also exist. This is not so. On the interval $(-1, 1] \times (-1, 1]$ in the plane, define f by

$$\text{(V)} \qquad f(u, v) = uv/(u^2 + v^2)^2 \quad \text{if} \quad (u, v) \neq (0, 0),$$
$$f(0, 0) = 0.$$

For each fixed u, the function $v \mapsto f(u, v)$ has the antiderivative $-u/2(u^2 + v^2)$, so

$$\int_{-1}^{+1} f(u, v)\, dv = 0.$$

Hence,

$$\int_{-1}^{1} \left\{ \int_{-1}^{1} f(u, v)\, dv \right\} du = 0.$$

In the same way,

$$\int_{-1}^{1} \left\{ \int_{-1}^{1} f(u, v)\, du \right\} dv = 0.$$

So both iterated integrals exist and have the same value. Nevertheless, f is not integrable over the square in the plane; we leave the proof as an exercise.

Another temptation is to define the multiple integral as *being* the iterated integral, whenever the latter exists. Of course, it is impossible to prove that a definition is wrong, but it is easy to prove that this definition would lead us into inconvenient situations. In the plane, there are examples in which one of the iterated integrals exists and the other does not, and there are examples in which both exist but have different values. Suppose we choose one of the two iterated integrals as the favored one and decree that the "double integral" is defined as being this iterated integral whenever it exists. Either example suffices to show that we would not even have the privilege of rotating axes by a right angle without changing the integral. These examples are given in Exercises 4-3 and 4-4.

EXERCISE 4-1 Find the integral of the function

$$f\colon (u, v) \mapsto u^3 v^2$$

over the set G of points (u, v) in the plane that are in the first quadrant, are inside the ellipse $u^2/25 + v^2/9 = 1$, and are outside the circle $u^2 + v^2 = 1$.

EXERCISE 4-2 Prove that the function defined in equation (V) is not integrable over the square $(-1, 1] \times (-1, 1]$. If it were, its integral over the squares $B_n = (1/n, 1] \times (1/n, 1]$ would be bounded for all positive integers n. This integral can be computed by iteration (why?) and is unbounded.

EXERCISE 4-3 Define f on $\mathscr{R}^2$ by

$$f(u, v) = v/u^3 \quad \text{if} \quad u > 0 \quad \text{and} \quad -u < v < u,$$
$$f(u, v) = 0 \quad \text{otherwise.}$$

Prove that of the two integrals

$$\int_0^1 \left\{ \int_{-1}^1 f(u,v)\,dv \right\} du, \qquad \int_{-1}^1 \left\{ \int_0^1 f(u,v)\,du \right\} dv,$$

the first is 0 and the second does not exist. By computing the integrals over the intervals $(1/n, 1] \times (0, 1]$, show that the double integral does not exist.

EXERCISE 4-4 Define f on $\mathscr{R}^2$ by

$$f(u,v) = 2(u-v)/(u+v)^3 \qquad \text{if} \quad u > 0 \quad \text{and} \quad v > 0,$$

$$f(u,v) = 0 \qquad \text{otherwise.}$$

(This is the function obtained from the one in the preceding exercise by a half-right-angle rotation of axes.) Prove that

$$\int_0^1 \left\{ \int_0^1 f(u,v)\,du \right\} dv = -1, \qquad \int_0^1 \left\{ \int_0^1 f(u,v)\,dv \right\} du = +1.$$

Show that on the set of (u, v) such that $1/n \leqq u < 1$ and $0 < v < u/2$ we have $f(u,v) > 4/27u^2$, so the integrals over these sets are unbounded, and the double integral cannot exist.

EXERCISE 4-5 Let f be positive valued and continuous on a bounded interval $[a, b]$ in R^1. Let C be the curve in the (x, y) plane that is the graph of $y = f(x)$. Let E be the subset of R^3 obtained by rotating about the x-axis the set of points in the (x, y)-plane that lie between the x-axis and the graph of f. Show that Corollary 4-2 can be applied to this to yield

$$m_L E = \int_a^b \pi f(x)^2\,dx.$$

EXERCISE 4-6 Use Exercise 4-5 to find the volume of a sphere of radius r.

EXERCISE 4-7 Let $A = [0, 1] \times [0, 1]$. On A define $f(u, v)$ to be 1 if $u = \frac{1}{2}$ and v is rational and to be 0 otherwise. Show that the hypotheses of Theorem 4-1 are satisfied, and verify equation (A). Show that the conclusions of the theorem are no longer valid if we interpret the integrals as Riemann integrals.

5. Change of Variables in Multiple Integrals

Substitution theory for multiple integrals is much more complicated than for simple integrals—reflecting the fact that even continuously differentiable mappings of sets G onto sets G^* in R^r can be more complicated when $r > 1$ than when $r = 1$. As in the case of the Fubini theorem, we shall establish two theorems. The one in this section is less general; it applies directly only to continuous integrands (and, by trickery, to certain discontinuous ones), but it is still general enough to be useful in ordinary applications of substitution theory. In Section 8 we shall prove another theorem in which the hypotheses on the integrand are very weak, amounting only to the obvious requirement that the integral we are computing should exist. This second theorem is more general, but because its proof requires extensions of the material in the later sections of Chapter II, it is not available to readers who have passed over those sections.

In stating and proving the substitution theorem, it will be convenient to make use of some of the elementary properties of the sets called *open sets.*

DEFINITION 5-1 *Let G be a set contained in R^r or in $\bar{R}^r$. A point x of R^r (or of $\bar{R}^r$) is an* ***interior point*** *of G if there exists a neighborhood of x that is contained in G. The set G is* ***open*** *if each point of G is an interior point of G. The set of all interior points of a set G is called the* ***interior*** *of G.*

An open interval A is an open set in the sense of this definition. For is x is in A, there is a neighborhood of x (namely A itself) that is contained in A. Likewise, if A is any interval, the set of all points interior to A is the largest open interval contained in A, and the use of the word *interior* in Definition 5-1 is consistent with its previous use.

The next lemma states a simple and frequently useful way of showing that a set is open.

LEMMA 5-2 *Let G be an open set in a space R^r and G^* an open set in a space R^s, and let f be a function that is defined and continuous on G and has its values in R^s. Then the inverse image $f^{-1}(G^*)$, consisting of all points x such that $f(x)$ is in G^*, is an open set.*

Let x' be a point of $f^{-1}(G^*)$. Then $f(x')$ is in the open set G^*, so it is interior to G^*, and there is a neighborhood V of $f(x')$ that is contained in G^*. Because f is continuous, there is a neighborhood $\gamma_1(x')$ such that $f(x)$ is in V whenever f is in $G \cap \gamma_1(x')$. Because G is open, there is a neighborhood $\gamma_2(x')$ contained in G. Then $\gamma(x') = \gamma_1(x') \cap \gamma_2(x')$ is a neighborhood of x', and it is contained in G because it is contained in $\gamma_2(x')$; and for every x in $\gamma(x')$, $f(x')$ is in V, which is contained in G^*, so every such x is in the inverse image $f^{-1}(G^*)$. Therefore, each point x' of $f^{-1}(G^*)$ has a neighborhood $\gamma(x')$ contained in $f^{-1}(G^*)$. This proves that $f^{-1}(G^*)$ is open, and the lemma is established.

In particular, suppose that f is real-valued and continuous on an open set G in R^r, and that c is a real number. The set $G^* = (c, \infty)$ is an open interval and is hence an open set, and by Lemma 5-2 its inverse image $f^{-1}(G^*)$ is open in R^r. This inverse image is the set of all x in G at which $f(x) > c$. Such sets were used in the preceding section.

We now establish a lemma and a corollary of a general nature that will be useful to us.

Lemma 5-3 *Let E be a set in R^r, and for each x in E let $\gamma(x)$ be a neighborhood of x. Then there exists a sequence $A_1, A_2, A_3, \ldots$ of pairwise disjoint left-open cubes and a sequence $\bar{x}_1, \bar{x}_2, \bar{x}_3, \ldots$ of points of E such that*

(i) *E is contained in the union of the A_i, and*
(ii) *for each positive integer i, $\bar{x}_i$ is in A_i and the closure A_i^- of A_i is contained in $\gamma(\bar{x}_i)$.*

For each x in E we choose a bounded neighborhood $\gamma_1(x)$, and we define $\gamma_2(x)$ to be $\gamma(x) \cap \gamma_1(x)$. We again use the sets $\mathcal{Q}_r[1], \mathcal{Q}_r[2], \ldots$ of intervals that were defined just before Theorem 3-1, and we arrange all these in one sequence: first the intervals in $\mathcal{Q}_r[1]$, then the intervals in $\mathcal{Q}_r[2]$, and so on. For convenience we denote this sequence by

$$\text{(A)} \qquad Q^*(1), Q^*(2), Q^*(3), \ldots .$$

We test these intervals successively, and we select those intervals $Q^*(n)$ that satisfy the following two conditions:

(B) there is a point x of E contained in $Q^*(n)$ such that $Q^*(n)^- \subset \gamma_2(x)$;

(C) $Q^*(n)$ is not contained in any $Q^*(n')$ that precedes it in the sequence (A) and has been selected.

We name the selected intervals $A_1, A_2, A_2, \ldots$. For each positive integer i, since A_i is one of the intervals $Q^*(n)$ that satisfies (B), there is a point x in E contained in A_i such that $A_i^- \subset \gamma_2(x)$. We select one such point and name it $\bar{x}_i$. Since A_i is contained in the bounded neighborhood $\gamma_1(\bar{x}_i)$ and all the intervals in the sets $Q_r[n]$ that are bounded are cubes, A_i is a left-open cube.

Suppose that A_i and $A_{i'}$ ($i' \neq i$) have a common point x. Since $A_i = Q(n,j)$ and $A_{i'} = Q(n',j')$ for some n, j, n', and j', we cannot have $n = n'$, for in each set $Q(n, 1), \ldots, Q(n, k(n))$ the intervals are pairwise disjoint. To be specific, suppose $n' > n$. By the construction of the $Q(n,j)$, the interval $Q(n',j')$ is either contained in $Q(n,j)$ or is disjoint from it. The latter cannot be the case, since they both contain x. So $Q(n',j')$ is contained in the selected cube $Q(n,j)$ and fails to satisfy (C). Therefore it could not have been selected as $A_{i'}$. This proves that the A_i are pairwise disjoint.

Let x be any point of E. The neighborhood $\gamma_2(x)$ is an open interval $(a^{(1)}, b^{(1)}) \times \cdots \times (a^{(r)}, b^{(r)})$, and all the $2r$ numbers $x^{(i)} - a^{(i)}$ and $b^{(i)} - x^{(i)}$ are positive. Let δ be the smallest of them. We can and do choose an integer n such that

$$2^n > \max\{|x^{(1)}|, \ldots, |x^{(r)}|\} \qquad \text{and} \qquad 2^{-n} < \delta.$$

The union of the intervals $Q(n, 1), \ldots, Q(n, k(n))$ is all of $\bar{R}^r$, so x is in one of them, say in $Q(n,j(n))$. This cannot be one of the unbounded intervals $Q(n,j)$ because all of these lie outside the interval $(-2^n, 2^n] \times \cdots \times (-2^n, 2^n]$ and cannot contain x. So $Q(n,j(n))$ is a left-open cube of edge-length 2^{-n}, which is less than δ. Since x is in $Q(n,j(n))$, $Q(n,j(n))^-$ is contained in $(x^{(1)} - \delta, x^{(1)} + \delta) \times \cdots \times (x^{(r)} - \delta, x^{(r)} + \delta)$, which is contained in $\gamma_2(x)$, and therefore $Q(n,j(n))$ satisfies (B). If it also satisfies (C), it is selected as one of the A_i, so in this case x is in the union of the A_i. If $Q(n,j(n))$ does not satisfy (C), it is contained in a previously selected cube $A_{i'}$. But in this case too x is in the union of the A_i. So (i) is satisfied, and the proof is complete.

This lemma has an interesting corollary that shows the closeness of the relationship between open sets and unions of intervals.

COROLLARY 5-4 *If G is an open set in R^r, it is the union of a sequence of pairwise disjoint left-open cubes $A_1, A_2, A_3, \ldots$ such that each closure A_i^- is contained in G.*

Each point x of G is interior to G, so we can and do select a neighborhood $\gamma(x)$ of x that is contained in G. Then the cubes $A_1, A_2, A_3, \ldots$ of Lemma 5-3 have the desired properties.

From Corollary 5-4 and Theorem III-10-8 it follows that every open set is measurable, and its measure is the sum of the measures of the cubes $A_1, A_2, A_3, \ldots$ of Corollary 5-4. Moreover, with the cubes A_i of Corollary 5-4, for each positive integer n the sum

$$S_n = 1_{A_1} + \cdots + 1_{A_n}$$

is a step-function on R^r, and it tends to 1_G at every point in R^r. Hence we can apply Theorem 4-1 to 1_G, and just as at the end of the preceding section we can obtain this result.

LEMMA 5-5 *If G is a bounded open set in R^r, and x in R^r is represented as (u, v) as was done in Theorem 4-1, then*

$$m_L G = \int_{RS} m_L G[u]\,du.$$

Another consequence of Lemma 5-2 is the following.

COROLLARY 5-6 *Let h be a continuous one-to-one mapping of an open set G in R^r onto an open set G^* in R^r, and let h have a continuous inverse. Then for every interval A contained in G, $h(A)$ is measurable.*

By Lemma 5-2 and the remark after it, for each i in the set $\{1, \ldots, r\}$ and each real number c the sets $\{x \text{ in } G : x^{(i)} < c\}$ and $\{x \text{ in } G : x^{(i)} > c\}$ are open, and because h is the inverse of the continuous mapping h^{-1}, by Lemma 5-2 the images $h\{x \text{ in } G: x^{(i)} < c\}$ and $h\{x \text{ in } G: x^{(i)} > c\}$ are open and hence are measurable. Since G^* is open and therefore measurable, by Theorem III-10-8 the differences

$$G^* \setminus h\{x \text{ in } G: x^{(i)} < c\}, \qquad G^* \setminus h\{x \text{ in } G: x^{(i)} > c\}$$

are measurable. These are the sets

$$h\{x \text{ in } G: x^{(i)} \geqq c\}, \qquad h\{x \text{ in } G: x^{(i)} \leqq c\},$$

respectively. If A is any interval contained in G, $h(A)$ is the intersection of at most $2r$ sets, each of which is of one of the four types that we have just proved measurable. So, $h(A)$ is measurable.

In the remainder of this section we shall simplify typography by omitting the parentheses around the superscripts that number the coordinates. Thus, what we have been calling $x^{(i)}$ will appear simply as x^i. No confusion should result, since we do not use exponents in any theorem in this section.

We shall assume that the reader has some slight familiarity with linear algebra and thus knows what a matrix is, what the determinant of a square matrix is, what the cofactor of an element of a square matrix is. The reader should know that the determinant of a square matrix can be computed by choosing any row or any column, multiplying each element of that row or column by its cofactor, and adding those products. If M is an $s \times r$ matrix (s rows, r columns), the entry in row i and column j will usually be denoted by the symbol M^i_j. Thus, the transpose of an $s \times r$ matrix M is the $r \times s$ matrix M^{T} such that

$$(M^{\mathrm{T}})^j_i = M^i_j \qquad (i = 1, \ldots, s; j = 1, \ldots, r).$$

We shall often use this notation for convenience. We have been writing points x in R^r in the form $(x^1, \ldots, x^r)$, and this can be regarded as the notation for a $1 \times r$ matrix. But in computations it is more useful to consider x as represented by an $r \times 1$ matrix that consists of a single column whose entries from top to bottom are $x^1, \ldots, x^r$. To simplify typography, we shall denote such a column by $(x^1, \ldots, x^r)^{\mathrm{T}}$. The row $(x^1, \ldots, x^r)$ is a $1 \times r$ matrix, and to take its transpose is to replace the row by the column that has the same elements in the same order; in other words, to set it up as a column.

There is one standard exception to the convention of denoting the elements of a matrix M by M^i_j. The identity matrix will be denoted by I, but its elements (1 on the principal diagonal, 0 elsewhere) will be denoted by the "Kronecker δ" δ^i_j, where

$$\delta^i_j = 1 \quad \text{if} \quad i = j,$$
$$\delta^i_j = 0 \quad \text{if} \quad i \neq j.$$

In this chapter we shall use the word *r-vector* to mean an $r \times 1$ matrix with a single column and r rows. The next chapter contains a more detailed discussion of vectors in which it will appear that we really should say that the $r \times 1$ matrix *represents* a vector, rather than that it *is* a vector. But for present purposes it is convenient and harmless to ignore this distinction. In particular, if M is an $s \times r$ matrix and x is an r-vector, by the standard definition of product of matrices, the product Mx will be the s-vector

$$\left(\sum_{j=1}^{r} M_j^1 x^j, \ldots, \sum_{j=1}^{r} M_j^s x^j \right)^{\mathrm{T}}.$$

If M is an $r \times r$ matrix and c is an r-vector, the mapping

(D) $$x \mapsto y = Mx + c,$$

or, in greater detail,

(E) $$y^i = \sum_{j=1}^{r} M_j^i x^j + c^i,$$

is called an **affine mapping** of R^r into itself. If M is nonsingular (that is, if the determinant $\det M$ is not 0), this can be solved for x as a function of y, and we thus find that the mapping maps R^r onto all of R^r, and it has an inverse that is also an affine mapping. The first part of our study of change of variables is to find what effect is produced on an integral by an affine change of variables.

LEMMA 5-7 *Let M be a nonsingular $r \times r$ matrix, c an r-vector, and f a function defined and continuous on R^r that vanishes outside a bounded interval. Then*

(F) $$\int_{R^r} f(y)\,dy = \int_{R^r} f(Mx + c)|\det M|\,dx.$$

If $r = 1$, M is a single real number, and $\det M = M$. If $M > 0$, $|\det M| = M$, and by Theorem I-11-1

$$\int_{-\infty}^{\infty} f(y)\,dy = \int_{-\infty}^{\infty} f(Mx + c)M\,dx,$$

which is (F). If $M < 0$, $|\det M| = -M$, and by Theorem I-11-1,

$$\int_{\infty}^{-\infty} f(y)\,dy = \int_{-\infty}^{\infty} f(Mx + c)M\,dx,$$

which implies (F). So the conclusion is valid when $r = 1$.

For larger values of r we proceed by induction. We first prove an auxiliary statement.

(G) If the conclusion of Lemma 5-7 holds when f is defined on $R, \ldots, R^{r-1}$, and the first column of M has only one nonzero entry, equation (F) is satisfied.

Let the one nonzero entry in the first column of M be the entry M_1^h in row h. For each x and y in R^r we define

(H) $$u = x^1, \qquad v = (x^2, \ldots, x^r)^T,$$
$$w = y^h, \qquad z = (y^1, \ldots, y^{h-1}, y^{h+1}, \ldots, y^r)^T,$$
$$d = c^h, \qquad e = (c^1, \ldots, c^{h-1}, c^{h+1}, \ldots, c^r)^T.$$

Then the equation $y = Mx + c$ is equivalent to the system

(I) $$w = Au + Cv + d, \qquad z = Bv + e,$$

where A is the 1×1 matrix consisting of the single number M_1^h, B is the $(r-1) \times (r-1)$ matrix obtained by deleting row h and column 1 from M, and C is the $1 \times (r-1)$ matrix $(M_2^h, \ldots, M_r^h)$.

The $r \times r$ matrix

(J) $$M' = \begin{bmatrix} A & C \\ 0 & B \end{bmatrix},$$

in which the 0 stands for the $(r-1) \times 1$ matrix $(0, \ldots, 0)^T$, is obtained from M by an interchange of rows. Therefore,

$$\det M' = \pm \det M.$$

The first column of M' is $(M_1^h, 0, \ldots, 0)^T$, and we compute the determinant of M' by expanding in cofactors of the first column. The result is

(K) $$|\det M| = |\det M'| = |M_1^h||\det B|.$$

Since $f(y)$ is determined by y, and y is determined by w and z via (H), $f(y)$ is determined by w and z, and we can write $f(y)$ as $F(w, z)$. Then

(L) $$f(Mx + c) = f(y) = F(w, z) = F(Au + Cv + d, Bv + e).$$

Let us denote the left member of (F) by J. Then by Theorem 4-1 (Fubini's theorem),

$$J = \int_R \left\{ \int_{R^{r-1}} F(w, z)\, dz \right\} dw.$$

For each fixed w in R we change the variable of integration in the inner integral from z to v by the last of equations (I). By the induction hypothesis, we can do this by equation (F), so that

(M) $$J = \int_R \left\{ \int_{R^{r-1}} F(w, Bv + e)|\det B|\, dv \right\} dw.$$

By Fubini's theorem, we can interchange the order of the two integrations in (M), thus obtaining

(N) $$J = |\det B| \int_{R^{r-1}} \left\{ \int_R F(w, Bv + e)\, dw \right\} dv.$$

For each fixed v in R^{r-1} we change the variable of integration in the inner integral in (N) by the first of equations (I). By hypothesis, this can be done by (F). This and Fubini's theorem yield

$$\text{(O)} \qquad J = |\det B| \int_{R^{r-1}} \left\{ \int_R F(Au + Cv + d, Bv + e) |\det A| \, du \right\} dv$$

$$= |\det B| |M_1^h| \int_{R^r} f(Mx + c) \, dx.$$

This and the definition of J imply that (G) is valid.

Now let M be an $r \times r$ matrix with nonzero determinant. In the first column there must be at least one nonzero entry; otherwise $\det M$ would be 0. If only M_1^1 is not 0, (F) holds by virtue of (G). Otherwise, let $M_1^h \neq 0$, where $h > 1$. Let P be the $r \times r$ matrix all of whose entries are 0 except those in column h; column h has the entries

$$(M_1^1/M_1^h, \ldots, M_1^{h-1}/M_1^h, 0, M_1^{h+1}/M_1^h, \ldots, M_1^r/M_1^h).$$

We easily compute that P^2 is the zero matrix. So if I is the $r \times r$ identity matrix,

$$(I + P)(I - P) = I - P^2 = I,$$

and $I + P$ and $I - P$ are reciprocals. By another elementary multiplication of matrices, the matrix

$$\text{(P)} \qquad N = (I - P)M$$

has a first column with a single nonzero entry, namely, the entry in row h. If we multiply both members of (P) by $I + P$, we obtain

$$\text{(Q)} \qquad M = (I + P)N.$$

For each x in R^r we define

$$z = Nx, \qquad y = (I + P)z + c.$$

Then by (Q)

$$y = (I + P)Nx + c = Mx + c.$$

Since $I + P$ and N each have a first column with a single nonzero entry, we can apply (G) twice and find

$$\int_{R^r} f(y) \, dy = \int_{R^r} f((I + P)x + c) |\det(I + P)| \, dz$$

$$= \int_{R^r} f((I + P)Nx + c) |\det(I + P)| |\det N| \, dx.$$

Since $(I + P)N = M$, $\det M$ is the product of the determinants of $I + P$ and N,

and

$$|\det(I + P)||\det N| = |\det M|.$$

So (F) follows from the preceding equation, and by induction it holds for all r. The proof is complete.

COROLLARY 5-8 *Let M be a nonsingular $r \times r$ matrix; let c be an r-vector; and let A be a bounded interval in R^r. Then*

$$m_L(MA + c) = |\det M| m_L A.$$

It is easy to construct a sequence of continuous functions $f_1, f_2, f_3, \ldots$, all with values in $[0, 1]$, all vanishing outside a bounded interval, and such that $f_n(x)$ tends to $1_A(x)$ for every x in R^r. For each y in R^r define

(R) $$g_n(y) = f_n(M^{-1}[y - c]).$$

Then for all x in R^r

(S) $$g_n(Mx + c) = f_n(x).$$

By Lemma 5-7, for $n = 1, 2, 3, \ldots$ we have

(T) $$\int_{R^r} g_n(y)\,dy = |\det M| \int_{R^r} g_n(Mx + c)\,dx$$

$$= |\det M| \int_{R^r} f_n(x)\,dx.$$

As n increases, f_n tends everywhere to 1_A. So by (S), $g_n(Mx + c)$ tends to $1_A(x)$, and $g_n(y)$ tends to $1_A(M^{-1}[y - c])$, which is the value at y of the indicator of $MA + c$. So by the dominated convergence theorem, (T) implies

$$\int_{R^r} 1_{MA+c}(y)\,dy = |\det M| \int_{R^r} 1_A(x)\,dx,$$

which is the conclusion of the corollary.

Now we turn our attention to more general mappings, not necessarily affine. A **continuously differentiable** mapping on an open set G in R^r is a function $x \mapsto h(x)$ on G with values in some space R^s such that each h^i is continuous, and at each x in G the partial derivatives of h^i exist, and these partial derivatives are continuous on G.

There are several notations for partial derivatives that are in common use. If a function f is defined and real-valued on a set E in R^r, and x_0 is an interior point of E, we shall denote the partial derivative of f with respect to the kth coordinate of the independent variable $x = (u, v)$ at the place x_0 by the symbol

$D_k f(x_0)$. Thus, if f is defined on R^2 by $x \mapsto f(x) = f(u, v) = u^3 \sin v$, we have

$$D_1 f(4, \pi) = 3[4^2] \sin \pi = 0,$$

$$D_2 f(4, \pi) = [4]^3(\cos \pi) = -64.$$

Let f be real-valued on a set X in R^r. If all the partial derivatives $D_1 f(x_0), \ldots, D_r f(x_0)$ exist at a point x_0 of X, and we write them in that order in a row, we obtain a $1 \times r$ matrix which we call $Df(x_0)$:

$$Df(x_0) = (D_1 f(x_0), \ldots, D_r f(x_0)).$$

More generally, we define the derivative-matrix of a vector-valued function f as follows.

DEFINITION 5-9 *Let X be a point-set in R^r. Let $x \mapsto f(x)$ (x in X) be a function whose values are vectors in R^s. Let x_0 be a point of X at which each component f^i of f has partial derivatives $D_1 f^i(x_0), \ldots, D_r f^i(x_0)$. Then $Df(x_0)$ is the $s \times r$ matrix*

$$Df(x_0) = \begin{bmatrix} D_1 f^1(x_0) & \cdots & D_r f^1(x_0) \\ D_1 f^2(x_0) & \cdots & D_r f^2(x_0) \\ \vdots & & \vdots \\ D_1 f^s(x_0) & \cdots & D_r f^s(x_0) \end{bmatrix},$$

where

$$D_j f^i(x_0) = \left. \frac{\partial f^i(x^1, \ldots, x^r)}{\partial x^j} \right|_{x = x_0}.$$

Thus, for example, if

$$\text{(U)} \qquad f^i(x) = \sum_{j=1}^{r} M^i_j x^j + c^i \qquad (i = 1, \ldots, s),$$

we find

$$D_j f^i(x_0) = D_j[M^i_1 x^1 + \cdots + M^i_r x^r]|_{x=x_0} = M^i_j,$$

and therefore

$$Df(x_0) = M \qquad (x_0 \text{ in } R^r).$$

So, with Definition 5-9, the elementary formula $D[ax + c] = a$ generalizes to affine transformations (U).

When $s = r$, the matrix $Df(x_0)$ in Definition 5-9 is square, with r rows and columns, so it has a determinant. Because the importance of this determinant first became clear in the work of K. G. J. Jacobi, it is called the "Jacobian."

DEFINITION 5-10 *Let f be a function defined on a subset X of R^r and with values that lie in R^r, and let x_0 be a point of X at which the derivative-matrix $Df(x_0)$ exists.*

*Then the **Jacobian** of f at x_0 is the determinant*

$$\det[Df(x_0)].$$

Suppose next that Y is a set in R^s and g a function on Y with values in R^t, and that X is a set in R^r and f a function on X whose values lie in Y. For each x in X, $f(x)$ has s coordinates, which we choose to write as a column in order from the top down. Likewise, we write the t coordinates of $g(y)$ in a column from the top down. Let x_0 be a point of X at which each of the functions f^j $(j = 1, \ldots, s)$ is differentiable (that is, has a differential), and let $y_0 = f(x_0)$ be a point of Y at which each of the functions g^h $(h = 1, \ldots, t)$ is differentiable. Then by the chain rule of the differential calculus, the composite function $F = g \circ f$, or $x \mapsto F(x) = g(f(x))$ (x in X), has t components, each of which is differentiable at x_0, and the partial derivative of F^h with respect to x^i satisfies

$$D_i F^h(x_0) = \sum_{j=1}^{s} D_j g^h(y_0)\, D_i f^j(x_0).$$

The right member is merely the explicit formula for the entry in row h and column i of the matrix product $Dg(y_0)\, Df(x_0)$, and the preceding equation can be written in the matrix form

$$DF(x_0) = Dg(y_0)\, Df(x_0).$$

Thus, we have proved the following extension of the chain rule.

LEMMA 5-11 *Let X be a set in R^r and Y a set in R^s. Let f be a function on X with values in Y, and let g be a function on Y with values in R^r. Let x_0 be a point of X at which each coordinate f^j of f is differentiable, and let $f(x_0)$ be a point of Y at which each coordinate g^h of g is differentiable. If the derivative-matrices are defined by Definition* 5-9,

$$D[g \circ f](x_0) = Dg(f(x_0))\, Df(x_0).$$

Of course, $Dg(f(x_0))$ means the value of the derivative-matrix Dg at the place $f(x_0)$.

Let h be a continuous one-to-one mapping of an open set G in R^r onto an open set G^* in R^r. Assume that h is continuously differentiable on G and that the inverse mapping g is continuous and continuously differentiable on G^*. Then for all x in G,

$$g(h(x)) = x.$$

The derivative-matrix of the right member is the identity matrix I, so by Lemma 5-11,

$$\text{(V)} \qquad Dg(h(x))\, Dh(x) = I \qquad (x \text{ in } G).$$

This implies that $Dh(x)$ and $Dg(h(x))$ are reciprocal matrices. But a matrix cannot have a reciprocal unless its determinant is different from 0, so we conclude that for each x in G, $Dh(x)$ has a nonzero determinant, and for each y in G^*, $Dg(y)$ has a nonzero determinant.

If h is differentiable at a point $\bar{x}$ of G, for all x near $\bar{x}$ the value of $h(x)$ is closely approximated by the linear expression $h(\bar{x}) + Dh(\bar{x})(x - \bar{x})$. If it were exactly the same as the linear expression, the mapping would be affine, and for each cube Q that contains $\bar{x}$ in its closure we would have

$$m_L h(Q) = |\det DH(\bar{x})| m_L Q.$$

We may therefore reasonably conjecture that if Q is a small cube whose closure contains $\bar{x}$, $m_L h(Q)$ will be nearly equal to $|\det Dh(\bar{x})| m_L Q$. This is in fact true. To prove it we would have to establish two inequalities. For each positive ε we would have to show, first, that for Q with small edge-lengths, $m_L h(Q)$ does not exceed $(1 + \varepsilon)|\det Dh(\bar{x})| m_L Q$ and, second, that it is not less than $(1 - \varepsilon)|\det Dh(\bar{x})| m_L Q$. The latter inequality is considerably harder to prove than the former. Fortunately, there is a device, thought up by Jacob T. Schwartz, that allows us to deduce a substitution theorem from the former inequality alone. This device will appear in the last part of the proof of Lemma 5-13.

LEMMA 5-12 *Let h be a continuously differentiable one-to-one mapping of an open set G in R^r onto an open set G^*, and let the inverse g of h also be continuously differentiable. Let ε be positive. Then for each $\bar{x}$ in G there is a neighborhood $\gamma_1(\bar{x})$ of $\bar{x}$ such that $\gamma_1(\bar{x}) \subset G$, and if Q is any cube such that $\bar{x}$ is in the closure Q^- and $Q^- \subset \gamma_1(\bar{x})$, then*

$$m_L h(Q) < (1 + \varepsilon)|\det Dh(\bar{x})| m_L Q.$$

We choose a positive number ε' such that

$$(1 + 2\varepsilon')^r < 1 + \varepsilon.$$

Let f be the composite mapping $x \mapsto f(x) = D(g(\bar{x}))h(x)$; that is,

$$f^i(x) = \sum_{k=1}^{r} D_k g^i(h(\bar{x}))h^k(x). \tag{W}$$

Then

$$D_j f^i(x) = \sum_{k=1}^{r} D_k g^i(h(\bar{x}))\, D_j h^k(x),$$

so, by (V), $Df(\bar{x})$ is the identity matrix. Since each partial derivative is continuous on G, there exists a neighborhood $\gamma_1(\bar{x})$ contained in G on which

$$\sum_{j=1}^{r} |D_j f^i(x) - \delta_j^i| < \varepsilon' \qquad (i = 1, \ldots, r). \tag{X}$$

Now let Q be a cube such that $\bar{x}$ is in the closure Q^- and $Q^- \subset \gamma_1(\bar{x})$. If $c = (c^1, \ldots, c^r)$ is the center of Q and e is its edge-length, Q^- is the interval

$$Q^- = [c^1 - e/2, c^1 + e/2] \times \cdots \times [c^r - e/2, c^r + e/2]. \tag{Y}$$

Let x be any point of Q^-. By the theorem of the mean, for each i in $\{1, \ldots, r\}$ there is a point $\tilde{x}_i$ on the line-segment joining $\bar{x}$ to x such that

$$\begin{aligned} f^i(x) &= f^i(\bar{x}) + \sum_{j=1}^{r} D_j f^i(\tilde{x}_i)(x^j - \bar{x}^j) \\ &= f^i(\bar{x}) + x^i - \bar{x}^i + \sum_{j=1}^{r} [D_j f^i(x_i) - \delta_j^i](x^j - \bar{x}^j). \end{aligned} \tag{Z}$$

In the last sum in the right member, each factor $(x^j - \bar{x}^j)$ has absolute value at most e, since x and $\bar{x}$ are both in the cube Q^- whose edge-length is e; and because x is in Q, x^i is in $[c^i - e/2,\ c^i + e/2]$. Therefore, by (Z),

$$f^i(x) \leqq f^i(\bar{x}) + (c^i + e/2) - \bar{x}^i + \varepsilon' e.$$

Likewise,

$$f^i(x) \geqq f^i(\bar{x}) + (c^i - e/2) - \bar{x}^i - \varepsilon' e.$$

So $f(x)$ is in the cube $Q_1 = Q_1^1 \times \cdots \times Q_1^r$, where Q_1^i is the interval

$$Q_1^i = (c^i + f^i(\bar{x}) - \bar{x}^i - e/2 - \varepsilon' e,\ c^i + f^i(\bar{x}) - \bar{x}^i + e/2 + \varepsilon' e)$$

whose length is $(1 + 2\varepsilon')e$. Therefore,

$$m_L Q_1 = (1 + 2\varepsilon')^r e^r < (1 + \varepsilon) m_L Q. \tag{AA}$$

Since $f(Q) \subset Q_1$, $Dh(\bar{x})f(Q) \subset Dh(\bar{x})Q_1$, and

$$m_L[Dh(\bar{x})f(Q)] \leqq m_L[Dh(\bar{x})Q_1]. \tag{BB}$$

But by (V) and (W),

$$Dh(\bar{x})f(Q) = Dh(\bar{x})[Dg(h(\bar{x})h(Q)] = h(Q),$$

and by Corollary 5-8,

$$m_L[Dh(\bar{x})Q_1] = |\det Dh(\bar{x})| m_L Q_1.$$

So (BB) implies

$$m_L h(Q) \leqq |\det Dh(\bar{x})| m_L Q_1.$$

This and (AA) complete the proof.

LEMMA 5-13 *Let G, G^*, h, and g be as in Lemma 5-12. Let* $\det Dh(x)$ *be bounded on G and* $\det Dg(y)$ *be bounded on G^*. Let $y \to f(y)$ be continuous, nonnegative, and bounded on G^*. Then*

$$\int_G f(h(x)) |\det Dh(x)|\, dx = \int_{G^*} f(y)\, dy. \tag{CC}$$

Let c be any number greater than the left member of (CC). We can and do choose a positive number ε such that

$$\text{(DD)} \qquad (1+\varepsilon)\int_G f(h(x))|\det Dh(x)|\,dx + (1+\varepsilon)\varepsilon m_1 G + \varepsilon m_1 G^* < c.$$

By Lemma 5-12, for each $\bar{x}$ in G there is a neighborhood $\gamma_1(\bar{x})$ contained in G such that if Q is any cube such that $\bar{x}$ is in Q^- and $Q^- \subset \gamma_1(\bar{x})$, then

$$\text{(EE)} \qquad m_L h(Q) < (1+\varepsilon)|\det Dh(\bar{x})| m_L Q.$$

Since the functions $x \mapsto f(h(x))$ and $x \mapsto f(h(x))|\det Dh(x)|$ are continuous on G, for each $\bar{x}$ in G there is a neighborhood $\gamma_2(\bar{x})$ such that, for every x in $\gamma_2(\bar{x}) \cap G$,

$$\text{(FF)} \qquad |f(h(x)) - f(h(\bar{x}))| < \varepsilon,$$
$$\big| f(h(x))|\det Dh(x)| - f(h(\bar{x}))|\det Dh(\bar{x})| \big| < \varepsilon.$$

We define $\gamma = \gamma_1 \cap \gamma_2$, and we choose a sequence of pairwise disjoint cubes $A_1, A_2, A_3, \ldots$ and a sequence of points $\bar{x}_1, \bar{x}_2, \bar{x}_3, \ldots$ of G such that

(i) G is contained in the union of the A_i, and
(ii) for each positive integer i, $\bar{x}_i$ is in A_i, and $A_i^- \subset \gamma(\bar{x}_i)$.

This is possible by Lemma 5-3. By (ii), the union of the A_i is contained in G, and by (i), it contains G, so it is G. Each $h(A_i)$ is measurable by Corollary 5-6. If y is in $h(A_i)$, then $y = h(x)$ for some x in A_i, and by (FF),

$$f(y) < f(h(\bar{x}_i)) + \varepsilon.$$

If x is in A_i, by (FF),

$$f(h(x))|\det Dh(x)| > f(h(\bar{x}_i))|\det Dh(\bar{x}_i)| - \varepsilon.$$

From these inequalities and (EE) we deduce

$$\begin{aligned}
\int_{G^*} f(y) 1_{h(A_i)}\,dy &= \int_{h(A_i)} f(y)\,dy \\
&\leqq \int_{h(A_i)} [f(h(\bar{x}_i)) + \varepsilon]\,dy \\
&= f(h(\bar{x}_i)) m_L h(A_i) + \varepsilon m_L h(A_i) \\
&\leqq f(h(\bar{x}_i))[(1+\varepsilon)|\det Dh(\bar{x}_i)|] m_L A_i + \varepsilon m_L h(A_i) \\
&= (1+\varepsilon)\int_{A_i} [f(h(\bar{x}_i))|\det Dh(\bar{x}_i)|]\,dx + \varepsilon m_L h(A_i) \\
&\leqq (1+\varepsilon)\int_{A_i} [f(h(x))|\det Dh(x)| + \varepsilon]\,dx + \varepsilon m_L h(A_i) \\
&= (1+\varepsilon)\int_G f(h(x))|\det Dh(x)| 1_{A_i}(x)\,dx \\
&\qquad + (1+\varepsilon)\varepsilon m_L A_i + \varepsilon m_L h(A_i).
\end{aligned}$$

We add these inequalities member by member (first and last members) for $i = 1, \ldots, n$ and then let n increase. Since the union of the A_i is G and the union of the $h(A_i)$ is G^*, by the monotone convergence theorem we obtain

$$\int_{G^*} f(y)\,dy \leqq (1+\varepsilon)\int_G f(h(x))|\det Dh(x)|\,dx + (1+\varepsilon)\varepsilon m_L G + \varepsilon m_L G^*.$$

But the right member is less than c, and c is any number greater than the left member of (CC), so we have established

$$\text{(GG)} \qquad \int_{G^*} f(y)\,dy \leqq \int_G f(h(x))|\det Dh(x)|\,dx.$$

The hypotheses of the lemma remain satisfied if we interchange G and G^* and also interchange g and h, so by what we have already proved we know that if ϕ is nonnegative, bounded, and continuous on G then, by (GG) with G and G^* and h and g interchanged,

$$\text{(HH)} \qquad \int_G \phi(x)\,dx \leqq \int_{G^*} \phi(g(y))|\det Dg(y)|\,dy.$$

We apply this with $\phi(x) = f(h(x))|\det Dh(x)|$. Then the integrand in the right member of (HH) is

$$f(h(g(y)))|\det Dh(g(y))|\,|\det Dg(y)|,$$

which is $f(y)$. So (HH) is the reverse inequality to (GG), and the members of (GG) are equal. The proof is complete.

We can now establish the substitution theorem toward which we have been working.

THEOREM 5-14 *Let h be a one-to-one continuously differentiable mapping of an open set G in R^r onto an open set G^*, with a continuously differentiable inverse g. Let f be continuous on G^*. Then f is integrable over G^* if and only if the function $x \mapsto f(h(x))|\det Dh(x)|$ is integrable over G, and in that case*

$$\text{(II)} \qquad \int_{G^*} f(y)\,dy = \int_G f(h(x))|\det Dh(x)|\,dx.$$

Suppose, first, that f is nonnegative. The function ϕ defined on G by

$$\phi(x) = \max\{|x^1|, \ldots, |x^r|, |h^1(x)|, \ldots, |h^r(x)|, f(h(x)), |\det Dh(x)|, |\det Dg(h(x))|\}$$

is continuous on G. For each positive integer n we define

$$G(n) = \{x \text{ in } G : \phi(x) < n\}.$$

By Lemma 5-2 this is open, and so is $G(n)^* = h(G(n))$. Evidently both are bounded, and f and $\det Dg$ are bounded on $G(n)^*$ and $\det Dh$ is bounded on $G(n)$.

By Lemma 5-13,

$$\text{(JJ)} \qquad \int_{G^*} f(y) 1_{G(n)^*}(y)\, dy = \int_G f(h(x)) |\det Dh(x)| 1_{G(n)}(x)\, dx.$$

As n increases, the integrands ascend. Every x in G is in $G(n)$ for all large n, and every y in G^* is in $G(n)^*$ for all large n. So the integrands in (JJ) tend to those in (II), and by the monotone convergence theorem, (II) holds. The conclusion is established for nonnegative f.

Suppose that f satisfies the hypotheses of the theorem. Then so do f^+ and f^-. If f is integrable over G^*, so are f^+ and f^-, and by the preceding part of this proof, both $(f^+)|\det Dh|$ and $(f^-)|\det Dh|$ are integrable over G, and

$$\int_{G^*} f^+(y)\, dy = \int_G f^+(h(x)) |\det Dh(x)|\, dx,$$

$$\int_{G^*} f^-(y)\, dy = \int_G f^-(h(x)) |\det Dh(x)|\, dx.$$

By subtraction, we find that the right member of (II) exists and that (II) is satisfied. Conversely, suppose that the function ϕ defined by

$$\phi(x) = f(h(x)) |\det Dh(x)|$$

is integrable over G. Then so are ϕ^+ and ϕ^-. But ϕ^+ and ϕ^- are, respectively, the functions satisfying

$$\phi^+(x) = f^+(h(x)) |\det Dh(x)|, \qquad \phi^-(x) = f^-(h(x)) |\det Dh(x)|,$$

so by the preceding proof, f^+ and f^- are integrable over G^*. Then so is their difference f. This completes the proof.

As an example, we shall consider the transformation from rectangular to polar coordinates in the plane. For points x in R^2 we shall write $x = (\rho, \theta)$, and for points y we shall write $y = (w, z)$. Then the transformation h defined by

$$y = h(x) = (\rho \cos\theta,\ \rho \sin\theta),$$

or

$$w = \rho\cos\theta, \qquad z = \rho\sin\theta,$$

is a continuously differentiable mapping of R^2 onto R^2, but it is not one-to-one. The matrix $Dh(x)$, or $Dh(\rho, \theta)$, is

$$\begin{bmatrix} \cos\theta & -\rho\sin\theta \\ \sin\theta & \rho\cos\theta \end{bmatrix},$$

so,

$$\det Dh(x) = \rho.$$

If we define B to be the union of the interval $(0, \infty) \times [0, 2\pi)$ with the single point $(0, 0)$, h maps B one-to-one onto R^2, but B is not open. But if we denote by W^+ the nonnegative w-axis,

$$W^+ = \{(w, z) \text{ in } R^2 : w \geqq 0,\ z = 0\},$$

$R^2 \setminus W^+$ is open, and h maps it one-to-one onto the interior

$$B^0 = (0, \infty) \times (0, 2\pi),$$

and the sets W^+ and $B \setminus B^0$ have measure 0. This lets us prove the following substitution formula.

(KK) Let E^* be an open set in R^2 and let f be continuous on E^*. Let E be the set of all points (ρ, θ) in the set

$$B - \{(0, \infty) \times [0, 2\pi)\} \cup \{(0, 0)\}$$

such that $(\rho \cos\theta, \rho \sin\theta)$ is in E^*. Then f is integrable over E^* if and only if the function

$$(\rho, \theta) \mapsto f(\rho\cos\theta, \rho\sin\theta)\rho$$

is integrable over E, and in that case

$$\int_{E^*} f(y)\, dy = \int_E f(\rho\cos\theta, \rho\sin\theta)\rho[d\rho\, d\theta].$$

Let $G^* = E^* \setminus W^+$ and $G = E \cap B^0$. These are open sets, and W^+ and $B \setminus B^0$ have measure 0, so

$$\int_{G^*} f(y)\, dy = \int_{E^*} f(y)\, dy,$$

$$\int_G f(\rho\cos\theta, \rho\sin\theta)\rho\, [d\rho\, d\theta] = \int_E f(\rho\cos\theta, \rho\sin\theta)\rho\, [d\rho\, d\theta].$$

By Theorem 5-14, the left members of these two equations are equal, and (KK) is established.

By use of (KK) we can prove that

$$\text{(LL)} \qquad \int_R \exp\left(-\frac{u^2}{2}\right) du = (2\pi)^{1/2}.$$

In Section II-11 we have already used this evaluation in anticipation of its proof. We see without difficulty that the integral has a finite value; we call it J. If we define f and g on R^2 by

$$g(w, z) = f(y) = \exp(-[w^2 + z^2]/2),$$

where, as before, $y = (w, z)$, we compute

$$\int_R g(w, z)\, dz = \int_R \exp\left(-\frac{w^2}{2}\right)\exp\left(-\frac{z^2}{2}\right) dz = J \exp\left(-\frac{w^2}{2}\right),$$

$$\int_R \left\{\int_R g(w, z)\, dw\right\} dz = \int_R J \exp\left(-\frac{w^2}{2}\right) dw = J^2.$$

Since f is obviously the limit of a sequence of step-functions, the hypotheses of Corollary 4-3 are satisfied. Therefore, f is integrable over R^2, and

$$\int_{R^2} f(y)\, dy = J^2.$$

With the same notation as before, we define

$$E^* = R^2 \setminus W^+, \qquad E = B^0 = (0, \infty) \times (0, 2\pi).$$

Since $R^2 \setminus E^*$ is the set W^+ of measure 0, by (KK),

$$J^2 = \int_{E^*} f(y)\, dy = \int_E f(\rho \cos\theta, \rho \sin\theta)\rho\, [d\rho\, d\theta].$$

The integrand in the last integral satisfies the hypotheses of Corollary 4-3, and

$$f(\rho\cos\theta, \rho\sin\theta)\rho = \exp\left(-\frac{\rho^2\cos^2\theta + \rho^2\sin^2\theta}{2}\right)\rho$$

$$= \rho \exp\left(-\frac{\rho^2}{2}\right) = D\left[-\exp\left(-\frac{\rho^2}{2}\right)\right].$$

So by Corollary 4-3 and the fundamental theorem,

$$J^2 = \int_{(0,\infty)} \left\{\int_{(0,2\pi)} \left[\exp\left(-\frac{\rho^2}{2}\right)\rho\, d\theta\right] d\rho\right.$$

$$= 2\pi \int_{(0,\infty)} D\left[-\exp\left(-\frac{\rho^2}{2}\right)\right] d\rho$$

$$= 2\pi \lim_{n\to\infty} \int_0^n D\left[-\exp\left(-\frac{\rho^2}{2}\right)\right] d\rho$$

$$= 2\pi \lim_{n\to\infty} \left[-\exp\left(-\frac{n^2}{2}\right) + \exp(0)\right]$$

$$= 2\pi.$$

Since $J \geqq 0$, this implies that $J = (2\pi)^{1/2}$, as was to be proved.

EXERCISE 5-1 In R^h, let $S(h, r)$ be the set

$$\{x \text{ in } R^h : (x^1)^2 + \cdots + (x^h)^2 < r^2\}.$$

(Thus, $S(2, r)$ is a circle of radius r and $S(3, r)$ is a ball of radius r.) Taking $u = x^1$ and $v = (x^2, \ldots, x^h)$, compute successively

$$m_L S(2, r) = \pi r^2, \qquad m_L S(3, r) = \tfrac{4}{3}\pi r^3, \qquad m_L S(4, r) = 8\pi r^4.$$

EXERCISE 5-2 Compute the volume of the set in R^3 that is bounded below by the (x^1, x^2)-plane and above by the surface

$$x^{(3)} = 4 - (x^{(1)})^2 - (x^{(2)})^2.$$

EXERCISE 5-3 Let $B = (-1, 1) \times (-1, 1)$. Compute

$$\int_B \frac{[du\,dv]}{(u^2 + v^2)^{1/2}}.$$

Suggestion: Corollary 4-2 applies to the region outside the circle $\{(u, v)$ in $R^2 : u^2 + v^2 \leqq \varepsilon^2\}$ and inside the triangle with vertices $(0, 0)$, $(1, 0)$, and $(1, 1)$. Use polar coordinates and let ε tend to 0.)

EXERCISE 5-4 Let S be the unit sphere $\{(x, y, z)$ in $R^3 : x^2 + y^2 + z^2 < 1\}$. For all real c define

$$J(c) = \int_S \frac{[dx\,dy\,dz]}{x^2 + y^2 + (z - c)^2}.$$

Show that if c is not 1, -1 or 0,

$$J(c) = \pi\{2 + (c - c^{-1})\log|c - 1| - (c - c^{-1})\log|c + 1|\}.$$

Suggestion: Integrate first with respect to $[dx\,dy]$; in the inner integral, change to polar coordinates. If $0 < c < 1$, use Exercise II-9-5.

EXERCISE 5-5 Using the notation of Exercise 5-1, compute $J(0)$ and show that $J(c)$ tends to $J(0)$ as c tends to 0.

EXERCISE 5-6 Evaluate

$$\int_H \frac{[dx\,dy]}{(1 + x^2 + y^2)^2},$$

where H is the regular hexagon of side 2 with center at the origin. *Suggestion*: There is an obvious way of subdividing the hexagon into six equilateral triangles. Use polar coordinates in each.

EXERCISE 5-7 Prove the following two statements:

(i) if G_1 and G_2 are open sets in R^r, their intersection $G_1 \cap G_2$ is open;
(ii) if $\mathcal{G}$ is a collection of open sets in R^r, the union of all the sets belonging to $\mathcal{G}$ is open.

6. Approximation of Sets by Unions of Intervals and of Integrable Functions by Limits of Step-Functions

Those readers who found it wise to pass over the latter sections of Chapter II will find it even wiser to pass over this section and the next two.

By Corollary 5-4 we see that if G is an open set, its measure can be computed as the sum of the measures of a sequence of intervals. If E is any set of finite measure, we can still approximate its measure arbitrarily closely by the sums of measures of a sequence of intervals.

★THEOREM 6-1 *Let E be a set of finite measure in R^r. Let ε be positive, and let G be an open set that contains E. Then there exists a sequence $A_1, A_2, A_3, \ldots$ of pairwise disjoint left-open cubes such that E is contained in the union of the A_i, and the closure of each A_i is contained in G, and*

$$\sum_{i=1}^{\infty} mA_i < mE + \varepsilon.$$

Let ε be positive and let G be an open set that contains E. Since the integral of 1_E is mE, there exists a gauge γ_1 on $\bar{R}^r$ such that for every γ_1-fine partition $\mathscr{P}$ of R^r,

$$\text{(A)} \qquad |S(\mathscr{P}; 1_E; m) - mE| < \varepsilon/6.$$

For each x in G, since G is open, there is a neighborhood $\gamma_2(x)$ contained in G; if x is not in G, we take $\gamma_2(x) = \bar{R}^r$. We define $\gamma = \gamma_1 \cap \gamma_2$. By Lemma 5-3, there is a sequence $A_1, A_2, A_3, \ldots$ of pairwise disjoint left-open cubes such that E is contained in the union of the A_i, and for each A_i there is a point $\bar{x}_i$ of E contained in A_i for which $A_i^- \subset \gamma(\bar{x}_i)$. Then $A_i^- \subset G$. By Corollary II-13-2, with (A), we find that, for each positive integer n,

$$\text{(B)} \qquad \sum_{i=1}^{n} \left| 1_E(\bar{x}_i) mA_i - \int_{A_i} 1_E(x)\, m(dx) \right| < 5\varepsilon/6.$$

Since $\bar{x}_i$ is in E, $1_E(\bar{x}_i) = 1$, and (B) implies

$$\sum_{i=1}^{n} mA_i < \sum_{i=1}^{n} \int_{A_i} 1_E(x)\, m(dx) + 5\varepsilon/6$$
$$\leqq \int_{R^r} 1_E(x)\, m(dx) + 5\varepsilon/6 = mE + 5\varepsilon/6.$$

We let n increase, and we obtain

$$\sum_{i=1}^{\infty} mA_i \leqq mE + 5\varepsilon/6 < mE + \varepsilon.$$

The proof is complete.

This theorem permits us to make a useful connection with the published literature of integration theory. Many authors *define* a set of measure 0 to be a set E that has the property

(C) for each positive ε there exists a sequence of intervals $A_1, A_2, A_3, \ldots$ whose union contains E and the sum of whose measures is less than ε.

Theorem 6-1 shows that every set of measure 0 has property (C). Conversely, by Theorem II-12-3, every set E with property (C) has measure 0. So if we encounter in some book or paper a set that is of measure 0 according to the definition based on property (C), we can safely understand it to be a set of measure 0 in our sense too.

From the point of view of integration, the functions next in simplicity to step-functions are those that are limits of ascending or descending sequences of step-functions. We shall call these U-functions and L-functions, respectively, although the names have previously been used with a slightly different meaning.

★DEFINITION 6-2 *A function u on R^r is a **U-function** if there exists an ascending sequence of step-functions, each with left-open intervals of constancy, whose limit is u. A function l on R^r is an **L-function** if there exists a descending sequence of step-functions, each with left-open intervals of constancy, whose limit is l.*

By Theorem II-12-6, every U-function u has an integral, finite or $+\infty$, and that integral is the limit of the integrals of any ascending sequence of step-functions with left-open intervals of constancy that converges to u. Likewise, every L-function l has an integral, finite or $-\infty$, and that integral is the limit of the integrals of any descending sequence of step-functions with left-open intervals of constancy that converges to l. The next lemma is trivial.

★LEMMA 6-3 (i) *A function u_1 is a U-function if and only if $-u_1$ is an L-function.*

(ii) *If $u_1, \ldots, u_n$ are U-functions, $u_1 + \cdots + u_n$ and $u_1 \wedge \cdots \wedge u_n$ are U-functions.*

Let $u_1, \ldots, u_n$ be U-functions. For each j in $\{1, \ldots, n\}$ there is an ascending sequence $s_{j,1}, s_{j,2}, s_{j,3}, \ldots$ of step-functions with left-open intervals of constancy that converges to u_j. Then for each positive integer k, $s_{1,k} + \cdots + s_{n,k}$ and $s_{1,k} \wedge \cdots \wedge s_{n,k}$ are step-functions with left-open intervals of constancy, and as k increases they ascend and approach the respective limits $u_1 + \cdots + u_n$ and $u_1 \wedge \cdots \wedge u_n$. This establishes conclusion (ii). Also, if u_1 is a U-function with the same sequence $s_{1,k}, s_{1,2}, s_{1,3}, \ldots$ that was used above, the sequence $-s_{1,1}$, $-s_{1,2}$, $-s_{1,3}, \ldots$ is a descending sequence of step-functions with left-open intervals of constancy that converges to $-u_1$, so $-u_1$ is an L-function. Similarly, we prove that if $-u_1$ is an L-function, u_1 is a U-function.

★THEOREM 6-4 *If f is integrable over R^r and $\varepsilon > 0$, there exist a U-function u and an L-function l such that $l \leqq f \leqq u$ and*

$$\text{(D)} \qquad \int_{R^r} u(x)\, m(dx) < \int_{R^r} f(x)\, m(dx) + \varepsilon,$$

$$\int_{R^r} l(x)\, m(dx) > \int_{R^r} f(x)\, m(dx) - \varepsilon.$$

For each positive integer n there is a gauge γ_n on $\bar{R}^r$ such that, for every γ_n-fine partition $\mathscr{P}$ of $\bar{R}^r$,

$$\left| S(\mathscr{P}; f_{R^r}; m) - \int_{R^r} f(x)\, m(dx) \right| < \frac{\varepsilon}{5 \cdot 2^{n+2}}.$$

We may suppose that $\gamma_n(x)$ is bounded whenever x is in R^r. Then by Theorem II-13-1, there exists a step-function s'_n with left-open intervals of constancy such that

$$\text{(E)} \qquad \int_{R^r} |f(x) - s'_n(x)|\, m(dx) < \frac{\varepsilon}{2^{n+2}}.$$

We define

$$s_n = s'_1 \vee \cdots \vee s'_n \qquad (n = 1, 2, 3, \ldots).$$

These form an ascending sequence of step-functions with left-open intervals of constancy, and

$$\begin{aligned} s_n - f &= [s'_1 - f] \vee \cdots \vee [s'_n - f] \\ &\leqq |s'_1 - f| \vee \cdots \vee |s'_n - f| \\ &\leqq |s'_1 - f| + \cdots + |s'_n - f|. \end{aligned}$$

Integration over R^r yields, with (E),

$$\text{(F)} \qquad \int_{R^r} [s_n(x) - f(x)]\, m(dx) \leqq \sum_{i=1}^{n} \frac{\varepsilon}{2^{i+2}}.$$

As n increases, the limit of the ascending sequence $s_1, s_2, s_3, \ldots$ is a function u' which by definition is a U-function. By (F) and the monotone convergence theorem,

$$\text{(G)} \qquad \int_{R^r} [u'(x) - f(x)]\, m(dx) \leqq \sum_{i=1}^{\infty} \varepsilon/2^{i+2} = \varepsilon/4.$$

For all n, $u' \geqq s_n \geqq s'_n$, so $f - f \wedge u' \leqq f - f \wedge s'_n \leqq |f - s'_n|$. If we integrate over R^r and recall (E), we obtain

$$\int_{R^r} [f(x) - f(x) \wedge u'(x)]\, m(dx) \leqq \int_{R^r} |f(x) - s'_n(x)|\, m(dx) < \varepsilon/2^{n+2}.$$

The first integral in this inequality is independent of n and is less than the arbitrarily small positive number $\varepsilon/2^{n+2}$, so it cannot be positive. The integrand is nonnegative, so the integral cannot be negative, and therefore it is 0. By Theorem II-12-4, the integrand must be 0 almost everywhere. That is, there exists a set N of measure 0 such that, except in N, $f(x) - f(x) \wedge u'(x) = 0$, which is the same as saying that except on N, $u'(x) \geqq f(x)$.

By Theorem 6-1, for each positive integer n there exists a sequence $A(n, 1)$, $A(n, 2)$, $A(n, 3), \ldots$ of pairwise disjoint left-open cubes such that their union contains N and the sum of their measures is less than $\varepsilon/2^{n+2}$. We define

$$\text{(H)} \qquad s_n'' = \sum_{i,j=1}^{n} 1_{A(i,j)}.$$

This is a step-function with left-open intervals of constancy. Integrating over R^r yields

$$\text{(I)} \qquad \int_{R^r} s_n''(x)\, m(dx) = \sum_{i,j=1}^{n} mA(i,j)$$
$$\leqq \sum_{i=1}^{n} \sum_{j=1}^{\infty} mA(i,j) < \sum_{i=1}^{n} \varepsilon/2^{i+2} < \varepsilon/4.$$

As n increases, the s_n'' ascend and approach a limit function, which by definition is a U-function. We denote it by u''. By the monotone convergence theorem, with (I),

$$\text{(J)} \qquad \int_{R^r} u''(x)\, m(dx) \leqq \varepsilon/4.$$

Let x be any point in N, and let k be any positive integer. For each i in the set $\{1, \ldots, k\}$ the union of the $A(i,j)$ $(j = 1, 2, 3, \ldots)$ contains x, so there is an integer $j(i)$ such that $A(i, j(i))$ contains x. Then for all n greater than the greatest of $j(1), \ldots, j(k)$ we have

$$\sum_{j=1}^{n} 1_{A(i,j)}(x) = 1 \qquad (i = 1, \ldots, k).$$

By (H), $s_n''(x) \geqq k$. Since k is an arbitrary integer, this implies that $s_n''(x)$ tends to ∞ as n increases, so

$$\text{(K)} \qquad u''(x) = \infty \qquad (x \text{ in } N).$$

We now define $u = u' + u''$. By Lemma 6-3, u is a U-function. By (G) and (J),

$$\int_{R^r} u(x)\, m(dx) \leqq \int_{R^r} f(x)\, m(dx) + \varepsilon/4 + \varepsilon/4 < \int_{R^r} f(x)\, m(dx) + \varepsilon.$$

For each x in R^r, either x is in N or x is in $R^r \setminus N$. If x is in N, $u''(x) = \infty$ by (K), so $u(x) = \infty \geqq f(x)$. If x is in $R^r \setminus N$, $u''(x) \geqq f(x)$ by definition of N, and

$u''(x) \geqq 0$, so in this case too we have $u(x) \geqq f(x)$. So u is a U-function such that $u \geqq f$, and the first inequality in (D) is satisfied.

Since f is integrable, so is $-f$. By the part of the proof already completed, there exists a U-function u_1 such that $u_1 \geqq -f$ and

$$\int_{R^r} u_1(x)\, m(dx) < \int_{R^r} [-f(x)]\, m(dx) + \varepsilon.$$

We define $l = -u_1$. By Lemma 6-3, l is an L-function and $l \leqq f$, and the second inequality in (D) is satisfied. The proof is complete.

A great step forward in mathematical analysis was the invention by Henri Lebesgue early in the twentieth century of a new form of integral. This was done first for real-valued functions on bounded intervals in R—the measure of an interval A being what we have called $m_L A$. (The "L" in m_L is in honor of Lebesgue.) This was extended to integrals over spaces of higher dimension and, in fact, to still more complicated spaces and also to measures other than m_L. The resulting integrals are sometimes called Radon–Lebesgue or Lebesgue–Stieltjes integrals and sometimes simply Lebesgue integrals. In 1918 P. J. Daniell published another way of defining an integral that could be shown to be equivalent to the Lebesgue integral. Starting with integrals of step-functions, the integrals of U-functions and of L-functions are defined as the obvious limits. (Of course, it has to be shown that these limits do not depend on the choice of sequence of step-functions but only on the U-function or L-function itself.) Then if there is a single number J that is a lower bound for the integrals of all U-functions $\geqq f$ and an upper bound for the integrals of all L-functions $\leqq f$, this number J is the integral (Daniell, or Lebesgue, or Lebesgue–Stieltjes) of f. Theorem 6-4 shows that if f has a gauge-integral over R^r, it has a Daniell (or Lebesgue) integral over R^r, and the two are equal. But by Lemma I-7-1, if f has a Daniell integral over R^r, it has a gauge-integral, and the two are equal. Therefore, the gauge-integral is equivalent to the Daniell integral. This opens the literature to us. If we encounter a theorem about the Lebesgue integral, we are at liberty to apply it to the gauge-integral, and vice versa.

The next corollary will be useful in the next two sections and elsewhere.

★COROLLARY 6-5 *Let f be integrable with respect to m over R^r. Then there exist functions g, h on R^r such that*

(i) *g is the limit of an ascending sequence of integrable L-functions, and h is the limit of a descending sequence of U-functions;*

(ii) *$g(x) \leqq f(x) \leqq h(x)$ for all x in R^r;*

(iii) *$g(x) = f(x) = h(x)$ for almost all x in R^r.*

For each positive integer n there is a U-function u'_n on R^r such that

$$u'_n \geqq f \tag{L}$$

and

$$\text{(M)} \qquad \int_{R^r} u_n'(x)\,m(dx) < \int_{R^r} f(x)\,m(dx) + 1/n,$$

by Theorem 6-4. We define $u_n = u_1' \wedge \cdots \wedge u_n'$ $(n = 1, 2, 3, \ldots)$. By Lemma 6-3, u_n is a U-function. By (L),

$$\text{(N)} \qquad u_n \geqq f.$$

Since $u_n \leqq u_n'$, by (M),

$$\text{(O)} \qquad \int_{R^r} u_n(x)\,m(dx) < \int_{R^r} f(x)\,m(dx) + 1/n.$$

The u_n form a descending sequence, so they approach a limit, which we call h. Then by (N),

$$\text{(P)} \qquad h \geqq f,$$

and by (O) and the monotone convergence theorem,

$$\text{(Q)} \qquad \int_{R^r} h(x)\,m(dx) \leqq \int_{R^r} f(x)\,m(dx).$$

Because of (P), equality must hold in (Q). Then the nonnegative function $h - f$ (understood to be 0 where $h(x) = f(x) = \pm\infty$) has integral 0. By Theorem II-12-4, it must be almost everywhere 0. This completes the proof of the statements concerning h.

Since $-f$ is also integrable over R^r, by the part of the proof already completed there is a descending sequence of integrable U-functions $u_1', u_2', u_3', \ldots$ with a limit h' such that $h' \geqq -f$ and $h'(x) = -f(x)$ for almost all x. Then $-u_1', -u_2', -u_3', \ldots$ is an ascending sequence of L-functions whose limit is $g = -h'$, and $g(x) = -h'(x) \leqq f(x)$ for all x in R^r, with equality holding for almost all x. The proof is complete.

Among the applications of Corollary 6-5 is a more powerful theorem on integration by substitution in R^1.

Theorem 6-6 *Let f be integrable over a closed interval $[a, b]$ in R^1, and let $\phi\colon y \mapsto \phi(y)$ $(\alpha \leqq y \leqq \beta)$ be a function with values in $[a, b]$ that is the indefinite integral of a function $\dot{\phi}$ that is either nonnegative or nonpositive on $[\alpha, \beta]$. Then*

$$\text{(R)} \qquad \int_\alpha^\beta f(\phi(y))\dot{\phi}(y)\,dy = \int_{\phi(\alpha)}^{\phi(\alpha)} f(x)\,dx.$$

To be specific we suppose that $\dot{\phi} \leqq 0$; the case $\dot{\phi} \geqq 0$ is similar but a trifle simpler. Then ϕ is nonincreasing but may have intervals of constancy.

Let $(c, d]$ be a nondegenerate subinterval of $[\phi(\beta), \phi(\alpha)]$. Let c' be the smallest y such that $\phi(y) = c$, and let d' and d'' be, respectively, the smallest and the largest y such that $\phi(y) = d$. Then

$$\alpha \leqq d' \leqq d'' < c' \leqq \beta.$$

It is easy to see that

$$c < \phi(y) < d \qquad \text{if and only if} \quad c' > y > d'',$$

$$\phi(y) = d \qquad \text{if and only if} \quad d' \leqq y \leqq d''.$$

These statements are equivalent to the equations

$$\text{(S)} \qquad 1_{(c,d)}(\phi(y)) = 1_{(d'',c')}(y),$$

$$\text{(T)} \qquad 1_{[d,d]}(\phi(y)) = 1_{[d',d'']}(y) \qquad (y \text{ in } R).$$

From (S) we deduce

$$\text{(U)} \qquad \int_{\phi(\alpha)}^{\phi(\beta)} 1_{(c,d)}(x)\,dx = -\int_{\phi(\beta)}^{\phi(\alpha)} 1_{(c,d)}(x)\,dx$$

$$= -\int_c^d 1\,dx = c - d = \phi(c') - \phi(d'')$$

$$= \int_{d''}^{c'} \dot{\phi}(y)\,dy$$

$$= \int_\alpha^\beta 1_{(d'',c')}(y)\dot{\phi}(y)\,dy$$

$$= \int_\alpha^\beta 1_{(c,d)}(\phi(y))\dot{\phi}(y)\,dy.$$

From (T),

$$\text{(V)} \qquad \int_{\phi(\alpha)}^{\phi(\beta)} 1_{[d,d]}(x)\,dx = 0 = \phi(d'') - \phi(d')$$

$$= \int_{d'}^{d''} \dot{\phi}(y)\,dy$$

$$= \int_\alpha^\beta 1_{[d',d'']}(y)\dot{\phi}(y)\,dy$$

$$= \int_\alpha^\beta 1_{[d,d]}(\phi(y))\dot{\phi}(y)\,dy.$$

By adding equations (U) and (V) member by member, we find that (R) is valid when f is the indicator $1_{(c,d]}$ of a left-open subinterval of $[\phi(\beta), \phi(\alpha)]$.

Next let f be a step-function on $[\phi(\beta), \phi(\alpha)]$ with left-open intervals of constancy. Then f is a sum of constant multiples of indicators of left-open subintervals of $[\phi(\beta), \phi(\alpha)]$. By applying the result (R) to each of these, multiplying by the constant factors, and adding, we find that (R) is valid for such step-functions.

If f is a U-function integrable over $(\phi(\beta), \phi(\alpha)]$, it is the limit of an ascending sequence of step-functions with left-open intervals of constancy, say of the step-functions $s_1, s_2, s_3, \ldots$. For each of these

$$\text{(W)} \qquad \int_{\alpha}^{\beta} s_n(\phi(y))\dot{\phi}(y)\,dy = \int_{\phi(\alpha)}^{\phi(\beta)} s_n(x)\,dx.$$

As n increases, by the monotone convergence theorem the right member of (W) tends to the right member of (R). Again by the monotone convergence theorem, the left member of (W) tends to the left member of (R), and so (R) holds for integrable U-functions. Likewise, it holds for L functions integrable over $[a, b]$.

If g is the limit of an ascending sequence of L-functions and is integrable over $[a, b]$, by the same argument as in the preceding paragraph, (R) holds for g. Likewise, (R) holds for every function h that is integrable and is the limit of a descending sequence of integrable U-functions.

Now let f be integrable over $[a, b]$ and let g and h be the functions of Corollary 6-5. By the preceding proof,

$$\text{(X)} \qquad \int_{\alpha}^{\beta} g(\phi(y))\dot{\phi}(y)\,dy = \int_{\phi(\alpha)}^{\phi(\beta)} g(x)\,dx,$$

$$\text{(Y)} \qquad \int_{\alpha}^{\beta} h(\phi(y))\dot{\phi}(y)\,dy = \int_{\phi(\alpha)}^{\phi(\beta)} h(x)\,dx.$$

By Corollary 6-5, the right members of (X) and (Y) are both equal to the right member of (R), so the left members are both equal to that number. Since $\dot{\phi} \leqq 0$,

$$h(\phi(y))\dot{\phi}(y) \leqq f(\phi(y))\dot{\phi}(y) \leqq g(\phi(y))\dot{\phi}(y)$$

for all y in $[\alpha, \beta]$. By Corollary I-7-3, this implies that (R) is valid, and the proof is complete.

EXERCISE 6-1 Prove that if M is an upper bound for f in Corollary 6-5, we can choose U-functions that descend and tend to f and that all have upper bound M. Prove the corresponding statement for lower bounds.

EXERCISE 6-2 Let g be a monotonic function that is an indefinite integral on an interval $[c, d]$. If E is a subset of $[c, d]$ with measure 0, $g(E)$ has measure 0.

EXERCISE 6-3 Let f_1 be a function integrable over R^s and f_2 a function integrable over R^t. Define f on R^r (where $r = s + t$) by

$$f(x) = f(u, v) = f_1(u)f_2(v).$$

Show that f is integrable over R^r, and

$$\int_{R^r} f(x)\,dx = \left[\int_{R^s} f_1(u)\,du\right]\left[\int_{R^t} f_2(v)\,dv\right].$$

Suggestion: This is easy for step-functions. For nonnegative U-functions f_1, f_2 and for nonnegative L-functions it holds by the monotone convergence theorem. Use Corollary 6-5.

EXERCISE 6-4 Let E be a set in the half-plane $\{(x, y): y \geqq 0\}$ such that $m_L E$ exists and is finite and the function $(x, y) \mapsto y$ is integrable over E. Let E^* be the set in (x, y, z)-space obtained by rotating E about the x-axis. Show that E^* has finite three-dimensional measure and that

$$m_L E^* = 2\pi \int_E y\,[dx\,dy].$$

7. A Second Form of Fubini's Theorem

We shall now prove a form of Fubini's theorem in which the conclusions are a little weaker than those in Theorem 4-1 but the hypotheses about the integrand f are as weak as possible; nothing is assumed about f beyond mere integrability. We shall use the same notation as in Section 4. Integration is over an interval A in R^r. This is represented as the Cartesian product of intervals $A(1)$ and $A(2)$ in R^s and R^t respectively, where s and t are positive integers with $s + t = r$. Points in R^r, R^s, and R^t will again be denoted by letters x, u, and v, respectively; a point x in R^r will also be written as (u, v) with u in R^s and v in R^t. Our improved Fubini theorem is the following.

THEOREM 7-1 *Let A be an interval in R^r that is the Cartesian product $A(1) \times A(2)$ of an interval $A(1)$ in R^s and an interval $A(2)$ in R^t. Let f be a function integrable over A. Then there exists a subset $N(1)$ of $A(1)$ with s-dimensional measure $m_L N(1)$ equal to 0, such that for all u in $A(1) \setminus N(1)$ the function $v \mapsto f(u, v)$ (v in $A(2)$) is integrable over $A(2)$, and the value of its integral defines a function integrable over $A(1) \setminus N(1)$, and*

$$\text{(A)} \qquad \int_{A(1)\setminus N(1)} \left\{\int_{A(2)} f(u, v)\,dv\right\} du = \int_A f(x)\,dx.$$

We first prove an auxiliary statement.

(B) Let $f_1, f_2, f_3, \ldots$ be an ascending or descending sequence of functions integrable over A whose limit f is integrable over A. Assume that for each positive integer n there exists a subset $N(1,n)$ of $A(1)$ with measure 0 such that for all u in $A(1) \setminus N(1,n)$ the function $v \mapsto f_n(u,v)$ is integrable over $A(2)$, and its integral defines a function integrable over $A(1) \setminus N(1,n)$, and

$$\text{(C)} \qquad \int_{A(1)\setminus N(1,n)} \left\{ \int_{A(2)} f_n(u,v)\,dv \right\} du = \int_A f_n(x)\,dx.$$

Then the conclusion of Theorem 7-1 holds for f.

By the monotone convergence theorem,

$$\text{(D)} \qquad \lim_{n\to\infty} \int_A f_n(x)\,dx = \int_A f(x)\,dx.$$

Let N' be the union of the sets $N(1,n)$ $(n = 1, 2, 3, \ldots)$. This has measure 0 and contains each $N(1,n)$, so for all u in $A(1) \setminus N'$ the integrals of the functions $v \mapsto f_n(u,v)$ over the interval $A(2)$ all exist, and (C) implies

$$\text{(E)} \qquad \int_{A(1)\setminus N'} \left\{ \int_{A(2)} f_n(u,v)\,dv \right\} du = \int_A f_n(x)\,dx.$$

By two more applications of the monotone convergence theorem,

$$\text{(F)} \qquad \lim_{n\to\infty} \int_{A(2)} f_n(u,v)\,dv = \int_{A(2)} f(u,v)\,dv$$

for all u in $A(1) \setminus N'$, and

$$\text{(G)} \qquad \lim_{n\to\infty} \int_{A(1)\setminus N'} \left\{ \int_{A(2)} f_n(u,v)\,dv \right\} du = \int_{A(1)\setminus N'} \left\{ \int_{A(2)} f(u,v)\,dv \right\} du.$$

The integral in the right member of (F) may be ∞ for some u in $A(1) \setminus N'$. But since the right member of (G) is finite, the subset N'' of $A(1) \setminus N$ on which the right member of (F) is ∞ must have measure 0. If we define $N(1) = N' \cup N''$, $N(1)$ has measure 0, and the right member of (G) is unchanged if we replace N' by $N(1)$. With this change, (D), (F), and (G) imply (A), and statement (B) is proved.

Suppose that f is an integrable U-function. Then there exists an ascending sequence of step-functions $f_1, f_2, f_3, \ldots$ that converges everywhere in R^r to f. By Theorem 4-1 (in fact, by the first part of the proof of Theorem 4-1), these f_n satisfy the hypotheses in statement (B), even with all $N(1,n)$ empty. By statement (B), the integrable U-function f satisfies the conclusions of the theorem. In a like manner, if f is an integrable L-function, the conclusions of the theorem hold for f.

Suppose next that h is integrable over A and is the limit of a descending sequence of integrable U-functions $f_1, f_2, f_3, \ldots$. As we have just shown, these f_n satisfy the hypotheses of statement (B), so h satisfies the conclusions of the theorem. Similarly, if g is an integrable function that is the limit of an ascending sequence of L-functions, g satisfies the conclusions of the theorem.

Now let f be integrable over A. Then f_A is integrable over R^r, and by Corollary 6-5 there are functions g, h on R^r such that

(i) g is the limit of an ascending sequence of integrable L-functions, and h is the limit of a descending sequence of integrable U-functions;
(ii) $g(x) \leqq f_A(x) \leqq h(x)$ for all x in R^r;
(iii) $g(x) = f_A(x) = h(x)$ for almost all x in R^r.

By (iii), f, g, and h have equal integrals over A. By the preceding part of the proof, g and h satisfy the conclusions of the theorem, so there is a set $N(1, g)$ of measure 0 such that if u is in $A(1) \setminus N(1, g)$, the integral

$$\text{(H)} \qquad \int_{A(2)} g(u, v)\, dv$$

exists, and its value defines a function integrable over $A(1) \setminus N(1, g)$, and

$$\text{(I)} \qquad \int_{A(1)\setminus N(1,g)} \left\{ \int_{A(2)} g(u, v)\, dv \right\} du = \int_A g(x)\, dx = \int_A f(x)\, dx.$$

Likewise, there exists a set $N(1, h)$ of measure 0 such that, for every u in $A(1) \setminus N(1, h)$, the integral

$$\text{(J)} \qquad \int_{A(2)} h(u, v)\, dv$$

exists, and it defines a function integrable over $A(1) \setminus N(1, h)$, and

$$\text{(K)} \qquad \int_{A(1)\setminus N(1,h)} \left\{ \int_{A(2)} h(u, v)\, dv \right\} du = \int_A f(x)\, dx.$$

By (I), the integral (H) has a finite integral over $A(1) \setminus N(1, g)$. Therefore, it must be finite-valued for almost all u in $A(1) \setminus N(1, g)$, say for all u except those in $N'(1, g)$. Likewise, the integral (J) is finite-valued for all u in $A(1) \setminus N(1, h)$ except those in a set $N'(1, h)$ of measure 0. Let $N'' = N(1, g) \cup N(1, h) \cup N'(1, g) \cup N'(1, h)$. This has measure 0, so the left members of (I) and (K) are unchanged if in them we replace $N(1, g)$ and $N(1, h)$ by N''. By (I) and (K),

$$\text{(L)} \qquad \int_{A(1)\setminus N''} \left\{ \int_{A(2)} h(u, v)\, dv - \int_{A(2)} g(u, v)\, dv \right\} du = 0.$$

The quantity in braces is nonnegative, so by Theorem II-12-4 it is 0 except on a set N^* of measure 0. We define $N(1) = N'' \cup N^*$. Since $g \leqq f \leqq h$ and (L) holds, by Corollary I-7-3 the function $v \mapsto f(u, v)$ is integrable over $A(2)$ whenever u is in $A(1) \setminus N(1)$, and for such u its integral over $A(2)$ is equal to that of $v \mapsto h(u, v)$. So the left member of (K) is unchanged if we replace $N(1, h)$ by $N(1)$ and $h(u, v)$ by $f(u, v)$. This completes the proof.

To show that this theorem can be extended to the case of n successive integrations, as in the latter part of Section 4, we exhibit the details for the case $n = 3$ only. Larger values of n can be treated similarly, with nothing worse than notational complications.

COROLLARY 7-2 *Let A be an interval in R^r that is the Cartesian product of intervals $A(1)$, $A(2)$, $A(3)$ in R^{r_1}, R^{r_2}, R^{r_3}, respectively. Let f be integrable over A. Then there exists a subset $N(1)$ of $A(1)$ with measure 0, and for each x_1 in $A(1) \setminus N(1)$ there exists a subset $N[x_1]$ of $A(2)$ with measure 0, such that all the integrals in the equation*

$$\text{(M)} \quad \int_{A(1)\setminus N(1)} \left\{ \int_{A(2)\setminus N[x_1]} \left[\int_{A(3)} f(x_1, x_2, x_3)\, dx_3 \right] dx_2 \right\} dx_1 = \int_A f(x)\, dx$$

exist, and the equation is valid.

Every x in A can be represented in the form $x = (x_1, x_2, x_3)$, with x_i in $A(i)$. We also use the notation v for (x_2, x_3). Since A is the Cartesian product of the intervals $A(1)$ and $A(2) \times A(3)$, by Theorem 7-1 there exists a subset $N(1)$ of $A(1)$ with measure 0 such that for every x_1 in $A(1) \setminus N(1)$ the integral

$$\text{(N)} \quad \int_{A(2)\times A(3)} f(x_1, v)\, dv$$

exists and is finite, and the function on $A(1) \setminus N(1)$ with values given by (N) is integrable over $A(1) \setminus N(1)$, and

$$\text{(O)} \quad \int_{A(1)\setminus N(1)} \left\{ \int_{A(2)\times A(3)} f(x_1, v)\, dv \right\} dx_1 = \int_A f(x)\, dx.$$

For each x_1 in $A(1) \setminus N(1)$ the function $v \mapsto f(x_1, v)$ is integrable over $A(2) \times A(3)$, so by Theorem 7-1 there is a subset $N[x_1]$ of $A(2)$ with measure 0 such that for all x_2 in $A(2) \setminus N[x_1]$ the function $x_3 \mapsto f(x_1, x_2, x_3)$ is integrable over $A(3)$, and

$$\int_{A(L)\setminus N[x_1]} \left[\int_{A(3)} f(x_1, x_2, x_3)\, dx_3 \right] dx_2 = \int_{A(2)\times A(3)} f(x_1, y)\, dy.$$

If we substitute this in (O), we obtain the conclusion of the corollary.

We continue to represent points x in R^r by the symbol (u, v), where u is in R^s and v is in R^t. In Definition 4-4 we defined the section $E[u]$ of a subset E of R^r to be

$$E[u] = \{v \text{ in } R^t : (u, v) \text{ in } E\}.$$

If E has finite measure in R^r, we can apply Theorem 7-1 to its indicator and obtain the following corollary.

COROLLARY 7-3 *If E is a set of finite measure in R^r, the section $E[u]$ has finite measure for all points u in R^s except those in a set N with $m_L N = 0$, and the integral*

$$\text{(P)} \qquad \int_{R^s \setminus N} m_L E[u]\, du$$

exists and has value $m_L E$.

COROLLARY 7-4 *If E is a set of measure 0 in R^r, then for almost all u in R^s the section $E[u]$ has measure 0 in R^t.*

If E has measure 0 in R^r, the integral (P) is 0. The integrand is nonnegative, so by Theorem II-12-4 it is 0 except on a set N' of R^s with measure 0. Then $N \cup N'$ is a subset of R^s with measure 0, and for all u in $R^s \setminus N \cup N'$ the section $E[u]$ has measure 0.

8. Second Form of the Substitution Theorem

We now prove a form of the theorem on change of variables in multiple integrals in which the substitutions allowed are the same as those in Theorem 5-14 but the hypotheses on the integrands are much weaker; we ask nothing more than the mere existence of one of the two integrals involved. We use the customary notation for the composite of two functions; if f is a function on a domain D_f and g is a function on D_g, $f \circ g$ is the function defined for all x in D_g for which $g(x)$ is in D_f, and for such x it has the value $f(g(x))$.

THEOREM 8-1 *Let h be a one-to-one continuously differentiable mapping of an open set G in R^r onto an open set G^*, with a continuously differentiable inverse g. Let f be a function on G^*. Then f is integrable over G^* if and only if $(f \circ h)|\det Dh|$ is integrable over G, and in that case*

$$\text{(A)} \qquad \int_{G^*} f(y)\, dy = \int_G f(h(x))|\det Dh(x)|\, dx.$$

Note that under the hypotheses, the composite $f \circ h$ is defined everywhere on G.

Suppose first that $f = 1_{G^* \cap B}$, where B is a bounded interval in R^r. Then $G^* \cap B^{\text{o}}$ is open (see Exercise 5-7) and 1 is integrable over it, so by Theorem 5-14,

$$\text{(B)} \qquad \int_{G^* \cap B^{\text{o}}} 1\, dy = \int_{g(G^* \cap B^{\text{o}})} 1\, |\det Dh(x)|\, dx.$$

If x is in G, $h(x)$ is in $G^* \cap B^O$ if and only if x is in $g(G^* \cap B^O)$, so $1_{g(G^* \cap B^O)}(x) = 1_{G^* \cap B^O}(h(x))$. Therefore, (B) can be written as

$$\text{(C)} \qquad \int_{G^*} 1_{G^* \cap B^O}(y)\, dy = \int_G 1_{G^* \cap B^O}(h(x))|\det Dh(x)|\, dx.$$

If B^- is the interval $[a^1, b^1] \times \cdots \times [a^r, b^r]$, it is the intersection of the open intervals

$$B(n) = (a^1 - 1/n, b^1 + 1/n) \times \cdots \times (a^r - 1/n, b^r + 1/n) \qquad (n = 1, 2, 3, \ldots).$$

Equation (C) holds for each $B(n)$, so

$$\text{(D)} \qquad \int_{G^*} 1_{G^* \cap B(n)}(y)\, dy = \int_G 1_{G^* \cap B(n)}(h(x))|\det Dh(x)|\, dx.$$

As n increases, the integrands in both members of (D) descend, so by the monotone convergence theorem,

$$\text{(E)} \qquad \int_{G^*} 1_{G^* \cap B^-}(y)\, dy = \int_G 1_{G^* \cap B^-}(h(x))|\det Dh(x)|\, dx.$$

The integrands in the left members of (C) and (E) differ only on $B^- \setminus B^O$, which has measure 0, so these left members are equal. Since $1_{G^* \cap B^O} \leqq 1_{G^* \cap B} \leqq 1_{G^* \cap B^-}$, by Corollary I-7-3, $(1_{G^* \cap B} \circ h)|\det Dh|$ is integrable over G, and it has the same integral as $(1_{G^* \cap B^O} \circ h)|\det Dh|$. From this and (C),

$$\int_{G^*} 1_{G^* \cap B}(y)\, dy = \int_G 1_{G^* \cap B}(h(x))|\det Dh(x)|\, dx,$$

and so (A) is satisfied by $f = 1_{G^* \cap B}$.

Suppose next that f is a step-function on R^r with values $c_1, \ldots, c_k$ on the respective pairwise disjoint bounded intervals $B(1), \ldots, B(k)$. By the preceding proof, for $j = 1, \ldots, k$ we have

$$\int_{G^*} 1_{G^* \cap B(j)}(y)\, dy = \int_G 1_{G^* \cap B(j)}(h(x))|\det Dh(x)|\, dx.$$

If we multiply both members by c_j and sum over $j = 1, \ldots, k$, we obtain (A). So (A) holds for all step-functions.

Suppose next that f is a U-function integrable over R^r. Then there exists an ascending sequence of step-functions $s_1, s_2, s_3, \ldots$ on R^r that converges everywhere to f. By the preceding paragraph,

$$\int_{G^*} s_n(y)\, dy = \int_G s_n(h(x))|\det Dh(x)|\, dx.$$

As n increases, the left members approach the integral of f over G^*. By the monotone convergence theorem, the limit of the integrands in the right member, which is $f(h(x))|\det Dh(x)|$, is integrable over G, and (A) is satisfied. By a similar proof, the conclusion of the theorem is valid if f is an integrable L-function.

Let f be a function integrable over R^r that is the limit of an ascending sequence of integrable L-functions $l_1, l_2, l_3, \ldots$. By the preceding paragraph, for each positive integer n we have

$$\int_{G^*} l_n(y)\,dy = \int_G l_n(h(x))|\det Dh(x)|\,dx.$$

As n increases, the integrand in the left member ascends and approaches $f(y)$. By the monotone convergence theorem, the left member converges to the integral of f over G^*. This implies that the integrals in the right member converge to a finite limit, so by the monotone convergence theorem they converge to the integral of the limit of the integrands, which is $f(h(x))|\det Dh(x)|$. Therefore (A) holds for this f also. By a similar proof, (A) holds if f is the limit of a descending sequence of U-functions.

Now assume that f is any function integrable over G^*. We apply Corollary 6-5, using the letters ϕ, ψ in place of h, g because h and g have already been used in this theorem with other meanings. By that corollary, there are functions ϕ, ψ on R^r such that

(i) ϕ is the limit of an ascending sequence of integrable L-functions, and ψ is the limit of a descending sequence of U-functions;

(ii) $\phi(y) \leqq f_{G^*}(y) \leqq \psi(y)$ for all y in R^r;

(iii) $\phi(y) = f_{G^*}(y) = \psi(y)$ for almost all y in R^r.

By the preceding paragraph,

$$\text{(F)} \qquad \int_{G^*} \phi(y)\,dy = \int_G \phi(h(x))|\det Dh(x)|,$$

$$\text{(G)} \qquad \int_{G^*} \psi(y)\,dy = \int_G \psi(h(x))|\det Dh(x)|\,dx.$$

By (iii), the left members of (F) and (G) are both equal to the left member of (A), so the right members of (F) and (G) are equal. For all x in G we have, by (ii),

$$\phi(h(x))|\det Dh(x)| \leqq f_{G^*}(h(x))|\det Dh(x)| \leqq \psi(h(x))|\det Dh(x)|,$$

so by Corollary I-7-3, the middle one of these three is integrable over G and its integral is equal to the integrals of the other two. By (F), this implies that (A) holds, and the proof is complete.

Even when the mapping $x \mapsto h(x)$ is affine and the integrand f is the indicator of a set, this theorem provides us with information that is important, although far from surprising. In any space R^r, if a set of finite measure is translated to another position, the translated set may reasonably be anticipated to have the same measure as the original set. To mathematicians of an earlier age, who regarded areas and volumes as basic ideas that needed no definition, this would have seemed a statement of the obvious. But to us, measure is a defined quantity, and we cannot honestly claim to have an intuitive perception of an unbounded set in R^{1000} whose indicator is everywhere discontinuous. To us, the invariance of measure under translation is not something to be accepted as "intuitively evident"; we have to prove it. Likewise, if we rotate a point-set about the origin, the rotated set ought to have the same measure as the original; but this has to be proved. More generally, if we map R^r onto itself by a mapping that leaves all distances unchanged, any set of finite measure should be mapped onto a set of the same measure. All these statements are special cases of Theorem 8-1, as we now show.

First, let $x \mapsto y = h(x) = x + c$ be a translation in R^r, and let E be a set in R^r. Since $\det Dh(x) = 1$ for all x, Theorem 8-1 informs us that if either of the integrals

$$\text{(H)} \qquad \int_{R^r} 1_{h(E)}(y)\,dy, \qquad \int_{R^r} 1_{h(E)}(h(x))1\,dx$$

exists, so does the other, and then they are equal. Since $1_{h(E)}(h(x)) = 1_E(x)$ for all x in R^r, the second of the integrals (H) is the integral of 1_E over R^r. So the conclusion of Theorem 8-1 is that if either $h(E)$ or E has finite measure, so has the other, and the two are equal. We have proved that translation leaves measure unaltered.

In R^2, a rotation about the origin through angle θ is given by $x \mapsto y = h(x)$, where

$$y^1 = x^1 \cos\theta - x^2 \sin\theta, \qquad y^2 = x^1 \sin\theta + x^2 \cos\theta.$$

This has the form $y = Ax$, where

$$A = \begin{bmatrix} \cos\theta & -\sin\theta \\ \sin\theta & \cos\theta \end{bmatrix}.$$

Then $\det A = 1$, so Theorem 8-1 informs us that if either of the integrals (H) exists, so does the other, and the two are equal. So, in the plane, measures are invariant under rotation.

We shall now accept a few elementary statements about vectors and linear algebra as known; all of them are well known and most of them will be proved again in Chapter V. If a vector x in R^r has components $x^1, \ldots, x^r$, its length is

$$|x| = \left[\sum_{i=1}^{r} (x^i)^2 \right]^{1/2}.$$

The distance between two points x, y in R^r is the length of the vector joining them; that is,

$$\operatorname{dist}(x, y) = \left[\sum_{i=1}^{r} (y^i - x^i)^2 \right]^{1/2}.$$

In order for a mapping $x \mapsto Ax$ of R^r onto R^r (A being an $r \times r$ matrix) to have the property that

$$|Ax| = |x|$$

for all x in R^r, it is necessary and sufficient that the transpose A^T of A be the inverse of A. The determinant of A^T is equal to the determinant of A, and the determinant of the product of two $r \times r$ matrices is the product of their determinants. This last implies

$$\text{(I)} \qquad 1 = \det AA^T = (\det A)(\det A^T) = (\det A)^2,$$

whence $|\det A| = 1$. As in the case of translations, we use this and Theorem 8-1 to prove that if E has finite measure, AE has the same measure, so the length-preserving mapping $x \mapsto Ax$ leaves measures unaltered.

EXERCISE 8-1 Let B be a set of finite measure in the (x, y)-plane, and let V be the point $(0, 0, h)$, where h is positive. The cone C with base B and vertex V is the set of all points $([h - w]u/h, [h - w]v/h, w)$ with (u, v) in B and $0 \leqq w \leqq h$. Show that C is measurable and $m_L C = hm_L B/3$. *Suggestion*: The mapping $(u, v, w) \mapsto (x, y, z)$ with

$$x = u[h - w]/h, \qquad y = v[h - w]/h, \qquad z = w$$

maps the set of points (u, v, w) with $w < h$ onto itself, and its inverse maps all of C but V onto $B \times [0, h)$. Apply Exercise 6-3 to $1_B(u, v)1_{[0,h)}(w)$.

9. Integration with Respect to Other Measures

In the Sections III-7 through III-12 we extended the concept of integration over sets in R to allow the use of measures of intervals in R that are different from the measure m_L, and we applied this integral to probability theory. In this section and in Sections 10–12 we shall show that this extension to other measures can also be made in spaces R^r with $r > 1$ and even in some infinite-dimensional spaces, and we shall develop deeper applications to probability theory. The extension of the results arrived at in Chapter III is easy, involving little more than a few changes in notation.

Suppose first that m is an extended-real-valued function defined on the class of all left-open intervals in R^r (r being any positive integer). If B is any set in R^r and f is any real-valued function on B, the integral of f with respect to m over B is defined as in Definition III-7-1, making only the trivial change of replacing all references to $\bar{R}$ with references to $\bar{R}^r$. We shall not continue to mention such trivial notational changes. This extension of the concept of integral is not only possible and easy; in many cases, integration in R^r is more natural than integration in R. For instance, let us consider a mass-distribution in R^3; we wish to find its moment of inertia about a line – say, the x^3-axis. If a particle of mass m is located at x, by elementary physics its moment of inertia about the x^3-axis is $m[(x^1)^2 + (x^2)^2]$. If a finite amount of matter is distributed in space so that the mass of the matter in each left-open interval A is mA, for each positive number ε and each $\bar{x}$ in R^3 we can find a neighborhood $\gamma(\bar{x})$ small enough so that on it the value of the function

$$x \mapsto f(x) = (x^{(1)})^2 + (x^{(2)})^2$$

differs by less than ε from $f(\bar{x})$. If A is a left-open interval whose closure is contained in $\gamma(\bar{x})$, moving all the material in A to the point $\bar{x}$ would change the moment of inertia of that part of the mass by less than εmA. The total mass of the matter is mR^3, and by hypothesis, this is finite. So if $\mathscr{P} = \{(\bar{x}_1, A_1), \ldots, (\bar{x}_k, A_k)\}$ is a γ-fine partition of R^3, the partition-sum $S(\mathscr{P}; f; m)$ will differ from the moment of inertia of the mass-distribution by less than εmR^3. That is to say, the moment of inertia is the gauge-limit of $S(\mathscr{P}; f; m)$ and is therefore equal to the integral

$$\text{(A)} \qquad \int_{R^3} [(x^{(1)})^2 + (x^{(2)})^2]\, m(dx).$$

In R, discrete distributions and distributions with densities were of particular importance; in fact, many texts on probability theory limit themselves to those two cases. There are also other kinds of measures in R, but they are somewhat difficult to describe. But in spaces R^r with $r > 1$ they can occur quite naturally. For example, let a one-dimensional wire with a mass of c grams per centimeter of length be bent along a smooth curve in R^3. For each interval A in R^3, mA will be c times the length of the piece of the curve that is in A. The discussion in the preceding paragraph shows that the moment of inertia is given by the integral (A). But this distribution is not discrete, because there are uncountably many points on the curve, and on the other hand it does not have a density.

The statements and proofs in Section III-8 extend to measures in R^r with no change except a trivial one in the proof of Lemma III-8-2. Let A_i be the Cartesian product $A_i = A_i^{(1)} \times \cdots \times A_i^{(r)}$ of r left-open intervals in R, and likewise let $B_j = B_j^{(1)} \times \cdots \times B_j^{(r)}$. For each k in $\{1, \ldots, r\}$ we denote by $c_1^{(k)}, \ldots, c_{n(k)}^{(k)}$ the end-points of all the intervals $A_i^{(k)}$ and $B_j^{(k)}$, arranged in increasing order. We define

$$C_n^{(k)} = (c_n^{(k)}, c_{n+1}^{(k)}] \qquad (n = 1, \ldots, n(k) - 1),$$

and we form all the Cartesian products

$$C_n = C_{j(1)}^{(1)} \times \cdots \times C_{j(r)}^{(r)}$$

in which each $C_{j(i)}^{(i)}$ is one of the intervals $C_1^{(i)}, \ldots, C_{n(i)-1}^{(i)}$. The rest of the proof of the extension is like that of Lemma III-8-2.

At the beginning of Section III-9 we briefly discussed pointwise densities. Although these continue to be unimportant for us, we shall say a few words about them because they are often mentioned in elementary texts. If E is any set in R^r for which $m_L E$ is finite and positive and mE has meaning, it is natural to define the mean density of m in E to be $mE/m_L E$, as we did in Chapter III. The pointwise density of m at a point $\bar{x}$ should be the limit of this mean density as E shrinks down to $\bar{x}$; but it is not instantly obvious what sets E should be used in this definition. It would be possible and not unreasonable to use intervals that contain $\bar{x}$, or cubes whose closures contain $\bar{x}$, or spheres centered at $\bar{x}$. For smooth distributions it is immaterial which class of sets E we choose to allow; but when the distribution is complicated, it can make a difference.

We choose to use cubes. Thus, the interval-function F has a pointwise density ρ at the point $\bar{x}$ if for each positive ε there exists a neighborhood U of $\bar{x}$ such that whenever E is a left-open cube whose closure E^- is contained in U and contains $\bar{x}$,

$$|F(A)/m_L A - \rho| < \varepsilon.$$

It is trivial to prove that if m has a density ρ in our sense (Definition III-9-1), then at each point $\bar{x}$ at which ρ is continuous, m has a pointwise density equal to $\rho(\bar{x})$. It is possible to prove a much stronger theorem. If m has a density ρ, then at almost all points $\bar{x}$ in R^r, m has a pointwise density equal to $\rho(\bar{x})$. But this is hard to prove and we do not need it, so we postpone its proof to Chapter VII.

Theorem III-9-2 extends to interval-functions in R^r without difficulty. All we need to do is to apply the construction of the intervals $(a_n, b_n]$ in the proof in Chapter III to each of the intervals $A^{(1)}, \ldots, A^{(r)}$ whose Cartesian product is A. Likewise, only trivial changes are needed in order to extend the proofs of Theorems III-9-3 and III-9-4 to higher dimensions.

The Fubini theorem 4-1 relates integrals over $A(1)$, over $A(2)$, and over A, where A, $A(1)$, and $A(2)$ are left-open intervals in R^r, R^s, and R^t, respectively and $s + t = r$. In order to extend this to other measures we must be able to integrate over intervals in R^r, R^s, and R^t, which means that we must start with interval-functions m, m_1, and m_2 on left-open intervals in R^r, R^s, and R^t, respectively. The conclusion of Theorem 4-1 should be replaced by an equation

$$\int_{A(1)} \left\{ \int_{A(2)} f(u, v)\, m_2(dv) \right\} m_1(du) = \int_A f(x)\, m(dx).$$

If this is to hold in the simple case in which f is the indicator of an interval $A = A(1) \times A(2)$, it must then be true that

$$m_1 A(1) m_2 A(2) = mA.$$

This is, in fact, the only peculiarity of the measure that is needed to allow us to extend Theorem 4-1, as follows.

THEOREM 9-1 *Let r, s, and t be positive integers with $r = s + t$. Let m, m_1, and m_2 be nonnegative additive regular functions of left-open intervals in R^r, R^s, and R^t,* *respectively, such that whenever A, $A(1)$, and $A(2)$ are left-open intervals in R^r, R^s, and R^t, respectively, such that $A = A(1) \times A(2)$, it is true that*

$$mA = [m_1 A(1)][m_2 A(2)].$$

Let A, $A(1)$, and $A(2)$ be bounded left-open intervals in R^r, R^s, and R^t, respectively, such that $A = A(1) \times A(2)$, and let f be a bounded function on A that is the limit of a sequence of step-functions with left-open intervals of constancy. Then for every u in $A(1)$ the function $v \mapsto f(u, v)$ (v in $A(2)$) is integrable over $A(2)$ with respect to m_2, and the integrals

$$\int_{A(1)} \left\{ \int_{A(2)} f(u, v)\, m_2(dv) \right\} m_1(du), \qquad \int_A f(x)\, m(dx)$$

both exist and are equal.

The proof is the same as that of Theorem 4-1, with only trivial notational changes. Likewise, the second form of Fubini's theorem (Theorem 7-1) can be generalized as follows.

THEOREM 9-2 *Let m, m_1, m_2, A, $A(1)$, and $A(2)$ be as in Theorem 9-1. Let f be a function integrable over A. Then there exists a subset $N(1)$ of $A(1)$ with $m_1 N(1) = 0$ such that for all u in $A(1) \setminus N(1)$ the function $v \mapsto f(u, v)$ is integrable with respect to m_2 over $A(2)$, and the value of its integral defines a function integrable with respect to m_1 over $A(1) \setminus N(1)$, and*

$$\int_{A(1)\setminus N(1)} \left\{ \int_{A(2)} f(u, v)\, m_2(dv) \right\} m_1(du) = \int_A f(x)\, m(dx).$$

Here too the proof is the same as that of Theorem 7-1 except for trivial notational changes.

EXERCISE 9-1 Let S consist of the two points $(0, 1)$, $(1, 0)$ of the plane, and for each interval A in the plane let mA be the number of points of S in A. Show that there cannot exist two interval functions m_1, m_2 in R^1 such that $m[A(1) \times A(2)] = m_1 A(1) m_2 A(2)$ for all left-open intervals $A(1)$, $A(2)$ in R^1. (Apply the equation to the three intervals $(-1, 1] \times (0, 1]$, $(0, 1] \times (-1, 1]$, $(0, 1] \times (0, 1]$ and show that the three resulting equations are incompatible.)

EXERCISE 9-2 Let m_1 and m_2 be functions of left-open intervals in R^s, R^t, respectively, and for each interval $A = A(1) \times A(2)$ with $A(1)$ in R^s and $A(2)$ in R^t define $mA = m_1 A(1) m_2 A(2)$. Prove:

(i) if m_1 and m_2 are nonnegative, so is m;
(ii) if m_1 and m_2 are additive, so is m;
(iii) if m_1 and m_2 are bounded, so is m;
(iv) if m_1 and m_2 are nonnegative, additive, bounded, and regular, so is m.

10. Applications to Probability Theory: Multivariate Distributions

Often, a chance occurrence results in the determination of r real numbers, in a definite order. The occurrence might be the result of r consecutive throws of a die, or it might be the age, height, and weight of an individual belonging to some population. The sort of chance occurrence that we can study mathematically is the type in which for each left-open interval A in R^r the ratio of the number of occurrences with outcomes in A to the total number of occurrences can be adequately approximated by some number $P(A)$, provided that the number of occurrences is large enough. This leads to the same kind of probability distribution that we met in Chapter III, except that now the probability measure P is defined for all left-open intervals in R^r.

The change from R to R^r causes no difficulty at all in the theorems we proved in Section III-11; nothing more is needed than to replace a few references to R and $\bar{R}$ by references to R^r and $\bar{R}^r$. However, to accord with custom in probability theory we shall usually give R^r the name Ω and denote its points by ω instead of by x. In the theorems in this section, no properties of events in Ω and of integrals with respect to P will be used except those that are proved in theorems marked by stars (★).

On a given Ω there are many random variables. It often happens that for some particular purpose, all the quantities of interest to us are determined by the values of a few of them, say $X_1, \ldots, X_k$. For example, if particles of a certain type have masses m, charges c, position-coordinates $x^{(1)}$, $x^{(2)}$, $x^{(3)}$ and velocity components $v^{(1)}$, $v^{(2)}$, $v^{(3)}$ such that $(m, c, x^{(1)}, x^{(2)}, x^{(3)}, v^{(1)}, v^{(2)}, v^{(3)})$ has a probability distribution in R^8, and we are interested only in the kinetic energy and functions of it, all the quantities of interest to us will be functions of the single random variable $X_1 = m[(v^{(1)})^2 + (v^{(2)})^2 + (v^{(3)})^2]$. If $k < r$, we may be able to simplify our work by mapping Ω into the lower-dimensional space R^k by the mapping

$$\omega \mapsto X(\omega) = (X_1(\omega), \ldots, X_k(\omega)).$$

As usual, if Ω_0 is any subset of Ω, we write $X(\Omega_0)$ for the set of all points $X(\omega)$ with ω in Ω_0, and for each set E in R^k we denote the inverse image $\{\omega \text{ in } \Omega : X(\omega) \text{ in } E\}$ by $X^{-1}(E)$.

If X_1 is a random variable and A is a left-open interval $(a, b]$ in R, then $X^{-1}(A) = \{X_1 \leqq b\} \setminus \{X_1 \leqq a\}$. Both these last-named sets are events, by Theorem III-10-12, so $X_1^{-1}(A)$ is an event by Theorem III-10-8. So, for every left-open interval A in R, $X_1^{-1}(A)$ is an event. If $X_1, \ldots, X_k$ are random variables on Ω, and $A = A^{(1)} \times \cdots \times A^{(k)}$ is a left-open interval in R^k, its inverse image $X^{-1}(A)$ is the intersection of the sets $\{X_1 \text{ in } A^{(1)}\}, \ldots, \{X_k \text{ in } A^{(k)}\}$. Each of these is an event, as we have just proved, so $X^{-1}(A)$ is an event. We define

$$\text{(A)} \qquad P_X(A) = P(X^{-1}(A))$$

for all left-open intervals A in R^k. Next we shall show that P_X is a probability distribution on $\bar{R}^k$; then we shall be justified in calling it the **joint distribution** of $X_1, \ldots, X_k$, or "the distribution of the random vector X."

Clearly, P_X is nonnegative. If A is a left-open interval in R^k that is the union of pairwise disjoint left-open intervals $A_1, \ldots, A_n$, the sets $\{X \text{ in } A_1\}, \ldots, \{X \text{ in } A_n\}$ are pairwise disjoint events in Ω, so

$$P_X(A) = P\{x \text{ in } A\} = \sum_{i=1}^{n} P(\{X \text{ in } A_i\}) = \sum_{i=1}^{n} P_X(A_i),$$

so P_X is an additive nonnegative function of left-open intervals in R^k. Let $A = A^{(1)} \times \cdots \times A^{(k)}$ be a left-open interval in R^k, and let the end-points of $A^{(i)}$ be $a^{(i)}$ and $b^{(i)}$. For each i, if $b^{(i)}$ is finite we define

$$A_n^{(i)} = A^{(i)} \cup (b^{(i)}, b^{(i)} + 1/n];$$

if $b^{(i)} = \infty$, we define $A_n^{(i)} = A^{(i)}$. In either case, $A_n^{(i)}$ is a left-open interval whose interior contains $A^{(i)}$, and as n increases the $A_n^{(i)}$ shrink and their intersection is $A^{(i)}$. We define

$$A_n = A_n^{(1)} \times \cdots \times A_n^{(k)}.$$

This is a left-open interval in R^k, and its interior contains A; and as n increases A_n shrinks, and the intersection of all the A_n is A. Therefore, the events $X^{-1}(A_n)$ shrink, and their intersection is $X^{-1}(A)$. By Corollary III-10-11,

$$\lim_{n \to \infty} P(X^{-1}(A_n)) = P(X^{-1}(A)).$$

So for each positive ε we can find a left-open interval A_n whose interior contains A and for which $P_X(A_n) < P_X(A) + \varepsilon$. By a similar proof, we can also find a bounded left-open interval F whose closure is contained in A and for which $P_X(F) > P_X(A) - \varepsilon$. So P_X is regular. Also,

$$P_X(R^k) = P(X^{-1}(R^k)) = P(\Omega) = 1,$$

so P_X is a probability distribution.

We have already defined

$$E(f) = \int_{\Omega} f(\omega)\, P(d\omega)$$

whenever f is a function on Ω for which the integral exists. Similarly, when f is a function on R^k we define

$$E_X(f) = \int_{R^k} f(x)\, P_X(dx)$$

whenever the integral exists.

When we are studying functions that depend only on the values of $X_1, \ldots, X_k$, it clearly can be advantageous to be able to work with the joint distribution of $X_1, \ldots, X_k$ instead of having to go back to Ω and the probability measure P. We now establish two theorems that allow us to do this under suitable conditions. The first is a special case of the second; it is included for the benefit of those readers who have not worked through Section 7.

THEOREM 10-1 *Let X be a random k-vector, and let f be a function on R^k that is the limit of a sequence of step-functions with left-open intervals of constancy. Then f is a random variable with respect to P_X, and $f \circ X$ is a random variable with respect to P, and if either of the expectations $E_X(f)$, $E(f \circ X)$ exists and is finite, the other also exists, and the two are equal.*

Suppose first that f is a step-function with the values $c_1, \ldots, c_n$ on the respective left-open intervals $A(1), \ldots, A(n)$. For $j = 1, \ldots, n$ we have, by (A) and Lemma I-3-2,

$$E_X(1_{A(j)}) = P_X(A(j)) = P(X^{-1}(A(j))). \tag{B}$$

But if ω is in $X^{-1}(A(j))$, $X(\omega)$ is in $A(j)$, and both members of the equation

$$1_{A(j)}(X(\omega)) = 1_{X^{-1}(A(j))}(\omega)$$

are 1, and otherwise both are 0; so the equation holds for all ω in Ω. This and (B) yield

$$E(1_{A(j)} \circ X) = E(1_{X^{-1}(A(j))}) = P(X^{-1}(A(j))) = E_X(1_{A(j)}).$$

We multiply both members by c_j and sum over $j = 1, \ldots, n$, obtaining

$$E(f \circ X) = E_X(f). \tag{C}$$

Suppose next that f is the limit of a sequence of step-functions $s_1, s_2, s_3, \ldots$ with left-open intervals of constancy. Let M be a positive number, and define

$$f_M = (f \wedge M) \vee (-M), \qquad s_{n,M} = (s_n \wedge M) \vee (-M).$$

These latter are step-functions with left-open intervals of constancy, and as n increases they tend to f_M, and their absolute values tend to $|f_M|$. They all remain less than M in absolute value, and M is integrable over R^k with respect to P_X and over Ω with respect to P. By the preceding paragraph, for $n = 1, 2, 3, \ldots$ the equations

$$E(s_{n,M} \circ X) = E_X(s_{n,M}), \tag{D}$$

$$E(|s_{n,M}| \circ X) = E_X(|s_{n,M}|) \tag{E}$$

are valid. As n increases, by the dominated convergence theorem the members of (D) converge to $E(f_M \circ X)$, $E_X(f_M)$, respectively, so

$$E(f_M \circ X) = E_X(f_M). \tag{F}$$

Likewise,

$$\text{(G)} \qquad E(|f_M| \circ X) = E_X(|f_M|).$$

As M increases through the positive integers, $|f_M| \circ X$ and $|f_M|$ ascend, and they approach $|f| \circ X$ and $|f|$, respectively; and f_M converges to f, and $f_M \circ X$ to $f \circ X$. The last statement implies that f is a random variable with respect to P_X and $f \circ X$ is a random variable with respect to P. If either of $E(f \circ X)$, $E_X(f)$ exists and is finite, the same holds for the absolute value, and the corresponding member of (G) is bounded. Then both members of (G) are bounded, and by the monotone convergence theorem both $E(|f| \circ X)$ and $E_X(|f|)$ are finite. Since

$$|f_M \circ X| \leqq |f| \circ X \qquad \text{and} \qquad |f_M| \leqq |f|,$$

by the dominated convergence theorem and (F) the expectations $E(f \circ X)$, $E_X(|f|)$ both exist, and they are equal.

The next corollary furnishes us with a useful formula for the moments of a random variable. In its proof we shall make use of the identity function on R, which is the function whose value at each point x in R is x. This function is often referred to as x, but this is not a good name for it since we have already used the letter x for points in R. We prefer to call the function $\mathbf{x}$, so that

$$\mathbf{x}(x) = x \qquad (x \text{ in } R).$$

COROLLARY 10-2 *Let X be a random variable and P_X its distribution, and let k be a positive integer. Then X has a finite kth moment if and only if the distribution P_X has a finite kth moment, and in that case the two are equal.*

Let f be $\mathbf{x}^k$; then $f(x) = x^k$ (x in R). For each positive integer n define f_n on R by

$$\begin{aligned} f_n(x) &= x^k && (-n < x \leqq n); \\ &= (2n - x)^k && (n < x \leqq 2n); \\ &= (-2n - x)^k && (-2n < x \leqq -n); \\ &= 0 && \text{elsewhere in } R. \end{aligned}$$

This is continuous and vanishes outside the bounded interval $[-2n, 2n]$, so it is integrable with respect to P_X over R. By Theorem 10-1,

$$\text{(H)} \qquad E_X(|f_n|) = E(|f_n| \circ X).$$

As n increases, both $|f_n|$ and $|f_n| \circ X$ ascend and converge to the respective limits $|\mathbf{x}|^k$, $|X|^k$. If X has a finite kth moment, $E(|X|^k)$ is finite; if the distribution has a finite kth moment, $E_X(|\mathbf{x}|^k)$ is finite. In either case the two members of (H) are bounded, and by the monotone convergence theorem the limit function $|\mathbf{x}|^k$ is integrable with respect to P_X over R, and the limit function $|X|^k$ is integrable with respect to P over Ω. For all n we have

$$|f_n(x)| \leqq |x|^k, \qquad |(f_n \circ X)(\omega)| \leqq |X(\omega)|^k,$$

so by (H) and the dominated convergence theorem,

$$\int_R x^k \, P_X(dx) = \int_\Omega X(\omega)^k \, P(d\omega).$$

This completes the proof.

COROLLARY 10-3 *Let X be a random variable and P_X its distribution. Then X has finite variance if and only if its distribution has finite variance, and in that case the two are equal.*

If X has finite variance, by Corollary III-11-5 it has finite first and second moments, so by Corollary 10-2, the same is true of the distribution P_X, and if $\mathbf{x}$ is the identity function on R,

$$E_X(\mathbf{x}^2) = E(X^2), \qquad E_X(\mathbf{x}) = E(X).$$

By Corollary III-11-7, the variance of the distribution is

$$E_X([\mathbf{x} - E_X(\mathbf{x})]^2) = E_X(\mathbf{x}^2) - [E_X(\mathbf{x})]^2 = E(X^2) - [E(X)]^2 = \sigma_X^2.$$

Conversely, if the distribution has finite variance, it has finite first and second moments. Then X has the same first and second moments, by Corollary 10-2, so X has finite variance.

It often happens that after we have selected a random vector X whose value determines all the quantities that interest us, we later wish to specialize still further and consider only a random vector X' formed of some of the components of X. Having already replaced Ω and P by R^k and P_X, we would like to determine $P_{X'}$ and $E_{X'}$ from P_X and E_X without having to go back to Ω, P, and E. Theorem 10-1 makes this possible. To simplify notation, we suppose that X' consists of the first h components of X ($h < k$). The function Π on R^k, defined by

$$\Pi(x^{(1)}, \ldots, x^{(k)}) = (x^{(1)}, \ldots, x^{(h)}),$$

maps R^k onto R^k; it is the operation of projecting each x in R^k onto the point in R^h whose coordinates are the first h coordinates of x. If $A = A^{(1)} \times \cdots \times A^{(h)}$ is a left-open interval in R^h, its inverse image under Π is the interval

$$\Pi^{-1}(A) = A^{(1)} \times \cdots \times A^{(h)} \times R \times \cdots \times R$$

in R^k, and this is left-open. Consequently, if f is a step-function on R^h with left-open intervals of constancy, $f \circ \Pi$ is a step-function on R^k with left-open intervals of constancy, and if a function f on R^h is the limit of a sequence of such step-functions on R^h, $f \circ \Pi$ is the limit on R^k of step-functions with left-open intervals of constancy. We can now apply Theorem 10-1 with X, R^k, Ω, P_X, and P replaced by Π, R^h, R^k, $P_{X'}$, and P_X, respectively. The result is the following corollary.

COROLLARY 10-4 *Let X be a random k-vector, and let X' consist of h of its components. Let Π project each x in R^k onto the point in R^h whose components are those components of x that are used in defining X'. Let f be a function on R^h that is the limit of a sequence of step-functions with left-open intervals of constancy. Then f is a random variable with respect to $P_{X'}$, and $f \circ \Pi$ is a random variable with respect to P_X, and $f \circ \Pi \circ X$ is a random variable with respect to P; and if any one of the three expectations $E_{X'}(f)$, $E_X(f \circ \Pi)$, $E(f \circ \Pi \circ X)$ exists and is finite, all three exist and have the same value.*

When X' is a single coordinate of X, say $X^{(j)}$, its distribution $P_{X'}$ is called the **marginal distribution** of $X^{(i)}$.

The functions to which Theorem 10-1 applies are numerous enough to include practically everything that one meets in uses of probability theory; in fact, we have already made good use of Theorem 10-1 in the three corollaries just proved. However, by use of the considerations in Section 6, we can establish a simple and powerful generalization.

THEOREM 10-5 *Let X be a random k-vector, and let P_X be its distribution. Then*

(i) *if f is a random variable on R^k with respect to P_X and $E_X(f)$ is finite, $f \circ X$ is a random variable on Ω with respect to P, and*

$$E(f \circ X) = E_X(f);$$

(ii) *if A is an event in R^k with respect to P_X, $X^{-1}(A)$ is an event in Ω with respect to P, and $P(X^{-1}(A)) = P_X(A)$;*

(iii) *if f is a random variable on R^k with respect to P_X, $f \circ X$ is a random variable on Ω with respect to P.*

Suppose, first, that $E_X(f)$ is finite. Let ε be positive. By Theorem 6-4, there exist a U-function u and an L-function l on R^k such that $l \leqq f \leqq u$ and

$$E_X(l) > E_X(f) - \varepsilon, \qquad E_X(u) < E_X(f) + \varepsilon.$$

By Theorem 10-1 and the definitions of U-function and L-function, $l \circ X$ and $u \circ X$ are integrable over Ω with respect to P, and

$$E(l \circ X) = E_X(l) > E_X(f) - \varepsilon, \qquad E(u \circ X) = E_X(u) < E_X(f) + \varepsilon.$$

Since $l \circ X \leqq f \circ X \leqq u \circ X$, by Lemma I-7-1 the preceding inequalities imply that $E(f \circ X)$ exists and is equal to $E_X(f)$. This establishes (i).

If A is an event in R^k with respect to P_X, 1_A is integrable with respect to P_X. By (i), $1_A \circ X$ is integrable with respect to P over Ω. But for every ω in Ω, $1_A(X(\omega))$

is 1 if ω is in $X^{-1}(A)$ and is 0 otherwise, so

$$1_A \circ X = 1_{X^{-1}(A)}.$$

Therefore, if A is an event, $1_{X^{-1}(A)}$ is integrable over Ω with respect to P, and its integral has the value

$$\int_{R^k} 1_A(x)\, P_X(dx).$$

This establishes (ii).

If f is a random variable with respect to P_X, there is a sequence $f_1, f_2, f_3, \ldots$ of real-valued functions on R^k that have finite expectations with respect to P_X and converge everywhere to f. By (i), the functions $f_j \circ X$ have finite expectations with respect to P, and they converge to $f \circ X$ at each point of Ω. So $f \circ X$ is a random variable with respect to P, and (iii) is established.

EXERCISE 10-1 In rolling two dice, the numbers X_1 and X_2 on the first and second die are determined by a point in a space of many dimensions, the coordinates of this point being the initial position and velocity of each die, the coefficient of friction of each square millimeter of the table top, the temperature, the air density, and many other such numbers. The set of all such points is a space Ω that can be subdivided into 36 subsets $E(i, j)$ $(i, j = 1, \ldots, 6)$; if ω is in $E(i, j)$, $X_1(\omega) = i$ and $X_2(\omega) = j$. We assume $P(E(i, j)) = \frac{1}{36}$ for each i and j. Find the distribution of the random vector (X_1, X_2) and of the random variable $X_1 + X_2$. Let $f(X_1, X_2) = 1$ if $X_1 + X_2 = 7$ let $f(X_1, X_2) = -3$ if $X_1 + X_2 = 2$ or 12, and $f(X_1, X_2) = 0$ otherwise. Find the expectation of f in three ways, using the original probability measure on Ω and the distributions of (X_1, X_2) and of $X_1 + X_2$.

11. Independence

Let A and B be two chance events with the respective likelihoods P_A, P_B of happening, and suppose also that the happening or nonhappening of A has no effect on the likelihood that B happens. Then in a large number N of trials, about $N_1 = NP_A$ occurrences of A can be expected. The fact that these N_1 trials are selected from the total set of N trials by the peculiarity that A took place should not alter the fact that of the N_1 trials, about $N_1 P_B$ may be expected to result in the happening of event B. So, the total number of trials in which both A and B happened may be expected to be about $N_1 P_B$, which is $NP_A P_B$. That is, if the happening of B is unaffected by the happening or nonhappening of A, the number of occurrences of both A and B in a large number of trials may be expected to be about $P_A P_B$ times the number of trials. This motivates the mathematical definition of *independence* (also called *stochastic independence* or *statistical independence*): if P is a probability distribution in any space, and A

and B are events in that space, they are independent if $P(A \cap B) = P(A)P(B)$. More generally,

Definition 11-1 *If $\mathscr{A}$ is any collection of events in a probability space, the events in $\mathscr{A}$ are* ***independent*** *if for every finite subset $A_1, \ldots, A_n$ of different events in the collection $\mathscr{A}$,*

$$P(A_1 \cap \cdots \cap A_n) = P(A_1) \cdots P(A_n).$$

By Theorem III-10-8, the intersection $A_1 \cap \cdots \cap A_n$ is an event. We shall now prove that if the events in $\mathscr{A}$ are independent and the collection $\mathscr{A}'$ is formed from $\mathscr{A}$ by replacing some or all of the members of $\mathscr{A}$ by their complements, the events in $\mathscr{A}'$ are also independent. Suppose that $A'_1, \ldots, A'_n$ is any finite set of different members of the collection $\mathscr{A}'$ and that, for each j in $\{1, \ldots, n\}$, A_j is a member of the collection $\mathscr{A}$ that is either A'_j or its complement. Consider first the case in which all the A'_j are in $\mathscr{A}$ except for a single one, say A'_1, which is the complement of an event A_1 in $\mathscr{A}$. Then $A_1 \cap A'_2 \cap \cdots \cap A'_n$ and $A'_1 \cap A'_2 \cap \cdots \cap A'_n$ are disjoint events whose union is $A'_2 \cap \cdots \cap A'_n$, and $A_1, A'_2, \ldots, A'_n$ all belong to $\mathscr{A}$, so

$$\begin{aligned} P(A'_1 \cap A'_2 \cap \cdots \cap A'_n) &= P(A'_2 \cap \cdots \cap A'_n) - P(A_1 \cap A'_2 \cap \cdots \cap A'_n) \\ &= P(A'_2)P(A'_3) \cdots P(A'_n) - P(A_1)P(A'_2) \cdots P(A'_n) \\ &= [1 - P(A_1)]P(A'_2) \cdots P(A'_n) \\ &= P(A'_1)P(A'_2) \cdots P(A'_n). \end{aligned}$$

We can continue and replace $A'_2, \ldots, A'_n$ by their complements without affecting the relation that the probability of the intersection is the product of the probabilities of the separate events.

It is trivially evident that if A is any event and B is either empty or the whole space, A and B are independent.

There is a similar relation of independence for random variables, and a slightly more general one for random vectors.

Definition 11-2 *Let $\mathscr{X}$ be any collection of random vectors. The vectors in collection $\mathscr{X}$ are* ***independent*** *if whenever $\{X_1, \ldots, X_n\}$ is a finite set consisting of different members of $\mathscr{X}$, and X_i is a random vector with values in $R^{r(i)}$ $(i = 1, \ldots, n)$, and $A_1, \ldots, A_n$ are left-open intervals in the respective spaces $R^{r(1)}, \ldots, R^{r(n)}$, then*

$$P(X_1 \text{ in } A_1, \ldots, X_n \text{ in } A_n) = P(X_1 \text{ in } A_1) \cdots P(X_n \text{ in } A_n).$$

The events in a collection $\mathscr{A}$ are independent if and only if their indicators are independent random variables. For if $A(1), \ldots, A(n)$ are different members of $\mathscr{A}$ and $B_1, \ldots, B_n$ are left-open intervals in R, the set $\{1_{A(j)} \text{ in } B_j\}$ is all of Ω if both 0 and 1 are in B_j; it is empty if neither is in B_j; it is $A(j)$ if 1 is in B_j but 0 is not;

and it is the complement of $A(j)$ if 0 is in B_j but 1 is not. Since Ω and $\varnothing$ are independent of all other sets, the sets $\{1_{A(j)} \text{ in } B_j\}$ ($j = 1, \ldots, n$) are independent if the indicators are independent. Conversely, if the indicators are independent, so are the sets $\{1_{A(j)} \text{ in } (\frac{1}{2}, \frac{3}{2}]\}$, which are the sets $A(j)$.

Until the beginning of the twentieth century, probability theory consisted almost entirely of the study of independent events and independent random variables. Today the theory has broadened to include many other situations, but the independent events and random variables continue to be highly important.

A particularly useful mathematical property of independent random variables is that if f_1 and f_2 are independent random variables and each has finite expectation, the expectation of the product $f_1 f_2$ is the product of the expectations $E(f_1)E(f_2)$. We shall deduce this from a lemma that has other uses too. The proof of the lemma in its full strength makes use of the ideas in Section 6. Those readers who have passed over that section will have to be content with the special case in which the f_i are limits of step-functions with left-open intervals of constancy, but as we have seen, such functions occur in many useful cases.

★LEMMA 11-3 *Let s and t be positive integers and r their sum. Let m, m_1, and m_2 be nonnegative additive regular functions of left-open intervals in R^r, R^s, and R^t, respectively, such that if $A(1)$ and $A(2)$ are left-open intervals in R^s and R^t, respectively, and $A = A(1) \times A(2)$, then*

(A) $$mA = (m_1 A(1))(m_2 A(2)).$$

Let f_1 be a function integrable over R^s with respect to m_1, and let f_2 be a function integrable over R^t with respect to m_2. Then the function f defined on R^r by

$$f(u, v) = f_1(u) f_2(v) \qquad (u \text{ in } R^s,\ v \text{ in } R^t)$$

is integrable over R^r, and

(B) $$\int_{R^r} f(x)\, m(dx) = \left\{\int_{R^s} f_1(u)\, m_1(du)\right\}\left\{\int_{R^t} f_2(v)\, m_2(dv)\right\}.$$

Let ε be positive; define

$$\varepsilon' = \tfrac{1}{5}\min\{\varepsilon, 1\}\left(\int_{R^s} |f_1(u)|\, m_1(du) + \int_{R^t} |f_2(v)|\, m_2(dv) + 1\right)^{-1}.$$

Since f_1 and $|f_1|$ are integrable with respect to m_1 over R^s, there is a gauge γ_1 on $\bar{R}^s$ such that if $\mathscr{P}_1$ is any γ_1-fine partition of R^s,

(C) $$\left| S(\mathscr{P}_1; f_1; m_1) - \int_{R^s} f_1(u)\, m_1(du) \right| < \varepsilon'$$

and

(D) $$\left| S(\mathscr{P}_1; |f_1|; m_1) - \int_{R^s} |f_1(u)|\, m_1(du) \right| < \varepsilon'.$$

Since f_2 is integrable with respect to m_2 over R^t, there is a gauge γ_2 on $\bar{R}^t$ such that if $\mathscr{P}_2$ is any γ_2-fine partition of R^t,

$$\text{(E)} \qquad \left| S(\mathscr{P}_2; f_2; m_2) - \int_{R^t} f_2(v)\, m_2(dv) \right| < \varepsilon'.$$

For each point $x = (u, v)$ of $\bar{R}^r$ we define

$$\gamma(x) = \gamma_1(u) \times \gamma_2(v).$$

This is a gauge on $\bar{R}^r$. Let $\mathscr{P}$ be any γ-fine partition of R^r; we shall prove

$$\text{(F)} \qquad \left| S(\mathscr{P}; f; m) - \left[\int_{R^s} f_1(u)\, m_1(du) \right] \left[\int_{R^t} f_2(v)\, m_2(dv) \right] \right| < \varepsilon.$$

Then, because (F) holds for every γ-fine partition $\mathscr{P}$ of R^r, by definition of the integral the conclusion of the lemma will be established.

Let $\mathscr{P}$ consist of the pairs $\{(\bar{x}_1, A_1), \ldots, (\bar{x}_{n^*}, A_{n^*})\}$. As in the proof of Lemma 2-1, we can find a "checkerboard partition" of R^r that consists of a collection of intervals

$$\text{(G)} \qquad A_{i,j} = A_i(1) \times A_j(2) \qquad (i = 1, \ldots, i^*; j = 1, \ldots, j^*)$$

in which the $A_i(1)$ are pairwise disjoint left-open intervals whose union is R^s and the $A_j(2)$ are pairwise disjoint left-open intervals whose union is R^t, such that each interval A_n of the partition $\mathscr{P}$ is the union of all the intervals $A_{i,j}$ contained in it. If A_n is the interval of $\mathscr{P}$ that contains $A_{i,j}$, we define

$$(u_{i,j}, v_{i,j}) = x_{i,j} = \bar{x}_n.$$

Then the set of pairs

$$\mathscr{P}^{\sim} = \{(x_{i,j}, A_{i,j}) : i = 1, \ldots, i^*; j = 1, \ldots, j^*\}$$

is also a γ-fine partition of R^r, and

$$\text{(H)} \qquad \begin{aligned} S(\mathscr{P}; f; m) &= S(\mathscr{P}^{\sim}; f; m) \\ &= \sum_{i=1}^{i^*} \sum_{j=1}^{j^*} f_1(u_{i,j}) f_2(v_{i,j}) m_1 A_i(1) m_2 A_j(2). \end{aligned}$$

Also,

$$(A_i(1) \times A_j(2))^{-} \subset A_n^{-} \subset \gamma(\bar{x}_n) = \gamma_1(u_{i,j}) \times \gamma_2(v_{i,j}),$$

which implies that

$$\text{(I)} \qquad A_i(1)^{-} \subset \gamma_1(u_{i,j}) \qquad \text{and} \qquad A_j(2)^{-} \subset \gamma_2(v_{i,j}).$$

We define for $i = 1, \ldots, i^*$ and $j = 1, \ldots, j^*$

$$D'_{i,j} = \int_{A_i(2)} f_1(u)\, m_1(du) - f_1(u_{i,j}) m_1 A_i(1),$$

$$D''_{i,j} = \int_{A_j(2)} f_2(v)\, m_2(dv) - f_2(v_{i,j}) m_2 A_j(2).$$

The intervals $A_i(1)$ $(i = 1, \ldots, i^*)$ are pairwise disjoint, and likewise the intervals $A_j(2)$ $(j = 1, \ldots, j^*)$. So Corollary II-13-2 yields the following conclusion.

(J) For all i in $\{1, \ldots, i^*\}$ and all j in $\{1, \ldots, j^*\}$,

$$\sum_{j=1}^{j^*} |D''_{i,j}| < 5\varepsilon' \qquad \text{and} \qquad \sum_{i=1}^{i^*} |D'_{i,j}| < 5\varepsilon'.$$

By (A), (G), and (I),

$$\text{(K)} \quad \left| S(\mathscr{P}; f; m) - \left[\int_{R^s} f_1(u)\, m_1(du)\right]\left[\int_{R^t} f_2(v)\, m_2(dv)\right]\right|$$

$$= \left| \sum_{i=1}^{i^*} \sum_{j=1}^{j^*} f_1(u_{i,j}) f_2(v_{i,j}) m_1 A_i(1) m_2 A_j(2) \right.$$

$$\left. - \sum_{i=1}^{i^*} \sum_{j=1}^{j^*} \left[\int_{A_i(1)} f_1(u)\, m_1(du)\right]\left[\int_{A_j(2)} f_2(v_{i,j})\, m_2(dv)\right]\right|$$

$$= \left| \sum_{i=1}^{i^*} \sum_{j=1}^{j^*} \left\{ f_1(u_{i,j}) m_1 A_i(1) \right\} \left\{ f_2(v_{i,j}) m_2 A_j(2) - \int_{A_j(2)} f_2(v)\, m_2(dv) \right\} \right.$$

$$+ \sum_{i=1}^{i^*} \sum_{j=1}^{j^*} \left\{ f_1(u_{i,j}) m_1 A_i(1) - \int_{A_i(1)} f_1(u)\, m_1(du) \right\}$$

$$\left. \times \left\{ \int_{A_j(2)} f_2(v)\, m_2(dv) \right\} \right|.$$

For each i in $\{1, \ldots, i^*\}$ we denote by U_i that one of the points $u_{i,1}, \ldots, u_{i,j^*}$ for which $|f_1(u_{i,j})|$ is greatest. The pairs $\{(U_1, A_1(1)), \ldots, (U_{i^*}, A_{i^*}(1))\}$ constitute a γ_1-fine partition of R^s, by (I). So by (D),

$$\sum_{i=1}^{i^*} |f_1(U_i)| m_1 A_i(1) < \int_{R^s} |f_1(u)|\, m_1(du) + \varepsilon'.$$

By this and (K),

$$\left| S(\mathscr{P}; f; m) - \left[\int_{R^s} f_1(u)\, m_1(du)\right]\left[\int_{R^t} f_2(v)\, m_2(dv)\right]\right|$$

$$\leqq \sum_{i=1}^{i^*} \sum_{j=1}^{j^*} |f_1(U_i)|\, m_m A_i(1) |D''_{i,j}| + \sum_{j=1}^{j^*} \sum_{i=1}^{i^*} |D'_{i,j}| \int_{A_j(2)} |f_2(v)|\, m_2(dv)$$

$$\leqq \sum_{i=1}^{i^*} |f_1(U_i) m_1 A_i(1)| (5\varepsilon') + \sum_{j=1}^{j^*} (5\varepsilon') \int_{A_j(2)} |f_2(v)|\, m_2(dv)$$

$$\leqq (5\varepsilon') \left\{ \int_{R^s} |f_1(u)|\, m_1(du) + \varepsilon' + \int_{R^t} |f_2(v)|\, m_2(dv) \right\}$$

$$< \varepsilon.$$

So (F) is valid, and the lemma is proved.

★COROLLARY 11-4 *Let $r(1), \ldots, r(n)$ be positive integers, and let r be their sum. Let $m, m_1, \ldots, m_n$ be nonnegative additive regular functions of left-open intervals in $R^r, R^{r(1)}, \ldots, R^{r(n)}$, respectively, with the property that if $A(1), \ldots, A(n)$ are left-open intervals in $R^{r(1)}, \ldots, R^{r(n)}$, respectively, and $A = A(1) \times \cdots \times A(n)$, then*

$$mA = (m_1 A(1)) \cdots (m_n A(n)).$$

For each i in $\{1, \ldots, n\}$ let f_i be a real-valued function on $R^{r(i)}$ that is integrable with respect to m_i over $R^{r(i)}$. Then the function f on R^r defined by

$$f(x_1, \ldots, x_n) = f_1(x_1) \cdots f_n(x_n)$$

is integrable with respect to m over R^r, and

$$\int_{R^r} f(x)\, m(dx) = \left\{\int_{R^{r(1)}} f_1(x_1)\, m_1(dx_1)\right\} \cdots \left\{\int_{R^{r(n)}} f_n(x_n)\, m_n(dx_n)\right\}.$$

We prove this by induction. It has been shown true for $n = 2$ in Lemma 11-3. Suppose it true for all integers less than n. With the notation of this corollary, define

$$r' = r(2) + \cdots + r(n), \qquad m'[A(2) \times \cdots \times A(n)] = m_2 A(2) \cdots m_n A(n),$$

$$v = (x_2, \ldots, x_n).$$

By the induction hypothesis, the function f' whose value at v is $f_2(x_2) \cdots f_n(x_n)$ is integrable over $R^{r'}$ with respect to m', and

$$\int_{R^{r'}} f'(v)\, m'(dv) = \left\{\int_{R^{r(2)}} f_2(x_2)\, m_2(dx_2)\right\} \cdots \left\{\int_{R^{r(n)}} f_n(x_n)\, m_n(dx_n)\right\}.$$

Again by the induction hypothesis, the function $f \mapsto f(x) = f(x_1, v)$ is integrable with respect to m over R^r, and

$$\int_{R^r} f(x)\, m(dx) = \left\{\int_{R^{r(1)}} f_1(x_1)\, m_1(dx_1)\right\}\left\{\int_{R^{r'}} f'(v)\, m'(dv)\right\}.$$

These two equations imply that the conclusion holds for n factors, and by induction the corollary holds for all n.

In the next theorem we use the same notation as in preceding proofs, and for easy reference we collect this notation here.

(L) First, P is a probability measure on a space Ω. $r(1), \ldots, r(n)$ are positive integers and r is their sum.

Second, for each i in $\{1, \ldots, n\}$, $X(i)$ is a random $r(i)$-vector that is measurable with respect to P on Ω.

Third, $P_{X(i)}$ is the distribution of $X(i)$, and $E_{X(i)}$ is the corresponding expectation that applies to real functions on $R^{r(i)}$.

And last, each x in R^r is also denoted by $(x_1, \ldots, x_n)$, with x_i in $R^{r(i)}$.

We can now state and prove an important theorem.

THEOREM 11-5 *Let $X(1), \ldots, X(2)$ be independent random vectors, $X(i)$ being an $r(i)$-vector. For each i in $\{1, \ldots, n\}$, let f_i be a real-valued function on $R^{r(i)}$ that has finite expectation $E_{X(i)}(f_i)$. Then the function f defined on R^r by*

$$f(x) = f_1(x_1) \cdots f_n(x_n)$$

has finite expectation, and

$$E_X(f) = [E_{X(1)}(f_1)] \cdots [E_{X(n)}(f_n)].$$

Let $A(i)$ be a left-open interval in $R^{r(i)}$ $(i = 1, \ldots, n)$, and define $A = A(1) \times \cdots \times A(n)$. For each ω in Ω, $X(\omega)$ is in A if and only if $X(i)(\omega)$ is in $A(i)$ for $i = 1, \ldots, n$, so

$$\{X \text{ in } A\} = \{X(1) \text{ in } A(1)\} \cap \cdots \cap \{X(n) \text{ in } A(n)\}.$$

Since the $X(i)$ are independent, this implies

$$P\{X \text{ in } A\} = [P\{X(1) \text{ in } A(1)\}] \cdots [P\{X(n) \text{ in } A(n)\}].$$

By the definition of P_X, etc., (equation (A) of Section 10), this can be written in the form

$$P_X(A) = [P_{X(1)}(A(1))] \cdots [P_{X(n)}(A(n))].$$

So if we take $m = P_X$ and $m_i = P_{X(i)}$ $(i = 1, \ldots, n)$, the hypotheses of Corollary 11-4 are satisfied. The conclusion of this theorem then follows at once from Corollary 11-4.

COROLLARY 11-6 *Let $X_1, \ldots, X_r$ be independent random variables on Ω, each with finite expectation. Then their product has finite expectation, and*

$$E(X_1 \cdots X_n) = E(X_1) \cdots E(X_n).$$

We apply Theorem 11-5 with all the $r(i)$ equal to 1 and all the f_i the identity function on R, so that $f_i(x) = x$. This is the limit of a sequence of step-functions with left-open intervals of constancy, so by Theorem 10-1, each $f_i \circ X_i$ is a random variable and

$$E(X_i) = E(f_i \circ X_i) = E_{X(i)}(f_i).$$

By Theorem 11-5, for the function f defined by $f(x) = x_1 \cdots x_r$ we have

$$E_X(f) = E_{X(1)}(f_1) \cdots E_{X(r)}(f_r).$$

We have just seen that the right member of this equation is the right member of the conclusion, and by Theorem 10-1,

$$E_X(f) = E(f \circ X) = E(X_1 \cdots X_r),$$

which is the left member of the conclusion. So the corollary is established.

This result is not only useful; it is remarkable. If the random variables X_1 and X_2 each have finite expectation but are not necessarily independent, their product may not have finite expectation, and if it does, the values of $E(X_1)$ and $E(X_2)$ give us no clue as to the value of $E(X_1 X_2)$. This contrasts with the good behavior of the product of independent random variables established in Corollary 11-6.

We know that if X_1 and X_2 are random variables with finite expectations, the expectation of their sum is the sum of their expectations. But even when X_1 and X_2 have finite variances, the variance of their sum has no such simple relationship to the variances of X_1 and X_2. For example, if X_1 has variance 1, then when $X_2 = X_1$ we have $\operatorname{Var} X_2 = 1$ and $\operatorname{Var}[X_1 + X_2] = 4$, whereas if $X_2 = -X_1$, we have $\operatorname{Var} X_2 = 1$ and $\operatorname{Var}[X_1 + X_2] = 0$. But when X_1 and X_2 are independent, we again have a simple relationship between the variances.

THEOREM 11-7 *If X_1 and X_2 are independent random variables with finite variance, $X_1 + X_2$ has finite variance, and*

$$\operatorname{Var}[X_1 + X_2] = \operatorname{Var} X_1 + \operatorname{Var} X_2.$$

Since X_1 and X_2 have finite variances, they have finite first and second moments, by Corollary III-11-5. Since $E(X_1 + X_2) = E(X_1) + E(X_2)$,

$$\text{(L)}\quad [(X_1 + X_2) - E(X_1 + X_2)]^2$$
$$= [(X_1 - E(X_1)) + (X_2 - E(X_2))]^2$$
$$= [X_1 - E(X_1)]^2 + 2[X_1 - E(X_1)][X_2 - E(X_2)] + [X_2 - E(X_2)]^2.$$

We define f_1 and f_2 on R by

$$f_i(x) = x - E(X_i).$$

Then for every set of real numbers a_1, b_1, a_2, b_2 the two sets

$$\{\omega \text{ in } \Omega : a_k < f_1(X_1(\omega)) \leqq b_1\}, \qquad \{\omega \text{ in } \Omega : a_2 < f_2(X_2(\omega)) \leqq b_2\}$$

are the same as the sets

$$\{\omega \text{ in } \Omega : a_1 + E(X_1) < X_1(\omega) \leqq b_1 + E(X_1)\},$$
$$\{\omega \text{ in } \Omega : a_2 + E(X_2) < X_2(\omega) \leqq b_2 + E(X_2)\},$$

and these are independent events because X_1 and X_2 are independent random variables. Their second moments are the variances of X_1 and X_2, which are finite, so by Corollary 11-6,

$$\begin{aligned} E(f_1(X_1)f_2(X_2)) &= E(f_1(X_1))E(f_2(X_2)) \\ &= [E(X_1 - E(X_1))][E(X_2 - E(X_2))] \\ &= [E(X_1) - E(E(X_1))][EX_2 - E(E(X_2))] \\ &= 0. \end{aligned}$$

So the middle term in the right member of (L) vanishes, and (L) reduces to the equation in the conclusion of the theorem.

A basic motivation in our definition of expectation was the idea that if an experiment is independently repeated many times and for each repetition we record the value of some function f determined by the outcome of the experiment, the average of the recorded numbers should be nearly equal to what we have called the *expectation* of the function. We shall now prove the weak law of large numbers. This law states that the foregoing is what happens, and for each positive number ε it gives a numerical estimate for the probability that the average of the several outcomes should be within ε of the (mathematically defined) expectation.

THEOREM 11-8 *Let $X_1, \ldots, X_n$ be independent random variables all with the same distribution and all with finite second moments. Then for every positive ε*

$$P(|[X_1 + \cdots + X_n]/n - E(X_1)| \geqq \varepsilon) \leqq [\operatorname{Var} X_1]/\varepsilon^2 n.$$

Let

$$Y = [X_1 + \cdots + X_n]/n - E(X_1).$$

Then

$$E(Y) = [E(X_1) + \cdots + E(X_n)]/n - E(X_1) = 0.$$

For $j = 1, \ldots, n$,

$$\operatorname{Var}[X_j/n] = n^{-2} \operatorname{Var} X_j = n^{-2} \operatorname{Var} X_1.$$

Since $E(Y) = 0$, by Theorem 11-7,

$$E(Y^2) = \operatorname{Var} Y = \sum_{j=1}^{n} \operatorname{Var}[X_j/n] = (\operatorname{Var} X_1)/n.$$

If we apply the Chebyshev inequality (Theorem III-11-8) with $c = \varepsilon$, we obtain the conclusion of this theorem.

Suppose that we are given r chance happenings, specified by numbers $X(1), \ldots, X(r)$, and that for each i we are willing to accept a certain probability distribution $P_{X(i)}$ as describing accurately enough the distribution of the values that occur. We consider that the chance of any one of these events happening is unaffected by the happening or nonhappening of any of the others, and we wish to construct a mathematical model. In this model there should be r independent random variables $X(1), \ldots, X(r)$, the distribution of $X(i)$ being the assigned $P_{X(i)}$. The constructions in the foregoing proofs show us how to do this. For each left-open interval $A = A(1) \times \cdots \times A(r)$ in R^r we define

$$\text{(M)} \qquad P(A) = P_{X(1)}(A(1)) \cdots P_{X(r)}(A(r)).$$

This is obviously nonnegative, and $P(R^r) = 1$. If A is subdivided by a hyperplane into two disjoint left-open intervals, these two subintervals must be produced by subdividing some one of the intervals $A(i)$ into two disjoint left-open intervals $A'(i), A''(i)$. Merely to simplify notation, we suppose that $A(1)$ is the interval that is subdivided. Then

$$\begin{aligned} A &= [A'(1) \cup A''(1)] \times A(2) \times \cdots \times A(r) \\ &= [A'(1) \times A(2) \times \cdots \times A(r)] \cup [A''(1) \times A(2) \times \cdots \times A(r)]. \end{aligned}$$

We denote by A', A'' the Cartesian products in the two brackets. Then since $P_{X(1)}$ is additive,

$$\begin{aligned} P(A) &= P_{X(1)}(A(1))P_{X(2)}(A(2)) \cdots P_{X(r)}(A(r)) \\ &= [P_{X(1)}(A'(1)) + P_{X(1)}(A''(1))]P_{X(2)}(A(2)) \cdots P_{X(r)}(A(r)) \\ &= PA' + PA''. \end{aligned}$$

Thus, P has the property stated for m_L in (A) of the proof of Lemma 2-1. From this statement we can deduce, as we did there, that P is additive. If $A = A(1) \times \cdots \times A(r)$ is left-open and $c > PA$, we can and do choose numbers $c_1, \ldots, c_r$ such that

$$c_i > P_{X(i)}A(i) \qquad \text{and} \qquad c_1 \cdots c_r < c.$$

Each $P_{X(i)}$ is regular, so for each i we can and do choose a left-open interval $B(i)$ such that $A(i) \subset B(i)^{\circ}$ and

$$P_{X(i)}(B(i)) < c_i.$$

Then $B = B(1) \times \cdots \times B(r)$ is a left-open interval in R^r whose interior contains A, and

$$P(B) < c_1 \cdots c_r < c.$$

Similarly, if $c < P(a)$ we can find a bounded left-open interval B whose closure is contained in A for which $P(B) > c$, and so P is regular. The coordinate functions $x \mapsto x^i$ are continuous on R^r, so they are random variables with respect to P. By (M) they are independent, and clearly the distribution of the ith coordinate function is $P_{X(i)}$.

For example, let us begin with the "Bernoulli distribution with parameter p," defined thus. First, p is a number in $(0, 1)$. Then, for each left-open interval A in R, $P_B(A) = p$ if A contains 1 but not 0; $P_B(A) = q = 1 - p$ if A contains 0 but not 1; $P_B(A) = 1$ if A contains both 0 and 1, and $P_B(A) = 0$ if A contains neither 0 nor 1. This was discussed in Section III-1, and it was shown there that for every real-valued function f on R

$$E(f) = f(0)q + f(1)p.$$

In particular, if f is the identity function $\mathbf{x}$ ($\mathbf{x}(x) = x$) on R,

$$E(\mathbf{x}) = 0q + 1p = p.$$

The variance of the distribution is the expectation of $(\mathbf{x} - p)^2$, which is

$$\operatorname{Var} X = (0 - p)^2 q + (1 - p)^2 p = p^2 q + q^2 p = pq(p + q) = pq.$$

To form the model for an r-vector whose components are independent Bernoulli with parameter p, we use (M). If $A = A(1) \times \cdots \times A(r)$ is a left-open interval in $\bar{R}^r$ that contains none of the points $(x^{(1)}, \ldots, x^{(r)})$ in which each $x^{(i)}$ is either 0 or 1, then for some i the interval $A(i)$ contains neither 0 nor 1. Then for this i, $P_{X(i)}(A(i)) = 0$, and by (M), $P(A) = 0$. Suppose that A contains exactly one of the points all of whose coordinates are 0 or 1, and that, for this point, j of the coordinates are 1 and $r - j$ of them are 0. Then in the right member of (M) there are j factors p and $r - j$ factors q, so

$$P(A) = p^j q^{r-j}.$$

It follows readily that for every left-open interval A, $P(A)$ is the sum of the numbers $p^j q^{r-j}$ for all the points with coordinates 0 and 1 that are contained in A.

It is obvious, but important enough to deserve explicit mention, that the joint distribution of r independent random variables is determined by the distributions of the several random variables and has nothing to do with any other property that the random variables may have in addition to their distribution. This follows from Definition 11-2 and is seen again in (M).

The *binomial distribution* with parameters r and p is defined to be the distribution of the sum of r independent random variables each of which has a Bernoulli distribution with parameter p. As just mentioned, this is enough to specify the binomial distribution. Suppose that Y_1 and Y_2 are two independent random variables with binomial distributions, Y_1 having parameters $r(1)$ and p, and Y_2 having parameters $r(2)$ and p. It is then easy to show that $Y_1 + Y_2$ has a binomial distribution with parameters $r(1) + r(2)$ and p. For let $r = r(1) + r(2)$, and let $X(1), \ldots, X(r)$ be r independent random variables with Bernoulli distributions with parameter p. The sum of the first $r(1)$ of them has binomial distribution with parameters $r(1)$ and p, which is the distribution of Y_1; and the sum of the rest of them has binomial distribution with parameters $r(2)$ and p, which is the distribution of Y_2. Since the two sums are independent, the distribution of their sum is the distribution of $Y_1 + Y_2$. But their sum is the sum of r independent random variables each having Bernoulli distribution with parameter p, so it is a binomial distribution with parameters r and p.

If Y is the sum of independent random variables $X(1), \ldots, X(r)$, each having Bernoulli distribution with parameter p, each $X(1)$ has expectation p and variance pq, as previously shown. So the expectation of Y is rp, and by Theorem 11-7, the variance of Y is rpq.

Let Y have binomial distribution with parameters r and p. We shall prove by induction that, for $k = 0, 1, \ldots, r$,

$$\text{(N)} \qquad P(Y = k) = \binom{r}{k} p^k q^{r-k}.$$

When this is established, it will follow at once that for every left-open interval A in R (and, in fact, for every set A in R), $P(A)$ is the sum of the numbers (N) for all k in A. Let Y be the sum of independent random variables $X(1), \ldots, X(r)$, each having Bernoulli distribution with parameter p, and let $S(r)$ be the statement "for $k = 0, 1, \ldots, r$, equation (N) is valid." Statement $S(1)$ is easily seen to be true; for with $r = 1$, if $k = 0$ both members of (N) are q and if $k = 1$ both members are p. Let h be a positive integer and assume that $S(h-1)$ is true; we shall prove that $S(h)$ is true. If $r = h$ and $k = 0$, both members of (N) have value q^r, for $Y = 0$ if and only if each $X(i) = 0$, and these are independent events each with probability q. If $k > 0$, the set $Y = k$ is the union of the two disjoint events

$$\{X(1) + \cdots + X(h-1) = k\} \cap \{X(h) = 0\},$$
$$\{X(1) + \cdots + X(h-1) = k-1\} \cap \{X(h) = 1\}.$$

Since the $X(1)$ are independent and $S(h-1)$ is true,

$$\begin{aligned} P(Y = k) &= P(X(1) + \cdots + X(h-1) = k) \cdot P(X(h) - 0) \\ &\quad + P(X(1) + \cdots + X(h-1) = k-1)P(X(h) = 1) \\ &= \left[\binom{h-1}{k} p^k q^{h-1-k}\right] q + \left[\binom{h-1}{k-1} p^{k-1} q^{h-k}\right] p \\ &= \left[\binom{h-1}{k} + \binom{h-1}{k-1}\right] p^k q^{h-k} \\ &= (h-1)! \left[\frac{1}{k!(h-1-k)!} + \frac{1}{(k-1)!(h-k)!}\right] p^k q^{h-k} \\ &= \binom{h}{k} p^k q^{h-k}, \end{aligned}$$

so (I is valid for $k = 1, \ldots, h$. This completes the proof that if $S(h-1)$ is true so is $S(h)$, and by induction, $S(r)$ is true for all positive integers r.

So far in this section we have avoided using the more advanced concepts presented in Section 6. In the next two theorems we shall make use of those concepts. Let $X_1, \ldots, X_n$ be independent random vectors in $R^{r(1)}, \ldots, R^{r(n)}$, respectively. By definition, when $A_1, \ldots, A_n$ are left-open intervals in $R^{r(1)}, \ldots, R^{r(n)}$, respectively, the sets $\{X_1 \text{ in } A_1\}, \ldots, \{X_n \text{ in } A_n\}$ are independent events. The next theorem shows that this remains true whenever the A_i are sets that are events with respect to $P_{X_1}, \ldots, P_{X_n}$, respectively.

THEOREM 11-9 *Let $X_1, \ldots, X_n$ be independent random vectors, X_i taking values in $R^{r(i)}$. Let $B(i)$ be a subset of $R^{r(i)}$ that is an event with respect to $P_{X(i)}$ $(i = 1, \ldots, n)$. Then the sets*

$$C(i) = \{\omega \text{ in } \Omega : X_i(\omega) \text{ in } B(i)\} \qquad (i = 1, \ldots, n)$$

are independent events in Ω.

Let $\{i(1), \ldots, i(k)\}$ be a subset of $\{1, \ldots, n\}$. We must show that the intersection of the sets $C(i(j))$ $(j = 1, \ldots, k)$ is an event whose P-measure is the product of the P-measures of the $C(i(j))$. To simplify notation, we suppose that $i(j) = j$ $(j = 1, \ldots, k)$; this can be brought about by renumbering the $C(i(j))$. For $j = 1, \ldots, k$ we define f_j by

$$f_j(x_j) = 1_{B(j)}(x_j) \qquad (x_j \text{ in } R^{r(j)});$$

for $j = k + 1, \ldots, n$ we define $f_j = 1$. Then for $j = 1, \ldots, k$, $f_j \circ X_j$ has the value 1 at each ω such that $X_j(\omega)$ is in $B(j)$, and it has the value 0 elsewhere, so by definition of $C(j)$,

$$\text{(O)} \qquad f_j \circ X_j = 1_{C(j)} \qquad (j = 1, \ldots, k).$$

By this and Theorem 10-5, for $j = 1, \ldots, k$

$$P(C(j)) = E(f_j \circ X_j) = E_{X(j)}(f_j).$$

If we define f on R^r by setting

$$f(x) = f_1(x_1) \cdots f_n(x_n),$$

by Theorems 10-5 and 11-5,

$$\begin{aligned} \text{(P)} \qquad E(f \circ X) &= E_X(f) = E_{X(1)}(f_1) \cdots E_{X(n)}(f_n) \\ &= P(C(1)) \cdots P(C(k)) \cdot 1 \cdots 1. \end{aligned}$$

By (O), we have for each ω in Ω

$$\begin{aligned} f(X(\omega)) &= f_1(X_1(\omega)) \cdots f_k(X_k(\omega)) \\ &= 1_{C(1)}(\omega) \cdots 1_{C(k)}(\omega) \\ &= 1_{C(1) \cap \cdots \cap C(k)}(\omega), \end{aligned}$$

so $f \circ X$ is the indicator of $C(1) \cap \cdots \cap C(k)$. If we substitute this in the left member of (P), we obtain

$$P(C(1) \cap \cdots \cap C(k)) = P(C(1)) \cdots P(C(k)),$$

which completes the proof.

COROLLARY 11-10 *For $i = i, \ldots, n$, let X_i be a random $r(i)$-vector on Ω, let the X_i be independent and let f_i be a random variable on $R^{r(i)}$ with respect to $P_{X(i)}$. Then the functions $f_i \circ X_i$ $(i = 1, \ldots, n)$ are independent random variables.*

Let $A(1), \ldots, A(n)$ be left-open intervals in $R^{r(1)}, \ldots, R^{r(n)}$, respectively. Then the sets

$$B(i) = \{x_i \text{ in } R^{r(i)} : f_i(x_i) \text{ in } A(i)\} \qquad (i = 1, \ldots, n)$$

are events with respect to $P_{X(1)}, \ldots, P_{X(n)}$, respectively. By Theorem 11-9, the sets

$$C(i) = \{\omega \text{ in } \Omega : X_i(\omega) \text{ in } B(i)\} \qquad (i = 1, \ldots, n)$$

are independent events. But $X_i(\omega)$ is in $B(i)$ if and only if $f_i(X_i(\omega))$ is in $A(i)$, so

$$C(i) = \{\omega \text{ in } \Omega : f_i(X_i(\omega)) \text{ in } A(i)\}.$$

Since these have been shown to be independent events, the functions $f_i \circ X_i$ are independent random variables.

If $A_1, A_2, A_3, \ldots$ is a sequence of sets, the **limit superior** of the A_n is defined to be the set of all points that belong to infinitely many of the A_n. A point x is in infinitely many of the A_n if and only if for every positive integer k it is in some A_n with $n \geqq k$ and hence in the union $A_k \cup A_{k+1} \cup A_{k+2} \cup \cdots$. We state this as a definition.

DEFINITION 11-11 *If $A_1, A_2, A_3, \ldots$ are sets,*

$$\limsup A_n = \bigcap_{k=1}^{\infty} \left\{ \bigcup_{n=k}^{\infty} A_n \right\}.$$

In the next two theorems we shall be concerned with infinite sequences of independent random variables. Given finitely many independent random variables, we have already seen (after Theorem 11-8) how we can construct a joint distribution for them on a finite-dimensional space. It is not so evident that we can construct a mathematical model for an infinite sequence of independent random variables with given distributions; it is not evident that such random variables exist on a set $\Omega = R^r$, and these are the only ones in which we have defined integration. It will be evident in the proofs that all we need to know about events and measure is that the events form a σ-algebra of subsets of some set Ω and that the measure P is countably additive on the set of events. But this still leaves us in an unsatisfactory state; that is all we need, but can we have even that much? We need a construction of an integral with respect to a measure P that will allow us to form a joint distribution for infinitely many random variables. This we shall do in detail in Section 14.

The next theorem is the *Borel–Cantelli lemma*, which is of great use in dealing with limits of sequences of random variables.

THEOREM 11-12 *Let P be a probability measure on a σ-algebra of events in a space Ω. Let $A_1, A_2, A_3, \ldots$ be a sequence of events. Then*

(i) *if $\Sigma_{n=1}^{\infty} P(A_n) < \infty$, $P(\limsup A_n) = 0$;*
(ii) *if the A_i are independent and $\Sigma_{n=1}^{\infty} P(A_n) = \infty$, then $P(\limsup A_n) = 1$.*

For (i), let ε be positive. There exists an integer k such that

$$\sum_{n=k}^{\infty} P(A_n) < \varepsilon.$$

By Definition 11-11, $\limsup A_n$ is contained in the union $A_k \cup A_{k+1} \cup A_{k+2} \cup \cdots$, so by Theorem II-12-3, $\limsup A_n$ has P-measure 0.

For (ii) we first observe that, as remarked just after Definition 11-1, the complements $\Omega \setminus A_n$ $(n = 1, 2, 3, \ldots)$ are a set of independent events. From this and Corollary III-10-11,

$$
\text{(Q)} \qquad
\begin{aligned}
P(\limsup A_n) &= \lim_{k\to\infty} P\left(\bigcup_{h=k}^{\infty} A_n\right) \\
&= \lim_{k\to\infty} P\left(\Omega \setminus \bigcap_{n=k}^{\infty} [\Omega \setminus A_n]\right) \\
&= \lim_{k\to\infty} \lim_{q\to\infty} P\left(\Omega \setminus \bigcap_{n=k}^{q} [\Omega \setminus A_n]\right) \\
&= \lim_{k\to\infty} \lim_{q\to\infty} \left\{1 - P\left(\bigcap_{n=k}^{q} [\Omega \setminus A_n]\right)\right\} \\
&= \lim_{k\to\infty} \lim_{q\to\infty} \left[1 - \prod_{n=k}^{q} P(\Omega \setminus A_n)\right] \\
&= \lim_{k\to\infty} \lim_{q\to\infty} \left[1 - \prod_{n=k}^{q} \{1 - P(A_n)\}\right].
\end{aligned}
$$

Since the second derivative of the exponential function $x \mapsto \exp x$ is positive, by Taylor's theorem

$$1 + x \leqq \exp x$$

for all real x. In particular,

$$\prod_{n=k}^{q} (1 - P(A_n)) \leqq \prod_{n=k}^{q} \exp(-P(A_n)) = \exp\left(-\sum_{n=k}^{q} P(A_n)\right).$$

As q increases, the sum in the right member increases without bound, so the right member of this inequality tends to 0, whereas the left member is nonnegative. So for all positive integers k,

$$\lim_{q\to\infty} \prod_{n=k}^{q} (1 - P(A_n)) = 0.$$

So the left member of (Q) is 1, and (ii) is established.

EXERCISE 11-1 Assume that when two dice are rolled, all six numbers on each die are equally likely and are unaffected by the number on the other die. Show that the model for this chance occurrence is that in Exercise 10-1.

EXERCISE 11-2 Let u be the latitude and v the longitude of a point on the earth's surface $(-\pi/2 \leqq u \leqq \pi/2,\ -\pi < v \leqq \pi)$. Suppose that a certain event can happen anywhere on the earth's surface with the same chance at each point.

Show that the model for this is a probability measure on $(-\pi/2, \pi/2] \times (-\pi, \pi]$ with probability density

$$p(u, v) = (\cos u)/4\pi.$$

(From elementary geometry, if two parallel planes intersect the surface of a sphere, the area of the part of the surface between them is proportional to the distance between the planes.) Show that the latitude of the event and its longitude are independent random variables.

EXERCISE 11-3 Let $X_1, \ldots, X_h, X_{h+1}, \ldots, X_k$ be independent random variables. Show that $(X_1, \ldots, X_h)$ and $(X_{h+1}, \ldots, X_k)$ are independent random vectors.

EXERCISE 11-4 Let X_1 and X_2 be independent random variables, each with a normal distribution with mean 0 and standard deviation 1. Let a and b be real numbers and let λ be positive. Show that the probability that $aX_1 + bX_2 < \lambda$ is the same as the probability that $X_1 < \lambda(a^2 + b^2)^{-1/2}$. *Suggestion*: Find the density of the distribution of (X_1, X_2). Rotate axes so that (a, b) goes to $([a^2 + b^2]^{1/2}, 0)$.

EXERCISE 11-5 Let $X(1)$ and $X(2)$ be independent normally distributed random variables with the respective means m_1, m_2 and the respective variances σ_1^2, σ_2^2. Show that $X(1) + X(2)$ is normally distributed with mean $m_1 + m_2$ and variance $\sigma_1^2 + \sigma_2^2$. *Suggestion*: Let $Y(i) = [X(i) - m_i]/\sigma_i$ $(i = 1, 2)$. Use preceding exercise.

EXERCISE 11-6 Show that the sum of any finite number of independent normally distributed random variables is normally distributed.

EXERCISE 11-7 Let $X(1), \ldots, X(n)$ be independent random variables whose distributions have the respective densities $p_1, \ldots, p_n$. Show that the function p on R^n defined by

$$p(x_1, \ldots, x_n) = p_1(x_1) \cdots p_n(x_n)$$

is a density for the joint distribution of the variables $X(1), \ldots, X(n)$.

EXERCISE 11-8 (Buffon's Needle Problem) A floor is marked with parallel east–west lines w units apart. A needle of length l $(< w)$ is dropped on the floor. Let x be the distance from the center of the needle to the nearest line south of it and θ be the angle of the needle from east. Assume that x is uniformly distributed over $(0, w]$ and θ is uniformly distributed over $(0, 2\pi]$, and that x and θ are independent. Find the probability that the needle will lie across a line. (This has been used to obtain an experimental estimate of π.)

EXERCISE 11-9 A particle has mass m, and the three components V_1, V_2, V_3 of its velocity are independent and normally distributed with mean 0 and variance σ. Find the distribution of the kinetic energy $(m/2)(V_1^2 + V_2^2 + V_3^2)$.

EXERCISE 11-10 Let $X(1)$ and $X(2)$ be random variables whose joint distribution has a continuous density p in the plane, and let Z be $X(1)/X(2)$ wherever $X(2) \neq 0$ and be any arbitrary real number where $X(2) = 0$. Show that Z has a density p_Z that satisfies

$$p_Z(z) = \int_{-\infty}^{\infty} p(uz, u)|u|\,du.$$

EXERCISE 11-11 Let the random variables $X(1)$, $X(2)$ have a joint distribution that is invariant under rotation, so that if a subset E of R^2 is an event and E_1 is obtained from E by rotating about the origin, E_1 has the same probability measure as E. Define Z as in Exercise 11-10. Prove that Z has the "Cauchy distribution" whose density p_Z is defined by

$$p_Z(z) = \pi/(z^2 + 1).$$

Verify that the integral of p_Z over R is 1.

EXERCISE 11-12 Let $X(1)$ and $X(2)$ be independent random variables, each normally distributed with mean 0 and variance 1. Define Z as in Exercise 11-10. Use Exercises 11-10 and 11-11 to obtain two proofs that Z has the Cauchy distribution.

12. Convolutions

In the final three sections of this chapter we shall study some parts of probability theory that are important but are less elementary than the material in foregoing sections. The first of these subjects is the convolution of functions and of measures. This is of great importance in many parts of analysis, and so we shall not restrict our attention to probability densities.

If f_1 and f_2 are real-valued functions on R, each integrable over R, then both $x \mapsto f_2(x)$ and $x \mapsto f_1(z - x)$ are integrable over R for each real z. But it does not follow that their product is integrable over R; the product of integrable functions is not necessarily integrable. When the integral of the product happens to exist for almost all z, it is called the *convolution* of f_1 and f_2 and is denoted by $f_2 * f_1$.

DEFINITION 12-1 *Let f_1 and f_2 be real-valued on R. If the integral*

(A) $$\int_{-\infty}^{\infty} f_1(z - x) f_2(x)\,dx$$

exists for all z in R except those in a set N of m_L-measure 0, the function whose value at each z in $R \setminus N$ is the integral (A) *and whose value on N is 0 is called the* ***convolution*** *of f_2 and f_1 and is denoted by $f_2 * f_1$:*

$$[f_2 * f_1](z) = \int_{-\infty}^{\infty} f_1(z-x) f_2(x)\, dx \qquad (z \text{ in } N),$$

$$[f_2 * f_1](z) = 0 \qquad (z \text{ in } R \setminus N).$$

In view of the fact that the product of two integrable functions is not necessarily integrable, it is rather surprising that whenever f_1 and f_2 are integrable, integral (A) exists for almost all z in R. This is one of the conclusions of the next theorem.

THEOREM 12-2 *Let f_1 and f_2 be real-valued functions integrable over R. Then for all z except those in a set N of measure 0, the integral* (A) *exists, and therefore the convolution $f_2 * f_1$ exists. Moreover, $f_2 * f_1$ is integrable over R, and if for each left-open interval B in R we define the* ***diagonal set*** *$D[B]$ to be the set $\{x \text{ in } R^2 : x_1 + x_2 \text{ in } B\}$, then for every left-open interval B in R*

$$\int_B [f_2 * f_1](u)\, du = \int_{D[B]} f_1(x_1) f_2(x_2)\, dx.$$

By Lemma 11-3 the function f defined by

$$f(x) = f_1(x_1) f_2(x_2) \qquad (x \text{ in } R^2)$$

is integrable over R^2. Suppose first that B is the interval (a, ∞) for some a in $\bar{R}$. This is open, and so is the diagonal set $D[B]$. So f is integrable over $D[B]$. We now make the substitution

$$x_1 = u_1 - u_2, \qquad x_2 = u_2.$$

This is a one-to-one map of R^2 onto itself, and its Jacobian is everywhere equal to 1. The image of $D[B]$ under this mapping is

$$\{u \text{ in } R^2 : u_1 > a\} = B \times R.$$

So, by Theorem 8-1,

$$\text{(B)} \qquad \int_{D[B]} f_1(x_1) f_2(x_2)\, dx = \int_{B \times R} f_1(u_1 - u_2) f_2(u_2)\, du.$$

In particular, by taking $a = -\infty$ we see that the integral in the right member exists when B is R. By the Fubini theorem (Theorem 7-1), the integral

$$\int_R f_1(u_1 - u_2) f_2(u_2)\, du_2$$

exists for all u_1 except those in a set N of measure 0, and

(C) $$\int_{B \times R} f_1(u_1 - u_2) f_2(u_2)\, du = \int_{B \setminus N} \left\{ \int_R f_1(u_1 - u_2) f_2(u_2)\, du_2 \right\} du_1$$
$$= \int_B [f_2 * f_1](u_1)\, du_1.$$

If B is the interval $(a, b]$, where $-\infty \leqq a < b < \infty$, we can apply (C) to the two intervals $B' = (a, \infty)$ and $B'' = (b, \infty)$. Since $B = B' \setminus B''$, by subtraction we obtain the equation in the conclusion of the theorem.

COROLLARY 12-3 *Let $X(1)$, $X(2)$ be independent random variables whose respective distributions $P_{X(1)}$, $P_{X(2)}$ have the densities p_1, p_2. Then the distribution of the sum $X(1) + X(2)$ has density $p_2 * p_1$.*

We first prove an auxiliary statement.

(D) The random vector $(X(1), X(2))$ has a distribution P_0 which has a density p_0 that satisfies

$$p_0(x) = p_1(x_1) p_2(x_2) \qquad (x \text{ in } R^2).$$

Let $A(1)$ and $A(2)$ be left-open intervals in R, and let A be $A(1) \times A(2)$. Since the events $\{X(1) \text{ in } A(1)\}$ and $\{X(2) \text{ in } A(2)\}$ are independent, their probabilities satisfy

$$P\{X(1) \text{ in } A(1) \text{ and } X(2) \text{ in } A(2)\} = P\{X(1) \text{ in } A(1)\} P\{X(2) \text{ in } A(2)\},$$

which implies

$$P_0(A) = P_{X(1)}(A(1)) P_{X(2)}(A(2)).$$

By definition of density, with Lemma 11-3 and the equation

$$1_A(x) = 1_{A(1)}(x_1) 1_{A(2)}(x_2),$$

this implies

$$P_0(A) = \left(\int_{A(1)} p_1(x_1)\, dx_1 \right) \left(\int_{A(2)} p_2(x_2)\, dx_2 \right)$$
$$= \left(\int_R p_1(x_1) 1_{A(1)}(x_1)\, dx_1 \right) \left(\int_R p_2(x_2) 1_{A(2)}(x_2)\, dx_2 \right)$$
$$= \int_{R^2} p_1(x_1) p_2(x_2) 1_{A(1)}(x_1) 1_{A(2)}(x_2)\, dx$$
$$= \int_{R^2} p_1(x_1) p_2(x_2) 1_A(x)\, dx$$
$$= \int_A p_1(x_1) p_2(x_2)\, dx.$$

So the function $x \mapsto p_1(x_1)p_2(x_2)$ is a density for P_0, which establishes statement (D).

Now let B be any left-open interval in R. Then $X(1) + X(2)$ is in B if and only if $(X(1), X(2))$ is in the diagonal set $D[B]$, so by Theorem 12-2,

$$P\{X(1) + X(2) \text{ in } B\} = P_0(D[B])$$
$$= \int_{D[B]} p_1(x_1)p_2(x_2)\,dx$$
$$= \int_B [p_2 * p_1](u)\,du.$$

This completes the proof.

COROLLARY 12-4 *Let f_1, f_2, and f_3 be integrable over R. Then for almost all points x in R*

(E) $$[f_2 * f_1](x) = [f_1 * f_2](x)$$

and

(F) $$([f_1 * f_2] * f_3)(x) = (f_1 * [f_2 * f_3])(x).$$

Also, for all real numbers a and b,

(G) $$[(af_2) * (bf_1)](x) = ab[f_2 * f_1](x)$$

and

(H) $$[(f_1 + f_2) * f_3](x) = [f_1 * f_2](x) + [f_2 * f_3](x).$$

Statements (G) and (H) are immediate consequences of Definition 12-1. For (E), let z be any point of the set $R \setminus N$ on which the integral

$$\int_{-\infty}^{\infty} f_2(z - x)f_1(x)\,dx$$

exists. In this integral we make the substitution

$$u = z - x.$$

By Theorem 6-6,

$$\int_{-\infty}^{\infty} f_2(z - x)f_1(x)\,dx = \int_{-\infty}^{\infty} f_2(u)f_1(z - u)\,du.$$

That is, $[f_2 * f_1](z) = [f_1 * f_2](z)$, and (E) is proved.

By Corollary 11-4, the function

$$f(x) = f_1(x_1)f_2(x_2)f_3(x_3) \qquad (x \text{ in } R^3)$$

is integrable over R^3. We make the substitution

$$x_1 = y_1, \qquad x_2 = y_2 - y_1 - y_3, \qquad x_3 = y_3.$$

This maps R^3 one-to-one onto itself, and its Jacobian is identically 1. So, by Theorem 8-1, the integral

$$\int_{R^3} f_1(y_1) f_2(y_2 - y_1 - y_3) f_3(y_3)\, dy$$

exists. By Fubini's theorem, there is a set N in R with measure 0 such that, for all y_2 in $R \setminus N$, the function

$$(y_1, y_3) \mapsto f_1(y_1) f_2(y_2 - y_1 - y_3) f_3(y_3)$$

is integrable over R^2. Let its value be denoted by $J(y_2)$. We apply Fubini's theorem to it in two ways. First, there is a subset N_2 of R with measure 0 such that for all y_1 in $R \setminus N_2$ the function

$$y_3 \mapsto f_1(y_1) f_2(y_2 - y_1 - y_3) f_3(y_3)$$

is integrable over R, and

$$J(y_2) = \int_{R \setminus N_2} \left\{ \int_R f_1(y_1) f_2(y_2 - y_1 - y_3) f_3(y_3)\, dy_3 \right\} dy_1.$$

The inner integral is $f_1(y_1)[f_3 * f_2](y_2 - y_1)$, and in the outer integral we can delete the N_2 without changing the integral, so

$$\text{(I)} \qquad J(y_2) = \int_R f_1(y_1)[f_3 * f_2](y_2 - y_1)\, dy_1 = [f_3 * f_2] * f_1(y_2).$$

Again, by Theorem 7-1 there is a set N_3 of measure 0 such that for all y_3 in $R \setminus N_3$ the function

$$y_1 \mapsto f_1(y_1) f_2(y_2 - y_1 - y_3) f_3(y_3)$$

is integrable over R, and

$$J(y_2) = \int_{R \setminus N_3} \left\{ \int_R f_1(y_1) f_2(y_2 - y_1 - y_3) f_3(y_3)\, dy_1 \right\} dy_3.$$

By the same argument as above,

$$\text{(J)} \qquad J(y_2) = \int_R [f_2 * f_1](y_2 - y_3) f_3(y_3)\, dy_3 = f_3 * [f_2 * f_1](y_2).$$

Equations (I) and (J) establish (F) and complete the proof.

As an example of a use of Corollary 12-3, we shall again prove that if two independent random variables have normal distributions, one with mean m_1 and variance σ_1^2 and the other with mean m_2 and variance σ_2^2, their sum is normally distributed with mean $m_1 + m_2$ and variance $\sigma_1^2 + \sigma_2^2$. If we use the notation

$$n_{m,\sigma}(x) = \frac{1}{(2\pi)^{1/2}\sigma} \exp\left[-\frac{(x-m)^2}{2\sigma^2} \right],$$

by Corollary 12-3, the sum of the random variables has a distribution with density at x equal to

$$\text{(K)} \qquad [n_{m_2,\sigma_2} * n_{m_1,\sigma_1}](x) = \frac{1}{2\pi\sigma_1\sigma_2} \int_{-\infty}^{\infty} \exp\left[-\frac{(x-m_1-u)^2}{2\sigma_1^2} - \frac{(u-m_2)^2}{2\sigma_2^2} \right] du.$$

If we define

$$\sigma = [\sigma_1^2 + \sigma_2^2]^{1/2}, \qquad m = m_1 + m_2,$$

we can without difficulty verify the identity

$$\frac{[(x-m)-(u-m_2)]^2}{\sigma_1^2} + \frac{(u-m_2)^2}{\sigma_2^2} = \frac{(x-m)^2}{\sigma^2} + \frac{[(u-m_2)\sigma^2-(x-m)\sigma_2^2]^2}{\sigma_1^2\sigma_2^2\sigma^2}.$$

We use this to change the form of the integrand in (K), and we then make the substitution

$$u = v + m_2 + (x-m)\sigma_2^2/\sigma^2.$$

The result is

$$\text{(L)} \quad [n_{m_2,\sigma_2} * n_{m_1,\sigma_1}](x) = \frac{1}{2\pi\sigma_1\sigma_2} \int_{-\infty}^{\infty} \exp\left[-\frac{(x-m)^2}{2\sigma^2} \right] \exp\left[-\frac{v^2\sigma^2}{2\sigma_1^2\sigma_2^2} \right] dv.$$

Since

$$\int_{-\infty}^{\infty} \exp\left[-\frac{v^2\sigma^2}{2\sigma_1^2\sigma_2^2} \right] dv = \frac{[2\pi]^{1/2}\sigma_1\sigma_2}{\sigma},$$

(L) implies

$$[n_{m_2,\sigma_2} * n_{m_1,\sigma_1}](x) = [1/(2\pi)^{1/2}\sigma] \exp[-(x-m)^2/2\sigma^2] = n_{m,\sigma}(x).$$

So the sum has normal distribution with mean m and variance σ.

There is a theorem analogous to Corollary 12-3 that furnishes the distribution of the sum of two independent random variables, even when they do not have densities. In it we use the notation $A - c$ to denote the set of all numbers x of the form $a - c$ with a in A. That is,

$$A - c = \{x \text{ in } R : x + c \text{ in } A\}.$$

We now define the convolution of two functions of intervals.

DEFINITION 12-5 *Let m_1 and m_2 be two real-valued functions of left-open intervals in R. Then the convolution $m_2 \star m_1$ is defined to be the interval-function whose value at the left-open interval B in R is*

$$[m_2 \star m_1](B) = \int_B m_1(B - x) m_2(dx),$$

provided that this integral exists for each such B.

THEOREM 12-6 *Let $X(1)$ and $X(2)$ be independent random variables with the respective distributions P_1, P_2. Then the convolution $P_2 \star P_1$ exists and is the distribution of $X(1) + X(2)$.*

Let P_0 be the interval-function defined for all left-open intervals in R^2 such that if $A = A(1) \times A(2)$ with $A(1)$ and $A(2)$ in R,

$$P_0 A = (P_1 A(1))(P_2 A(2)).$$

Then P_0 is the distribution of the vector $(X(1), X(2))$. If B is a left-open interval in R and, as before, $D[B]$ is the diagonal set consisting of all x in R^2 with $x_1 + x_2$ in B, the set of all ω with $X(1) + X(2)$ in B is the set of all ω with $(X(1), X(2))$ in $D[B]$. It thus has probability measure $P_0(D[B])$. If we denote $D[B]$ by C, for each real number x the section $C[x]$ of C at x is the set of all y such that (x, y) is in $D[B]$, which is $B - x$. If $B = (a, b]$, $B - x = (a - x, b - x]$, and

$$P_1(B - x) = P_1(-\infty, b - x) - P_1(-\infty, a - x).$$

This is the difference between two bounded monotone functions, so it is integrable with respect to P_2 over R. By Corollary 7-3,

$$P_0(D[B]) = \int_R P_1(B - x) P_2(dx) = [P_2 \star P_1](B).$$

Since the left member is the probability that $X(1) + X(2)$ is in B, the theorem is proved.

COROLLARY 12-7 *Let P_1, P_2, and P_3 be distribution functions. Then for all left-open intervals B in R*

$$[P_2 \star P_1](B) = [P_1 \star P_2](B),$$

$$[P_1 \star P_2] \star P_3(B) = P_1 \star [P_2 \star P_3](B).$$

In the first of these equations, the left member is the probability that $X(1) + X(2)$ is in B and the right is the probability that $X(2) + X(1)$ is in B, so they are equal. In the second equation, the left member is the probability that $X(1) + [X(2) + X(3)]$ is in B, and the right member is the probability that $[X(1) + X(2)] + X(3)$ is in B, so they are equal.

As an example, consider two independent random variables $X(1)$, $X(2)$, both taking on only nonnegative integral values. $X(1)$ takes value i ($i = 1, 2, 3, \ldots$) with probability p_i', and $X(2)$ takes value i with probability p_i''. Suppose that B is a left-open interval in R that contains no nonnegative integer. Then $B - u$ also contains no such integer whenever u is a nonnegative integer; for if v is a nonnegative integer in $B - u$, $u + v$ is a nonnegative integer in B. Since

$$\text{(M)} \qquad \int_R P_1(B - u)\, P_2(du) = P_1(B - 0)p_0'' + P_1(B - 1)p_1'' + P_1(B - 2)p_2'' + \cdots,$$

when B contains no nonnegative integer, the left member of (M) has value 0. If B contains a single nonnegative integer n, $B - u$ contains $n - u$ for $u = 0, 1, 2, \ldots, n$ and contains no nonnegative integer for other integers u. The left member of (M) then has value

$$\text{(N)} \qquad p_n' p_0'' + p_{n-1}' p_1'' + \cdots + p_0' p_n''.$$

So the distribution of $X(1) + X(2)$ is the measure that assigns measure (N) to the point n for $n = 0, 1, 2, \ldots$ and assigns measure 0 to all sets that contain no nonnegative integer.

This suggests extension of the notation for convolutions to sequences. If S' is the sequence with terms a_0', a_1', $a_2', \ldots$ and S'' is the sequence with terms a_0'', a_1'', $a_2'', \ldots$, we define

$$S'' \star S' = (a_0' a_0'', a_1' a_0'' + a_0' a_1'', a_2' a_0'' + a_1' a_1'' + a_0' a_2'', \ldots).$$

By an easy computation, if f_1 is the sum of a power series in which the coefficients form a sequence $S_1 = (a_0', a_1', a_2', \ldots)$ and f'' is the sum of a power series whose coefficients form a sequence $S_2 = (a_0'', a_1'', a_2'', \ldots)$, the product $f_1(x)f_2(x)$ is the sum of a power series in x whose coefficients form the sequence $S_2 \star S_1$. In particular, if $f_2 = 1 + x$, $f_1(x)f_2(x)$ will have the sequence of coefficients

$$\text{(O)} \qquad a_0', a_0' + a_1', a_1' + a_2', a_2' + a_3', \ldots .$$

Let us denote the coefficient of x^j in the expansion of $(1 + x)^n$ by $\binom{n}{j}$. Then by (O),

$$\binom{n+1}{j} = \binom{n}{j} + \binom{n}{j-1}.$$

This can be expressed as follows. Write the coefficients for the expansion of $(1 + x)^n$ in the nth line, the first coefficient in each line being vertically below the first coefficient in the preceding line. Then each coefficient in line $n + 1$ is the sum of the coefficient above it and the coefficient next left of the one above it.

This gives the diagram

$$\begin{array}{cccccccc}
1 & 0 & 0 & 0 & 0 & 0 & \cdots \\
1 & 1 & 0 & 0 & 0 & 0 & \cdots \\
1 & 2 & 1 & 0 & 0 & 0 & \cdots \\
1 & 3 & 3 & 1 & 0 & 0 & \cdots \\
1 & 4 & 6 & 4 & 1 & 0 & \cdots \\
\vdots & \vdots & \vdots & \vdots & \vdots & \vdots & \cdots
\end{array}$$

This, with the zeroes omitted, is known as *Pascal's triangle* for the binomial coefficients. But in fact, it was known to Omar Khayyam, centuries before Pascal.

EXERCISE 12-1 Starting with a Bernoulli distribution in which 1 has probability p and 0 has probability $q = 1 - p$, use (O) to find the binomial distribution that is the distribution of the sum of four independent Bernoulli-distributed random variables.

EXERCISE 12-2 Show that if P_1 and P_2 are distributions, then if $P_1(-\infty, x]$ is a continuous function of x, so is $[P_2 \star P_1](-\infty, x]$, and if $P_1(-\infty, x]$ has a bounded derivative on R, so has $[P_2 \star P_1](-\infty, x]$.

EXERCISE 12-3 Let P be the uniform distribution with density $p = 1$ on the interval $(-\frac{1}{2}, \frac{1}{2}]$. Compute and graph $p * p$, $p * p * p$, and $p * p * p * p$. Note that $p(-\infty, x]$ is discontinuous, $[p * p](-\infty, x]$ is continuous but has a discontinuous derivative, and $[p * p * p](-\infty, x]$ has a continuous derivative and a second derivative with finitely many discontinuities. Note also that the shape of the graph more closely resembles that of the normal density as the number of factors p increases.

EXERCISE 12-4 Tables of numerical data are sometimes "smoothed" by replacing each entry a_i by an average, such as $(a_{i-1} + a_i + a_{i+1})/3$. Show that this is a convolution. Apply it to the sequence $1, -1, 1, -1, 1, \ldots$ to see that it does smooth the entries.

EXERCISE 12-5 Let U be a distribution with a continuously differentiable density u that vanishes outside $(-\delta, \delta)$. Show that for every distribution P and every function f such that $|f(x_1) - f(x_2)| < \varepsilon$ whenever $|x_1 - x_2| < \delta$, the distribution $P \star U$ has a continuously differentiable density and

$$\left| \int_R f(x)\,[P \star U](dx) - \int_R f(x)\,P(dx) \right| < \varepsilon.$$

(The replacement of P by $P \star U$ is sometimes called "smearing.")

EXERCISE 12-6 Prove that the operation of convolution has the following strong continuity property. For each pair P_1, P_2 of probability distributions on R, let the distance $\rho(P_1, P_2)$ be the infimum of numbers ε such that whenever $-\infty < a < b < \infty$,

$$P_1((a - \varepsilon, b + \varepsilon]) + \varepsilon \geqq P_2((a, b])$$

and

$$P_2((a - \varepsilon, b + \varepsilon]) + \varepsilon \geqq P_1((a, b]).$$

Then whenever P_1, P_2, P'_1, and P'_2 are probability distributions on R,

$$\rho(P_1 \star P_2, P'_1 \star P'_2) \leqq \rho(P_1, P'_1) + \rho(P_2, P'_2).$$

Suggestion: If $c_1 > \rho(P_1, P'_1)$ and $c_2 > \rho(P_2, P'_2)$, and $-\infty < a < b < \infty$,

$$[P'_2 \star P'_1]((a, b]) \leqq \int_R P_1((a - c_1 - x, b + c_1 - x])\, P'_2(dx) + \int_R c_1\, P'_2(dx).$$

Use Corollary 12-7, then replace P'_2 by its estimate in terms of P_2.

13. The Central Limit Theorem

The central limit theorem of probability theory has several forms with different degrees of generality. One useful form states, roughly, that if a random variable S is the sum of many independent random variables, no one being much larger than the others in the sense that the variance of any one is much smaller than the sum of the variances of all the others, then the sum S has a distribution that is nearly normal. This accounts for the frequent applicability of the normal distribution in experimental situations. For example, let the measurement of some fixed quantity be repeated many times. Each measurement will have an error, and this error is the sum of the contributions made by many independent small sources of error. Then the total error will be nearly normally distributed. We are about to prove a form of the central limit theorem, but one that does not reach as much generality as we have been describing; we shall consider only sums of many independent random variables, each with the same distribution. This situation we have met before; for example, the binomial distribution is the distribution of the sum of several independent random variables, each with the same Bernoulli distribution.

If each of n identically distributed independent random variables has mean m and variance σ^2, the sum of the n variables will have mean nm and variance $n\sigma^2$. As n increases, the distribution of the sum will spread out thinly over R. To keep it in bounds, we "center" it by subtracting nm from the sum, and we "scale" it by multiplying it by $1/n^{1/2}\sigma$, which reduces the variance to 1. The resulting random variable is called the *scaled centered sum*. This is what converges to the standard normal distribution.

THEOREM 13-1 *Let $X_1, X_2, X_3, \ldots$ be a sequence of independent random variables, each with the same distribution P that has a finite mean m and a finite variance σ^2. Then the scaled centered sum*

$$S_n^* = [X_1 + \cdots + X_n - nm]/\sigma n^{1/2}$$

has a distribution P_n^* that approaches the standard normal distribution, in the sense that for each positive number ε there is an integer $m(\varepsilon)$ such that for all n greater than $n(\varepsilon)$ and all left-open intervals B in R,

$$\text{(A)} \qquad \left| P_n^*(B) - (2\pi)^{-1/2} \int_B \exp\left(-\frac{x^2}{2}\right) dx \right| < \varepsilon.$$

We lose no generality in assuming that $m = 0$ and $\sigma = 1$. For if the conclusion is established in this case, we can apply it to the independent random variables $Y_i = (X_i - m)/\sigma$, which all have the same distribution with mean 0 and variance 1; the scaled centered sum for the Y_i is the same as for the X_i. We shall prove the theorem by first establishing three lemmas.

LEMMA 13-2 *Define, for all real x,*

$$u_1(x) = x^3(1 - x)^3 1_{(0,1]}(x) \Big/ \int_0^1 y^3(1 - y)^3\, dy,$$

and for each positive ε

$$\text{(A)} \qquad u_\varepsilon(x) = u_1(x/\varepsilon)/\varepsilon.$$

Then u_ε and its first and second derivatives are defined and continuous on R, and u_ε is positive on $(0, \varepsilon)$ and 0 elsewhere, and its integral over R is 1, and there exist constants C_2, C_3 such that for all x in R

$$\text{(B)} \qquad |Du(x)| \leqq C_2/\varepsilon^2,$$

$$\text{(C)} \qquad |D^2u(x)| \leqq C_3/\varepsilon^3.$$

We write $p(x)$ for $x \vee 0$. The for all x in R and all positive integers q

$$p(x)^{q+1} = (q + 1) \int_0^x p(y)^q\, dy;$$

for if $x \leqq 0$, both members are 0, and if $x > 0$, the equation takes the familiar form

$$x^{q+1} = (q + 1) \int_0^x y^q\, dy.$$

By this and the fundamental theorem, p^2 has the continuous derivative $2p$, and p^3 has the second derivative $6p$. Then $p(1-x)^3$ also has continuous first and second derivatives, and so have $p(x)^3p(1-x)^3$, $u_1(x)$, and $u_\varepsilon(x)$. By the substitution $z = \varepsilon x$, we find

$$\int_R u_\varepsilon(z)\,dz = \int_{-\infty}^{\infty} u_1 \frac{z/\varepsilon}{\varepsilon}\,dz = \int_{-\infty}^{\infty} u_1(x)\,dx = 1.$$

If we define

$$C_2 = \sup\{|Du_1(x)| : 0 \leqq x \leqq 1\},$$
$$C_3 = \sup\{|D^2u_1(x)| : 0 \leqq x \leqq 1\},$$

we have

$$|Du_\varepsilon(x)| = |\varepsilon^{-2}Du_1(x/\varepsilon)| \leqq \varepsilon^{-2}C_2,$$
$$|D^2u(x)| = |\varepsilon^{-3}D^2u_1(x/\varepsilon)| \leqq \varepsilon^{-3}C_3.$$

The proof is complete.

The function u_ε defined in Lemma 13-2 is the density of a probability distribution U such that for every left-open interval B in R

$$\text{(D)} \qquad U(B) = \int_B u_\varepsilon(x)\,dx.$$

There exists a random variable X_0 with distribution U such that the random variables $X_0, X_1, X_2, \ldots$ are independent.

We define convolution-powers just as we define ordinary powers; if P is any distribution,

$$P^{1\star} = P, \qquad P^{2\star} - P\star P, \ldots, \qquad P^{(n+1)\star} = P\star P^n, \ldots$$

LEMMA 13-3 *Let $A_x = (-\infty, x]$. Then for every pair of distributions P_1, P_2, every positive integer n, and every x in R*

$$\text{(E)} \qquad |[P_1^{n\star}\star U](A_x) - [P_2^{n\star}\star U](A_x)| \leqq n\sup\{|[P_1\star U](A_x) - [P_2\star U](A_x)| : x \text{ in } R\}.$$

This is evidently true for $n = 1$. Suppose it is true for n equal to a positive integer $k \geqq 1$. For every x in R,

$$\text{(F)} \qquad |[P_1^{k\star}\star U](A_x) - [P_2^{k\star}\star U](A_x)|$$
$$\leqq |[P_1\star P_1^{(k-1)\star}\star U](A_x) - [P_2\star P_1^{(k-1)\star}\star U](A_x)|$$
$$+ |[P_2\star P_1^{(k-1)\star}\star U](A_x) - [P_2\star P_2^{(k-1)\star}\star U](A_x)|$$
$$= |[P_1^{(k-1)\star}\star(P_1 - P_2)\star U](A_x)|$$
$$+ |[P_2\star\{P_1^{(k-1)\star} - P_2^{(k-1)\star}\}\star U](A_x)|.$$

If we define

$$C = \sup\{|[(P_1 - P_2) \star U](A_x)| : x \text{ in } R\}$$

and observe that

$$A_x - y = A_{x-y} \qquad (y \text{ in } R),$$

then since $P_1^{(k-1)\star}$ is a probability distribution, the first term in the right member of (F) is

$$\text{(G)} \quad \left| \int_R [(P_1 - P_2) \star U](A_{x-y}) \, P_1^{(k-1)\star}(dy) \right| \leqq \int_R C \, P_1^{(k-1)\star}(dy) = C.$$

Since (E) is assumed to be valid for $n = k - 1$, the second term in the right member of (F) is

$$\text{(H)} \quad \left| \int_R \{[P_1^{(k-1)\star} \star U](A_{x-y}) - [P_2^{(k-1)\star} \star U](A_{x-y})\} \, P_2(dy) \right|$$

$$\leqq \int_R (k-1) C \, P_2(dy) \leqq (k-1)C.$$

Inequalities (F), (G), and (H) imply that (E) holds for $n = k$, so by induction it holds for all positive integers n.

We define P_n to be the distribution of the random variable $X_1/n^{1/2}$. Since X_1 has mean 0 and variance 1,

$$\text{(I)} \quad \int_R P_n(dy) = 1, \qquad \int_R y \, P_n(dy) = 0, \qquad \int_R y^2 \, P_n(dy) = \frac{1}{n}.$$

We next prove an auxiliary statement.

(J) If f is nonnegative and continuous on R,

$$\int_R f(y) \, P_n(dy) = \int_R f(n^{-1/2} z) \, P_1(dz).$$

If f is the indicator of a left-open interval $(a, b]$, the function $z \mapsto f(n^{-1/2}z)$ is the indicator of $(n^{1/2}a, n^{1/2}b]$. So the left member of the equation in (J) is the probability that $X_1/n^{1/2}$ is in $(a, b]$, and the right member is the probability that X_1 is in $(n^{1/2}a, n^{1/2}b]$. These are evidently equal. So the equation in (J) holds when f is the indicator of a left-open interval. Every step-function with left-open intervals of constancy is a linear combination of such indicators, so the equation holds for all such step-functions. Every nonnegative continuous function is the limit of an ascending sequence of step-functions with left-open intervals of constancy, so by the monotone convergence theorem, statement (J) is correct.

LEMMA 13-4 *If P_n is the distribution of $X_1/n^{1/2}$, and $A_x = (-\infty, x]$, and U is as defined in* (D), *then*

$$\lim_{n\to\infty} \sup\{|n([P_n \star U](A_x) - U(A_x)) - Du_\varepsilon(x)/2| : x \text{ in } R\} = 0.$$

We write $V(x)$ for $U(A_x)$. Then

$$D^{k+1}V(x) = D^k u_\varepsilon(x) \qquad (k = 0, 1, 2),$$

and by (B) and (C),

$$\text{(K)} \qquad |D^2 V(x)| \leqq C_2/\varepsilon^2, \qquad |D^3 V(x)/6| \leqq C_3/\varepsilon^3$$

for all x in R. By Taylor's theorem, for each pair of real numbers x, y there exist numbers x^*, x^{**} between x and y such that

$$\text{(L)} \qquad V(x-y) - V(x) + y\,DV(x) - (y^2/2)\,D^2V(x) = -(y^3/6)\,D^3V(x^*),$$
$$V(x-y) - V(x) + y\,DV(x) = (y^2/2)\,D^2V(x^{**}).$$

From the first of these, with (K),

$$\text{(M)} \qquad |V(x-y) - V(x) + y\,DV(x) - (y^2/2)\,D^2V(x)| \leqq C_3|y|^3/\varepsilon^3,$$

and from the second of equations (L), with (K),

$$\text{(N)} \qquad |V(x-y) - V(x) + y\,DV(x) - (y^2/2)\,D^2V(x)|$$
$$\leqq (y^2/2)[|D^2V(x)| + |D^2V(x^{**})|] \leqq C_2 y^2/\varepsilon^2.$$

Inequalities (M) and (N) imply

$$\text{(O)} \qquad |n[V(x-y) - V(x) + yu_\varepsilon(x) - (y^2/2)\,Du_\varepsilon(x)]|$$
$$\leqq n[(C_3|y|^3\varepsilon^{-3}) \wedge (C_2 y^2 \varepsilon^{-2})].$$

The left member of (O) is a continuous function of y, and by (O) it does not exceed $nC_2y^2\varepsilon^{-2}$, which is integrable with respect to P_n. So, the left member of (O) is integrable with respect to P_n over R. Recalling (J),

$$\text{(P)} \qquad \left|\int_R \left[nV(x-y) - nV(x) + nyu_\varepsilon(x) - n\frac{y^2}{2}Du_\varepsilon(x)\right]P_n(dy)\right|$$
$$\leqq \int_R n[(C_3|y|^3\varepsilon^{-3}) \wedge (C_2y^2\varepsilon^{-2})]\,P_n(dy)$$
$$\leqq \int_R n[(C_3|z|^3\varepsilon^{-3}n^{-3/2}) \wedge (C_2z^2\varepsilon^{-2}n^{-1})]\,P_1(dz).$$

The integrand in the right member is continuous, and it is integrable because it does not exceed the integrable function $C_2z^2\varepsilon^{-2}$, and as n increases it tends everywhere to 0 because $\varepsilon^{-3}C_3z^3n^{-1/2}$ does so. By the dominated convergence theorem, the right member of (P) tends to 0 as n increases. Since by (I) the left member of (P) can be written in the form

$$|n\{[P_n \star U](A_x) - U(A_x)\} - Du_\varepsilon(x)/2|,$$

the conclusion of the lemma is established.

We now turn to the proof of Theorem 13-1. Let ε be positive, and let U and u_ε be defined as in (D) and Lemma 13-2. By Lemma 13-4, there exists an integer $n'(\varepsilon)$ such that if $n > n'(\varepsilon)$,

$$\text{(Q)} \qquad |n\{[P_n \star U](A_x) - U(A_x)\} - Du_\varepsilon(x)/2\}| < \varepsilon/4$$

for all real x. Let $N_{m,\sigma}$ denote the normal distribution with mean m and variance σ^2. Lemma 13-4 also applies to the case in which $X(1)$ has distribution $N_{0,1}$, in which case P_n is $N_{0,1/n}$. Then by Lemma 13-4 there exists an integer $n''(\varepsilon)$ such that for all $n > n''(\varepsilon)$,

$$|n\{[N_{0,1/n} \star U](A_x) - U(A_x)\} - Du_\varepsilon(x)/2| < \varepsilon/4$$

for all real x. Let $n(\varepsilon)$ be the greater of $n'(\varepsilon)$ and $n''(\varepsilon)$; then for all $n > n(\varepsilon)$ both the preceding inequalities hold, whence

$$|n\{[P_n \star U](A_x) - N_{0,1/n} \star U](A_x)\}| < \varepsilon/2.$$

By Lemma 13-3, for every interval $A_x = (-\infty, x]$ and every integer $n > n(\varepsilon)$,

$$\text{(R)} \qquad |[P_n^{n\star} \star U](A_x) - [N_{0,1/n}^{n\star} \star U](A_x)| < \varepsilon/2.$$

The independent random variables $X_0, X_1/n^{1/2}, \ldots, X_n/n^{1/2}$ have the respective distributions $U, P_n, \ldots, P_n$, so $P_n^{n\star} \star U$ is the distribution of the sum

$$X_1/n^{1/2} + \cdots + X_n/n^{1/2} + X_0 = S_n^* + X_0.$$

The convolution-power $N_{0,1/n}^{n\star}$ is the distribution of the sum of n independent random variables, each with mean 0 and variance $1/n$, so it is a normal distribution with mean 0 and variance 1. That is,

$$N_{0,1/n}^{n\star} = N_{0,1}.$$

Let Y be a random variable independent of $X_0, X_1, X_2, \ldots$ and having normal distribution with mean 0 and variance 1. Then $N_{0,1/n}^{n\star} \star U$ is the distribution of $Y + X_0$.

For the standard normal distribution $N_{0,1}$ we have for all intervals $(a, b]$

$$\text{(S)} \qquad N_{0,1}(a, b] = (2\pi)^{-1/2} \int_a^b \exp\left(-\frac{x^2}{2}\right) dx \leqq \frac{b-a}{2}.$$

We have already introduced the name P_n^* for the distribution of the sum S_n^*. The random variable X_0 takes on values only in the interval $[0, \varepsilon]$. So for every real x

we have by (R) and (S)

$$\begin{aligned}
P_n^*(A_x) &= P(S_n^* \leqq x) \\
&\leqq P(S_n^* + X_0 \leqq x + \varepsilon) \\
&= [P_n^{n\star} \star U](A_{x+\varepsilon}) \\
&< [N_{0,1/n}^{n\star} \star U](A_{x+\varepsilon}) + \varepsilon/2 \\
&= [N_{0,1} \star U](A_{x+\varepsilon}) + \varepsilon/2 \\
&= P(Y + X_0 \leqq x + \varepsilon) + \varepsilon/2 \\
&\leqq P(Y \leqq x + \varepsilon) + \varepsilon/2 \\
&= N_{0,1}(A_{x+\varepsilon}) + \varepsilon/2 \\
&= N_{0,1}(A_x) + N_{0,1}(x, x + \varepsilon] + \varepsilon/2 \\
&\leqq N_{0,1}(A_x) + \varepsilon.
\end{aligned}$$

On the other hand,

$$\begin{aligned}
P_n^*(A_x) &= P(S_n^* \leqq x) \\
&\geqq P(S_n^* + X_0 \leqq x) \\
&= [P_n^{n\star} \star U](A_x) \\
&> [N_{0,1/n}^{n\star} \star U](A_x) - \varepsilon/2 \\
&= [N_{0,1} \star U](A_x) - \varepsilon/2 \\
&= P(Y + X_0 \leq x) - \varepsilon/2 \\
&\geqq P(Y + \varepsilon \leqq x) - \varepsilon/2 \\
&= N_{0,1}(A_{x-\varepsilon}) - \varepsilon/2 \\
&= N_{0,1}(A_x) - N_{0,1}(x - \varepsilon, x] - \varepsilon/2 \\
&\geqq N_{0,1}(A_x) - \varepsilon.
\end{aligned}$$

These two inequalities establish the theorem.

EXERCISE 13-1 A fair coin is tossed six times. Find the distribution of the number of "heads." Show that Chebyshev's inequality assures us that the number of heads is in $(\frac{3}{2}, \frac{9}{2})$ with probability at least $\frac{1}{3}$; but that by the central limit theorem, this probability is about 0.78. Show that it is 0.78125.

EXERCISE 13-2 Let S be the sum of 1000 Bernoulli trials each assigning probability 0.999 to 0 and 0.001 to 1. Show that if we apply the central limit theorem, we find that the probability that $S < 0$ is about 0.16, which is ridiculous. *Conclusion*: Although the sum of n independent identically distributed random variables is nearly normally distributed if n is large, with some distributions of the summands the number n may have to be very large indeed before the distribution of the sum is anywhere near its limit.

EXERCISE 13-3 Assume that when each number in a list is rounded to the nearest integer, the rounding error is uniformly distributed over $(-\frac{1}{2}, \frac{1}{2}]$. Show that the error in a sum of 100 numbers produced by rounding can be anywhere from -50 to $+50$ but that the central limit theorem informs us that it is between -5 and 5 with probability about 0.9858. What does Theorem 11-8 tell us?

EXERCISE 13-4 Let P_1, P_2 be two distributions with mean 0 and finite variance. Let S be the sum of n_1 random variables with distribution P_1 and n_2 random variables with distribution P_2, all being independent. Show that if n_1 and n_2 are large, S has nearly a normal distribution.

14. Distributions in Some Infinite-Dimensional Spaces

The types of multivariate distributions studied in the preceding sections include many important special cases, but there are some quite elementary cases that cannot be handled by their use. For example, if the experiment consists of tossing a fair coin until two consecutive tosses yield "heads," what is the expectation of the number of tosses needed to end the trial? Here there is no a priori limit on the number of tosses that will be needed; if we specify a number N, then if N is large it will be unlikely that more than N tosses will be needed to get two consecutive "heads," but it is not impossible. We can think of each sequence of tosses as the beginning of an infinite sequence of tosses; to each such infinite sequence s we assign a number $k(s)$, namely, the smallest integer k such that tosses $k-1$ and k both give "heads." This is a function on the set of all sequences of tosses, and we wish to know its expectation. But for this purpose we need a probability distribution on the space of all infinite sequences of tosses, and this is an infinite-dimensional space. If we define $X(j)$ to be 1 if the jth toss is "heads" and 0 if the jth toss is "tails," we know the joint distribution of $(X(1), \ldots, X(n))$ for every integer n. Our problem is to construct a probability distribution on the space of all sequences $(x_1, x_2, x_3, \ldots)$ such that the distribution of the first n coordinates $x_1, \ldots, x_n$ agrees with the given distribution of the random variables $X(1), \ldots, X(n)$. Fortunately, this can be done in almost exactly the same way as in the finite-dimensional case. It can even be done when T is an uncountably infinite set. But the uncountable case requires the use of more advanced ideas from topology than does the countable case, so to save trouble we shall restrict our attention to the case of countable sets T.

Suppose that T is a nonempty set and that to each t in T there corresponds a nonempty set $A(t)$. The Cartesian product

$$\text{(A)} \qquad A = \mathop{\times}_{t \text{ in } T} A(t)$$

is defined to be the set of all functions $x: t \mapsto x(t)$ on T such that for each t in T, $x(t)$ is in $A(t)$. Thus if T has r members and each $A(t)$ is R, the Cartesian product defined in (A) is the set of all ordered r-tuples of real numbers, which is R^r; and if T is the set of all positive integers and each $A(t)$ is $\bar{R}$, the Cartesian product is the set of all sequences of extended real numbers. When each factor-set $A(t)$ is R, the Cartesian product is denoted by R^T; when each factor-set $A(t)$ is $\bar{R}$, the Cartesian product is denoted by $\bar{R}^T$.

An **interval** in $\bar{R}^T$ is defined to be a Cartesian product (A) in which each $A(t)$ is an interval in $\bar{R}$ and for all but finitely many values of t, $A(t)$ is the whole of $\bar{R}$. Likewise, an interval in R^T is a Cartesian product (A) in which each $A(t)$ is an interval in R and for all but finitely many values of t is R itself. The interval A is **open** or **closed** or **left-open** according as all $A(t)$ are open or closed or left-open. It should be remembered that in $\bar{R}$, $\bar{R}$ is an interval that is open, closed, and left-open, and likewise in R, R is an interval that is open, closed, and left-open.

Just as in $\bar{R}^r$, a **neighborhood** of a point x of $\bar{R}^T$ is an open interval in $\bar{R}^T$ that contains x. A **gauge** on $\bar{R}^T$ is a function γ on $\bar{R}^T$ such that for each x in $\bar{R}^T$, $\gamma(x)$ is a neighborhood of x. An **allotted partition** of a set C in $\bar{R}^T$ is a finite set of pairs

$$\text{(B)} \qquad \mathscr{P} = \{(\bar{x}_1, A_1), \ldots, (\bar{x}_k, A_k)\}$$

in which each $\bar{x}_i$ is a point of $\bar{R}^T$ and the A_i are pairwise disjoint left-open intervals whose union is C. If γ is a gauge on $\bar{R}^T$ and $\mathscr{P}$ is an allotted partition – for which we use the notation of (B) – $\mathscr{P}$ is a **γ-fine partition** if for each i in $\{1, \ldots, k\}$ the closure A_i^- is contained in $\gamma(\bar{x}_i)$.

If f is a real-valued function defined on $\bar{R}^T$, and m is an extended-real-valued function defined on the set of all left-open intervals in R^T, and $\mathscr{P}$ is an allotted partition of R^T, the partition-sum corresponding to $\mathscr{P}$, f, and m is

$$S(\mathscr{P}; f; m) = \sum_{i=1}^{k} f(\bar{x}_i) m A_i,$$

provided that this sum exists. If B is a subset of R^T, and f is a real-valued function on B, and f_B is the function on $\bar{R}^T$ that coincides with f on B and is 0 on $\bar{R}^T \setminus B$, the integral of f over B with respect to m is defined to be the gauge-limit of $S(\mathscr{P}; f_B; m)$ (as defined in Definition I-1-7), provided that that gauge-limit exists.

However, as in earlier chapters, this definition would be useless if we could not prove that for each gauge γ on $\bar{R}^T$ there exists a γ-fine partition of R^T. This is, in fact, true even for uncountable T; but to avoid difficulties, we shall prove it only for the case of countable T.

THEOREM 14-1 *Let T be a countable set. Let γ be a gauge on $\bar{R}^T$, and let B be a left-open interval in R^T. Then there exists a γ-fine partition*

$$\mathscr{P} = \{(\bar{x}_1, A_1), \ldots, (\bar{x}_k, A_k)\}$$

of B such that for $i = 1, \ldots, k$, $\bar{x}_i$ is in the closure A_i^- of A_i in $\bar{R}$.

Since T is countable, its points can be listed in a sequence $t_1, t_2, t_3, \ldots$. For each positive integer n we first form the same $2 \cdot 4^n + 2$ intervals that we used in the proof of Theorem I-4-2, namely,

$$\text{(C)} \qquad [-\infty, -2^n],$$

$$(-2^n + (j-1)2^{-n}, -2^n + j2^{-n}] \qquad (j = 1, \ldots, 2 \cdot 4^n),$$

$$(2^n, \infty].$$

Then we form the set of all intervals Q in $\bar{R}^T$ such that for each t in the set $\{t_1, \ldots, t_n\}$, $Q(t)$ is one of the intervals in the list (C), and for all t not in the set $\{t_1, \ldots, t_n\}$, $Q(t)$ is $\bar{R}$. There are $(2 \cdot 4^n + 2)^n$ such intervals. We call them $Q(n,j)$, $j = 1, \ldots, k(n)$, where $k(n) = (2 \cdot 4^n + 2)^n$. Each interval $Q(n+1, j)$ is contained in one of the intervals $Q(n, i)$.

Temporarily, for each left-open interval B^* in R^T we shall define a "special" partition of B^* to be a γ-fine partition $\mathscr{P}$ of B^* such that for each pair (x, A) in $\mathscr{P}$, A is the intersection of B^* with the interval $Q(n,j)$ for some n and j, and x is in the closure A^- in $\bar{R}$. As in the proof of Theorem 3-1, if B had no special partition there would be a number j_1 in the set $\{1, \ldots, k(1)\}$ such that the interval $B \cap Q(1, j_1)$ had no special partition. We choose such a j_1 and denote $B \cap Q(1, j_1)$ by B_1. Since B_1 has no special partition, there is a number j_2 in the set $\{1, \ldots, k(2)\}$ such that the interval $B_1 \cap Q(2, j_2)$ has no special partition. We choose such a j_2 and denote $B_1 \cap Q(2, j_2)$ by B_2. Continuing thus, we obtain a sequence of intervals $B_0 = B, B_1, B_2, \ldots$ such that for each positive integer n,

$$B_n = B_{n-1} \cap Q(n, j_n)$$

for some j_n in the set $\{1, \ldots, k(n)\}$, and no B_n has a special partition.

As usual, we denote B_n by

$$B_n = B_n(t_1) \times B_n(t_2) \times B_n(t_3) \times \cdots.$$

For each t_j in T, the intervals $B_n(t_j)$ shrink as n increases, and they are never empty. By Theorem 2-2 in the Introduction, there is a point of $\bar{R}$ contained in the intersection of all $B_n(t_j)^-$. We choose such a point and call it $\bar{x}(t_j)$. We have thus defined a function $t_j \mapsto \bar{x}(t_j)$ on T, and by definition this is a point $\bar{x}$ of $\bar{R}^T$. Then the neighborhood $\gamma(\bar{x})$ is an open interval G in $\bar{R}$ that contains $\bar{x}$. We write $G = G(t_1) \times G(t_2) \times G(t_3) \times \cdots$. By definition of open interval, there is an integer h such that for $j > h$, $G(t_j) = \bar{R}$. For $j = 1, \ldots, h$, $G(t_j)$ is an open interval in $\bar{R}$ that contains $\bar{x}(t_j)$, so as in the proof of Theorem I-4-2, for all large n the closure $B_n(t_j)^-$ of $B_n(t_j)$ in $\bar{R}$ is contained in $G(t_j)$. We choose an n for which this is true for $j = 1, \ldots, h$. Then the single pair $(\bar{x}, B_n)$ is a special partition of the interval B_n. But B_n was constructed so as to have no special partitions. So the assumption that the theorem is false leads to a contradiction, and the theorem is proved.

The definitions of additivity and nonnegativeness are as in Definition III-8-1 except for the trivial change of replacing $\bar{R}$ by $\bar{R}^T$, which we shall not continue to mention. For such functions m, all the theorems, definitions, etc., whose names are preceded by stars remain valid for all $\bar{R}^T$ with countable T. (They are valid for uncountable T also, but we are not presenting the proofs.) However, as we saw earlier in this chapter, in order to show that for every left-open interval A, mA is the integral with respect to m of the indicator of A, we have to assume more than additivity and nonnegativeness for m. In R, this added property was called *regularity*. We shall extend it to interval-functions in R^T, but we cannot keep the wording of Definition III-8-3 unchanged because in R^T with infinite T there are no bounded intervals. To get around this difficulty, we first observe that whenever S is a subset of T, to each left-open interval

$$A = \bigtimes_{t \text{ in } S} A(t)$$

in the space R^S there corresponds a left-open interval in R^T which we denote by $A^\dagger$ and which is defined as

$$A^\dagger = \bigtimes_{t \text{ in } T} A^\dagger(t),$$

where

$$A^\dagger(t) = A(t) \qquad (t \text{ in } S)$$

$$A^\dagger(t) = R \qquad (t \text{ in } T \setminus S).$$

On the set of left-open intervals in R^S, we define a function m_S by setting

$$m_S A = mA^\dagger \qquad (A \text{ a left-open interval in } R^S).$$

(If T is finite and m is a probability distribution in R^t, m_S is what we have previously called the *marginal distribution* on R^S.)

With the help of this concept we can extend the definition of regularity.

★DEFINITION 14-2 *Let m be an extended-real-valued function on the family of left-open intervals in R^T. Then m is an* ***additive nonnegative regular function of intervals*** *if for each finite subset T_0 of T there is a finite set S such that $T_0 \subset S \subset T$ and that m_S is an additive nonnegative regular function on the family of left-open intervals in R^S, as defined in Definition* III-8-3.

★THEOREM 14-3 *Let m be a regular nonnegative additive function on the family of left-open intervals in R^T, and let A be a left-open interval in R^T. Then 1_A has an* *integral over R^T, and*

$$\int_{R^T} 1_A(x)\, m(dx) = mA.$$

Let A be a left-open interval in R^T of the form

$$A = \bigtimes_{t \text{ in } T} A(t).$$

By definition, there is a finite subset T_0 of T such that $A(t) = R$ whenever t is not in T_0. Let S be a finite subset of T such that $S \supset T_0$ and the measure m_S defined above is regular on R^S. It is convenient to introduce some notation. If x is in $\bar{R}^T$, by x_S we shall mean its "restriction to S," which is the function $t \mapsto x(t)$ (t in S). Likewise, by A_S we shall mean the interval

$$\bigtimes_{t \text{ in } S} A(t),$$

which is in R^S. Suppose first that A_S is bounded. Let ε be positive. By Definition III-8-3, we can and do select left-open intervals F and G in R^S such that

$$F^- \subset A_S \subset G^O$$

and

$$m_S G < m_S A_S + \varepsilon = mA + \varepsilon,$$

$$m_S F > m_S A_S - \varepsilon = mA - \varepsilon.$$

If x is in A, x_S is in A_S, which is contained in G^O. So $[G^O]^\dagger$ is an open interval in $\bar{R}^T$ that contains x. We choose it for $\gamma(x)$. If x is not in A, x_S is not in F^-. Just as in the second paragraph of Section 2, there is a neighborhood U of x_S in $\bar{R}^S$ that is disjoint from F^-. The interval $U^\dagger$ is an open interval in $\bar{R}^T$ that contains x and is disjoint from $F^\dagger$. The rest of the proof is a mere repetition of the proof of Lemma I-3-2.

The extension to intervals A for which A_S is unbounded is effected just as in the proof of Corollary III-8-4.

Now, as before, we define a measurable set to be one whose indicator has an integral, finite or infinite. As in the case of R^r, these sets form a σ-algebra, and if for each measurable set B we define

$$mB = \int_{R^T} 1_B(x)\, m(dx),$$

the function thus defined is an extension to the family of all measurable sets of the original function m on the family of left-open intervals. In particular, if $mR^T = 1$, m is called a *probability measure*, and the measurable sets are called *events*, as before. Since the whole of probability theory can be worded in terms of events and the countably additive measure m (with $mR^T = 1$) on them, we now have a complete basis for probability theory on spaces R^T with countable T.

The example at the beginning of this section is a typical example of the way in which distributions on spaces R^T can be encountered. A trial of some sort is repeated infinitely often. For each finite set of repetitions, the joint distribution of the outcomes is known. We wish to construct a distribution on R^T that specializes to each of the possible finite subsets of trials. Suppose, then, that for each set S in a collection of finite subsets of a countable set T we are given a distribution P_S on the space R^S. We are looking for a distribution P on R^T that includes all of them, in the sense that whenever A is an event in R^S with respect to distribution P_S, and we "pad it out" to a set $A^\dagger$ in R^T consisting of all x in R^T for which the restriction

$$x_S : t \mapsto x(t) \qquad (t \text{ in } S)$$

belongs to A, $P_S(A)$ is equal to $P(A^\dagger)$. If this is to be possible, the distributions P_S must have a certain consistency property. Suppose that S and S' are finite subsets of T with $S \subset S'$. If A is a set in R^S that is P_S-measurable, and B is the set of all points x in $R^{S'}$ whose restriction $t \mapsto x(t)$ (t in S) is in A, then the "padded out" set $B^\dagger$ (consisting of all x in R^T for which the restriction $x_{S'}$ is in B) is the same as the set $A^\dagger$ (consisting of all x in R^T with x_S in A). Then if there is a distribution P on R^T that specializes down to P_S and to $P_{S'}$, both $P_S A$ and $P_{S'} B$ must be equal to $PA^\dagger$, and therefore they must be equal to each other. This consistency is the key hypothesis in the next theorem, which is a special case of an extension theorem due to N. Kolmogorov.

THEOREM 14-4 *Let T be a nonempty countable set, and let $\mathscr{S}$ be a collection of finite subsets of T such that every finite subset of T is contained in some set S that belongs to the family $\mathscr{S}$. For each S in $\mathscr{S}$, let P_S be a probability distribution on R^S. Assume that whenever S and S' belong to $\mathscr{S}$, and $S \subset S'$, and A is a left-open interval in R^S, and A' is the left-open interval in $R^{S'}$ for which*

$$A'(t) = A(t) \qquad (t \text{ in } S)$$

$$A'(t) = R \qquad (t \text{ in } S' \setminus S),$$

it is true that

$$P_S A = P_{S'} A'.$$

Then there exists a probability distribution P on R^T such that whenever S is a member of $\mathscr{S}$, and A is a left-open interval in R^S, and $A^\dagger$ is the interval in R^T for which $A^\dagger(t) = A(t)$ whenever t is in S and $A^\dagger(t) = R$ whenever t is in $T \setminus S$, then

$$P(A^\dagger) = P_S(A).$$

Moreover, if S is in $\mathscr{S}$ and f is a real-valued function on R^S, and $f^\dagger$ is the function on R^T defined by $f^\dagger(x) = f(x_S)$, then if f is P_S-measurable, $f^\dagger$ is P-measurable; and if f is integrable with respect to P_S, $f^\dagger$ is integrable with respect to P, and

$$\text{(D)} \qquad \int_{R^T} f^\dagger(x)\, P(dx) = \int_{R^S} f(x)\, P_S(dx).$$

If A is a left-open interval in R^T, there is a finite subset T_0 of T such that if t is in $T \setminus T_0$, $A(t) = R$. Let S be any member of the set $\mathscr{S}$ that contains T_0; define

$$\text{(E)} \qquad P(A) = P_S(A_S),$$

where, as before, A_S is the Cartesian product

$$A_S = \bigtimes_{t \text{ in } S} A(t).$$

In spite of the freedom of choice of S, this definition is unambiguous. For let S and S' be two members of the family $\mathscr{S}$ that contain T_0. By hypothesis, there is a member S'' of $\mathscr{S}$ that contains the (finite) union $S \cup S'$. Also by hypothesis, both $P_S A_S$ and $P_{S'} A_{S'}$ are equal to $P_{S''} A_{S''}$, so they are equal to each other, and the definition of $P(A)$ is unambiguous.

If $A_1, \ldots, A_k$ are pairwise disjoint left-open intervals in R^T whose union is a left-open interval A in R^T, there are finite subsets $T_1, \ldots, T_k$ of T such that if t is in $T \setminus T_i$, $A_i(t) = R$. Let S be a member of $\mathscr{S}$ that contains the union of the T_i. Then $A_{1,S}, \ldots, A_{k,S}$, A_S are all left-open intervals in R^S, and $A_{1,S}, \ldots, A_{k,S}$ are pairwise disjoint, and their union is A_S. Since P_S is a probability distribution,

$$P_S(A_S) = P_S(A_{1,S}) + \cdots + P_S(A_{k,S}).$$

But by definition of P, this implies

$$P(A) = P(A_1) + \cdots + P(A_k),$$

so P is finitely additive. It is obviously nonnegative and satisfies $P(R^T) = 1$, and since each P_S is regular, so is P. So P is a probability distribution on R^T that specializes in the desired way to each P_S.

If f is the indicator of a left-open interval in R^S, equation (D) follows by Theorem 14-3 from definition (E) of P. If f is a step-function on R^S, it is a linear combination of indicators of intervals in R^S, and (D) follows for f from its validity for each of the indicators. If f is an integrable U-function on R^S, it is the limit of a rising sequence of step-functions. If we write (D) for each of the step-functions and apply the monotone convergence theorem, we find that (D) holds for integrable U-functions f. Likewise, it holds for integrable L-functions. In the same way, (D) holds whenever f is the limit of a descending sequence of U-functions or of an ascending sequence of L-functions. If f is integrable over R^S with respect to P_S, there are, by Corollary 6-5, functions g, h on R^S such that $g \leqq f \leqq h$, and g is the limit of an ascending sequence of L-functions and h the limit of a descending sequence of U-functions, and $g = f = h$ except on a set of P_S-measure 0. By the part of the proof already completed,

$$\int_{R^T} g^\dagger(x)\, P(dx) = \int_{R^S} g(x)\, P_S(dx) = \int_{R^S} h(x)\, P_S(Dx) = \int_{R^T} h^\dagger(x)\, P(dx).$$

This, with the obvious inequality $g^\dagger(x) \leqq f^\dagger(x) \leqq h^\dagger(x)$ and Corollary I-7-3, implies that (D) is valid and completes the proof.

As an easy example, we apply Theorem 14-4 to the case of an infinite sequence of tosses of a fair coin. Here we can take S to be the family of all sets $\{1, \ldots, n\}$. For this S, P_S is the distribution that assigns probability 2^{-n} to each ordered n-tuple of zeros and ones. By Theorem 14-4, there is a distribution on R^T ($T = \{1, 2, 3, \ldots\}$) that specializes to each of these distributions. For each positive integer n, the event that the game described at the beginning of this section continues to at least $n + 1$ tosses is the set C_n of all sequences $x = (x(1), x(2), x(3), \ldots)$ such that the ordered n-tuple $(x(1), \ldots, x(n))$ contains no two consecutive ones. This we subdivide into the two events H_n, consisting of all x in C_n with $x(n) = 1$, and the event T_n, consisting of all x in C_n with $x(n) = 0$. We leave it to the reader to prove by induction that

$$P(H_n) = 2^{-n}F_n, \qquad P(T_n) = 2^{-n}F_{n+1},$$

where F_n is the nth Fibonacci number. (The Fibonacci numbers are $1, 1, 2, 3, 5, 8, 13, \ldots$, each number after the second being the sum of the two that immediately precede it.) The probability that the game ends at the nth toss is $P(H_{n-1})/2$, which is $2^{-n}F_{n-1}$. The probability that it lasts beyond the nth toss is

$$P(C_n) = P(H_n) + P(T_n) = 2^{-n}F_{n+2}.$$

It is easy to prove by induction that

$$F_n < (\tfrac{3}{2})^n,$$

so $P(C_n)$ tends to 0 as n increases. Our model matches our anticipation that the probability of an unending game is 0. If f is a real-valued function on a set that contains the positive integers, and the player wins an amount $f(n)$ if the game ends at the nth toss, the expectation of f is

$$E(f) = \sum_{n=2}^{\infty} f(n)F_{n-1}/2^n.$$

In particular, the expected length of the game is

$$\sum_{n=2}^{\infty} nF_{n-1}/2^n$$

tosses.

For another example, we start with a homogeneous regular icosahedron two of whose faces are marked "0," two marked "1," and so on. Each experiment consists of an infinite sequence of independent tosses of the icosahedron, resulting in an infinite sequence of digits $d_1, d_2, d_3, \ldots$. This is a point in R^T, where T is the set of positive integers. For each interval $(a, b]$ in R we define $P_1(a, b]$ to be 0.1 times the number of the digits $0, 1, \ldots, 9$ contained in $(a, b]$; that is, if

$$D = \{0, 1, \ldots, 9\},$$

then

$$P_1(a, b] = \text{number of points in} \quad D \cap (a, b].$$

For each positive integer n we define $S(n)$ to be $\{1, \ldots, n\}$, and for each left-open interval $A = A(1) \times \cdots \times A(n)$ in $R^{S(n)}$ we define

$$P_{S(n)}(A) = P_1(A(1))P_1(A(2)) \cdots P_1(A(n)).$$

By Theorem 14-4 there is a probability distribution P on R^T that specializes to each of these. In particular, if A is a set in R^T such that the nth term $x(n)$ in each sequence belonging to A is not a digit, $P(A) = 0$; and by adding for all positive integers n, the set of all sequences (points of R^T) in which not all terms are digits is a set of P-measure 0.

Let k be a positive integer, and let N_k be the subset of R^T such that for each x in N_k, $x(j) = 9$ for $j \geqq k$. For each integer n greater than k, we define f_n to be the function on $R^{S(n)}$ (where $S(n) = \{1, \ldots, n\}$) and we define $f_n^\dagger$ to be the function on R^T such that $f_n(x)$ and $f_n^\dagger(x)$ are 1 if $x(k) = x(k+1) = \cdots = x(n) = 9$ and are 0 otherwise. Then $f_n \geqq 1_{N_k}$, and

$$\int_{R^T} f_n^\dagger(x)\, P(dx) = \int_{R^{S(n)}} f_n(x)\, P_{S(n)}(dx) = 10^{-(n-k+1)}.$$

By Lemma I-7-1 the integral of 1_{N_k} over R^T with respect to P is 0, so $P(N_k) = 0$. The union of all N_k $(k = 1, 2, 3, \ldots)$ also has P-measure 0, so the set of all sequences that are 9 from a certain point on has P-measure 0. We define Ω_0 to be the set of all sequences of digits that are not constantly equal to 9 after some point. Then $P(\Omega_0) = 1$.

Every number y in $[0, 1)$ has a decimal representation that is unique if wc agree not to use decimals that are ultimately 9s. Hence there is a one-to-one correspondence between Ω_0 and $[0, 1)$. The set of all real numbers whose decimal expansions begin with n assigned digits $a_1, a_2, \ldots, a_n$ is an interval whose lower end-point is $.a_1 a_2 \cdots a_n$ and whose upper end-point is $.a_1 a_2 \cdots a_n + 10^{-n}$. The length 10^{-n} of this interval equals the probability of the event that the sequence x in Ω_0 determines a real number in the interval. Every interval in $[0, 1)$ whose end-points are terminating decimals is the union of finitely many pairwise disjoint intervals of the type just described, so its length is equal to the P-measure of the set of sequences in Ω_0 that determine points in the interval. By an easy limiting process, this remains true for every subinterval of $[0, 1)$. If $y(x)$ is the real number determined by the point x of Ω_0, and f is the indicator of an interval in $[0, 1)$, what we have just proved is

$$\int_0^1 f(u)\, du = \int_{\Omega_0} f(y(x))\, P(dx).$$

This extends at once to step-functions and then, by an argument we have often used before, it extends to all U-functions and all L-functions, to all limits of ascending sequences of integrable L-functions and all limits of descending sequences of integrable U-functions, and finally to all functions f that are integrable with respect to m_L over $[0, 1)$. So for this particular probability measure we have reduced integration over R^T to an ordinary integration with respect to m_L over an interval in R.

There are some other measure in R^T that can by one device or another be reduced to integrals with respect to m_L over an interval in R. Such reductions were once very important, in the time when the theory of integration with respect to m_L in R was well advanced but integration in more general spaces was as yet undeveloped. To us it is much less important since, as we have seen, the bulk of the theory of integration generalizes in a perfectly straightforward way from one dimension to infinitely many.

Given any sequence of probability distributions $P_1, P_2, P_3, \ldots$ in R, it is now easy to construct a probability distribution in R^T such that the coordinate variables X_i defined by $X_i(x) = x(i)$ are independent random variables with X_i having distribution P_i. Theorem 14-4 shows us how to do this. By use of this measure we can discuss a stronger form of the law of large numbers than we proved in Theorem 11-8. There we showed that if $X_1, X_2, X_3, \ldots$ is a sequence of independent random variables, all having the same distribution with finite variance, the average $[X_1 + \cdots + X_n]/n$ of the outcomes of n experiments differs from the expectation $E(X_1)$ by less than an arbitrary positive ε except on a set of ω whose probability measure is near 0 when n is large. But this does not imply that for every possible sequence of outcomes the average will tend to $E(X_1)$ as a limit. Nevertheless, this can be proved true if we omit a set of sequences of probability 0. This is called the "strong law of large numbers." It can be proved for all sequences of independent identically distributed random variables with finite expectation, but it is easier to prove it when the random variables have finite variance, and this is the only case that we shall consider.

★THEOREM 14-5 *Let $X_1, X_2, X_3, \ldots$ be independent random variables, all having the same distribution P_1 with finite variance. Let P be the probability measure on the space Ω on which all the X_i are defined. (If we constructed Ω by Theorem* 14-4, *it is R^T with $T = \{1, 2, 3, \ldots\}$.) Then for all ω in Ω except those in a set of P-measure* 0

$$\text{(F)} \qquad \lim_{n \to \infty} [X_1(\omega) + \cdots + X_n(\omega)]/n = E(X_1).$$

Let V denote the variance of X_1, and for each positive integer j define

$$B_j = \{\omega \text{ in } \Omega : |[X_1(\omega) + \cdots + X_{j^4}(\omega)]/j^4 - E(x_1)| \geqq j^{-1}\}.$$

By Theorem 11-8,

$$P(B_j) \leqq V/j^2.$$

So, the sum of the $P(B_j)$ converges, and by the Borel-Cantelli lemma (Theorem 11-12), the set N of points that belong to infinitely many sets B_j has P-measure

0. If ω is not in N, it is missing from B_j for all j greater than a certain j', which means that

$$|[X_1(\omega) + \cdots + X_{j^4}(\omega)]/j^4 - E(X_1)| < j^{-1}$$

for all j greater than j'. This implies that

$$\text{(G)} \qquad \lim_{j \to \infty} [X_1(\omega) + \cdots + X_{j^4}(\omega)]/j^4 = E(X_1).$$

If the X_i take on only nonnegative values, for each positive integer n we define $j(n)$ to be the largest integer j for which $j^4 \leqq n$. Then, since the X_i are nonnegative, we have at each ω

$$\begin{aligned} &\{[X_1(\omega) + \cdots + X_{j(n)^4}(\omega)]/j(n)^4\}\{j(n)^4/[j(n) + 1]^4\} \\ &\leqq [X_1(\omega) + \cdots + X_n(\omega)]/n \\ &\leqq \{[X_1(\omega) + \cdots + X_{[j(n)+1]^4}(\omega)]/[j(n) + 1]^4\}\{[j(n) + 1]^4\}/j(n). \end{aligned}$$

In the first and last expressions in this inequality the first factor tends to $E(X_1)$ by (G) if ω is not in N. The second factor tends to 1, so (F) is established under the supplementary hypothesis that the X_1 are nonnegative.

To remove this restriction, we need only notice that if the X_1 satisfy the hypotheses, so do $X_1^+, X_2^+, X_3^+, \ldots$ and so do $X_1^-, X_2^-, X_3^-, \ldots$. By the part of the proof already completed,

$$\lim_{n \to \infty} [X_1^+(x) + \cdots + X_1^+(x)]/n = E(X_1^+)$$

for all x except those in a set N_1 with $P(N_1) = 0$, and

$$\lim_{n \to \infty} [X_1^-(x) + \cdots + X_n^-(x)]/n = E(X_1^-)$$

for all x except those in a set N_2 with $P(N_2) = 0$. The set $N = N_1 \cup N_2$ has P-measure 0, and except for x in it both these equations hold, and by subtraction so does (F).

As an example, let X_j have the value 1 if $x_j = 6$ and the value 0 otherwise. With the distribution P in R^T corresponding to the icosahedron-tossing experiment, these are independent and identically distributed random variables, each with expectation 0.1 and finite variance. The number

$$[X_1(x) + \cdots + X_n(x)]/n$$

is the ratio of the number of digits 6 among the first n digits of x to the number n; it is called the *relative frequency* of digit 6 among the first n digits of x. By Theorem 14-5, as n increases this tends to 0.1 for all x except those in a set of P-measure 0. If we map R^T onto $[0, 1)$, as we did in the second example after Theorem 14-4, we find that there is a subset N_6 of $[0, 1)$ with $m_L N_6 = 0$ such that for all x in $[0, 1) \setminus N_6$ the relative frequency of the digit 6 among the first n digits of x tends to 0.1 as n increases. The same argument applied to 6 applies also to each other digit, so there is a set $N = N_0 \cup \cdots \cup N_9$ with $m_L N = 0$ such that for all x in $[0, 1) \setminus N$, the relative frequency of each digit among the first n digits of x tends to 0.1 as n increases.

EXERCISE 14-1 Let m be a function of left-open intervals in R^T with T finite. Show that m is additive, nonnegative, and regular by Definition 14-2 if and only if it is additive, nonnegative, and regular by Definition III-8-3, as extended to spaces R^r.

EXERCISE 14-2 Show that when m is additive, nonnegative, and regular on a space R^T, there may exist a finite subset S of T such that m_S (as defined just before Definition 14-2) fails to be regular. (Take $T = \{1, 2\}$, $S = \{2\}$ and for each left-open interval A in R^T define mA to be the length of the arc of the graph of $x_2 = 1/x_1$ $(x_1 > 0)$ that is contained in A.)

EXERCISE 14-3 The game described at the beginning of this section either continues beyond the nth toss or ends at the jth toss for some $j \leqq n$. This implies

$$2^{-2}F_1 + 2^{-3}F_2 + \cdots + 2^{-n}F_{n-1} + 2^{-n}F_{n+2} = 1.$$

Verify this directly from the definition of the Fibonacci numbers.

EXERCISE 14-4 A typewriter has 44 keys plus a shift key and a space bar. The well-known monkeys are trained to type 14 symbols at random from the 89 possibilities; the carriage then returns to start and each monkey repeats the process at 5 symbols per second. Show that there is probability 1 that at some time the sentence *I'm overworked* will appear, but that if there are 10^9 monkeys working day and night, there is only about 0.5 probability of success in 1.2×10^{12} years.

EXERCISE 14-5 Let $a_1, a_2, a_3, \ldots$ be a sequence of numbers in the interval $(0, 1)$. Prove the following statements.

(i) On the space R^T (T the set of positive integers) there is a probability measure P such that the sets

$$A_j = \{x \text{ in } R^T : 0 < x_j \leqq a_j\} \qquad (j = 1, 2, 3, \ldots)$$

are independent events, and $P(A_j) = a_j$. *Suggestion*: Use Theorem 14-4, each one-dimensional distribution being uniform on $(0, 1]$.

(ii) The limit

$$p_0 = \lim_{n \to \infty} \prod_{j=1}^{n} (1 - a_j)$$

exists and is in $[0, 1)$.

(iii) Let $\mathscr{S}$ be the family of all finite subsets of T, and for each S in $\mathscr{S}$ let B_S be the set of all x in R^T such that for all t in T, x_t is in $[0, a_t)$ if and only if t is in S. Then

$$P(B_S) = p_0 \prod_{t \text{ in } S} \frac{a_t}{1 - a_t},$$

the product in the right membeR being understood to be 1 if S is empty.

(iv) If S_∞ is the set of all x in R^T such that $0 \leqq x_n \leqq a_n$ for infinitely many n, S_∞ is an event, and

$$P(S_\infty) + \sum_{S \text{ in } \mathscr{S}} P(B_S) = 1.$$

(v) If Σa_n diverges, $p_0 = 0$; if Σa_n converges, $p_0 > 0$, and

$$p_0^{-1} = \sum_{S \text{ in } \mathscr{S}} \prod_{t \text{ in } S} \frac{a_t}{1 - a_t}.$$

(vi) In particular, if $q_1, q_2, q_3, \ldots$ are the prime numbers $2, 3, 5, \ldots$ in increasing order, and $u > 1$,

$$\left\{ \prod_{n=1}^{\infty} (1 - q_n^{-u}) \right\}^{-1} = \sum_{S \text{ in } \mathscr{S}} \prod_{t \text{ in } S} \frac{1}{q_t^u - 1}.$$

(The left member of the last equation is the Riemann zeta function. Observe that, as in Exercise 14-3, we have used probability theory to establish an equation that has no obvious connection with probability. This problem is taken from a note by F. Stern (1978), in *American Mathematical Monthly* **85**, 363.)

Line Integrals and Areas of Surfaces

1. Geometry in r-Dimensional Space

More than two millenia after the Greek mathematicians had begun the systematic study of plane and solid geometry, René Descartes introduced, early in the seventeenth century, the idea of bringing numbers to the aid of geometry by means of coordinate systems. This is now familiar even to beginners in mathematics. In three-space we choose a point O for "origin," and we construct three lines through O; these we call the first, second, and third axes. For simplicity we shall assume that these are perpendicular to each other. Each is subdivided by O into two half-axes, one of which we name the positive half-axis and the other the negative half-axis. Let P be any point of space, and through P pass a plane parallel to the second and third axes. This will meet the first axis at a point P'. If P' is on the positive half-axis, we define x^1 to be the distance dist OP'; otherwise we define x^1 to be $-$dist OP'. Similarly, we define x^2 and x^3. Thus, P determines an ordered number-triple (x^1, x^2, x^3), which is a member of R^3. Conversely, each ordered number-triple corresponds to a point of space, and it is easy to see that the mapping of space into R^3 is continuous and so is its inverse.

But this correspondence is not merely a means of identifying points. All the fundamental ideas of solid geometry, such as distance and angle, are expressible in terms of the numbers in the triples that correspond to points, and we can conduct the study of solid geometry by working with the numbers. The situation in the plane is similar but a little simpler; we need only two axes, and each point P in the plane corresponds to an ordered number-pair (x^1, x^2) that is a member of R^2.

Descartes introduced this numerical representation of points in order to use the pairs or triples of numbers as aids in the study of geometry. But more recently, ordered r-tuples of numbers have proved useful in their own right. For example, if the girders of a bridge meet in N joints, then in terms of some selected system of axes in space, the kth joint will have certain coordinates (x_k, y_k, z_k).

The $3N$-tuple

$$(x_1, y_1, z_1, x_2, y_2, z_2, x_3, \ldots, x_N, y_N, z_N)$$

will locate every joint and thus describe the state of deformation of the bridge. Each shape of the bridge corresponds to a point in R^{3N}. If we had developed a geometry in R^{3N}, we could use it to gain information about the states of the bridge. The spaces R^2 and R^3 have geometries because they correspond to the plane and to three-dimensional space. The task we undertake here is to invent a geometry that applies to every R^r and that takes the familiar form of plane geometry when $r = 2$ and of solid geometry when $r = 3$.

The first step is to introduce an idea of distance into R^r. Suppose that in the plane we choose and fix a pair of perpendicular axes. If P' and Q' are points of the plane, there will be, corresponding to P', a number-pair (x^1, x^2) that is itself a member P of R^2, and, corresponding to Q', a number-pair (y^1, y^2) that is itself a member Q of R^2. By use of the theorem of Pythagoras, it is easy to prove the well-known theorem that the distance from P' to Q' is

$$\text{(A)} \qquad \operatorname{dist}(P', Q') = [(x^1 - y^1)^2 + (x^2 - y^2)^2]^{1/2}.$$

We have not yet defined the distance between pairs of points in R^2. But it is natural to define the distance between points P and Q of R^2 to be the same as the distance between points P' and Q' of the plane to which the number-pairs P and Q correspond. Likewise, if P' and Q' are two points of three-dimensional space and we choose and fix three mutually perpendicular axes in that space, P' and Q' will correspond, respectively, to two points

$$P = (x^1, x^2, x^3), \qquad Q = (y^1, y^2, y^3)$$

of R^3. Again, we can prove by the theorem of Pythagoras that

$$\text{(B)} \qquad \operatorname{dist}(P', Q') = [(x^1 - y^1)^2 + (x^2 - y^2)^2 + (x^3 - y^3)^2]^{1/2}.$$

We have not yet defined distance in R^3, but it is natural to define the distance between points P and Q of R^3 to be the same as the distance between points P' and Q' of space to which P and Q correspond. This suggests the following generalization.

DEFINITION 1-1 *If $P = (x^1, \ldots, x^r)$ and $Q = (y^1, \ldots, y^r)$ belong to R^r, the* ***distance*** *from P to Q is*

$$\operatorname{dist}(P, Q) = [(x^1 - y^1)^2 + \cdots + (x^r - y^r)^2]^{1/2}.$$

We have already met expressions similar to this, although in different notation, in our study of probability theory in Section II-11.

When r is 2 or 3, the distance thus defined will have all the familiar properties of distance in the plane or in space, for then P and Q correspond to points P' and Q' in the plane or in space, and by (A) or (B) their distance will be equal to the distance $\operatorname{dist}(P', Q')$. But for values of r greater than 3 the points of R^r will not correspond to anything in elementary geometry, and all the properties of distance will have to be proved from Definition 1-1 by using our knowledge of the properties of numbers. There are four fundamental properties of the distance defined in Definition 1-1, and we now state and prove them.

LEMMA 1-2 *Let P, Q, and R be points of R^r. Then*

(i) $\text{dist}(P, Q) = \text{dist}(Q, P)$;
(ii) $\text{dist}(P, Q) \geqq 0$;
(iii) $\text{dist}(P, Q) = 0$ *if and only if* $P = Q$;
(iv) $\text{dist}(P, R) \leqq \text{dist}(P, Q) + \text{dist}(Q, R)$.

Let

$$P = (x^1, \ldots, x^r), \qquad Q = (y^1, \ldots, y^r), \qquad R = (z^1, \ldots, z^r).$$

The first three conclusions are trivially easy to prove. The fourth, called the *triangle inequality*, states that if P, Q, and R are the vertices of a triangle, the length of the side from P to Q is, at most, equal to the sum of the lengths of the other two sides. This is a familiar theorem of plane geometry. We need to show that it follows from properties of real numbers, without calling on the axioms of geometry. To do this we first establish an inequality so useful that it is astonishing that it was not proved before the early nineteenth century.

LEMMA 1-3 (Cauchy's Inequality) *If $(v_1, \ldots, v_r)$ and $(w_1, \ldots, w_r)$ are two r-tuples of real numbers, then*

$$\text{(C)} \qquad \left|\sum_{i=1}^{r} v_i w_i\right| \leqq \sum_{i=1}^{r} |v_i|\,|w_i| \leq \left[\sum_{i=1}^{r} v_i^2\right]^{1/2} \left[\sum_{i=1}^{r} w_i^2\right]^{1/2}.$$

Equality holds if and only if there are numbers a, b not both 0 such that

$$\text{(D)} \qquad av_i + bw_i = 0 \qquad (i = 1, \ldots, r).$$

The first of the two inequalities in (C) is obvious. The second holds if either all the v_i are 0 or all the w_i are 0. Suppose, then, that neither all the v_i nor all the w_i are 0; define

$$t = \left[\sum_{i=1}^{r} |w_i|^2 \Big/ \sum_{i=1}^{r} |v_i|^2\right]^{1/4}.$$

Then

$$0 \leqq \sum_{i=1}^{r} \frac{(t|v_i| - t^{-1}|w_i|)^2}{2}$$

$$= \frac{t^2}{2} \sum_{i=1}^{r} |v_i|^2 - \sum_{i=1}^{r} |v_i|\,|w_i| + \frac{t^{-2}}{2} \sum_{i=1}^{r} |w_i|^2.$$

If we transpose the middle term in the last expression to the left member and substitute the value of t as defined, we obtain (C).

If (D) is satisfied, one of the two statements

$$w_i = -(a/b)v_i, \qquad v_i = -(b/a)w_i \qquad (i = 1, \ldots, r)$$

is meaningful and correct, and by substitution we find that equality holds in (C). If equality holds in (C), we consider two cases. If all the v_i are 0 or all the w_i are 0, (D) holds with one of the numbers a, b being 0 and the other being 1. If neither all the v_i nor all the w_i are 0, let s be 1 if $v_1 w_1 + \cdots + v_r w_r > 0$ and be -1 otherwise. We define t as before. Then

$$\begin{aligned} \sum_{i=1}^{r} (tv_i - st^{-1}w_i)^2 &= t^2 \sum_{i=1}^{r} v_i^2 - 2s \sum_{i=1}^{r} v_i w_i + t^{-2} \sum_{i=1}^{r} w_i^2 \\ &= 2\left(\sum_{i=1}^{r} v_i^2\right)^{1/2}\left(\sum_{i=1}^{r} w_i^2\right)^{1/2} - 2\left|\sum_{i=1}^{r} v_i w_i\right| \\ &= 0. \end{aligned}$$

Every term in the first sum in these equations must be 0, so (D) holds with $a = t$ and $b = -st^{-1}$. The proof of Lemma 1-3 is complete.

With the notation for P, Q, and R in Lemma 1-2, by Definition 1-1,

$$\begin{aligned} [\operatorname{dist}(P, R)]^2 &= \sum_{i=1}^{r} (z^i - x^i)^2 \\ &= \sum_{i=1}^{r} [(z^i - y^i) + (y^i - x^i)]^2 \\ &= \sum_{i=1}^{r} (z^i - y^i)^2 + 2 \sum_{i=1}^{r} (z^i - y^i)(y^i - x^i) + \sum_{i=1}^{r} (y^i - x^i)^2. \end{aligned}$$

By this and the Cauchy inequality (Lemma 1-3),

$$\begin{aligned} [\operatorname{dist}(P, R)]^2 &\leqq \sum_{i=1}^{r} (z^i - x^i)^2 + 2\left[\sum_{i=1}^{r} (z^i - y^i)^2\right]^{1/2}\left[\sum_{i-1}^{r} (y^i - x^i)^2\right]^{1/2} \\ &\quad + \sum_{i=1}^{r} (y^i - x^i)^2 \\ &= [\operatorname{dist}(Q, R)]^2 + 2 \operatorname{dist}(Q, R) \operatorname{dist}(P, Q) + [\operatorname{dist}(P, Q)]^2 \\ &= [\operatorname{dist}(Q, R) + \operatorname{dist}(P, Q)]^2. \end{aligned}$$

The triangle inequality, (iv) of Lemma 1-2, follows immediately, and the proof of Lemma 1-2 is complete.

COROLLARY 1-4 *If P, Q, and R are points of R^r, then*

$$\text{(E)} \qquad |\operatorname{dist}(P, Q) - \operatorname{dist}(Q, R)| \leqq \operatorname{dist}(P, R).$$

By two applications of conclusion (iv) of Lemma 1-2,

$$\text{dist}(P, Q) \leqq \text{dist}(P, R) + \text{dist}(R, Q),$$

$$\text{dist}(R, Q) \leqq \text{dist}(R, P) + \text{dist}(P, Q).$$

In each of these we transpose the last term of the right member to the left member. Since $\text{dist}(R, Q) = \text{dist}(Q, R)$ and $\text{dist}(P, R) = \text{dist}(R, P)$, one of the two inequalities that we obtain is inequality (E).

If P_0 is a point of R^r and c is a positive number, the **ball** with center P_0 and radius c is defined to be the set

$$\{P \text{ in } R^r : \text{dist}(P, P_0) < c\}.$$

These can be used in place of the open intervals as the neighborhoods of P_0 in R^r with no change in the meaning of limit, continuity, etc., because of the following lemma.

LEMMA 1-5 *Let P_0 be a point of R^r. Then every ball with center P_0 contains an open interval that contains P_0, and every open interval that contains P_0 contains a ball with center P_0.*

Let B be the ball with center $P_0 = (x_0^1, \ldots, x_0^r)$ and radius c. Define B to be the open interval

$$(x_0^1 - cr^{-1/2}, x_0^1 + cr^{-1/2}) \times \cdots \times (x_0^r - cr^{-1/2}, x_0^r + cr^{-1/2}).$$

If $P = (x^1, \ldots, x^r)$ is a point of the interval,

$$|x^i \quad x_0^i| < cr^{-1/2} \qquad (i = 1, \ldots, r).$$

By Definition 1-1,

$$\begin{aligned}\text{dist}(P, P_0) &= [(x^1 - x_0^1)^2 + \cdots + (x^r - x_0^r)^2]^{1/2} \\ &< [c^2r^{-1} + \cdots + c^2r^{-1}]^{1/2} = c,\end{aligned}$$

so P is in the ball B. Conversely, let

$$A = (a^1, b^1) \times \cdots \times (a^r, b^r)$$

be an open interval that contains P_0. Then the numbers

$$x_0^i - a^i,\ b^i - x_0^i$$

are all positive. Let c be the smallest of them, and let B be the ball with center P_0 and radius c. By Definition 1-1, if $P = (x^1, \ldots, x^r)$ is a point of B and i is in $\{1, \ldots, r\}$,

$$x_0^i - x^i \leqq \text{dist}(P, P_0) < c \leqq x_0^i - a^i,$$

whence $x^i > a^i$. Similarly, we prove $x^i < b^i$, so x is in the interval A.

In plane geometry and in solid geometry we have a concept of line-segment and a concept of line. Whether or not a set of points is a line-segment or a line can be tested by means of distances and the concept of *betweenness*. A point R is "between" points P and Q of the plane (or of space) – that is to say, R is on the line-segment joining P and Q – if and only if

$$\text{(F)} \qquad \operatorname{dist}(P, Q) = \operatorname{dist}(P, R) + \operatorname{dist}(R, Q).$$

(Note that we have somewhat stretched the usual meaning of the word "between," since by our definition both P and Q are "between" P and Q.) When r is a positive integer and P, Q, and R are points of R^r, we shall accept (F) as *defining* the meaning of the statement that R is between P and Q. Likewise, in plane or solid geometry, R is on the line through the (distinct) points P and Q if and only if R is between P and Q, or Q is between R and P, or P is between R and Q. We accept this as the definition of the line through P and Q in R^r also. This leads us to a method of identifying the points of the segment with ends P and Q (or of the line through P and Q) by means of the numbers that constitute the r-tuples P and Q.

LEMMA 1-6 *Let*

$$P = (x^1, \ldots, x^r), \qquad Q = (y^1, \ldots, y^r)$$

be distinct points of R^r. *Then*

(i) *the line-segment with ends* P *and* Q *consists of all points* $Z = (z^1, \ldots, z^r)$ *in* R^r *such that for some* t *in* $[0, 1]$,

$$\text{(G)} \qquad z^i = x^i + t(y^i - x^i);$$

(ii) *the line through* P *and* Q *consists of all points* $Z = (z^1, \ldots, z^r)$ *such that* (G) *holds for some real* t.

If the numbers $z^1, \ldots, z^r$ satisfy (G) with some t in $[0, 1]$,

$$z^i - x^i = t(y^i - x^i), \qquad y^i - z^i = (1 - t)(y^i - x^i) \qquad (i = 1, \ldots, r).$$

Therefore

$$\operatorname{dist}(P, Z) = t \operatorname{dist}(P, Q), \qquad \operatorname{dist}(Z, Q) = (1 - t) \operatorname{dist}(P, Q),$$

whence

$$\operatorname{dist}(P, Z) + \operatorname{dist}(Z, Q) = \operatorname{dist}(P, Q),$$

and by definition Z is between P and Q. Conversely, if Z is between P and Q,

$$\text{(H)} \qquad [\operatorname{dist}(P, Q) - \operatorname{dist}(P, Z)]^2 = [\operatorname{dist}(Z, Q)]^2.$$

The left member of (H) is

$$\sum_{i=1}^{r} (x^i - y^i)^2 - 2\left[\sum_{i-1}^{r} (x^i - y^i)^2\right]^{1/2} \left[\sum_{i=1}^{r} (x^i - z^i)^2\right]^{1/2} + \sum_{i=1}^{r} (x^i - z^i)^2;$$

the right member of (H) is

$$\sum_{i=1}^{r} (z^i - y^i)^2 = \sum_{i=1}^{r} [(z^i - x^i) - (y^i - x^i)]^2$$

$$= \sum_{i=1}^{r} (z^i - x^i)^2 - 2\sum_{i=1}^{r} (z^i - x^i)(y^i - x^i) + \sum_{i=1}^{r} (y^i - x^i)^2.$$

By (H), these last two expressions are equal, so

$$\text{(I)} \qquad \sum_{i=1}^{r} (z^i - x^i)(y^i - x^i) = \left[\sum_{i=1}^{r} (y^i - x^i)^2\right]^{1/2} \left[\sum_{i=1}^{r} (z^i - x_i)^2\right]^{1/2}.$$

By Lemma 1-3, there are numbers a, b not both 0 such that

$$a(z^i - x^i) + b(y^i - x^i) = 0 \qquad (i = 1, \ldots, r).$$

Here a cannot be 0, since $P \neq Q$, so if we define $t = -b/a$, we find

$$z^i - x^i = t(y^i - x^i) \qquad (i = 1, \ldots, r).$$

If we substitute this in (I), we obtain

$$t\sum_{i=1}^{r} (y^i - x^i)^2 = |t| \sum_{i=1}^{r} (y^i - x^i)^2,$$

so $t \geqq 0$. On the other hand, if $t > 1$, the distance from P to Z is t times the distance from P to Q, so the equation

$$\text{dist}(P, Q) = \text{dist}(P, Z) + \text{dist}(Z, Q)$$

could not hold. So $0 \leqq t \leqq 1$, and the conclusion about line-segments is established.

If Z is a point of the line through P and Q, either Z is between P and Q, or P is between Z and Q, or Q is between P and Z. In the first case, (G) holds with some t in $[0, 1]$. In the second case, by conclusion (i) there is a number t' in $[0, 1]$ such that

$$x^i - z^i = t'(y^i - z^i) \qquad (i = 1, \ldots, r).$$

Here $t' \neq 1$, since $P \neq Q$, so the preceding equation implies

$$z^i - x^i = [t'/(t' - 1)](y^i - x^i) \qquad (i = 1, \ldots, r).$$

If we define $t = t'/(t' - 1)$, this is (G). In the remaining case, in which Q is between P and Z, there is a number t' in $[0, 1]$ for which

$$y^i - x^i = t'(z^i - x^i) \qquad (i = 1, \ldots, r).$$

Since $P \neq Q$, $t' \neq 0$, so

$$z^i - x^i = (1/t')(y^i - x^i) \qquad (i = 1, \ldots, r).$$

If we define $t = 1/t'$, this is (G).

The *length* of a line-segment is, of course, defined to be the distance between its end-points.

A *translation* of the plane into itself can be defined as a mapping $P \mapsto T(P)$ such that the line-segment from P to $T(P)$ has the same length and the same direction for all points P in the plane, and similarly in three-dimensional space. But we cannot adopt this unchanged in R^r because we have not defined *direction* in R^r. A way out of this difficulty is suggested by the usual representation of the plane by R^2 and of space by R^3. Let us choose two perpendicular axes in the plane, as in the first paragraph of this section. Then, as in that paragraph, to each point P' in the plane there corresponds a number-pair P in R^2. Suppose that T is a translation of the plane by which each P' is mapped on a point $T(P')$ such that the line through P' and $T(P')$ is parallel to the first axis. The origin will move to a point $T(0) = (c^1, 0)$. Let P' be any point of the plane; it will correspond to a point $P = (x^1, x^2)$ of R^2, and its transform $T(P')$ will correspond to a point (y^1, y^2) of R^2. To determine x^2 we construct a line through P' parallel to the first axis; this will intersect the second axis at a point that bears a number-label which, by definition, is x^2. To determine y^2 we follow the same procedure. But the line through $T(P')$ parallel to the first axis is the same as the line through P' parallel to the first axis, so $y^2 = x^2$. To determine x^1 we construct a line through P' parallel to the second axis. This will meet the first axis at a point Q' that bears a number-label which, by definition, is x^1. Similarly, a parallel to the second axis through $T(P')$ will meet the first axis at a point Q'' whose number-label is y^1. From the construction it is evident that the figure $P'Q'Q''T(P')$ is a rectangle. Therefore the opposite sides $P'T(P')$ and $Q'Q''$ have the same length and the same direction. But by definition of translation in the plane, the segment $OT(O)$ also has the same length and same direction. The point $T(O)$ is $|c^1|$ units from O and is in the direction of the positive or negative first axis according as c^1 is positive or negative. So Q'' is $|c^1|$ units from Q' and is in the direction of the positive or the negative first axis according as c^1 is positive or negative. This implies that the number-label y^1 of Q'' is c^1 plus the number-label x^1 of Q'; that is,

$$y^1 = x^1 + c^1.$$

So if P corresponds to (x^1, x^2), $T(P)$ corresponds to $(x^1 + c^1, x^2)$. In the same way, if the translation is in the direction of the positive or negative second axis, there is a number c^2 such that if P corresponds to (x^1, x^2), $T(P)$ corresponds to $(x^1, x^2 + c^2)$. Every translation in the plane can be obtained by first translating in the direction of the first axis and then in the direction of the second axis. So if T is any translation in the plane, there exist two real numbers c^1, c^2 such that when P corresponds to (x^1, x^2), $T(P)$ corresponds to $(x^1 + c^1, x^2 + c^2)$.

By a similar discussion, if T is any translation in three-space, there exist three real numbers c^1, c^2, c^3 such that when P is the point that corresponds to (x^1, x^2, x^3), $T(P)$ is the point that corresponds to $(x^1 + c^1, x^2 + c^2, x^3 + c^3)$.

This suggests the following definition.

DEFINITION 1-7 *Let r be a positive integer. A mapping T of R^r onto itself is a* ***translation*** *if there are real numbers $c^1, \ldots, c^r$ such that for each point P of R^r, if $P = (x^1, x^2, \ldots, x^r)$, then $T(P) = (x^1 + c^1, x^2 + c^2, \ldots, x^r + c^r)$.*

We can now define parallelism in a way that agrees with the standard meaning in the plane and in space. A line l_1 in R^r is **parallel** to a line l_2 in R^r if there exists a translation of R^r onto itself that maps l_1 onto l_2.

Suppose that r and s are positive integers, that L is an $s \times r$ matrix

$$L = (L^i_j : i = 1, \ldots, s; j = 1, \ldots, r)$$

of real numbers, and that $c^1, \ldots, c^s$ are real numbers. We can define a mapping M of R^r into R^s by setting $M(x) = y$, where if $x = (x^1, \ldots, x^r)$ the functional value $M(x) = \tilde{x} = (\tilde{x}^1, \ldots, \tilde{x}^s)$ is given by

$$\text{(J)} \qquad \tilde{x}^i = \sum_{j=1}^{r} L^i_j x^j + c^i \qquad (i = 1, \ldots, s).$$

Such mappings have the following important property.

(K) If l is a line in R^r, the image $M(l)$ of l in R^s is either a single point or a line in R^s.

For, let there exist two points P, Q on l that have different images $M(P)$, $M(Q)$. If Z is on l, and

$$P = (x^1, \ldots, x^r), \qquad Q = (y^1, \ldots, y^r), \qquad Z = (z^1, \ldots, z^r),$$

by Lemma 1-6 there exists a real number t such that

$$\text{(L)} \qquad z^j = x^j + t(y^j - x^j) \qquad (j = 1, \ldots, r).$$

Then for the images

$$\tilde{P} = (\tilde{x}^1, \ldots, \tilde{x}^s) = M(P), \qquad \tilde{Q} = (\tilde{y}^1, \ldots, \tilde{y}^s) = M(Q),$$
$$\tilde{Z} = (\tilde{z}^1, \ldots, \tilde{z}^s) = M(Z)$$

we have by (J),

$$\begin{aligned}
\tilde{z}^i &= \sum_{j=1}^{r} L^i_j z^j + c^i \\
&= \sum_{j=1}^{r} L^i_j [x^j + t(y^j - x^j)] + c^i \\
&= \left[\sum_{j=1}^{r} L^i_j x^j + c^i\right] + t\left[\left\{\sum_{j=1}^{r} L^i_j y^j + c^j\right\} - \left\{\sum_{j=1}^{r} L^i_j x^j + c^j\right\}\right] \\
&= \tilde{x}^i + t(\tilde{y}^i - \tilde{x}^i).
\end{aligned}$$

By Lemma 1-5, $\tilde{Z}$ is on the line through $M(P)$ and $M(Q)$. Conversely, if $\tilde{Z}$ is on the line through $M(P)$ and $M(Q)$, there is a t such that $\tilde{z}^i = \tilde{x}^i + t(\tilde{y}^i - \tilde{x}^i)$ for $i = 1, \ldots, s$, and by the above computation $\tilde{Z}$ is the image of the Z that is defined by (L). This establishes statement (K).

Because of (L), mappings M defined by (J) are called **linear** mappings. As a very special case, when $s = r$ and L is the identity matrix, equation (J) reduces to

$$\tilde{x}^i = x^i + c^i \qquad (i = 1, \ldots, r),$$

and this mapping is a translation.

Up to this point we have carefully avoided any use of the word "coordinate" in this section. The reason is that the ideas of length, betweenness, etc., are intrinsic ideas of the spaces themselves and do not need any introduction of coordinate systems to define and discuss them. Nevertheless, coordinates can be convenient devices, so we now introduce them into the plane, three-space, and R^r.

Suppose that S is a set of points in the plane, or in three-space, or in some space R^s, or, for that matter, in any space in which neighborhoods are defined for each point. Suppose also that there is a set G in a space R^r that is mapped one-to-one onto S by a function ϕ that is continuous and has a continuous inverse. Then ϕ is called a **coordinate system** in S, and each r-tuple x in G is the **set of coordinates** of the point $\phi(x)$ in S. For example, let S be all the surface of the earth except the meridian of longitude 180°, and let G be the rectangle $(-90, +90) \times (-180, +180)$. To each ordered pair (A, B) in G there corresponds a point of S that has latitude A degrees and longitude (from Greenwich) B degrees. The correspondence between (A, B) and the point on earth is continuous, and so is its inverse. The coordinate systems of greatest use to us in this chapter are those in which both S and G are all of a space R^r. To avoid the confusion that could be caused by having both points of S and points of G represented by rows of r numbers, we adopt a typographical distinction. The points of R^r, which are the principal objects of study, consist of ordered r-tuples, and these we have been writing as horizontal rows of numbers with the first element at the left, as is natural for users of any Western alphabet. But for the points of G, which are of interest only because they specify points of R^r, we use the classical Chinese custom; the ordered r-tuple that is a point of G will be written as a column, with the first element at the top. Thus New Orleans, located at latitude 30° N and longitude 90° W, would have coordinates $\binom{30}{-90}$. However, although this convention may help to avoid confusion, columns of numbers are space-consuming and expensive to set in type. So, we use a device familiar from elementary matrix theory. An r-row, or ordered row of r numbers, is the same as a $1 \times r$ matrix, which has a single row and r columns. To transpose any matrix, we change its rows into columns and its columns into rows, so the transpose of the $1 \times r$ matrix $(x^1, \ldots, x^r)$ is the $r \times 1$ matrix whose one column consists of the numbers $x^1, \ldots, x^r$ from the top down,

$$\text{(M)} \qquad (x^1, x^2, \ldots, x^r)^{\mathrm{T}} = \begin{pmatrix} x^1 \\ x^2 \\ \vdots \\ x^r \end{pmatrix}.$$

There is no difference in meaning between the two members of (M). In writing on scratch paper, the reader would be well advised to use the right member of (M). But here, for reasons of economy, we shall use the left member.

We have given the name *r-row* to the $1 \times r$ matrix $(x^1, \ldots, x^r)$. Similarly, we shall give the name *r-column* to an $r \times 1$ matrix such as the right member of (M). As an aid to memory, we shall continue to use capital letters for points of R^r, the plane, or three-space. For r-columns we shall use lower-case italic letters, and we shall denote the elements in the column by placing superscripts $1, \ldots, r$ on the name of the r-column. Thus, a point P of R^r may have coordinate r-column x, and then x is the r-column in the right member of (M).

The next lemma is quite trivially easy, but it is still useful.

LEMMA 1-8 *Let $P_0, P_1, P_2, \ldots$ be a sequence of points of R^r or of the plane or of three-space, P_n having coordinate r-column $x_n = (x_n^1, \ldots, x_n^r)^{\mathrm{T}}$. Then P_n converges to P_0 as n increases if and only if, for each i in $\{1, \ldots, r\}$, x_n^i converges to x_0^i. A similar statement holds for all other kinds of convergence, such as gauge-limits of functions of allotted partitions that have values in R^r.*

As in the definition of coordinates, we use the name G for the space R^r when R^r is used as the space of coordinate r-tuples. Let B be either an open interval in R^r that contains P_0, or a ball with center P_0. Let ϕ be the mapping of G into R^r. By definition of coordinates, ϕ and its inverse are continuous. So (using Lemma 1-5 if B is a ball), there is an open interval $A = A^1 \times A^2 \times \cdots \times A^r$ in G that contains x_0 and has $\phi(x)$ in B for all x in A. If x_n^i converges to x_0^i for each i, there is an n_0 such that for all n greater than n_0, x_n^i is in the neighborhood A^i of x_0^i for all n greater than n_0. Then for such n, P_n is in B, and so P_n converges to P_0. Conversely, suppose that P_n converges to P_0. Let i be one of the numbers $1, \ldots, r$, and let A^i be a neighborhood of x_0^i in R. Let A be $R \times \cdots \times R \times A^i \times R \times \cdots \times R$, with A^i in ith place. This is a neighborhood of x_0 in G, and ϕ^{-1} is continuous, so there is a neighborhood B of P_0 in R^r such that $\phi^{-1}(B)$ is contained in A. For all sufficiently large n, P_n is in B, so $\phi^{-1}(P_n)$, which is the r-tuple $(x_n^1, \ldots, x_n^r)$, is in A, and x_n^i is in A^i. This completes the proof.

The most natural of coordinate systems in R^r is that in which the coordinates of each point $(x^1, \ldots, x^r)$ of R^r are the same numbers arranged in a column, as in the right member of (M). This will be called the *natural* or *original* coordinate system in R^r. If S is the plane and in it we choose and fix two perpendicular axes, to each point P' of the plane there will correspond a point (x^1, x^2) of R^2, as in the first paragraph of this section. The two-column $(x^1, x^2)^{\mathrm{T}}$ will be called the original coordinate system in the plane. (It is "original" for us because this mapping of plane onto R^2 is our entry into the geometry of the plane, and it is "original" in the sense that it is the one that Descartes first thought of. It is not particularly "natural.") Likewise, in three-space we choose and fix three mutually perpendicular axes. Then each point P' of space corresponds to a triple

(x^1, x^2, x^3), and the three-column $(x^1, x^2, x^3)^T$ will be called the original coordinates of P'.

Among the simplest coordinate systems are those in which the mapping ϕ of the coordinate-set $(x^1, \ldots, x^r)^T$ onto the corresponding point $y = \phi(x)$ is linear, as in (J); that is, there exist numbers L^i_j, c^i $(i, j = 1, \ldots, r)$ such that if the point y of R^r has coordinates $(x^1, \ldots, x^r)^T$, then

$$\text{(N)} \qquad y^i = \sum_{j=1}^{r} L^i_j x^j + c^i \qquad (i = 1, \ldots, r).$$

In order that this be a coordinate system, the matrix L must have an inverse, so that just one r-column x corresponds to each point y. Such coordinate systems are called **linear**. An analog of Lemma 1-6 holds for them.

LEMMA 1-9 *Let P and Q be distinct points of R^r, and let them have the respective coordinates $(x^1, \ldots, x^r)^T$, $(y^1, \ldots, y^r)^T$ in a linear coordinate system. Then*

(i) *the line-segment with ends P and Q consists of all points Z whose coordinates $(z^1, \ldots, z^r)^T$ satisfy*

$$\text{(O)} \qquad z^i = x^i + t(y^i - x^i) \qquad (i = 1, \ldots, r)$$

with some t in $[0, 1]$;

(ii) *the line through P and Q consists of all points Z whose coordinates $(z^1, \ldots, z^r)^T$ satisfy* (O) *with some real t.*

By (K), Z is on the line through P and Q if and only if the image $(z^1, \ldots, z^r)$ is on the line through $(x^1, \ldots, x^r)$ and $(y^1, \ldots, y^r)$, and by Lemma 1-6 this is true if and only if (O) holds. The statement about line-segments is proved by repeating the proof of (K) with the restriction $0 \leqq t \leqq 1$.

It is obvious that if we map the space of coordinates onto itself by a translation, the space R^r of points corresponding to the coordinates is also mapped onto itself by a translation if the coordinate system is linear.

EXERCISE 1-1 Show that in R^2, if P, Q, and Z have the respective coordinate two-columns x, y, and z, Z is in the line through P and Q if and only if

$$\det \begin{bmatrix} x^1 & y^1 & z^1 \\ x^2 & y^2 & z^2 \\ 1 & 1 & 1 \end{bmatrix} = 0.$$

EXERCISE 1-2 The triangle with vertices P_0, P_1, and P_2 in R^r can be defined to be the set of points P such that P is between P_0 and some point Q that is itself between P_1 and P_2. Show that if P_0, P_1, and P_2 have the respective coordinate r-columns x_0, x_1, and x_2, a point with coordinate r-column x is in the triangle with vertices P_0, P_1, and P_2 if and only if there exist nonnegative numbers p_0, p_1, and p_2 such that $p_0 + p_1 + p_2 = 1$ and

$$x = p_0 x_0 + p_1 x_1 + p_2 x_2.$$

EXERCISE 1-3 Let $P_1, \ldots, P_5$ be points in R^r. Show that if P_2 and P_4 are both between P_1 and P_5, and P_3 is between P_2 and P_4, then P_3 is between P_1 and P_5.

2. Vectors

In physics one encounters quantities whose description involves both a magnitude and a direction. These are called *vector quantities*. It is customary and convenient to picture them as directed line segments in some space R^r – a directed line segment being a line segment for which one end-point has been designated the *initial* or *beginning* point and the other the *terminal* or *end* point. This representation does not merely give us a picture for visualizing the quantity; it also allows us to perform operations, namely, addition and multiplication by real numbers, that have physical significance.

If a directed line segment represents a vector quantity, any other directed line segment with the same length and direction will equally represent the same quantity, since all we are interested in is length and direction, not position in the space. However, we here meet a small difficulty: we have not defined "direction." But it is in accord with our primitive notions of direction to say that if a directed line segment $\overrightarrow{AB}$ can be translated so that its beginning point A is mapped onto P and its terminal point B is mapped on Q, then the directed line segments $\overrightarrow{AB}$ and $\overrightarrow{PQ}$ have the same length (this we have proved) and the same direction. So if $\overrightarrow{AB}$ and $\overrightarrow{CD}$ are directed line segments in the plane, or in three-space, or in some R^r, they are **equivalent** (or have the same length and direction) provided that there exists a translation of the space under which A is mapped on C and B is mapped on D. To justify the use of the word "equivalent," we have to prove that each directed line-segment is equivalent to itself; that if directed line-segment $\overrightarrow{AB}$ is equivalent to $\overrightarrow{CD}$, then $\overrightarrow{CD}$ is equivalent to $\overrightarrow{AB}$; and that if $\overrightarrow{AB}$ is equivalent to $\overrightarrow{CD}$ and $\overrightarrow{CD}$ is equivalent to $\overrightarrow{EF}$, then $\overrightarrow{AB}$ is equivalent to $\overrightarrow{EF}$. But this is trivially easy, and we omit the proof. Since we have no reason for favoring one directed line-segment over any other equivalent to it, for each directed line-segment $\overrightarrow{AB}$ we lump together all the directed line-segments equivalent to $\overrightarrow{AB}$, and we call the resulting class a **vector**. If all the line-segments in the class belong to the plane, we should call the class a "vector whose members are in the plane," but we shall consistently misuse the English language and call it "a vector in the plane." Likewise, if the members are directed line-segments in three-space, the vector is miscalled a "vector in three-space," and if they all belong to a space R^r the vector is miscalled a "vector in R^r."

In this chapter we shall use lower-case boldface letters to denote vectors in the plane, in three-space, or in R^r, and we shall denote the class of all vectors in R^r by V^r. If $\mathbf{v}$ is a vector that belongs to V^r, it is a class of directed line-segments lying in R^r, and any one of those directed line-segments is said to **represent** $\mathbf{v}$.

It is obvious that if $\mathbf{v}$ is any vector in V^r and A is any point of R^r, there is just one point B in R^r such that the directed line-segment $\overrightarrow{AB}$ represents $\mathbf{v}$. A similar statement holds for vectors in the plane or in three-space.

All the directed line-segments that represent a vector $\mathbf{v}$ have the same length. We define the length of $\mathbf{v}$ to be the length of any directed line-segment that represents $\mathbf{v}$, and we denote the length of $\mathbf{v}$ by $|\mathbf{v}|$.

In particular, if we have chosen one point O of the space to be the origin, there is a one-to-one correspondence between the points P of the space and the vectors $\mathbf{v}$ in V^r—the vector $\mathbf{v}$ corresponding to P being the vector represented by the directed line-segment $\overrightarrow{OP}$. This vector is called the **position-vector** of the point P.

There is a traditional way of defining the sum $\mathbf{v}_1 + \mathbf{v}_2$ of two vectors in the plane, or in three-space, or in R^r, that is designed to suit the needs of the physical applications of vectors but that also serves the needs of pure mathematics. To be specific, suppose that $\mathbf{v}_1$ and $\mathbf{v}_2$ are vectors in R^r. Let A be any point of R^r. There is just one point B in R^r such that $\overrightarrow{AB}$ represents $\mathbf{v}_1$, and there is just one point C of R^r such that $\overrightarrow{BC}$ represents $\mathbf{v}_2$. Then the directed line-segment $\overrightarrow{AC}$ represents a vector $\mathbf{v}_3$, and by definition this vector $\mathbf{v}_3$ is $\mathbf{v}_1 + \mathbf{v}_2$. When $\mathbf{v}_1$ and $\mathbf{v}_2$ do not have the same or opposite directions, this can be stated in a different way that is preferred by some people; we form a parallelogram $ABCD$ such that $\overrightarrow{AB}$ represents $\mathbf{v}_1$ and $\overrightarrow{AD}$ represents $\mathbf{v}_2$; then the diagonal $\overrightarrow{AC}$ represents $\mathbf{v}_1 + \mathbf{v}_2$.

From this definition it is not difficult to prove that addition of vectors is commutative and associative. Since it will be even easier after we introduce coordinates, we postpone the proof.

There is a vector in V^r which we call $\mathbf{0}$, that consists of all directed line-segments $\overrightarrow{AA}$ that end where they begin. By our definition of addition, for every vector $\mathbf{v}$ in V^r it is true that

$$\mathbf{0} + \mathbf{v} = \mathbf{v} + \mathbf{0} = \mathbf{v},$$

so $\mathbf{0}$ has the defining property of a zero for addition. Also, if $\mathbf{v}$ is a vector and $\overrightarrow{AB}$ represents $\mathbf{v}$, the vector $\overrightarrow{BA}$ has the property that

$$\overrightarrow{AB} + \overrightarrow{BA} = \overrightarrow{BA} + \overrightarrow{AB} = \mathbf{0},$$

so $\overrightarrow{BA}$ represents the negative of the vector $\mathbf{v}$ and will be denoted by the symbol $-\mathbf{v}$.

We can easily use the idea of betweenness to define the concepts of *same direction* and *opposite directions* for vectors $\mathbf{v}_1$ and $\mathbf{v}_2$ both in V^r, or both in the plane, or both in three-space. To be specific, let $\mathbf{v}_1$ and $\mathbf{v}_2$ be vectors in R^r. We choose an origin O in R^r; then there is a unique point A in R^r such that $\overrightarrow{OA}$ represents $\mathbf{v}_1$, and there is a unique point B such that $\overrightarrow{OB}$ represents $\mathbf{v}_2$. Then $\mathbf{v}_1$ and $\mathbf{v}_2$ have **opposite directions** if O is between A and B.

In particular, let us choose any linear coordinate system in R^r. Then O will have coordinates $(0, \ldots, 0)^T$, and A will have certain coordinates $(x^1, \ldots, x^r)^T$. As we saw in the preceding paragraph, $\overrightarrow{AO}$ will represent $-\mathbf{v}_1$. The translation that maps each point Q of R^r, with coordinates $(y^1, \ldots, y^r)^T$, onto the point with coordinates $(y^1 - x^1, \ldots, y^r - x^r)^T$ will map A onto O and O onto the point B with coordinates $(-x^1, \ldots, -x^r)^T$. Since O is between the points with coordinates $(x^1, \ldots, x^r)^T$ and $(-x^1, \ldots, -x^r)^T$, the vectors $\mathbf{v}_1$ and $-\mathbf{v}_1$ have opposite directions.

A vector $\mathbf{w}$ has the ***same direction*** as a vector $\mathbf{v}$ if it has the direction opposite to that of $-\mathbf{v}$.

If $\mathbf{v}$ is a vector belonging to V^r and b is a real number, the product $b\mathbf{v}$ is defined to be the vector in V^r that has length $|b|\,|v|$ and has the same direction as $\mathbf{v}$ if $b \geqq 0$ and the opposite direction to that of $\mathbf{v}$ if $b < 0$.

The geometric representation of vectors is fundamental, but there is a numerical representation that is usually much more convenient in computing. Let us introduce a linear coordinate system ϕ into R^r as in (J) of Section 1. To the coordinate r-column $x = (x^1, \ldots, x^r)^T$ there corresponds the point $(\tilde{x}^1, \ldots, \tilde{x}^r)$ of R^r, where

$$\text{(A)} \qquad \tilde{x}^i = \sum_{j=1}^{r} L^i_j x^j + \tilde{c}^i \qquad (i = 1, \ldots, r).$$

To the coordinate column $(0, \ldots, 0)^T$ there corresponds the point

$$\text{(B)} \qquad O = (\tilde{c}^1, \ldots, \tilde{c}^r),$$

which we call the *origin*. If $\mathbf{v}$ is a vector that belongs to V^r, there is just one point

$$P = (\tilde{v}^1, \ldots, \tilde{v}^r)$$

in R^r such that $\overrightarrow{OP}$ represents $\mathbf{v}$. This point P has a column

$$v = (v^1, \ldots, v^r)^T$$

of coordinates, and this r-column v is the numerical representation of $\mathbf{v}$ **in** or **induced by** the coordinate system ϕ in R^r.

By definition of coordinate system, equations (A) can be solved for the x^i in terms of the $\tilde{x}^i$:

$$\text{(C)} \qquad x^j = \sum_{i=1}^{r} M^j_i(\tilde{x}^i - \tilde{c}^i) \qquad (i = 1, \ldots, r).$$

Here M is the matrix that is the inverse of matrix L. Let

$$C = (\tilde{y}^1, \ldots, \tilde{y}^r), \qquad D = (\tilde{z}^1, \ldots, \tilde{z}^r)$$

be two points of R^r such that $\overrightarrow{CD}$ also represents $\mathbf{v}$. The translation T in R^r that maps O onto C is given by

$$(T\tilde{w})^i = \tilde{w}^i + (\tilde{y}^i - \tilde{c}^i) \qquad (i = 1, \ldots, r).$$

Since CD is a translate of OP, T must also map P onto D, so that

(D) $$\tilde{z}^i = \tilde{v}^i + (\tilde{y}^i - \tilde{c}^i).$$

By (C) and (D), the coordinate r-columns v, y, and z of P, C, and D respectively satisfy for $j = 1, \ldots, r$

$$z^j - y^j = \sum_{i=1}^{r} [M_i^j(\tilde{z}^i - \tilde{c}^i) - M_i^j(\tilde{y}^i - \tilde{c}^i)]$$

$$= \sum_{i=1}^{r} M_i^j(\tilde{z}^i - \tilde{y}^i) = \sum_{i=1}^{r} M_i^j(\tilde{v}^i - \tilde{c}^i) = v^i.$$

We have thus proved the following statement.

(E) Let $\mathbf{v}$ belong to V^r, and let ϕ be a linear coordinate system on R^r. If $\overrightarrow{CD}$ is any representation of $\mathbf{v}$, and y and z are the coordinate r-columns of C and D, respectively, the (numerical) representation of $\mathbf{v}$ in, or induced by, the ϕ-system is the r-column $z - y$.

In particular, $\overrightarrow{DC}$ represents $-\mathbf{v}$, so the r-column that represents $-\mathbf{v}$ is $y - z$, which is $-v$.

The space of r-columns is a space of ordered r-tuples, so in it we can continue to use the same neighborhoods as always before; a neighborhood of an r-column is an open interval that contains it. In V^r we have not yet defined neighborhoods, but there is a natural definition. There is a one-to-one correspondence between the points P of R^r and the vectors $\mathbf{v}$ that are their position-vectors. Let $\mathbf{v}_0$ belong to V^r, and let it be the position-vector of a point P_0 in R^r. A set G in V^r is a neighborhood of $\mathbf{v}_0$ if and only if G consists of the position-vectors of the points in a neighborhood G' of P_0 in R^r. Now, the correspondence between the vectors in V^r and the r-columns of numbers is one-to-one and is obviously continuous, so this mapping of r-columns onto the vectors that they represent is a coordinate system in the space V^r. Moreover, we have the following test.

LEMMA 2-1 *Let $\mathbf{v}_0, \mathbf{v}_1, \mathbf{v}_2, \ldots$ be vectors belonging to V^r, and in a linear coordinate system let $\mathbf{v}_n$ be represented by $(v_n^1, \ldots, v_n^r)^{\mathrm{T}}$. Then $\mathbf{v}_n$ converges to $\mathbf{v}_0$ if and only if, for each i in $\{1, \ldots, r\}$, v_n^i converges to v_0^i. A similar statement holds for other kinds of convergence.*

For $n = 0, 1, 2, 3, \ldots$, let $\mathbf{v}_n$ be the position-vector of a point P_n. By the above definition of neighborhood of $\mathbf{v}_0$, $\mathbf{v}_n$ converges to $\mathbf{v}_0$ if and only if P_n converges to P_0. But the r-column $(v_n^1, \ldots, v_n^r)^{\mathrm{T}}$ is the coordinate representation of P_n, so by Lemma 1-8, P_n converges to P_0 if and only if for $i = 1, \ldots, r$ it is true that v_n^i converges to v_0^i. This completes the proof.

The importance of these r-columns is that they do not merely represent the vectors in V^r but also give a means of calculating.

(F) If **v** and **w** are vectors belonging to V^r, and in a linear coordinate system they have the respective representations

$$v = (v^1, \ldots, v^r)^{\mathrm{T}}, \qquad w = (w^1, \ldots, w^r)^{\mathrm{T}},$$

the sum $\mathbf{v} + \mathbf{w}$ has representation

$$v + w = (v^1 + w^1, \ldots, v^r + w^r)^{\mathrm{T}}.$$

Let O be the point with coordinates $(0, \ldots, 0)^{\mathrm{T}}$, and let A and B be points such that $\overrightarrow{OA}$ and $\overrightarrow{OB}$ represent **v** and **w**, respectively. There is a point C such that $\overrightarrow{AC}$ represents **w**. By definition, $\overrightarrow{OC}$ represents $\mathbf{v} + \mathbf{w}$. Let $(y^1, \ldots, y^r)$ be the coordinates of C. Since $\overrightarrow{AC}$ represents **w**, by (E) the r-column w that represents $\overrightarrow{AC}$ must satisfy

$$w^i = y^i - v^i \qquad (i = 1, \ldots, r).$$

It follows at once that $y = v + w$, which is the conclusion of (F).

The addition of r-columns of real numbers is commutative and associative, so by (F) the addition of vectors in V^r is also commutative and associative.

(G) If **v** belongs to V^r and b is a real number, and in a linear coordinate system ϕ in R^r the numerical representation of **v** is the r-column $v = (v^1, \ldots, v^r)^{\mathrm{T}}$, then the representation of $b\mathbf{v}$ is $bv = (bv^1, \ldots, bv^r)^{\mathrm{T}}$.

Again let the coordinate system be given by (A). Let O and P be the points of R^r with respective coordinate r-columns $(0, \ldots, 0)^{\mathrm{T}}$ and v; then the directed line-segment $\overrightarrow{OP}$ represents the vector **v**. Let Q be the point with coordinates $(bv^1, \ldots, bv^r)^{\mathrm{T}}$. If P and Q are, respectively, the r-rows $(\tilde{v}^1, \ldots, \tilde{v}^r)$ and $(\tilde{w}^1, \ldots, \tilde{w}^r)$, we have by (A),

$$\tilde{v}^i = \sum_{j=1}^{r} L^i_j v^j + \tilde{c}^i, \qquad \tilde{w}^i = \sum_{i=1}^{r} L^i_j bv^j + \tilde{c}^i.$$

Since O is the point $(\tilde{c}^1, \ldots, \tilde{c}^r)$ of R^r,

$$\begin{aligned}
|\mathbf{v}| &= \operatorname{dist}(O, P) \\
&= \left[\left(\sum_{j=1}^{r} L^1_j v^j + \tilde{c}^1 - \tilde{c}^1\right)^2 + \cdots + \left(\sum_{j=1}^{r} L^r_j v^j + \tilde{c}^r - \tilde{c}^r\right)^2\right]^{1/2}, \\
\operatorname{dist}(O, Q) &= \left[\left(\sum_{i=1}^{r} L^1_j bv^j + \tilde{c}^1 - \tilde{c}^1\right)^2 + \cdots + \left(\sum_{j=1}^{r} L^r_j bv^j + \tilde{c}^r - \tilde{c}^r\right)^2\right]^{1/2} \\
&= |b|\,|\mathbf{v}|.
\end{aligned}$$

If $b < 0$, we define $t = 1/(1 + |b|)$. This is in the interval $[0, 1]$, and for $i = 1, \ldots, r$

$$\left(\sum_{j=1}^{r} L^i_j v^j + \tilde{c}^i\right) + t\left[\left(\sum_{j=1}^{r} L^i_j bv^j + \tilde{c}^i\right) - \left(\sum_{j=1}^{r} L^i_j v^j + \tilde{c}^i\right)\right] = \tilde{c}^i.$$

Therefore the point O is between P and Q, and $\overrightarrow{OQ}$ has the direction opposite to that of $\overrightarrow{OP}$. By definition, $\overrightarrow{OQ}$ represents $b\mathbf{v}$, so $b\mathbf{v}$ is numerically represented by the r-column $(bv^1, \ldots, bv^r)^{\mathrm{T}}$ of coordinates of Q.

If b is positive, we notice that $b\mathbf{v} = (-b)(-\mathbf{v})$. Since $-b$ is negative, we can apply the preceding paragraph and conclude that the r-column $((-b)(-v^1), \ldots, (-b)(-v^r))^{\mathrm{T}}$ represents a vector with length $|-b||-v|$ and direction opposite to $-\mathbf{v}$. That is, bv represents a vector with length $|b|\,|\mathbf{v}|$ and direction the same as that of $\mathbf{v}$, which by definition is $b\mathbf{v}$. The proof of (G) is complete.

There are many other collections of objects in which it is useful to define operations of addition and multiplication by real numbers that have the fundamental properties of those operations in V^r. Such spaces are called (real) vector spaces. We define them thus.

DEFINITION 2-2 *Let $(V, +, \cdot)$ be a triple such that for each pair of members v_1, v_2 of V, $v_1 + v_2$ is a member of V, and for each real number b and each member v of V, $b \cdot v$ is a member of V, and the following conditions are satisfied:*

(i) *For all v_1, v_2, and v_3 in V, $v_1 + v_2 = v_2 + v_1$ and $(v_1 + v_2) + v_3 = v_1 + (v_2 + v_3)$;*

(ii) *there is a member 0 of V such that $v + 0 = v$ for all v in V, and for each v in V there is a member $-v$ of V such that $v + (-v) = 0$;*

(iii) *for all v in V, $0 \cdot v = 0$ and $1 \cdot v = v$;*

(iv) *for all real b_1 and b_2 and all v in V*

$$b_1(b_2 v) = (b_1 b_2)v \text{ and } (b_1 + b_2)v = b_1 v + b_2 v;$$

(v) *for all real b and all v_1 and v_2 in V,*

$$b(v_1 + v_2) = bv_1 + bv_2 .$$

Then the triple is called a (**real**) **vector space**.

Usually, when the operations $+$ and $\cdot$ are clearly understood, we say that "V is a (real) vector space."

A real-valued function $v \mapsto |v|$ on a vector space is called a **norm**, or **length**, if the following definition is satisfied.

DEFINITION 2-3 *A function $v \mapsto |v|$ defined on a (real) vector space is a* **norm** *(or* **length***) if*

(i) *for each v in V, $|v| \geqq 0$, and $|v| = 0$ if and only if $v = 0$;*

(ii) *for all v_1 and v_2 in V, $|v_1 + v_2| \leqq |v_1| + |v_2|$;*

(iii) *for all v in v and all real b,*

$$|bv| = |b|\,|v|.$$

A vector space on which a norm is defined is called a **normed vector space**.

For all positive integers r, the space V^r, with addition and multiplication by real numbers defined as in this section and with the length of v as $|v|$, satisfies all the requirements in Definitions 2-2 and 2-3; most of them have already been proved, and the others are trivial consequences of (F) and (G). So V^r (with these operations) is a normed vector space. So are the Euclidean plane and three-space, since these correspond to R^2 and R^3 by the original coordinate systems.

EXERCISE 2-1 Prove that if a vector v belonging to V^r has representation $(v^1, \ldots, v^r)^T$ in a rectangular coordinate system,

$$v \leqq r^{1/2} \sum_{i=1}^{r} |v^i|.$$

Suggestion: Let $w^i = 1$ if $v^i \geqq 0$ and $w^i = -1$ if $v^i < 0$. Apply Lemma 1-3 to the r-tuples v and w.

EXERCISE 2-2 Prove that for each fixed n the real polynomials on R of degree at most n form a real vector space.

3. Inner Products and Length-Preserving Maps

In plane geometry it is proved that if we have two triangles with the sides of the one equal to the sides of the other, the angles in the one are equal to the corresponding angles in the other. We have not defined the concept of *angle* in higher-dimensional spaces, but we shall base its definition on the requirement that if ABC is a triangle in a space R^r and $A'B'C'$ is a triangle in a space R^s, and the lengths of the sides AB, BC, CA are equal, respectively, to the lengths of the sides $A'B'$, $B'C'$, $C'A'$, then the angles A, B, C are equal to the corresponding angles A', B', C'. Suppose, then, that ABC is a triangle in R^r. For each angle θ in the interval $[0, \pi]$ we construct the triangle in the plane whose vertices O, P, Q have the respective original coordinates

$$(0,0)^T, \qquad (\text{dist}(A,B), 0)^T, \qquad (\text{dist}(A,C)\cos\theta, \text{dist}(A,C)\sin\theta)^T.$$

Then

$$\text{(A)} \qquad \begin{aligned} \text{dist}(O,P) &= \text{dist}(A,B), \\ \text{dist}(O,Q) &= [\text{dist}(A,C)^2\cos^2\theta + \text{dist}(A,C)^2\sin^2\theta]^{1/2} \\ &= \text{dist}(A,C), \\ \text{dist}(P,Q) &= [(\text{dist}(A,B) - \text{dist}(A,C)\cos\theta)^2 \\ &\qquad + \text{dist}(A,C)^2\sin^2\theta]^{1/2} \\ &= [(\text{dist}(A,B))^2 + (\text{dist}(A,C))^2 \\ &\qquad - 2(\text{dist}(A,B))(\text{dist}(A,C))\cos\theta]^{1/2}. \end{aligned}$$

When $\theta = 0$, the right member of the last equation in (A) is

$$|\text{dist}(A, B) - \text{dist}(A, C)|,$$

which by Corollary 1-4 is at most equal to dist(B, C). When $\theta = \pi$ radians, the right member of the last equation is dist(A, B) + dist(A, C), which by Lemma 1-2 is at least dist(B, C). Since the expression in the right member of that equation is a continuous function of θ, there is a value of θ in $[0, \pi]$ such that

(B) $$(\text{dist}(B, C))^2 = (\text{dist}(A, B))^2 + (\text{dist}(A, C))^2 - 2(\text{dist}(A, B))(\text{dist}(A, C)) \cos \theta.$$

If neither dist(A, B) nor dist(A, C) is 0, there is just one such θ, and this θ we define to be the angle between AB and AC. If either dist(A, B) or dist(A, C) is 0, equation (B) holds for every value of θ, and we consider the angle between AB and AC to be any number in the interval $[0, \pi]$.

From (B) we at once obtain the corresponding formula for vectors. Let $\mathbf{v}_1$ and $\mathbf{v}_2$ be vectors in R^r. Let A be any point of R^r; then there are points B, C such that AB represents $\mathbf{v}_1$ and AC represents $\mathbf{v}_2$. Then (B) takes the form

(C) $$|\mathbf{v}_2 - \mathbf{v}_1|^2 = |\mathbf{v}_1|^2 + |\mathbf{v}_2|^2 - 2|\mathbf{v}_1||\mathbf{v}_2| \cos \theta.$$

The quantity multiplied by -2 in the right member of (C) is of considerable usefulness, so we assign it a name and a symbol. It is called the **inner product** of v_1 and v_2, and it is denoted by $\langle v_1, v_2 \rangle$.

DEFINITION 3-1 *If* $\mathbf{v}_1$ *and* $\mathbf{v}_2$ *are vectors that belong to* V^r, *the* ***inner product*** *of* $\mathbf{v}_1$ *and* $\mathbf{v}_2$ *is the number*

(D) $$\langle \mathbf{v}_1, \mathbf{v}_2 \rangle = |\mathbf{v}_1||\mathbf{v}_2| \cos \theta,$$

where θ *is the angle between* $\mathbf{v}_1$ *and* $\mathbf{v}_2$.

If neither $\mathbf{v}_1$ nor $\mathbf{v}_2$ is 0, from (D) we obtain

$$\cos \theta = \langle \mathbf{v}_1, \mathbf{v}_2 \rangle / |\mathbf{v}_1||\mathbf{v}_2|.$$

As a special case, if $\mathbf{v}_1 = \mathbf{v}_2$, the angle θ is 0 and $\cos \theta = 1$, so

$$\langle \mathbf{v}_1, \mathbf{v}_1 \rangle^{1/2} = |\mathbf{v}_1|.$$

The simplest and most useful properties of the inner product are gathered in the next lemma.

LEMMA 3-2 *Let* $\mathbf{u}$, $\mathbf{v}$, *and* $\mathbf{w}$ *be vectors that belong to* V^r, *and let* c *be a real number. Then*

(i) $\langle \mathbf{u}, \mathbf{v} \rangle = \langle \mathbf{v}, \mathbf{u} \rangle$,
(ii) $\langle c\mathbf{u}, \mathbf{v} \rangle = c\langle \mathbf{u}, \mathbf{v} \rangle$,
(iii) $\langle \mathbf{u} + \mathbf{v}, \mathbf{w} \rangle = \langle \mathbf{u}, \mathbf{w} \rangle + \langle \mathbf{v}, \mathbf{w} \rangle$,
(iv) $|\langle \mathbf{u}, \mathbf{v} \rangle| \leqq |\mathbf{u}||\mathbf{v}|$.

Let A be the point $(0, \ldots, 0)$ of R^r. There are points

$$B = (u^1, \ldots, u^r), \qquad C = (v^1, \ldots, v^r)$$

of R^r such that $\overrightarrow{AB}$ represents $\mathbf{u}$ and $\overrightarrow{AC}$ represents $\mathbf{v}$. Then by (D) and (C),

$$\text{(E)} \quad \langle \mathbf{u}, \mathbf{v} \rangle = \tfrac{1}{2}[|\mathbf{u}|^2 + |\mathbf{v}|^2 - |\mathbf{v} - \mathbf{u}|^2]$$

$$= \tfrac{1}{2}[(u^1)^2 + \cdots + (u^r)^2 + (v^1)^2 + \cdots + (v^r)^2 - (v^1 - u^1)^2 - \cdots - (v^r - u^r)^2]$$

$$= u^1 v^1 + \cdots + u^r v^r.$$

Evidently, conclusions (i), (ii), and (iii) follow at once from this. Inequality (iv) follows from (D), since $|\cos \theta| \leqq 1$.

If $P = (x^1, \ldots, x^r)$ and $Q = (y^1, \ldots, y^r)$ are points of R^r, we have defined their "original" coordinates to be the r-columns $(x^0, \ldots, x^r)^T$, $(y^1, \ldots, y^r)^T$, and the distance from P to Q is given by Definition 1-1 to be

$$\text{dist}(P, Q) = [(x^1 - y^1)^2 + \cdots + (x^r - y^r)^2]^{1/2}.$$

However, there are other coordinate systems in which the distance from P to Q is given by this same simple formula, and they are all convenient to use in computation. We call such systems *rectangular* coordinate systems.

DEFINITION 3-3 *A linear coordinate system in a space F^r is* **rectangular** *if whenever P and Q are points of R^r and their coordinate r-columns are x and y, respectively,*

$$\text{dist}(P, Q) = [(x^1 - y^1)^2 + \cdots + (x^r - y^r)^2]^{1/2}.$$

This formula for the distance was all we needed in proving (E), so the same proof establishes the next lemma.

LEMMA 3-4 *Let* $\mathbf{u}$ *and* $\mathbf{v}$ *be vectors in V^r, and let u, v be the r-columns that represent* $\mathbf{u}$, $\mathbf{v}$, *respectively, in a rectangular coordinate system. Then*

$$\langle \mathbf{u}, \mathbf{v} \rangle = u^1 v^1 + \cdots + u^r v^r.$$

COROLLARY 3-5 *If* $\mathbf{u}$ *and* $\mathbf{v}$ *have representations u, v in a rectangular coordinate system, they are perpendicular if and only if $u^1 v^1 + \cdots + u^r v^r = 0$.*

For if neither $\mathbf{u}$ nor $\mathbf{v}$ is $\mathbf{0}$, by (D) and Lemma 3-4 the equation holds if and only if $\cos \theta = 0$; and if, say, $\mathbf{u} = \mathbf{0}$, the equation holds and by convention, $\mathbf{u}$ is perpendicular to all vectors.

It is customary to represent the elements of the identity $r \times r$ matrix by $\delta^i{}_j$, so that

(F)
$$\delta^i{}_j = 1 \quad \text{if} \quad i = j,$$
$$\delta^i{}_j = 0 \quad \text{if} \quad i \neq j \qquad (i, j = 1, \ldots, r).$$

The ith column of this matrix is δ_i, where

(G)
$$\delta_i = (\delta^1{}_i, \delta^2{}_i, \ldots, \delta^*{}_i)^{\mathrm{T}} = (0, \ldots, 0, 1, 0, \ldots, 0)^{\mathrm{T}},$$

the "1" in the last expression being in ith place. If the coordinate system is rectangular, this column δ_i represents a vector, called $\mathbf{e}_i$, that has length 1 by Definition 3-3. It is called the "ith coordinate unit vector" and is the vector of length 1 that points in the direction of the positive ith axis. For $i, j = 1, \ldots, r$ the inner product of $\mathbf{e}_i$ and $\mathbf{e}_j$ is given by Lemma 3-4 to be

(H)
$$\langle \mathbf{e}_i, \mathbf{e}_j \rangle = \sum_{k=1}^{r} \delta^k{}_i \delta^k{}_j = \delta^i{}_j.$$

In particular, if $i \neq j$, the vectors $\mathbf{e}_i$ and $\mathbf{e}_j$ are perpendicular, so the axes in a rectangular system are perpendicular to each other, which explains the name.

We can use the $\mathbf{e}_i$ or any other set of mutually perpendicular unit vectors to obtain a convenient expression for inner products, as follows.

LEMMA 3-6 *Let $\mathbf{u}_1, \ldots, \mathbf{u}_r$ be mutually perpendicular unit vectors belonging to V^r. Then for every pair of vectors $\mathbf{u}$, $\mathbf{v}$ belonging to V^r,*

(i) $\mathbf{u} = \langle \mathbf{u}, \mathbf{u}_1 \rangle \mathbf{u}_1 + \cdots + \langle \mathbf{u}, \mathbf{u}_r \rangle \mathbf{u}_r$,
(ii) $\langle \mathbf{u}, \mathbf{v} \rangle = \langle \mathbf{u}, \mathbf{u}_1 \rangle \langle \mathbf{v}, \mathbf{u}_1 \rangle + \cdots + \langle \mathbf{u}, \mathbf{v}_r \rangle \langle \mathbf{v}, \mathbf{u}_r \rangle$.

In the original coordinate system, the vectors $\mathbf{u}, \mathbf{u}_1, \ldots, \mathbf{u}_r$ are represented by respective r-columns $u, u_1, \ldots, u_r$. These $r + 1$ r-columns cannot be linearly independent, so there are numbers $a_0, a_1, \ldots, a_r$ not all 0 such that

$$a_0 u + a_1 u_1 + \cdots + a_r u_r = 0.$$

The left member represents the vector $a_0 \mathbf{u} + a_1 \mathbf{u}_1 + \cdots + a_r \mathbf{u}_r$, so

(I)
$$a_0 \mathbf{u} + a_1 \mathbf{u}_1 + \cdots + a_r \mathbf{u}_r = 0.$$

For each i in $\{1, \ldots, r\}$ we take the inner product of both members of (I) with $\mathbf{u}_i$. Because the $\mathbf{u}_i$ are mutually perpendicular unit vectors,

(J)
$$\langle \mathbf{u}_i, \mathbf{u}_j \rangle = \delta^i{}_j,$$

so from (I) we obtain

$$a_0 \langle \mathbf{u}, \mathbf{u}_i \rangle + a_i = 0.$$

This implies, first, that $a_0 \neq 0$; otherwise all the a_i would be 0, contrary to the way they were chosen. Second, it implies

$$a_i/a_0 = -\langle \mathbf{u}, \mathbf{u}_i \rangle.$$

If we substitute this in (I), we obtain conclusion (i). If we apply (i) to $\mathbf{u}$ and $\mathbf{v}$, we obtain with the help of (J)

$$\begin{aligned} \langle \mathbf{u}, \mathbf{v} \rangle &= \langle \langle \mathbf{u}, \mathbf{u}_1 \rangle \mathbf{u}_1 + \cdots + \langle \mathbf{u}, \mathbf{u}_r \rangle \mathbf{u}_r, \langle \mathbf{v}, \mathbf{u}_1 \rangle \mathbf{u}_1 + \cdots + \langle \mathbf{v}, \mathbf{u}_r \rangle \mathbf{u}_r \rangle \\ &= \langle \mathbf{u}, \mathbf{u}_1 \rangle \langle \mathbf{v}, \mathbf{v}_1 \rangle + \cdots + \langle \mathbf{u}, \mathbf{u}_r \rangle \langle \mathbf{v}, \mathbf{u}_r \rangle, \end{aligned}$$

establishing (ii).

From Lemma 3-6 we may deduce, correctly, that systems of mutually perpendicular unit vectors can be useful, and it is desirable to have a device for constructing such systems. Such a device is the "Gram–Schmidt process," described in the next lemma.

Lemma 3-7 *Let $\mathbf{v}_1, \ldots, \mathbf{v}_k$ ($k \leqq r$) be vectors belonging to V^r, with $\mathbf{v}_1 \neq \mathbf{0}$. Then there exist mutually perpendicular unit vectors $\mathbf{u}_1, \ldots, \mathbf{u}_r$ in V^r such that for $j = 1, \ldots, k$, $\mathbf{v}_j$ is a linear combination of $\mathbf{u}_1, \ldots, \mathbf{u}_j$.*

Suppose first that $\mathbf{v}_1, \ldots, \mathbf{v}_k$ are linearly independent. If $k < r$, we can adjoin vectors $\mathbf{v}_{k+1}, \ldots, \mathbf{v}_r$ such that the r vectors $\mathbf{v}_1, \ldots, \mathbf{v}_r$ are all linearly independent. For $\mathbf{u}_1$ we choose $\mathbf{v}_1/|\mathbf{v}_1|$; then $\mathbf{u}_1$ is a unit vector, and $\mathbf{v}_1$ is a linear combination of $\mathbf{u}_1$, namely, $\mathbf{v}_1 = |\mathbf{v}_1|\mathbf{u}_1$. We proceed by induction. Suppose that for some number j in $\{2, \ldots, r\}$ there exist mutually perpendicular unit vectors $\mathbf{u}_1, \ldots, \mathbf{u}_{j-1}$ such that

$$\begin{aligned} \mathbf{v}_1 &= c_{1,1}\mathbf{u}_1, \\ \mathbf{v}_2 &= c_{2,1}\mathbf{u}_1 + c_{2,2}\mathbf{u}_2, \\ &\vdots \\ \mathbf{v}_{j-1} &= c_{j-1,1}\mathbf{u}_1 + \cdots + c_{j-1,j-1}\mathbf{u}_{j-1}. \end{aligned}$$

If any $c_{i,i}$ were 0, the i vectors $\mathbf{v}_1, \ldots, \mathbf{v}_i$ would all be linear combinations of the $i - 1$ vectors $\mathbf{u}_1, \ldots, \mathbf{u}_{i-1}$ and could not be linearly independent. So all $c_{i,i}$ are different from 0, and the preceding equations can be solved for each $\mathbf{u}_i$ ($i = 1, \ldots, j-1$) as a linear combination of $\mathbf{v}_1, \ldots, \mathbf{v}_i$. We define

$$\text{(K)} \qquad \mathbf{w}_j = \mathbf{v}_j - \langle \mathbf{v}_j, \mathbf{u}_1 \rangle \mathbf{u}_1 - \cdots - \langle \mathbf{v}_j, \mathbf{u}_{j-1} \rangle \mathbf{u}_{j-1}.$$

If $\mathbf{w}_j$ were $\mathbf{0}$, $\mathbf{v}_j$ would be a linear combination of $\mathbf{v}_1, \ldots, \mathbf{v}_{j-1}$, contrary to the hypothesis that the $\mathbf{v}_i$ are linearly independent. So $\mathbf{w}_j \neq \mathbf{0}$. We define

$$\mathbf{u}_j = \mathbf{w}_j/|w_j|.$$

Then $\mathbf{u}_j$ is a unit vector, and (K) expresses $\mathbf{v}_j$ as a linear combination of

$\mathbf{u}_1, \ldots, \mathbf{u}_j$. Also, for $i = 1, \ldots, j-1$,

$$\begin{aligned}\langle \mathbf{u}_j, \mathbf{u}_i \rangle &= |\mathbf{w}_j|^{-1} \langle [\mathbf{v}_j - \langle \mathbf{v}_j, \mathbf{u}_1 \rangle \mathbf{u}_1 - \cdots - \langle \mathbf{v}_j, \mathbf{u}_{j-1} \rangle \mathbf{u}_{j-1}], \mathbf{u}_i \rangle \\ &= |w_j|^{-1} [\langle \mathbf{v}_j, \mathbf{u}_i \rangle - \langle \mathbf{v}_j, \mathbf{u}_1 \rangle \langle \mathbf{u}_1, \mathbf{u}_i \rangle - \cdots - \langle \mathbf{v}_j, \mathbf{u}_{j-1} \rangle \langle \mathbf{u}_{j-1}, \mathbf{u}_i \rangle] \\ &= 0,\end{aligned}$$

so all the vectors $\mathbf{u}_1, \ldots, \mathbf{u}_j$ are mutually perpendicular. By induction, a set $\mathbf{u}_1, \ldots, \mathbf{u}_r$ exists whose members are mutually perpendicular unit vectors with each $\mathbf{v}_j$ a linear combination of $\mathbf{u}_1, \ldots, \mathbf{u}_j$.

If the $\mathbf{v}_i$ are linearly dependent, we first discard those $\mathbf{v}_i$ that are linear combinations of preceding vectors $\mathbf{v}_j$. Let the remaining vectors be $\mathbf{v}_{j(1)}, \ldots, \mathbf{v}_{j(h)}$; then $j(i) \geqq i$. By the preceding proof, there are mutually perpendicular unit vectors $\mathbf{u}_1, \ldots, \mathbf{u}_r$ such that for each i in $1, \ldots, h$, $\mathbf{v}_{j(i)}$ is a linear combination of $\mathbf{u}_1, \ldots, \mathbf{u}_i$ and therefore of $\mathbf{u}_1, \ldots, \mathbf{u}_{j(i)}$. If $\mathbf{v}_n$ is not one of the $\mathbf{v}_{j(i)}$, it is a linear combination of vectors $\mathbf{v}_{j(1)}, \ldots, \mathbf{v}_{j(i)}$ with $j(i) < n$. Therefore it is a linear combination of the vectors $\mathbf{u}_1, \ldots, \mathbf{u}_{j(i)}$, which are among the vectors $\mathbf{u}_1, \ldots, \mathbf{u}_n$. The proof is complete.

Let a linear coordinate system ϕ be defined by (J) of Section 1, so that the coordinate r-column x maps onto the point $P = (\tilde{x}^1, \ldots, \tilde{x}^r)$ of R^r for which

$$\text{(L)} \qquad \tilde{x}^i = \sum_{j=1}^{r} L^i{}_j x^j + \tilde{c}^i \qquad (i = 1, \ldots, r).$$

In the natural coordinate system, the coordinate r-column of P is simply $(\tilde{x}^1, \ldots, \tilde{x}^r)^{\mathrm{T}}$, and this we shall denote by $\tilde{x}$. Then if we denote by L the $r \times r$ matrix with elements $L^i{}_j$ and by x and $\tilde{c}$ the r-columns $(x^1, \ldots, x^r)^{\mathrm{T}}$ and $(\tilde{c}^1, \ldots, \tilde{c}^r)^{\mathrm{T}}$, respectively, equation (L) can be abbreviated to

$$\text{(M)} \qquad \tilde{x} = Lx + \tilde{c}.$$

It is useful to know the conditions on L and $\tilde{c}$ that characterize rectangular coordinate systems. This will follow as a corollary from the following computation.

LEMMA 3-8 *Let L be an $r \times r$ matrix and $\tilde{c}$ an r-column. In order that the function $x \mapsto Lx + \tilde{c}$ on the set of r-columns shall have the property that for all r-columns x and y*

$$\text{(N)} \qquad \sum_{i=1}^{r} [(Lx + \tilde{c})^i - (Ly + \tilde{c})^i]^2 = \sum_{i=1}^{r} (x^i - y^i)^2,$$

it is necessary and sufficient that $L^{\mathrm{T}} = L^{-1}$.

If we write v^i for $x^i - y^i$, we see that (N) holds for all r-columns x and y if and only if

$$\text{(O)} \qquad \sum_{i=1}^{r} \left[\sum_{j=1}^{r} L^i{}_j v^j \right]^2 = \sum_{i=1}^{r} [v^i]^2$$

for all r-columns v. Suppose that this is true. Let h, k be two different members of the set $\{1, \ldots, r\}$ and let v have l in hth place, a real number t in kth place, and 0 in all other places. Then by (O),

$$\sum_{i=1}^{r} [L^i{}_h + tL^i{}_k]^2 = 1 + t^2.$$

Since this holds for all real t, the coefficients of t and of t^2 in the right and left members must be equal, hence

$$\sum_{i=1}^{r} L^i{}_h L^i{}_k = 0, \qquad \sum_{i=1}^{r} [L^i{}_k]^2 = 1.$$

These can be combined into

$$\text{(P)} \qquad \sum_{i=1}^{r} L^i{}_h L^i{}_k = \delta^h{}_k,$$

where $\delta^h{}_k$ was defined in (F). The left member is the element in row h, column k of the matrix product $L^{\mathrm{T}}L$, and the right member is the element in row h, column k of the identity matrix 1, so $L^{\mathrm{T}}L = 1$. This implies $L^{\mathrm{T}} = L^{-1}$.

Conversely, suppose $L^{\mathrm{T}} = L^{-1}$. Then (P) holds, and

$$\begin{aligned}
\sum_{i=1}^{r} [L^i{}_j v^j]^2 &= \sum_{i=1}^{r} \left[\sum_{h=1}^{r} L^i{}_h v^h \right] \left[\sum_{k=1}^{r} L^i{}_k v^k \right] \\
&= \sum_{h,k=1}^{r} \left[v^h \left(\sum_{i=1}^{r} L^i{}_h L^i{}_k \right) v^k \right] \\
&= \sum_{h,k=1}^{r} v^h \delta^h{}_k v^k = \sum_{h=1}^{r} (v^h)^2.
\end{aligned}$$

So (O) holds, which implies that (N) is satisfied.

DEFINITION 3-9 *An $r \times r$ matrix L is* ***orthogonal*** *if $L^{-1} = L^{\mathrm{T}}$.*

COROLLARY 3-10 *A linear coordinate system ϕ, defined by*

$$\text{(Q)} \qquad \tilde{x}^i = \sum_{j=1}^{r} L^i{}_j x^j + \tilde{c}^i,$$

is rectangular if and only if the matrix L is orthogonal.

The natural coordinates of $P = (\tilde{x}^1, \ldots, \tilde{x}^r)$ are the r-column $\tilde{x} = (\tilde{x}^1, \ldots, \tilde{x}^r)^{\mathrm{T}}$, and (Q) can be written in the form

$$x \mapsto \tilde{x} = Lx + \tilde{c}.$$

Let $P = (\tilde{x}^1, \ldots, \tilde{x}^r)$ and $Q = (\tilde{y}^1, \ldots, \tilde{y}^r)$. Then

$$\text{(R)} \qquad [\operatorname{dist}(P, Q)]^2 = (\tilde{x}^1 - \tilde{y}^1)^2 + \cdots + (\tilde{x}^r - \tilde{y}^r)^2$$

$$= \sum_{i=1}^{r} [(Lx + c)^i - (Ly + c)^i]^2.$$

By definition, the mapping ϕ is rectangular if and only if

$$[\operatorname{dist}(P, Q)]^2 = \sum_{i=1}^{r} (x^i - y^i)^2$$

for all P and Q; by (R) this is true if and only if (N) holds for all r-columns x and y, and by Lemma 3-8 and Definition 3-9 this is true if and only if L is orthogonal.

COROLLARY 3-10 *Let L and M be orthogonal matrices. Then*

(i) $\det L = \pm 1$;
(ii) L^{T} *and* L^{-1} *are orthogonal*;
(iii) LM *is orthogonal.*

To prove (i) we have to make use of the well-known theorems that the determinant of the product of two square matrices is the product of their determinants, and that the determinant of a square matrix is equal to the determinant of its transpose. Since L is orthogonal, $L^{\mathrm{T}}L = L^{-1}L = 1$, so

$$1 = \det L^{\mathrm{T}}L = (\det L^{\mathrm{T}})(\det L) = (\det L)^2,$$

which implies conclusion (i). Also, from $L^{\mathrm{T}}L = 1$ we deduce

$$1 = L^{\mathrm{T}}L = (L^{\mathrm{T}})(L^{\mathrm{T}})^{\mathrm{T}},$$

so that $(L^{\mathrm{T}})^{\mathrm{T}}$ is the inverse of L^{T}, and by Definition 3-9, L^{T} is orthogonal. So is L^{-1}, which is the same as L^{T}.

By an easy calculation, the transpose of the product of two matrices is the product of their transposes in the reverse order. If L and M are orthogonal,

$$(LM)^{\mathrm{T}}(LM) = (M^{\mathrm{T}}L^{\mathrm{T}})(LM) = (M^{-1}L^{-1})(LM) = M^{-1}1M = 1,$$

so $(LM)^{\mathrm{T}}$ is the inverse of LM, and LM is orthogonal.

EXERCISE 3-1 In the plane, let the first axis be horizontal. Let $\mathbf{u}$ have length 2 and angle 15° with the positive first axis, and let $\mathbf{v}$ have length 3 and angle 45° with the first axis. Find $\langle \mathbf{u}, \mathbf{v} \rangle$ by Definition 3-1. Introduce rectangular axes and compute $\langle \mathbf{u}, \mathbf{v} \rangle$ by Lemma 3-4.

EXERCISE 3-2 Let P, Q, R have coordinates $(0, 0, 0)^{\mathrm{T}}$, $(1, 1, 1)^{\mathrm{T}}$, $(1, 0, 1)^{\mathrm{T}}$ in the original coordinate system in R^3. Find a rectangular coordinate system in which they all have their third coordinates equal to 0.

EXERCISE 3-3 Show that for any fixed θ, the rotation in R^2 defined by

$$\text{(S)} \qquad y^1 = x^1 \cos\theta - x^2 \sin\theta, \qquad y^2 = x^1 \sin\theta + x^2 \cos\theta$$

is length-preserving, and so is

$$\text{(T)} \qquad y^1 = x^1, \qquad y^2 = -x^2.$$

Transformation (T) is called a *reflection*. Why? Show that the matrix of coefficients in (S) has determinant 1, whereas that in (T) has determinant -1. Is the transformation

$$y^1 = -x^1 \cos\theta + x^2 \sin\theta, \qquad y^2 = x^1 \sin\theta + x^2 \cos\theta$$

length-preserving? a rotation? a reflection? a combination of both?

EXERCISE 3-4 Prove that a linear mapping L of R^r onto itself preserves angles if and only if there is a positive number p such that for all vectors $\mathbf{v}$, $|L\mathbf{v}| = p|\mathbf{v}|$.

Before stating the next exercise, we need to explain some details of the *Mercator projection*, which has been widely used in mapmaking for the past four centuries. Each point on the earth's surface has a longitude v^1 and a latitude v^2; the meridian of longitude 0 is the half-great-circle that passes through the poles and the cross-hair of the meridian telescope at Greenwich Observatory. We shall measure all angles in radians. In three-space we use a rectangular coodinate system with origin at the earth's center; the positive first, second, and third axes meet the earth's surface at the point with longitude 0 and latitude 0, the point with longitude $\pi/2$ and latitude 0, and the North Pole, respectively. We regard the earth as a sphere with radius a. The rectangle $(-\pi, \pi] \times (-\pi/2, \pi/2)$ is mapped one-to-one on all the earth's surface except the poles by $v \mapsto X(v) = (a \cos v^1 \cos v^2,\ a \sin v^1 \cos v^2,\ a \sin v^2)$. If c is a positive number, the same rectangle is mapped one-to-one onto the strip $M = (-c, c] \times (-\infty, \infty)$ by

$$v \mapsto u = \phi(v) = ([c/\pi]v^1, [c/\pi]\log(\sec v^2 + \tan v^2)).$$

The mapping between the earth's surface and the strip M given by

$$u \mapsto X(\phi^{-1}(u)) \qquad (-c < u^1 \leqq c, u^2 \text{ in } R)$$

is the **Mercator projection.**

EXERCISE 3-5 Let u_0 be a point of the strip M and $\mathbf{t}$ a vector in the plane. Then there exists a curve $s \mapsto u(s)$ (s in R) that is in M for s in some closed interval that contains 0 and has $u(0) = u_0$ and $Du(0) = \mathbf{t}$. This curve defines another curve

$$s \mapsto X \circ \phi^{-1} \circ u(s) = X(\phi^{-1}(u(s)))$$

on the earth's surface. Let $\mathbf{w}$ be the tangent vector

$$\mathbf{w} = DX \circ \phi^{-1} \circ u(0).$$

Show that the mapping of vectors **t** onto vectors **w** is linear. Show that if **t** points right (that is, in the direction of the positive u^1 axis), **w** points due east. Show that the ratio $|\mathbf{w}|/|\mathbf{t}|$ depends on u_0 but not on **t**. Show that if **t** makes angle θ with the positive u^1 axis, **w** makes angle θ with the direction due east. (This last was important before longitudes could be accurately determined. If on the Mercator map the straight line from port A to port B was θ degrees clockwise from vertically up, then by leaving port A and traveling steadily on a course θ degrees clockwise from due north, a ship would reach port B.)

4. Covectors

Suppose that a particle is moving in three-space and is acted on by a force that is the same wherever the particle is. If $\overrightarrow{CD}$ is a translate of $\overrightarrow{AB}$, in moving from C to D the particle goes the same distance in the same direction against the same force as in going from A to B, so the work done on the particle by the force is the same in both motions. That is, the work done on the particle is determined by the vector that specifies the motion of the particle. The work done in going from A to C is independent of the path followed in going from A to C, so the work done in traversing the directed line-segment $\overrightarrow{AB}$ and then the directed line segment $\overrightarrow{BC}$ is the same as the work done in traversing the directed line-segment $\overrightarrow{AC}$, which by definition is $\overrightarrow{AB} + \overrightarrow{BC}$. If we denote by $W(\mathbf{v})$ the work done on the particle when it traverses a directed line-segment that represents the vector $\mathbf{v}$, we thus have

$$W(\mathbf{v} + \mathbf{w}) = W(\mathbf{v}) + W(\mathbf{w}).$$

It follows that for each positive integer n,

$$W(n\mathbf{v}) = nW(\mathbf{v}).$$

If $\mathbf{w} = (m/n)\mathbf{v}$, with m and n positive integers, then

$$W(n[\mathbf{v}/n]) = nW(\mathbf{v}/n), \qquad W(m[\mathbf{v}/n]) = mW(\mathbf{v}/n),$$

which implies

$$W(\mathbf{w}) = (m/n)W(\mathbf{v}).$$

If W is a continuous function of the position of the end-point of v, as physics tells us it must be, this extends to all real numbers, and we obtain

$$W(c\mathbf{v}) = cW(\mathbf{v})$$

for all positive c. Since, clearly, $W(-\mathbf{v}) = -W(\mathbf{v})$, this holds for all real c. So the function W (whose value for each displacement $\overrightarrow{AB}$ is the work done on the particle by the force in moving from A to B) is a linear function on the space V^3 of vectors in three-space, in the sense of the following definition.

Definition 4-1 *Let r be a positive integer. A function* $\mathbf{F}$ *on* V^r *is a* ***linear function*** *(on* V^r*) if it is defined and real-valued on* V^r*, and for each pair of vectors* $\mathbf{v}_1, \mathbf{v}_2$ *in* V^r *and each pair of real numbers* c_1, c_2 *it is true that*

$$\mathbf{F}(c_1\mathbf{v}_1 + c_2\mathbf{v}_2) = c_1\mathbf{F}(\mathbf{v}_1) + c_2\mathbf{F}(\mathbf{v}_2).$$

Such functions have many applications besides the one sketched in the first paragraph of this section. In this chapter we shall use boldface capital letters to denote linear functions on a space V^r.

We have already defined addition of functions (on the same domain) and multiplication of functions by real numbers. As always, if $\mathbf{F}_1$ and $\mathbf{F}_2$ are linear functions on V^r and c is a real number, by $\mathbf{F}_1 + \mathbf{F}_2$ and $c\mathbf{F}_1$ we shall understand the functions whose values at each $\mathbf{v}$ in V^r are $\mathbf{F}_1(\mathbf{v}) + \mathbf{F}_2(\mathbf{v})$ and $c\mathbf{F}_1(\mathbf{v})$, respectively. These are evidently linear functions on V^r, and the requirements in Definition 2-2 are clearly satisfied, so the set of all linear functions on V^r with the definition just given for addition and for multiplication by real numbers is a vector space in the sense of Definition 2-2. It is called the space **adjoint** to V^r and is denoted by either of the symbols $(V^r)^*$, V_r. Both are in common use. In this chapter we shall most frequently use V_r; in the next we shall prefer $(V^r)^*$. The members of V_r can properly be called vectors, but to help keep in mind the important fact that they are linear functions on V^r, we shall give them the special name **covectors**. In this chapter we shall, as we have already said, use boldface capital letters for covectors.

For each $\mathbf{F}$ in V_r we define $|\mathbf{F}|$ to be the least upper bound of $|\mathbf{F}(\mathbf{v})|$ on the unit ball, consisting of all $\mathbf{v}$ of length at most 1. Thus,

$$\text{(A)} \qquad |\mathbf{F}| = \sup\{\mathbf{F}(\mathbf{v}) : \mathbf{v} \text{ in } V^r, |\mathbf{v}| \leqq 1\}.$$

This is obviously nonnegative. Soon we shall show that it is finite. If $\mathbf{F}_1$ and $\mathbf{F}_2$ belong to V_r, for each $\mathbf{v}$ with $\mathbf{v} \leqq 1$ we have

$$|[\mathbf{F}_1 + \mathbf{F}_2](\mathbf{v})| = |\mathbf{F}_1(\mathbf{v}) + \mathbf{F}_2(\mathbf{v})| \leqq |\mathbf{F}_1| + |\mathbf{F}_2|,$$

so $|\mathbf{F}_1 + \mathbf{F}_2|$ is an upper bound for the absolute values of $\mathbf{F}_1 + \mathbf{F}_2$ on the unit ball and is not less than the least upper bound. That is,

$$|\mathbf{F}_1 + \mathbf{F}_2| \leqq |\mathbf{F}_1| + |\mathbf{F}_2|.$$

If $b \neq 0$, for all $\mathbf{v}$ in the unit ball

$$|b\mathbf{F}(\mathbf{v})| = |b|\,|\mathbf{F}(\mathbf{v})| \leqq |b|\,|\mathbf{F}|,$$

so the least upper bound of the left member is not more than the right member:

$$\text{(B)} \qquad |bF| \leqq |b|\,|\mathbf{F}|.$$

Likewise,

$$|\mathbf{F}(\mathbf{v})| = |b^{-1}b\mathbf{F}(\mathbf{v})| = |b|^{-1}|b\mathbf{F}(\mathbf{v})| \leqq |b|^{-1}|b\mathbf{F}|,$$

so $|\mathbf{F}| \leqq |b|^{-1}|b\mathbf{F}|$, which with (B) implies

$$|b\mathbf{F}| = |b|\,|\mathbf{F}|.$$

This evidently is still valid if $b = 0$, so all the requirements in Definition 2-3 are satisfied, and the function $\mathbf{F} \mapsto |\mathbf{F}|$ is a norm on V_r. Thus, V_r is a normed vector space.

COROLLARY 4-2 *If F is in V_r and v is in V^r, then*

$$|\mathbf{F}(\mathbf{v})| \leqq |\mathbf{F}|\,|\mathbf{v}|.$$

If $\mathbf{v} = 0$, this is trivial. Otherwise, the vector $\mathbf{u} = \mathbf{v}/|\mathbf{v}|$ has length 1, and so $|F(u)| \leqq |\mathbf{F}|$. But

$$|\mathbf{F}(\mathbf{u})| = |\mathbf{F}(\mathbf{v})|/|\mathbf{v}|,$$

from which the conclusion follows.

So far, we have had no need of coordinate systems in connection with covectors. But if we have chosen a linear coordinate system in V^r, we can define a numerical representation of each covector $\mathbf{F}$ in such a way as to make it especially easy to calculate the value of $\mathbf{F}(\mathbf{v})$. As in the paragraph containing (F) and (G) of Section 3, we define $\delta^i{}_j$ to be 1 if $i = j$ and 0 if $i \neq j$, and we define

$$\delta_i = (\delta^1{}_i, \delta^2{}_i, \ldots, \delta^r{}_i)^{\mathrm{T}} = (0, \ldots, 0, 1, 0, \ldots, 0)^{\mathrm{T}},$$

where in the last expression the "1" is in ith place. If ϕ is a linear coordinate system in V^r, the r-column δ_i represents a vector, which we call $\mathbf{e}_i$, in V^r. Let $\mathbf{v}$ belong to V^r; it has a representation $v = (v^1, \ldots, v^r)^{\mathrm{T}}$ in the coordinate system ϕ. For each i, $v^i\mathbf{e}_i$ is represented by $(0, \ldots, 0, v^i, 0, \ldots, 0)^{\mathrm{T}}$, where the v^i is in ith place, so

$$\mathbf{v} = v^1\mathbf{e}_1 + \cdots + v^r\mathbf{e}_r.$$

Let $\mathbf{F}$ belong to V_r. Then

$$\mathbf{F}(\mathbf{v}) = \mathbf{F}(v^1\mathbf{e}_1 + \cdots + v^r\mathbf{e}_r) = v^1\mathbf{F}(\mathbf{e}_1) + \cdots + v^r\mathbf{F}(\mathbf{e}_r).$$

If we define

$$F_j = \mathbf{F}(\mathbf{e}_j) \qquad (j = 1, \ldots, r),$$

this takes the form

(C) $$\mathbf{F}(\mathbf{v}) = F_1 v^1 + \cdots + F_r v^r.$$

We take the $1 \times r$ matrix, or r-row,

(D) $$F = (F_1, \ldots, F_r)$$

to be the numerical representative of the covector $\mathbf{F}$. This is in fact a coordinate system in V_r; but whereas in V^r we have written the coordinates of points as r-columns, we choose to write the coordinates of covectors as r-rows, as in (D). The representation of $\mathbf{F}$ by the r-row F as in (D) is called the representation of $\mathbf{F}$ **induced** by the coordinate system ϕ in R^r, just as the representation of the vector **v** by the r-column of differences of coordinates of beginning and end points of a directed segment representing **v** was called the representation of **v** induced by ϕ. The correspondence between covectors and r-rows just established is the coordinate system in V_r **induced** by the coordinate system ϕ in R^r.

Since F is a $1 \times r$ matrix $(F_1, \ldots, F_r)$ and v is an $r \times 1$ matrix $(v^1, \ldots, v^r)^{\mathrm{T}}$, by the standard rule of matrix multiplication the product Fv is a 1×1 matrix whose one and only element is the right member of (C). Strictly speaking, the 1×1 matrix with the single element Fv is not the same as the number Fv, just as the person named John Doe is not the same as a parade in which the one and only marcher is John Doe. (The latter would presumably require a city license to go along a street, whereas John Doe would not.) Nevertheless, when convenient we shall ignore the distinction and abbreviate (C) to the form

$$F(v) = Fv. \tag{E}$$

LEMMA 4-3 *Let* **F** *belong to* V_r. *Let* ϕ *be a rectangular coordinate system in* R^r, *and let* $F = (F_1, \ldots, F_r)$ *be the representation of* **F** *induced by* ϕ. *Then*

$$|\mathbf{F}| = [F_1^2 + \cdots + F_r^2]^{1/2}. \tag{F}$$

This is trivial if $\mathbf{F} = 0$, so we suppose $\mathbf{F} \neq 0$. If **v** is any vector in the unit ball in V^r, then by Lemma 3-4,

$$|\mathbf{v}| = [(v^1)^2 + \cdots + (v^r)^2]^{1/2} \leqq 1.$$

Then by (C) and the Cauchy inequality Lemma 1-3,

$$\begin{aligned}|\mathbf{F}(\mathbf{v})| &= |F_1 v^1 + \cdots + F_r v^r| \\ &\leqq [F_1^2 + \cdots + F_r^2]^{1/2}[(v^1)^2 + \cdots + (v^r)^2]^{1/2}.\end{aligned}$$

The last factor is at most 1, so the right member of (F) is an upper bound for $|\mathbf{F}(\mathbf{v})|$ on the unit ball and must be at least equal to the least upper bound $|\mathbf{F}|$:

$$|\mathbf{F}| \leqq [F_1^2 + \cdots + F_r^2]^{1/2}.$$

On the other hand, if **v** is the vector represented by $(v^1, \ldots, v^r)^{\mathrm{T}}$, where

$$v^i = F_i/[F_1^2 + \cdots + F_r^2]^{1/2},$$

we compute easily that $|\mathbf{v}| = 1$, and by (C),

$$\begin{aligned}|\mathbf{F}(\mathbf{v})| &= |F_1(F_1/[F_1^2 + \cdots + F_r^2]^{1/2}) + \cdots + F_r(F_r/[F_1^2 + \cdots + F_r^2]^{1/2})| \\ &= [F_1^2 + \cdots + F_r^2]^{1/2}.\end{aligned}$$

So the right member of (F) is the least upper bound for $\mathbf{F}(\mathbf{v})$ on the unit ball, and Lemma 4-3 is established.

If ϕ and $\bar{\phi}$ are two linear coordinate systems on R^r, there are two $r \times r$ matrices M, N possessing inverses and two r-columns a, b such that if a point

$P = (\tilde{x}^1, \ldots, \tilde{x}^r)$ has coordinates $x = (x^1, \ldots, x^r)^T$ in the ϕ-system and coordinates $\bar{x} = (\bar{x}^1, \ldots, \bar{x}^r)^T$ in the $\bar{\phi}$-system, then

$$\tilde{x}^i = \sum_{j=1}^{r} M^i{}_j x^j + a^i, \qquad \tilde{x}^i = \sum_{j=1}^{r} N^i{}_j \bar{x}^j + b^j \qquad (i = 1, \ldots, r).$$

Therefore, the coordinates of P in the two systems are related by

$$Mx + a = N\bar{x} + b,$$

or

$$\bar{x} = N^{-1}(Mx + a - b).$$

If we define $L = N^{-1}M$ and $c = N^{-1}(a - b)$, this takes the form

(G) $$\bar{x} = Lx + c.$$

The same change of coordinate systems in R^r, from the ϕ-system to the $\bar{\phi}$-system, will produce changes in the induced representation of vectors in V^r and of covectors in V_r. The next lemma states what those changes are.

LEMMA 4-4 *Let ϕ and $\bar{\phi}$ be two linear coordinate systems in R^r, and let them be so related that for each P in R^r, if P has coordinate r-column x in the ϕ-system and coordinate r-column $\bar{x}$ in the $\bar{\phi}$-system, equation* (G) *holds. Let* $\mathbf{v}$ *be a vector in V^r and* $\mathbf{F}$ *a covector in V_r. If these have the respective representations v, F induced in the ϕ-system and the respective representations $\bar{v}$, $\bar{F}$ induced in the $\bar{\phi}$-system, then*

(H) $$\bar{v} = Lv,$$

(I) $$\bar{F} = FL^{-1}.$$

Let $\mathbf{v}$ be a vector, and let $\overrightarrow{AB}$ be a directed line-segment that represents $\mathbf{v}$. Let A and B have the respective coordinate representations $(x^1, \ldots, x^r)^T$, $(y^1, \ldots, y^r)^T$ in the ϕ-system and the respective representations $(\bar{x}^1, \ldots, \bar{x}^r)^T$, $(\bar{y}^1, \ldots, \bar{y}^r)^T$ in the $\bar{\phi}$-system. Then by (E) of Section 2 and (G),

$$\bar{v} = \bar{y} - \bar{x} = (Ly + c) - (Lx + c) = L(y - x) = Lv,$$

and (H) is established.

The r-rows F, $\bar{F}$ that represent $\mathbf{F}$ in the two systems must satisfy

$$\mathbf{F}(\mathbf{v}) = Fv = \bar{F}\bar{v}$$

for every $\mathbf{v}$ in V^r, v and $\bar{v}$ being the representations of $\mathbf{v}$ in the two systems. By (H), this implies

$$\sum_{i=1}^{r} \bar{F}_i \left[\sum_{j=1}^{r} L^i{}_j v^j \right] = \sum_{j=1}^{r} F_j v^j.$$

This holds for all $v^1, \ldots, v^r$ if and only if

$$\sum_{i=1}^{r} \bar{F}_i L^i{}_j = F_j \qquad (j = 1, \ldots, r),$$

which is the same as the matrix equation

$$\bar{F}L = F.$$

If we multiply both members on the right by L^{-1}, we obtain (I).

EXERCISE 4-1 Let g denote the gravitational acceleration, which is considered constant. The work done by gravity on a particle of mass m that moves from A to B is $-mg$(height of B − height of A). Let **F** be the force on the particle.

(i) Show that if the coordinates are rectangular with the third axis vertically upward, **F** is represented by $(0, 0, -mg)$.

(ii) Change to coordinates ρ, θ, ϕ by

$$x^1 = \rho\cos\theta\cos\phi, \qquad x^2 = \rho\sin\theta\cos\phi, \qquad x^3 = \rho\sin\theta.$$

Find the representation of F at (ρ, θ, ϕ) in this system.

(iii) In exterior ballistics, some use has been made of a coordinate system in which the first axis points in the direction of the initial velocity of a projectile and the second coordinate is the height of the projectile minus the height of the point on the x^1-axis vertically above or below the projectile. (The motion is assumed to be in a plane.) Find the representation of **F** in this system.

5. Differentiation and Integration of Vector-Valued Functions

Many of the definitions and theorems in preceding sections have used only the properties of real numbers and their absolute values that are listed in Definitions 2-2 and 2-3 of a normed vector space. These definitions and theorems will therefore have immediate generalizations from real-valued functions to functions whose values lie in a normed vector space. For example, let f be a function defined on an interval A and having values in V^r or in any other normed vector space. Then f has a derivative $Df(x_0)$ at a point x_0 of A if to each positive ε there corresponds a neighborhood $\gamma(x_0)$ of x_0 such that whenever x is a point of A that is in $\gamma(x_0)$ and is different from x_0,

$$|[f(x) - f(x_0)]/(x - x_0) - Df(x_0)| < \varepsilon.$$

There is no need for coordinates in defining the derivative, and the proof of the statement, "whenever f_1 and f_2 are functions with values in a normed vector space that are both differentiable at x_0, and a_1 and a_2 are real numbers, the function $a_1 f_1 + a_2 f_2$ has derivative $a_1\,Df_1(x_0) + a_2\,Df_2(x_0)$," is the same as it was for real-valued functions. However, in computations it is often desirable to make use of the coordinates of the vector f, which are real numbers. The next lemma is therefore useful.

LEMMA 5-1 *Let* $\mathbf{f}$ *be a function defined on an interval* A *and having values in* V^r. *Let* $\mathbf{f}(x)$ *have coordinate representation* $(f^1(x),\ldots,f^r(x))^{\mathrm{T}}$ *in a linear coordinate system. Then* $\mathbf{f}$ *has a derivative at a point* x_0 *of* A *if and only if each* f^i $(i = 1,\ldots,r)$ *has a derivative at* x_0, *and in that case* $D\mathbf{f}(x_0)$ *has coordinates* $(Df^1(x_0),\ldots,Df^r(x_0))^{\mathrm{T}}$.

Let x be a point of A different from x_0. Then the vector

$$\text{(A)}\qquad [\mathbf{f}(x) - \mathbf{f}(x_0)]/(x - x_0)$$

is represented by the r-column

$$\text{(B)}\qquad ([f^1(x) - f^1(x_0)]/(x - x_0),\ldots,[f^r(x) - f^r(x_0)]/(x - x_0))^{\mathrm{T}}.$$

By Lemma 2-1, the vector (A) has a limit as x tends to x_0 if and only if each of the r numbers in the column (B) has a limit; that is, if and only if each f^i has a derivative at x_0; and in that case the limit $D\mathbf{f}(x_0)$ of vector (A) is represented by the r-column

$$\left(\lim_{x\to x_0}[f^1(x) - f^1(x_0)]/(x - x_0),\ldots,\lim_{x\to x_0}[f^r(x) - f^r(x_0)]/(x - x_0)\right)^{\mathrm{T}}$$
$$= (Df^1(x_0),\ldots,Df^r(x_0))^{\mathrm{T}}.$$

This completes the proof.

Similarly, in any of the successively more general definitions of the integral (Definitions I-2-1, III-7-1, and IV-1-1), we can replace the real-valued integrands by integrands with values in a normed vector space with no other changes. For example, Definition IV-1-1 generalizes to this form.

DEFINITION 5-2 *Let* m *be a real-valued function of left-open intervals in a space* R^q. *Let* X *be a normed vector space (for example, the space* V^r *for some integer* r*). Let* B *be a set contained in* R^q, *and let* f *be a function that is defined on a subset* D *of* R^q *and has its values in* X. *Then* f *is* ***gauge-integrable*** *over* B *if* B *is contained in* D *and the gauge-limit of* $S(\mathscr{P}; f_B; m)$ *exists. If* J *is a vector in the space* X *such that*

$$J = \textit{gauge-limit of } S(\mathscr{P}; f_B; m),$$

we denote J *by the symbol*

$$\int_B f(x)\,m(dx).$$

Here, as before, f_B is the function such that $f_B(x)$ is $f(x)$ if x is in B and is 0 otherwise, and if $\mathscr{P}$ is the allotted partition

(C) $$\{(\bar{x}_1, A_1), \ldots, (\bar{x}_k, A_k)\},$$

$S(\mathscr{P}; f_B; m)$ is the sum

(D) $$\sum_{j=1}^{k} f_B(\bar{x}_j) m A_j.$$

As we proved Theorem I-5-3, we can prove that if f_1 and f_2 are functions with values in a normed vector space, and they are both integrable over B, and a_1 and a_2 are real numbers, the function $a_1 f_1 + a_2 f_2$ is integrable over B, and

$$\int_B [a_1 f_1(x) + a_2 f_2(x)]\, m(dx) = a_1 \int_B f_1(x)\, m(dx) + a_2 \int_B f_2(x)\, m(dx).$$

Other similar elementary statements generalize just as easily. But when the vector space is V^r, we have another way of proving most of these theorems because, as in Lemma 5-1, we can perform the integration by using the coordinates, thus.

THEOREM 5-3 *Let* $\mathbf{f}$ *be a function defined on a set* B *in* R^q *and assuming values in a space* V^r. *Let* m *be a function real-valued on the set of left-open intervals in* R^q. *In a linear coordinate system, let the vector* $\mathbf{f}(x)$ *be represented by the r-column* $(f^1(x), \ldots, f^r(x))^{\mathrm{T}}$. *Then* $\mathbf{f}$ *is integrable with respect to* m *over* B *if and only if for each* i *in* $\{1, \ldots, r\}$, *the real-valued function* f^i *is integrable with respect to* m *over* B, *and in that case the integral of* $\mathbf{f}$ *over* B *is represented by*

(E) $$\left(\int_B f^1(x)\, m(dx), \ldots, \int_B f^r(x)\, m(dx)\right)^{\mathrm{T}}.$$

Let $\mathscr{P}$ be an allotted partition, with notation (C). For each j in $\{1, \ldots, k\}$ the vector $\mathbf{f}(\bar{x}_j)$ has representation $(f^1(\bar{x}_j), \ldots, f^r(\bar{x}^j))^{\mathrm{T}}$, so by multiplying by mA_j and adding, we find that $S(\mathscr{P}; \mathbf{f}_B; m)$ has the representation

(F) $$(S(\mathscr{P}; f^1_B; m), \ldots, S(\mathscr{P}; f^r_B; m))^{\mathrm{T}}.$$

The integral of $\mathbf{f}$ over B exists if and only if $S(\mathscr{P}; \mathbf{f}_B; m)$ has a gauge-limit, which by Lemma 2-1 is true if and only if each of the r numbers in (F) has a gauge-limit, which is true if and only if each f^i is integrable with respect to m over B. Moreover, by Lemma 2-1, if the limit exists it is represented by the r-column (E). This completes the proof.

The definition of absolute integrability (Definition II-2-2) extends without change to functions with values in any normed vector space. But the theorem that a function is integrable if and only if it is absolutely integrable cannot be proved for functions with values in an arbitrary normed vector space. However, the situation is by no means hopeless; the theorem is true for all functions with values in a space V^r, as we now prove.

THEOREM 5-4 *Let m be a nonnegative additive function of left-open intervals in a space R^q; let B be a subset of R^q, and let $\mathbf{f}$ be a function defined on B and having its values in a space V^r. Then $\mathbf{f}$ is integrable with respect to m over B if and only if it is absolutely integrable with respect to m over B.*

To prove that if $\mathbf{f}$ is absolutely integrable, it is integrable, we need only repeat the first paragraph of the proof of Theorem II-2-4. Suppose, then, that $\mathbf{f}$ is integrable with respect to m over B. We choose any rectangular coordinate system in V^r, and as before we denote by $\mathbf{e}_i$ $(i = 1, \ldots, r)$ the vector represented by the r-column δ_i that has 1 in ith place and 0 everywhere else. If a vector $\mathbf{v}$ has representation $v = (v^1, \ldots, v^r)^{\mathrm{T}}$ in the coordinate system, the vector $v^i\mathbf{e}_i$ is represented by an r-column with v^i in ith place and 0 elsewhere, and the sum of these for $i = 1, \ldots, r$ is the r-column v. Hence,

$$\mathbf{v} = v^1\mathbf{e}_1 + \cdots + v^r\mathbf{e}_r.$$

By the triangle inequality, since $v^i\mathbf{e}_i$ has length $|v^i|1$,

$$\text{(G)} \qquad |\mathbf{v}| \leqq |v^1| + \cdots + |v^r|.$$

Let $\mathbf{f}(x)$ have representation $(f^1(x), \ldots, f^r(x))^{\mathrm{T}}$ in the chosen coordinate system. Since $\mathbf{f}$ is integrable, by Theorem 5-3 each f^i is integrable with respect to m over B. By Theorem II-2-4, which extends to integrals of real-valued functions over sets B in R^q, each f^i is absolutely integrable with respect to m over B. Now let ε be any positive number. For $i = 1, \ldots, r$ there is a gauge γ_i on $\bar{R}^q$ such that whenever

$$\text{(H)} \qquad \mathscr{P}' = \{(x_1', A_1'), \ldots, (x_h', A_h')\} \qquad \text{and}$$
$$\mathscr{P}'' = \{(x_1'', A_1''), \ldots, (x_k'', A_k'')\}$$

are γ_i-fine partitions of R^q,

$$\text{(I)} \qquad \sum_{j=1}^{h} \sum_{n=1}^{k} |f^i(x_j') - f^i(x_n'')| m(A_j' \cap A_n'') < \varepsilon/r.$$

Let γ be the gauge $\gamma_1 \cap \gamma_2 \cap \cdots \cap \gamma_r$. If P' and P'' are γ-fine partitions of R^q, with the notation in (H), all the inequalities (I) are satisfied, so by (G),

$$\begin{aligned}\sum_{j=1}^{h} \sum_{n=1}^{k} &|\mathbf{f}(x_j') - \mathbf{f}(x_n'')| m(A_j' \cap A_n'') \\ &\leqq \sum_{j=1}^{h} \sum_{n=1}^{k} \sum_{i=1}^{r} |f^i(x_j') - f^i(x_n'')| m(A_j' \cap A_n'') \\ &< \varepsilon.\end{aligned}$$

So f is absolutely integrable with respect to m over B, and the proof is complete.

The definition of Lipschitz continuity in Section II-3 extends immediately to any spaces in which distance has been defined. Let g be a function defined on a set D in a space X on which a distance has been defined, and let the values of g lie in a space Y on which a distance has been defined. Then g is **Lipschitzian** on D if there exists a number L such that for all x and x' in D

$$\operatorname{dist}(g(x), g(x')) \leqq L \operatorname{dist}(x, x').$$

It is then easy to extend Theorem II-3-1, as follows.

Theorem 5-5 *Let* $\mathbf{f}$ *be a function with values in a space* V^r *that is defined and integrable with respect to a nonnegative additive interval function m over a subset B of* R^q. *Let g be real-valued and Lipschitzian on a set D in* V^r *that contains all the points* $\mathbf{f}(x)$ *for x in B and also contains* $(0, \ldots, 0)^T$ *and has* $g((0, \ldots, 0)^T) = 0$. *Then the composite function* $g \circ f$ *is integrable with respect to m over B.*

To prove this, we repeat the proof of Theorem II-3-1 unchanged.

Obviously, this can be written in a different notation. If $f^1, \ldots, f^r$ are real-valued functions that are integrable with respect to m over B, and g is Lipschitzian on a set in R^r that contains all the points $(f^1(x), \ldots, f^r(x))$ with x in B and also contains $(0, \ldots, 0)$, then the function $x \mapsto g(f^1(x), \ldots, f^r(x))$ is integrable with respect to m over B.

Corollary 5-6 *If* $\mathbf{f}$ *is a function defined on a set B in* R^q, *with values in* V^r *and integrable over B with respect to a nonnegative additive interval function m, then* $|\mathbf{f}|$ *is integrable with respect to m over B, and*

$$\text{(J)} \qquad \left| \int_B \mathbf{f}(x)\, m(dx) \right| \leqq \int_B |\mathbf{f}(x)|\, m(dx).$$

If $\mathbf{v}$ and $\mathbf{v}'$ belong to V^r and O is the origin in R^r, there are points P, P' in R^r such that $\overrightarrow{OP}$ represents $\mathbf{v}$ and $\overrightarrow{OP'}$ represents $\mathbf{v}'$. Then $\overrightarrow{PP'}$ represents $\mathbf{v}' - \mathbf{v}$, and

$$\operatorname{dist}(P, P') = |\mathbf{v}' - \mathbf{v}|.$$

By Corollary 1-4, with P, O, P' in place of P, Q, R,

$$||\mathbf{v}| - |\mathbf{v}'|| \leqq |\mathbf{v}' - \mathbf{v}|,$$

so the function $\mathbf{v} \mapsto |\mathbf{v}|$ is Lipschitzian, and by Theorem 5-5, $|\mathbf{f}|$ is integrable over B.

In R^r we introduce rectangular coordinates. Then each f^i is integrable over B by Theorem 5-3; let its integral be J^i. If all J^i are 0, inequality (J) is obvious.

Otherwise, by Cauchy's inequality (Lemma 1-3),

$$
\begin{aligned}
|\mathbf{J}|^2 = \sum_{i=1}^{r} (J^i)^2 = \sum_{i=1}^{r} J^i \int_B f^i(x)\, m(dx) \\
&= \int_B \sum_{i=1}^{r} J^i f^i(x)\, m(dx) \\
&\leqq \int_B \left[\sum_{i=1}^{r} (J^i)^2 \right]^{1/2} \left[\sum_{i=1}^{r} f^i(x)^2 \right]^{1/2} m(dx) \\
&= |\mathbf{J}| \int_B |\mathbf{f}(x)|\, m(dx).
\end{aligned}
$$

Dividing by $|\mathbf{J}|$ gives us inequality (J).

EXERCISE 5-1 Show that it is possible for a function $x \mapsto f(x)$ $(a \leqq x \leqq b)$ with values in V^r to have a continuous derivative on $[a, b]$ and yet for there to be no $\bar{x}$ in $[a, b]$ for which

$$f(b) - f(a) = (b - a)\, Df(\bar{x}).$$

Suggestion: Take $r = 2$, $a = 0$, $b = 2\pi$, $f(x) = (\cos x, \sin x)^{\mathrm{T}}$.

EXERCISE 5-2 A set H in R^r is a **hyperplane** if there exists a nonzero linear function F with $F(0) = 0$ and a real number c such that

$$H = \{x \text{ in } R^r : F(x) = c\}.$$

This divides R^r into two (closed) half-spaces

$$\{x \text{ in } R^r : F(x) \leqq c\}, \qquad \{x \text{ in } R^r : F(x) \geqq c\}.$$

Let $\mathbf{f}(x)$ be the position-vector of the point $f(x)$. Show that if $\mathbf{f}$, m, and B are as in Definition 5-2 and all values of $f(x)$ (x in B) are in one of the half-spaces, and m is nonnegative and $mB = 1$, and $\mathbf{f}$ is integrable with respect to m over B, then the point

$$\left(\int_B f^1(x)\, m(dx), \ldots, \int_B f^r(x)\, m(dx) \right)^{\mathrm{T}}$$

is in the same half-space.

EXERCISE 5-3 Let C be a nonempty collection of real-valued functions on a set D in R^r, all satisfying a Lipschitz condition with the same constant L. For each x in D, let $F(x)$ be the supremum of $f(x)$ for all f in C. Show that if F has a finite value at some x in D, it is finite at all x in D, and then on D it satisfies a Lipschitz condition of constant L.

6. Curves and Their Lengths

The word *curve* unfortunately has (at least) two quite different meanings in mathematics. One is static, as in elementary geometry. A curve in this sense is a set of points of a particular nature – for example, the set of all points at a given distance r from a given point C. The other meaning is, in essence, the history of a moving point. By a **representation** we shall mean a continuous function $t \mapsto P(t)$ $(a \leqq t \leqq b)$ on an interval in R with values $P(t)$ in a space R^r. (We really mean this to be a representation of a curve, but for a short while we avoid this expression until we have defined what a curve is.) Let $u \mapsto P'(u)$ $(c \leqq u \leqq d)$ be another representation. If these are regarded as records of voyages, it is not enough that each value of P should also be a value of P'; if a traveler goes on a one-track railroad from town A to town B and another traveler goes from A to B and back again, we regard them as having made different voyages, even though all the points through which the first traveler passes are also on the route of the second traveler, and vice versa. On the other hand, if we say that the two travelers made the same trip, we do not mean that they went equally fast, or even that each stopped wherever the other stopped. What we mean is that if the first traveler went through a place A and later through a place B and still later through a place C, then the second traveler went through A and later went through B and still later went through C. To this idea of "making the same trip" we shall give the name *equivalence* of representations, and we define it precisely thus.

DEFINITION 6-1 *Let $t \mapsto P(t)$ $(a \leqq t \leqq b)$ and $u \mapsto Q(u)$ $(c \leqq u \leqq d)$ be continuous functions with values in a space R^r. These functions are* ***equivalent representations*** *if whenever*

(A) $$t_0 = a \leqq t_1 \leqq t_2 \leqq \cdots \leqq t_k = b$$

is a nondecreasing sequence of numbers in $[a, b]$, there exists a nondecreasing sequence

(B) $$u_0 = c \leqq u_1 \leqq u_2 \leqq \cdots \leqq u_k = d$$

of numbers in $[c, d]$ such that

(C) $$P(t_j) = Q(u_j) \qquad (j = 1, \ldots, k),$$

and whenever $u_0, \ldots, u_k$ is a nondecreasing sequence (C) *of numbers in $[c, d]$, there exists a nondecreasing sequence* (A) *in $[a, b]$ such that* (C) *holds.*

We leave it as an exercise to show that if $u \mapsto t(u)$ $(c \leqq u \leqq d)$ is a nondecreasing continuous function with $t(c) = a$ and $t(d) = b$, the representation $u \mapsto P(t(u))$ $(c \leqq u \leqq d)$ is equivalent to the representation $t \mapsto P(t)$ $(a \leqq t \leqq b)$, even if $u \mapsto t(u)$ lacks an inverse.

Two continuous functions P and Q may fail to be equivalent even when every value taken by $P(t)$ is a value of $Q(u)$, and vice versa. For example, if we choose rectangular coordinates in the plane and define $P(t)$ and $Q(t)$ by the coordinate representations

$$P: f(t) = (-\cos t, \sin t)^{\mathrm{T}} \qquad (0 \leqq t \leqq \pi),$$

$$Q: g(u) = (-\cos u, |\sin u|)^{\mathrm{T}} \qquad (0 \leqq u \leqq 2\pi),$$

the curves represented by these functions are not equivalent, even though the points on the one curve are the same as the points on the other. For $f(\pi)$ and $g(2\pi)$ are different, whereas if the two representations were equivalent, these would have to be equal.

Sometimes we find it convenient to use the position-vectors of points rather than the points themselves. If $\mathbf{x}(t)$ is the position-vector of $P(t)$, we consider the function $t \mapsto \mathbf{x}(t)$ as merely another notation for $t \mapsto P(t)$.

So far we have spoken of representations of curves, but we have not said what a curve is. We define a **curve** to be an equivalence class of representations. (Compare this with the definition of **vector** in Section 1.) That is, if $t \mapsto P(t)$ is a continuous function on a bounded closed interval in R, and the values of P are in R^r, the set of all functions that are equivalent to $t \mapsto P(t)$ is called a curve, and each of the functions in the class is called a *representation* of the curve.

Suppose that $\mathscr{C}$ is a curve in R^r with a representation $t \mapsto P(t)$ $(a \leqq t \leqq b)$. In any given coordinate system, $P(t)$ will have a coordinate representation $f(t)$ by an r-column. Then the function

$$t \mapsto f(t) \qquad (a \leqq t \leqq b)$$

is a representation of $\mathscr{C}$ in the given coordinate system. A curve Π is a **polygon** if in some linear coordinate system it has a representation $t \mapsto p(t)$ $(a \leqq t \leqq b)$ in which p is continuous, and there are finitely many points

$$t_0 = a \leqq t_1 \leqq t_2 \leqq \cdots \leqq t_n = b$$

such that $p(t)$ is linear on each subinterval $[t_{j-1}, t_j]$ $(j = 1, \ldots, n)$. (Of course, we could discard all t_j such that $t_j = t_{j-1}$.) The points $p(t_0), \ldots, p(t_n)$ are called the **vertices** of the polygon Π. The polygon Π is **inscribed** in the curve $\mathscr{C}$ if in a linear coordinate system they have the respective representations

$$t \mapsto p(t), \qquad t \mapsto f(t) \qquad (a \leqq t \leqq b)$$

such that for some sequence $t_0 = a \leqq t_1 \leqq \cdots \leqq t_n = b$, p is linear on each $[t_{j-1}, t_j]$ and

$$p(t_j) = f(t_j) \qquad (j = 0, 1, \ldots, n).$$

This seems to depend on the representation chosen for $\mathscr{C}$, but it does not. If $u \mapsto g(u)$ $(c \leqq u \leqq d)$ is another representation of $\mathscr{C}$, by Definition 6-1 there are points $u_0 = c \leqq u_1 \leqq \cdots \leqq u_n = d$ such that $g(u_j) = f(t_j)$ $(j = 0, 1, \ldots, n)$. Then the function that coincides with $g(u_j)$ at each u_j and is linear on each interval $[u_{j-1}, u_j]$ is obviously another representation of Π.

Since the portion $t \mapsto p(t)$ $(t_{j-1} \leqq t \leqq t_j)$ is a line-segment of length

$$\text{dist}(p(t_j), p(t_{j-1})),$$

it is reasonable to define the **length of** Π to be

$$\text{(D)} \qquad \text{length } \Pi = \sum_{j=1}^{r} \text{dist}(f(t_{j-1}), f(t_j));$$

and we do this. We wish to define the length of $\mathscr{C}$ in accordance with the ideas of elementary geometry. In particular, in order to extend to all curves the property that "a straight line is the shortest distance between two points," the length we ascribe to the portion $t \mapsto f(t)$ $(t_{j-1} \leqq t \leqq t_j)$ of $\mathscr{C}$ must be at least as great as the length $\text{dist}(f(t_j), f(t_{j-1}))$ of the line-segment with the same ends; so, by adding, the length of $\mathscr{C}$ should not be less than the length of any inscribed polygon. Taking this as a suggestion, we adopt the next definition.

DEFINITION 6-2 *If $\mathscr{C}$ is a curve in R^r, the* ***length*** *$L(\mathscr{C})$ of $\mathscr{C}$ is the supremum of the lengths of all polygons inscribed in $\mathscr{C}$.*

For polygons Π this gives us two definitions of length – the original definition in (D) and the new one in Definition 6-2. In the next theorem we shall show that for every polygon the two definitions give equal values for the length. Until this is proved, we shall observe the precaution of using the expression "length of Π" and avoiding the symbol $L(\Pi)$ for the length defined in (D), and we shall use the expression "length $L(\Pi)$ of Π" for the quantity defined in Definition 6-2.

Suppose that Π is a polygon that has a representation $t \mapsto P(t)$ $(a \leqq t \leqq b)$, and that in a linear coordinate system the coordinates $f(t)$ of $P(t)$ are linear on subintervals $[t_{j-1}, t_j]$ $(j = 1, \ldots, n)$, where $t_0 = a \leqq t_1 \leqq \cdots \leqq t_n = b$. Let $\mathbf{f}(t)$ be the position-vector of $P(t)$. Then $D\mathbf{f}(t)$ has a constant value $\mathbf{c}_j$ on each interval (t_{j-1}, t_j). Let $\dot{\mathbf{f}}$ be any function on $[a, b]$ such that $\dot{\mathbf{f}}(t) = D\mathbf{f}(t)$ on each interval (t_{j-1}, t_j); at $t_0, \ldots, t_n$ we assign $\dot{\mathbf{f}}(t)$ any value. By Theorem 5-3,

$$\begin{aligned}
|\mathbf{f}(t_j) - \mathbf{f}(t_{j-1})| &= \left| \int_{t_{j-1}}^{t_j} \dot{\mathbf{f}}(t)\,dt \right| \\
&= \left| \int_{t_{j-1}}^{t_j} \mathbf{c}_j\,dt \right| \\
&= |(c_j^1[t_j - t_{j-1}], \ldots, c_j^r[t_j - t_{j-1}])^{\mathrm{T}}| \\
&= |(c_j^1, \ldots, c_j^r)^{\mathrm{T}}|(t_j - t_{j-1}) \\
&= |\mathbf{c}_j|(t_j - t_{j-1}) \\
&= \int_{t_{j-1}}^{t_j} |\dot{\mathbf{f}}(t)|\,dt.
\end{aligned}$$

Adding for $j = 1, \ldots, n$ yields

$$\text{(E)} \qquad \text{length } \Pi = \int_a^b |\dot{\mathbf{f}}(t)|\, dt.$$

It is a reasonable conjecture that if $\mathscr{C}$ has a representation in which the f^i have continuous derivatives, equation (E) will still be valid. For then we can cut $\mathscr{C}$ into many small arcs, each nearly straight, so $\mathscr{C}$ is in a loose sense nearly a polygon. But we are now about to show that (E) holds under much weaker hypotheses. All that we need to know about the function $\mathbf{f}$ is that it is the indefinite integral of some function $\dot{\mathbf{f}}$; this last does not have to be bounded or to have even a single point of continuity. Later, in Chapter VII, we shall prove that this is the strongest possible theorem on lengths; no weaker hypotheses will yield the conclusion.

THEOREM 6-3 *Let $\mathscr{C}$ be a continuous curve in R^r that has a representation $t \mapsto P(t)$ $(a \leqq t \leqq b)$ such that the position-vector $\mathbf{f} = \overrightarrow{OP}$ is the indefinite integral of a (vector-valued) function $\dot{\mathbf{f}}$:*

$$\text{(F)} \qquad \mathbf{f}(t) = \mathbf{f}(a) + \int_a^t \dot{\mathbf{f}}(u)\, du \qquad (a \leqq t \leqq b).$$

Then $\mathscr{C}$ has finite length, and

$$\text{(G)} \qquad L(\mathscr{C}) = \int_a^b |\dot{\mathbf{f}}(t)|\, dt.$$

Let Π be a polygon inscribed in $\mathscr{C}$. Then there are points

$$\text{(H)} \qquad t_0 = a < t_1 < \cdots < t_n = b$$

such that the continuous function that is linear on each interval $[t_{j-1}, t_j]$ and coincides with $f(t)$ at each t_j is a representation of Π. By Definition (D) and Corollary 5-6,

$$\begin{aligned} \text{length } \Pi &= \sum_{j=1}^n |f(t_j) - f(t_{j-1})| = \sum_{j=1}^n \left| \int_{t_{j-1}}^{t_j} \dot{f}(t)\, dt \right| \\ &\leqq \sum_{j=1}^n \int_{t_{j-1}}^{t_j} |\dot{f}(t)|\, dt = \int_a^b |\dot{f}(t)|\, dt. \end{aligned}$$

So the integral of $|\dot{f}|$ is an upper bound for the lengths of polygons inscribed in $\mathscr{C}$ and by Definition 6-2,

$$\text{(I)} \qquad L(\mathscr{C}) \leqq \int_a^b |\dot{f}(t)|\, dt.$$

To prove the reversed inequality, let ε be any positive number. For each i in $\{1, \ldots, r\}$ the function $\dot{f}^i$ is integrable from a to b, so we can and do choose a gauge γ_i on $\bar{R}$ such that if $\mathscr{P}$ is any γ_i-fine partition of R,

$$\text{(J)} \qquad \left| S(\mathscr{P}; f^i_{(a,b]}; m_L) - \int_a^b f^i(t)\,dt \right| < \frac{\varepsilon}{10r}.$$

Let $\gamma = \gamma_1 \cap \cdots \cap \gamma_r$. We can and do choose γ-fine partitions of $(a, b]$, of $(-\infty, a]$, and of (b, ∞) and form their union. This union is a γ-fine partition

$$\mathscr{P} = \{(\bar{t}_1, A_1), \ldots, (\bar{t}_q, A_q)\}$$

of R. Without loss of generality we may assume that the intervals $A_1, \ldots, A_n$ are the ones that are in $(a, b]$ and that they are numbered from left to right. Thus there are numbers as listed in (H) for which

$$A_j = (t_{j-1}, t_j] \qquad (j = 1, \ldots, n).$$

Let s be the vector-valued step-function that on each interval A_j has the constant value

$$\text{(K)} \qquad s(t) = \left(\int_{A_j} \dot{f}(t)\,dt \right) \Big/ m_L A_j \qquad (t \text{ in } A_j;\, j = 1, \ldots, q).$$

Since (J) holds for every γ_i-fine partition and this partition $\mathscr{P}$ is γ_i-fine, by Corollary II-13-3

$$\int_R |\dot{f}^i_R(t) - s^i(t)|\,dt < \frac{\varepsilon}{r}.$$

By (G) of Section 5, this implies

$$\int_a^b |\dot{f}(t) - s(t)|\,dt \leqq \int_R |\dot{f}_R(t) - s(t)|\,dt < \varepsilon.$$

Therefore,

$$\int_a^b |s(t)|\,dt \geqq \int_a^b |\dot{f}(t)|\,dt - \varepsilon.$$

If in this we substitute the value of $|s(t)|$ from K, we obtain

$$\text{(L)} \qquad \sum_{j=1}^{n} \left| \int_{A_j} \dot{f}(t)\,dt \right| \geqq \int_a^b |\dot{f}(t)|\,dt - \varepsilon.$$

By hypothesis,

$$\int_{t_{j-1}}^{t_j} \dot{f}(t)\,dt = f(t_j) - f(t_{j-1}),$$

so the left member of (L) is the sum of the vector-lengths $|f(t_j) - f(t_{j-1})|$. By (D), this is the length of the polygon inscribed in $\mathscr{C}$ with vertices at $f(t_0), f(t_1), \ldots, f(t_n)$, so by Definition 6-2 it is not less than $L(\mathscr{C})$. Therefore

(L) implies

$$L(\mathscr{C}) \geqq \int_a^b |\dot{f}(t)|\, dt - \varepsilon.$$

This holds for every positive ε, so it continues to hold if we replace ε by 0. That is, the reversed inequality to (I) is satisfied, and therefore equality must hold in (I). The proof is complete.

Suppose now that Π is a polygon, with a representation $t \mapsto f(t)$ $(a \leqq t \leqq b)$ that is linear on intervals $[t_0, t_1], \ldots, [t_{n-1}, t_n]$, as in (H). Then by (E) the length of Π as defined in (D) is

$$\text{length } \Pi = \int_a^b |\dot{f}(t)|\, dt.$$

By Theorem 6-3,

$$L(\Pi) = \int_a^b |\dot{f}(t)|\, dt.$$

So,

(M) the length of Π as defined in (D) and the length $L(\Pi)$ of Π as defined in Definition 6-2 are equal, and we no longer need to distinguish between them.

If for each t' in $[a, b]$ we define $s(t')$ to be the length of the portion

$$t \mapsto P(t) \qquad (a \leqq t \leqq t')$$

of the curve $\mathscr{C}$, we see by Theorem 6-3 that under the hypotheses of that theorem, $|\dot{\mathbf{f}}|$ is a density for s. That is, for each subinterval $[c, d]$ of $[a, b]$, $s(d) - s(c)$ is the integral of $|\dot{\mathbf{f}}|$ over $[c, d]$. It is therefore consistent with our notation to write

$$\dot{s} = |\dot{\mathbf{f}}|.$$

Also, since

(N) $$s(t') = \int_a^{t'} |\dot{\mathbf{f}}(t)|\, dt,$$

at each point t' at which $\dot{\mathbf{f}}$ is continuous we have

(O) $$Ds(t') = |\dot{\mathbf{f}}(t')|.$$

In particular, if the parameter t is arc length, so that $s(t') = t'$ for all t' in $[0, L(\mathscr{C})]$, then at each point of continuity of $\dot{\mathbf{f}}$ we have

$$1 = |\dot{\mathbf{f}}(t)|.$$

In the notation of differentials,

$$df^{(i)} = \dot{f}^{(i)}(t)\, dt, \qquad ds = \dot{s}(t)\, dt$$

at all points where the derivative exists. So in this notation, (O) takes the form

$$ds = [(df^{(1)})^2 + \cdots + (df^{(q)})^2]^{1/2}.$$

This expression is frequently used. However, it should be kept in mind that it is meaningful only when the $f^{(i)}$ have derivatives at all but a few points and are the indefinite integrals of those derivatives, whereas (N) holds under the much less restrictive hypotheses of Theorem 6-3.

Suppose that the hypotheses of Theorem 6-3 hold and that c is in $[a, b]$. Let $\mathscr{C}_{a,c}$, $\mathscr{C}_{c,b}$ denote the curves represented by

$$x = f(t) \qquad (a \leqq t \leqq c), \qquad x = f(t) \qquad (c \leqq t \leqq b),$$

respectively. From the conclusion of Theorem 6-3 it follows at once that

$$L(\mathscr{C}) = L(\mathscr{C}_{a,c}) + L(\mathscr{C}_{c,b}).$$

This holds, in fact, in full generality, but since we do not need it we shall merely give a brief sketch of the proof. Let ε be positive; let Π_1, Π_2, Π_3 be polygons inscribed in $\mathscr{C}$, $\mathscr{C}_{a,c}$, $\mathscr{C}_{c,b}$, respectively, such that

$$L(\Pi_1) > L(\mathscr{C}) - \varepsilon, \qquad L(\Pi_2) > L(\mathscr{C}_{a,c}) - \varepsilon/2, \qquad L(\Pi_3) > L(\mathscr{C}_{c,b}) - \varepsilon/2.$$

Let Π be inscribed in $\mathscr{C}$ and have all the vertices of Π_1, Π_2, and Π_3, and let $\Pi_{a,c}$ and $\Pi_{c,b}$ be the parts of Π inscribed in $\mathscr{C}_{a,c}$ and $\mathscr{C}_{c,b}$, respectively. Then Π_1, Π_2, and Π_3 are inscribed in Π, $\Pi_{a,c}$, and $\Pi_{c,b}$, respectively, and

$$L(\mathscr{C}) \geqq L(\Pi) \geqq L(\Pi_1) > L(\mathscr{C}) - \varepsilon,$$

$$L(\mathscr{C}_{a,c}) \geqq L(\Pi_{a,c}) \geqq L(\Pi_2) > L(\mathscr{C}_{a,c}) - \varepsilon/2,$$

$$L(\mathscr{C}_{c,b}) \geqq L(\Pi_{c,b}) \geqq L(\Pi_3) > L(\mathscr{C}_{c,b}) - \varepsilon/2.$$

The conclusion can be deduced from these inequalities.

EXERCISE 6-1 Define

$$f(t) = (-1, 0) \qquad (-2 \leqq t < -1),$$

$$f(t) = (t, [1 - t^2]^{1/2}) \qquad (-1 \leqq t \leqq 1)$$

and

$$g(u) = (-\cos u, \sin u) \qquad (0 \leqq u \leqq \pi),$$

$$g(u) = (1, 0)(\pi < u \leqq 2\pi).$$

Show that these are equivalent and therefore represent the same curve although there is no continuous one-to-one mapping $u \mapsto t(u)$ of $[0, 2\pi]$ onto $[-2, 1]$ such that $f(t(u)) = g(u)$.

EXERCISE 6-2 Let $\mathscr{C} : t \mapsto P(t)$ $(a \leqq t \leqq b)$ be a curve in R^r with finite length, and for each t' in $[a, b]$ let $s(t')$ be the length of the arc $t \mapsto P(t)$ $(a \leqq t \leqq t')$. Show that $t' \mapsto s(t')$ is continuous on $[a, b]$. *Suggestion*: For each t^* in $[a, b]$ and each positive ε, let (H) define the vertices of a polygon Π with $L(\Pi) > L(\mathscr{C}) - \varepsilon/3$, with t^* one of the t_i (say, t_n) and with $\text{dist}(P(t_{n-1}), P(t_n))$ and $\text{dist}(P(t_n), P(t_{n+1})) < \varepsilon/3$; modifications at ends are obvious. The two arcs of $\mathscr{C}$ corresponding to $a \leqq t \leqq t_{n-1}$ and $t_{n+1} \leqq t \leqq b$ have total length $> L(\mathscr{C}) - \varepsilon$, so $s(t^*) - \varepsilon < s(t) < s(t^*) + \varepsilon$ for t in (t_{n-1}, t_{n+1}).

EXERCISE 6-3 Show that with the notation of Exercise 6-2, for s' in $[0, L(\mathscr{C})]$ there is at least one t in $[a, b]$ with $s(t) = s'$. For all of these, $P(t)$ has the same value; call it $P^*(s')$. Then $s \mapsto P^*(s)$ $(0 \leqq s \leqq L(\mathscr{C}))$ is a representation of $\mathscr{C}$, and the parameter s is arc length.

EXERCISE 6-4 Define

$$f(t) = (t, t \sin t^{-2}) \qquad (0 < t \leqq \pi^{-1/2}),$$
$$f(0) = (0, 0).$$

Show that f is continuous but represents a curve of infinite length.

In the next two exercises we shall refer to the *Fréchet distance* between two curves. Let $\mathscr{C}_1$, $\mathscr{C}_2$ be curves with the respective representations

$$t \mapsto P(t) \qquad (a \leqq t \leqq b), \qquad u \mapsto Q(u) \qquad (c \leqq u \leqq d).$$

To each increasing function $u \mapsto t(u)$ that has a continuous inverse and satisfies $t(c) = a$, $t(d) = b$, there corresponds a supremum

$$\sup \text{dist}\{(P(t(u)), Q(u)) : c \leqq u \leqq d\}.$$

The infimum of this quantity for all mappings t is the Fréchet distance between $\mathscr{C}_1$ and $\mathscr{C}_2$.

EXERCISE 6-5 Show that for every curve $\mathscr{C}$ in R^r and every number c less than $L(\mathscr{C})$ there is a positive δ such that if $\mathscr{C}'$ is a curve whose Fréchet distance from C is less than δ, then

$$L(\mathscr{C}) > c.$$

Suggestion: There is a polygon Π inscribed in $\mathscr{C}$ with $L(\Pi) > c$. If the vertices of Π are at $t_0, \ldots, t_n$, take $\delta = (L(\Pi) - c)/2(n + 1)$. Suppose $t_j = t(u_j)$. The sum of the lengths of the segment with ends $f(t_{j-1})$ and $g(u_{j-1})$, the arc $u \mapsto g(u)$ $(u_{j-1} \leqq u \leqq u_j)$, and the segment with ends $g(u_j)$ and $f(t_j)$ is greater than the length of the side of the polygon that joins $f(t_{j-1})$ and $f(t_j)$.

EXERCISE 6-6 Show that there is no finite upper bound for the lengths of curves near a given curve, in the sense of the Fréchet distance.

EXERCISE 6-7 Let a curve $\mathscr{C}$ have representation $t \mapsto f(t)$ $(a \leqq t \leqq b)$ in a linear coordinate system. Show that for every number c less than $L(\mathscr{C})$ there is a positive δ such that whenever Π is a polygon with vertices $f(t_0), \ldots, f(t_p)$ inscribed in $\mathscr{C}$ and having $t_j - t_{j-1} < \delta$ $(j = 1, \ldots, p)$, it is true that $L(\Pi) > c$.

EXERCISE 6-8 Let $\mathscr{C}$ be a curve with representation $t \mapsto f(t)$ $(a \leqq t \leqq b)$. For each allotted partition $\mathscr{P} = \{(\bar{t}_1, A_1), \ldots, (\bar{t}_k, A_k)\}$ of $(a, b]$ with $A_1 = (t_0, t_1]$, $\ldots, A_k = (t_{k-1}, t_k]$, and $t_0 = a$, $t_k = b$, let

$$F(\mathscr{P}) = \sum_{j=1}^{k} \operatorname{dist}(f(t_{j-1}), f(t_j)).$$

Show that $L(\mathscr{C})$ is the gauge-limit of $F(\mathscr{P})$. *Suggestion*: For $\varepsilon > 0$, choose Π inscribed in $\mathscr{C}$ with $L(\Pi) > L(\mathscr{C}) - \varepsilon$. If the vertices of Π are at $t = t_0, \ldots, t_k$, $L(\Pi) = F(\mathscr{P})$. Let $\gamma(t)$ contain none of $\{t_0, \ldots, t_k\}$ if t is not in that set and contain t_j only if $t = t_j$. Then if $\mathscr{P}$ is γ-fine, $F(\mathscr{P}) \geqq L(\Pi)$.

7. Line Integrals

Suppose, first, that a particle moves in three-dimensional space along a curve and is acted on by a force that is the same at all points of the space. If it moves from P_0 to P_1, we use the symbol ΔP to denote the vector represented by the directed line-segment from P_0 to P_1. We have seen in the beginning of Section 4 that the work done on the particle by the force is a linear function of the displacement ΔP and therefore is a covector $\mathbf{F}$ belonging to V_r. The work done is then $\mathbf{F}(\Delta P)$.

Suppose next that the force that acts on the particle when it is at place P is not constant throughout space but is a continuous covector-valued function $P \mapsto \mathbf{F}(P)$. Let the particle move along a curve $\mathscr{C}$ that has a representation

$$\text{(A)} \qquad t \mapsto P(t) \qquad (a \leqq t \leqq b).$$

For each point P_0 of space and each positive number ε there is a neighborhood $\gamma(P_0)$ of P_0 such that if P is in $\gamma(P_0)$

$$\text{(B)} \qquad |\mathbf{F}(P) - \mathbf{F}(P_0)| < \varepsilon.$$

Since P is a continuous function, for each $\bar{t}$ in $[a, b]$ there is a neighborhood $\gamma_1(\bar{t})$ such that for all t in $\gamma_1(\bar{t})$, $P(t)$ is in $\gamma(P(\bar{t}))$. By Theorem I-4-2, there is a γ_1-fine partition

$$\text{(C)} \qquad \mathscr{P} = \{(\bar{t}_1, A_1), \ldots, (\bar{t}_k, A_k)\}$$

of $[a, b]$; without loss of generality we may suppose that the intervals A_i are numbered from left to right, so that A_i is $(t_{i-1}, t_i]$, with $t_0 = a$ and $t_k = b$. For each i in $\{1, \ldots, k\}$, the closed interval A_i^- is in $\gamma_1(\bar{t}_i)$, so the arc

$$\text{(D)} \qquad C[A_i]: t \mapsto P(t) \qquad (t \text{ in } A_i^-)$$

lies in $\gamma(P(\bar{t}))$, and therefore for all t in A_i^-

(E) $$|\mathbf{F}(P(t)) - \mathbf{F}(P(\bar{t}_i))| < \varepsilon.$$

Let $\mathbf{\Delta}P(A_i)$ denote the vector represented by the directed line-segment that goes from the beginning of the arc $\mathscr{C}[A_i]$ to its end-point. If along $\mathscr{C}[A_i]$ the force were constantly equal to that represented by $\mathbf{F}(P(\bar{t}_i))$, the work done on the particle in traversing $\mathscr{C}[A_i]$ would have been

$$\mathbf{F}(P(\bar{t}_i))(\mathbf{\Delta}P(A_i)).$$

Since (E) holds, for small ε this should be close to the work done by the actual force on the particle in traversing the arc, so the sum

(F) $$\sum_{i=1}^{k} \mathbf{F}(P(\bar{t}_i))(\mathbf{\Delta}P(A_i))$$

should be close to the work done in traversing all of $\mathscr{C}$.

We have practically duplicated the steps in past definitions of the integral, and we now formalize an extension of integration that includes the one just discussed informally. First, let $\mathscr{C}$ be a curve in R^r, and let (A) be a representation of $\mathscr{C}$. To each allotted partition $\mathscr{P}$ of $[a, b]$, with notation (C), there corresponds a set of pairs

(G) $$\mathscr{P}_{\mathscr{C}} = \{(P(\bar{t}_1), \mathscr{C}[A_1]), \ldots, (P(\bar{t}_k), \mathscr{C}[A_k])\}$$

in which each $P(\bar{t}_i)$ is a point of $\mathscr{C}$ and each $\mathscr{C}[A_i]$ is an arc of $\mathscr{C}$. (We are assuming that in (C), all $\bar{t}_i$ are in $[a, b]$.) This set of pairs is called an **allotted partition of** $\mathscr{C}$. If γ is a gauge on $\bar{R}^r$, and for $i = 1, \ldots, k$ the arc $\mathscr{C}[A_i]$ is contained in the neighborhood $\gamma(P(\bar{t}_i))$, the allotted partition $\mathscr{P}_{\mathscr{C}}$ is a **γ-fine partition of** $\mathscr{C}$. For each A_i the vector $\mathbf{\Delta}P(A_i)$ is defined to be the vector represented by the directed line-segment that begins at the initial point of $\mathscr{C}[A_i]$ and ends at the end-point of $\mathscr{C}[A_i]$. If

$$P \mapsto \mathbf{F}(P)$$

is a function defined on a set that contains all the points of $\mathscr{C}$, and the values of $\mathbf{F}$ are in V_r, we define the **partition-sum** corresponding to the function $\mathbf{F}$ and the allotted partition $\mathscr{P}_{\mathscr{C}}$ to be

(H) $$S(\mathscr{P}_{\mathscr{C}}; \mathbf{F} \circ P; \mathbf{\Delta}P) = \sum_{i=1}^{k} \mathbf{F}(P(\bar{t}_i))(\mathbf{\Delta}P(A_i)).$$

Then, using this notation, we make the following definition.

DEFINITION 7-1 *The V_r-valued $\mathbf{F}$ has a **line-integral** along the curve $\mathscr{C}$ if there exists a number J such that to each positive number ε there corresponds a gauge γ on $\bar{R}^r$ such that for every γ-fine partition $\mathscr{P}_{\mathscr{C}}$ of $\mathscr{C}$,*

$$|S(\mathscr{P}_{\mathscr{C}}; \mathbf{F} \circ P; \mathbf{\Delta}P) - J| < \varepsilon.$$

The number J is said to be the value of the integral of **F** *along (or over) C and is denoted by*

$$\int_{\mathscr{C}} \mathbf{F}(P)\,\Delta P.$$

This obviously does not depend on any choice of coordinate system in R^r, since no coordinates are even mentioned. However, there is a mention of a representation of $\mathscr{C}$, namely (A). Still, the existence and value of the integral defined in Definition 7-1 do not depend on the choice of representation of $\mathscr{C}$. For let

$$\text{(I)} \qquad u \mapsto Q(u) \qquad (c \leqq u \leqq d)$$

be another representation of $\mathscr{C}$. In (C) there is no loss of generality in assuming that we have numbered the intervals $A_1, \ldots, A_k$ from left to right. Then there are numbers

$$t_0 = a \leqq t_1 \leqq t_2 \leqq \cdots \leqq t_k = b$$

such that $A_i = (t_{i-1}, t_i]$. Since (I) is another representation of $\mathscr{C}$, there are numbers

$$u_0 = c \leqq u_1 \leqq u_2 \leqq \cdots \leqq u_k = d$$

such that

$$P(t_i) = Q(u_i) \qquad (i = 0, 1, \ldots, k),$$

and there are numbers $\bar{u}_i$ $(i = 1, \ldots, k)$ such that

$$P(\bar{t}_i) = Q(\bar{u}_i) \qquad (i = 1, \ldots, k).$$

Then if we write B_i for $(u_{i-1}, u_i]$, we see that

$$u \mapsto Q(u) \qquad (u \text{ in } B_i^-)$$

is another representation of $\mathscr{C}[A_i]$, so $\Delta Q(B_i) = \Delta P(A_i)$, and the partition of $\mathscr{C}$ defined by representation (I) and the u_i and $\bar{u}_i$ is the same as the partition of $\mathscr{C}$ defined by the representation (A) and the t_i and $\bar{t}_i$. So the partition-sums take the same set of values, independent of the representation of $\mathscr{C}$, and the existence and value of the integral is independent of the representation.

Since computations are usually carried out more conveniently with the help of coordinates, it is useful to introduce another notation, in terms of coordinates. Let a linear coordinate system be chosen in R^r. The point $P(t)$ of the curve $\mathscr{C}$ will have for coordinates an r-column, which we denote by $x(t)$. At the place P with coordinate r-column x, the value of the function **F** will be a covector $\mathbf{F}(P)$, and this will be represented by an r-row. This r-row is determined by the r-column x, so we denote it by $F(x) = (F_1(x), \ldots, F_r(x))$. If A_i is $(t_{i-1}, t_i]$, the coordinates of the beginning and end points of $\mathscr{C}[A_i]$ will be $x(t_{i-1})$ and $x(t_i)$, respectively, so the vector $\mathbf{\Delta}P(A_i)$ represented by the directed line-segment joining them is numerically represented by the r-column $x(t_i) - x(t_{i-1})$. This we denote by $\Delta x(A_i)$.

The value of the covector $\mathbf{F}(P(\bar{t}_i))$ at the vector $\mathbf{\Delta}P(A_i)$ is then given by the matrix product $F(x(\bar{t}_i)\,\Delta x(A_i)$, and

$$S(\mathscr{P}_{\mathscr{C}}; \mathbf{F} \circ P; \mathbf{\Delta}P) = \sum_{i=1}^{k} F(x(\bar{t}_i))\,\Delta x(A_i). \tag{J}$$

Accordingly, we adopt the alternative symbol

$$\int_{\mathscr{C}} F(x)\,dx$$

for the value of the integral. In the future we shall use whichever is more convenient.

As an easy example, if F has a constant value on R^r, then for every curve $\mathscr{C}$, with representation (A), the integral of $\mathbf{F}$ over $\mathscr{C}$ exists, and

$$\int_{\mathscr{C}} \mathbf{F}(P)\,\mathbf{\Delta}P = \mathbf{F}\,\mathbf{\Delta}P(C) = (F)(x(b) - x(a)).$$

For then in (J) the right member is

$$(F)\left(\sum_{i=1}^{k} \Delta x(A_i)\right) = (F)(x(b) - x(a)),$$

no matter what the allotted partition of $\mathscr{C}$.

The line integral has the same sort of linearity as all the integrals in preceding chapters, as follows.

THEOREM 7-2 *Let $\mathscr{C}$ be a curve in R^r; let $\mathbf{F}_1$ and $\mathbf{F}_2$ be functions defined on a set of points that contains all points of $\mathscr{C}$, and let them have values in V_r; and let c_1 and c_2 be real numbers. If $\mathbf{F}_1$ and $\mathbf{F}_2$ are integrable over $\mathscr{C}$, so is $c_1\mathbf{F}_1 + c_2\mathbf{F}_2$, and in that case*

$$\int_{\mathscr{C}} [c_1F_1(x) + c_2F_2(x)]\,dx = c_1\int_{\mathscr{C}} F_1(x)\,dx + c_2\int_{\mathscr{C}} F_2(x)\,dx.$$

The proof is just like the proofs of all preceding theorems of linearity (for example, Theorem I-5-3) and there is no point in writing it out once again.

There is a rough but useful estimate of the magnitude of the line-integral.

THEOREM 7-3 *Let $\mathbf{F}$ be integrable over a curve $\mathscr{C}$ of finite length. Then*

$$\left|\int_{\mathscr{C}} F(x)\,dx\right| \leqq [\sup\{|\mathbf{F}(x)| : x \text{ on } \mathscr{C}\}]L(\mathscr{C}). \tag{K}$$

Let J denote the left member of (K), and let M denote the supremum of $|\mathbf{F}(x)|$ on $\mathscr{C}$. Suppose the theorem false. Then $|J| - ML(\mathscr{C})$ is positive. Let $t \mapsto x(t)$ $(a \leqq t \leqq b)$ represent $\mathscr{C}$ in a linear coordinate system. Since $\mathbf{F}$ is integrable over $\mathscr{C}$, there exists an allotted partition $\mathscr{P} = \{(\bar{t}_1, A_1), \ldots, (\bar{t}_k, A_k)\}$ of $(a, b]$ that produces a partition $\mathscr{P}_{\mathscr{C}}$ of $\mathscr{C}$ for which

(L) $$|S(\mathscr{P}_{\mathscr{C}}; F \circ x; \Delta x) - J| < |J| - ML(\mathscr{C}).$$

But if A_i is the interval $(t_{i-1}, t_i]$,

(M) $$\begin{aligned}|S(\mathscr{P}_{\mathscr{C}}; F \circ x; x)| &= \left|\sum_{i=1}^{k} F(x(\bar{t}_i))(x(t_i) - x(t_{i-1}))\right| \\ &\leqq \sum_{i=1}^{k} |F(x(\bar{t}_i))|\,|x(t_i) - x(t_{i-1})| \\ &\leqq M \sum_{i=1}^{k} |x(t_i) - x(t_{i-1})|.\end{aligned}$$

The last-named sum is the length of a polygon inscribed in $\mathscr{C}$, so it cannot exceed $L(\mathscr{C})$, and (M) implies

$$|S(\mathscr{P}_{\mathscr{C}}; F \circ x; \Delta x)| \leqq ML(\mathscr{C}).$$

This contradicts (L), and the theorem is established.

For the rest of this section we shall save work by restricting our attention to the special but especially important case of continuous integrands $\mathbf{F}$.

Theorem 7-4 *Let $\mathscr{C}$ be a curve of finite length in R^r, and let $\mathbf{F}$ be a function with values in V_r that is defined on a set D of points that contains all points of $\mathscr{C}$ and that is continuous at all points of $\mathscr{C}$. Then $\mathbf{F}$ is integrable over $\mathscr{C}$.*

We choose a linear coordinate system in R^r. Let $t \mapsto x(t)$ $(a \leqq t \leqq b)$ be a representation of $\mathscr{C}$, and let ε be positive. For each point x_0 on C there is a neighborhood $\gamma(x_0)$ of x_0 such that if x is in $\gamma(x_0) \cap D$,

$$|F(x) - F(x_0)| < \varepsilon/(2L(\mathscr{C}) + 2).$$

For x_0 in $\bar{R}^r$ but not on $\mathscr{C}$ we define $\gamma(x_0)$ to be $\bar{R}^r$. Then γ is a gauge on $\bar{R}^r$. Let

$$\mathscr{P}' = \{(t'_1, A'_1), \ldots, (t'_h, A'_h)\}, \qquad \mathscr{P}'' = \{(t''_1, A''_1), \ldots, (t''_k, A''_k)\}$$

be two allotted partitions of $(a, b]$ that produce γ-fine partitions of C. We suppose the intervals numbered from left to right. For $i = 1, \ldots, h$ the intervals $A'_i \cap A''_1, A'_i \cap A''_2, \ldots, A'_i \cap A''_k$ are subintervals of A'_i numbered from left to right, and their union is A'_i. Clearly,

$$\Delta x(A'_i) = \sum_{j=1}^{k} \Delta x(A'_i \cap A''_j).$$

We multiply each member of this equation by $F(x(t_i'))$ and add; the result is

$$S(\mathscr{P}'_{\mathscr{C}}; F \circ x; \Delta x) = \sum_{i=1}^{h} \sum_{j=1}^{k} F(x(t_i'))\, \Delta x(A_i' \cap A_j'').$$

In the same way

$$S(\mathscr{P}''_{\mathscr{C}}; F \circ x; \Delta x) = \sum_{i=1}^{h} \sum_{j=1}^{k} F(x(t_j''))\, \Delta x(A_i' \cap A_j'').$$

So

(N) $$|S(\mathscr{P}'_{\mathscr{C}}; F \circ x; \Delta x) - S(\mathscr{P}''_{\mathscr{C}}; F \circ x; \Delta x)|$$
$$= \left| \sum_{i=1}^{h} \sum_{j=1}^{k} [F(x(t_i')) - F(x(t_j''))]\, \Delta x(A_i' \cap A_j'') \right|$$
$$\leqq \sum_{i=1}^{h} \sum_{j=1}^{k} |F(x(t_i')) - F(x(t_j''))|\, |\Delta x(A_i' \cap A_j'')|.$$

For each i in $\{1, \ldots, h\}$ and each j in $\{1, \ldots, k\}$, if $A_i' \cap A_j''$ is not empty it contains a point t^*, and $x(t^*)$ is in $\gamma(x(t_i'))$ because t^* is in A_i and $\mathscr{P}'_{\mathscr{C}}$ is γ-fine. So

$$|F(x(t^*)) - F(x(t_i'))| < \varepsilon/(2L(\mathscr{C}) + 2).$$

Similarly,

$$|F(x(t^*)) - F(x(t_j''))| < \varepsilon/(2L(\mathscr{C}) + 2),$$

so

$$|F(x(t_i')) - F(x(t_j''))|\, |\Delta x(A_i' \cap A_j'')| \leqq [\varepsilon/(L(\mathscr{C}) + 1)]|\Delta x(A_i' \cap A_j'')|.$$

This is also true if $A_i' \cap A_j''$ is empty, since then both members are 0. So by (N),

(O) $$|S(\mathscr{P}'_{\mathscr{C}}; F \circ x; \Delta x) - S(\mathscr{P}''_{\mathscr{C}}; F \circ x; \Delta x)|$$
$$\leqq \frac{\varepsilon}{L(\mathscr{C}) + 1} \sum_{i=1}^{h} \sum_{j=1}^{k} |\Delta x(A_i' \cap A_j'')|.$$

For each i in $\{1, \ldots, h\}$ the intersections $A_i' \cap A_1'', A_i' \cap A_2'', \ldots, A_i' \cap A_k''$ are pairwise disjoint intervals (some empty) whose union is A_i', so the whole collection of intersections $A_i' \cap A_j''$ is pairwise disjoint and has union $(a, b]$. So the last-named sum in (O) is the length of a polygon inscribed in $\mathscr{C}$ and cannot be greater than $L(\mathscr{C})$. That is,

$$|S(\mathscr{P}'_{\mathscr{C}}; F \circ x; \Delta x) - S(\mathscr{P}''_{\mathscr{C}}; F \circ x; \Delta x)| < \varepsilon.$$

Here, just as in the proof of Theorem II-1-1, the fact that this holds for all γ-fine partitions $\mathscr{P}'$ and $\mathscr{P}''$ is the Cauchy test for convergence. The gauge limit of $S(\mathscr{P}_{\mathscr{C}}; F \circ x; \Delta x)$ exists, and the theorem is proved.

The line integrals of continuous functions **F** also have the additivity property that if a curve is cut into finitely many arcs, the integral over the whole curve is the sum of the integrals over the parts.

THEOREM 7-5 *Let $\mathscr{C}$ be a curve in R^r with finite length, and in a linear coordinate system in R^r let $\mathscr{C}$ have representation $t \mapsto x(t)$ $(a \leqq t \leqq b)$. Let $B_1, \ldots, B_k$ be intervals $(b_0, b_1]$, $(b_1, b_2], \ldots, (b_{k-1}, b_k]$, where $a = b_0 < b_1 < b_2 < \cdots < b_k = b$. Let* **F** *be a V_r-valued function defined on a set of points that contains the points of $\mathscr{C}$, and let* **F** *be continuous at the points of $\mathscr{C}$. Then*

$$\int_{\mathscr{C}} F(x)\,dx = \sum_{j=1}^{k} \int_{\mathscr{C}[B_j]} F(x)\,dx.$$

Let ε be positive. There is a gauge γ_0 on $\bar{R}^r$ such that if $\mathscr{P}_{\mathscr{C}}$ is a γ_0-fine partition of $\mathscr{C}$,

$$\text{(P)} \qquad \left| S(\mathscr{P}_{\mathscr{C}}; F \circ x; \Delta x) - \int_{\mathscr{C}} F(x)\,dx \right| < \frac{\varepsilon}{2},$$

and for each j in $\{1, \ldots, k\}$ there is a gauge γ_j on $\bar{R}^r$ such that if $\mathscr{P}_{\mathscr{C}[B_j]}$ is a γ-fine partition of the arc $\mathscr{C}[B_j]$,

$$\text{(Q)} \qquad \left| S(\mathscr{P}_{\mathscr{C}[B_j]}; F \circ x; \Delta x) - \int_{\mathscr{C}[B_j]} F(x)\,dx \right| < \varepsilon/2k.$$

Define $\gamma = \gamma_0 \cap \gamma_1 \cap \cdots \cap \gamma_k$, and for $j = 1, \ldots, k$ let $\mathscr{P}_j$ be an allotted partition of $(a, b]$ that produces a γ-fine partition $\mathscr{P}_{\mathscr{C}[B_j]}$ of $\mathscr{C}[B_j]$. The union $\mathscr{P}_0$ of $\mathscr{P}_1, \ldots, \mathscr{P}_k$ is an allotted partition of $(a, b]$ that produces a γ-fine partition $\mathscr{P}_{\mathscr{C}}$ of $\mathscr{C}$. Then (P) and (Q) are satisfied, and

$$S(\mathscr{P}_{\mathscr{C}}; F \circ x; \Delta x) = \sum_{j=1}^{k} S(\mathscr{P}_{\mathscr{C}[B_j]}; F \circ x; \Delta x).$$

From this, with (P) and (Q), we deduce

$$\begin{aligned}
&\left| \int_{\mathscr{C}} F(x)\,dx - \sum_{j=1}^{k} \int_{\mathscr{C}[B_j]} F(x)\,dx \right| \\
&\quad = \left| \left\{ \int_{\mathscr{C}} F(x)\,dx - S(\mathscr{P}_{\mathscr{C}}; F \circ x; \Delta x) \right\} \right. \\
&\qquad \left. - \sum_{j=1}^{k} \left\{ \int_{\mathscr{C}[B_j]} F(x)\,dx - S(\mathscr{P}_{\mathscr{C}[B_j]}; F \circ x; \Delta x) \right\} \right| \\
&\quad < \varepsilon.
\end{aligned}$$

The first member of this inequality is a fixed nonnegative number, and it is less than the arbitrary positive number ε, so it is 0. The proof is complete.

It is a reasonable and correct conjecture that if $\mathbf{F}$ is continuous and a sequence of curves $\mathscr{C}_n$ tends to a curve $\mathscr{C}_0$ both in position and in direction everywhere, the integral of $\mathbf{F}$ over $\mathscr{C}_n$ should approach its integral over $\mathscr{C}_0$. The next theorem is interesting and useful because it states a condition for convergence of the integral over $\mathscr{C}_n$ to the integral over $\mathscr{C}_0$ without mentioning the direction or derivative or tangent of any curve.

THEOREM 7-6 *Let $\mathscr{C}_0, \mathscr{C}_1, \mathscr{C}_2, \ldots$ be curves in R^r such that in a linear coordinate system, $\mathscr{C}_n$ has a representation $t \mapsto x_j(t)$ $(a \leqq t \leqq b)$, $(j = 0, 1, 2, \ldots)$. Let $\mathbf{F}$ be a function with values on V_r that is defined on a set that contains all the points of all the $\mathscr{C}_j$, and let $\mathbf{F}$ be continuous at all points of all curves $\mathscr{C}_0, \mathscr{C}_1, \mathscr{C}_2, \ldots$. Let the functions x_j converge to x_0 uniformly on $[a, b]$ as j increases, and let there exist a number M such that all the $\mathscr{C}_j$ have lengths less than M. Then*

$$\lim_{j \to \infty} \int_{\mathscr{C}_j} F(x)\, dx = \int_{\mathscr{C}_0} F(x)\, dx.$$

Let ε be positive. To each point x_0 on $\mathscr{C}_0$ there corresponds a neighborhood $\gamma(x_0)$ such that for all x in $\gamma(x_0)$ that are on some $\mathscr{C}_j$,

$$|\mathbf{F}(x) - \mathbf{F}(x_0)| < \varepsilon/3M. \tag{R}$$

To each point $\bar{x}$ of $\bar{R}^r$ that is not on $\mathscr{C}_0$ we assign $\bar{R}^r$ as $\gamma(\bar{x})$. Then γ is a gauge on $\bar{R}^r$. Let

$$\mathscr{P} = \{(\bar{t}_1, A_1), \ldots, (\bar{t}_k, A_k)\}$$

be an allotted partition of $(a, b]$ that produces a γ-fine partition of $\mathscr{C}_0$. We may assume that the A_i are numbered in order from left to right and that A_i is $(t_{i-1}, t_i]$, where

$$t_0 = a < t_1 < \cdots < t_k = b.$$

Let i be one of the numbers $1, \ldots, k$. The arc $t \mapsto x_0(t)$ $(t_{i-1} \leqq t \leqq t_i)$ is in $\gamma(x_0(\bar{t}_i))$, so the distance from $x_0(t)$ to the complement of that neighborhood is a function that is positive-valued and continuous on $[t_{i-1}, t_i]$. It therefore has a positive lower bound δ_i on $[t_{i-1}, t_i]$. For all large j, the distance from $x_j(t)$ to $x_0(t)$ is less than δ_i on all of $[a, b]$, so the arc $\mathscr{C}_j[A_i]$ defined by $t \mapsto x_j(t)$ $(t_{i-1} \leqq t \leqq t_i)$ is also in $\gamma(x_0(\bar{t}_i))$. We choose an n_0 such that for $j > n_0$ this is true for all i in $\{1, \ldots, k\}$.

For all i in $\{1, \ldots, k\}$ and all nonnegative integers j,

$$\begin{aligned} \int_{\mathscr{C}_j[A_i]} F(x_0(\bar{t}_i))\, dx &= F(x_0(\bar{t}_i))[x_j(t_i) - x_j(t_{i-1})] \\ &= F(x_0(\bar{t}_i))\, \Delta x_j(A_i). \end{aligned}$$

From this and Theorem 7-5,

$$\text{(S)}\quad \int_{\mathscr{C}_j} F(x)\,dx - \int_{\mathscr{C}_0} F(x)\,dx$$

$$= \sum_{i=1}^{k} \int_{\mathscr{C}_j[A_i]} F(x)\,dx - \sum_{i=1}^{k} \int_{\mathscr{C}_0[A_i]} F(x)\,dx$$

$$= \sum_{i=1}^{k} \int_{\mathscr{C}_j[A_i]} \{F(x) - F(x_0(\bar{t}_i))\}\,dx + \sum_{i=1}^{k} F(x_0(\bar{t}_i))\,\Delta x_j(A_i)$$

$$- \sum_{i=1}^{k} \int_{\mathscr{C}_0[A_i]} \{F(x) - F(x_0(\bar{t}_i))\}\,dx + \sum_{i=1}^{k} F(x_0(\bar{t}_i))\,\Delta x_0(A_i).$$

Since all points x of the arc $\mathscr{C}[A_j]$ are in the neighborhood $\gamma(x(\bar{t}_j))$, by (R) the integrand $F(x) - F(x(\bar{t}_j))$ has norm $|F(x) - F(x(\bar{t}_j))|$ at most $\varepsilon/3M$ on $\mathscr{C}[A_j]$, so by Theorem 7-3,

$$\left| \sum_{i=1}^{k} \int_{\mathscr{C}[A_j]} \{F(x) - F(x(\bar{t}_j))\}\,dx \right| \leqq \sum_{i=1}^{k} \frac{\varepsilon}{3M} L(\mathscr{C}[A_j]) < \frac{\varepsilon}{3}.$$

In a like manner, the other sum of integrals in the last member of (S) has absolute value less than $\varepsilon/3$. For $i = 1, \ldots, k$, as j increases, the difference $\Delta x_j(A_i) = x_j(t_i) - x_j(t_{i-1})$ converges to $\Delta x_0(A_i) = x_0(t_i) - x_0(t_{i-1})$, so the sum of all the remaining terms in (S) also tends to 0, and for all large j it has absolute value less than $\varepsilon/3$. So for all large j the absolute value of the right member is less than ε, and the proof is complete.

There is considerable computational advantage in reducing the line-integrals to ordinary integrals with respect to measure m_L as, for example, we did for integrals with respect to dm in Theorem III-9-3. A similar theorem holds here also.

THEOREM 7-7 *Let $\mathscr{C}$ be a curve in R^r that has a representation $t \mapsto x(t)$ $(a \leqq t \leqq b)$ such that there exists an r-column of real-valued functions $t \mapsto \dot{x}(t)$ $(a \leqq t \leqq b)$ with which*

$$x(t) = x(a) + \int_a^t \dot{x}(u)\,du \qquad (a \leqq t \leqq b).$$

Let **F** *be a function with values in V_r that is defined on a set of points containing the points of $\mathscr{C}$ and that is continuous at all the points of $\mathscr{C}$. Then*

$$\int_{\mathscr{C}} F(x)\,dx = \int_a^b F(x(t))\dot{x}(t)\,dt.$$

Let ε be positive. By Theorem 7-4, $\mathbf{F}$ is integrable over $\mathscr{C}$, so there exists a gauge γ_1 on R^r such that for every γ_1-fine partition $\mathscr{P}_{\mathscr{C}}$ of $\mathscr{C}$,

$$\text{(T)} \qquad \left| S(\mathscr{P}_{\mathscr{C}}; F \circ x; \Delta x) - \int_{\mathscr{C}} F(x)\,dx \right| < \varepsilon/2.$$

Since F is continuous at the points of $\mathscr{C}$, to each point $\bar{x}$ on $\mathscr{C}$ there corresponds a neighborhood $\gamma_2(\bar{x})$ such that for all x on $\mathscr{C}$ that are in $\gamma_2(\bar{x})$,

$$|F(x) - F(\bar{x})| < \varepsilon/(2L(\mathscr{C}) + 2).$$

For $\bar{x}$ in $\bar{R}^r$ but not on $\mathscr{C}$, we choose $\gamma_2(\bar{x})$ to be $\bar{R}^r$. Then γ_2 is a gauge on $\bar{R}^r$, and so is $\gamma = \gamma_1 \cap \gamma_2$.

Now let $\mathscr{P} = \{(\bar{t}_i, A_1), \ldots, (\bar{t}_k, A_k)\}$ be an allotted partition of $(a, b]$ that produces a γ-fine partition $\mathscr{P}_{\mathscr{C}}$ of $\mathscr{C}$. Without loss of generality we may assume that the A_i are numbered from left to right, so that A_i is $(t_{i-1}, t_i]$, with $t_0 = a < t_i < \cdots < t_k = b$. Then

$$\text{(U)} \qquad F(x(\bar{t}_i))\,\Delta x(A_i) = F(x(\bar{t}_i))[x(t_i) - x(t_{i-1})] = \int_{t_{i-1}}^{t_i} F(x(\bar{t}_j))\dot{x}(t)\,dt.$$

For all t in A_i^-, $x(t)$ is on $\mathscr{C}[A_i]$, hence in $\gamma_2(x(\bar{t}_i))$, so

$$\text{(V)} \qquad |F(x(t)) - F(x(\bar{t}_i))| < \varepsilon/(2L(\mathscr{C}) + 2).$$

Now, by Corollaries 4-2 and I-6-4 and Theorem 6-3,

$$\begin{aligned}
&\left| \int_a^b F(x(t))\dot{x}(t)\,dt - S(\mathscr{P}_{\mathscr{C}}; F \circ x; \Delta x) \right| \\
&\quad = \left| \sum_{j=1}^{k} \int_{t_{i-1}}^{t_i} [F(x(t))\,dt - F(x(\bar{t}_j))]\dot{x}(t)\,dt \right| \\
&\quad \leqq \sum_{j=1}^{k} \int_{t_{j-1}}^{t_j} |F(x(t)) - F(x(\bar{t}_j))|\,|\dot{x}(t)|\,dt \\
&\quad \leqq \sum_{j=1}^{k} \frac{\varepsilon}{2L(\mathscr{C}) + 2} \int_{t_{j-1}}^{t_j} |\dot{x}(t)|\,dt \\
&\quad = \frac{\varepsilon}{2L(\mathscr{C}) + 2} \int_a^b |\dot{x}(t)|\,dt \\
&\quad < \frac{\varepsilon}{2}.
\end{aligned}$$

This and (T) imply

$$\left| \int_{\mathscr{C}} F(x)\,dx - \int_a^b F(x(t))\dot{x}(t)\,dt \right| < \varepsilon.$$

The left member is a fixed nonnegative number, which has just been shown to be less than an arbitrary positive number ε, so it is 0. The theorem is proved.

Among line-integrals an especially well-behaved subset consists of those in which the value of the integral along $\mathscr{C}$ depends only on the location of the beginning and end of $\mathscr{C}$, not on its course between them. For instance, in physics it often happens that a force field is described by a covector-valued function $\mathbf{F}$ such that the integral of $\mathbf{F}$ along $\mathscr{C}$ depends only on the beginning and end of $\mathscr{C}$. Such force fields are called **conservative.** The next theorem furnishes a test by which this global property (equality of the integrals) can be deduced from a local test involving the derivatives of $\mathbf{F}$ at individual points.

Suppose that G is an open set in R^r, that P_0 and P_1 are points of G, and that $\mathscr{C}_0$ and $\mathscr{C}_1$ are two curves lying in G, both having beginning-point P_0 and end-point P_1. The geometric picture of deforming $\mathscr{C}_0$ into $\mathscr{C}_1$ without ever leaving G or moving the end-points can be given an analytic expression thus. There is a continuous function

$$(t, u) \mapsto x(t, u) \qquad ((t, u) \text{ in } [0, 1] \times [0, 1])$$

with values in G such that $x(0, u)$ is P_0 for all u in $[0, 1]$; $x(1, u)$ is P_1 for all u in $[0, 1]$; the function $t \mapsto x(t, 0)$ $(0 \leqq t \leqq 1)$ represents $\mathscr{C}_0$; and the function $t \mapsto x(t, 1)$ $(0 \leqq t \leqq 1)$ represents $\mathscr{C}_1$. With this understanding we can state and prove the following theorem.

THEOREM 7-8 *Let G be an open set in R^r, and let $\mathbf{F}$ be a function with values in V_r that is defined and continuously differentiable on G. In a linear coordinate system, let $\mathbf{F}(x)$ be represented by the r-row $(F_1(x), \ldots, F_r(x))$. Then these two statements are equivalent:*

(i) *whenever $\mathscr{C}_0$ and $\mathscr{C}_1$ are curves of finite length lying in G and such that $\mathscr{C}_0$ can be deformed into $\mathscr{C}_1$ in G without moving the end-points,*

$$\int_{\mathscr{C}_1} F(x)\, dx = \int_{\mathscr{C}_0} F(x)\, dx;$$

(ii) *for each i and j in $\{1, \ldots, r\}$ the partial derivatives of F_i and F_j with respect to the jth and ith coordinates satisfy*

$$\text{(W)} \qquad D_i F_j(x) = D_j F_i(x)$$

at all points of G.

Suppose, first, that (i) is satisfied. Let x_0 be any point of G. Since G is open, there is a positive δ such that the cube

$$(x_0^1 - \delta, x_0^1 + \delta) \times \cdots \times (x_0^r - \delta, x_0^r + \delta)$$

is contained in G. As usual, we denote by δ_i the r-column with 1 in ith place and 0 elsewhere. If $0 \leqq e < \delta$, the two-sided polygon C_0 with vertices x_0, $x_0 + e\delta_i$,

$x_0 + e\delta_i + e\delta_j$ and the polygon with vertices $x_0, x_0 + e\delta_j, x_0 + e\delta_j + e\delta_i$ can be deformed into each other in G without moving the end-points; all we need to do is to move the middle vertex continuously along the diagonal of the square. The polygon C_0 consists of the two sides $t \mapsto x_0 + t\delta_i, t \mapsto x_0 + e\delta_i + t\delta_j$ $(0 \leqq t \leqq e)$, and C_1 consists of the two sides $t \mapsto x_0 + t\delta_j, t \mapsto x_0 + e\delta_j + t\delta_i$ $(0 \leqq t \leqq e)$, so the equality of the integrals of $\mathbf{F}$ over the two polygons, together with Theorem 7-7 and the observation that $F\delta_i = F_i$ and $F\delta_j = F_j$, yields

$$\int_0^e \{F_i(x_0 + t\delta_i) + F_j(x_0 + e\delta_i + t\delta_j) - F_j(x_0 + t\delta_j) - F_i(x_0 + e\delta_j + t\delta_i)\}\, dt = 0.$$

By the fundamental theorem,

$$-\int_0^e \int_0^e D_j F_i(x_0 + t\delta_i + s\delta_j)\, ds\, dt + \int_0^e \int_0^e D_i F_j(x_0 + s\delta_i + t\delta_j)\, ds\, dt = 0.$$

In the last integral we can interchange s and t without changing the value of the integral. Then the integral over $[0, e] \times [0, e]$ of the continuous function

$$D_i F_j(x_0 + t\delta_i + s\delta_j) - D_j F_i(x_0 + t\delta_i + s\delta_j)$$

is 0, so the function has to vanish at some point (t', s') in that square. Since the function is continuous and vanishes at a point in an arbitrarily small square that contains x_0, it vanishes at x_0, and (ii) is satisfied.

Conversely, suppose that (ii) is satisfied. Let $\mathscr{C}_0$ and $\mathscr{C}_1$ be curves of finite length that lie in G and can be deformed into each other in G without moving the end-points. Then there is a continuous function

$$(t, u) \mapsto x(t, u) \qquad ((t, u) \text{ in } [0, 1] \times [0, 1])$$

with values in G such that $x(0, u)$ is $x(0, 0)$ and $x(1, u)$ is $x(1, 0)$ for all u in $[0, 1]$, and the functions $t \mapsto x(t, 0)$ and $t \mapsto x(t, 1)$ $(0 \leqq t \leqq 1)$ represent $\mathscr{C}_0$ and $\mathscr{C}_1$, respectively. Let n be a positive integer. We first subdivide the unit square $[0, 1] \times [0, 1]$ into n^2 congruent squares and then subdivide each of these into four triangles by drawing the diagonals. Now let x_n be the function on the unit square that on each of the $4n^2$ triangles is linear and coincides with x at the vertices. Then x_n converges uniformly to x as n increases, so for all large n (we discard the others) $x_n(t, u)$ is in G. For each fixed u in $[0, 1]$, the function $t \mapsto x_n(t, u)$ $(0 \leqq t \leqq 1)$ represents a polygon in G with the same beginning and end as $\mathscr{C}_0$ and $\mathscr{C}_1$. We call this $\Pi(n, u)$.

As in the statement of the theorem, for any function of several variables, D_i denotes partial differentiation with respect to the ith independent variable. For example, $D_2 x(t, u)$ means what is often written $\partial x(t, u)/\partial u$. By Theorem 7-7,

$$\int_{\Pi(n,u)} F(x)\, dx = \int_0^1 \sum_{i=1}^r F_i(x_n(t, u))\, D_1 x_n^i(t, u)\, dt.$$

By Theorem II-11-1, we can differentiate under the integral sign with respect to u and obtain

$$D\int_{\Pi(n,u)} F(x)\,dx = \int_0^1 \left[\sum_{i,j=1}^{r} D_jF_i(x_n(t,u))\,D_2x_n^j(t,u)\,D_1x_n^i(t,u) + \sum_{i=1}^{r} F_i(x_n(t,u))\,D_2\,D_1x_n^i(t,u) \right] dt.$$

In the last term in the integrand we interchange the order of the partial differentiations, as we may because the x_n^i are piecewise linear, and we integrate by parts. The result is

$$D\int_{\Pi(n,u)} F(x)\,dx = \int_0^1 \sum_{i,j=1}^{r} [D_jF_i(x_n(t,u)) - D_iF_j(x_n(t,u))] \times [D_2x_n^j(t,u)][D_1x_n^i(t,u)]\,dt.$$

But the integrand is identically 0 by hypothesis, so the integral of F over $\Pi(nu)$ is constant. In particular,

$$\int_{\Pi(n,1)} F(x)\,dx = \int_{\Pi(n,0)} F(x)\,dx.$$

The polygons $\Pi(n,0)$ and $\Pi(n,1)$ have lengths not greater than $\max\{L(\mathscr{C}_0), L(\mathscr{C}_1)\}$, and their representing functions $t \mapsto x_n(t,0)$, $t \mapsto x_n(t,1)$ converge uniformly to the functions that represent $\mathscr{C}_0$ and $\mathscr{C}_1$, respectively. By Theorem 7-6, the integrals of **F** over the $\Pi(n,0)$ converge to the integral of F over $\mathscr{C}_0$, and the integrals of **F** over the $\Pi(n,1)$—which are equal to those over the $\Pi(n,0)$—converge to the integral over $\mathscr{C}_1$. So the integrals of **F** over $\mathscr{C}_0$ and $\mathscr{C}_1$ are equal, and the proof is complete.

EXERCISE 7-1 Let G be an open set in R^r such that any two points in G can be joined by a polygon in G. Let **F** be a continuous covector-valued function on G. Assume that for any two polygons $\mathscr{C}_1, \mathscr{C}_2$ in G with the same beginning and the same end,

$$\int_{\mathscr{C}_1} \mathbf{F}(x)\,d\mathbf{x} = \int_{\mathscr{C}_2} \mathbf{F}(x)\,d\mathbf{x}.$$

Prove that there exists a real-valued function Φ on G such that

$$F_j(x) = D_j\Phi(x) \qquad (j = 1, \ldots, r;\ x \text{ in } G).$$

Suggestion: Fix 0 in G; let $\Phi(x_n)$ be the value of the integral of **F** along any polygon in G from 0 to the point with coordinates x_0. If x_0 is in G and ε is a small positive number, the segment from $x = (x_0^1 - \varepsilon, x_0^2, \ldots, x_0^r)$ to x_0 is in G. Let $t \mapsto P(t)$ $(0 \leqq t \leqq 2)$ be a polygon in G with $P(t) = (x_0^1 - \varepsilon(2 - t), x_0^2, \ldots, x_0^r)$ $(1 \leqq t \leqq 2)$. By Theorem 7-7, compute $D_1\Phi(P(t))$ at $t = 2$.

EXERCISE 7-2 The unit vector at a point $x \neq O$ in R^3 that points to the origin has rectangular components $v^i = -x^i/|x|$, where $|x| = [(x^1)^2 + (x^2)^2 + (x^3)^2]^{1/2}$. A mass M at the origin attracts a particle of mass m at x with a gravitational force pointing to the origin and of magnitude $GMm|x|^{-2}$, where G is a constant. Find the components of the gravitational force covector $\mathbf{F}$. Show that its components satisfy equations (W). Show that $F_i = D_i[GMm/|x|]$. Show that for every curve $\mathscr{C}$ that satisfies the hypotheses of Theorem 7-7,

$$\int_{\mathscr{C}} \mathbf{F}(x)\, d\mathbf{x} = GMm[|x(b)|^{-1} - |x(a)|^{-1}].$$

EXERCISE 7-3 A long straight wire along the x^3-axis carries a current. A second wire carries a current, is parallel to the x^3-axis, and passes through the point $Q = (x^1, x^2, 0)$. The current in the first wire produces a magnetic field that produces a force on the second wire that can be represented by a vector in the x^1x^2-plane whose length is a constant multiple of $|OQ|^{-1}$ and whose direction is one right angle ahead of the vector $\overrightarrow{OQ}$. Find the components of the force and show that they satisfy (W). Let $\mathscr{C}_0$ and $\mathscr{C}_1$ be two paths in the x^1x^2-plane that are halves of the same circle with center O and with the same beginning and the same end, but with one clockwise and the other counterclockwise. Show that two translations of the second wire, one so that its intersection with the x^1x^2-plane follows $\mathscr{C}_0$ and the other so that the intersection follows $\mathscr{C}_1$, yield different values for the integral of the force along the path. (That is why electric motors work.)

8. The Behavior of Inscribed Polyhedra

If f is a function on an interval B in R^1 and has a derivative at a point x_0 of B, and $[a_n, b_n]$ is a sequence of subintervals of B such that each one contains x_0 and $b_n - a_n$ tends to 0, and f_n is a linear function that coincides with f at a_n and b_n, then Df_n will tend to $Df(x_0)$ as n increases. This fact lies at the basis of the proof of Theorem 6-3. Suppose now that f is defined on an interval B in R^2 and is differentiable at a point x_0 of B, and that $T_1, T_2, T_3, \ldots$ is a sequence of triangles contained in B that contain x_0 and have edge-lengths tending to 0. If f_n is the linear function that coincides with f at the vertices of T_n, one might optimistically hope that D_1f_n and D_2f_n converge to $D_1f(x_0)$ and $D_2f(x_0)$, respectively. But this is false. For example, let B be the square $[-1, 1] \times [-1, 1]$, and on B let f be defined by

$$f(x^1, x^2) = [x^1]^2.$$

For each positive integer n, the triangle T_n with vertices $(0, 0)$, $(-1/n, 1/n^4)$, and $(1/n, 1/n^4)$ lies in B, contains $(0, 0)$, and has its longest edge of length $2/n$, which tends to 0 as n increases. The linear function f_n that coincides with f at the vertices of T_n is defined by

$$f_n(x^1, x^2) = n^2[x^2].$$

So $D_2 f_n = n^2$, which tends to ∞ with n, whereas $D_2 f = 0$.

The trouble arises because the triangles T_n are thin in proportion to their lengths. The smallest square that contains T_n has area $(2/n)^2$, and the area of $T_n = 1/n^4$, and the ratio of $1/n^4$ to $(2/n)^2$ tends to 0 as n increases. With this as a guide, we define the **thickness ratio** of any bounded measurable set B in R^2 to be the ratio of mB to the measure (area) of the smallest square that contains B. The next three lemmas indicate that if in defining inscribed polyhedra we use only triangles with thickness ratio bounded away from 0, the kind of trouble shown in the example will not arise. These lemmas could be extended to higher-dimensional spaces with no trouble except for slightly more complicated notation.

We begin with a simple estimate.

LEMMA 8-1 *If S is a (closed) triangle in R^2 contained in a square of edge e, and $mS \geqq \tau e^2$ for some positive τ, and g is a linear function on R^2 whose values at the vertices of S lie in an interval $[M_1, M_2]$, then the two partial derivatives of g satisfy*

$$|D_i g(x)| \leqq (M_2 - M_1)/\tau e \qquad (i = 1, 2).$$

Let S be contained in the square

$$Q = [a^1, a^1 + e] \times [a^2, a^2 + e].$$

For each v in the interval $[a^2, a^2 + e]$ we denote by $S[v]$ the set of all u in R such that (u, v) is in S. Since S is a triangle, $S[v]$ is an interval. By Corollary IV-4-3 or Lemma IV-5-5, we know that

$$m_L S = \int_{[a^2, a^2+e]} m_L S[v]\, du.$$

(The m_L in the left member is two-dimensional measure; that in the right member is length of $S[v]$.) By hypothesis, the left member is at least τe^2. If $m_L S[v]$ were less than τe for all v in $[a^2, a^2 + e]$, the right member would be less than $\tau e m_L[a^2, a^2 + e]$, which is τe^2. So there is some v_0 in $[a^2, a^2 + e]$ such that

$$m_L S[v_0] \geqq \tau e.$$

Let $S[v_0]$ be $[a, b]$; then $b - a \geqq \tau e$, and the points

$$x_1 = (a, v_0), \qquad x_2 = (b, v_0)$$

belong to S. Since g is linear, $D_1 g$ is constant, and

$$|D_1 g| = |[g(x_2) - g(x_1)]/(b - a)| \leqq (M_2 - M_1)/\tau e.$$

In the same way we prove that

$$|D_2 g| \leqq (M_2 - M_1)/\tau e.$$

From this lemma we deduce two useful corollaries.

LEMMA 8-2 *Let f be defined and satisfy a Lipschitz condition of constant K on a set B in R^2. Let S be a (closed) triangle in R^2 that is contained in B and has thickness ratio $\tau > 0$, and let g be the linear function that coincides with f at the vertices of S. Then the two partial derivatives of g satisfy*

(A) $$|D_i g| \leqq 2^{3/2} K/\tau.$$

Let the vertices of S be A_0, A_1, and A_2, and let

$$Q = [a^1, a^1 + e] \times [a^2, a^2 + e]$$

be a square with the shortest possible edge that contains S. Then A_1 and A_2 have distance at most $2^{1/2}e$ from A_0, so for $i = 1, 2$,

$$f(A_0) - K2^{1/2}e \leqq f(A_i) \leqq f(A_0) + K2^{1/2}e.$$

The hypotheses of Lemma 8-1 are satisfied with

$$M_1 = f(A_0) - K2^{1/2}e, \qquad M_2 = f(A_0) + K2^{1/2}e,$$

so (A) is valid by Lemma 8-1.

LEMMA 8-3 *Let F be defined on a set B in R^2 and be differentiable at a point x_0 of B. Let $\tau, \varepsilon_1, \varepsilon_2, \varepsilon_3, \ldots$ be positive numbers such that ε_n tends to 0 as n increases. For each positive integer n, let S_n be a triangle with thickness ratio at least τ such that x_0 is in S_n, and the vertices of S_n are in B, and S_n is contained in a square of edge-length less than ε_n. Let f_n be the linear function on R^2 that coincides with f at the vertices of S_n. Then*

(B) $$\lim_{n \to \infty} D_i f_n = D_i f(x_0) \qquad (i = 1, 2).$$

Let Q_n be a square

$$Q_n = [a_n^1, a_n^1 + e_n] \times [a_n^2, a_n^2 + e_n]$$

of least possible edge-length that contains S_n. Since S_n is contained in some square of edge-length less than ε_n, we must have $e_n < \varepsilon_n$, and therefore

(C) $$\lim_{n \to \infty} e_n = 0.$$

Let ε be any positive number. Since f has a differential at x_0, there is a positive δ such that for every point x of B with $\operatorname{dist}(x, x_0) < \delta$,

(D) $$\left| f(x) - f(x_0) - \sum_{i=1}^{2} D_i f(x_0)(x^i - x_0^i) \right| \leqq \frac{\tau\varepsilon}{2^{3/2}} \operatorname{dist}(x, x_0).$$

The function g_n defined by

$$\text{(E)} \qquad g_n(x) = f_n(x) - f(x_0) - \sum_{i=1}^{2} D_i f(x_0)(x^i - x_0^i)$$

is linear. Let the vertices of the triangle S_n be denoted by $y_{n,0}$, $y_{n,1}$, and $y_{n,2}$. Since these points and the point x_0 are all in Q_n, which has edge-length e_n,

$$\text{(F)} \qquad \operatorname{dist}(y_{n,j}, x_0) \leqq 2^{1/2} e_n \qquad (j = 0, 1, 2),$$

and by (C), the right member of this inequality is less than δ for all large n. Since f_n coincides with f at the vertices of S_n, for such n we have by (F) and (D), for $j = 1, 2, 3$,

$$\begin{aligned} |g_n(y_{n,j})| &= \left| f(y_{n,j}) - f(x_0) - \sum_{i=1}^{2} D_i f(x_0)(y_{n,j}^i - x_0^i) \right| \\ &\leqq \frac{\tau \varepsilon}{2^{3/2}} 2^{1/2} e_n = \frac{\tau \varepsilon e_n}{2}. \end{aligned}$$

By Lemma 8-1, this implies

$$|D_i f_n - D_i f(x_0)| = |D_i g_n| \leqq 2(\tau \varepsilon e_n / 2)/\tau e_n = \varepsilon.$$

Since for each positive ε this holds for all large n, (B) is valid. The proof is complete.

9. Areas of Surfaces

In discussing curves, we began with representations and then defined a curve as a class of equivalent representations. We shall begin our discussion of surfaces by defining representations. A surface will be an equivalence class of representations. Our discussion of equivalence, however, will not be exhaustive, and we postpone it to the end of this section.

By a **representation of a surface** in R^r we shall mean a function with values in R^r that is defined and continuous on the closure of a bounded open set in R^2. This definition is more inclusive than is customary. For example, we allow the open set to consist of several parts with disjoint closures, and it would be more customary to think of a continuous R^r-valued function on such a set as representing several surfaces, not just one. But for present purposes there is no disadvantage in allowing the larger class of representations.

We shall define the area of a surface by a limit process, starting with the areas of polyhedra that consist of collections of triangles inscribed in the surface. So we have to start by investigating the areas of triangles in R^r. In fact, we have not even defined a triangle in R^r for $r > 2$; so for completeness we shall say that if P_0, P_1, and P_2 are points in R^r, the triangle with vertices P_0, P_1, and P_2 consists of all points P that are between P_0 and a point Q that is itself between P_1 and P_2. The area of a triangle in R^2 is simply the ordinary measure $m_L T$ of the point-set T as a set in R^2. The definition of the area of a triangle in R^r $(r > 2)$ is unambiguously determined if we make the requirement, suggested by elementary geometry, that if two triangles, even in different spaces, have the three sides of the one equal in length to the three sides of the other, the two triangles have equal areas. It follows at once that if P_0, P_1, and P_2 are points in R^r and θ is the angle between the directed line-segments P_0P_1 and P_0P_2, then

$$\text{(A)} \qquad \text{area}\, \Delta P_0P_1P_2 = \tfrac{1}{2}(\text{dist}\, P_0P_1)(\text{dist}\, P_0P_2)(\sin\theta),$$

this is true for a triangle in the plane that has sides equal to those of triangle $P_0P_1P_2$ and therefore has angle θ between the sides, by (C) of Section 3.

From (A) and Definition 3-1,

$$\text{(B)} \qquad \begin{aligned}[2\,\text{area}\, \Delta P_0P_1P_2]^2 &= |\overrightarrow{P_0P_1}|^2|\overrightarrow{P_0P_2}|^2(1 - \cos^2\theta) \\ &= |\overrightarrow{P_0P_1}|^2|\overrightarrow{P_0P_2}|^2 - \langle \overrightarrow{P_0P_1}, \overrightarrow{P_0P_2}\rangle^2.\end{aligned}$$

Now suppose that we choose a rectangular coordinate system in R^2. In R^r we do not choose a coordinate system, but we do choose an origin, so that each point of R^r has a position-vector $\mathbf{x}$. Let $\mathbf{x}$ be a linear mapping of R^2 into R^r, the point of R^2 with coordinates $(u^1, u^2)^{\mathrm{T}}$ being mapped onto the point with position vector $\mathbf{x}(u^1, u^2)$. As usual, if g is a function of several real variables, we denote by $D_j g$ the partial derivative of g with respect to the jth independent variable. Because $\mathbf{x}$ is linear, the partial derivatives $D_1\mathbf{x}$, $D_2\mathbf{x}$ are constant, and for all points u,

$$\text{(C)} \qquad \mathbf{x}(u^1, u^2) = (D_1\mathbf{x})u^1 + (D_2\mathbf{x})u^2 + \mathbf{x}(0, 0).$$

Let A_0, A_1, A_2 be points of R^2 with the respective coordinate two-columns u_0, u_1, u_2, and let P_i be the point of R^r with position vector $\mathbf{x}(u_i)$ $(i = 0, 1, 2)$. For convenience, we use the abbreviations $(s, t)^{\mathrm{T}}$ for $(u_1^1 - u_0^1, u_1^2 - u_0^2)^{\mathrm{T}}$; (v, w) for $(u_2^1 - u_0^1, u_2^2 - u_0^2)^{\mathrm{T}}$; $\mathbf{y}$ for $\mathbf{x}(u_1) - \mathbf{x}(u_0)$; and $\mathbf{z}$ for $\mathbf{x}(u_2) - \mathbf{x}(u_0)$. Then by (C),

$$\text{(D)} \qquad \mathbf{y} = (D_1\mathbf{x})s + (D_2\mathbf{x})t, \qquad \mathbf{z} = (D_1\mathbf{x})v + (D_2\mathbf{x})w,$$

and (B) takes the form

$$\text{(E)} \qquad [2\ \text{area}\, \Delta P_0P_1P_2]^2 = \langle \mathbf{y}, \mathbf{y}\rangle\langle \mathbf{z}, \mathbf{z}\rangle - \langle \mathbf{y}, \mathbf{z}\rangle^2.$$

By (D) and (E),

$$\begin{aligned}[2\ \text{area}\, \Delta P_0P_1P_2]^2 &= \langle s\, D_1\mathbf{x} + t\, D_2\mathbf{x}, s\, D_1\mathbf{x} + t\, D_2\mathbf{x}\rangle \\ &\quad \times \langle v\, D_1\mathbf{x} + w\, D_2\mathbf{x}, v\, D_1\mathbf{x} + w\, D_2\mathbf{x}\rangle \\ &\quad - \langle s\, D_1\mathbf{x} + t\, D_2\mathbf{x}, v\, D_1\mathbf{x} + w\, D_2\mathbf{x}\rangle^2.\end{aligned}$$

We define

$$\text{(F)} \qquad E = \langle D_1\mathbf{x}, D_1\mathbf{x}\rangle, \qquad F = \langle D_1\mathbf{x}, D_2\mathbf{x}\rangle, \qquad G = \langle D_2\mathbf{x}, D_2\mathbf{x}\rangle.$$

After multiplying out the parentheses in (E) and performing a little tedious collecting of terms, we find

$$\text{(G)} \qquad [2 \text{ area } \Delta P_0P_1P_2]^2 = (EG - F^2)(sw - tv)^2 = (EG - F^2)\det\begin{pmatrix} s & v \\ t & w \end{pmatrix}^2.$$

As a special case, let r be 2 and let $\mathbf{x}$ be the identity map. Then $E = G = 1$ and $F = 0$, and if in (G) we substitute the meanings of the abbreviations s, t, v, and w, we obtain the formula, familiar from analytic geometry,

$$\text{(H)} \qquad \text{area } \Delta A_0A_1A_2 = \frac{1}{2}\left|\det\begin{pmatrix} u_1^1 - u_0^1 & u_2^1 - u_0^1 \\ u_1^2 - u_0^2 & u_2^2 - u_0^2 \end{pmatrix}\right|.$$

Returning with this to (G) yields

$$\text{(I)} \qquad \text{area } \Delta P_0P_1P_2 = (EG - F^2)^{1/2}(\text{area } \Delta A_0A_1A_2) = \int_{\Delta A_0A_1A_2} (EG - F^2)^{1/2}\, du.$$

It should be observed that this is independent of the choice of coordinates in R^r. It was in fact proved without mentioning any coordinate system in R^r. We did choose an origin, but change of origin would change only the constant $\mathbf{x}(0, 0)$ in (C) and would leave E, F, and G unchanged.

If we choose rectangular coordinates in R^r, the inner products take their familiar simple forms, given in Lemma 3-4, and (F) becomes

$$\text{(J)} \qquad E - \sum_{i=1}^{r} (D_1x^i)^2,$$

$$F = \sum_{i=1}^{r} (D_1x^i)(D_2x^i),$$

$$G = \sum_{i=1}^{r} (D_2x^i)^2.$$

With rectangular coordinates, the following method of calculating $EG - F^2$ is sometimes convenient.

(K) If the coordinate system is rectangular,

$$EG - F^2 = \sum_{1 \leq i < j \leq r} \left[\det\begin{pmatrix} D_1x^i & D_2x^i \\ D_1x^j & D_2x^j \end{pmatrix}\right]^2 = \tfrac{1}{2}\sum_{i,j=1}^{r} \left[\det\begin{pmatrix} D_1x^i & D_2x^i \\ D_1x^j & D_2x^j \end{pmatrix}\right]^2.$$

For convenience, we introduce the symbol

(L) $$S = \sum_{1 \leqq i < j \leqq r} \left[\det \begin{pmatrix} D_1 x^i & D_2 x^i \\ D_1 x^j & D_2 x^j \end{pmatrix} \right]^2 .$$

Interchanging the names i and j does not change the sum. Interchanging the rows of the determinant changes its sign but does not affect its square. So,

$$S = \sum_{1 \leqq j < i \leqq r} \left[\det \begin{pmatrix} D_1 x^i & D_2 x^i \\ D_1 x^j & D_2 x^j \end{pmatrix} \right]^2 .$$

We add the last two equations member by member and to the right member add the sum of the squares of the determinants with $j = i$, each of which is 0. The result is

(M) $$2S = \sum_{i,j=1}^{r} \left[\det \begin{pmatrix} D_1 x^i & D_2 x^i \\ D_1 x^j & D_2 x^j \end{pmatrix} \right]^2 .$$

Expanding the determinants in the right member yields

$$\begin{aligned} 2S &= \sum_{i,j=1}^{r} [(D_1 x^i)^2 (D_2 x^j)^2 - 2(D_1 x^i)(D_2 x^j)(D_1 x^j)(D_2 x^i) + (D_1 x^j)^2 (D_2 x^i)^2] \\ &= 2\left[\sum_{i=1}^{r} (D_1 x^i)^2 \right]\left[\sum_{j=1}^{r} (D_2 x^j)^2 \right] - 2\left[\sum_{i=1}^{r} D_1 x^i D_2 x^i \right]\left[\sum_{j=1}^{r} D_1 x^j D_2 x^j \right] \\ &= 2EG - 2F^2. \end{aligned}$$

This, with (L) and (M), establishes (K).

By a **polyhedron** in R^r we shall mean a finite set of triangles in R^r. The **area** of the polyhedron is defined to be the sum of the areas of the triangles that constitute the polyhedron. If

$$u \mapsto \mathbf{x}(u) \qquad (u \text{ in } G^-)$$

is a representation of a surface in R^r, the polyhedron consisting of triangles $S_1, \ldots, S_k$ in R^r is **inscribed** in the surface if there exist triangles $T_1, \ldots, T_k$ in G^- with pairwise disjoint interiors such that for $j = 1, \ldots, k$, if $u_{j,0}, u_{j,1}, u_{j,2}$ are the coordinates of the vertices of T_j, the vertices of S_j have position-vectors $\mathbf{x}(u_{j,0})$, $\mathbf{x}(u_{j,1})$, $\mathbf{x}(u_{j,2})$. It is tempting to imitate the definition of the length of a curve and to define the area of a surface to be the supremum of the areas of all polyhedra inscribed in the surface, but this leads to nonsense. For example, let a surface in R^3 have the representation $u \mapsto \mathbf{x}(u)$, where in rectangular coordinates

$$x^1(u) = u^1, \qquad x^2(u) = u^2, \qquad x^3(u) = (u^1)^2 \qquad (u^1 \text{ and } u^2 \text{ in } [-1, 1]).$$

As in the beginning of Section 8, let T_n be the triangle in R^2 with the vertices

$$u_0 = (0, 0), \qquad u_1 = (-1/n, 1/n^4), \qquad u_2 = (1/n, 1/n^4).$$

We leave it as an exercise (Exercise 9-3) to show that the triangle in R^3 with vertices $\mathbf{x}(u_0)$, $\mathbf{x}(u_1)$, $\mathbf{x}(u_2)$ has area greater than $1/n^3$. By translating T_n parallel to the u^2-axis by multiples of $1/n^4$, we can obtain in $[-1, 1] \times [-1, 1]$ a set of $2n^4$ translates of T_n with pairwise disjoint interiors, and the corresponding inscribed polyhedron of $2n^4$ faces has area greater than $2n$, which tends to ∞ with n.

The existence of such examples forces us to put rather complicated restrictions on the surfaces whose areas we define and on the inscribed polyhedra that we use in defining the area, as in the next lemma and definition. In the lemma we define area for a class of surfaces almost, but not quite, large enough to be adequate for applications.

LEMMA 9-1 *Let a rectangular coordinate system be chosen in R^2 and an origin be chosen in R^r. Let G be a bounded open set in R^2, and let*

$$u \mapsto \mathbf{x}(u) \qquad (u \text{ in } G^-)$$

be a representation of a surface in R^r such that $\mathbf{x}$ is Lipschitzian on G^- [and is differentiable at all points of G except those in a set N with $m_L N = 0$]. Let $\tau, \varepsilon_1, \varepsilon_2, \varepsilon_3, \ldots$ be positive numbers such that ε_n tends to 0 as n increases. For each positive integer n, let $\mathscr{T}_n$ be a finite set

$$\mathscr{T}_n = \{T_{n,1}, T_{n,2}, \ldots, T_{n,p(n)}\}$$

of triangles contained in G^-, each $T_{n,j}$ in $\mathscr{T}_n$ having thickness ratio at least τ and edges all of length less than ε_n. Assume that all points of G except those in a set N_1 of measure 0 belong to the union of interiors

$$U_n = T^0_{n,1} \cup T^0_{n,2} \cup \cdots \cup T^0_{n,p(n)}$$

for all but finitely many n. Let the vertices of $T_{n,j}$ be $u_{n,j,0}$, $u_{n,j,1}$, and $u_{n,j,2}$, and let $\Pi(\mathscr{T}_n)$ be the inscribed polyhedron consisting of the $p(n)$ triangles with vertices $\mathbf{x}(u_{n,j,0})$, $\mathbf{x}(u_{n,j,1})$, and $\mathbf{x}(u_{n,j,2})$.

Then as n increases, the area of the inscribed polyhedron $\Pi(\mathscr{T}_n)$ will converge to the limit

$$\int_{A \setminus N} [EG - F^2]^{1/2}\, du.$$

The hypothesis that is enclosed in square brackets is superfluous; if $\mathbf{x}$ is Lipschitzian, it is almost everywhere differentiable. But we have not proved this, so we state the differentiability as a hypothesis.

Neither the hypotheses nor the conclusion of the lemma involves any coordinate system in R^r, but as a help in the proof it is convenient to put a rectangular coordinate system in R^r. For each positive integer n and each j in $\{1, \ldots, p(n)\}$, let $\mathbf{x}_{n,j}$ be the linear function on R^2 that coincides with $\mathbf{x}$ at each of the three vertices of $T_{n,j}$. If we define

$$E_{n,j} = \sum_{i=1}^{r} (D_1 x^i_{n,j})^2,$$

$$F_{n,j} = \sum_{i=1}^{r} (D_1 x^i_{n,j})(D_2 x^i_{n,j}),$$

$$G_{n,j} = \sum_{i=1}^{r} (D_2 x^i_{n,j})^2$$

we know by (I) and (J) that the area of the triangle with vertices $\mathbf{x}(u_{n,j,0})$, $\mathbf{x}(u_{n,j,1})$, and $\mathbf{x}(u_{n,j,2})$ is the integral of $[E_{n,j}G_{n,j} - F^2_{n,j}]^{1/2}$ over the triangle $T_{n,j}$. But the boundary of $T_{n,j}$ has measure 0, so this integral is the same as the integral over the interior of $T_{n,j}$, and therefore

$$\text{(N)} \qquad \text{area } \Delta(\mathbf{x}(u_{n,j,0}), \mathbf{x}(u_{n,j,1}), \mathbf{x}(u_{n,j,2})) = \int_{T^0_{n,j}} [E_{n,j}G_{n,j} - F^2_{n,j}]^{1/2}\, du.$$

Now let g_n be the function that on each interior $T^0_{n,j}$ has the value $[E_{n,j}G_{n,j} - F^2_{n,j}]^{1/2}$ and outside of the union U_n of those interiors has the value 0. If we integrate this over G, we obtain the sum of the integrals of the $[E_{n,j}G_{n,j} - F^2_{n,j}]^{1/2}$ over all the $T^0_{n,j}$, and by (N),

$$\text{(O)} \qquad \int_G g_n(u)\, du = \text{area } \Pi(\mathscr{T}_n).$$

Let u be a point of G, and let n be a positive integer. If u is not in U_n, by definition we have $g_n(u) = 0$. If u is in U_n, it is interior to a triangle $T_{n,j}$ of the set $\mathscr{T}_n$. The Lipschitz constant K for the vector-valued function $\mathbf{x}$ is also a Lipschitz constant for each component x^i, and $x^i_{n,j}$ coincides with x^i at each vertex of $T_{n,j}$, so by Lemma 8-2 we have for each i in $\{1, \ldots, r\}$,

$$|D_1 x^i_{n,j}(u)| \leqq 2^{3/2}K/\tau, \qquad |D_2 x^i_{n,j}(u)| \leqq 2^{3/2}K/\tau.$$

By (J),

$$E \leqq 8rK^2/\tau^2, \qquad G \leqq 8rK^2/\tau^2,$$

so

$$|g_n(u)| \leqq 8rK^2/\tau^2.$$

Therefore the g_n all have the bound $8rK^2/\tau^2$.

If u is in $G \setminus (N \cup N_1)$, it is in U_n for all large n, and x is differentiable at u. By Lemma 5-1, each component function x^i is differentiable at u. For each n such that u is in U_n, there is a number $j(n)$ in the set $\{1, \ldots, p(n)\}$ such that u is in $T^0_{n,j(n)}$. Then by Lemma 8-3,

$$\lim_{n \to \infty} D_\alpha x^i_{n,j(n)} = D_\alpha x_n(u) \qquad (\alpha = 1, 2).$$

By (J), this implies

$$\lim_{n\to\infty} E_{n,j}(u) = E(u), \qquad \lim_{n\to\infty} F_{n,j}(u) = F(u), \qquad \lim_{n\to\infty} G_{n,j}(u) = G(u).$$

Therefore, at all points of $G \setminus N$ except the set N_1 of measure 0, it is true that $g_n(u)$ converges to $[EG - F^2]^{1/2}$. Since the g_n all have the same bound, by the dominated convergence theorem we have

$$\lim_{n\to\infty} \int_{G \setminus N} g_n(u)\, du = \int_{G \setminus N} [EG - F^2]^{1/2}\, du.$$

Since N has measure 0, this and (O) imply the conclusion of the lemma.

DEFINITION 9-2 *Let $u \mapsto x(u)$ (u in G^-) be a representation of a surface in R^r that satisfies the conditions in Lemma* 9-1. *Then the area of the surface is defined to be the number*

$$\text{(P)} \qquad \int_{G \setminus N} [EG - F^2]^{1/2}\, du$$

that is the limit of the areas of inscribed polyhedra that satisfy the conditions in Lemma 9-1.

There are, however, numerous examples in which the $\mathbf{x}$ is not Lipschitzian, but if we cut out of G a suitably chosen subset of arbitrarily small measure, $\mathbf{x}$ is Lipschitzian on what is left of G. This suggests the following generalization.

THEOREM 9-3 *Let G be a bounded open set in R^2, and let $u \mapsto \mathbf{x}(u)$ (u in G^-) be a representation of a surface in R^r. Assume that there exists a sequence of open subsets $G_1, G_2, G_3, \ldots$ of G such that $G_1 \subset G_2 \subset G_3 \subset \cdots$ and*

$$\lim_{n\to\infty} m_L G_n = m_L G.$$

Assume also that on each G_n^- the function $u \mapsto x(u)$ (u in G_n^-) satisfies the hypotheses of Lemma 9-1 [*and that $\mathbf{x}$ is differentiable at all points of G except those in a set N with $m_L N = 0$*]. *Then as n increases, the area of the surface represented by $u \mapsto x(u)$ (u in G_n) tends to*

$$\int_{G \setminus N} [EG - F^2]^{1/2}\, du.$$

The union U of the G_n satisfies $G_n \subset U \subset G$ for every n, so

$$m_L G_n \leqq m_L U \leqq m_L G.$$

The first of these three measures is arbitrarily close to the third, so $m_L U$ must be equal to $m_L G$, and $G \setminus U$ is a set of measure 0. By Definition 9-2, the area of the surface represented by $u \mapsto x(u)$ (u in G_n^-) is

$$\int_{G_n \setminus N} [EG - F^2]^{1/2}\, du = \int_{G \setminus N} [EG - F^2]^{1/2}\, 1_{G_n}(u)\, du.$$

As n increases, the last integrand ascends at all points u of $G \setminus N$ and tends to $[EG - F^2]^{1/2}$ at all points of $U \setminus N$, so the limit of the areas of the surfaces is

$$\int_{U \setminus N} [EG - F^2]^{1/2}\, du.$$

Since $G \setminus U$ has measure 0, the last integral is the same as the integral (P), and the proof is complete.

DEFINITION 9-4 *If the representation $u \mapsto x(u)$ (u in G^-) satisfies the conditions in Theorem 9-3, the area of the surface represented by $u \mapsto x(u)$ (u in G^-) is defined to be the value of the integral* (P).

A useful special case is that in which $r = 3$ and the surface is represented in "nonparametric form." That is, there exists a bounded open set G in R^2 and a continuous real-valued function $f(x)$ (x in G^-) whose graph is the surface S. In this case the representation of S can be written as $u \mapsto x(u)$ (u in G^-), where

$$\text{(Q)} \qquad x^1(u) = u^1, \qquad x^2(u) = u^2, \qquad x^3(u) = f(u^1, u^2) \qquad (u \text{ in } G^-).$$

If S is represented by (Q) and the function f has a differential at a point u of G, we readily compute

$$D_1 x(u) = (1, 0, D_1 f(u))^{\mathrm{T}}, \qquad D_2 x(u) = (0, 1, D_2 f(u))^{\mathrm{T}},$$

$$E = 1 + (D_1 f(u))^2, \qquad F = D_1 f(u)\, D_2 f(u), \qquad G = 1 + (D_2 f(u))^2,$$

$$EG - F^2 = 1 + (D_1 f(u))^2 + (D_2 f(u))^2.$$

From this and Theorem 9-3 and Definition 9-4 we obtain this corollary.

COROLLARY 9-5 *Let G be a bounded open set in R^2 and f a continuous real-valued function on G^-. Assume [that f is differentiable at all points of G except those in a set N of measure* 0, *and] that there exists a sequence of open subsets $G_1, G_2, G_3, \ldots$ of G such that $G_1 \subset G_2 \subset G_3 \subset \cdots$ and the limit as n increases of $m_{\mathrm{L}} G_n$ is $m_{\mathrm{L}} G$. Assume that f is Lipschitzian on each G_n^-. Then the area of the surface that is the graph of the function f is*

$$\int_{G \setminus N} [1 + (D_1 f(u))^2 + (D_2 f(u))^2]^{1/2}\, du.$$

For an example, we let G be the interior of a circle of radius r and center $(0, 0)$ in R^2. For convenience, we denote the coordinates (in a rectangular system) by u and v instead of by u^1 and u^2. The surface S that is the graph of

$$(u, v) \mapsto f(u, v) = [r^2 - u^2 - v^2]^{1/2} \qquad ((u, v) \text{ in } G^-)$$

is a hemisphere of radius r. On G^- the function f is not Lipschitzian; its partial derivatives are unbounded near the circumference of G. But if we define G_n to be the disk

$$\{(u, v) \text{ in } R^2 : u^2 + v^2 < (1 - 1/n)r^2\},$$

these sets expand as n increases, and their union is all of G, and f is continuously differentiable (hence Lipschitzian) on each of them. By Corollary 9-5, the area of S is

$$\text{(R)} \qquad \text{area } S = \int_G [1 + (D_1 f(u, v))^2 + (D_2 f(u, v))^2]^{1/2}\, du\, dv.$$

Since

$$D_1 f(u, v) = -u[r^2 - u^2 - v^2]^{-1/2},$$
$$D_2 f(u, v) = -v[r^2 - u^2 - v^2]^{-1/2},$$

we find

$$[1 + (D_1 f(u, v))^2 + (D_2 f(u, v))^2]^{1/2} = r[r^2 - u^2 - v^2]^{-1/2}.$$

From (R) and Fubini's theorem,

$$\begin{aligned} \text{area } S &= \int_G r[r^2 - u^2 - v^2]^{-1/2}\, du\, dv \\ &= \int_{-r}^{r} \left\{ \int_{-(r^2 - u^2)^{1/2}}^{(r^2 - u^2)^{1/2}} r[r^2 - u^2 - v^2]^{-1/2}\, dv \right\} du. \end{aligned}$$

For each fixed u in $(-r, r)$,

$$(r^2 - u^2 - v^2)^{-1/2} = D \arcsin[v/(r^2 - u^2)^{1/2}],$$

so from the preceding equation we obtain

$$\text{area } S = 2\pi r^2,$$

in agreement with elementary geometry.

We have postponed until now the question of equivalence of representations of surfaces. Let us define a **homeomorphism** between sets A and B, or a **homeomorphic mapping** of A onto B, as a function $u \mapsto T(u)$ (u in A) that establishes a one-to-one correspondence between the points of A and those of B and is continuous and has a continuous inverse. If $u \mapsto \mathbf{x}(u)$ (u in $[a, b]$) and $v \mapsto \mathbf{y}(v)$ (v in $[c, d]$) represent curves, and there is a homeomorphism $u \mapsto T(u)$ of $[a, b]$ onto $[c, d]$ such that $\mathbf{y}(T(u)) = \mathbf{x}(u)$ for all u in $[a, b]$, and $T(a) = c$ and $T(b) = d$, the functions $\mathbf{x}$ and $\mathbf{y}$ are equivalent in the sense of Section 6. (See the remark after Definition 6-1.) If we were interested only in the lengths of curves, the requirement $T(a) = c$ and $T(b) = d$ could be dropped because it serves to distinguish between curve $\mathscr{C}$ and its reverse, and they have equal lengths. The requirement that $T(a) = c$ and $T(b) = d$ has an analog in R^2, but it is complicated, and since all that we are interested in at present is area, we simply omit it. So, as a first step, we could say that when two representations

$$u \mapsto \mathbf{x}(u) \qquad (u \text{ in } G^-), \qquad v \mapsto \mathbf{y}(v) \qquad (v \text{ in } H^-)$$

(G and H being bounded open sets in R^2) have the property that there exists a homeomorphic mapping $u \mapsto T(u)$ of G onto H for which

$$\mathbf{y}(T(u)) = \mathbf{x}(u) \qquad (u \text{ in } G),$$

then the two representations are equivalent.

This has an immediate and adequate generalization. Suppose that $\mathbf{x}$ and $\mathbf{y}$ are as in the preceding paragraph, and for each positive integer n there exists a homeomorphism T_n of G onto H such that as n increases, the function $u \mapsto \mathbf{y}(T_n(u))$ (u in G) converges to $\mathbf{x}(u)$ uniformly on G. We shall then say that the two representations are **equivalent.**

Having stated this satisfactorily general definition of equivalence, we do nothing at all with it because we have not developed the analytical machinery needed to cope with it. Our definitions and theorems have all involved derivatives, and we are not in a position to prove that our definition of area gives the same number for all representations because we have not even stated the definition for representations that are not almost everywhere differentiable. We have to go in the opposite direction. First, instead of considering all representations, we shall allow only representations that satisfy the hypotheses of Theorem 9-3. Second, in defining equivalence, we shall allow only changes of parameter $u \mapsto T(u)$ (u in G) that are homeomorphisms and that are differentiable and have differentiable inverses on open subsets of G that expand and fill almost all of G.

If $u \mapsto T(u)$ (u in G^-) is a representation that satisfies the hypotheses of Theorem 9-3, the area is defined in Definition 9-4. Strictly speaking, we should not have called this the area of the surface, but the area of the surface in its $\mathbf{x}$-representation. After the next theorem is proved, we shall at last know that the value of the integral is the same for all such representations, so it is really the area of the surface. However, the expression $[EG - F^2]^{1/2}$ for the area-integrand in Lemma 9-1 is inconvenient to work with, so we introduce a new symbol for it. If $\mathbf{x}$ is defined on G and is differentiable at a point u of G, we define

$$\text{(S)} \quad A_{\mathbf{x}}(u) = [\langle D_1\mathbf{x}(u), D_1\mathbf{x}(u)\rangle\langle D_2\mathbf{x}(u), D_2\mathbf{x}(u)\rangle - \langle D_1\mathbf{x}(u), D_2\mathbf{x}(u)\rangle^2]^{1/2}.$$

Similarly, if the representation is $v \mapsto y(v)$ (v in H), and $\mathbf{y}$ is differentiable at a point v of H, we define the area-integrand $A_{\mathbf{y}}(v)$ at v by replacing $\mathbf{x}(u)$ with $\mathbf{y}(v)$ in (S).

We can now state and prove the theorem that if two representations satisfy the hypotheses of Theorem 9-3 and are equivalent in the strong sense defined just above, the values of the area-integral in the two representations are equal. Thus, the area is a property of the surface and does not depend on the representation, provided that the hypotheses of Theorem 9-3 are satisfied.

THEOREM 9-6 *Let G and H be bounded open sets in R^2. Let*

$$u \mapsto x(u) \qquad (u \text{ in } G^-), \qquad v \mapsto y(v) \qquad (v \text{ in } H^-)$$

be representations of surfaces that are differentiable on the respective sets $G \setminus N_1$, $H \setminus N_2$, where N_1 and N_2 have measure 0. Let

$$u \mapsto T(u) = (T^1(u), T^2(u)) \qquad (u \text{ in } G)$$

be a homeomorphic mapping of G onto H such that for every u in G,

$$\text{(T)} \qquad \mathbf{y}(T(u)) = \mathbf{x}(u).$$

Assume that there exists a sequence $G(1), G(2), G(3), \ldots$ of open subsets of G such that $G(1) \subset G(2) \subset G(3) \subset \cdots$, and for each positive integer n the function T is continuously differentiable on $G(n)$ and its inverse is continuously differentiable on $T(G(n))$, and

$$\lim_{n \to \infty} m_{\mathrm{L}} G(n) = m_{\mathrm{L}} G, \qquad \lim_{n \to \infty} m_{\mathrm{L}} T(G(n)) = m_{\mathrm{L}} H.$$

Then

$$\text{(U)} \qquad \int_{G \setminus N_1} A_{\mathbf{x}}(u)\, du = \int_{H \setminus N_2} A_{\mathbf{y}}(v)\, dv.$$

We introduce rectangular coordinates in R^r. By (K), at each point u of $G \setminus N_1$

$$\text{(V)} \qquad A_{\mathbf{x}}(u)^2 = \sum_{i,j=1}^{r} \left[\det \begin{pmatrix} D_1 x^i(u) & D_2 x^i(u) \\ D_1 x^j(u) & D_2 x^j(u) \end{pmatrix} \right]^2,$$

and at each point v of $H \setminus N_2$

$$\text{(W)} \qquad A_{\mathbf{y}}(v)^2 = \sum_{i,j=1}^{r} \left[\det \begin{pmatrix} D_1 y^i(v) & D_2 y^i(v) \\ D_1 y^j(v) & D_2 y^j(v) \end{pmatrix} \right]^2.$$

For each positive integer n and each u in $G(n) \setminus N_1$ we obtain by applying the chain rule to identity (T),

$$\text{(X)} \qquad \begin{aligned} D_1 x^i(u) &= D_1 y^i(T(u))\, D_1 T^1(u) + D_2 y^i(T(u))\, D_1 T^2(u), \\ D_2 x^i(u) &= D_1 y^i(T(u))\, D_2 T^1(u) + D_2 y^i(T(u))\, D_2 T^2(u). \end{aligned}$$

For brevity we write $D_\alpha y^i$ for $D_\alpha y^i(T(u))$ and $D_\alpha T^i$ for $D_\alpha T^i(u)$. Then if we substitute (X) in (V), we obtain

$$\begin{aligned} A_x(u)^2 &= \sum_{i,j=1}^{r} \{[D_1 y^i D_1 T^1 + D_2 y^i D_1 T^2][D_1 y^j D_2 T^1 + D_2 y^j D_2 T^2] \\ &\qquad - [D_1 y^j D_1 T^1 + D_2 y^j D_1 T^2][D_1 y^i D_2 T^1 + D_2 y^i D_2 T^2]\}^2 \\ &= \sum_{i,j=1}^{r} \{D_1 y^i D_2 y^j D_1 T^1 D_2 T^2 + D_2 y^i D_1 y^j D_1 T^2 D_2 T^1 \\ &\qquad - D_1 y^j D_2 y^i D_1 T^1 D_2 T^2 - D_2 y^j D_1 y^i D_1 T^2 D_2 T^1\}^2 \\ &= \left[\det\begin{pmatrix} D_1 T^1 & D_2 T^1 \\ D_1 T^2 & D_2 T^2 \end{pmatrix}\right]^2 A_{\mathbf{y}}(T(u)). \end{aligned}$$

From this and Theorem IV-8-1 we deduce that for each positive integer n,

$$\int_{G(n)\setminus N_1} A_{\mathbf{x}}(u)\,du = \int_{T(G(n))\setminus N_2} A_{\mathbf{y}}(v)\,dv,$$

which is the same as

$$\int_{G\setminus N_1} A_{\mathbf{x}}(u) 1_{G(n)}(u)\,du = \int_{H\setminus N_2} A_{\mathbf{y}}(v) 1_{T(G(n))}(v)\,dv.$$

As n increases, the integrands in both members ascend and tend to those in (U) except on sets of measure 0 – namely, the set of points of G that are not in any $G(n)$, and the set of points in H that are not in any $T(G(n))$. By the monotone convergence theorem, (U) is valid, and the theorem is established.

EXERCISE 9-1 A triangle in R^4 has vertices $P_0 = (1, -1, 2, 0)$, $P_1 = (1, -1, 3, 1)$, and $P_2 = (0, -2, 2, 0)$. Find its area by (B) and verify that it agrees with the area given by the formula from elementary trigonometry

$$\text{area} = [s(s - P_0P_1)(s - P_0P_2)(s - P_1P_2)]^{1/2},$$

where

$$s = [P_0P_1 + P_0P_2 + P_1P_2]/2.$$

EXERCISE 9-2 Show that the expression for $EG - F^2$ just before (and used in) Corollary 9-5 can be obtained slightly more easily by use of (K).

EXERCISE 9-3 Show that the triangle in R^3 with vertices $(0, 0, 0)$, $(-1/n, 1/n^4, 1/n^2)$, $(1/n, 1/n^4, 1/n^2)$ has area greater than $1/n^3$. *Suggestion*: Use (B).

EXERCISE 9-4 Show that the graph of

$$(x^1, x^2) \mapsto hr^{-1}[(x^1)^2 + (x^2)^2]^{1/2} \qquad ((x^1)^2 + (x^2)^2 \leq r^2)$$

is a circular cone with radius of base equal to r and altitude equal to h. Find its area.

EXERCISE 9-5 Verify that the representation

$$\text{(Y)} \qquad x^1 = u^1 \cos u^2, \qquad x^2 = u^1 \sin u^2,$$
$$x^3 = hu^1/r \qquad (0 \leqq u^1 \leqq r,\ 0 < u^2 \leqq 2\pi)$$

is equivalent to that in Exercise 9-4. Find the area using representation (Y).

EXERCISE 9-6 For convenience, write (u, v) for (u^1, u^2). Find the area of the surface represented by

$$x^1 = u^2 - v^2, \qquad x^2 = \log(u + v) + \log(u - v),$$
$$x^3 = \cos u^2 \cos v^2 + \sin u^2 \sin v^2 \qquad (1 \leqq u \leqq 2,\ 0 \leqq v \leqq 2).$$

Explain the (perhaps surprising) result.

EXERCISE 9-7 Verify that the surface in R^3 represented by

$$(u^1, u^2) \mapsto (r \cos u^1 \cos u^2, r \cos u^1 \sin u^2, r \sin u^1)$$
$$(-\pi/2 \leqq u^1 \leqq \pi/2,\ -\pi \leqq u^2 \leqq \pi)$$

is a sphere of radius r. Find its area by Theorem 9-3.

EXERCISE 9-8 Show that the surface represented by

$$(u^1, u^2) \mapsto (r \cos u^1 \cos u^2, r \cos u^1 \sin u^2, r \sin u^1)$$
$$(0 \leqq u^1 \leqq \pi/2,\ -\pi \leqq u^2 \leqq \pi)$$

is a hemisphere. Find its area by Theorem 9-3. Verify that it is equivalent to the representation

$$(u^1, u^2) \mapsto (u^1, u^2, [r^2 - (u^1)^2 - (u^2)^2]^{1/2} \qquad ((u^1)^2 + (u^2)^2 \leqq r^2)$$

that was investigated just after Corollary 9-5.

VI

Vector Spaces, Orthogonal Expansions, and Fourier Transforms

1. Complex Vector Spaces

In Chapter V we saw that it was convenient to represent vectors in the space V^r by $1 \times r$ matrices $(x^1, \ldots, x^r)^T$ of real numbers. Such a matrix can be regarded as a real-valued function $i \mapsto x^i$ on the set $\{1, \ldots, r\}$. The latter wording suggests an important generalization, which we now study.

If to each i in the set $\{1, \ldots, r\}$ there corresponds a real number x^i, these numbers can be regarded as the coordinates of a single point P in R^r, or as the components of a vector $\mathbf{x}$ in V^r. (The superscript i merely identifies the ith component; it is not an exponent.) In Chapter V we saw that by regarding the numbers not as individuals but as joined together as a function $i \mapsto x^i$ (i in $\{1, \ldots, r\}$) to represent one vector, we gained both in clarity of ideas and in ease of manipulation.

Suppose, then, that to each t in a set B there corresponds a real number $f(t)$. We can think of these individually, as when we construct a table for the sine-function. Or we can think of all of them together, to form a single function $t \mapsto f(t)$ (t in B). From this point of view, the mental picture corresponding to the sine-function would be a graph in the form of a wave oscillating in height between -1 and 1. This function $t \mapsto f(t)$ has a close resemblance to the function $i \mapsto x^i$ that represents a vector in V^r.

In Definition V-2-2 we introduced the modern extended use of the word *vector*. In order to refer to the real-valued functions on B as vectors, in the wider sense, we need to define addition of functions and multiplication of functions by real numbers in such a way that the conditions in Definition V-2-2 are satisfied. This is nothing new; we have already used such a definition.

DEFINITION 1-1 *If f_1 and f_2 are real-valued functions on a set B and b is a real number,*

(i) *$f_1 + f_2$ is the function whose value at each t in B is $f_1(t) + f_2(t)$,*
(ii) *bf_1 is the function whose value at each t in B is $bf_1(t)$.*

It is now evident that the statements in Definition V-2-2 are valid and that the class of all real-valued functions on B is a vector space, or linear space.

There are considerable advantages, however, both in theory and in applications, if we extend from real-valued to complex-valued functions and also if, instead of considering the class of all complex-valued functions on B, we consider various subclasses chosen to have some useful properties.

Merely to connect complex numbers with vectors, we recall that one way of defining the space $\mathscr{C}$ of complex numbers is to say that $\mathscr{C}$ is the vector space V^2 with a multiplication defined by

$$\text{(A)} \qquad \begin{pmatrix} a \\ b \end{pmatrix}\begin{pmatrix} c \\ d \end{pmatrix} = \begin{pmatrix} ac - bd \\ ad + bc \end{pmatrix} \qquad (a, b, c, d \text{ in } R).$$

This multiplication is obviously commutative. By expanding by (A), we readily compute that

$$\left[\begin{pmatrix} a \\ b \end{pmatrix}\begin{pmatrix} c \\ d \end{pmatrix}\right]\begin{pmatrix} e \\ f \end{pmatrix} = \begin{pmatrix} a \\ b \end{pmatrix}\left[\begin{pmatrix} c \\ d \end{pmatrix}\begin{pmatrix} e \\ f \end{pmatrix}\right]$$

and

$$\begin{pmatrix} a \\ b \end{pmatrix}\left[\begin{pmatrix} c \\ d \end{pmatrix} + \begin{pmatrix} e \\ f \end{pmatrix}\right] = \begin{pmatrix} a \\ b \end{pmatrix}\begin{pmatrix} c \\ d \end{pmatrix} + \begin{pmatrix} a \\ b \end{pmatrix}\begin{pmatrix} e \\ f \end{pmatrix},$$

so multiplication is associative and distributive.

The two coordinate vectors are

$$\text{(B)} \qquad \mathbf{e}_1 = \begin{pmatrix} 1 \\ 0 \end{pmatrix}, \qquad \mathbf{e}_2 = \begin{pmatrix} 0 \\ 1 \end{pmatrix},$$

and every vector $(a, b)^{\mathrm{T}}$ can be written as

$$\text{(C)} \qquad (a, b)^{\mathrm{T}} = a\mathbf{e}_1 + b\mathbf{e}_2.$$

It is easy to verify that the set of complex numbers $a\mathbf{e}_1$ (a real) behave in all respects like the real numbers; for example, $(a \cdot \mathbf{e}_1)(c \cdot \mathbf{e}_1) = (ac) \cdot \mathbf{e}_1$. It is customary to simplify notation by simply omitting the factor $\mathbf{e}_1$ in (C). It is also customary to use i as another notation for $\mathbf{e}_2$. Then (C) takes the form

$$\text{(D)} \qquad \begin{pmatrix} a \\ b \end{pmatrix} = a + bi = a + ib.$$

From (A) we obtain

$$\text{(E)} \qquad i^2 = -1;$$

this, with the commutative, associative, and distributive laws, allows us to reconstruct (A) as

(F) $$(a + ib)(c + id) = (ac - bd) + i(ad + bc).$$

If $z = a + ib$, a is called the **real part** of z and is denoted by $\operatorname{Re} z$, and b is called the **imaginary part** of z and is denoted by $\operatorname{Im} z$. The latter name comes to us from a time when complex numbers were felt to be more a product of the human imagination than real numbers are.

As usual, we define the **conjugate** of z to be

(G) $$\bar{z} = \operatorname{Re} z - i \operatorname{Im} z.$$

It is also written z^{-}

The length of the vector $z = (a, b)^{\mathrm{T}}$ has already been defined in Section V-1 to be

$$|z| = [a^2 + b^2]^{1/2}.$$

By a trivial calculation,

(H) $$|z|^2 = z\bar{z} = |\bar{z}|^2.$$

It is also easy to establish that if z, z_1, and z_2 are complex numbers,

(I) $$(1/z)^{-} = 1/\bar{z} \qquad \text{if} \quad z \neq 0,$$

(J) $$(z_1 + z_2)^{-} = \bar{z}_1 + \bar{z}_2,$$

(K) $$(z_1 z_2)^{-} = \bar{z}_1 \bar{z}_2.$$

For by (H), if $z = a + ib \neq 0$,

$$\begin{aligned}(1/z)^{-} &= (\bar{z}/|z|^2)^{-} \\ &= (a/|z|^2 - ib/|z|^2)^{-} \\ &= a/|z|^2 + ib/|z|^2 \\ &= z/|z|^2 \\ &= 1/\bar{z};\end{aligned}$$

and if $z_1 = a_1 + ib_1$ and $z_2 = a_2 + ib_2$, then

$$\begin{aligned}(z_1 + z_2)^{-} &= (a_1 + a_2 + i[b_1 + b_2])^{-} \\ &= a_1 + a_2 - i(b_1 + b_2) \\ &= \bar{z}_1 + \bar{z}_2, \\ (z_1 z_2)^{-} &= ([a_1 a_2 - b_1 b_2] + i[a_1 b_2 + a_2 b_1])^{-} \\ &= a_1 a_2 - b_1 b_2 - i(a_1 b_2 + a_2 b_1) \\ &= (a_1 - ib_1)(a_2 - ib_2) \\ &= \bar{z}_1 \bar{z}_2.\end{aligned}$$

Definition V-2-2 can be extended thus.

DEFINITION 1-2 *A* ***complex vector space***, *or* ***complex linear space***, *is a triple* $(V, +, \cdot)$ *such that* V *is a set and*

(i) $+$ *is a function on* $V \times V$ *with values in* V;
(ii) $\cdot$ *is a function on* $\mathscr{C} \times V$ *with values in* V;

and for all v_1, v_2, v_3 *in* V *and all complex numbers* c_1, c_2

(iii) $v_1 + v_2 = v_2 + v_1$,
(iv) $(v_1 + v_2) + v_3 = v_1 + (v_2 + v_3)$,
(v) *there is a member* 0 *of* V *such that* $v + 0 = v$ *for all* v *in* V;
(vi) *to each* v *in* V *corresponds a member* $(-v)$ *of* V *such that* $v + (-v) = 0$;
(vii) $c_1(c_2 v_1) = (c_1 c_2) v_1$;
(viii) $0 v_1 = 0$, $1 v_1 = v_1$;
(ix) $(c_1 + c_2) v_1 = c_1 v_1 + c_2 v_1$;
(x) $c_1(v_1 + v_2) = c_1 v_1 + c_1 v_2$.

Although the name *complex vector* (*or linear*) *space* applies to the triple $(V, +, \cdot)$, when the meanings of $+$ and $\cdot$ are obvious we shall often speak of *the complex vector* (*or linear*) *space* V."

If we replace the word *complex* by *real* everywhere in Definition 1-2, we obtain the definition of a real vector space, as in Section V-2.

When we are discussing a complex vector space, it is customary to use the word **scalar** to denote a complex number; when we are discussing a real vector space, the word **scalar** is used to denote a real number. So, in either case multiplication of vectors by scalars is always possible—the product being a vector.

COROLLARY 1-3 *Let* V *be a set of functions all defined and complex-valued on a set* B. *Let* t *and* $\cdot$ *be defined by Definition* 1-1, *where* b *is allowed to be any complex number. If for every* f_1 *and* f_2 *in* V *and every complex number* b *both* $f_1 + f_2$ *and* bf_1 *are in* V, *then* V *is a complex vector space.*

By hypothesis, (i) and (ii) in Definition 1-2 are satisfied. All the other requirements in Definition 1-2 follow at once from Definition 1-1 because of known properties of complex numbers.

The set of real-valued functions $i \to x^i$ (i in $1, \ldots, r$) is the set of representations of V^r; it is a real vector space R^r. The set of all complex-valued functions $i \to z^i$ (i in $\{1, \ldots, r\}$) is a complex vector space; we shall call it $\mathscr{C}^r$.

As an exercise, the reader should verify that the set of all bounded complex-valued functions on any set B is a complex vector space, and if B is a set in R^r, the set of all complex-valued functions continuous on B is a complex vector space.

EXERCISE 1-1 Test each of the following sets of complex-valued functions $t \mapsto f(t)$ (t in R) to find if it is or is not a vector-space:

(i) all polynomials;
(ii) all functions with continuous tenth derivatives;

(iii) all continuous functions f on R such that $|f|$ is integrable over R;
(iv) all continuous functions that do not vanish on R;
(v) all functions such that $\lim_{t \to \infty} f(t) = 0$.

2. The Spaces $\mathscr{L}_1$ and $\mathscr{L}_2$

In manipulating vectors in V^r, it was often essential to sum the values of some functions of i over $i = 1, \ldots, r$. For example, the definition of the length $|x|$ of a vector in V^r represented by $(x^1, \ldots, x^r)^{\mathrm{T}}$ involves such a sum. In manipulating vectors in a space of complex functions on a set B in R^r, we need a substitute for a sum over $\{1, \ldots, r\}$, and a natural substitute is an integral over B. Suppose, then, that m is a function of left-open intervals in R^r. (Any reader who prefers not to be bothered with such measures can replace m by the elementary measure m_{L}.) If B is a set in R^r and f is a complex-valued function on B, we have already defined $\int_B f(x)\, m(dt)$. For complex numbers are vectors belonging to V^2, and in Definition V-5-2 we have defined the integral of a vector-valued function. Moreover, as a special case of Theorem V-5-3 we have the following.

THEOREM 2-1 *Let m be an additive nonnegative function of left-open intervals in R^r. Let D be a subset of R^r and let f be a complex-valued function on D. Then f is integrable with respect to m over D if and only if* Re f *and* Im f *are integrable with respect to m over D, and in that case*

$$\int_D f(t)\, m(dt) = \int_D \operatorname{Re} f(t)\, m(dt) + i \int_D \operatorname{Im} f(t)\, m(dt).$$

COROLLARY 2-2 *Let m be an additive nonnegative function of left-open intervals in R^r and let D be a subset of R^r. Let f_1 and f_2 be complex-valued functions integrable with respect to m over D, and let c be a complex number. Then $f_1 + f_2$, cf_1, $|f_1|$, and $\bar{f}_1$ are integrable with respect to m over D, and if one of the functions f_1, f_2 is bounded, then $f_1 f_2$ is integrable with respect to m over D, and*

(i) $$\int_D [f_1(t) + f_2(t)]\, m(dt) = \int_D f_1(t)\, m(dt) + \int_D f_2(t)\, m(dt),$$

(ii) $$\int_D c f_1(t)\, m(dt) = c \int_D f_1(t)\, m(dt),$$

(iii) $$\int_D |f_1(t)|\, m(dt) \geqq \left| \int_D f_1(t)\, m(dt) \right|,$$

(iv) $$\int_D \bar{f}_1(t)\, m(dt) = \left[\int_D f_1(t)\, m(dt) \right]^{-}.$$

By Theorem V-5-3 and Corollary V-5-6, (i) and (iii) are valid. Since f_1 is integrable with respect to m over D, by Theorem 2-1 so are $\operatorname{Re} f_1$ and $\operatorname{Im} f_1$. Then

$$\begin{aligned}\int_D \operatorname{Re} f_1(t)\, m(dt) + i\int_D [-\operatorname{Im} f_1(t)]\, m(dt) &= \left(\int_D \operatorname{Re} f_1(t)\, m(dt) + i\int_D \operatorname{Im} f_1(t)\, m(dt)\right)^- \\ &= \left(\int_D f_1(t)\, m(dt)\right)^- .\end{aligned}$$

By Theorem 2-1, this implies (iv).

If f_1 and f_2 are integrable, so are $\operatorname{Re} f_1$, $\operatorname{Re} f_2$, $\operatorname{Im} f_1$, and $\operatorname{Im} f_2$. If one of them, say f_2, is bounded, then $\operatorname{Re} f_2$ and $\operatorname{Im} f_2$ are bounded, and by Theorem II-5-4, the four products

$$(\operatorname{Re} f_1)(\operatorname{Re} f_2),\ (\operatorname{Re} f_1)(\operatorname{Im} f_2),\ (\operatorname{Im} f_1)(\operatorname{Re} f_2),\ (\operatorname{Im} f_1)(\operatorname{Im} f_2)$$

are all integrable with respect to m over D. Since

$$\text{(A)}\qquad \begin{aligned} f_1 f_2 = {} & \{(\operatorname{Re} f_1)(\operatorname{Re} f_2) - (\operatorname{Im} f_1)(\operatorname{Im} f_2)\} \\ & + \{(\operatorname{Re} f_1)(\operatorname{Im} f_2) + (\operatorname{Im} f_1)(\operatorname{Re} f_2)\}\end{aligned}$$

and the real-valued functions in the curly braces are integrable with respect to m over D, so is $f_1 f_2$. If f_2 is a constant c, equation (A) holds with $\operatorname{Re} c$ and $\operatorname{Im} c$ in place of $\operatorname{Re} f_2$ and $\operatorname{Im} f_2$, and all terms in the curly braces are integrable by Theorem I-5-3. Then cf_1 is integrable, and by integrating both members of (A) and using Theorem 2-1,

$$\begin{aligned}\int_D cf_1(t)\, m(dt) &= \int_D \{(\operatorname{Re} c)(\operatorname{Re} f_1(t)) - (\operatorname{Im} c)(\operatorname{Im} f_1(t))\}\, m(dt) \\ &\quad + i\int_D \{(\operatorname{Re} c)(\operatorname{Im} f_1(t)) + (\operatorname{Im} c)(\operatorname{Re} f_1(t))\}\, m(dt) \\ &= \left\{(\operatorname{Re} c)\operatorname{Re}\int_D f_1(t)\, m(dt) - (\operatorname{Im} c)\operatorname{Im}\int_D f_1(t)\, m(dt)\right\} \\ &\quad + i\left\{(\operatorname{Re} c)\operatorname{Im}\int_D f_1(t)\, m(dt) + (\operatorname{Im} c)\operatorname{Re}\int_D f_1(t)\, m(dt)\right\} \\ &= c\int_D f_1(t)\, m(dt),\end{aligned}$$

which is conclusion (ii). The proof is complete.

Our definition of measurable functions is an immediate extension of Definition III-10-1.

DEFINITION 2-3 *Let m be an additive nonnegative function of left-open intervals in R^r, and let D be a subset of R^r. Let f be a complex-valued function on D. Then f is* ***m*-measurable** *on D if there exists a sequence of functions $f_1, f_2, \ldots$ integrable with respect to m over D such that*

$$\lim_{n \to \infty} f_n(x) = f(x)$$

for all x in D.

COROLLARY 2-4 *Let m be an additive nonnegative function of left-open intervals in R^r, D a subset of R^r, f and g complex-valued functions on D, and c a complex number. Then*

(i) *f is m-measurable on D if and only if* $\mathrm{Re}\, f$ *and* $\mathrm{Im}\, f$ *are m-measurable on D;*
(ii) *if f and g are m-measurable on D, so are $f + g$, fg, cf, f, and $\bar{f}$.*

Suppose that f and g are m-measurable on D. We choose two sequences f_1, $f_2, \ldots$ and $g_1, g_2, \ldots$ of functions, all integrable with respect to m over D and converging to f and g, respectively, at each point of D. We may assume the f_n bounded; otherwise we replace f_n by $n \wedge [f_n \vee (-n)]$. Then the functions

(B) $$f_n + g_n,\ f_n g_n,\ cf_n,\ |f_n|,\ \bar{f}_n$$

are all integrable with respect to m over D by Corollary 2-2, and they converge, respectively, to

(C) $$f + g,\ fg,\ cf,\ |f|,\ \bar{f}$$

at each point of D. So, by definition, the functions in the list (C) are all m-measurable on D, and (ii) is established.

If f is m-measurable on D, by (ii) so are the functions

$$\bar{f},\ f + \bar{f},\ f - \bar{f},\ (f + \bar{f})/2,\ (f - \bar{f})/2,$$

and the last two are $\mathrm{Re}\, f$ and $\mathrm{Im}\, f$. Conversely, if $\mathrm{Re}\, f$ and $\mathrm{Im}\, f$ are m-measurable on D, by (ii) so are $i(\mathrm{Im}\, f)$ and $(\mathrm{Re}\, f) + i(\mathrm{Im}\, f)$, which completes the proof of (i).

From Corollary 2-4 it follows at once that the set of all complex-valued functions m-measurable on D is a (complex) vector space.

For each $p \geqq 1$ we define a very important class of functions, thus.

DEFINITION 2-5 *Let m be a nonnegative additive function of left-open intervals in R^r; let D be a set in R^r and p a number in $[1, \infty)$. Then (complex) $\mathscr{L}_p[m, D]$ is the set of all m-measurable complex-valued functions f on D such that $|f|^p$ is integrable with respect to m over D. Also, real $\mathscr{L}_p[m, D]$ is the set of all m-measurable real-valued functions f on D such that $|f|^p$ is integrable with respect to m over D.*

An interesting special case is that in which Z is a subset of the integers (positive, negative, or 0) and for each interval A in R, $m_Z A$ is the number of integers of set Z that are in A. We have discussed this before. Every function on Z is m-measurable, and a complex-valued function f on Z is integrable with respect to m_Z if and only if

$$\sum_{j \text{ in } Z} |f(j)|$$

is finite, and in that case

$$\int_Z f(x)\, m_Z(dx) = \sum_{j \text{ in } Z} f(j).$$

With this Z and m_Z, the class $\mathscr{L}_p[m_Z, Z]$ is the set of all complex-valued functions on Z such that

$$\sum_{j \text{ in } Z} |f(j)|^p < \infty.$$

This particular $\mathscr{L}_p[m_Z, Z]$ is usually denoted by $l_p[Z]$. Historically, $l_2[Z]$ was the first *Hilbert space* that was defined and used (by David Hilbert, of course).

With m and D as in Definition 2-5, if f is integrable it is measurable and $|f|$ is integrable, so f is in $\mathscr{L}_1[m, D]$. Conversely, by Corollary III-10-4, if f is measurable and $|f|$ is integrable, f is integrable with respect to m over D. So $\mathscr{L}_1[m, D]$ is the set of all complex-valued functions integrable with respect to m over D.

It can be shown that, for all $p \geqq 1$, $\mathscr{L}_p[m, D]$ is a complex vector space, and real $\mathscr{L}_p[m, D]$ is a real vector space. But $\mathscr{L}_1$ and $\mathscr{L}_2$ are especially important spaces, and fortunately they are especially convenient to work with, so we shall consider them only. Moreover, we shall give proofs for complex $\mathscr{L}_p[m, D]$ ($p = 1$ or 2) only; the case of real $\mathscr{L}_p[m, D]$ is an obvious simplification.

COROLLARY 2-6 *Let m and D be as in Definition* 2-3. *Then $\mathscr{L}_1[m, D]$ is a complex vector space.*

This is an immediate consequence of Corollary 2-2.

LEMMA 2-7 *Let m and D be as in Definition* 2-3. *If f_1 and f_2 belong to $\mathscr{L}_2[m, D]$, $f_1 f_2$ and $f_1 \bar{f}_2$ are integrable with respect to m over D.*

Since f_1 and f_2 are m-measurable over D, so are $\bar{f}_2$, $f_1 f_2$, and $f_1 \bar{f}_2$, by Corollary 2-4. Also, for each t in D

$$|f_1(t)\overline{f_2(t)}| = |f_1(t) f_2(t)| \leqq \tfrac{1}{2}\{|f_1(t)|^2 + |f_2(t)|^2\},$$

and $|f_1|^2$ and $|f_2|^2$ are integrable with respect to m over D by hypothesis, so by Corollary III-10-4, $f_1 f_2$ and $f_1 \bar{f}_2$ are integrable with respect to m over D.

COROLLARY 2-8 *Let m be an additive nonnegative function of left-open intervals in R^r, and let D be a subset of R^r. Then $\mathscr{L}_2[m, D]$ is a vector space.*

Let f and g belong to $\mathscr{L}_2[m, D]$, and let c be a complex number. Then f and g are m-measurable, so $f + g$ and cf are m-measurable. Also, for each t in D

$$\begin{aligned}|f(t) + g(t)|^2 &= [f(t) + g(t)][f(t)^- + g(t)^-] \\ &= f(t)f(t)^- + f(t)g(t)^- + g(t)f(t)^- + g(t)g(t)^-,\end{aligned}$$

and all four terms in the right member are integrable with respect to m over D by Lemma 2-7, and

$$|cf|^2 = |c|^2|f|^2,$$

which is integrable with respect to m over D. So $f + g$ and cf both belong to $\mathscr{L}_2[m, D]$, and $\mathscr{L}_2[m, D]$ is a linear space.

3. Normed Vector Spaces

Along with the elementary operations on vectors in finite-dimensional spaces, we have often needed the idea of the length of a vector. The generalization of the concept of length to a large class of vector (or linear) spaces is usually called the *norm* of the vector.

DEFINITION 3-1 *A function $v \mapsto \|v\|$ on a linear space V is called a* **norm** *on V if it satisfies the conditions*

(i) $\|0\| = 0$;
(ii) $\|v\| > 0$ *for all v in V other than* 0;
(iii) *for all v_1 and v_2 in V,* $\|v_1 + v_2\| \leqq \|v_1\| + \|v_2\|$;
(iv) *for all v in V and all scalars c,* $\|cv\| = |c| \cdot \|v\|$.

It is called a **pseudo-norm** *on V if it satisfies* (i), (iii), *and* (iv) *and instead of* (ii) *satisfies the weaker condition*

(ii′) *for all v in V,* $\|v\| \geqq 0$.

If $v \mapsto \|v\|$ is a norm on V, the pair $\{V, \|\cdot\|\}$ is called a **normed linear space**, *or* **normed vector space**.

Given a norm on V, we can at once define a distance on V by defining the distance between u and v to be $\|u - v\|$. This has the familiar properties of distance: the distance from u to v is 0 if $u = v$ and greater than 0 if $u \neq v$; the distance from u to v equals the distance from v to u; and the distance from u to w cannot exceed the sum of the distance from u to v and the distance from v to w. From a pseudo-norm we can define a pseudo-distance in the same way; it differs from a true distance only in that two different points u, v can have pseudo-distance 0.

For every f in the class $\mathscr{L}_1[m, D]$ we define

(A) $$\|f\|_1 = \int_D |f(x)|\, m(dx).$$

This clearly satisfies (i) of Definition 3-1, and (iv) and (iii) follow by integration from the relations

(B) $$|cf(x)| = |c|\, |f(x)|,$$
$$|f_1(x) + f_2(x)| \leqq |f_1(x)| + |f_2(x)|,$$

valid for all x in D. Also, condition (ii′) is obviously satisfied. However, (ii) is not satisfied. The integral of $|f|$ is 0 whenever $f(x) = 0$ except on a set of measure 0; in fact, the integral of $|f|$ is 0 if and only if $f(x) = 0$ almost everywhere in D, as we showed in Theorems II-12-2 and II-12-4. We can convert this pseudo-norm into a true norm by lumping together with each function f integrable over D all the other functions equivalent to f; that is, equal to f except on a set with m-measure 0. Then every function f in $\mathscr{L}_1[m, D]$ belongs to exactly one equivalence-class, and all members of that class are almost everywhere equal to f. The set of equivalence-classes is called $L_1[m, D]$. We can specify an equivalence-class by naming any function in it; if an equivalence-class contains a function f, we can and shall denote it by the symbol $[f]$. More generally, we adopt the following definition.

Definition 3-2 *For every nonnegative additive function m of left-open intervals in R^r, every set D contained in R^r, and every number $p \geqq 1$, we define $L_p[m, D]$ to be the set of all equivalence-classes of functions belonging to $\mathscr{L}_p[m, D]$; each class $[f]$ in $L_p[m, D]$ contains, along with any one member f, all functions equivalent to f.*

If f' is in a class $[f]$ and g' is in a class $[g]$, the sum $f' + g'$ determines an equivalence-class $[f' + g']$. If from $[f]$ we had selected any other member f'' and from $[g]$ any other member g'', then these would have differed only on sets of measure 0 from f' and g', respectively, so $f'' + g''$ would have been equivalent to $f' + g'$. That is, $[f'' + g'']$ would have been the same equivalence-class as $[f' + g']$. The equivalence-class of the sum is determined uniquely by the equivalence-classes $[f]$ and $[g]$ and does not depend on the representatives we choose. Thus, the operation of adding representatives furnishes for each pair of classes $[f]$, $[g]$ a unique class, which we can call $[f] + [g]$. In the same way, multiplying any representative of the class $[f]$ by a scalar c furnishes a uniquely determined equivalence-class, which we can call $c[f]$, and replacing any representative f of the class $[f]$ by its conjugate $\bar{f}$ furnishes a uniquely determined equivalence-class, which we call $[f]^-$. We have thus provided the set

$L_p[m, D]$ with operations of addition and scalar multiplication that satisfy the requirements in Definition 1-2, so $L_p[m, D]$ (with these operations) is a linear space. Also, the right member of equation (A) is unchanged if we replace f by any function equivalent to f, so the integral is determined by the equivalence-class to which f belongs. We adopt this value as the norm of the equivalence-class $[f]$ in $L_1[m, D]$; if $[f]$ belongs to $L_1[D]$, we define $\|[f]\|_1$ to be the integral of the absolute value of any one of the functions that belong to the class $[f]$. With this definition, $L_1[m, D]$ becomes a normed linear space.

The sentences in the preceding paragraph were long and intricate because of the necessity of distinguishing between the functions that are members of $\mathscr{L}_1[m, D]$ and the equivalence-classes of such functions – the equivalence-classes being the members of $L_1[m, D]$. One who is well versed in the use of these spaces can safely use the solecism "f belongs to L_1" instead of the correct statement "f belongs to $\mathscr{L}_1$"; if the speaker and listener are expert, they will both recognize what the substitute stands for. But until one is adept, it is safer to avoid such shortcuts and keep the distinction explicit.

We could use a procedure somewhat like that in the preceding paragraphs to introduce a pseudo-norm in $\mathscr{L}_2[m, D]$ and a norm in $L_2[m, D]$. But we shall not do this, because we can do better. We shall introduce a generalization to spaces $\mathscr{L}_2[m, D]$ and $L_2[m, D]$ of the inner product that we have met in finite-dimensional vector spaces, in Section V-3.

DEFINITION 3-3 *If V is a complex vector space, an* ***inner product*** *on V is a function on $V \times V$ that assigns to each pair of vectors v and w in V a complex number $\langle v, w\rangle$ such that*

(i) $\langle 0, 0\rangle = 0$,
(ii) $\langle v, v\rangle > 0$ *if* $v \neq 0$,
(iii) *for all u, v, and w in V,* $\langle u + v, w\rangle = \langle u, w\rangle + \langle v, w\rangle$,
(iv) *for all complex numbers c and all v and w in V,* $\langle cv, w\rangle = c\langle v, w\rangle$,
(v) *for all v and w in V,* $\langle v, w\rangle = \langle w, v\rangle^-$.

If the function $(u, v) \mapsto \langle u, v\rangle$ satisfies (i), (iii), (iv), *and* (v), *but instead of* (ii) *satisfies the weaker condition*

(ii′) $\langle v, v\rangle \geqq 0$ *for all v in V,*

the function is called a ***pseudo-inner-product***.

Let f and g both belong to $\mathscr{L}_2[D]$. We define

$$\langle f, g\rangle = \int_D f(x)\overline{g(x)}\, m(dx). \tag{C}$$

This exists, by Lemma 2-7. Conditions (i), (iii), (iv), and (v) of Definition 3-3 are trivially easy to verify. Moreover, for all f we have $f(x)f(x)^- \geqq 0$, so (ii′) is satisfied, and (C) defines a pseudo-inner-product on $L_2[m, D]$. It is not an inner product, since $\langle f, f\rangle = 0$ whenever $f(x) = 0$ at almost all points x of D.

If $[f]$ and $[g]$ are two equivalence-classes that are members of the space $L_2[m, D]$, and f is in the equivalence-class $[f]$, and g is in the class $[g]$, we define

$$\text{(D)} \qquad \langle [f], [g]\rangle = \langle f, g\rangle = \int_D f(x)g(x)^- \, m(dx).$$

This seems at a glance to depend on the choice of the representatives f, g of the classes $[f]$ and $[g]$. But it does not. Suppose that f' is another member of $[f]$ and g' is another member of $[g]$. Then $f'(x)g'(x) = f(x)g(x)$ except on a set of measure 0, so the right member of (D) is unchanged if we replace f by f' and g by g'. The function $([f], [g]) \mapsto \langle [f], [g]\rangle$ defined in (D) is, in fact, a function on $L_2[D] \times L_2[D]$. The zero element in $L_2[m, D]$ is, of course, the class of functions equivalent to 0, so (i) is satisfied. If f is in $[f]$, then

$$\langle [f], [f]\rangle = \int_D |f(x)|^2 \, m(dx).$$

This is not negative, and by Theorem II-12-4 it is 0 only if $|f(x)|^2 = 0$ for almost all x, that is, it is 0 only if f is in the equivalence-class $[0]$. So (ii) is satisfied. Obviously, (iii) and (iv) hold. For (v),

$$\begin{aligned}
\langle [f], [g]\rangle &= \int_D f(x)g(x)^- \, m(dx) \\
&= \int_D [g(x)f(x)^-]^- \, m(dx) \\
&= \left\{\int_D g(x)f(x)^- \, m(dx)\right\}^- \\
&= \langle [g], [f]\rangle^-.
\end{aligned}$$

Therefore, equation (C) defines an inner product on $L_2[m, D]$.

Definition 3-3 has an immediate consequence that we shall henceforth use without explicit mention, namely,

$$\langle u, cv\rangle = \bar{c}\langle u, v\rangle.$$

This follows from (iv) and (v) of Definition 3-2 by

$$\langle u, cv\rangle = \langle cv, u\rangle^- = [c\langle v, u\rangle]^- = \bar{c}\langle v, u\rangle^- = \bar{c}\langle u, v\rangle.$$

From now on, whenever convenient, we shall omit the square brackets and write $\langle f, g\rangle$ in place of $\langle [f], [g]\rangle$ for $[f]$ and $[g]$ in $L_2[m, D]$.

The next theorem is important for use in spaces with inner products or pseudo-inner-products. It is often called the **Cauchy–Schwarz inequality**.

THEOREM 3-4 *Let V be a complex linear space, and let $(u, v) \mapsto \langle u, v\rangle$ be a pseudo-inner-product on $V \times V$. Then for all u and v in V,*

(E) $$|\langle u, v\rangle| \leqq \langle u, u\rangle^{1/2}\langle v, v\rangle^{1/2}.$$

We prove this theorem only for complex linear spaces. To obtain the proof for real linear spaces, we merely ignore the conjugate signs.

If $\langle u, v\rangle = 0$, the conclusion is evident, so we consider the case $\langle u, v\rangle \neq 0$. We define

$$c = \langle u, v\rangle/|\langle u, v\rangle|.$$

Then

$$|c| = 1 \qquad \text{and} \qquad 1/c = \bar{c},$$

and if we define $w = cv$, we obtain

$$\langle u, w\rangle = \bar{c}\langle u, v\rangle = |\langle u, v\rangle|,$$

$$\langle w, w\rangle = \langle cv, cv\rangle = c\bar{c}\langle v, v\rangle = \langle v, v\rangle,$$

$$\langle w, u\rangle = \langle u, w\rangle^{-} = |\langle u, v\rangle|.$$

For each positive real number t, $tu - t^{-1}w$ is in V, so by hypothesis,

$$\begin{aligned} 0 &\leqq \langle tu - t^{-1}w, tu - t^{-1}w\rangle \\ &= t^2\langle u, u\rangle - \langle u, w\rangle - \langle w, u\rangle + t^{-2}\langle w, w\rangle \\ &= t^2\langle u, u\rangle - 2|\langle u, v\rangle| + t^{-2}\langle v, v\rangle, \end{aligned}$$

hence

(F) $$|\langle u, v\rangle| \leqq (t^2/2)\langle u, u\rangle + (1/2t^2)\langle v, v\rangle.$$

If $\langle u, u\rangle$ were 0, the right member of (E) could be made smaller than any preassigned positive ε by choosing t large. This contradicts the hypothesis that $\langle u, v\rangle \neq 0$; so $\langle u, u\rangle > 0$. Likewise, if $\langle v, v\rangle$ were 0, by choosing t near 0 we could make the right member of (E) arbitrarily close to 0, contradicting the hypothesis $\langle u, v\rangle \neq 0$, so $\langle v, v\rangle \neq 0$. Thus, neither $\langle u, u\rangle$ nor $\langle v, v\rangle$ is 0. We choose $t = [\langle v, v\rangle/\langle u, u\rangle]^{1/4}$, and (F) takes the form (E). So (E) holds in all cases.

In any space V in which an inner product is defined, we define a norm by setting

(G) $$\|u\| = [\langle u, u\rangle]^{1/2}.$$

Properties (i) and (ii) in Definition 3-1 follow from properties (i) and (ii) in Definition 3-3. If c is complex,

$$\|cu\|^2 = \langle cu, cu\rangle = c\bar{c}\langle u, u\rangle = |c|^2\|u\|^2,$$

and by taking square roots we obtain (iv) of Definition 3-1. For the triangle inequality, which is (iii) of Definition 3-1, we observe that by Theorem 3-4

$$\begin{aligned}\|u + v\|^2 &= \langle u + v, u + v\rangle \\ &= \langle u, u\rangle + \langle v, u\rangle + \langle u, v\rangle + \langle v, v\rangle \\ &\leqq \langle u, u\rangle + 2[\langle u, u\rangle]^{1/2}[\langle v, v\rangle]^{1/2} + \langle v, v\rangle \\ &= \|u\|^2 + 2\|u\| \cdot \|v\| + \|v\|^2 \\ &= [\|u\| + \|v\|]^2,\end{aligned}$$

and (iii) of Definition 3-1 follows by taking square roots. The function $u \mapsto \|u\|$ defined in (F) is therefore a norm on the space V.

In particular, if the space V is $L_2[m, D]$, the norm defined in (G) is often designated by $\|u\|_2$. Thus, if $\lfloor f \rfloor$ is in $L_2[m, D]$,

(II) $$\|\lfloor f \rfloor\|_2 = \left\{\int_D |f(x)|^2 \, dx\right\}^{1/2}.$$

We use the same symbol for $\|f\|$ in $\mathscr{L}_2[D]$. Thus, the statement that $\lfloor f_n \rfloor$ tends to $\lfloor f \rfloor$ in L_2-norm means that $\|\lfloor f_n \rfloor - \lfloor f \rfloor\|_2$ tends to 0, and the statement that f_n tends to f in $\mathscr{L}_2$-pseudo-norm means that $\|f_n - f\|_2$ tends to 0. But these expressions are a bit tedious, and instead we shall use the technically incorrect (but not confusing) shortening "f_n tends to f in L_2-norm" for both of them. We also use the expression "the L_2-distance between f and g" to mean $\|f - g\|_2$.

Theorem 3-5 *Let V be a linear space with an inner product. Then $\langle u, v\rangle$ is continuous in both variables.*

Let u_0 and v_0 be in V, and let ε be positive. Define

$$\delta = \varepsilon/(\|u_0\| + \|v_0\| + 1 + \varepsilon).$$

Let u and v be points of V such that

$$\|u - u_0\| < \delta \qquad \text{and} \qquad \|v - v_0\| < \delta.$$

Then by Theorem 3-4,

(I) $$\begin{aligned}|\langle u, v\rangle - \langle u_0, v_0\rangle| &= |\langle u - u_0, v\rangle + \langle u_0, v - v_0\rangle| \\ &\leqq |\langle u - u_0, v\rangle| + |\langle u_0, v - v_0\rangle| \\ &\leqq \|u - u_0\| \, \|v\| + \|u_0\| \, \|v - v_0\|.\end{aligned}$$

Since $\delta < 1$, by the triangle inequality,

$$\|v\| \leqq \|v_0\| + \|v - v_0\| < \|v_0\| + 1,$$

so

$$\|u - u_0\| \, \|v\| + \|u_0\| \, \|v - v_0\| < \delta(\|v_0\| + 1) + \|u_0\|\delta < \varepsilon.$$

By this and (I),

$$|\langle u, v\rangle - \langle u_0, v_0\rangle| < \varepsilon,$$

completing the proof.

EXERCISE 3-1 Let D be a set in R^r. Show that if f is in $\mathscr{L}_2[D]$, so are f, f^-, $\operatorname{Re} f$, $\operatorname{Im} f$, $(\operatorname{Re} f)^+$, $(\operatorname{Re} f)^-$, $(\operatorname{Im} f)^+$, and $(\operatorname{Im} f)^-$.

EXERCISE 3-2 With $l_p(Z)$ defined as in Section 2, if Z is an infinite set of integers, $l_p(Z)$ consists of all functions f on Z such that $\sum_{z \text{ in } Z} |f(z)|^p$ is finite. For $Z = \{1, 2, 3, \ldots\}$ and $p = 2$, what is the form of the Cauchy–Schwarz inequality? Show that if D is the union of all intervals $(z_i, z_i + 1]$ with z_i in Z, there is a one-to-one length-preserving mapping of $l_p(Z)$ into $L_p[m_L, D]$.

EXERCISE 3-3 Show that if f is nonnegative on an m-measurable set D and f^2 is integrable over D, f is in $\mathscr{L}_2[m, D]$. (Let B be a bounded interval in R^r. For each positive integer n, $f^2 + n^{-2}$ is integrable over $D \cap B$. Define $g(y) = y^{1/2}$ for $y \geqq n^{-2}$, $g(y) = ny$ for $y < n^{-2}$. Then $x \mapsto g(f(x)^2 + n^{-2})$ is integrable over $D \cap B$, and it is $[f(x)^2 + n^{-2}]^{1/2}$ on $D \cap B$. These functions descend and converge to f on $D \cap B$.)

EXERCISE 3-4 Prove that for f defined on D, if any one of the following three statements is true, all three are true:

(i) f is in $\mathscr{L}_2[m, D]$;
(ii) the squares of $(\operatorname{Re} f)^+$, $(\operatorname{Re} f)^-$, $(\operatorname{Im} f)^+$, and $(\operatorname{Im} f)^-$ are all integrable over D;
(iii) there exist nonnegative functions f_1, f_2, f_3, f_4 on D, all having squares integrable over D, such that $f = f_1 - f_2 + if_3 - if_4$.

EXERCISE 3-5 Prove that if f is complex-valued on an m-measurable set D contained in R^r, f is in $\mathscr{L}_2[m, D]$ if and only if $f|f|$ is integrable over D. (If $f|f|$ is integrable, so is $[f|f|]^+$, which is $[f^+]^2$. By Exercise 3-3, f^+ is in $\mathscr{L}_2[m, D]$. Similarly, f^- is in $\mathscr{L}_2[m, D]$.)

EXERCISE 3-6 Prove that if F is a real or complex-valued linear function on a normed vector space V that is continuous at 0, there exists a positive number B such that for all v in V, $|F(v)| \leqq B\|v\|$. (There exists a positive δ such that if $\|v\| < \delta$, $|F(v) - F(0)| < 1$. For each $v \neq 0$ in V, $|F(v)| = [2\|v\|/\delta]|F([\delta/2\|v\|]v)|$.) *Suggestion*: Do not try to prove that every linear function on a normed vector space is continuous, as it was in R^r. It is not so.

EXERCISE 3-7 Let m be a regular nonnegative additive function of left-open intervals in R^r, and let B be a set in R^r. Let f be a function on B with values in a normed linear space V. Show that the definition of the gauge-integral can be extended to define

$$\int_B f(x)\, m(dx);$$

only trivial changes are required. By use of Exercise 3-6, prove that if f is such an integrable function and F is a real-valued or complex-valued function linear and continuous on V, then the composite function $x \mapsto F(f(x))$ (x in B) is integrable over B, and

$$\int_B F(f(x))\, m(dx) = F\left(\int_B f(x)\, m(dx)\right).$$

4. Completeness of Spaces $\mathscr{L}_1$, $\mathscr{L}_2$, L_1, and L_2

If $v, v_1, v_2, v_3, \ldots$ belong to a normed linear space V, the statement

$$\lim_{n \to \infty} v_n = v$$

has the obvious meaning that

$$\lim_{n \to \infty} \|v_n - v\| = 0.$$

If this is the case, the sequence certainly satisfies the **Cauchy condition**:

(A) to each positive ε there corresponds a positive integer N such that if $m > N$ and $n > N$, then $\|v_m - v_n\| < \varepsilon$.

For we need only choose N large enough so that if $m > N$, $\|v_m - v\| < \varepsilon/2$. Then if m and n both exceed N,

$$\|v_m - v_n\| \leqq \|v_m - v\| + \|v - v_n\| < \varepsilon.$$

But there are normed linear spaces in which sequences exist that satisfy the Cauchy condition but do not converge to any limit. For example, for each bounded complex-valued function f on $D = [-1, 1]$ let us define

$$\|f\| = \sup\{|f(x)| : x \text{ in } D\}.$$

The class of all continuously differentiable functions on D, with norm $\|\cdot\|$, forms a normed vector space. For each x in D we define

$$f_\infty(x) = |x|, \qquad f_n(x) = [x^2 + n^{-2}]^{1/2} \qquad (n = 1, 2, 3, \ldots).$$

Then if $1 \leqq m \leqq n \leqq \infty$,

$$|x| \leqq f_n(x) \leqq f_m(x) \leqq |x| + 1/m.$$

Therefore the f_n form a Cauchy sequence of members of V, and $\|f_m - f_\infty\|$ tends to 0 as m increases. The f_m converge to f_∞ and cannot converge to any other function; in particular, they cannot converge to any member of V.

Investigations that involve limits in some space can often be carried out only if it can be shown that in that space every sequence that satisfies the Cauchy condition necessarily has a limit in the space. The space $L_2[D]$ has the linearity properties of the finite-dimensional spaces, and it has an inner product that resembles the inner product in finite-dimensional spaces, so we may hope that many of the theorems that are familiar to us in finite-dimensional space will have analogs in $L_2[m, D]$. But unless we can prove that every Cauchy sequence in $L_2[m, D]$ converges to some point of $L_2[m, D]$, the resemblance is not close enough to allow us to carry over theorems involving limits. Fortunately, we can prove that this is true both in $L_1[m, D]$ and in $L_2[m, D]$.

The property that we have been discussing is important enough to have a name.

DEFINITION 4-1 *A pseudo-normed linear space V is* ***complete*** *if for each sequence $v_1, v_2, v_3, \ldots$ in V that satisfies the Cauchy condition* (A), *there is a v in V such that* $\lim_{n\to\infty} v_n = v$.

As a simple and useful application of this concept, we show that the comparison test for convergence of series extends to complete normed linear spaces.

COROLLARY 4-2 *If $v_1 + v_2 + v_3 + \cdots$ is a series of vectors in a complete pseudo-normed linear space V, and there is a convergent series $p_1 + p_2 + p_3 + \cdots$ of nonnegative numbers such that for all n,*

$$\|v_n\| \leqq p_n,$$

then the series $v_1 + v_2 + v_3 + \cdots$ converges to some v_0 in V.

Let

$$s_n = v_1 + \cdots + v_n \qquad (n = 1, 2, 3, \ldots).$$

For each positive ε there is an $n(\varepsilon)$ such that if $n > m > n(\varepsilon)$,

$$p_{m+1} + \cdots + p_n < \varepsilon.$$

Then

$$\|s_n - s_m\| = \|v_{m+1} + \cdots + v_n\| \leqq \|v_{m+1}\| + \cdots + \|v_n\|$$
$$\leqq p_{m+1} + \cdots + p_n < \varepsilon.$$

So $s_1, s_2, s_3, \ldots$ satisfies the Cauchy condition. Since V is complete, s_n converges in norm or in pseudo-norm to some v_0 in V.

To prove that the spaces L_1 and L_2 (of equivalence-classes) are complete, we first prove a theorem stronger than the statement that the spaces $\mathscr{L}_1$ and $\mathscr{L}_2$ (of functions) are complete.

THEOREM 4-3 *Let m be an additive nonnegative regular function of left-open intervals in R^r. Let D be a subset of R^r, and let p be 1 or 2. If $f_1, f_2, f_3, \ldots$ is a Cauchy sequence in real or complex $\mathscr{L}_p[m, D]$, there exists an f in $\mathscr{L}_p[m, D]$ such that f_n converges to f in L_p-norm, and there exists a subsequence $f_{n(1)}, f_{n(2)}, f_{n(3)}, \ldots$ that converges to $f(x)$ at all points x of D except those in a set of m-measure 0.*

We first show that we can choose an almost-everywhere-convergent subsequence of the f_i. Since the Cauchy condition is satisfied, we can and do choose a positive integer $n(1)$ such that

$$\text{if } n \geqq n(1),\ \|f_n - f_{n(1)}\|_p < 2^{-1};$$

and then successively, having chosen $n(1), n(2), \ldots, n(j-1)$, we can and do choose $n(j)$ such that

$$\text{(B)} \qquad n(j) > n(j-1)$$

and

$$\text{(C)} \qquad \text{if } n \geqq n(j),\ \|f_n - f_{n(j)}\|_p < 2^{-j}.$$

For convenience, we introduce the symbol

$$\text{(D)} \qquad g_j = f_{n(j)}.$$

By the triangle inequality and (C), for each pair of positive integers j and k we have

$$\begin{aligned}\text{(E)} \quad & \| |g_{j+1} - g_j| + |g_{j+2} - g_{j+1}| + \cdots + |g_{j+k} - g_{j+k-1}| \|_p \\ & \leqq \|g_{j+1} - g_j\|_p + \|g_{j+2} - g_{j+1}\|_p + \cdots + \|g_{j+k} - g_{j+k-1}\|_p \\ & \leqq 2^{-j} + 2^{-j-1} + \cdots + 2^{-j-k+1} < 2^{1-j}.\end{aligned}$$

By the definition of pseudo-norm in $\mathscr{L}_p[m, D]$, this implies

$$\text{(F)} \quad \int_D \{|g_{j+1}(x) - g_j(x)| + \cdots + |g_{j+k}(x) - g_{j+k-1}(x)|\}^p\, m(dx) < 2^{(1+j)p}.$$

For each x in D and each positive integer j, the limit

$$\text{(G)} \qquad R_j(x) = \lim_{k\to\infty} \sum_{i=j}^{k} |g_{i+1}(x) - g_i(x)|$$

exists, finite or ∞. As k increases, the integrand in (F) ascends and converges everywhere to $R_j(x)^p$. So by (F) and the monotone convergence theorem, R_j^p is integrable with respect to m over D, and

$$\text{(H)} \qquad \int_D R_j(x)^p\, m(dx) \leqq 2^{(1-j)p}.$$

The integrability of R_1^p implies that it is finite at all x in D except those in a set N with $mN = 0$. From this and (G), at all points x in $D \setminus N$ the series

$$\sum_{i=1}^{\infty} [g_{i+1}(x) - g_i(x)]$$

is absolutely convergent and is hence convergent. So the limit

$$\lim_{n \to \infty} \sum_{i=1}^{n-1} [g_{i+1}(x) - g_i(x)] = \lim_{n \to \infty} [g_n(x) - g_1(x)]$$

exists and is finite; hence $g_n(x)$ converges to a finite limit.

We define a function f on D by setting

$$\text{(I)} \qquad f(x) = \lim_{n \to \infty} g_n(x) \qquad (x \text{ in } D \setminus N)$$

$$f(x) = 0 \qquad (x \text{ in } N).$$

Then f is the limit everywhere in D of the m-integrable functions $g_n 1_{D \setminus N}$, so by definition it is m-measurable. If j and k are any positive integers,

$$|g_{j+k}(x) - g_j(x)|^p \leqq \{|g_{j+1}(x) - g_j(x)| + \cdots + |g_{j+k}(x) - g_{j+k-1}(x)|\}^p$$
$$\leqq \left\{ \sum_{i=j}^{\infty} |g_{i+1}(x) - g_i(x)| \right\}^p = R_j(x)^p.$$

The last function is integrable over $D \setminus N$, and for all x in $D \setminus N$,

$$\lim_{k \to \infty} |g_{j+k}(x) - g_j(x)|^p = |f(x) - g_j(x)|^p,$$

so by the dominated convergence theorem, $|f - g_j|^p$ is m-integrable over $D \setminus N$, and (since $mN = 0$ and (H) is satisfied)

$$\text{(J)} \qquad \int_{D \setminus N} |f(x) - g_j(x)|^p \, m(dx) = \lim_{k \to \infty} \int_{D \setminus N} |g_{j+k}(x) - g_j(x)|^p \, m(dx)$$
$$\leqq \int_{D \setminus N} R_j(x)^p \, m(dx) \leqq 2^{(1-j)p}.$$

Since $f - g_j$ is measurable and $|f - g_j|^p$ is m-integrable over $D \setminus N$ (and therefore over D), $f - g_j$ is in $\mathscr{L}_p[m, D]$; and since g_j is in $\mathscr{L}_p[m, D]$, so is the sum $f = [f - g_j] + g_j$. So by (D) and (I), the subsequence $f_{n(1)}, f_{n(2)}, f_{n(3)}, \ldots$ converges to the member f of $\mathscr{L}_p[m, D]$ at all points in $D \setminus N$, where $mN = 0$. Since $mN = 0$, (J) and (D) imply

$$\text{(K)} \qquad \|f - f_{n(j)}\|_p \leqq 2^{1-j}.$$

Now let ε be positive. We can and do choose a positive integer j so large that

$$2^{1-j} < \varepsilon/2.$$

Let n be greater than $n(j)$. By (C), and (K),

$$\|f - f_n\|_p \leqq \|f - f_{n(j)}\|_p + \|f_{n(j)} - f_n\|_p < 2^{1-j} + 2^{-j} < \varepsilon.$$

So f_n converges to f in $L_p[m, D]$-norm, and the proof is complete.

COROLLARY 4-4 *If m is a regular nonnegative additive function of left-open intervals in R^r and p is 1 or 2, $L_p[m, D]$ is complete.*

Let $[f_1], [f_2], [f_3], \ldots$ be a Cauchy sequence in $L_p[m, D]$. From each class $[f_j]$ we choose a function $x \to f_j(x)$ (x in D); this is in $\mathscr{L}_p$, and the sequence $f_1, f_2, f_3, \ldots$ satisfies the Cauchy condition. By Theorem 4-3 there is a function f in $\mathscr{L}_p[m, D]$ such that

$$\lim_{j \to \infty} \|f_j - f\|_p = 0.$$

Then the class $[f]$ is in L_p and is the limit of $[f_j]$ in L_p-norm.

A vector in V^r is a single entity; we can picture it as represented by a directed line segment. But for computational purposes it is often convenient to choose a coordinate system and then to represent a vector v by its components $v^1, \ldots, v^r$. These are one-by-one the values of the function $i \mapsto v^i$ (i in $\{1, \ldots, r\}$). In V^r, convergence of a sequence of vectors $\mathbf{v}_1, \mathbf{v}_2, \ldots$ to a vector $\mathbf{v}$ was defined in Section V-2 in terms of the vectors themselves. But it is a computational convenience that by Lemma V-2-1 $\mathbf{v}_n$ tends to $\mathbf{v}$ if and only if for each i in $\{1, \ldots, r\}$ the component v_n^i tends to v^i. Analogously, a vector in $\mathscr{L}_p[m, D]$ is a single entity that we can visualize as the graph of a function. This is not a mere mathematical trick. A sound wave (at a given place) can be represented by a function that shows air pressure as a function of time. What one hears depends on this function as a whole, not on its value at some particular moment. In quantum mechanics, the state of an atom is specified by a wave-function, which is a function in a space $\mathscr{L}_2[m_L, R^r]$ for some r. To predict the result of an experiment performed on the atom, we need to know this function as a whole, not merely its value at some spot.

A vector $\mathbf{v}$ in V^r is specified by the functional values v^i ($i = 1, \ldots, r$). Analogously, a vector v in $\mathscr{L}_p[m, D]$ is specified by the functional values $v(x)$ (x in D). In $\mathscr{L}_p[m, D]$, convergence of vectors v_n to a vector v is defined in terms of the vectors themselves, just as it was in V^r; it means

$$\lim_{n \to \infty} \|v_n - v_0\|_p = 0.$$

By analogy with the spaces V^r, we would naturally hope that v_n tends to v in $\mathscr{L}_p[m, D]$ if and only if for each x in D the component $v_n(x)$ tends to $v(x)$. Inconveniently, the situation is not quite so simple.

The examples at the beginning of Section II-4 show that f_n can converge pointwise to f on D without having f_n converge to f in L_1-norm or in L_2-norm.

On the other hand, if we form the sequence of subintervals $A_1, A_2, A_3, \ldots$ of $D = (0, 1]$, defined successively as

$$(0,1]; (0,\tfrac{1}{2}], (\tfrac{1}{2},1]; (0,\tfrac{1}{4}], (\tfrac{1}{4},\tfrac{2}{4}], (\tfrac{2}{4},\tfrac{3}{4}], (\tfrac{3}{4},1]; (0,\tfrac{1}{8}], \ldots$$

(the pattern is obvious), and define f_n to be the indicator of A_n, we find readily that f_n tends to 0 in L_1-norm and in L_2-norm. But if x is in D, there are infinitely many A_n that contain x and infinitely many that do not. So $f_n(x)$ takes each of 0 and 1 as functional value for infinitely many n and does not converge to any limit.

Nevertheless, there are useful, even though incomplete, correspondences between the two kinds of limit. If f_n converges to f everywhere in D (or even almost everywhere in D), and in addition there is an integrable g such that $|f_n(x)| \leqq g(x)$ for all x in D, then f_n converges to f in L_1-norm, by the dominated convergence theorem. A similar theorem holds for L_2-convergence; we leave its easy proof as an exercise. In the other direction, if f_n converges to f in L_1-norm, or in L_2-norm, $f_n(x)$ may not converge to $f(x)$ anywhere, but there is a subsequence of the f_n that converges to $f(x)$ almost everywhere. This is part of the content of the next theorem.

THEOREM 4-5 *Let $g, f_1, f_2, f_3, \ldots$ be complex-valued functions on a set D contained in R^r. Then*

(i) *if $g, f, f_1, f_2, \ldots$ are in $\mathscr{L}_p[m, D]$ ($p = 1$ or 2) and converge to f in L_p-norm, there is a subsequence $f_{n(1)}, f_{n(2)}, f_{n(3)}, \ldots$ that converges to f at almost all points x of D;*

(ii) *if f_n converges to f in L_1-norm, in L_2-norm, or almost everywhere, and also f_n converges to g in L_1-norm, in L_2-norm, or almost everywhere, then $f(x) = g(x)$ for almost all x in D.*

We first prove the special case of (ii) in which f_n converges both to f and to g in L_1-norm, or to both in L_2-norm, or to both almost everywhere. If f_n tends to f and to g in L_1-norm, for all n we have by the triangle inequality,

$$0 \leqq \|f - g\|_1 \leqq \|f_n - f\|_1 + \|f_n - g\|_1.$$

The right member is arbitrarily near 0 for large n, so the left member is a nonnegative number that is less than any positive number; it therefore has to be 0. Then the integral of $|f - g|$ over D is 0, and by Theorem II-12-4 it must be 0 almost everywhere in D. A similar proof applies if f_n converges both to f and to g in L_2-norm. If $f_n(x)$ converges to $f(x)$ on $D \setminus N_1$, where $m_L N_1 = 0$, and $f_n(x)$ converges to $g(x)$ on $D \setminus N_2$, where $m_L N_2 = 0$, then $f_n(x)$ converges to both $f(x)$ and $g(x)$ on $D \setminus [N_1 \cup N_2]$. But $N_1 \cup N_2$ has measure 0, and the limit of a sequence of complex numbers is unique, so $f(x) = g(x)$ at all points of D except those in the set $N_1 \cup N_2$ of measure 0.

Suppose next that p is 1 or 2, and that $f, f_1, f_2, f_3, \ldots$ are all in $\mathscr{L}_p$, and that f_n converges to f in L_p-norm. Then the sequence satisfies the Cauchy condition, so by Theorem 4-3 there exists a member g of $\mathscr{L}_p$ and a subsequence $f_{n(1)}, f_{n(2)}, f_{n(3)}, \ldots$ that converges to g in L_p-norm, and moreover $f_{n(i)}(x)$ converges to $g(x)$ almost everywhere in D, say at all points x of D except those in a set N_1 of measure 0. But by the preceding paragraph, since f_n tends both to f and to g in L_p-norm, $f(x) = g(x)$ for all x in D except those in a subset N_2 of measure 0. But then the subsequence $f_{n(i)}$ $(i = 1, 2, 3, \ldots)$ converges to $f(x)$ at all points x of D except those in the set $N_1 \cup N_2$ of measure 0. This completes the proof of (i).

Returning to (ii), let f_n converge to f in any one of the three modes. Then there exists a subsequence $f_{n(i)}$ $(i = 1, 2, 3, \ldots)$ that converges to f almost everywhere in D. The sequence $f_1, f_2, f_3, \ldots$ converges to g in one of the three modes, so the subsequence $f_{n(i)}$ $(i = 1, 2, 3, \ldots)$ converges to g in the same mode. By part (i) of this theorem, there is a subsequence of $f_{n(i)}$ $(i = 1, 2, 3, \ldots)$ that converges to $g(x)$ for almost all x in D. But this last subsequence, being a subsequence of $f_{n(i)}$ $(i = 1, 2, 3, \ldots)$, still converges to $f(x)$ for almost all x in D. The proof is complete.

Suppose that, with D and m as before, we encounter a complex-valued function

$$f: (x, t) \mapsto f(x, t) \qquad (x \text{ in } D;\ a < t < b).$$

If t_0 and $t_0 + h$ are two points of (a, b), we can form the difference quotient

$$\text{(L)} \qquad [f(x, t_0 + h) - f(x, t_0)]/h \qquad (x \text{ in } D).$$

If at a point x of D this tends to a limit as h tends to 0, that limit is the ordinary partial derivative of f with respect to t at (x, t_0) and is denoted by any of several symbols, such as $D_{r+1}f(x, t_0)$ or $\partial f(x, t)/\partial t$. But it can happen that for each t in (a, b) the function $x \mapsto f(x, t)$ belongs to $\mathscr{L}_p$, and its purpose is to represent a point of L_p. This is the case, for example, in quantum mechanics, where the function f is at each time t a member of $\mathscr{L}_2$ that is the state-function of some system, such as an atom or molecule. In this case the appropriate idea of derivative is the idea of the derivative of a vector, which we have already met in Chapter V. In this sense, the difference quotient (L) will have a limit g in $\mathscr{L}_p$ if, when we denote the function $x \mapsto f(x, t)$ by $f(\,\cdot\,, t)$,

$$\text{(M)} \qquad \lim_{h \to 0} \|[f(\cdot, t_0 + h) - f(\cdot, t_0)]/h - g\|_p = 0.$$

Then the limit g is called the "derivative in $\mathscr{L}_p$ sense," as we now state formally.

DEFINITION 4-6 *Let D be a subset of R^r and m a nonnegative additive regular function of left-open intervals in R^r. For each t in (a, b) let the function $f(\cdot, t)$, or*

$$x \mapsto f(x, t) \qquad (x \text{ in } D),$$

belong to $\mathscr{L}_p[m, D]$, and let t_0 be in (a, b). If the difference quotient (L) *converges in L_p-norm to a function g in $\mathscr{L}_p[m, D]$ as h tends to 0, f is said to have a* ***derivative***

***with respect to** t **at** t_0 **in the** L_p **sense**, and g is called a **version** of that derivative and is denoted by placing the symbol (L_p) before any of the customary symbols for the derivative with respect to t, such as $(L_p)\, D_{r+1}f(x, t_0)$ or $(L_p)\, \partial f(x, t)/\partial t$.*

As an important special case, suppose that D is all of R^r and that F belongs to $\mathscr{L}_p[m, D]$. If we define f on R^{r+1} by

$$(x, t) \mapsto f(x, t) = F(x^1, \ldots, x^{j-1}, x^j + t, x^{j+1}, \ldots, x^r),$$

the pointwise limit of the difference quotient (L) at $t_0 = 0$ is then the ordinary partial derivative of F with respect to x^j. If the difference quotient (L) tends in L_2-norm to a limit function g in $\mathscr{L}_2[m, R^r]$, this g is the L_2 partial derivative of F, and we write

$$\text{(N)} \qquad (L_2)\, D_j F(x) = \frac{(L_2)\, \partial F(x)}{\partial x^j} = g.$$

Although the ordinary, or pointwise, partial derivative and the derivative in L_p sense are conceptually different, fortunately they are closely related, so that with well-behaved functions we can rely on the familiar partial derivative to let us compute the derivative in L_p sense. This is shown in the next theorem.

THEOREM 4-7 *Let D be a subset of R^r, m a nonnegative additive regular function of left-open intervals in R^r, (a, b) an open interval in R, and t_0 a point of (a, b). For each t in (a, b), let*

$$x \mapsto f(x, t) \qquad (x \text{ in } D)$$

be a function in $\mathscr{L}_p[m, D]$, where $p = 1$ or 2. Assume that for each x in $D \setminus N$, where $mN = 0$, the partial derivative $D_{r+1}f(x, t_0)$ of f with respect to t at t_0 exists. Then

(i) if there exists a positive number δ and a function M integrable with respect to m over D such that whenever $0 < |h| < \delta$, then $t_0 + h$ is in (a, b) and

$$|f(x, t_0 + h) - f(x, t_0)|^p/|h|^p \leqq M(x),$$

the derivative of f with respect to t at t_0 exists in the L_p sense;

(ii) if the derivative of f with respect to t at t_0 exists in the L_p sense, and g is any version of that derivative,

$$g(x) = D_{r+1}f(x, t_0)$$

for almost all x in $D \setminus N$.

Let ϕ denote the function on D that is equal to the partial derivative $D_{r+1}f(x, t_0)$ for x in $D \setminus N$ and is 0 on N. Let $h(1), h(2), h(3), \ldots$ be any sequence of nonzero numbers tending to 0 such that for each n, $t_0 + h(n)$ is in (a, b). For each positive integer n the difference quotient

$$\text{(O)} \qquad Q_n(x) = [f(x, t_0 + h(n)) - f(x, t_0)]/h(n)$$

is a measurable function on D, and it tends almost everywhere on D to $\phi(x)$, so ϕ is measurable on D. Also, by the hypothesis in (A)

$$\text{(P)} \qquad |Q_n(x)|^p \leqq M(x) \qquad (x \text{ in } D),$$

and the left member tends almost everywhere to $|\phi(x)|^p$, so by the dominated convergence theorem, $|\phi|^p$ is integrable, and ϕ belongs to $\mathscr{L}_p$, and

$$\text{(Q)} \qquad |\phi(x)|^p \leqq M(x) \qquad (x \text{ in } D).$$

By (P) and (Q),

$$\begin{aligned}|Q_n(x) - \phi(x)|^p &\leqq [|Q_n(x)| + |\phi(x)|]^p \\ &\leqq [M(x)^{1/p} + M(x)^{1/p}]^p \\ &= 2^p M(x),\end{aligned}$$

which is integrable. The left member tends to 0 almost everywhere in D, so by the dominated convergence theorem,

$$\lim_{n\to\infty} \int_D |Q_n(x) - \phi(x)|^p \, m(dx) = 0.$$

That is, $\|Q_n - \phi\|_p$ tends to 0; and since this is true for every sequence $h(n)$ of nonzero numbers tending to 0, it is true that

$$\lim_{h\to 0} \|[f(x, t_0 + h) - f(x, t_0)]/h - \phi(x)\|_p = 0.$$

So ϕ is a version of the derivative of f at t_0 in the L_p sense, and the proof of conclusion (i) is complete.

For (ii), let $h(1), h(2), h(3), \ldots$ be as above. Since g is a version of the derivative in the L_p sense, the ratio $Q_n(x)$ defined in (O) tends in L_p-norm to $g(x)$. By hypothesis, it tends to $D_{r+1}f(x, t_0)$ on $D \setminus N$. By Theorem 4-5, $g(x)$ and $D_{r+1}f(x, t_0)$ are equal almost everywhere in $D \setminus N$, and (ii) is established.

It is never commendable to harbor mental confusion. But Theorem 4-7 shows that if we fail to distinguish between the L_2 derivative and the ordinary partial derivative, and also have the good fortune to meet only functions that satisfy the rather mild condition in (i) of Theorem 4-7, the mental confusion will not produce any incorrect results.

EXERCISE 4-1 Show that the function f defined by

$$f(x) = x^{-1} \sin x^2 \qquad (x \neq 0), \qquad f(0) = 0$$

is in $L_2[m_L, (-\infty, \infty)]$, and Df is continuous on $(-\infty, \infty)$, but $(L_2)\,Df$ does not exist.

EXERCISE 4-2 Prove that if f is the indicator of the rational numbers. Df does not exist, but $(L_2)Df = 0$.

EXERCISE 4-3 Let $Z = \{1, 2, 3, \ldots\}$, and for each j in Z let $e_j = (0, 0, \ldots, 0, 1, 0, \ldots)$ with 1 in the jth place. Prove that for each v in $l_2(Z)$, $\langle v, e_j \rangle$ tends to 0 as j increases but e_j has no limit in $l_2(Z)$.

5. Hilbert Spaces and Their Geometry

Whenever a linear space has an inner product or a pseudo-inner-product defined on it, we shall assume without further mention that the norm and the distance in the space are those corresponding to that inner product or pseudo-inner-product so that

$$\|v\| = \langle v, v \rangle^{1/2}, \qquad \operatorname{dist}(v, w) = \|v - w\|.$$

By a **complete pseudo-inner-product space** we shall, of course, mean a vector space in which a pseudo-norm is defined and which is complete in the pseudo-metric corresponding to that pseudo-inner-product. The definition of a **complete inner-product space** is self-suggesting; we merely omit the "pseudo." But we do not use the expression "complete inner-product space" because such spaces are always referred to as *Hilbert spaces*. We are going to prove a number of theorems about Hilbert spaces. But all of the proofs apply without change to complete pseudo-inner-product spaces, and these are not negligible objects, because they include the spaces $\mathscr{L}_2[m, D]$. So the theorems in this section will all be stated and proved for complete pseudo-inner-product spaces, which will make them applicable to Hilbert spaces too. Hilbert spaces and complete pseudo-inner-product spaces retain many of the properties of finite-dimensional spaces, and we shall now investigate some of them.

In any pseudo-normed linear space, the neighborhoods of a point v_0 of the space are defined to be the (open) balls with center v_0 and positive radius; that is, for each $r > 0$ the set $B(v_0, r) = \{v \text{ in } V : \|v - v_0\| < r\}$ is a neighborhood of v_0. As before, a set G in V is **open** if for each v_0 in G there is a neighborhood of v_0 that is contained in G; and a set F in V is **closed** if its complement $V \setminus F$ is open, which is the case if and only if for every point v_0 of V such that every neighborhood of v_0 contains at least one point of F, v_0 is itself a point of F.

In V^2 and V^3 it is a familiar fact that two vectors v_1, v_2 are orthogonal (perpendicular) if and only if their inner product is 0. We adopt this as the definition of orthogonality in all spaces with pseudo-inner-products.

DEFINITION 5-1 *Let W be a space in which a pseudo-inner-product is defined. If v_1 and v_2 are in W, they are **orthogonal** if $\langle v_1, v_2 \rangle = 0$.*

If v and w are two orthogonal vectors in a pseudo-inner-product space W, they are the legs of a right triangle whose hypotenuse has length $\|v - w\|$. So the next statement is a generalization of the theorem of Pythagoras.

THEOREM 5-2 *If v and w are orthogonal vectors in a space W with a pseudo-inner-product,*

$$\|v - w\|^2 = \|v\|^2 + \|w\|^2$$

and

$$\|v + w\|^2 = \|v\|^2 + \|w\|^2.$$

For then

$$\begin{aligned}\|v \pm w\|^2 &= \langle v \pm w, v \pm w\rangle \\ &= \langle v, v\rangle \pm \langle w, v\rangle \pm \langle v, w\rangle + \langle w, w\rangle,\end{aligned}$$

and the second and third terms in the right member are 0.

In elementary geometry there is a theorem that states that the sum of the squares of the lengths of the diagonals of a parallelogram is equal to the sum of the squares of the lengths of the four sides. The next theorem extends this to Hilbert spaces.

THEOREM 5-3 *For any two members v and w of a space with a pseudo-inner-product,*

$$\|v + w\|^2 + \|v - w\|^2 = 2\|v\|^2 + 2\|w\|^2.$$

For

$$\begin{aligned}\|v + w\|^2 + \|v - w\|^2 &= \langle v + w, v + w\rangle + \langle v - w, v - w\rangle \\ &= 2\langle v, v\rangle + 2\langle w, w\rangle.\end{aligned}$$

A subset W_0 of a linear space W is a **subspace** (or **linear subspace**) of W if for every two members w_1, w_2 of W_0 and every scalar c, $w_1 + w_2$ and cw_1 also belong to W_0. If W is a normed linear space, a **closed subspace** of W is a (linear) subspace of W that is a closed set. In elementary geometry a frequently used construction was to find the foot of a perpendicular from a point v_0 to a subspace (a line, or in R^3 a plane). The foot of the perpendicular was also the point of the subspace nearest to v_0. We can prove an analog for Hilbert spaces; to each v_0 there corresponds a nearest point in a closed subspace, and that point is the foot of the perpendicular from v_0 to the subspace. But with no more effort, we can prove a stronger result. A **convex** set K in a linear space W is a set such that whenever v and w are in K, so is every point of the line-segment, consisting of all points $tv + (1 - t)w$ $(0 \leqq t \leqq 1)$, that joins v to w.

THEOREM 5-4 *Let v be a point of a complete pseudo-inner-product space W and let K be a closed convex set in W. Then there is a point w in K nearest to v; that is, $\|w - v\| = d$, where d is the greatest lower bound of distances $\|u - v\|$ for all u in K.*

For each positive integer n there is a point u_n of K such that $\|u_n - v\| < d + 1/n$. Let ε be positive, and fix a positive number N so large that

$$d/N + 1/2N^2 < \varepsilon^2/8. \tag{A}$$

Let m and n be integers greater than N. Since both u_n and u_m are in the convex set K, so is $u_n/2 + u_m/2$, and therefore

$$\|u_n/2 + u_m/2 - v\| \geqq d. \tag{B}$$

We apply Theorem 5-3 to $(u_n - v)/2$ and $(u_m - v)/2$ and obtain, using (B),

$$\left\|\frac{u_n}{2} - \frac{u_m}{2}\right\|^2 = 2\left\|\frac{u_n - v}{2}\right\|^2 + 2\left\|\frac{u_m - v}{2}\right\|^2 - \left\|\frac{u_n - v}{2} + \frac{u_m - v}{2}\right\|^2$$

$$\leqq \frac{[d + 1/n]^2}{2} + \frac{[d + 1/m]^2}{2} - d^2 = \frac{d}{n} + \frac{1}{2n^2} + \frac{d}{m} + \frac{1}{2m^2} < \frac{\varepsilon^2}{4}.$$

This implies that

$$\|u_n - u_m\| < \varepsilon.$$

So the sequence $u_1, u_2, u_3, \ldots$ satisfies the Cauchy condition and must converge to a point w of W. Since K is closed, w is in K, and by Theorem 3-5,

$$\|v - w\| = \lim_{n \to \infty} \|v - u_n\| = d.$$

So w is the point sought.

Subspaces are evidently convex, so the conclusion holds for all closed subspaces. The next theorem relates the minimum-distance property to perpendicularity.

THEOREM 5-5 *Let W_0 be a subspace (not necessarily closed) of a space W with a pseudo-inner-product, and let v be in W. A point w of W_0 is the point of W_0 nearest to v if and only if $v - w$ is orthogonal to every w' in W_0.*

Let $v - w$ be orthogonal to every w' in W_0. For every u in W_0, $u - w$ is in W_0 and so is orthogonal to $v - w$. By Theorem 5-2,

$$\|v - u\|^2 = \|v - w\|^2 + \|w - u\|^2 \geqq \|v - w\|^2,$$

so w is the point of W_0 at least distance from v.

Conversely, let w be the point of W_0 nearest to V, and let w' be any point of W_0. Then for some real number θ we have

$$\langle v - w, w' \rangle = |\langle v - w, w' \rangle| e^{i\theta}.$$

This implies

$$|\langle v - w, w' \rangle| = \langle v - w, w' e^{i\theta} \rangle = \langle w' e^{i\theta}, v - w \rangle. \tag{C}$$

For every real number t, $w + te^{i\theta}w'$ is in W_0, so its distance from v is at least as great as the distance of w from v. Therefore the square of the distance, which is

$$\|v - w - te^{i\theta}w'\|^2 = \langle v - w, v - w\rangle - t\langle v - w, e^{i\theta}w'\rangle$$
$$- t\langle e^{i\theta}w', v - w\rangle + t^2\langle w', w'\rangle,$$

has its least value when $t = 0$. Its derivative therefore vanishes at $t = 0$. By (C), this yields $|\langle v - w, w'\rangle| = 0$, as was to be proved.

A set of nonzero vectors in a linear space with a pseudo-inner-product is called an **orthogonal set** if any two different elements of the set are perpendicular. If, in addition, each of the vectors in the set is a unit vector, the set is called **orthonormal.** In this chapter we shall discuss orthogonal sets with finitely many members or with countably infinitely many members. The discussions could be extended without serious difficulty to the case of spaces (and there are such spaces) in which there are uncountable orthogonal sets.

In Section V-3 we saw that sets of r mutually perpendicular vectors in V^r were convenient objects to work with. Given an origin, each such set corresponded to a choice of a rectangular coordinate system; if $\mathbf{e}_1, \ldots, \mathbf{e}_r$ are mutually perpendicular unit vectors, for each j the end-points of the set of multiples $t\mathbf{e}_j$ (t in R) are the points of the jth coordinate axis. By Lemma V-3-4, given such a set, every vector $\mathbf{u}$ in V^r is the sum of vectors of the form $\langle \mathbf{u}, \mathbf{e}_j\rangle\mathbf{e}_j$, which are the components of $\mathbf{u}$ along the several axes.

Hilbert spaces and, more generally, complete pseudo-inner-product spaces are so closely related to finite-dimensional spaces that very similar results can be obtained. If $e_1, e_2, \ldots$ are mutually perpendicular unit vectors, we can define the component of a vector v along e_j to be $\langle v, e_j\rangle e_j$, just as in V^r. But even in V^r the sum of these components is not v unless there are enough of the e_j; if $\mathbf{e}_1$ and $\mathbf{e}_2$ are perpendicular vectors in V^3, not every vector $\mathbf{v}$ in V^3 can be written as $\langle \mathbf{v}, \mathbf{e}_1\rangle\mathbf{e}_1 + \langle \mathbf{v}, \mathbf{e}_2\rangle\mathbf{e}_2$. In V^r it is easy to settle whether we have a complete set of $\mathbf{e}_j$; if we have r of them in V^r, that is enough. In infinite-dimensional spaces it is not quite so simple; if we have a set with enough members and leave one out, we are left with infinitely many, but not with a large enough set. We shall now study such representations of vectors as sums of components. However, one small generalization is useful: instead of orthonormal sets, we shall work with orthogonal sets $v_1, v_2, v_3, \ldots$. The v_j may not be unit vectors, but from them we easily obtain a set of orthogonal unit vectors

$$e_j = v_j/\|v_j\| \qquad (j = 1, 2, 3, \ldots).$$

The components of a vector w along the jth axis was previously given by the expression $\langle w, e_j\rangle e_j$; in view of the definition of e_j just given, this can be written as

(D) $$[\langle w, v_j\rangle/\|v_j\|^2]v_j.$$

The expression $\langle w, v_j\rangle/\|v_j\|^2$ is called the **expansion coefficient** of w along v_j. We shall now see what we can say about the relationship between a vector in a pseudo-inner-product space and its components along the various axes.

THEOREM 5-6 *Let V be a space with an inner product, and let $v_1, v_2, v_3, \ldots$ be a finite set or infinite sequence of nonzero members of V that form an orthogonal set. For each w in V and each positive integer, define $\hat{w}(j)$ to be the expansion coefficient $\langle w, v_j\rangle/\|v_j\|^2$. Then*

$$\sum_j |\hat{w}(j)|^2\|v_j\|^2 \leqq \|w\|^2.$$

This inequality is known as **Bessel's inequality**.

Let $v_1, \ldots, v_n$ be a finite subset of the orthogonal set. Then

$$\text{(E)} \qquad 0 \leqq \langle w - \hat{w}(1)v_1 - \cdots - \hat{w}(n)v_n, w - \hat{w}(1)v_1 - \cdots - \hat{w}(n)v_n\rangle$$

$$= \langle w, w\rangle - \sum_{i=1}^{n} \hat{w}(i)\langle v_i, w\rangle - \sum_{j=1}^{n} \hat{w}(j)^-\langle w, v_j\rangle$$

$$+ \sum_{i=1}^{n}\sum_{j=1}^{n} \hat{w}(i)\hat{w}(j)^-\langle v_i, v_j\rangle.$$

In the first and second sums we have $\langle v_i, w\rangle = \langle w, v_i\rangle^- = \hat{w}(i)^-\|v_i\|^2$. In the last sum, for each i the factor $\langle v_i, v_j\rangle$ is 0 unless $j = i$, in which case it is $\|v_i\|^2$. So (E) reduces to

$$\text{(F)} \quad 0 \leqq \|w\|^2 - \sum_{i=1}^{n} \hat{w}(i)\hat{w}(i)^-\|v_i\|^2 - \sum_{j=1}^{n} \hat{w}(j)^-\hat{w}(j)\|v_j\|^2 + \sum_{i=1}^{n} \hat{w}(i)\hat{w}(i)^-\|v_i\|^2$$

$$= \|w\|^2 - \sum_{i=1}^{n} |\hat{w}(i)|^2\|v_i\|^2.$$

If the orthogonal set is finite, this completes the proof. If it is an infinite sequence, by (F) the sum of the first n terms of the infinite series

$$\sum_{i=1}^{\infty} |\hat{w}(i)|^2\|v\|^2$$

is at most $\|w\|^2$, so the series converges and its sum is at most $\|w\|^2$.

THEOREM 5-7 *Let $v_1, v_2, v_3, \ldots$ be an orthogonal set in a complete pseudo-inner-product space W. Let c_1, c_2, c_3 be a sequence of scalars such that the series*

$$|c_1|^2\|v_1\|^2 + |c_2|^2\|v_2\|^2 + |c_3|^2\|v_3\|^2 + \cdots$$

is convergent. Then the series

$$\sum_{j=1}^{\infty} c_j v_j$$

converges to a vector in W.

For each positive integer n define

$$s_n = c_1 v_1 + \cdots + c_n v_n.$$

The vectors $c_i v_i$ are orthogonal to each other, since the v_i are, and the length of $c_i v_i$ is $|c_i|\,\|v_i\|$.

Let ε be positive. Since $\Sigma |c_i|^2 \|v_i\|^2$ converges, there is an N such that if $m > n > N$,

$$\sum_{j=n}^{m} |c_j|^2 \|v_j\|^2 < \varepsilon^2.$$

The Pythagorean theorem 5-2 extends to finite sums of orthogonal vectors, so

$$\begin{aligned} \|s_m - s_n\|^2 &= \|c_{n+1} v_{n+1} + \cdots + c_m v_m\|^2 \\ &= |c_{n+1}|^2 \|v_{n+1}\|^2 + \cdots + |c_m|^2 \|v_m\|^2 < \varepsilon^2. \end{aligned}$$

The partial sums s_n satisfy the Cauchy condition, and since W is complete, they have a limit in W.

Definition 5-8 *Let V be a normed linear space and E a set contained in V. The **linear span of** E is the set of all linear combinations $c_1 u_1 + \cdots + c_k u_k$ in which the u_i are vectors belonging to E and the c_i are scalars. The **closed linear span of** E is the set of all vectors v in V such that every neighborhood of v contains at least one point of the linear span of E.*

We shall omit the easy proofs that the linear span of E is a linear subspace of V and that the closed linear span of E is a closed linear subspace of V.

If Π is a plane in R^3 that passes through the origin O, the position vectors $v = \overrightarrow{OP}$ of points P in Π form a linear subspace V of V^3. We can choose two mutually perpendicular unit vectors u_1, u_2 in V. Then V is the linear span of the set $E = \{u_1, u_2\}$. Since V is closed, it is also the closed linear span of E. If $\overrightarrow{OA}$ is the representation of any vector x in V^3, the problems of (i) finding the point of the plane P nearest to A, (ii) dropping a perpendicular from A to P, and (iii) resolving x into a component in V and a component perpendicular to V are three variant forms of the same problem. It is clearly useful to have a formula for the solution of this problem in V^3. In the next theorem we show that not only in V^3 but even in all complete pseudo-inner-product spaces there is a simple formula for the solution.

Theorem 5-9 *Let W be a complete pseudo-inner-product space, let $V = \{v_1, v_2, \ldots\}$ be a finite or countably infinite orthogonal set in W, and let V^{CL} be the closed linear span of V. Let w be any point of W, and for each v_j in V define*

$$\hat{w}(j) = \langle w, v_j \rangle / \|v_j\|^2. \qquad \text{(G)}$$

Then

(i) *the series*

$$\sum_{v_j \text{ in } V} \hat{w}(j) v_j$$

converges;

(ii) *the sum p of the series in* (i) *is a point of* V^{CL} *and is the point of* V^{CL} *closest to* w;

(iii) $w - p$ *is orthogonal to* V^{CL}.

By Bessel's inequality, Theorem 5-6, the series $\Sigma\, |\hat{w}(j)|^2 \|v_j\|^2$ converges. By Theorem 5-7, the series $\Sigma\, \hat{w}(j) v_j$ converges to a point of W. (If v is a finite set, this is trivial.) The partial sums

$$s_n = \sum_{j=1}^{n} \hat{w}(j) v_j$$

are in the linear span of V. If V is finite, this completes the proof that the sum p is in V^{CL}. If V is an infinite sequence, the sum p of the series $\Sigma\, w(j) v_j$ is the limit of the points s_n of the linear span of V, so p is in V^{CL}.

By the continuity of the inner product (Theorem 3-5), for each v_k in V we have

$$\begin{aligned} \langle w - p, v_k \rangle &= \lim_{n \to \infty} \langle w - s_n, v_k \rangle \\ &= \lim_{n \to \infty} \left[\langle w, v_k \rangle - \sum_{j=1}^{n} \hat{w}(j) \langle v_j, v_k \rangle \right]. \end{aligned}$$

The first term in the right member is $\hat{w}(k)\|v_k\|^2$. If $n > k$, all the factors $\langle v_j, v_k \rangle$ are 0 except the one with $j = k$, which is $\|v_k\|^2$. So the right member of the equation is 0, and $\langle w - p, v_k \rangle = 0$ for each v_k in V. If v' is any point of the linear span of V, v' has the form $c_1 v_1 + \cdots + c_h v_h$, so

$$\langle w - p, v' \rangle = \sum_{j=1}^{h} c_j^{-} \langle w - p, v_j \rangle = 0.$$

If v'' is in V^{CL}, it is the limit of a sequence of points $v'_1, v'_2, \ldots$ of the linear span of V, and each inner product $\langle w - p, v'_j \rangle$ is 0, so by the continuity of the inner product, $\langle w - p, v'' \rangle = 0$. Conclusion (iii) is established. By Theorem 5-5, conclusion (ii) follows from this.

By far the most interesting case is that in which the sum $\Sigma\, \hat{w}(j) v_j$ is equal to w. This happens when the orthonormal set V has a property called **completeness**, which can be expressed in several equivalent ways (as we shall see in Theorem 5-14).

DEFINITION 5-10 *Let W be a space in which a pseudo-inner-product is defined. A set* V_0 *of mutually orthogonal vectors in W is* **complete** *if the closed linear span of*

V_0 is W; that is, if for every w in W and every positive ε there exists a finite linear combination

$$v = c_1 v_1 + \cdots + c_k v_k$$

of members of V_0 such that

$$\langle w - v, w - v \rangle < \varepsilon^2.$$

This kind of completeness has nothing to do with the completeness of the vector space itself, as defined in Definition 4-1.

In the discussion following Theorem 5-5, we brought out that a main use of orthogonal sets of vectors was to express an arbitrary vector as the sum of its components along the axes, and that to do this we needed to have enough axes; that is, enough orthogonal vectors. We now show that the kind of completeness we defined in Definition 5-10 is the right one for this purpose by proving two especially important theorems concerning vectors and their expansions.

THEOREM 5-11 *Let W be a complete inner-product-space, and let $v_1, v_2, v_3, \ldots$ be a countably infinite orthogonal set that is complete in W. Then for every w in W*

$$w = \sum_{j=1}^{\infty} \hat{w}(j) v_j,$$

where, as before,

$$\hat{w}(j) = \langle w, v_j \rangle / \|v_j\|^2.$$

For by Theorem 5-9, the infinite series is convergent and its sum p is the point of V^{CL} nearest to w. But $V^{\mathrm{CL}} = W$, and the point of W nearest to w is w.

THEOREM 5-12 *Let W be a complete pseudo-inner-product space; let $v_1, v_2, v_3, \ldots$ be a complete orthogonal set in W, and let $c_1, c_2, c_3, \ldots$ be a sequence of complex numbers such that the series*

$$\sum_{j=1}^{\infty} |c_j|^2 \|v_j\|^2$$

converges. Then the series of vectors

$$\sum_{j=1}^{\infty} c_j v_j$$

converges to a vector w in W, and the expansion coefficients of this function w are

$$\hat{w}(j) = c_j \qquad (j = 1, 2, 3, \ldots).$$

By Theorem 5-7, the finite sums

$$s_n = c_1 v_1 + \cdots + c_n v_n$$

converge to a vector w in W. By the continuity of the inner product (Theorem 3-5), for each positive integer j

$$\begin{aligned}\hat{w}(j) &= \langle w, v_j\rangle/\|v_j\|^2 \\ &= \lim_{n\to\infty} \langle c_1 v_1 + \cdots + c_n v_n, v_j\rangle/\|v_j\|^2 \\ &= \lim_{n\to\infty} \sum_{i=1}^{n} c_i\langle v_i, v_j\rangle/\|v_j\|^2.\end{aligned}$$

For all n greater than j, all inner products $\langle v_i, v_j\rangle$ are 0 except the one with $i = j$. So the right member is c_j for all $n > j$, and the proof is complete.

In V^r, a vector $\mathbf{v}$ has representation $(v^1, \ldots, v^r)^{\mathrm{T}}$ if and only if (in the language of this chapter) it has expansion coefficients $v^1, \ldots, v^r$ in terms of the coordinate vectors $\mathbf{e}_1, \ldots, \mathbf{e}_r$. In Lemma V-3-4 we found that when vectors $\mathbf{u}$ and $\mathbf{v}$ have the respective expansion coefficients $u^1, \ldots, u^r$ and $v^1, \ldots, v^r$, their inner product is

$$\langle \mathbf{u}, \mathbf{v}\rangle = u^1 v^1 + \cdots + u^r v^r.$$

This useful formula generalizes to orthogonal sets in complete pseudo-inner-product spaces, as is shown in the following theorem, known as **Parseval's theorem**.

THEOREM 5-13 *Let $v_1, v_2, v_3, \ldots$ be a complete orthogonal set in a complete pseudo-inner-product space W. Let v and w belong to W, and for each positive integer j define*

$$\hat{v}(j) = \langle v, v_j\rangle/\|v_j\|^2, \qquad \hat{w}(j) = \langle w, v_j\rangle/\|v_j\|^2.$$

Then

$$\text{(H)} \qquad \langle v, w\rangle = \sum_{j=1}^{\infty} [\hat{v}(j)\|v_j\|][\hat{w}(j)\|v_j\|]^-.$$

In particular,

$$\text{(I)} \qquad \|w\|^2 = \sum_{j=1}^{\infty} |\hat{w}(j)|^2\|v_j\|^2.$$

Because the v_j form an orthogonal set, for every positive integer n we have

$$\text{(J)} \qquad \begin{aligned}\left\langle \sum_{i=1}^{n} \hat{v}(i)v_i, \sum_{j=1}^{n} \hat{w}(j)v_j \right\rangle &= \sum_{i=1}^{n}\sum_{j=1}^{n} \hat{v}(i)\hat{w}(j)^-\langle v_i, v_j\rangle \\ &= \sum_{i=1}^{n} \hat{v}(i)\hat{w}(i)^-\|v_i\|^2.\end{aligned}$$

As n increases, the two sums in the left member of (J) converge to v and w, respectively, by Theorem 5-11. By the continuity of the inner product, the left member of (J) then converges to $\langle v, w\rangle$. Therefore the right member of (J) also converges to $\langle v, w\rangle$ as n increases. By definition, the infinite series $\Sigma\, \hat{v}(i)\hat{w}(i)\|v_i\|^2$ converges, and its sum is also $\langle v, w\rangle$. This is conclusion (H). If we set $v = w$, we obtain conclusion (I).

The earlier theorems in this chapter show that Hilbert spaces have many of the properties familiar to us in Euclidean solid geometry. The last three show that some of the basic ideas of coordinate geometry also carry over with small change. Given a complete orthogonal set of vectors in the Hilbert space, every vector in the space can be represented as the sum of its components along the axes (Theorem 5-11), and every sequence of numbers $c_1, c_2, c_3, \ldots$ for which $\Sigma\, |c_i|^2\|v_i\|^2$ converges is the set of expansion coefficients of some vector in the space (Theorem 5-12), and these expansion coefficients can be used to compute inner products by a formula like that in the finite-dimensional case. This not only shows that Hilbert spaces are very direct generalizations of finite-dimensional spaces; it also shows that the property of completeness of orthogonal sets is an important one. It is therefore useful to know several different ways of expressing it.

If A is a subset of a normed linear space and E is a subset of A, the statement that E is **dense** in A means that for every vector a in A and every positive ε, the ball $B(a, \varepsilon)$ contains at least one point of E.

THEOREM 5-14 *Let V be an orthogonal set in a complete pseudo-inner-product space W. The following three statements are equivalent*:

(i) *V is complete*;
(ii) *the linear span of V is dense in W*;
(iii) *there is no nonzero vector in W that is orthogonal to all the vectors v in V.*

The equivalence of (i) and (ii) is trivial. The set V is complete if and only if its closed linear span is W, which is to say that for every w in W each ball centered at w contains a point of the linear span of V; and this is the meaning of (ii).

Let V^{CL} be the closed linear span of V. Suppose that there is a nonzero vector w in W orthogonal to all v in V. Every vector v' in the linear span of V can be written in the form $v' = c_1 v_1 + \cdots + c_k v_k$, where the v_j are in V and the c_j are scalars. Then

$$\langle w, v'\rangle = c_1^{-}\langle w, v_1\rangle + \cdots + c_k^{-}\langle w, v_k\rangle = 0.$$

Every v'' in V^{CL} is the limit of a sequence of vectors v_j' in the linear span of V, so by the continuity of the inner product, w is orthogonal to v''. We apply Theorem 5-5 with $W_0 = V^{\mathrm{CL}}$ and find that 0 is the point of V^{CL} nearest to w. That is, for all v in V^{CL} we have

$$\|w - v\| \geqq \|w - 0\| > 0,$$

so no point of V^{CL} is in the ball $B(w, \|w\|)$. Therefore V is not complete, and so (i) implies (iii).

Conversely, suppose that V is not complete. Then V^{CL} is not all of W, and there is a vector h in $W \setminus V^{CL}$. Let d be the greatest lower bound of the norms $\|h - v\|$ for all v in V^{CL}. By Theorem 5-4, there is a w in V^{CL} such that $\|h - w\| = d$. Since h is not in V^{CL} and w is, $\|h - w\|$ cannot be 0. By Theorem 5-5, $h - w$ is orthogonal to every vector v in V^{CL}; in particular, it is orthogonal to every vector v_i in V. So (iii) implies (i), and the proof is complete.

EXERCISE 5-1 Let K be a closed set in a complete normed vector space. Show that if K has any one of the following three properties, it has all three of them:

(i) K is convex;
(ii) if A and B are points of K, the midpoint of the segment AB is in K;
(iii) if A and B are distinct points of K, some point of the segment AB different from A and B belongs to K.

Suggestion: If K is not convex, there are distinct points A, B of K and there is a point C of the segment AB that is not in K. Let D be the point of $AC \cap K$ nearest to C, and E the point of $CB \cap K$ nearest C. Consider CD.

EXERCISE 5-2 If K is a closed convex set in a complete real pseudo-inner-product space W, and w_0 is not in K, show that there is a continuous linear function L on W and a real number c such that $L(v) \leqq c$ for all v in K and $L(w_0) > c$. *Suggestion*: Let y_0 be the point of K nearest to w_0. Define, for each w in W,

$$L(w) = \langle w, w_0 - y_0 \rangle, \qquad c = \langle y_0, w_0 - y_0 \rangle.$$

If w is in K, so is $y_0 + t(w - y_0)$ for $0 \leqq t \leqq 1$, and $\|y_0 + t(w - y_0) - w_0\|^2$ has its least value at $t = 0$.

EXERCISE 5-3 Let m be a regular additive nonnegative function of left-open intervals in R^r, let D be a set in R^r with $0 < mD < \infty$, and let f be a function on D with values in a complete pseudo-inner-product space W such that f is integrable with respect to m over D. Define the *integral mean* of f to be

$$Mf = [mD]^{-1} \int_D f(t)\, m(dt).$$

Show that if K is a closed convex set in W that contains $f(t)$ for all t in D, it also contains Mf. *Suggestion*: If Mf is not in K, let L be a linear function such that $L(v) \leqq c$ for v in K and $L(Mf) > c$. Since $Lf(t) \leqq c$ for all t in D, its integral over D is, at most, cmD.

EXERCISE 5-4 Let K be a closed convex set in R^s. A real-valued function ϕ on K is **convex** if whenever x_1 and x_2 are in K and $0 \leqq t \leqq 1$,

$$\phi((1 - t)x_1 + tx_2) \leqq (1 - t)\phi(x_1) + t\phi(x_2).$$

Show that this is equivalent to the condition that the set K^+ of points (x, z) in R^{s+1} such that x is in K and $z \geqq \phi(x)$ is a convex set. Show that if f is a function on a set D in R^r, and m is a regular additive nonnegative function of left-open intervals in R^r, and f takes values in K and is integrable over D, and ϕ is convex and $0 < mD < \infty$, then

$$[mD]^{-1} \int_D \phi(f(x))\, m(dx) \leqq \phi\left([mD]^{-1} \int_D f(x)\, m(dx) \right).$$

This is known as **Jensen's inequality**.

6. Approximation by Step-Functions and by Differentiable Functions

Often it is desirable to study a function in a class $\mathscr{L}_p$ by approximating it by functions with some other desirable properties. In this section we shall show that such approximation is possible by means of step-functions and by means of arbitrarily smooth functions. We first show that functions in $\mathscr{L}_p$ can be arbitrarily closely approximated, not merely by step-functions but by the special kind of step-functions that we call rational step-functions. A function s on R^r is a **rational step-function** if it can be represented as a sum

$$s = \sum_{j=1}^{k} c_j 1_{B(j)}$$

in which each c_j is a complex number with rational real and imaginary parts and each $B(j)$ is a rational interval; that is, each $B(j)$ is a bounded left-open interval and all the coordinates of each vertex of $B(j)$ are rational numbers.

THEOREM 6-1 *Let m be an additive nonnegative regular function of left-open intervals in R^r. If f is a function on R^r that belongs to $\mathscr{L}_{p(h)}$ for each $p(h)$ in a finite set $\{p(1), \ldots, p(h^*)\}$ of numbers each $\geqq 1$, then for each positive ε there exists a rational step-function s on R^r such that*

(A) $$\|f - s\|_{p(h)} < \varepsilon \qquad (h = 1, \ldots, h^*).$$

Define

(B) $$g_1 = (\operatorname{Re} f)^+, \qquad g_2 = (\operatorname{Re} f)^-, \qquad g_3 = (\operatorname{Im} f)^+, \qquad g_4 = (\operatorname{Im} f)^-.$$

By Corollary 2-4, these are measurable, and since

$$|g_\alpha(x)|^{p(h)} \leqq |f(x)|^{p(h)} \qquad (\alpha = 1, 2, 3, 4;\ h = 1, \ldots, h^*;\ x \text{ in } R^r)$$

they all belong to $\mathscr{L}_{p(h)}[m, R^r]$ for $h = 1, \ldots, h^*$. Since for each α in $\{1, 2, 3, 4\}$

and each h in $\{1, \ldots, h^*\}$ the function g_α is nonnegative and $[g_\alpha]^{p(h)}$ is integrable over R^r, to each positive ε there corresponds a gauge $\gamma(\alpha, h)$ on $\bar{R}^r$ such that $\gamma(\alpha, h)(x)$ is bounded whenever x is in R^r, and whenever

$$\text{(C)} \qquad \mathscr{P} = \{(x_1, A(1)), \ldots, (x_{k'}, A(k'))\}$$

is a $\gamma(\alpha, h)$-fine partition of R^r,

$$\text{(D)} \qquad \left| S(\mathscr{P}; [(g_\alpha)^{p(h)}]_{R^r}; m) - \int_{R^r} g_\alpha(x)^{p(h)}\, m(dx) \right| < \frac{\varepsilon^{p(h)}}{5 \cdot 8^{p(h)}}.$$

Define γ to be the infimum

$$\gamma = \gamma(1,1) \cap \cdots \cap \gamma(4, h^*)$$

of the $4h^*$ gauges $\gamma(\alpha, h)$. By Theorem IV-3-1, there exists a partition $\mathscr{P}$ that is γ-fine and is such that each interval of the partition that is bounded is a rational cube. We choose such a partition and denote it by the notation (C), choosing the numbering so that $A(1), \ldots, A(k)$ are bounded intervals and $A(k+1), \ldots, A(k')$ are unbounded. We define s'_α to be the function that on each interval $A(j)$ $(j = 1, \ldots, k')$ has the constant value $(g_\alpha)_{R^r}(x_j)$. If $A(j)$ is unbounded, this value is 0, so

$$s'_\alpha = \sum_{j=1}^{k} g_\alpha(x_j) 1_{A(j)}.$$

Therefore s'_α is a step-function constant on rational cubes, and by Theorem II-13-1,

$$\text{(E)} \quad \int_{R^r} |g_\alpha(x)^{p(h)} - s'_\alpha(x)^{p(h)}|\, m(dx) < \left(\frac{\varepsilon}{8}\right)^{p(h)} \qquad (\alpha = 1, 2, 3, 4;\ h = 1, \ldots, h^*).$$

We next show that if a and b are nonnegative numbers and $p \geqq 1$,

$$\text{(F)} \qquad |b - a|^p \leqq |b^p - a^p|.$$

There is no loss of generality in assuming $0 \leqq a \leqq b$; otherwise we interchange names of a and b. Then the function defined by

$$(t + b - a)^p - t^p \qquad (t \geqq 0)$$

has derivative

$$p(t + b - a)^{p-1} - pt^{p-1},$$

which is nonnegative on $[0, \infty)$. So its value at $t = 0$ is not greater than its value at $t = a$, which is to say

$$(0 + b - a)^p \leqq b^p - a^p,$$

establishing (F). We apply this to $g_\alpha(x)$ and $s'_\alpha(x)$ and obtain

$$|g_\alpha(x) - s'_\alpha(x)|^{p(h)} \leqq |g_\alpha(x)^{p(h)} - s'_\alpha(x)^{p(h)}|.$$

Integrating over R^r yields, with (E),

$$\int_{R^r} |g_\alpha(x) - s'_\alpha(x)|^{p(h)}\, m(dx) < \left(\frac{\varepsilon}{8}\right)^{p(h)},$$

or

$$\text{(G)} \qquad \|g_\alpha - s'_\alpha\|_{p(h)} < \varepsilon/8 \qquad (\alpha = 1, 2, 3, 4;\ h = 1, \ldots, h^*).$$

Without loss of generality we may suppose that $p(1)$ is the smallest of the $p(h)$. We denote the union $A(1) \cup \cdots \cup A(k)$ by U. We can and do choose rational numbers $c_{\alpha,1}, \ldots, c_{\alpha,k}$ such that

$$|c_{\alpha,j} - g_\alpha(x_j)| < \varepsilon/8(1 + mU)^{1/p(1)} \qquad (j = 1, \ldots, k;\ \alpha = 1, 2, 3, 4).$$

Then the functions

$$s_\alpha = \sum_{j=1}^{k} c_{\alpha,j} 1_{A(j)}$$

are rational step-functions, and for $h = 1, \ldots, h^*$

$$\begin{aligned}
\|s_\alpha - s'_\alpha\|_{p(h)} &= \left\{\int_{R^r} |s_\alpha(x) - s'_\alpha(x)|^{p(h)}\, m(dx)\right\}^{1/p(h)} \\
&\leqq \left\{\int_U \left[\frac{\varepsilon}{8(1 + mU)^{1/p(1)}}\right]^{p(h)} m(dx)\right\}^{1/p(h)} \\
&= \left[\frac{\varepsilon}{8(1 + mU)^{1/p(1)}}\right](mU)^{1/p(h)} < \frac{\varepsilon}{8}.
\end{aligned}$$

This and (G) imply

$$\|s_\alpha - g_\alpha\|_{p(h)} < \varepsilon/4 \qquad (\alpha = 1, 2, 3, 4;\ h = 1, \ldots, h^*).$$

If we now define

$$s = s_1 - s_2 + is_3 - is_4,$$

we compute that for $h = 1, \ldots, h^*$

$$\begin{aligned}
\|f - s\|_{p(h)} &= \|(g_1 - s_1) - (g_2 - s_2) + i(g_3 - s_3) - i(g_4 - s_4)\|_{p(h)} \\
&\leqq \sum_{\alpha=1}^{4} \|g_\alpha - s_\alpha\|_{p(h)} < \varepsilon.
\end{aligned}$$

Since s is a rational step-function, this completes the proof.

One reason for the usefulness of Theorem 6-1 is that the set of rational step-functions is denumerably infinite. In many proofs it is fairly easy to manage countably many functions. In such cases we can often control the situation by first working with the rational step-functions and then extending the results to all of $\mathscr{L}_p$ by using the fact that the rational step-functions are dense in $\mathscr{L}_p$.

To avoid verbosity, we now introduce three commonly used expressions. A function f on a domain D has **bounded support** if the set $\{x \text{ in } D : f(x) \neq 0\}$ is bounded. It is **of class** C^n (n a positive integer) if f and all its derivatives of order $\leqq n$ are defined and continuous on D. It is **of class** C^∞ if f and all its derivatives of all orders are defined and continuous on D.

LEMMA 6-2 *Let m be a regular additive nonnegative function of left-open intervals in R^r; let s be a step-function on R^r with left-open intervals of constancy, and let q be a positive integer. Then there exists a sequence $g_1, g_2, g_3, \ldots$ of functions of class C^q on R^r, all vanishing outside the same bounded interval and all with values in the interval* $[\inf s, \sup s]$, *such that*

(i) *for every x in R^r,*

$$\lim_{n\to\infty} g_n(x) = s(x);$$

(ii) *for every number $p \geqq 1$,*

$$\lim_{n\to\infty} \|g_n - s\|_p = 0.$$

By a trivial modification of the proof of Lemma IV-13-2, we show that the function

$$z \mapsto h(z) = z^{q+1}(1-z)^{q+1} 1_{(0,1]}(z) \Big/ \int_0^1 z^{q+1}(1-z)^{q+1}\, dz$$

is defined and continuous together with its derivatives $Dh, \ldots, D^q h$ on R, and its integral over R is 1. By the substitution $z = nx - ny$, we obtain for every positive integer n and every real x

$$\text{(H)} \qquad \int_R nh(nx - ny)\, dy = 1,$$

and the function $y \mapsto nh(nx - ny)$ is of class C^q on R. We define for all x in R^r

$$\text{(I)} \qquad g_n(x) = \int_{R^r} [n^r h(nx^1 - ny^1) \cdots h(nx^r - ny^r)] s(y)\, dy.$$

The factor in square brackets in the right member of (I) vanishes for y outside the interval

$$\text{(J)} \qquad W(x, n) = (x^1 - 1/n, x^1] \times \cdots \times (x^r - 1/n, x^r],$$

and we can and do choose a number c such that the factor $s(y)$ vanishes for y outside the interval

$$W[c] = (-c, c] \times \cdots \times (-c, c].$$

The partial derivatives with respect to $x^1, \ldots, x^r$ of the integrand in (I) of order

$\leqq q$ are continuous and bounded on $W[c]$ and vanish outside $W[c]$. So we can use the theorem on differentiation under the integral sign (Theorem II-11-1) repeatedly to find that g_n is continuous, and all its partial derivatives of order $\leqq q$ exist and are continuous, on R^r.

The factor in square brackets in the right member of (I) is positive for y in $W(x, n)$ and is 0 elsewhere, so in (I) we may integrate over $W(x, n)$ instead of over R^r. By (H) and Fubini's theorem,

$$\text{(K)} \qquad \int_{R^r} n^r[h(nx^1 - ny^1) \cdots h(nx^r - ny^r)]\, dy = 1.$$

By this and the first theorem of the mean for integrals (Corollary I-6-4), we obtain

$$\text{(L)} \qquad \inf\{s(y) : y \text{ in } W(x, n)\} \leqq g_n(x) \leqq \sup\{s(y) : y \text{ in } W(x, n)\}.$$

This implies, first,

$$\text{(M)} \qquad |g_n(x)| \leqq \sup\{|s(y)| : y \text{ in } R^r\}$$

for all x in R^r. Second, if x does not belong to the interval

$$W[c + 1] = (-c - 1, c + 1] \times \cdots \times (-c - 1, c + 1],$$

no point of $W(x, n)$ is in $W[c]$, so s vanishes identically on $W(x, n)$, and by (L),

$$\text{(N)} \qquad g_n(x) = 0 \qquad (x \text{ in } R^r \setminus W[c + 1]).$$

Third, we can subdivide $W[c + 1]$ into pairwise disjoint left-open intervals $B_1, \ldots, B_p$ on each of which s is constant. If x is in $W[c]$, it is in a certain one B_i of these intervals. We can represent B_i as

$$B_i = (a_1, b_1] \times \cdots \times (a_r, b_r].$$

Then, since x is in B_i,

$$a_j < x^j \leqq b_j \qquad (j = 1, \ldots, r).$$

If

$$\text{(O)} \qquad n > \max\{1/(x^j - a_j) : j = 1, \ldots, r\},$$

we have

$$a_j < x^j - 1/n < x^j \leqq b_j,$$

so

$$W(x, n) \subset B_i.$$

But s is constant on B_i and has value $s(x)$ at the point x of B_i, so $s(y) = s(x)$ for all y in $W(x, n)$. By (L),

$$g_n(x) = s(x)$$

for all n that satisfy (O). By this and (N),

$$\text{(P)} \qquad \lim_{n\to\infty} g_n(x) = s(x)$$

for all x in R^r.

If $p \geqq 1$, since (M) holds and both g_n and s vanish outside $W[c+1]$,

$$|g_n(x) - s(x)|^p \leqq (2 \sup|s|)^p 1_{W[c+1]}(x) \qquad (x \text{ in } R^r).$$

The right member represents a function integrable over R^r, so by (P) and the dominated convergence theorem,

$$\lim_{n\to\infty} \int_{R^r} |g_n(x) - s(x)|^p \, m(dx) = 0.$$

This completes the proof.

COROLLARY 6-3 *Let m be a regular nonnegative additive function of left-open intervals in R^r, and let D be a left-open interval in R^r. If f is a function that belongs to $\mathscr{L}_{p(h)}[m, D]$ for each number $p(h)$ in a collection of numbers $\{p(1), \ldots, p(h^*)\}$ each $\geqq 1$, for each positive ε there is a function g of class C^4 on R^r of bounded support such that the values of g lie in $[\inf f, \sup f]$ and*

$$\|g - f\|_{p(h)} < \varepsilon \qquad (h = 1, \ldots, h^*).$$

By Theorem 6-1, there exists a step-function s with left-open intervals of constancy that takes values in $[\inf f, \sup f]$ such that

$$\|f - s\|_{p(h)} < \varepsilon/2 \qquad (h = 1, \ldots, h^*).$$

By Lemma 6-2 there is a function g of class C^4 that has bounded support, takes values in $[\inf f, \sup f]$ and is such that

$$\|g - s\|_{p(h)} < \varepsilon/2 \qquad (h = 1, \ldots, h^*).$$

EXERCISE 6-1 Show that the functions that vanish outside a bounded interval and have continuous derivatives of all orders are dense in $\mathscr{L}_p$. *Suggestion*: In place of the function h used in the proof of Lemma 6-2, use the function $x \mapsto \phi(x)\phi(1-x)$, where

$$\phi(x) = \exp(-1/x^2) \qquad (x > 0),$$

$$\phi(x) = 0 \qquad (x \leqq 0)$$

The properties of ϕ were discussed in Exercise II-6-5.

EXERCISE 6-2 Prove that if f is bounded and integrable over an interval $[\alpha, \beta]$, and ϕ is a function on $[a, b]$ with values in $[\alpha, \beta]$ that is the indefinite integral of a function $\dot{\phi}$, then

$$\int_{\phi(a)}^{\phi(b)} f(x)\,dx = \int_a^b f(\phi(y))\dot{\phi}(y)\,dy.$$

Suggestion: If f is continuous and has bounded support, this holds by Theorem II-9-6. Use the dominated convergence theorem to extend to all step-functions, then to U-functions and L-functions, then to the functions g, h in Corollary IV-6-5. Then with this g and h,

$$\int_a^x g(\phi(y))\dot{\phi}(y)\,dy = \int_a^x h(\phi(y))\dot{\phi}(y)\,dy,$$

where $a \leqq x \leqq b$. By the fundamental theorem, the integrands are almost everywhere equal, and $f(\phi(y))\dot{\phi}(y)$ is between them.

7. Fourier Series

If for a function

$$x \mapsto f(x) \qquad (x \text{ in } R)$$

there exists a positive number p such that

$$f(x + p) = f(x)$$

for all x in R, the function f is said to have **period** p. If we know the values of such an f on an interval $(a, b]$ with $b = a + p$, we can find its values for all x, since every x in R has the form $x = np + x'$, where n is an integer and x' is in $(a, b]$; and by periodicity,

$$f(x) = f(np + x') = f(x').$$

If we imagine the real axis rolled around a circle of circumference p, then to all the points x of the axis that fall on a given point of the circumference will correspond the same value of $f(x)$, so the function f can equally well be thought of as a function on the circumference of the circle. The circle of circumference p has radius

$$T = p/2\pi,$$

and it is often more convenient to use this T than to use the period p.

The most familiar functions of period p are those with values

$$\text{(A)} \qquad 1, \cos x/T, \sin x/T, \ldots, \cos nx/T, \sin nx/T, \ldots \qquad (n = 1, 2, 3, \ldots).$$

As early as 1748, Euler asserted that "every" function of period p can be represented as the sum of an infinite series, each term of which is a constant multiple of one of the functions in the list (A). But in his time mathematicians had not felt it necessary to state just what they meant by a function, and the idea of convergence of a series had not yet been stated. It was not until the beginning of the nineteenth century that J. B. J. Fourier introduced some exactness into the study of such series, which he used in studying the flow of heat. Such series are now called *Fourier series.*

Today Fourier series continue to be of great importance both in mathematical theory and in applications, but most people who work with them prefer to use complex exponentials instead of the trigonometric functions (A). For each complex number z the value of the exponential, cosine, and sine functions at z are defined by the same power series that we met in Section II-6:

$$\exp z = \sum_{j=0}^{\infty} \frac{z^j}{j!}, \qquad \cos z = \sum_{j=0}^{\infty} \frac{(-1)^j z^{2j}}{(2j)!}, \qquad \sin z = \sum_{j=0}^{\infty} \frac{(-1)^j z^{2j+1}}{(2j+1)!},$$

where, as usual, $0! = 1$. These converge for all z and converge uniformly on the set of all z with $|z| \leqq c$ for every positive number c; the proofs are the same as for the real case, since in these proofs only absolute values of numbers were used. In particular, for each complex number z we have, recalling Examples II-6-11 and II-6-13,

$$\text{(B)} \qquad \begin{aligned} \exp(iz) &= 1 + (iz)/1! + (iz)^2/2! + (iz)^3/3! + (iz)^4/4! + \cdots \\ &= [1 - z^2/2! + z^4/4! - \cdots] + i[z/1! - z^3/3! + \cdots] \\ &= \cos z + i \sin z. \end{aligned}$$

From this we readily compute

$$\text{(C)} \qquad \begin{aligned} \cos z &= [\exp(iz) + \exp(-iz)]/2, \\ \sin z &= [\exp(iz) - \exp(-iz)]/2i. \end{aligned}$$

So any statement involving the functions $\cos nx/T$ and $\sin nx/T$ can easily be changed into one involving the functions defined by

$$\text{(D)} \qquad x \mapsto \exp(inx/T) \qquad (x \text{ in } R;\ n = 0, \pm 1, \pm 2, \ldots).$$

In using these exponential functions, in the real case it was important to know that

$$\exp(x + y) = (\exp x)(\exp y).$$

Conveniently, this useful formula is valid for the complex case also, as we now prove. Let z and w be complex numbers; let D denote differentiation with respect to t $(-\infty < t < \infty)$. The equation

$$D(z + tw)^n = n(z + tw)^{n-1} w \qquad (t \text{ in } R)$$

is established exactly as in the real case. Then

$$D(z + tw)^n/n! = w(z + tw)^{n-1}/(n-1)! \qquad (n = 1, 2, 3, \ldots)$$

is the nth term of the power series whose sum is $w\exp(z + tw)$, and the convergence is uniform for $0 \leqq t \leqq 1$. Therefore the integral of $w\exp(z + tw)$ over every subinterval $[0, t']$ of $[0, 1]$ is the sum of the integrals of the terms of the series, so that

$$\begin{aligned} w\int_0^{t'} \exp(z + tw)\,dt &= \sum_{n=1}^{\infty} \int_0^{t'} \frac{D(z + tw)^n}{n!}\,dt \\ &= \sum_{n=1}^{\infty} \frac{(z + t'w)^n}{n!} - \frac{(z + 0w)^n}{n!} \\ &= \exp(z + t'w) - \exp z. \end{aligned}$$

By the fundamental theorem, the left member is differentiable with respect to t', and its derivative is (writing t in place of t')

$$w\exp(z + tw) = D\exp(z + tw).$$

We utilize this to differentiate the product

$$\exp(z + tw)\exp(t[-w]).$$

The result is

$$\begin{aligned} D\{\exp(z + tw)\exp(-tw)\} &= [w\exp(z + tw)]\exp(-tw) \\ &\quad + w\exp(z + tw)[-w\exp(-tw)] \\ &= 0. \end{aligned}$$

By the theorem of mean value, the product has the same value at 1 and at 0, so

$$\text{(E)} \qquad \exp(z + w)\exp(-w) = (\exp z)(\exp 0).$$

In particular, when $z = 0$, this yields

$$\text{(F)} \qquad (\exp w)(\exp[-w]) = 1.$$

So $\exp w$ is never 0, and $\exp(-w)$ is its reciprocal. Now (E) yields

$$\text{(G)} \qquad \exp(z + w) = (\exp z)(\exp w),$$

as was to be proved.

In the rest of this section we shall consider only functions on a fixed interval $(a, b]$; this can be extended to all of R as a function with period

$$b - a = 2\pi T.$$

Since we shall keep this notation, there is no need to continue to repeat it. Also, the only measure that we shall use is m_{L}, and there is no need to keep repeating that. Therefore we shall usually simplify notation by abbreviating $\mathscr{L}_p[m_{\mathrm{L}}, (a, b]]$ and $L_p[m_{\mathrm{L}}, (a, b]]$ to $\mathscr{L}_p$ and L_p, respectively.

We define the functions v_n ($n = 0, \pm 1, \pm 2, \ldots$) to be the same ones already listed in (D); that is,

$$v_n(x) = \exp(inx/T) \qquad (x \text{ in } [a, b];\ n = 0, \pm 1, \pm 2, \ldots).$$

It is obvious that each v_n belongs to $\mathscr{L}_p$ for all $p \geqq 1$, and that for all x in $[a, b]$

(H) $$v_{-n}(x) = v_n(x)^-,$$

(I) $$v_m(x)v_n(x) = v_{m+n}(x),$$

(J) $$Dv_n(x) = (in/T)v_n(x).$$

The v_n all have the same value at b as at a. By (H) and (I), for all integers m and n

(K) $$\langle v_m, v_n \rangle = \int_a^b v_m(x)v_n(x)^- \, dx = \int_a^b v_{m-n}(x) \, dx.$$

If $m \neq n$, by (J) and the fundamental theorem,

(L) $$\langle v_m, v_n \rangle = (T/i[m - n])[v_{m-n}(b) - v_{m-n}(a)] = 0.$$

The v_n are, therefore, an orthogonal set. If $m = n$, since $v_0 = 1$ and $b - a = 2\pi T$, we deduce from (I) and (K)

(M) $$\langle v_n, v_n \rangle = \int_a^b 1 \, dx = 2\pi T.$$

We now prove that the v_n form a complete orthogonal set.

THEOREM 7-1 *The functions*

$$v_n : x \mapsto v_n(x) = \exp(inx/T) \qquad (a \leqq x \leqq b),$$

where $T = (b - a)/2\pi$, form a complete orthogonal set in $\mathscr{L}_2[m_L, (a, b]]$.

We have already proved them orthogonal. We next prove

(N) If F is real-valued and continuous on $[a, b]$ and $\langle F, v_n \rangle = 0$ for all integers n, then F is nowhere positive on $[a, b]$.

Suppose this false; then there exists an x_0 in (a, b) at which $F(x_0) > 0$. There is then a positive δ such that

$$a < x_0 - \delta < x + \delta < b$$

and

(O) $$F(x) > 0 \qquad (x_0 - \delta \leqq x \leqq x_0 + \delta).$$

We define

(P) $$g(x) = 1 - \cos(\delta/T) + \cos[(x - x_0)/T] \qquad (a \leqq x \leqq b).$$

Then

$$\text{(Q)}\quad g(x) > 1 \qquad (x_0 - \delta < x < x_0 + \delta)$$

$$-1 < -\cos\delta/T \leqq g(x) \leqq 1 \qquad (a \leqq x \leqq x_0 - \delta \text{ or } x_0 + \delta \leqq x \leqq b).$$

Since g is real-valued, for every positive integer j,

$$\text{(R)}\qquad \langle F, g^j\rangle = \int_{(a,x_0-\delta]} F(x)g(x)^j\,dx + \int_{(x_0-\delta,x_0+\delta)} F(x)g(x)^j\,dx + \int_{[x_0+\delta,b]} F(x)g(x)^j\,dx.$$

Since F is continuous, $F(x)$ has a finite upper bound M on $[a, b]$. So by (Q), the integrands in the first and third integrals in the right member of (R) are never below $-M$, and the sum of these integrals is at least $-M(b - a)$. In the second integral in the right member, by (O) and (Q) the integral increases and tends to ∞, so by the monotone convergence theorem this integral tends to ∞. So

$$\text{(S)}\qquad \lim_{j\to\infty}\int_a^b F(x)g(x)^j\,dt = \infty.$$

On the other hand, by (P), (C), and (G), we have

$$g(x) = [1 - \cos(\delta/T)] + v_1(x)[v_1(-x_0)/2] + v_{-1}(x)[v_{-1}(-x_0)/2],$$

so if we raise both members to the jth power and recall (I), we obtain

$$g(x)^j = \sum_{h=-j}^{j} c_h v_h(x),$$

where the c_h are complex numbers whose values we do not need to know. If we multiply both members by $F(x)$ and integrate from 0 to T, we obtain

$$\int_a^b F(x)g(x)^j\,dx = \sum_{h=-j}^{j} c_h \int_a^b F(x)v_h(x)\,dx = 0.$$

This contradicts (S), so the assumption that (N) is false has led to a contradiction, and (N) is established.

Now let f be a member of $\mathscr{L}_2$ that is orthogonal to all the v_n. Then for all integers n

$$\begin{aligned}\langle \bar{f}, v_n\rangle &= \int_a^b \bar{f}(x)\bar{v}_n(x)\,dx\\ &= \left(\int_a^b f(x)v_n(x)\,dx\right)^{-}\\ &= \langle f, \bar{v}_n\rangle^{-} = \langle f, v_{-n}\rangle^{-} = 0,\end{aligned}$$

so $\bar{f}$ is also orthogonal to all v_n. For each x in $[a, b]$ the indicator $1_{(a,x]}$ is in $\mathscr{L}_2$. By Lemma 2-7, the integral

$$F(x) = \int_a^b [f(u) + \bar{f}(u)] 1_{(a,x]}(u)\, du$$

exists. Since $f + \bar{f}$ is orthogonal to v_0 and $\bar{v}_0 = v_0 = 1$, we have

$$F(b) = \int_a^b [f(u) + \bar{f}(u)] v_0(u)\, du = 0.$$

By integration by parts, with (H) and (J), we find that for each nonzero integer n

$$\begin{aligned} \langle F, v_n \rangle &= \int_a^b F(x) v_{-n}(x)\, dx \\ &= F(b)[-T/ni] v_{-n}(b) - F(a)[-T/ni] v_{-n}(a) \\ &\quad + \int_a^b [f(x) + \bar{f}(x)][T/ni] v_{-n}(x)\, dx \\ &= 0. \end{aligned}$$

By (N), F is nowhere positive in $[a, b]$, and by applying (N) to $-F$ we find that $-F$ is nowhere positive, so F vanishes identically on $[a, b]$. By Theorem II-13-4, $f + \bar{f}$ is 0 almost everywhere in $[a, b]$. That is, $\operatorname{Re} f = 0$ almost everywhere. But if f is orthogonal to all v_n, so is if, and by the preceding proof, $\operatorname{Re}(if) = 0$ almost everywhere. Since $\operatorname{Re}(if) = -\operatorname{Im} f$, we now know that if f is orthogonal to all v_n, both $\operatorname{Re} f$ and $\operatorname{Im} f$ are almost everywhere 0, so that f is the zero member of $\mathscr{L}_2$. By Theorem 5-14, the set of v_n is complete.

Just before Theorem 5-6 we defined expansion coefficients; the expansion coefficients of w in terms of any orthogonal set are the numbers

(T) $$\langle w, v_j \rangle / \|v_j\|^2.$$

When f is in $\mathscr{L}_2[m_L, (a, b]]$ and the v_n are the functions in Theorem 7-1, the expansion coefficients are called **Fourier coefficients**. By (T) and (M), these are the numbers $\hat{f}(n)$ defined by

(U) $$\begin{aligned} \hat{f}(n) &= (2\pi T)^{-1} \int_a^b f(x) v_{-n}(x)\, dx \\ &= (b - a)^{-1} \int_a^b f(x) v_{-n}(x)\, dx. \end{aligned}$$

Thus, $\hat{f}(n)$ is the average of $f v_{-n}$ over $(a, b]$.

Although we are primarily concerned with functions in $\mathscr{L}_2$, it should be pointed out that the integral in (U) exists whenever f is integrable over $(a, b]$, and (U) is taken as the definition of the Fourier coefficient $\hat{f}(n)$ for all such f.

For f in $\mathscr{L}_2$, Bessel's inequality (Theorem 5-6) holds, but Theorem 5-13 gives a much better result. For by (I) of that theorem, with (M),

(V) $$\|f\|^2 = 2\pi T \sum_{j=-\infty}^{\infty} \hat{f}(j)^2,$$

and by (H) of that theorem, if f and g are in L_2,

(W) $$\langle f, g\rangle = 2\pi T \sum_{j=-\infty}^{\infty} \hat{f}(j)\hat{g}(j)^-.$$

From Theorem 5-11 we deduce the most important expansion theorem.

THEOREM 7-2 *If f is in $\mathscr{L}_2[m_L, (0, T]]$, then*

$$f = \sum_{j=-\infty}^{\infty} \hat{f}(j)v_j;$$

that is, for each positive δ there is an n such that if $j' > n$ and $j'' > n$,

$$\left\| f - \sum_{j=-j'}^{j''} \hat{f}(j)v_j \right\|_2 < \delta.$$

Theorem 7-2 gives an exact result, using the entire set of v_j. However, sometimes we are limited (for example, by the computer we are using) to a finite set of v_j, say, those v_j with j in a finite set Z of integers. We want to form a combination

$$\sum_{j \text{ in } Z} c_j v_j$$

that is in some sense as close as possible to a function f in $\mathscr{L}_2$. To measure the discrepancy between f and the approximation it is often reasonable to use the L_2-norm of their difference. The question is, how should we choose the c_j to make this as small as possible? Conveniently, the answer is simple: choose the Fourier coefficients. This, and more, is stated in the following corollary.

COROLLARY 7-3 *Let Z be a finite or infinite set of integers, and let V_Z consist of all g in L_2 of the form*

(X) $$g = \sum_{j \text{ in } Z} c_j v_j$$

(with complex coefficients c_j) that are either finite sums (if Z is finite) or are series that are convergent in L_2-norm (if Z is infinite). Let f be in L_2. Then the series

(Y) $$\sum_{j \text{ in } Z} \hat{f}(j)v_j$$

converges if Z is infinite, and whether Z is finite or infinite, the sum (Y) represents the point of V_Z nearest to f in L_2-norm.

This follows at once from Theorem 5-9.

For brevity, when Z is the set of all integers, we shall denote the space $l_2[Z]$ by l_2. Each member $[f]$ of $L_2[m_L, [0, T]]$ determines a set of Fourier coefficients $\hat{f}(j)$ $(j = 0, \pm 1, \pm 2, \ldots)$ that can be represented as the values of a two-way sequence, or function on Z, denoted, according to our custom, by $\hat{f}$; thus $\hat{f}$ is the function

$$j \mapsto \hat{f}(j) \qquad (j = 0, \pm 1, \pm 2, \ldots).$$

By Theorem 5-13 and equation (V), this $\hat{f}$ belongs to l_2, and its l_2-norm is

$$\text{(Z)} \qquad \|\hat{f}\|_2 = \left\{ \sum_{j=-\infty}^{\infty} |\hat{f}(j)|^2 \right\}^{1/2} = \frac{\|f\|_2}{2\pi T}.$$

In the next theorem, known as the *Riesz–Fischer theorem*, we shall show that this mapping of L_2 into l_2 is one-to-one. It is thus a one-to-one mapping of each space into the other that is length preserving except for the scale-factor $2\pi T$ in (Z). This gives us great freedom in working with either space; we can go back and forth between them by means of the one-to-one correspondence and freely choose whichever representation – the member $[f]$ of L_2 or its image $\hat{f}$ in l_2 – is most convenient.

THEOREM 7-4 *If $j \mapsto c_j$ $(j = 0, \pm 1, \pm 2, \ldots)$ is a two-way sequence of complex numbers such that the series*

$$\sum_{j=-\infty}^{\infty} |c_j|^2$$

converges, the series

$$\sum_{j=-\infty}^{\infty} c_j v_j$$

converges in L_2-norm to a function f in $\mathscr{L}_2[m_L, [a, b]]$, and the Fourier coefficients of f are the c_j:

$$\hat{f}(j) = c_j \qquad (j = 0, \pm 1, \pm 2, \ldots).$$

Since the functions v_j $(j = 0, \pm 1, \pm 2, \ldots)$ are a complete orthogonal set in L_2 by Theorem 7-1, this is an immediate consequence of Theorem 5-12.

EXERCISE 7-1 From (B) and (G), deduce the addition formulas for the sine and cosine.

EXERCISE 7-2 Show that the Fourier series for the function

$$f(x) = -1 \qquad (x < 0),$$

$$f(x) = +1 \qquad (x \geqq 0)$$

on the interval $[-\pi, \pi]$ is

$$\sum_{n \text{ odd}} \frac{2}{n\pi i} e^{inx},$$

or

$$\frac{4}{\pi}\left(\frac{\sin x}{1} + \frac{\sin 3x}{3} + \cdots + \frac{\sin(2n+1)x}{2n+1} + \cdots\right).$$

EXERCISE 7-3 Show that the Fourier series for the function $x \mapsto |x|$ on $[-\pi, \pi]$ is

$$\frac{\pi}{2} - \sum_{n \text{ odd}} \frac{2}{n^2 \pi} e^{inx},$$

or

$$\frac{\pi}{2} - \frac{4}{\pi}\left(\frac{\cos x}{1^2} + \frac{\cos 3x}{3^2} + \frac{\cos 5x}{5^2} + \cdots\right).$$

Show that this converges uniformly to $|x|$ on $[-\pi, \pi]$. (It converges uniformly to something, by Lemma II-6-6; use Theorems 7-2 and 4-5.)

EXERCISE 7-4 Let c_n $(n = 0, \pm 1, \pm 2, \ldots)$ be a two-way sequence of complex numbers such that for some constant C and some $q > 1$,

$$c_n \leqq C|n|^{-q} \qquad (n = \pm 1, \pm 2, \ldots).$$

Show that the series

$$\sum_{-\infty}^{\infty} c_n e^{inx}$$

converges uniformly on $[-\pi, \pi]$ to a function f that belongs to $\mathscr{L}_2[-\pi, \pi]$, and it is the Fourier series of that function.

EXERCISE 7-5 Show that the Fourier series for the function $x \mapsto x$ $(-\pi \leqq x \leqq \pi)$ is

$$\sum_{n \neq 0} \frac{(-1)^{n+1}}{in} e^{inx},$$

or

$$2(\sin x - \tfrac{1}{2}\sin 2x + \tfrac{1}{3}\sin 3x - \cdots).$$

EXERCISE 7-6 Show that the Fourier series for the function $x \mapsto x^2$ $[-\pi \leqq x \leqq \pi]$ is

$$\frac{\pi^2}{3} + \sum_{n \neq 0} \frac{2\pi(-1)^n}{n^2} e^{inx},$$

or

$$\frac{\pi^2}{3} + 4\left(\frac{-\cos x}{1^2} + \frac{\cos 2x}{2^2} - \frac{\cos 3x}{3^2} + \cdots\right).$$

Show that this converges uniformly to x^2 on $[-\pi, \pi]$.

8. Indefinite Integrals and the Weierstrass Approximation Theorem

In this section we continue to use the functions v_n, defined by

$$v_n(x) = \exp(inx/T) \qquad (n = 0, \pm 1, \pm 2, \ldots),$$

that we introduced in the preceding section.

If f is in $\mathscr{L}_2$, we know by Theorem 7-2 that the partial sums

$$\text{(A)} \qquad S_n(t) = \sum_{j=-n}^{n} \hat{f}(j)v_j(x)$$

converge in L_2-norm to f. By Theorem 4-5, a subsequence $S_{n(1)}(x), S_{n(2)}(x), \ldots$ converges almost everywhere to a limit, and this limit is almost everywhere $f(x)$. But even if f is continuous, it may not be true that S_n converges almost everywhere, and still less that it converges uniformly to f. In order to attain this, we have to strengthen the assumptions on f.

In many cases of interest, f is a function of period $2\pi T$ that is continuous and that is "piecewise smooth"; that is, there are points $a = x_0 < x < \cdots < x_n = b = a + 2\pi T$ such that on each interval $[x_{j-1}, x_j]$, f coincides with a continuously differentiable function. If on $(x_{j-1}, x_j]$ we define $\dot{f}(x)$ to be the derivative of that function, f is the indefinite integral of $\dot{f}$. Being an indefinite integral is the key hypothesis in the next lemma and theorem; it is a weaker requirement than piecewise smoothness.

LEMMA 8-1 *Let F be a function on $[a, b]$ that is the indefinite integral of a function f. Assume also that $F(b) = F(a)$. Then the Fourier coefficients of F and f satisfy*

$$\text{(B)} \qquad \hat{F}(n) = [T/ni]\hat{f}(n) \qquad (n = \pm 1, \pm 2, \ldots).$$

By integration by parts, with (J) of Section 7,

$$\begin{aligned}\hat{F}(n) &= \int_a^b F(x)(2\pi T)^{-1}v_{-n}(x)\,dx \\ &= -\frac{T}{ni}\Big\{F(b)(2\pi T)^{-1}v_{-n}(b) - F(a)(2\pi T)^{-1}v_{-n}(a) \\ &\qquad - \int_a^b f(x)(2\pi T)^{-1}v_{-n}(x)\,dx\Big\}.\end{aligned}$$

Since $v_{-n}(b) = v_{-n}(a) = 1$, and the last integral is $-\hat{f}(n)$, this implies (B).

THEOREM 8-2 *Let F be a function on an interval $[a, b]$ that is the indefinite integral of a function f belonging to $\mathscr{L}_2[m_L, [a, b]]$ and that satisfies $F(b) = F(a)$. Then the partial sums*

$$S_n(x) = \sum_{j=-n}^{n} \hat{F}(j)v_j(x) \qquad (n = 1, 2, 3, \ldots)$$

of the Fourier series for F converge absolutely and uniformly to $F(x)$ on $[a, b]$.

For each positive integer n, let $Z(n)$ denote the set of all integers $-n, \ldots, n$ except 0. By Lemma 8-1 and the Cauchy inequality (Lemma V-1-3), for each positive integer n,

$$\sum_{j \text{ in } Z(n)} |\hat{F}(j)| = \sum_{j \text{ in } Z(n)} \left|\frac{T}{j}\right| |\hat{f}(j)|$$

$$\leqq T \left[\sum_{j \text{ in } Z(n)} \frac{1}{j^2} \right]^{1/2} \left[\sum_{j \text{ in } Z(n)} |\hat{f}(j)|^2 \right]^{1/2}.$$

The quantity in the first bracket in the right member is bounded because $1/j^2$ is a convergent series, and the quantity in the second bracket is bounded by Bessel's inequality (Theorem 5-6). So the left member is bounded for all n, and the series with terms $\hat{F}(j)$ is absolutely convergent. Since

$$|\hat{F}(j)v_j(x)| = |\hat{F}(j)| \qquad (x \text{ in } R),$$

the Fourier series for F is absolutely convergent. By Lemma II-6-6, it is uniformly convergent. Temporarily, denote its sum by G. Then G is continuous on $[a, b]$, being the uniform limit of continuous functions (the partial sums S_n). These S_n converge uniformly to G, and by Theorem 7-2 they converge in L_2-norm to F. By Theorem 4-5, F and G are equal at almost all points of $[a, b]$. Suppose that the set of measure $b - a$ on which they are equal is called M. Then every nondegenerate subinterval of $[a, b]$ must contain points of M; otherwise mM would be less than $b - a$. If x is any point of $[0, T]$, we can and do choose points $x_1, x_2, x_3, \ldots$ of M that converge to x. Since the t_n are in M,

$$F(x_n) = G(x_n) \qquad (n = 1, 2, 3, \ldots).$$

Since x_n tends to x and both F and G are continuous, this implies

$$F(x) = G(x),$$

and the proof is complete.

We can use this theorem to prove the frequently useful approximation theorem of Weierstrass; every function continuous on a bounded closed interval in R can be uniformly approximated as closely as desired by polynomials.

THEOREM 8-3 *If f is a complex-valued continuous function on a bounded closed interval $[c, d]$, and ε is positive, there is a polynomial p with complex* *coefficients such that*

$$|f(x) - p(x)| < \varepsilon \qquad (c \leqq x \leqq d).$$

If f is real-valued, the polynomial can be chosen to have real coefficients.

We consider the complex case first. We define

$$a = c, \qquad b = 2d - c,$$

and we first extend the definition of f to $[a, b]$ by setting

$$f(x) = f(2c - x) \qquad (c \leqq x \leqq b).$$

This is continuous on $[a, b]$ and has $f(b) = f(a)$. We extend it to all of R with period $b - a$; the extended function is then continuous on R. It is continuous, and hence uniformly continuous, on $[a - 1, b + 1]$. So, if ε is positive, we can and do choose a positive δ (we take it less than 1) such that if x' and x'' are in $[a - 1, b + 1]$ and $|x' - x''| \leqq \delta$, then

$$|f(x') - f(x'')| < \varepsilon/3.$$

We define a function g in R by setting

$$g(x) = \delta^{-1} \int_x^{x+\delta} f(u)\, du.$$

By the fundamental theorem, this has derivative

$$Dg(x) = [f(x + \delta) - f(x)]/\delta,$$

which is a continuous function of x; so g is the indefinite integral of its own derivative on $[a - 1, b + 1]$. By Corollary I-6-4, for each x in R there is an x^* in $[x, x + \delta]$ such that

$$g(x) = \delta^{-1}\{f(x^*)\delta\}.$$

Since $|x^* - x| < \delta$, for all x in $[a, b]$ we have

$$\text{(C)} \qquad |f(x) - g(x)| < \varepsilon/3.$$

Since g is continuously differentiable and $g(b) = g(a)$, by Theorem 8-2 the partial sums

$$S_n(x) = \sum_{j=-n}^{n} g(j)v_j$$

converge uniformly to $g(x)$ as n increases. Therefore we can and do choose an integer n such that

$$\text{(D)} \qquad |S_n(x) - g(x)| < \varepsilon/3$$

for all x in R.

Next we define

$$M = n \max\{|a|, |b|\}.$$

By Example II-6-11, the power series for $\exp iu$ converges uniformly for $-M \leqq u \leqq M$. (The proof in Chapter II was for real exponentials, but the theorem holds for the complex case too, since in the proof only the absolute values of the independent variable x were used.) Therefore we can and do choose an integer q such that for the partial sum

$$P(u) = \sum_{j=0}^{q} \frac{(iu)^j}{j!}$$

it is true that

$$|\exp(iu) - P(u)| < \varepsilon/(3 + 3|\hat{g}(-n)| + \cdots + 3|\hat{g}(n)|).$$

For each x in $[a, b]$ and each j in $\{-n, \ldots, n\}$, jx is in $[-M, M]$, so

$$\begin{aligned} |v_j(x) - P(jx)| &= |\exp(ijx) - P(jx)| \\ &\leqq \varepsilon/(3 + 3|\hat{g}(-n)| + \cdots + 3|\hat{g}(n)|). \end{aligned}$$

This implies

$$\text{(E)} \quad \begin{aligned} \left| S_n(x) - \sum_{j=-n}^{n} \hat{g}(j) P(jx) \right| &= \left| \sum_{j=-n}^{n} \hat{g}(j)[v_j(x) - P(jx)] \right| \\ &\leqq \sum_{j=-n}^{n} |\hat{g}(j)|\,|v_j(x) - P(jx)| \\ &\leqq \sum_{j=-n}^{n} |\hat{g}(j)|[\varepsilon/(3 + 3|\hat{g}(-n)| + \cdots + 3|\hat{g}(n)|)] \\ &< \varepsilon/3. \end{aligned}$$

Comparing (C), (D), and (E) yields

$$\left| f(x) - \sum_{j=-n}^{n} \hat{g}(j) P(jx) \right| < \varepsilon \qquad (a \leqq t \leqq b).$$

Since the sum in the left member is a polynomial in x, the theorem is proved for complex f.

If f is real-valued, we can by the preceding proof approximate it to within ε by a polynomial

$$x \mapsto p(x) = \sum_{j=0}^{n} c_j x^j$$

on $[a, b]$, the c_j being complex. Let p' be the polynomial

$$p'(x) = \sum_{j=0}^{n} (\operatorname{Re} c_j) x^j.$$

Then

$$|f(x) - p'(x)| = |\mathrm{Re}[f(x) - p(x)]|$$
$$\leqq |f(x) - p(x)| < \varepsilon,$$

and the proof is complete.

EXERCISE 8-1 By use of Lemma 8-1, deduce the conclusions of Exercise 7-3 and 7-6 from the statements of Exercises 7-2 and 7-5.

EXERCISE 8-2 Show that for each bounded interval B the set of polynomials with rational coefficients is denumerable and is dense in $L_p[m, B]$ $(p \geqq 1)$. (Use Corollary 6-3 and Theorem 8-3.)

EXERCISE 8-3 Show that the sum $F(x)$ of the series

$$\sum_{n=1}^{\infty} n^{-1.6} \cos nx \qquad (-\pi \leqq x \leqq \pi)$$

is the indefinite integral of a function f, and that $F(\pi) = F(-\pi)$.

EXERCISE 8-4 Prove the Riemann–Lebesgue theorem: if f is integrable over $[a, b]$, its Fourier coefficients $\hat{f}(j)$ tend to 0 as j tends to ∞ or to $-\infty$. *Suggestion*: If $\varepsilon > 0$, there is a step-function s such that the integral of $|f - s|$ is less than $(b - a)\varepsilon/2$. Then $|\hat{f}(j) - \hat{s}(j)| < \varepsilon/2$. If s has k intervals of constancy, $\hat{s}(j)$ is the sum of k integrals, each over an interval of constancy, that tend to 0 as $|j|$ increases.

9. Legendre Polynomials

The Legendre polynomials are the polynomials defined in the interval $[-1, 1]$ by

(A) $$P_0(x) = 1, \qquad P_n(x) = D^n(x^2 - 1)^n/2^n n! \qquad (n = 1, 2, 3, \ldots),$$

where, as usual, D^n is the differentiation operator

$$D^0 f = f \quad \text{and} \quad D^n f = D(D^{n-1} f) \qquad (n = 1, 2, 3, \ldots).$$

Obviously, P_n is a polynomial of degree exactly n, so it belongs to $\mathscr{L}_2[m_{\mathrm{L}}, [-1, 1]]$. To show that these P_n form an orthogonal set, we first observe that for $j \leqq n$,

(B) $$D^j(x^2 - 1)^n = Q_j(x)(x^2 - 1)^{n-j},$$

where $Q_j(x)$ is a polynomial in x. This is obvious for $j = 0$. If it holds for an integer j, then

$$\begin{aligned} D^{j+1}(x^2-1)^n &= D[Q_j(x)(x^2-1)^{n-j}] \\ &= [DQ_j(x)](x^2-1)^{n-j} + (n-j)Q_j(x)(x^2-1)^{n-j-1}(2x) \\ &= Q_{j+1}(x)(x^2-1)^{n-j-1}, \end{aligned}$$

where Q_{j+1} is the polynomial

$$(x^2-1)\,DQ_j(x) + 2(n-j)xQ_j(x).$$

So, by induction, (B) holds for all nonnegative integers n.

We next prove

(C) If j and k are positive integers, and $j \leqq n$, then

$$\int_{-1}^{1} D^j(x^2-1)^n\, D^k(x^2-1)^n\, dx = -\int_{-1}^{1} D^{j-1}(x^2-1)^n\, D^{k+1}(x^2-1)^n\, dx.$$

By integration by parts, the left member of this equation is equal to

$$D^{j-1}(x^2-1)^n\, D^k(x^2-1)^n \Big|_{-1}^{+1} - \int_{-1}^{+1} D^{j-1}(x^2-1)^n\, D^{k+1}(x^2-1)^n\, dx.$$

By (B), $D^{j-1}(x^2-1)^n$ is the product of a polynomial and a factor $(x^2-1)^{n-j+1}$. Since $n-j+1 \geqq 1$, this last factor vanishes at -1 and at 1, and (C) is proved.

Now let m and n be integers with $0 \leqq m \leqq n$. By (A) and (C) (the latter applied n times),

(D)
$$\begin{aligned} \langle P_n, P_m\rangle &= \frac{1}{2^{m+n}m!n!}\int_{-1}^{+1} D^n(x^2-1)^n\, D^m(x^2-1)^m\, dx \\ &= (-1)^n \frac{1}{2^{m+n}m!n!}\int_{-1}^{+1} D^0(x^2-1)^n\, D^{m+n}(x^2-1)^m\, dx. \end{aligned}$$

If $n > m$, the factor $D^{m+n}(x^2-1)^m$ is identically 0, so

(E)
$$\langle P_n, P_m\rangle = 0 \qquad (0 \leqq m < n).$$

So the Legendre polynomials form an orthogonal set.

If $m = n$, we notice that $(x^2-1)^n$ is a polynomial whose leading term is x^{2n}, all others being of lower degree, so

$$D^{2n}(x^2-1)^n = (2n)!.$$

From this and (D),

(F)
$$\langle P_n, P_n\rangle = (-1)^n \frac{1}{2^{2n}(n!)^2}\int_{-1}^{+1} (x^2-1)^n (2n)!\, dx.$$

To finish the computation, all we need is to establish

$$\text{(G)} \quad \int_{-1}^{+1} (x^2 - 1)^n \, dx = 2(-1)^n \cdot \frac{2}{3} \cdot \frac{4}{5} \cdot \frac{6}{7} \cdots \frac{2n}{2n+1} \qquad (n = 1, 2, 3, \ldots).$$

This is an easy elementary calculation if $n = 1$. Suppose that it holds for a positive integer n. By integration by parts,

$$\int_{-1}^{+1} (x^2 - 1)^{n+1} \, dx = (x^2 - 1)^{n+1} x \Big|_{-1}^{+1} - \int_{-1}^{+1} [(n+1)(x^2 - 1)^n (2x)] x \, dx$$

$$= -2(n+1) \int_{-1}^{+1} [(x^2 - 1)^{n+1} + (x^2 - 1)^n] \, dx.$$

Transposing the first term in the right member yields

$$(2n+3) \int_{-1}^{+1} (x^2 - 1)^{n+1} \, dx = -(2n+2) \int_{-1}^{+1} (x^2 - 1)^n \, dx.$$

In the right member, we substitute the value of the integral from (G); then we find that (G) holds for $n + 1$ also, and by induction (G) holds for all positive integers n.

If we substitute (G) in (F) and do a little simplification, we obtain

$$\text{(H)} \qquad \langle P_n, P_n \rangle = 2/(2n+1).$$

Since each P_n has degree exactly n, every polynomial of degree k is a linear combination of $P_0, \ldots, P_k$. Let ε be positive, and let f be any member of $\mathscr{L}_2[m_L, [-1, 1]]$. We may suppose it extended to all of R by setting $f = 0$ outside $[-1, 1]$. By Corollary 6-3, there is a continuous function ϕ on $[-1, 1]$ such that

$$\|f - \phi\|_2 < \varepsilon/2.$$

By the Weierstrass approximation theorem (Theorem 8-3), there is a polynomial Q such that

$$\|Q(x) - \phi(x)\| < \varepsilon/4 \qquad (-1 \leqq x \leqq 1).$$

Then

$$\|Q - \phi\|_2 = \left\{ \int_{-1}^{+1} |Q(x) - \phi(x)|^2 \, dx \right\}^{1/2} < \varepsilon/2,$$

so by the triangle inequality,

$$\|f - Q\|_2 < \varepsilon.$$

The polynomial Q is a linear combination of the Legendre polynomials P_0, $P_1, \ldots$, so the combinations of these polynomials are dense in $\mathscr{L}_2[m_L, [-1, 1]]$, and they form a complete set of orthogonal functions.

These polynomials are useful in finding the closest least-squares approximation to a given function by polynomials of a given degree. Let f belong to $\mathscr{L}_2[m_L, [-1, 1]]$. Its expansion in Legendre polynomials is, recalling (H),

$$\text{(I)} \qquad \sum_{j=0}^{\infty} [(2n + 1)/2]\langle f, P_j\rangle P_j;$$

this series converges in L_2-norm to f. By conclusion (ii) of Theorem 5-9, if we stop this with the term $j = k$, we obtain that combination of $P_0, \ldots, P_k$ that is closest to f in L_2-norm. But the combinations of $P_0, \ldots, P_k$ are all the polynomials of degree k or less, so the sum of the first $k + 1$ terms of the expansion (I) gives us the polynomial of degree k that is closest to f in L_2-norm; that is, it is the approximation in the sense of least-square error. A virtue of this procedure is that if we find the best approximation of a given degree and then decide that we want an approximation of higher degree, we do not have to discard the work we have done; we need only compute one more term of (I) and add it on.

The first few Legendre polynomials are

$$\text{(J)} \qquad \begin{aligned} P_0 &= 1, \\ P_1(x) &= x, \\ P_2(x) &= (3x^2 - 1)/2, \\ P_3(x) &= (5x^3 - 3x)/2, \\ P_4(x) &= (35x^4 - 30x^2 + 3)/8. \end{aligned}$$

EXERCISE 9-1 Show that P_n contains only even powers of x if n is even, and only odd powers of x if n is odd. (This follows readily from (A).)

EXERCISE 9-2 Show that if f is an odd function, so that $f(-x) = -f(x)$ on $[-1, 1]$, all its even-numbered Legendre expansion coefficients c_n are 0; and if it is an even function, so that $f(-x) = f(x)$ in $[-1, 1]$, all its odd-numbered Legendre expansion coefficients are 0.

EXERCISE 9-3 Show that if f is the indicator of $(0, \infty)$, its first five Legendre expansion coefficients $c_0, \ldots, c_4$ are

$$\tfrac{1}{2}; \tfrac{3}{4}; 0; -\tfrac{7}{16}; 0.$$

Show that c_n is 0 if n is positive and even. *Suggestion*: $f - \frac{1}{2}$ differs from an odd function at only one point.

EXERCISE 9-4 Expand $x \mapsto x^4$ in Legendre polynomials. Show that $c_n = 0$ if $n > 4$. What cubic polynomial is the best approximation to this function?

EXERCISE 9-5 Find the first five terms in the expansion of $x \mapsto \sin \pi x$ in Legendre polynomials. Why does this differ from the sum of the first five terms of the power series for $\sin \pi x$?

10. The Hermite Polynomials and the Hermite Functions

There is a certain sequence of polynomials on R whose study would be forced on us by the needs of quantum mechanics even if they had no intrinsic mathematical interest – which they do have. But we do not wish to go deeply into physics, so we shall present a motivation that may seem rather artificial. The Legendre polynomials form an orthogonal set on $[-1, 1]$. But polynomials do not belong to $\mathscr{L}_2[m_L, R]$, so if we wish to make up a Hilbert space that contains the polynomials, we need to replace m_L by some other measure with respect to which all polynomials are integrable. In Section II-11 we found such a measure, one that involves the function $\exp(-x^2/2)$ that is of importance in probability theory. We shall use this measure, with a trivial change; to make some formulas simpler, we change scale on the x-axis and introduce a measure m_H, defined for every left-open interval B in R by

$$m_H B = \int_B \exp(-x^2)\, dx.$$

Then by Theorem III-9-4, a function F on R is integrable with respect to m_H over R if and only if the function f defined by

$$f(x) = F(x)\exp(-x^2)$$

is integrable over R with respect to m_L, and in that case

$$\text{(A)} \qquad \int_R F(x)\, m_H(dx) = \int_R F(x)\exp(-x^2)\, dx = \int_R f(x)\, dx.$$

It is easy to see that if either of the functions F, $x \mapsto F(x)\exp(-x^2)$ is measurable with respect to either m_H or m_L, both are measurable with respect to both measures. The inner product in $\mathscr{L}_2[m_H, R]$ will be denoted by $\langle \cdot , \cdot \rangle_H$. Thus, if both F and G belong to $\mathscr{L}_2[m_H, R]$, their inner product is

$$\text{(B)} \qquad \langle F, G\rangle_H = \int_R F(x)\overline{G(x)}\, m_H(dx).$$

One straightforward way of obtaining an orthogonal set of polynomials would be to start with the sequence $1, x, x^2, x^3, \ldots$ and orthogonalize these functions by the Gram–Schmidt procedure described in the proof of Lemma

V-3-7. But we can reach the same goal with less effort if we use a trick bequeathed to us by our predecessors. For each nonnegative integer n we define the **Hermite polynomial** H_n by

(C) $$H_n(x) = (-1)^n(\exp x^2)\,D^n \exp(-x^2),$$

where $D^n \exp(-x^2)$ is the nth derivative of $\exp(-x^2)$ if $n > 0$, and is $\exp(-x^2)$ itself if $n = 0$. It requires only a few seconds to compute that

(D) $$H_0(x) = 1, \qquad H_3(x) = 8x^3 - 12x,$$
$$H_1(x) = 2x, \qquad H_4(x) = 16x^4 - 48x^2 + 12,$$
$$H_2(x) = 4x^2 - 2, \qquad H_5(x) = 32x^5 - 160x^3 + 120x.$$

These expressions lead to the following conjecture.

(E) For each nonnegative integer n, H_n is a polynomial, and its term of highest degree is $2^n x^n$.

This is true for $n = 0, 1, \ldots, 5$, by (D). Suppose it true for an integer n. By (C),

(F) $$DH_n(x) = (-1)^n\{2x(\exp x^2)\,D^n \exp(-x^2) + (\exp x^2)\,D^{n+1}\exp(-x^2)\}$$
$$= 2xH_n(x) - H_{n+1}(x).$$

This implies

(G) $$H_{n+1}(x) = 2xH_n(x) - DH_n(x).$$

The right member is a polynomial in x whose term of highest degree is $2^{n+1}x^{n+1}$, so by induction, (E) holds for all n.

For every positive integer j, every polynomial P, and every positive number k, we have, by integration by parts from $-k$ to k,

$$\int_{-k}^{k} P(x)\,D^j \exp(-x^2)\,dx$$
$$= P(x)\,D^{j-1}\exp(-x^2)\Big|_{-k}^{k} - \int_{-k}^{k} DP(x)\,D^{j-1}\exp(-x^2)\,dx.$$

Each of the terms evaluated at k and at $-k$ is the product of a polynomial in k with $\exp(-k^2)$ and therefore tends to 0 as k increases. Therefore,

(H) $$\int_{-\infty}^{\infty} P(x)\,D^j \exp(-x^2)\,dx = -\int_{-\infty}^{\infty} DP(x)\,D^{j-1}\exp(-x^2)\,dx.$$

If n is a positive integer and m is an integer such that $0 \leqq m \leqq n$, by (A), (B), and (C),

(I) $$\langle H_m, H_n\rangle_{\mathrm{H}} = \int_R H_m(x)(-1)^n D^n \exp(-x^2)\,dx.$$

If $m < n$, we apply (H) $m + 1$ times to the integral in (I), obtaining

$$\langle H_m, H_n \rangle_{\mathrm{H}} = (-1)^{n+m+1} \int_R [D^{m+1} H_m(x)]\, D^{n-m-1} \exp(-x^2)\, dx.$$

The first factor in the integrand is 0 because by (E), H_m is a polynomial of degree m in x. So

(J) if $0 \leqq m < n$, $\langle H_m, H_n \rangle_{\mathrm{H}} = 0$.

Therefore the H_n form an orthogonal set in $\mathscr{L}_2[m_{\mathrm{H}}, R]$. In fact, they form a complete orthogonal set. But this fact will be much easier to prove after we have established some properties of the Fourier transform, so we postpone its proof to Section 14 of this chapter.

If $m = n$, we apply (H) n times to the right member of (I). By (E),

$$D^n H_n(x) = 2^n n!,$$

so the result is

$$\text{(K)} \qquad \langle H_n, H_n \rangle_{\mathrm{H}} = (-1)^{2n} \int_R (2^n n!) \exp(-x^2)\, dx.$$

In equation (LL) of Section IV-5 we showed that

$$\int_R \exp\left(-\frac{u^2}{2}\right) du = (2\pi)^{1/2}.$$

If we make the substitution $u = 2^{1/2} x$, we obtain

$$\text{(L)} \qquad \int_R \exp(-x^2)\, dx = \pi^{1/2}.$$

By substituting this in (K), we obtain

$$\text{(M)} \qquad \langle H_n, H_n \rangle_{\mathrm{H}} = 2^n n! \pi^{1/2} \qquad (n = 0, 1, 2, \ldots).$$

The equation

$$\text{(N)} \qquad D^{n+1} \exp(-x^2) + 2x\, D^n \exp(-x^2) + 2n\, D^{n-1} \exp(-x^2) = 0$$

holds when $n = 1$, as is easily verified. If it holds for a positive integer n, by differentiation

$$D^{n+2} \exp(-x^2) + 2x\, D^{n+1} \exp(-x^2) + 2(n + 1)\, D^n \exp(-x^2) = 0,$$

which is (N) with $n + 1$ in place of n. By induction, (N) holds for all positive integers n. If we multiply both members of (N) by $(-1)^{n+1} \exp x^2$, we obtain

$$\text{(O)} \qquad H_{n+1}(x) - 2x H_n(x) + 2n H_{n-1}(x) = 0.$$

Starting with $H_0 = 1$ and $H_1 = 2x$, we can quickly compute successive $H_n(x)$ by means of this recursion formula.

If we substitute the value of H_{n+1} from (O) in (F), we obtain the neat differential equation

(P) $$DH_n(x) = 2nH_{n-1}(x).$$

The **Hermite functions** $h_0, h_1, h_2, \ldots$ are defined by

(Q) $$h_n(x) = \exp(-x^2/2)H_n(x).$$

All the preceding statements about the Hermite polynomials can be transformed into statements about the Hermite functions. By (A), the space $\mathscr{L}_2[m_{\mathrm{H}}, R]$ consists of all functions F on R such that the function f defined by

$$f(x) = F(x)\exp(-x^2/2)$$

belongs to $\mathscr{L}_2[m_{\mathrm{L}}, R]$. We shall denote the inner product of functions in $\mathscr{L}_2[m_{\mathrm{L}}, R]$ by $\langle \cdot, \cdot \rangle_{\mathrm{L}}$. Then if F and G are in $\mathscr{L}_2[m_{\mathrm{H}}, R]$ and

$$f(x) = F(x)\exp(-x^2/2), \qquad g(x) = G(x)\exp(-x^2/2),$$

we have

$$\langle f, g \rangle_{\mathrm{L}} = \langle F, G \rangle_{\mathrm{H}}.$$

Likewise, the norm $\|F\|_{\mathrm{H}}$ of F in $\mathscr{L}_2[m_{\mathrm{H}}, R]$ is equal to the norm $\|f\|_{\mathrm{L}}$ of f in $\mathscr{L}_2[m_{\mathrm{L}}, R]$. In particular, by (M),

(R) $$\langle h_n, h_n \rangle_{\mathrm{L}} = \langle H_n, H_n \rangle_{\mathrm{H}} = 2^n n! \pi^{1/2}.$$

As we have seen, if f is in $\mathscr{L}_2[m_{\mathrm{L}}, R]$, the function defined by $F(x) = f(x)\exp(x^2/2)$ is in $\mathscr{L}_2[m_{\mathrm{H}}, R]$. Since the Hermite polynomials form a complete orthogonal set in $\mathscr{L}_2[m_{\mathrm{H}}, R]$, for each positive ε there is a linear combination $c_0H_0 + \cdots + c_kH_k$ such that

$$\|F - [c_0H_0 + \cdots + c_kH_k]\|_{\mathrm{H}} < \varepsilon.$$

This is the same as

$$\int_R |F(x) - [c_0H_0(x) + \cdots + c_kH_k(x)]|^2 \, m_{\mathrm{H}}(dx) < \varepsilon^2,$$

which by (A) is equivalent to

$$\begin{aligned}\varepsilon^2 &> \int_R |F(x) - [c_0H_0(x) + \cdots + c_kH_k(x)|^2 \exp(-x^2)\,dx \\ &= \int_R |f(x) - [c_0h_0(x) + \cdots + c_kh_k(x)|^2 \,dx.\end{aligned}$$

So the linear combinations of the h_n are dense in $\mathscr{L}_2[m_{\mathrm{L}}, R]$, and they form a complete orthogonal sequence in $\mathscr{L}_2[m_{\mathrm{L}}, R]$. Consequently, every f in $\mathscr{L}_2[m_{\mathrm{L}}, R]$ can be expanded in a series of multiples of the h_n, the series converging to f in the norm of $\mathscr{L}_2[m_L, R]$. By (Q) and (F),

$$\text{(S)}\qquad \begin{aligned} Dh_n(x) &= -x[\exp(-x^2/2)]H_n(x) + [\exp(-x^2/2)]\,DH_n(x) \\ &= -xh_n(x) + 2xh_n(x) - h_{n+1}(x) \\ &= xh_n(x) - h_{n+1}(x). \end{aligned}$$

If we combine equations (F) and (P), we obtain

$$\begin{aligned} D^2H_n(x) &= D[DH_n(x)] \\ &= D[2xH_n(x) - H_{n+1}(x)] \\ &= 2H_n(x) + 2x\,DH_n(x) - DH_{n+1}(x) \\ &= 2H_n(x) + 2x\,DH_n(x) - 2(n+1)H_n(x) \\ &= 2x\,DH_n(x) - 2nH_n(x). \end{aligned}$$

In this we substitute $\exp(x^2/2)h_n(x)$ for $H_n(x)$; after some elementary computation and simplification, we find

$$\text{(T)}\qquad D^2h_n(x) + [2n + 1 - x^2]h_n(x) = 0 \qquad (x \text{ in } R;\ n = 0, 1, 2, 3, \ldots).$$

EXERCISE 10-1 Show that for each real x the function f defined on R by

$$y \mapsto f(x, y) = \exp(2xy - y^2)$$

has the power-series expansion

$$f(x, y) = \sum_{n=0}^{\infty} \frac{H_n(x)}{n!} y^n.$$

Suggestion: $f(x, y) = (\exp x^2)\phi(x - y)$, where $\phi(z) = \exp(-z^2)$. By expansion in power series (Taylor's series),

$$\begin{aligned} \phi(x - y) = \phi(x) + (-y)\,D\phi(x) + [(-y)^2/2!]\,D^2\phi(x) + \cdots \\ + [(-y)^n/n!]\,D^n\phi(x) + \cdots. \end{aligned}$$

Compare with (C).

11. The Schrödinger Equation for the Harmonic Oscillator

In this section we shall use theorems about expansions in orthogonal functions to solve one of the simplest equations of quantum mechanics, the Schrödinger equation for the harmonic oscillator. In doing this, we shall show the rationale of the procedure known as *separation of the variables*.

In quantum mechanics the state of a system (an atom or molecule or ion) at a given time t is expressed by means of a *wave function* or *state function* $\phi(t)$ that belongs to a space $\mathscr{L}_2[m_L, R^r]$ and has unit length in that space. That is, ϕ is a complex-valued function

$$(x, t) \mapsto \phi(x, t) \qquad (x \text{ in } R^r;\ -\infty < t < \infty)$$

that for each t is an m_L-measurable function on R^r and has

$$\int_{R^r} |\phi(x, t)|^2\, dx = 1.$$

The *Schrödinger equation* in quantum mechanics asserts that the time derivative of the state-function, regarded as a moving point in $\mathscr{L}_2[m_L, R^r]$, is equal to a certain function in $\mathscr{L}_2[m_L, R^r]$ that is determined by x, ϕ, and the partial derivatives of ϕ with respect to the x^i. For the harmonic oscillator, $r = 1$, and if we choose the units of length and time properly, the Schrödinger equation is

$$\text{(A)} \qquad i\{(L_2)\partial\phi(x, t)/\partial t\} = (L_2)\, D^2\phi(x, t) - x^2\phi(x, t),$$

where D denotes differentiation with respect to x.

We are not going to derive this, or to explain the physics that leads to it. All we plan to do is to solve it.

If $g_0, g_1, g_2, \ldots$ is any orthogonal sequence complete in $\mathscr{L}_2[m_L, R]$, any solution ϕ of (A) can be expanded at each time t in a series

$$\text{(B)} \qquad \phi(x, t) = \sum_{j=0}^{\infty} c_j(t) g_j(x),$$

the right member converging in L_2-norm. Unless the g_j have some useful peculiarity, this is not much help. But it is a reasonable conjecture, verified by experience, that the series expansion (B) will be easier to work with if each individual term in the series is a solution of (A). Suppose, then, that $c_j(t)g_j(x)$ satisfies (A). The left member of (A) is then

$$i(L_2) \lim_{h \to 0} [c_j(t + h)g_j(x) - c_j(t)g_j(x)]/h,$$

which is the same as

$$i[dc_j(t)/dt]g_j(x).$$

So (A) takes the form

$$\text{(C)} \qquad i[dc_j(t)/dt]g_j(x) = c_j(t)[(L_2)\, D^2 g_j(x) - x^2 g_j(x)].$$

The ratio $-i[dc_j(t)/dt]/c_j(t)$ depends on t only and not on x. For the moment we call it C. Then (C) takes the form

$$\text{(D)} \qquad (L_2)\, D^2 g_j(x) + [C - x^2]g_j(x) = 0.$$

By (T) of the preceding section, the Hermite function h_n satisfies the equation

(E) $$D^2h_n(x) + (2n + 1 - x^2)h_n(x) = 0 \qquad (-\infty < x < \infty).$$

In this, the D^2 is the ordinary second derivative. But the equation

$$Dh_n(x) = \lim_{u \to 0} [h_n(x + u) - h_n(x)]/u$$

remains valid if the limit is understood to be in L_2-norm, because of the factor $\exp(-x^2/2)$ in h_n; and likewise for the second derivative. That is, h_n satisfies (D) with $2n + 1$ in place of C. Therefore we choose the Hermite functions for the orthogonal functions g_j in which we expand ϕ, and (B) becomes

(F) $$\phi(x, t) = \sum_{j=0}^{\infty} c_j(t)h_j(x).$$

We simplify notation by dropping the subscript L from the symbol for the inner product in $\mathscr{L}_2[m_L, R]$; we denote it simply by $\langle \cdot, \cdot \rangle$. We are seeking only solutions of the Schrödinger equation in which for each t all the functions

$$x \mapsto D\phi(x, t), \qquad x \mapsto D^2\phi(x, t), \qquad x \mapsto x^2\phi(x, t)$$

belong to $\mathscr{L}_2[m, R]$, and the first two are the same as the L_2-derivatives $(L_2)\, D\phi$, $(L_2)\, D^2\phi$.

If $A < 0 < B$, by integration by parts we obtain

(G) $$\int_A^B [D^2\phi(x, t)]h_j(x)\, dx = [D\phi(B, t)]h_j(B) - [D\phi(A, t)]h_j(A)$$
$$- \int_A^B [D\phi(x, t)][Dh_j(x)]\, dx.$$

Since $D\phi(x, t)$ belongs to $\mathscr{L}_2[m_L, R]$, it cannot tend to ∞ in absolute value as x tends to ∞. Therefore we can select positive numbers $B(1), B(2), B(3), \ldots$ tending to ∞ such that the numbers $D\phi(B(n), t)$ remain bounded as n increases. Then the product $D\phi(B(n), t)h_j(B(n))$ tends to 0 as n increases. Likewise, there are negative numbers $A(1), A(2), A(3), \ldots$ tending to $-\infty$ such that $D\phi(A(n), t)h_j(A(n))$ tends to 0 as n increases. We substitute $A(n)$ for A and $B(n)$ for B in (G) and let n increase; by (G),

(H) $$\int_{-\infty}^{\infty} [D^2\phi(x, t)]h_j(x)\, dx = -\int_{\infty}^{\infty} [D\phi(x, t)][Dh_j(x)]\, dx.$$

We repeat the argument, integrating by parts a second time. The result is

(I) $$\int_{-\infty}^{\infty} [D^2\phi(x, t)]h_j(x)\, dx = \int_{-\infty}^{\infty} \phi(x, t)\, D^2h_j(x)\, dx.$$

Let us define $H^*\phi(t)$ to be the member of $\mathscr{L}_2[m_L, R]$ defined by

(J) $$H^*\phi(t) \colon x \mapsto D^2\phi(x, t) - x^2\phi(x, t).$$

Then, since h_j is real-valued, (I) and (E) imply

$$\text{(K)} \qquad \begin{aligned} \langle H^*\phi(t), h_j\rangle &= \int_{-\infty}^{\infty} [D^2\phi(x,t) - x^2\phi(x,t)]h_j(x)\,dx \\ &= \int_{-\infty}^{\infty} \phi(x,t)[D^2h_j(x) - x^2h_j(x)]\,dx \\ &= \int_{-\infty}^{\infty} \phi(x,t)(-2j-1)h_j(x)\,dx \\ &= -(2j+1)\langle\phi(\cdot,t), h_j\rangle. \end{aligned}$$

If we take the inner product of both members of (A) with h_j and use the notation (J), we obtain from (K)

$$\text{(L)} \qquad \langle i(L_2)\,\partial\phi/\partial t, h_j\rangle = \langle H^*\phi(t), h_j\rangle = -(2j+1)\langle\phi(\cdot,t), h_j\rangle.$$

The left member of (A) is the limit in L_2-norm of the difference quotient

$$\text{(M)} \qquad i[\phi(\cdot, t+y) - \phi(\cdot,t)]/y$$

as y tends to 0, where again $\phi(\cdot,t)$ means the function $x \mapsto \phi(x,t)$. By the continuity of the inner product, the convergence in L_2-norm of quotient (M) to $i(L_2)\,\partial\phi/\partial t$ implies

$$\text{(N)} \qquad \lim_{y\to 0} \langle i[\phi(\cdot,t+y) - \phi(\cdot,t)]/y, h_j\rangle = \langle i(L_2)\,\partial\phi/\partial t, h_j\rangle.$$

The left member of this equation is

$$\lim_{y\to 0} \{\langle i\phi(\cdot,t+y), h_j\rangle - \langle i\phi(\cdot,t), h_j\rangle\}/y,$$

which by definition is

$$\text{(O)} \qquad i\,d\langle\phi(\cdot,t), h_j\rangle/dt.$$

By (A), the right member of (N) is $\langle H^*\phi(t), h_j\rangle$. So by (K) and (O),

$$i\,d\langle\phi(\cdot,t), h_j\rangle/dt = -(2j+1)\langle\phi(\cdot,t), h_j\rangle.$$

This equation is easy to solve; its solution is

$$\langle\phi(\cdot,t), h_j\rangle = \exp[(2j+1)it]\langle\phi(\cdot,0), h_j\rangle.$$

Therefore, the series whose sum is the solution of (A) is (recalling (R) of the preceding section)

$$\begin{aligned} \phi(x,t) &= \sum_{j=0}^{\infty} \frac{\langle\phi(\cdot,t), h_j\rangle}{\langle h_j, h_j\rangle} h_j(x) \\ &= \sum_{j=0}^{\infty} \frac{\langle\phi(\cdot,0), h_j\rangle}{n!2^n\pi^{1/2}} \exp[(2j+1)it]h_j(x). \end{aligned}$$

Given the initial value of the state-function ϕ at time $t = 0$, this expansion furnishes $\phi(x, t)$ at all other t.

12. The Fourier Transform for Certain Smooth Functions

In previous sections we discussed several kinds of expansions of functions into series of orthogonal functions. Among these, the Fourier series was of especial importance for several reasons. For instance, there are many systems, such as electric circuits, whose response to a sine-wave input is easily calculated, and the response to any other input representable as the sum of a series of sine-wave inputs may then be computed by adding the responses to the several sine-wave components of the input. But the Fourier series is limited to periodic functions, or to functions on a bounded interval $(a, b]$ (which can be made into periodic functions by repeating the values of the function with period $b - a$). This suggests that it might be highly desirable to find a way of expressing nonperiodic functions on the whole real axis R as some combination of pure sine waves or, equivalently, as combinations of complex exponentials. In (O) of Section III-12 we already had an indication of the kind of combination we could use. There we saw that the sum of an infinite series can be written as the value of an integral with respect to a certain measure that assigned nonzero measure to some integers and zero measure to all the rest of R. This suggests that the representation that we are looking for might have the form of the integral of a pure complex exponential $x \mapsto \exp(iyx)$ with respect to some more general kind of measure. This is, in fact, the kind of representation that we shall arrive at. However, we shall not reach for the greatest possible generality; we shall stop with measures that have densities.

In order to find the right way to express functions as combinations of complex exponentials, we shall start with functions of a rather simple type, namely, the functions of class C^2 that have bounded support. For these we can deduce the integral representation on all of R by a fairly simple passage to the limit from their expansions as Fourier series. This representation of C^2 functions with bounded support is much more than a device for finding what form the expansion should have; it is the germ from which the general theory grows.

In this and the following sections we shall often encounter functions defined by some formula. Heretofore such a function has been denoted by a symbol $x \mapsto f(x)$, where $f(x)$ is some quantity computable somehow from x. This is somewhat unwieldy, and we shall introduce an alternative symbol. The function $x \mapsto f(x)$ will be denoted by $f(\mathbf{x})$; that is, we omit the "$x \mapsto$" and change the letter x to boldface type. For example, the sine function has been expressed by $x \mapsto \sin x$. We shall also represent it by $\sin \mathbf{x}$. This use of boldface letters should not cause any confusion with their previous use to denote vectors because in the remainder of this chapter we shall not use boldface type for vectors.

Suppose, then, that f is a function of class C^2 on R that vanishes outside a bounded interval B. Let T be any integer greater than 4 that is large enough so that the interval $[-T\pi, T\pi]$ contains B. We could expand the function f on the interval

$$[a, b] = [-T\pi, T\pi]$$

in a Fourier series as in Section 7, but we prefer to make one small notational change. Instead of the functions

$$v_n(x) = \exp(inx/T) \qquad (n = 0, \pm 1, \pm 2, \ldots)$$

on $[a, b]$ that we used in Section 7, we shall use the functions

(A) $$u_n = v_n/\|v_n\|.$$

Then

(B) $$\|u_n\| = 1$$

for all integers n. By (M) of Section 7, (A) implies

(C) $$u_n(x) = (2\pi T)^{-1/2} \exp(inx/T).$$

Since the u_n are merely positive multiples of the v_n, they too form a complete orthogonal sequence of functions. The function f is the indefinite integral of its own derivative, and it vanishes at a and at b. So by Theorem 8-2, with (B), the equation

(D) $$f(x) = \sum_{j=-\infty}^{\infty} \langle f, u_n \rangle u_n(x)$$

is valid, the series in the right member converging uniformly on $[-T\pi, T\pi]$ to $f(x)$.

We now introduce two new symbols whose only justification at this moment is that they provide a convenient way of rewriting equation (D). Later we shall see that they have much more importance than that.

DEFINITION 12-1 *Let f be a function integrable over R. Then the* ***Fourier transform*** *of f is the function $\hat{f}$ (also called $f^\wedge$) defined by*

$$\hat{f}(\mathbf{y}) = (2\pi)^{-1/2} \int_R f(x) \exp(-i\mathbf{y}x)\, dx,$$

and the ***conjugate Fourier transform*** *of f is the function $\check{f}$ (also called $f^\vee$) defined by*

$$\check{f}(\mathbf{x}) = (2\pi)^{-1/2} \int_R f(y) \exp(i\mathbf{x}y)\, dy.$$

Since f is integrable over R and $\exp(-iyx)$ is continuous and bounded for each y in R, the integrals that define $\hat{f}$ and $\check{f}$ surely exist.

The use of the letters x and y in Definition 12-1 accords more or less with custom. It really would make no mathematical difference if we replaced x by any other letter and **y** by any other boldface letter in the definition of $\hat{f}$ and similarly in the definition of $\check{f}$. But usually, in discussing Fourier transforms and conjugate Fourier transforms, one letter is preferred to express the independent variable in the function and another to express the independent variable in the transform. For example, if the function is $x \mapsto f(x)$, its Fourier transform is $y \mapsto \hat{f}(y)$. This is not mere whimsy. When Fourier transforms are used in physical applications, the independent variable in f will have one interpretation, such as time, and the independent variable in $\hat{f}$ will have another, such as frequency.

For the particular function f under consideration, f vanishes outside $[-T\pi, T\pi]$, so

$$\text{(E)} \qquad \hat{f}(y) = (2\pi)^{-1/2} \int_{-T\pi}^{T\pi} f(x) \exp(-iyx)\, dx.$$

By Theorem II-11-1, $\hat{f}$ is continuous on R, and so are its derivatives of all orders. With this notation, recalling (C),

$$\text{(F)} \qquad \langle f, u_n \rangle = (2\pi T)^{-1/2} \int_{-T\pi}^{T\pi} f(x) \exp(-inx/T)\, dx = T^{-1/2} \hat{f}(n/T).$$

Now (D) can be rewritten as

$$\text{(G)} \quad f(x) = \sum_{j=-\infty}^{\infty} T^{-1/2} \hat{f}\left(\frac{n}{T}\right) u_n(x) = (2\pi)^{-1/2} \sum_{j=-\infty}^{\infty} \hat{f}\left(\frac{n}{T}\right) \exp\left(\frac{inx}{T}\right) T^{-1}.$$

We also define

$$\text{(H)} \qquad y_n = n/T \qquad (n = 0, \pm 1, \pm 2, \ldots).$$

Then (G) implies that for all x in $[-T\pi, T\pi]$

$$\text{(I)} \qquad f(x) = (2\pi)^{-1/2} \sum_{j=-\infty}^{\infty} \hat{f}(y_n) \exp(iy_n x)[y_{n+1} - y_n].$$

The sum in the right member of (I) resembles a Riemann sum for the integral of $\hat{f}(y)\exp(iyx)$, and it is reasonable to conjecture that it will converge to that integral as T increases and the intervals $(y_n, y_{n+1}]$ become shorter. To prove that this is actually true, we need an estimate for $|\hat{f}(y)|$.

Since f and its derivatives vanish at $T\pi$ and at $-T\pi$, by integrating by parts twice we deduce from (E) that for $y \neq 0$

$$\begin{aligned} \hat{f}(y) &= (2\pi)^{-1/2} \frac{1}{iy} \int_{-T\pi}^{T\pi} Df(x) \exp(-iyx)\, dx \\ &= (2\pi)^{-1/2} \left(-\frac{1}{y^2}\right) \int_{-T\pi}^{T\pi} D^2 f(x) \exp(-iyx)\, dx. \end{aligned}$$

So for $y \neq 0$

$$\text{(J)} \qquad |\hat{f}(y)| \leqq \left\{(2\pi)^{-1/2} \int_{-T\pi}^{T\pi} |D^2 f(x)|\, dx\right\} y^{-2}.$$

Clearly, from (E),

$$\text{(K)} \qquad |\hat{f}(y)| \leqq (2\pi)^{-1/2} \int_{-T\pi}^{T\pi} |f(x)|\, dx.$$

So if we define

$$C = \frac{1}{2\pi} \max\left\{\int_{-T\pi}^{T\pi} |D^2 f(x)|\, dx, \int_{-T\pi}^{T\pi} |f(x)|\, dx\right\},$$

and define the function M on $(-\infty, \infty)$ by

$$M(y) = \begin{cases} Cy^{-2} & (|y| \geqq 1), \\ C & (|y| < 1), \end{cases}$$

the inequality

$$\text{(L)} \qquad (2\pi)^{-1/2} |\hat{f}(y)| \leqq M(y)$$

holds for all real y, by (J) if $|y| \geqq 1$ and by (K) if $|y| < 1$.

For each x in $[-T\pi, T\pi]$, let $F_{T,x}$ be the step-function that on each interval $(y_n, y_{n+1}]$ has the constant value

$$\text{(M)} \qquad F_{T,x}(y) = (2\pi)^{-1/2} \hat{f}(y_n) \exp(iy_n x).$$

Then (I) implies that

$$\text{(N)} \qquad f(x) = \int_{-\infty}^{\infty} F_{T,x}(y)\, dy \qquad (-T\pi \leqq x \leqq T\pi).$$

Recall that T is an integer greater than 4. Let y be any real number; for some n, $y_n < y \leqq y_{n+1}$. If $n \geqq T$, the interval $(y_n, y_{n+1}]$ is contained in $(1, \infty)$, on which M coincides with the decreasing function $y \mapsto Cy^{-2}$. So by (M) and (L),

$$\begin{aligned} |F_{T,x}(y)| &\leqq M(y_n) \\ &= Cy_n^{-2} \\ &= CT^2/n^2 \\ &= [CT^2/(n+1)^2](1 + 1/n)^2 \\ &\leqq [Cy_{n+1}^{-2}](1 + 1/4)^2 \\ &< 2M(y_{n+1}) \\ &\leq 2M(y). \end{aligned}$$

If $n < T$, the whole interval (y_n, y_{n+1}) lies in $(-\infty, 1]$, on which M is nondecreasing, so by (M) and (L)

$$|F_{T,x}(y)| \leqq M(y_n) \leqq M(y).$$

This and the preceding inequality imply that

$$\text{(O)} \qquad |F_{T,x}(y)| \leqq 2M(y)$$

holds for all y.

For each y in R, the y_n in the right member of (M) and the y in the left member are both in the interval $[y_n, y_{n+1}]$, which has length $1/T$. So, by the continuity of $\hat{f}$,

$$\text{(P)} \qquad \lim_{T \to \infty} F_{T,x}(y) = (2\pi)^{-1/2} \hat{f}(y) \exp(iyx).$$

The function M is integrable over R, so by (O), (P), and the dominated convergence theorem,

$$\lim_{T \to \infty} \int_R F_{T,x}(y)\, dy = (2\pi)^{-1/2} \int_R \hat{f}(y) \exp(iyx)\, dy.$$

By (N), the integral in the left member of this equation has the value $f(x)$ for all T large enough so that x is in $[-T\pi, T\pi]$, so the limit of the integral is $f(x)$, and therefore

$$\text{(Q)} \qquad f(x) = (2\pi)^{-1/2} \int_R \hat{f}(y) \exp(iyx)\, dy.$$

This has the advantage over (G) that it is valid for all x in R, whereas (G) furnishes a representation of $f(x)$ only for x in $[-T\pi, T\pi]$. For all functions f that are of class C^2 and have bounded support, (Q) provides the representation of f as a combination of complex exponentials. The individual exponentials $\exp(iyx)$, instead of being multiplied by Fourier coefficients and added as in the case of Fourier series, are multiplied by a factor $(2\pi)^{-1/2}\hat{f}(y)$ and combined by integration, the coefficient function $\hat{f}$ being the Fourier transform of f defined in Definition 12-1. This is the form of analysis of f into pure sine waves, or complex exponentials, that we have been seeking. But the class of functions to which it applies is as yet much too small. We summarize what we have proved, and state some facts yet to be proved, in the following theorem.

THEOREM 12-2 *Let f be a complex-valued function on R that is of class C^2 and has bounded support. Then its Fourier transform $\hat{f}$ and its conjugate Fourier transform $\check{f}$ are of class C^∞, are bounded, and belong both to $\mathscr{L}_1$ and to $\mathscr{L}_2$. The mappings $f \mapsto \hat{f}$ and $f \mapsto \check{f}$ are linear on the class of all functions of class C^2 with bounded support, and for every such function f*

$$\text{(R)} \qquad (\hat{f})\check{} = f \quad \textit{and} \quad (\check{f})\hat{} = f,$$

and

(S) $$\|\hat{f}\|_2 = \|\check{f}\|_2 = \|f\|_2 .$$

By Theorem II-11-1, $\hat{f}$ has continuous derivatives of all orders. For all integrable functions f the equations

(T) $$\check{f}(y) = [\bar{f}\,\hat{}\,(y)]^-, \qquad \hat{f}(y) = [\bar{f}\,\check{}\,(y)]^-$$

are obvious consequences of Definition 12-1. So if f is of class C^2 with bounded support, so is $\bar{f}$; $(\bar{f})\hat{}$ is of class C^∞, as has already been proved; and so is $[(\bar{f})\hat{}\,]^-$, which by (T) is $\check{f}$. We have seen that $\hat{f}$ is bounded; by (T), so is $\check{f}$. We have proved in (Q) the first of equations (R). We apply it to $\bar{f}$ and use (T); the result is

$$f^{--\vee--\wedge-} = f^-,$$

or

$$f^{\vee\wedge-} = f^-,$$

which implies the second of equations (R).

The linearity of the mapping $f \mapsto \hat{f}$ is evident; if f_1 and f_2 are of class C^2 and have bounded supports, and c_1 and c_2 are complex numbers, by Definition 12-1

$$\begin{aligned}(c_1 f_1 + c_2 f_2)\hat{}\,(y) &= (2\pi)^{-1/2} \int_R (c_1 f_1(x) + c_2 f_2(x)) \exp(-iyx)\, dx \\ &= (2\pi)^{-1/2} c_1 \int_R f_1(x) \exp(-iyx)\, dx \\ &\quad + (2\pi)^{-1/2} c_2 \int_R f_2(x) \exp(-iyx)\, dx \\ &= c_1 \hat{f}_1(y) + c_2 \hat{f}_2(y).\end{aligned}$$

To prove that the mapping leaves L_2-norms unchanged, we note that by Parseval's theorem (Theorem 5-13), with (F) and (H), for each f of class C^2 with bounded support

(U) $$\|f\|_2^2 = \sum_{j=-\infty}^{\infty} |\langle f, u_j \rangle|^2 = \sum_{j=-\infty}^{\infty} |\hat{f}(y_n)|^2 (y_{n+1} - y_n).$$

By definition (M), the equation

$$2\pi |F_{T,x}(y)|^2 = |\hat{f}(y_n)|^2$$

holds on $(y_n, y_{n+1}]$, so (U) yields

(V) $$\|f\|_2^2 = 2\pi \int_R |F_{T,x}(y)|^2\, dy.$$

By (O), for all large T

$$|F_{T,x}(y)|^2 \leqq 4M(y)^2,$$

the right member being integrable. This and (P) and (V), with the dominated convergence theorem, imply

$$\|f\|_2^2 = \int_R |\hat{f}(y)|^2\,dy = \|\hat{f}\|_2^2,$$

which implies one of equations (S). For the other, by (T) and the first of (S),

$$\|\check{f}\|_2 = \|f^{-\wedge-}\|_2 = \|f^{-\wedge}\|_2 = \|f^{-}\|_2 = \|f\|.$$

This completes the proof.

EXERCISE 12-1 Prove that if f is integrable and even on R, then $\hat{f}$ and $\check{f}$ are even, and

$$\hat{f}(y) = \check{f}(y) = \left(\frac{2}{\pi}\right)^{1/2} \int_0^\infty f(x)(\cos yx)\,dx;$$

and if f is integrable and odd, then $\hat{f}$ and $\check{f}$ are odd, and

$$\hat{f}(y) = -\check{f}(y) = -\left(\frac{2}{\pi}\right)^{1/2} \int_0^\infty f(x)(\sin yx)\,dx.$$

EXERCISE 12-2 Prove that if f is integrable and c is real, and $g(x) = \exp(icx)f(x)$ for all x in R, then for all y

$$\hat{g}(y) = \hat{f}(y - c), \qquad \check{g}(y) = \check{f}(y + c).$$

EXERCISE 12-3 Prove that if f is integrable over R, and a and b are real numbers with $a > 0$, and g is the function defined by

$$g(x) = f(ax + b) \qquad (x \text{ in } R),$$

then

$$\hat{g}(y) = (1/a)\exp(iyb/a)\hat{f}(y/a).$$

(In the integral defining $\hat{g}(y)$, make the substitution $u = ax + b$.)

EXERCISE 12-4 Prove that if f is integrable and $g(x) = f(-x)$ for all x, then

$$\hat{g}(y) = \check{f}(y) \qquad \text{and} \qquad \check{g}(y) = \hat{f}(y)$$

for all real y, and also

$$\hat{g}(y) = \hat{f}(-y) \qquad \text{and} \qquad \check{g}(y) = \check{f}(-y).$$

EXERCISE 12-5 Prove that if f is of class C^2 and has bounded support,

$$f^{\wedge\wedge}(x) = f^{\vee\vee}(x) = f(-x)$$

for all x in R. (Use Exercise 12-4 and Theorem 12-2.)

EXERCISE 12-6 Let f be the indicator of $[-a, a]$. Show that

$$\hat{f}(y) = (2/\pi)^{1/2} y^{-1} \sin ya \qquad (y \neq 0),$$

$$\hat{f}(0) = (2/\pi)^{1/2} a.$$

EXERCISE 12-7 From Exercises 12-6 and 12-3, compute $\hat{f}$ when $f = 1_{(a,b]}$.

EXERCISE 12-8 Let f be defined by

$$f(x) = (1 - |x|/2a)^+,$$

where $a > 0$. Compute $\hat{f}$.

13. The Fourier–Plancherel Transformation

The class of functions to which Theorem 12-2 applies is too small to be of much use. We could extend the mapping $f \mapsto \hat{f}$ to the space $\mathscr{L}_2$, but there are advantages in regarding the mapping as one defined on the space $L_2[R]$ of equivalence-classes of functions belonging to $\mathscr{L}_2$ and transforming the members of $L_2[R]$ into members of $L_2[R]$. This mapping we shall call the **Fourier–Plancherel** transformation. In making this distinction we are insisting on more precision in language than is usual. Experienced mathematicians usually speak of "functions belonging to $L_2[R]$," although in fact only equivalence-classes of functions can belong to $L_2[R]$, and they use the same name, "Fourier transform," for the transform $\hat{f}$ defined in Definition 12-1 and for the Fourier–Plancherel transform we are about to define. This is the kind of laxity of language that is harmless and convenient among experts. The speaker and the hearer both know that the language is inexact, and both can without effort replace the abbreviated and inexact statement by the more precise one that it stands for. But until one has become adept in the use of spaces $\mathscr{L}_p$ and L_p and of Fourier and Fourier–Plancherel transforms, the cautious procedure is to use the full and accurate statement instead of the abbreviation. We shall do so. If and when the reader feels sufficiently at home in the subject, he should feel free to abandon our somewhat extreme caution.

As in Section 12, the only measure that we shall use in this section is m_{L}, and the only space is R, so we shall write $\mathscr{L}_p$ and L_p in place of $\mathscr{L}_p[m_{\mathrm{L}}, R]$ and $L_p[m_{\mathrm{L}}, R]$, respectively.

The definition of the Fourier transform $\hat{f}$ in Definition 12-1 cannot be applied to arbitrary functions f in $\mathscr{L}_2$ because such functions may not be integrable over R, so that the integral in the definition may fail to exist. But we can extend the idea of the Fourier transform to such functions by an easy limit process, starting with the functions that we learned to handle in Section 12.

LEMMA 13-1 *Let $[f]$ belong to L_2. Then there exist unique members $[g']$, $[g'']$ of L_2 such that whenever $f_1, f_2, f_3, \ldots$ is a sequence of functions, each of class C^2 and with bounded support, that converges in L_2-norm to a function in the class $[f]$, the sequence $[\hat{f}_1], [\hat{f}_2], [\hat{f}_3], \ldots$ of members of L_2 converges in L_2-norm to $[g']$ and the sequence $[\check{f}_1], [\check{f}_2], [\check{f}_3], \ldots$ converges in L_2-norm to $[g'']$.*

We shall discuss only the sequence $[\hat{f}_j]$; the sequence $[\check{f}_j]$ can be treated in just the same way.

Let $f_1, f_2, f_3, \ldots$ be a sequence of functions, each of class C^2 and with bounded support, that converges in L_2-norm to some function f in the class $[f]$. Then it is a Cauchy sequence in $\mathscr{L}_2$. By Theorem 12-2, for each pair of positive integers i, j

$$\|\hat{f}_i - \hat{f}_j\|_2 = \|f_i - f_j\|_2,$$

so the sequence $\hat{f}_1, \hat{f}_2, \hat{f}_3, \ldots$ is also a Cauchy sequence in $\mathscr{L}_2$. By Theorem 4-3, there exists a function g' in $\mathscr{L}_2$ such that $\hat{f}_j$ converges in L_2-norm to g'. Then the sequence $[\hat{f}_1], [\hat{f}_2], [\hat{f}_3], \ldots$ converges in L_2 to $[g']$.

Let $h_1, h_2, h_3, \ldots$ be another sequence of functions, each of class C^2 and with bounded support, that converges in L_2-norm to some function f' that belongs to $[f]$. Then

$$\lim_{n \to \infty} \|h_n - f_n\|_2 = \|f - f'\|_2 = 0.$$

By Theorem 12-2,

$$\|\hat{h}_n - \hat{f}_n\|_2 = \|h_n - f_n\|_2,$$

so

$$\lim_{n \to \infty} \|\hat{h}_n - \hat{g}_n\|_2 = 0.$$

We have already shown that

$$\lim_{n \to \infty} \|\hat{f}_n - g'\|_2 = 0,$$

and by the triangle inequality,

$$\|\hat{h}_n - g'\|_2 \leqq \|\hat{h}_n - \hat{f}_n\|_2 + \|\hat{f}_n - g'\|_2,$$

so

$$\lim_{n \to \infty} \|\hat{h}_n - g'\|_2 = 0.$$

So h_n converges to g' in L_2-norm in the space $\mathscr{L}_2$, and therefore $[h_n]$ converges to $[g']$ in the space L_2, and the limit $[g']$ is uniquely determined.

By Corollary 6-3, for every member $[f]$ of L_2 there exist sequences $f_1, f_2, f_3, \ldots$ of functions of class C^2 with bounded support that converge in L_2-norm to a function f that belongs to $[f]$.

DEFINITION 13-2 *Let $[f]$ belong to L_2. Then the unique member $[g']$ of L_2 that is the limit of the sequence $[\hat{f}_1], [\hat{f}_2], [\hat{f}_3], \ldots$ whenever $f_1, f_2, f_3, \ldots$ is a sequence of functions of class C^2 with bounded support that converges in L_2-norm to some function f belonging to the class $[f]$ is called the* **Fourier–Plancherel transform** *of $[f]$ and is denoted by $[f]\hat{}$; and the unique member $[g'']$ of L_2 that is the limit of the sequence $[\check{f}_1], [\check{f}_2], [\check{f}_3], \ldots$ for such sequences $f_1, f_2, f_3, \ldots$ is called the* **conjugate Fourier–Plancherel transform** *of $[f]$ and is denoted by $[f]\check{}$.*

From the definition there follow several useful properties of the transform.

THEOREM 13-3 (i) *The Fourier–Plancherel transform and the conjugate Fourier–Plancherel transform are one-to-one linear length-preserving maps of L_2 onto (all of) L_2.*

(ii) *If f is in both $\mathscr{L}_1$ and $\mathscr{L}$, $f\hat{}$ is in $\mathscr{L}_2$, and*

$$[f]\hat{} = [\qquad \text{and} \qquad [f]\check{} = [\check{f}].$$

(iii) *If $[f]$ is in L_2,*

$$[f]\check{} = \{([f]^{-})\hat{}\}^{-}.$$

(iv) *The Fourier–Plancherel transformation and the conjugate Fourier–Plancherel transformation are inverses of each other; for all $[f]$ in L_2,*

$$([f]\hat{})\check{} = [f] \qquad \textit{and} \qquad ([f]\check{})\hat{} = [f].$$

Let $[f]$ and $[g]$ belong to L_2, and let a and b be complex numbers. We can and do choose sequences $f_1, f_2, f_3, \ldots, g_1, g_2, g_3, \ldots$ of functions of class C^2 with bounded support such that f_n converges to f and g_n converges to g in L_2-norm, where f is some member of $[f]$ and g is some member of $[g]$. Then $af_n + bg_n$ is of class C^2, has bounded support, and converges to $af + bg$ in L_2-norm. By definition,

$$\text{(A)} \qquad [af + bg]\hat{} = \lim_{n \to \infty} [(af_n + bg_n)\hat{}],$$

the limit (and all other limits) being in L_2-norm. By Theorem 12-2,

$$(af_n + bg_n)\hat{} = a\hat{f}_n + b\hat{g}_n,$$

so in L_2

$$[(af_n + bg_n)\hat{}] = a[\hat{f}_n] + b[\hat{g}_n].$$

We take the limits as n increases, applying Definition 13-2 to the right member and (A) to the left member; the result is

$$[af + bg]^{\wedge} = a[f]^{\wedge} + b[g]^{\wedge}.$$

So the Fourier–Plancherel transform is linear. Similarly, the conjugate transform is linear. With the same notation, we have by definition,

$$\lim_{n \to \infty} [f_n] = [f] \qquad \text{and} \qquad \lim_{n \to \infty} [\hat{f}] = [f]^{\wedge},$$

and by Theorem 12-2,

$$\| [\hat{f}_n] \|_2 = \| [f_n] \|_2$$

for all positive integers n, so

$$\| [f]^{\wedge} \|_2 = \| [f] \|_2,$$

and the Fourier–Plancherel transform is length preserving. Similarly, we prove that the conjugate Fourier–Plancherel transform is length preserving.

Suppose that f belongs both to $\mathscr{L}_1$ and to $\mathscr{L}_2$. By Corollary 6-3, we can and do choose a sequence $f_1, f_2, f_3, \ldots$ of functions, each of class C^2 and with bounded support, such that f_n converges to f both in L_1-norm and in L_2-norm. By Lemma 13-1, the Fourier transforms $\hat{f}_n$ converge in L_2-norm to a function g' in L_2, which by Definition 13-2 is a member of $[f]^{\wedge}$. By Definition 12-1, for each real number y

$$|\hat{f}(y) - \hat{f}_n(y)| = \left| (2\pi)^{-1/2} \int_R [f(x) - f_n(x)] \exp(-iyx)\, dx \right|$$
$$\leqq (2\pi)^{-1/2} \int_R |f(x) - f_n(x)|\, dx,$$

and the last integral tends to 0 because f_n tends to f in L_1-norm. So the functions $\hat{f}_n$ converge pointwise to $\hat{f}$ and converge in L_2-norm to g'. By Theorem 4-5, $f^{\wedge}$ and g' are equal almost everywhere, and therefore $f^{\wedge}$ belongs to the class $[g']$, which is $[f]^{\wedge}$ by Definition 13-2. That is,

$$[\hat{f}] = [f]^{\wedge}.$$

Similarly,

$$[\check{f}] = [f]^{\vee},$$

and conclusion (ii) is established.

Let $[f]$ belong to L_2, and let $f_1, f_2, f_3, \ldots$ be a sequence of functions of class C^2 with bounded support that converges in L_2-norm to some function f in the class $[f]$. We have seen in Section 12 that for each n, both f_n and $\hat{f}_n$ belong both to $\mathscr{L}_1$ and to $\mathscr{L}_2$. By conclusion (ii), for each n

(B) $$[(\hat{f}_n)^{\vee}] = [\hat{f}_n]^{\vee}.$$

The left member is $[f_n]$, by Theorem 12-2, so it converges in L_2-norm to $[f]$. By Definition 13-2, $\hat{f}_n$ converges in L_2-norm to a function g' that belongs to $[f]^{\wedge}$, so $[\hat{f}_n]$ converges in L_2 to $[f]^{\wedge}$. The conjugate Fourier–Plancherel transformation is continuous, being length preserving, so in L_2-norm

$$\lim_{n\to\infty} [\hat{f}_n]^{\vee} = ([f]^{\wedge})^{\vee}.$$

Now the left member of (B) converges in L_2-norm to $[f]$ and the right member converges in L_2-norm to $([f]^{\wedge})^{\vee}$, so these are equal. In a similar way,

$$[f] = ([f]^{\vee})^{\wedge},$$

and conclusion (iv) is established.

This implies that every $[f]$ in L_2 is the Fourier–Plancherel transform of a member g of L_2, namely, of $g = [f]^{\vee}$, so the Fourier–Plancherel transformation maps L_2 onto all of L_2. A similar proof applies to the conjugate Fourier–Plancherel transform, and the proof of (i) is complete.

If $[f]$ is in L_2, and $f_1, f_2, f_3, \ldots$ is a sequence of functions of class C^2 with bounded support that converges in L_2-norm to a function f belonging to $[f]$, then in L_2 we see that $[f_n]$ converges to $[f]$, $[f_n]^{-}$ converges to $[f]^{-}$, $([f_n]^{-})^{\wedge}$ converges to $([f]^{-})^{\wedge}$, and $\{([f_n]^{-})^{\wedge}\}^{-}$ converges to $\{([f]^{-})^{\wedge}\}^{-}$. But because both f_n and f_n^{-} belong to $\mathscr{L}_1$ and to $\mathscr{L}_2$,

$$[f_n]^{-} = [f_n^{-}], \qquad ([f_n]^{-})^{\wedge} = [f_n^{-}]^{\wedge} = [f_n^{-\wedge}],$$

$$\{([f_n]^{-})^{\wedge}\}^{-} = [f_n^{-\wedge}]^{-} = [f_n^{-\wedge-}] = [f_n^{\vee}].$$

The left member of the last equation converges to $\{([f]^{-})^{\wedge}\}^{-}$ and the right member converges to $[f]^{\vee}$, so conclusion (iii) is established. The proof of the theorem is complete.

Calculating a Fourier transform by Definition 12-1 calls for integration; calculating a Fourier–Plancherel transform by Definition 13-2 calls for a sequence of integrations followed by a passage to the limit in $\mathscr{L}_2$, which is much less convenient. It is therefore convenient that in the many examples in which f belongs both to $\mathscr{L}_1$ and to $\mathscr{L}_2$, $[f]$ and $[f]$ can be calculated as easily as by Definition 12-1. For in that case, we first calculate $f^{\wedge}$ and $f^{\vee}$ by Definition 12-1. Then we already have the Fourier–Plancherel transforms $[f]^{\wedge}$ and $[f]^{\vee}$; for by conclusion (ii) of Theorem 13-3,

$$[f]^{\wedge} = [\hat{f}], \qquad [f]^{\vee} = [\check{f}].$$

To emphasize that there really are many examples of functions that belong both to $\mathscr{L}_1$ and to $\mathscr{L}_2$, we prove the following remark.

(C) If f is integrable and bounded over R, it belongs to $\mathscr{L}_p$ for all $p \geqq 1$.

For if f is integrable, it is measurable. Therefore $|f|$ is measurable, and if $p \geqq 1$, so is $|f|^p$. If M is an upper bound for $|f|$,

$$|f|^p = |f|^{p-1}|f| \leqq M^{p-1}|f|,$$

so $|f|^p$ is integrable, and f is in $\mathscr{L}_p$.

EXAMPLE 13-4 In Exercise 12-6 we computed that if f is the indicator of an interval $[-a, a]$, its Fourier transform is given by

$$\hat{f}(y) = (2/\pi)^{1/2}[\sin ya]/y \qquad (y \text{ in } R,\ y \neq 0),$$

$$\hat{f}(y) = (2/\pi)^{1/2}\, a \qquad (y = 0).$$

By Theorem 13-3, the Fourier–Plancherel transform is $[f]\hat{\ } = [\hat{f}]$. But $\hat{f}$ does not have a Fourier transform because it is not integrable. We return to this in Example 13-9.

Since the Fourier–Plancherel transformation is length preserving in L_2,

$$\left\{\frac{2}{\pi}\int_{-\infty}^{\infty} \frac{\sin^2 ya}{y^2}\, dy\right\}^{1/2} = \left\{\int_{-\infty}^{\infty} 1_{[-a,a]}(x)^2\, dx\right\}^{1/2},$$

hence

(D) $$\int_0^{\infty} y^{-2}\sin^2 ay\, dy = \pi a/2.$$

EXAMPLE 13-5 Let c be a complex number with negative real part, and let f be the function on R defined by

$$f(x) = \exp cx \qquad (x \geqq 0),$$

$$f(x) = 0 \qquad (x < 0).$$

This is continuous on $[0, \infty)$, and

$$|f(x)| = \exp[(\operatorname{Re} c)x],$$

which is integrable over $[0, \infty)$. Hence, f is bounded and integrable over R and belongs to $\mathscr{L}_1$ and to $\mathscr{L}_2$. By an elementary calculation,

$$\hat{f}(y) = (2\pi)^{-1/2}\int_0^{\infty} \exp(cx - iyx)\, dx$$

$$= (2\pi)^{-1/2}(iy - c)^{-1}.$$

This belongs to $\mathscr{L}_2$ but not to $\mathscr{L}_1$, so we cannot apply the conjugate Fourier transformation to it. We return to this in Example 13-8.

When f is in $\mathscr{L}_2$ but not in $\mathscr{L}_1$, the following corollary may be helpful.

COROLLARY 13-6 *Let f belong to $\mathscr{L}_2$. If there exist two sequences $a_1, a_2, a_3, \ldots, b_1, b_2, b_3, \ldots$ of positive numbers, both tending to ∞, such that the functions g_n defined by*

$$g_n(y) = (2\pi)^{-1/2} \int_{-a_n}^{b_n} f(x) \exp(-iyx)\, dx$$

converge for almost all y to a limit $g(y)$, then $[g] = [f]\hat{}$. A similar statement holds for $[f]\check{}$.

Define

$$f_n(x) = f(x) \qquad (-a_n \leqq x \leqq b_n),$$

$$f_n(x) = 0 \qquad (-\infty < x < a_n \text{ and } b_n < x < \infty).$$

This is the product of f and the indicator of $[-a_n, b_n]$, both of which belong to $\mathscr{L}_2$, so by Lemma 2-7, f_n is integrable. Since it is measurable and

$$|f_n|^2 \leqq |f|^2,$$

which is integrable, f_n belongs to $\mathscr{L}_2$. By Theorem 13-3(ii),

$$\text{(E)} \qquad [g_n] = [f_n]\hat{}.$$

As n increases, $|f_n - f|^2$ converges everywhere to 0, remaining at most equal to $|f|^2$, which is integrable. So f_n converges to f in L_2-norm. By Theorem 13-3, $[f_n]\hat{}$ converges to $[f]\hat{}$ in L_2-norm. This and (E) imply that g_n converges in L_2-norm to some function h in the class $[f]\hat{}$. But g_n converges to g almost everywhere, by hypothesis. So by Theorem 4-5, $g = h$ almost everywhere, and g is a member of the equivalence-class $[h] = [f]\hat{}$. This completes the proof.

Corollary 13-6 can be used sometimes to invert the Fourier transform, even when $\hat{f}$ is not integrable, as Examples 13-8 and 13-9 will show. In preparation we prove a lemma.

LEMMA 13-7 *Let g and h be complex-valued functions on an interval $[a, \infty)$ such that*

(i) *g is an indefinite integral, $g(x)$ tends to 0 as $x \to \infty$, and* Re g *and* Im g *are monotonic;*

(ii) *there is a positive M such that for all x in $[a, \infty)$,*

$$\left| \int_a^x h(u)\, du \right| \leqq M.$$

Then the limit

$$\lim_{x \to \infty} \int_a^x g(u) h(u)\, du$$

exists and is finite.

Suppose first that g and h are real-valued. Let ε be positive. There exists an x_0 such that if $x \geqq x_0$,

$$|g(x)| < \varepsilon/4M.$$

Define

$$H(x) = \int_a^x h(u)\,du \qquad (x \geqq a).$$

Let x_1 and x_2 $(x_2 > x_1)$ be two points in $[x_0, \infty)$. By the second theorem of the mean (Exercise II-9-2), there exists a number $\bar{x}$ in $[x_1, x_2]$ such that

$$\int_{x_1}^{x_2} g(u)h(u)\,du = g(x_1)\int_{x_1}^{\bar{x}} h(u)\,du + g(x_2)\int_{\bar{x}}^{x_2} h(u)\,du.$$

Then

$$\left|\int_a^{x_2} g(u)h(u)\,du - \int_a^{x_1} g(u)h(u)\,du\right|$$

$$\leqq |g(x_1)[H(\bar{x}) - H(x_1)]| + |g(x_2)[H(x_2) - H(\bar{x})]|$$

$$\leqq \frac{\varepsilon}{4M}2M + \frac{\varepsilon}{4M}2M = \varepsilon.$$

So the function

$$x \mapsto \int_a^x g(u)h(u)\,du \qquad (x \geqq a)$$

satisfies the Cauchy condition and therefore tends to a finite limit as x increases.

For complex g and h we write

$$\int_a^x g(u)h(u)\,du = \int_a^x \operatorname{Re} g(u) \operatorname{Re} h(u)\,du - \int_a^x \operatorname{Im} g(u) \operatorname{Im} h(u)\,du$$

$$+ i\int_a^x \operatorname{Re} g(u) \operatorname{Im} h(u)\,du + i\int_a^x \operatorname{Im} g(u) \operatorname{Re} h(u)\,du.$$

By the preceding paragraph, each of the four integrals in the right member tends to a finite limit as x increases. So the left member tends to a finite limit, and the proof is complete.

EXAMPLE 13-8 With f and $\hat{f}$ as in Example 13-5 and c real and $\neq 0$, for almost all real x,

(F) $$f(x) = \lim_{u\to\infty} (2\pi)^{-1/2}\int_{-u}^{u} \hat{f}(y)\exp(iyx)\,dy.$$

For all y,

$$\hat{f}(y) = (2\pi)^{-1/2}\{-c/(c^2 + y^2) - iy/(c^2 + y^2)\}.$$

Both $\mathrm{Re}\hat{f}$ and $\mathrm{Im}\hat{f}$ are continuously differentiable and hence are indefinite integrals on $[0, \infty)$, and both tend to 0 as y increases. The indefinite integral of $\exp(iyx)$ is bounded for all x except 0, so by Lemma 13-7, for all nonzero x the limit

$$\lim_{u\to\infty} \int_0^u \hat{f}(y)\exp(iyx)\,dy$$

exists and is finite. The same applies to the integral from $-u$ to 0, so

$$\lim_{u\to\infty} (2\pi)^{-1/2} \int_{-u}^{u} \hat{f}(y)\exp(iyx)\,dy$$

exists and is finite for all nonzero x. By Corollary 13-6, it belongs to the class $[f^{\wedge}]^{\vee}$, which by Theorem 13-3 is $[f]$.

EXAMPLE 13-9 With f and $\hat{f}$ as in Example 13-4, equation (F) is satisfied. By Example 13-4, for $y \neq 0$,

$$\begin{aligned} f^{\wedge}(y)\exp(iyx) = {} & (2\pi)^{-1/2} iy^{-1} \exp[iy(x-a)] \\ & - (2\pi)^{-1/2} iy^{-1} \exp[iy(x+a)]. \end{aligned}$$

By Lemma 13-7, the limits as u increases of the integrals

$$\int_1^u \hat{f}(y)\exp(iyx)\,dy, \qquad \int_{-u}^{-1} \hat{f}(y)\exp(iyx)\,dy$$

exist and are finite; and $\hat{f}$ is continuous on $[-1, 1]$, so

$$\int_{-1}^{1} \hat{f}(y)\exp(iyx)\,dy$$

exists. We add the three integrals and find that the integral of $\hat{f}(y)\exp(iyx)$ over $(-u, u]$ tends to a finite limit as $u \to \infty$. By Corollary 13-6, equation (F) is satisfied.

EXERCISE 13-1 If f is in $\mathscr{L}_2$ and $g(x) = f(-x)$ for all x, then

$$[f]^{\wedge\wedge} = [f]^{\vee\vee} = [g].$$

Suggestion: Approximate f in L_2-norm by functions of class C^2 with bounded support. Use Exercise 12-5 and Theorem 13-3(i).

EXERCISE 13-2 Prove that if f is continuous and in $\mathscr{L}_2$, and both f and $\hat{f}$ are integrable, then

$$f^{\wedge\vee}(x) = f(x)$$

for all x in R. (By Theorem 13-3, $[f^{\wedge\vee}] = [f^{\wedge}]^{\vee} = [f]^{\wedge\vee} = [f]$, so the two continuous functions $f, f^{\wedge\vee}$ are almost everywhere equal.)

14. The Fourier Transformation and the Fourier–Plancherel Transformation

Although the Fourier transformation, defined for integrable functions in Definition 12-1, does not have all the virtues of the Fourier–Plancherel transformation, it has some properties of its own that make it worth studying. The first theorem concerns continuity and uniqueness.

THEOREM 14-1 *Let C_0 denote the class of bounded continuous complex-valued functions g on R such that*

$$\lim_{y\to\infty} g(y) = \lim_{y\to-\infty} g(y) = 0.$$

Then

(i) *the Fourier transformation $f \mapsto \hat{f}$ and the conjugate Fourier transformation $f \mapsto \check{f}$ are linear mappings of $\mathscr{L}_1$ into C_0, and*

$$|\hat{f}(y)| \leqq (2\pi)^{-1/2}\|f\|_1, \qquad |\check{f}(y)| \leqq (2\pi)^{-1/2}\|f\|_1$$

for all y in R;

(ii) *if f_1 and f_2 are in $\mathscr{L}_1$, then $\hat{f}_1(y) = \hat{f}_2(y)$ for all y in R if and only if $f_1(x) = f_2(x)$ for almost all x in R.*

The linearity of both maps is evident from Definition 12-1. Also, for all real y,

$$\text{(A)} \qquad |\hat{f}(y)| \leqq (2\pi)^{-1/2}\int_R |f(x)\exp(-iyx)|\,dx$$

$$= (2\pi)^{-1/2}\int_R |f(x)|\,dx = (2\pi)^{-1/2}\|f\|_1,$$

and likewise with $\check{f}$ in place of $\hat{f}$.

Let f be integrable. By Corollary 6-3, there exists a sequence $g_1, g_2, g_3, \ldots$ of functions of class C^2 with bounded support such that

$$\lim_{n\to\infty} \|f - g_n\|_1 = 0.$$

By Theorem 12-2, each Fourier transform $\hat{g}_n$ is infinitely differentiable, and by (A) they converge uniformly to $\hat{f}$, so $\hat{f}$ is continuous. By (L) of Section 12, each $\hat{g}_n(y)$ tends to 0 as y tends to ∞ or to $-\infty$, so the same is true of the uniform limit $\hat{f}$. The same discussion applies to $\check{f}$, or if we prefer we can handle it by (T) of Section 12, and conclusion (i) is established.

If $f_1(x) = f_2(x)$ for almost all x, and f_1 is integrable, by Definition 12-1 and Theorem II-12-2 we have

$$\hat{f}_1(y) = \hat{f}_2(y) \qquad (y \text{ in } R).$$

Suppose, conversely, that $\hat{f}_1 = \hat{f}_2$. Define

$$f = f_1 - f_2; \tag{B}$$

then

$$\hat{f}(y) = \hat{f}_1(y) - \hat{f}_2(y) = 0 \qquad (y \text{ in } R).$$

Let g be any function on R of class C^2 with bounded support. By Theorem 12-2, $\check{g}$ is integrable. By Lemma IV-11-3 (which extends readily to complex-valued functions), the product

$$(x, y) \mapsto f(x)\check{g}(y)$$

is integrable over R^2. Since the function $(x, y) \mapsto \exp(-iyx)$ is continuous and bounded on the plane, the product

$$(x, y) \mapsto f(x)\check{g}(y)\exp(-iyx)$$

is integrable over the plane. By Fubini's theorem (Theorem IV-7-1), there is a set N in R with $mN = 0$ such that the iterated integrals in the equation

$$\begin{aligned}(2\pi)^{-1/2}\int_{R\setminus N}\left\{\int_R f(x)\check{g}(y)\exp(-iyx)\,dy\right\}dx \\ = (2\pi)^{-1/2}\int_{R\setminus N}\left\{\int_R f(x)\check{g}(y)\exp(-iyx)\,dx\right\}dy\end{aligned} \tag{C}$$

both exist and are both equal to the integral over R^2. By Theorem 12-2, the left member of (C) is

$$\int_{R\setminus N} f(x)g(x)\,dx. \tag{D}$$

The right member of (C) is equal to

$$\int_{R\setminus N} \check{g}(y)\hat{f}(y)\,dy, \tag{E}$$

and this is 0 because $\hat{f}(y)$ is identically 0. Since the integrand in (D) is defined for all x and $mN = 0$, this implies

$$\int_R f(x)g(x)\,dx = 0. \tag{F}$$

Let $(a, b]$ be any bounded left-open interval in R. There exists a sequence of functions $g_1, g_2, g_3, \ldots$ of class C^2 on R, vanishing outside $[a - 1, b + 1]$ and all having values in $[0, 1]$, such that $g_n(x)$ tends to $1_{(a,b]}(x)$ for all x in R. Then by the dominated convergence theorem and (F),

$$\int_R f(x)1_{(a,b]}(x)\,dx = \lim_{n\to\infty}\int_R f(x)g_n(x)\,dx = 0.$$

So the integral of f over every bounded interval is equal to 0, and by Theorem II-13-4, $f(x) = 0$ for almost all x. By (B), this completes the proof.

In Section IV-12 we showed (as we shall show again) that if f and g are integrable over R, for almost all x in R the integral

$$\text{(G)} \qquad \int_R f(x-y)g(y)\,dy$$

exists, and we defined the convolution $f*g$ to be the function that has the value (G) where the integral exists and the value 0 elsewhere in R. This can be extended without trouble to complex-valued functions, since we can express f as $\operatorname{Re} f + i\operatorname{Im} f$, and g likewise. This operation of convolution is transformed by the Fourier transformation into the simpler operation of multiplication.

THEOREM 14-2 *If f and g are integrable functions,*

$$(f*g)\hat{} = (2\pi)^{1/2}\hat{f}\hat{g}.$$

By Lemma IV-11-3, the function $x\mapsto f(x^1)g(x^2)$ is integrable over the plane R^2. Since for each real number y the functions $x\mapsto\exp(-iyx^1)$ and $x\mapsto\exp(-iyx^2)$ are continuous and bounded on R^2, the function

$$x\mapsto(2\pi)^{-1/2}f(x^1)g(x^2)\exp(-iyx^1)\exp(-iyx^2)$$

is integrable over R^2 for each real y. We integrate it over R^2 and make the substitution

$$x^1 = u^1 - u^2, \qquad x^2 = u^2$$

in the integral. The Jacobian of this transformation is 1, so by Theorem IV-8-1,

$$\text{(H)} \qquad (2\pi)^{-1/2}\int_{R^2} f(u^1-u^2)g(u^2)\exp(-iyu^1)\,du$$

$$= (2\pi)^{-1/2}\int_{R^2} f(x^1)g(x^2)\exp(-iyx^1)\exp(-iyx^2)\,dx.$$

The integral in the left member of (H) exists for all real y, and in particular for $y = 0$. So by Fubini's theorem (Theorem IV-7-1), there is a set N of measure 0 in R such that if u^1 is not in N, the function

$$u^2\mapsto f(u^1-u^2)g(u^2)$$

is integrable over R. Since $\exp(-iyu^1)$ is continuous and bounded, the function

$$u^2 \mapsto f(u^1 - u^2)g(u^2)\exp(-iyu^1)$$

is also integrable over R whenever u^1 is not in N. By Fubini's theorem, and recalling that $mN = 0$,

$$\text{(I)} \qquad \int_{R^2} f(u^1 - u^2)g(u^2)\exp(-iyu^1)\,du$$
$$= \int_{R\setminus N}\left\{\int_R f(u^1 - u^2)g(u^2)\exp(-iyu^1)\,du^2\right\}du^1$$
$$= \int_{R\setminus N}[f*g](u^1)\exp(-iyu^1)\,du^1$$
$$- \int_R [f*g](u^1)\exp(-iyu^1)\,du^1$$
$$= (2\pi)^{1/2}[f*g]\hat{\ }(y).$$

Applying Fubini's theorem to the right member of (H), we obtain

$$\text{(J)} \qquad \int_{R^2} f(x^1)g(x^2)\exp(-iyx^1)\exp(-iyx^2)\,dx$$
$$= \int_R\left\{\int_R f(x^1)g(x^2)\exp(-iyx^1)\exp(-iyx^2)\,dx^1\right\}dx^2$$
$$= 2\pi\hat{f}(y)\hat{g}(y).$$

(In the right member, the outer integral can be taken over all of R because the inner integral exists for all x^2 in R.) Substituting (I) and (J) in (H) yields the conclusion of the theorem.

Fourier transforms are useful in solving differential equations, and their usefulness rests chiefly on the next two theorems.

THEOREM 14-3 *If the functions $x \mapsto f(x)$ and $x \mapsto xf(x)$ are both integrable over R, the derivative of $\hat{f}$ is*

$$D\hat{f} = (-i\mathbf{x}f(\mathbf{x}))\hat{\ }.$$

To prove this we need a simple estimate; for all real r,

$$\text{(K)} \qquad |\exp(ir) - 1| \leqq |r|.$$

The point $\exp(ir) = \cos r + i\sin r$ is on the unit circle and is the point located by starting at 1 and moving a distance $|r|$ in one direction or the other along the circumference. The distance $\exp(ir) - 1$ between beginning and end points of the arc cannot exceed the distance $|r|$ traveled along the circumference.

Let y be any real number and $y_1, y_2, y_3, \ldots$ any sequence of real numbers different from y and tending to y. Then

$$\text{(L)} \qquad \frac{\hat{f}(y_n) - \hat{f}(y)}{y_n - y} = (2\pi)^{-1/2} \int_R f(x) \frac{\exp(-iy_n x) - \exp(-iyx)}{y_n - y} dx.$$

Since by (K),

$$|\exp(-iy_n x) - \exp(-iyx)| = |[\exp(-i[y_n x - yx]) - 1]\exp(-iyx)|$$
$$\leqq |y_n x - yx|1 = |x|\,|y_n - y|,$$

the integrand in the right member cannot exceed $|f(x) \cdot x|$, which is integrable. As y_n tends to y, the ratio

$$[\exp(-iy_n x) - \exp(-iyx)]/(y_n - y)$$

tends to the derivative at y of the function $y \mapsto \exp(-iyx)$, which is $-ix\exp(-iyx)$. So by the dominated convergence theorem, the right member of (L) tends to

$$\text{(M)} \qquad (2\pi)^{-1/2} \int_R f(x)(-ix)\exp(-iyx)\,dx.$$

To simplify notation, we define

$$g(x) = -ixf(x).$$

Then (M) is $\hat{g}(y)$. So the left member of (L) tends to $\hat{g}(y)$ whenever $y_1, y_2, y_3, \ldots$ is a sequence of numbers different from y and tending to y. That is, $\hat{g}(y)$ is the limit of the ratio

$$[\hat{f}(y') - \hat{f}(y)]/[y' - y]$$

as $y' \neq y$ tends to y, which is the conclusion of the theorem.

THEOREM 14-4 *Let F and f be functions integrable over R such that for all real x,*

$$F(x) = \int_{-\infty}^{x} f(u)\,du.$$

Then

$$\text{(N)} \qquad \hat{f}(y) = iy\hat{F}(y) \qquad (y \text{ in } R).$$

If a and b are positive and $y \neq 0$, by integration by parts,

$$\text{(O)} \qquad \int_{-a}^{b} F(x)\exp(-iyx)\,dx = F(b)(-iy)^{-1}\exp(-iyb)$$
$$- F(-a)(-iy)^{-1}\exp(iya)$$
$$- \int_{-a}^{b} f(x)(-iy)^{-1}\exp(-iyx)\,dx.$$

Since F is integrable over R, it cannot remain bounded away from 0 on any interval (n, ∞), so we can pick a sequence of values $b(1), b(2), \ldots$ tending to ∞ such that $F(b(n))$ tends to 0. Likewise, we can select a sequence $a(1), a(2), \ldots$ tending to ∞ such that $F(-a(n))$ tends to 0. We substitute these in (O) and let n increase, and we obtain

$$\int_R F(x) \exp(-iyx)\, dx = (1/iy) \int_R f(x) \exp(-iyx)\, dx,$$

which establishes the theorem.

We have seen in Section III-12 that the density of the normal probability distribution with mean m and variance σ^2 is

$$p_{m,\sigma}(x) = (2\pi)^{-1/2}\sigma^{-1} \exp(-[x - m]^2/2\sigma^2).$$

We shall now compute its Fourier transform.

EXAMPLE 14-5 If $p_{m,\sigma}(x) = (2\pi)^{-1/2}\sigma^{-1} \exp(-[x - m]^2/2\sigma^2)$, then

$$\hat{p}_{m,\sigma}(y) = (2\pi)^{-1/2} \exp(-iym) \exp(-\sigma^2 y^2/2).$$

To simplify notation, we first prove this when $\sigma = 1$ and $m = 0$; we write $p(x)$ for $p_{0,1}(x)$. At the end of Section IV-5 we showed that

$$\text{(P)} \qquad \int_R p(x)\, dx = 1.$$

By Theorem 14-3,

$$D\hat{p}(y) = (2\pi)^{-1/2} \int_R (-ix)\left[(2\pi)^{-1/2} \exp\left(-\frac{x^2}{2}\right)\right] \exp(-iyx)\, dx$$

$$= \lim_{n \to \infty} \int_{-n}^{n} \frac{i}{2\pi}\left[D \exp\left(-\frac{x^2}{2}\right)\right] \exp(-iyx)\, dx.$$

By integration by parts,

$$D\hat{p}(y) = \lim_{n \to \infty} \frac{i}{2\pi}\left\{\exp\left(-\frac{n^2}{2}\right)\exp(-iyn) - \exp\left(-\frac{n^2}{2}\right)\exp(iyn)\right.$$

$$\left. - \int_{-n}^{n}\left[\exp\left(-\frac{x^2}{2}\right)\right](-iy)\exp(-iyx)\, dx\right\}$$

$$= -y\hat{p}(y).$$

This differential equation for $\hat{p}$, with the initial value $(2\pi)^{-1/2}$ at $y = 0$ given by (P), is easily solved; the solution is

$$\text{(Q)} \qquad \hat{p}(y) = (2\pi)^{-1/2} \exp(-y^2/2).$$

For arbitrary positive σ and real m we make the substitution

$$u = [x - m]/\sigma$$

in the right member of the equation

$$\hat{p}_{m,\sigma}(y) = (2\pi)^{-1/2} \int_R (2\pi)^{-1/2}\sigma^{-1} \exp\left[-\frac{(x-m)^2}{2\sigma^2}\right] \exp(-iyx)\, dx$$

and obtain

$$\hat{p}_{m,\sigma}(y) = (2\pi)^{-1} \int_R \exp\left(-\frac{u^2}{2}\right) \exp(-iyu\sigma)\exp(-iym)\, du.$$

By (Q), this implies

$$\hat{p}_{m,\sigma}(y) = (2\pi)^{-1/2} \exp(-iym)\hat{p}(y\sigma),$$

which was to be proved.

In Section IV-12 of Chapter IV we stated the following corollary.

COROLLARY 14-6 *If X_1 and X_2 are independent normally distributed random variables with the respective means m_1, m_2 and the respective variances σ_1^2, σ_2^2, then $X_1 + X_2$ is normally distributed, with mean $m_1 + m_2$ and variance $\sigma_1^2 + \sigma_2^2$.*

In Chapter IV we sketched a proof of this by elementary computations. We now show that by use of Fourier transforms the proof becomes trivial. We define

$$m_3 = m_1 + m_2, \qquad \sigma_3 = [\sigma_1^2 + \sigma_2^2]^{1/2},$$

and we simplify notation by writing

(R) $$p_i = p_{m_i,\sigma_i} \qquad (i = 1, 2, 3).$$

By Example 14-5,

$$\hat{p}_i(y) = (2\pi)^{-1/2} \exp(-iym_i)\exp(-\sigma_i^2 y^2/2) \qquad (i = 1, 2, 3),$$

so that

(S) $$\hat{p}_3(y) = (2\pi)^{1/2}\hat{p}_1(y)\hat{p}_2(y).$$

By Corollary IV-12-3, the distribution of $X_1 + X_2$ has density $p_1 * p_2$, and by Theorem 14-2,

$$(p_1 * p_2)\hat{}\,(y) = (2\pi)^{1/2}\hat{p}_1(y)\hat{p}_2(y).$$

This and (S) show that the distribution $p_1 * p_2$ of $X_1 + X_2$ and the normal distribution p_3 have the same Fourier transform, so by Theorem 14-1, they are almost everywhere equal, and the density of the normal distribution with mean $m_1 + m_2$ and variance $\sigma_1^2 + \sigma_2^2$ is the density of the distribution of the sum $X_1 + X_2$.

Example 14-7 The Fourier transform of the Hermite function h_n $(n = 0, 1, 2, \ldots)$ is

$$\text{(T)} \qquad \hat{h}_n(y) = (-i)^n h_n(y).$$

This holds for $n = 0$, by Example 14-5 with $m = 0$ and $\sigma = 1$. Suppose that it holds for a nonnegative integer n. By equation (S) of Section 10,

$$\text{(U)} \qquad h_{n+1}(x) = -Dh_n(x) + xh_n(x)$$

for all x. Since the h_n belong to $\mathscr{L}_1[m_L, R]$ and to $\mathscr{L}_2[m_L, R]$, we can take Fourier transforms and apply Theorems 14-3 and 14-4, obtaining

$$\hat{h}_{n+1}(y) = -iy\hat{h}_n(y) + i\,D\hat{h}_n(y).$$

By the induction hypothesis, this and (U) imply

$$\begin{aligned} \hat{h}_{n+1}(y) &= [(-i)^{n+1} yh_n(y) - (-i)^{n+1}\,Dh_n(y)] \\ &= (-i)^{n+1} h_{n+1}(y). \end{aligned}$$

By induction, (T) holds for all nonnegative integers n.

We now have the material needed to prove, as asserted in Section 10 of this chapter, that the Hermite polynomials form a complete orthogonal set in $\mathscr{L}_2[m_H, R]$. Let f be a real-valued function in $\mathscr{L}_2[m_H, R]$ that is orthogonal to all the H_n. We now prove that

$$\text{(V)} \qquad \int_R f(x)x^n\, m_H(dx) = 0 \qquad (n = 0, 1, \ldots).$$

This is true for $n = 0$ and $n = 1$, since $H_0 = 1$ and $H_1(x) = 2x$. If it holds for $0, 1, \ldots, n$, by (E) of Section 10

$$2^{n+1}x^{n+1} = H_{n+1}(x) + \sum_{j=0}^{n} c_j x^j,$$

where the c_j are real numbers. We multiply by $f(x)$ and integrate with respect to m_H over R. The first term in the right is then 0 because f is orthogonal to all H_n, and the other terms are 0 by the induction hypothesis, so (V) holds for $n + 1$. It then holds for all n by induction.

For each fixed real number y,

$$\text{(W)} \qquad \exp(-x^2)\exp(-iyx) = \sum_{n=0}^{\infty} \frac{(-iy)^n}{n!} x^n \exp(-x^2).$$

The function $x \mapsto x^n \exp(-x^2)$ has an L_2-norm that can be computed by Lemma II-11-3 with $\sigma = \frac{1}{2}$:

$$\begin{aligned} \|\mathbf{x}^n \exp(-x^2)\|_2^2 &= \int_R x^{2n} \exp(-2x^2)\, dx \\ &= (2\pi)^{1/2}(2^{-2n-1})(1 \cdot 3 \cdot 5 \cdots [2n-1]). \end{aligned}$$

So the ratio of the norm of the term in (W) with $n = j + 1$ to that of the term with $n = j$ is

$$[|y|/(j+1)]2^{-1}[2j+1]^{1/2},$$

which tends to 0 as j increases. By the ratio test, the series whose terms are the norms of the terms in series (W) is convergent. By Corollary 4-2, for fixed y the series in the right member of (W), regarded as a series of members of $\mathscr{L}_2[m_L, R]$, converges in L_2-norm to some member of $\mathscr{L}_2[m_L, R]$. But the right member of (W) converges pointwise to the left member of (W), so by Theorem 4-5 the left member of (W) represents the function in $\mathscr{L}_2[m_L, R]$ that is the sum of the series in the right member; that is,

$$\text{(X)} \qquad \lim_{j\to\infty} \left\| \sum_{n=0}^{j} \frac{(iy)^n}{n!} x^n \exp(-x^2) - \exp(-x^2)\exp(iyx) \right\|_2 = 0.$$

We take the inner product of both members of (W) with f. Since the inner product is continuous, (X) implies

$$\lim_{j\to\infty} \sum_{n=0}^{j} \int_R f(x) \frac{(-iy)^n}{n!} x^n \exp(-x^2)\,dx = \int_R f(x)\exp(-iyx)\exp(-x^2)\,dx.$$

Each term in the left member is 0 by (V), with (A) of Section 10, so the right member is also 0. This implies

$$[f(\mathbf{x})\exp(-\mathbf{x}^2)]\hat{\ }(y) = 0$$

for all y. By Theorem 14-1,

$$f(x)\exp(-x^2) = 0$$

for almost all x. But $\exp(-x^2)$ is never 0, so $f(x) = 0$ for almost all x. The only functions orthogonal to all H_n are almost everywhere 0, and the H_n form a complete set.

EXERCISE 14-1 Use the facts

(i) there is an $\bar{x}$ between 0 and r such that

$$|\sin r| = |0 + r\cos\bar{x}| \leq |r|,$$

(ii) $|\exp(ir) - 1| = |\exp(-ir/2)[\exp(ir) - 1]| = 2|\sin r/2|$

to obtain another proof of (K).

EXERCISE 14-2 If f is in $\mathscr{L}_2$, it is the sum of four functions g_0, g_1, g_2, g_3 in $\mathscr{L}_2$ such that

$$(g_n)\hat{\ } = (-i)^n g_n \qquad (n = 0, 1, 2, 3).$$

Suggestion: Expand in Hermite series

$$f = \sum_{j=0}^{\infty} c_j h_j.$$

Let

$$g_n = \sum_{j=0}^{\infty} c_{4j+n} h_{4j+n}.$$

EXERCISE 14-3 Show that if $0 < a \leqq b$, the function

$$f = 1_{[-a,a]} * 1_{[-b,b]}$$

is continuous, is constantly 0 on $(-\infty, -a - b)$ and on $(a + b, \infty)$, is constantly $2a$ on $(-b + a, b - a)$, and is linear in the remaining two intervals. By Theorem 14-2 and Example 13-4, compute $\hat{f}$.

EXERCISE 14-4 Use Exercise 14-3 to compute $\hat{g}$ when $g(x) = (1 - |x|/2a)^+$.

15. Applications to Differential Equations

We shall first apply the Fourier transformation to solve an ordinary differential equation with given conditions at a certain starting time.

Consider the simple electrical circuit of Fig. VI-1, where R is the resistance, L is the inductance, and C is the capacity of the condenser. If $f(t)$ is the current

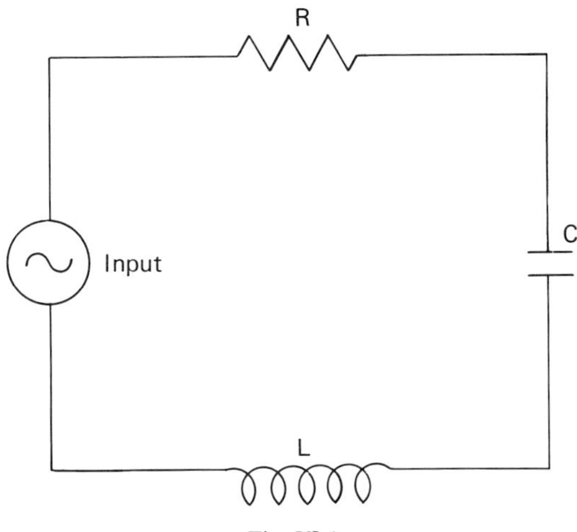

Fig. VI-1

at time t, the charge $q(t)$ on the condenser at time t (> 0) is

$$q(t) = q(0) + \int_0^t f(u)\,du.$$

The voltage drop across the condenser is $q(t)/C$, the voltage drop across the resistance is $Rf(t)$, and the voltage drop across the inductance is $L\,Df(t)$, provided that f is differentiable. The sum of all three must be equal to the impressed voltage $v(t)$, so

$$\text{(A)} \qquad L\,Df(t) + Rf(t) + C^{-1}\left[q(0) + \int_0^t f(u)\,du\right] = v(t).$$

We shall assume that v is continuously differentiable. Then solving (A) is equivalent to solving

$$\text{(B)} \qquad L\,D^2f(t) + R\,Df(t) + C^{-1}f(t) = w(t),$$

where we have written w for Dv. We wish to solve this on some interval $[0, T]$, given the values of f and Df at time $t = 0$. We know by Theorem III-2-1 that equation (B) has a solution on $[0, T+1]$. We know that functions of class C^2 exist that are constantly 1 on $(-\infty, T]$ and are constantly 0 on $[T+1, \infty)$. We multiply f by such a function and obtain a function of class C^2 on $[0, \infty)$ that coincides with f on $[0, T]$ and vanishes for $t > T + 1$. If we replace f in the left member of (B) by this function, the left member will be equal to w on $[0, T]$ and will vanish on $[T+1, \infty)$. We have thus replaced (B) by an equation in which f and w are the same as in the original equation on $[0, T]$ and vanish on $[T+1, \infty)$; the new f and w may differ from the original for $t > T$, but this does not interest us. For $t < 0$ we define f and w to be 0.

We cannot apply Theorem 14-4 to f and Df because f is discontinuous at 0 and is not an indefinite integral. By integration by parts, recalling that f and w are 0 for $t < 0$,

$$\begin{aligned}
(Df)\hat{}(y) &= (2\pi)^{-1/2}\int_0^\infty Df(t)\exp(-iyt)\,dt \\
&= (2\pi)^{-1/2}\left\{f(t)\exp(-iyt)\Big|_0^\infty - \int_0^\infty f(t)(-iy)\exp(-iyt)\,dt\right\} \\
&= -(2\pi)^{-1/2}f(0) + iy\hat{f}(y)
\end{aligned}$$

$$\begin{aligned}
(D^2f)\hat{}(y) &= (2\pi)^{-1/2}\int_0^\infty D^2f(t)\exp(-iyt)\,dt \\
&= (2\pi)^{-1/2}\left\{Df(t)\exp(-iyt)\Big|_0^\infty - \int_0^\infty Df(t)(-iy)\exp(-iyt)\,dt\right\} \\
&= -(2\pi)^{-1/2}\,Df(0) + iy\,(Df)\hat{}(y) \\
&= -(2\pi)^{-1/2}\,Df(0) - (2\pi)^{-1/2}(iy)f(0) - y^2\hat{f}(y).
\end{aligned}$$

So if we take the Fourier transforms of both members of (B), we obtain

(C) $$[-Ly^2 + Riy + C^{-1}]\hat{f}(y) - (2\pi)^{-1/2}[L\,Df(0) + (iyL + R)f(0)] = \hat{w}(y).$$

We now confine our attention to the case in which

(D) $$R^2 < 4LC^{-1}.$$

If we define

(E) $$a = R/2L, \qquad b = (4LC^{-1} - R^2)^{1/2}/2L,$$

we readily calculate

$$-Ly^2 + iRy + C^{-1} = L(iy + a + ib)(iy + a - ib),$$

hence

(F) $$1/[-Ly^2 + Riy + C^{-1}] = (1/2iLb)[(iy + a - ib)^{-1} - (iy + a + ib)^{-1}].$$

This and (C) imply

(G) $$\hat{f}(y) = (1/2iLb)[(iy + a - ib)^{-1} - (iy + a + ib)^{-1}] \times [\hat{w}(y) + (2\pi)^{-1/2}L\,Df(0) + (2\pi)^{-1/2}(iyL + R)f(0)].$$

Let us define

(H) $$\begin{aligned}\phi(t) &= (2iLb)^{-1}[\exp(-a + ib)t - \exp(-a - ib)t]1_{[0,\infty)}(t)\\ &= (Lb)^{-1}\exp(-at)(\sin bt)1_{[0,\infty)}(t).\end{aligned}$$

By definition,

(I) $$\begin{aligned}\hat{\phi}(y) &= (2\pi)^{-1/2}(2iLb)^{-1} \times \int_0^\infty [\exp(-a + ib)t - \exp(-a - ib)t]\exp(-iyt)\,dt\\ &= (2\pi)^{-1/2}(2iLb)^{-1}[(-a + ib - iy)^{-1}\exp(-a + ib - iy)t + (a + ib + iy)^{-1}\exp(-a - ib - iy)t]_0^\infty\\ &= (2\pi)^{-1/2}(2iLb)^{-1}[(a - ib + iy)^{-1} - (a + ib + iy)^{-1}].\end{aligned}$$

Evidently both ϕ and $D\phi$ are integrable over R, and

$$\int_{-\infty}^{t} D\phi(u)\,du = \phi(t)$$

for all real t. So by (I) and (G), with Theorems 14-2 and 14-4,

(J) $$\begin{aligned}\hat{f}(y) &= (2\pi)^{1/2}\hat{\phi}(y)\hat{w}(y) + \hat{\phi}(y)[L\,Df(0) + Rf(0)] + \hat{\phi}(y)(iy)Lf(0)\\ &= (\phi * w)\hat{\ }(y) + [L\,Df(0) + Rf(0)]\hat{\phi}(y) + [Lf(0)](D\phi)\hat{\ }(y).\end{aligned}$$

So by Theorem 13-3,

$$\text{(K)}\quad f(t) = \int_{-\infty}^{\infty} w(u)\phi(t-u)\,du + [L\,Df(0) + Rf(0)]\phi(t) + Lf(0)\,D\phi(t).$$

The terms with factors $\phi(t)$ or $D\phi(t)$ tend rapidly to 0 as t increases because of the factor $\exp(-at)$. They are called *transients*. The first term in the right member of (K) persists. It has an interpretation that we shall describe rather sloppily. Although the integration is from $-\infty$ to ∞, one factor in the integrand is 0 if $u < 0$ and the other is 0 if $u > t$, so we need only integrate from 0 to t. We can think of the integral as approximately the sum of many terms, each of the form $\phi(t-u)[w(u)\,du]$. The second factor represents the increment of voltage delivered to the system in the time-interval u to $u + du$; the first factor represents the effect of a unit increment of voltage $t - u$ time-units after the increment happened. The term is the delayed effect of the differential input.

We could not have reached the solution by applying the conjugate Fourier transform to the members of (G), since the last term in (G) is not integrable unless $f(0) = 0$.

Our second application is to a well-known partial differential equation, the wave equation. This is

$$\text{(L)}\qquad \frac{\partial^2 u(x,t)}{\partial t^2} = \frac{\partial^2 u(x,t)}{\partial x^2} \qquad (t > 0,\ x \text{ in } R),$$

with the boundary conditions

$$\text{(M)}\qquad \lim_{t\to 0+} u(x,t) = f(x),$$

$$\text{(N)}\qquad \lim_{t\to 0+} \frac{\partial u(x,t)}{\partial t} = g(x),$$

where f is of class C^2 and g of class C^1 on R. We first solve the problem under the extra hypotheses

(O) f and g are integrable over R,

(P) the left members of (M) and (N) converge to the right members in L_1-norm,

and we seek a solution u such that in the equation

$$\hat{u}(y,t) = (2\pi)^{-1/2} \int_R u(x,t)\exp(-iyx)\,dx$$

we can compute the partial derivatives $\partial\hat{u}/\partial t$ and $\partial^2\hat{u}/\partial t^2$ by differentiating under the integral sign. Then by (L),

$$\frac{\partial^2 \hat{u}(y,t)}{\partial t^2} = (2\pi)^{-1/2} \int_R \frac{\partial^2 u(x,t)}{\partial t^2} \exp(-iyx)\,dx$$
$$= (2\pi)^{-1/2} \int_R \frac{\partial^2 u(x,t)}{\partial x^2} \exp(-iyx)\,dx.$$

By two applications of Theorem 14-4 to the right member,

$$\frac{\partial^2 \hat{u}(y,t)}{\partial t^2} = -y^2 \hat{u}(y,t).$$

The solution is given for each y by the expression

(Q) $$\hat{u}(y,t) = A(y)\cos yt + B(y)\sin yt.$$

If (P) holds, it is easy to verify that for all y in R,

$$\lim_{t\to 0+} \hat{u}(y,t) = \hat{f}(y),$$
$$\lim_{t\to 0+} [\partial \hat{u}(x,t)/\partial t](y) = \lim_{t\to 0+} [\partial \hat{u}(y,t)/\partial t] = g(y).$$

From this and (Q),

$$A(y) = f(y), \qquad yB(y) = g(y).$$

So for all $y \neq 0$ we have by (Q),

(R) $$\hat{u}(y,t) = [\cos yt]\hat{f}(y) + [y^{-1}\sin yt]\hat{g}(y)$$
$$= \tfrac{1}{2}\exp(iyt)\hat{f}(y) + \tfrac{1}{2}\exp(-iyt)\hat{f}(y) + [y^{-1}\sin yt]\hat{g}(y).$$

By Exercise 12-3,

$$\exp(iyt)\hat{f}(y) = [f(\mathbf{x}+t)]\hat{\ }(y),$$
$$\exp(-iyt)\hat{f}(y) = [f(\mathbf{x}-t)]\hat{\ }(y),$$

and by Exercise 12-6,

$$y^{-1}\sin yt = (\pi/2)^{1/2}[1_{(-t,t]}]\hat{\ }(y).$$

So (R) implies, with Theorem 14-2,

$$\hat{u}(t,y) = \tfrac{1}{2}f(\mathbf{x}+t)\hat{\ }(y) + \tfrac{1}{2}f(\mathbf{x}-t)\hat{\ }(y) + \tfrac{1}{2}[1_{(-t,t]} * g]\hat{\ }(y).$$

By the uniqueness theorem (Theorem 14-1),

(S) $$u(t,x) = \tfrac{1}{2}[f(x+t) + f(x-t)] + \tfrac{1}{2}[1_{(-t,t]} * g](x)$$
$$= \tfrac{1}{2}[f(x+t) + f(x-t)] + \tfrac{1}{2}\int_{-\infty}^{\infty} 1_{(-t,t]}(x-u)g(u)\,du$$
$$= \tfrac{1}{2}[f(x+t) + f(x-t)] + \tfrac{1}{2}\int_{x-t}^{x+t} g(u)\,du.$$

Now we could remove the extra hypotheses (O) and (P) by appropriate approximation techniques, but that would be a waste of time. Having the formula (S), we can verify by straightforward differentiation that whenever f is of class C^2 and g of class C^1, the $u(x, t)$ given by (S) satisfies the differential equation (L) and the boundary conditions (M) and (N).

EXERCISE 15-1 Show that if (C) holds, and the input voltage is 0, the solution of (B) (which is the transient alone) changes sign every $2\pi L(4LC^{-1} - R^2)^{-1/2}$ time-units. Show also that if $R^2 > 4LC^{-1}$, the solution with $u = 0$ always keeps the same sign and tends to 0.

EXERCISE 15-2 Discuss the behavior as t tends to ∞ of the solution (K) corresponding to $w(t) = \sin bt$.

EXERCISE 15-3 Let the temperature at place x and time $t \geqq 0$ of an infinite homogeneous rod be $u(t, x)$. Accept from physics the statement that (with properly chosen units) u satisfies the heat equation

$$\partial u/\partial t = \tfrac{1}{2}\partial^2 u/\partial x^2 \qquad (t > 0).$$

Show that if the initial condition

$$\lim_{t \to 0, t > 0} u(x, t) = f(x)$$

is satisfied with f continuous and integrable, then

$$u(t, x) = (2\pi t)^{-1/2} \int_{-\infty}^{\infty} f(y) \exp\left(-\frac{[y - x]^2}{2t}\right) dy.$$

Suggestion: Under assumptions like those used in discussing the wave equation,

$$\partial \hat{u}(t, y)/\partial t = (-y^2/2)\hat{u}(t, y).$$

Solve this and use Example 14-5 and Theorem 14-2:

$$\begin{aligned} \hat{u}(t, \mathbf{y}) &= \hat{u}(0, \mathbf{y}) \exp(-\mathbf{y}^2 t/2) \\ &= \hat{u}(0, \mathbf{y})[t^{-1/2} \exp(-\mathbf{x}^2/2t)]\hat{\ } \\ &= (2\pi t)^{-1/2}[f(\mathbf{x}) * \exp(-\mathbf{x}^2/2t)]\hat{\ }. \end{aligned}$$

VII

Measure Theory

1. σ-Algebras and Measurable Functions

The development of integration theory in the preceding chapters is only one of many ways of approaching the subject. It was set forth in the belief that it is especially easy for a beginner to comprehend and is well suited for teaching a student of physics, chemistry, or engineering enough integration theory to be of clear benefit. But one who intends to go further into mathematics needs to know other approaches too, for several reasons. One is the need to read published papers involving integration, which will require, at the least, knowledge of the concepts involved in various approaches and of the relations (usually equivalence) between the various kinds of integrals. Another reason is that no one approach is superior in all ways to all others. Knowing several different methods of defining the integral will give a mathematician a choice of several ways of thinking about a proposition involving integrals; sometimes one of these and sometimes another will lead most smoothly to the desired result.

In our development of integration theory we defined the integral first, and we deduced theorems about measure from theorems about integrals. But in many modern papers involving integration, the reversed order is used, as indeed it was in Lebesgue's definition of the integral at the very beginning of the twentieth century. There are several ways of defining measure and several ways of using measure to define integral, and a reader of any paper on the subject would be well advised to look up exactly what the author means by each of the familiar words. Nevertheless, though the word *measure* has several definitions with different degrees of generality, they agree with each other in fundamental respects, and a reader who learns any one of them will usually find it easy to make the adjustment to any other.

The various kinds of measure that we have developed in previous chapters have several essential features in common. First, each is defined on a σ-algebra of sets. We here repeat the definition (Definition III-10-7) of σ-algebra and add something to it.

DEFINITION 1-1 (i) *Let X be any nonempty set. A nonempty collection $\mathscr{A}$ of subsets of X is a* ***σ-algebra*** *(more specifically, a* ***σ-algebra of subsets of*** *X) if whenever A is in $\mathscr{A}$ so is $X \setminus A$, and whenever $A_1, A_2, A_3, \ldots$ is a finite or countably infinite collection of sets belonging to $\mathscr{A}$, the union of the A_i also belongs to $\mathscr{A}$;*

(ii) *A* ***measurable space*** *is a pair $(X, \mathscr{A})$ in which X is a nonempty set and $\mathscr{A}$ is a σ-algebra of subsets of X.*

For brevity, whenever $(X, \mathscr{A})$ is a measurable space, the statement "A belongs to the σ-algebra $\mathscr{A}$" will usually be abbreviated to "A is an $\mathscr{A}$-set."

It is convenient to list some operations on sets belonging to a σ-algebra that leave us still in the σ-algebra.

LEMMA 1-2 *Let $(X, \mathscr{A})$ be a measurable space. Then:*

(i) *X is in $\mathscr{A}$;*

(ii) *the empty set $\varnothing$ is in $\mathscr{A}$;*

(iii) *if A_1 and A_2 are in $\mathscr{A}$, so is $A_1 \setminus A_2$;*

(iv) *if $A_1, A_2, A_3, \ldots$ is a finite or countably infinite collection of sets belonging to $\mathscr{A}$, the intersection $A_1 \cap A_2 \cap A_3 \cap \cdots$ belongs to $\mathscr{A}$.*

We can and do choose a set A that belongs to $\mathscr{A}$. By Definition 1-1, $X \setminus A$ belongs to the σ-algebra $\mathscr{A}$. Therefore so does the union $A \cup [X \setminus A]$, which is X, and (i) is proved. Since X is in $\mathscr{A}$, so is $X \setminus X$ by Definition 1-1, and (ii) is proved. If $A_1, A_2, A_3, \ldots$ belong to $\mathscr{A}$, by Definition 1-1 so do $X \setminus A_1$, $X \setminus A_2$, $X \setminus A_3, \ldots$; so does the union $[X \setminus A_1] \cup [X \setminus A_2] \cup [X \setminus A_3] \cup \cdots$; and so does

$$X \setminus ([X \setminus A_1] \cup [X \setminus A_2] \cup [X \setminus A_3] \cup \cdots).$$

This last set is $A_1 \cap A_2 \cap A_3 \cap \cdots$, so (iv) is proved. Finally, if A_1 and A_2 belong to $\mathscr{A}$, so does $X \setminus A_2$ by Definition 1-1, and so does $A_1 \cap [X \setminus A_2]$ by (iv). This last set is $A_1 \setminus A_2$, so (iii) is proved.

All the measures that we have defined in preceding chapters satisfy the requirements of the following definition.

DEFINITION 1-3 *Let $(X, \mathscr{A})$ be a measurable space. A* ***measure*** *on $(X, \mathscr{A})$ is a function m on $\mathscr{A}$ that takes values in $[0, \infty]$, is countably additive, and satisfies $m\varnothing = 0$; that is, whenever $A_1, A_2, A_3, \ldots$ is a finite or countable collection of pairwise disjoint sets all belonging to $\mathscr{A}$,*

$$m(A_1 \cup A_2 \cup A_3 \cup \cdots) = mA_1 + mA_2 + mA_3 + \cdots.$$

The alternative procedure for defining an integral is to postulate that we are given a measure on some measurable space and then to use this measure to construct an integral. This means, of course, that several of the key theorems will have to be proved again, from a different set of assumptions. But in addition to this, there is an important point to remember. It will be somewhat deceptively easy to prove statements about integrals from postulates about measure; it was easy to prove theorems about measure from theorems about integrals. The theories of measure and of integration are so closely related that postulating properties of measures is quite close to postulating properties of integrals. This can produce a spurious appearance of simplicity. What must not be forgotten is that whenever such a development of integration theory is to be used in any specific situation, it is necessary to show that the postulates of the theory are satisfied in that situation. Also, in order to be sure that the theory has any content at all, it is necessary to exhibit some system (preferably, some interesting system) that satisfies the postulates. To give a specific example, we could use the measurable space $(X, \mathscr{A})$ in which X is the real number system R and $\mathscr{A}$ is the collection of all subsets of R, and on this space we could postulate a measure m defined and nonnegative for all subsets of R, mB being the length of B when B is an interval, and m being invariant under translation. That is, whenever A and B are sets such that for some real c, x is in A if and only if $x + c$ is in B, it is true that $mA = mB$. With this measure many proofs can be simplified, but all are nonsense. It can be proved that no such measure m can exist, and all statements about it are statements about nothing at all.

Measurability of a function was defined in Definition III-10-1. But this form of definition is not usable here because it uses the idea of integral and we have not yet defined integrals. So instead we use a definition suggested by Theorem III-10-12.

Definition 1-4 *Let $\mathscr{A}$ be a collection of sets and f an extended-real-valued function on a set A_0 that belongs to $\mathscr{A}$. Then f is* ***measurable*** *with respect to $\mathscr{A}$ on A_0 if for each y in $\bar{R}$ the set $\{x \text{ in } A_0 : f(x) \leqq y\}$ belongs to $\mathscr{A}$.*

The statement that f is measurable with respect to $\mathscr{A}$ will usually be abbreviated to "f is $\mathscr{A}$-measurable" or even to "f is measurable."

Lemma 1-5 *Let $(X, \mathscr{A})$ be a measurable space, and let f be a function defined and extended-real-valued on a set A_0 that belongs to $\mathscr{A}$. Then the following four statements are equivalent:*

(i) *f is measurable on A_0 with respect to $\mathscr{A}$;*
(ii) *for each y in $\bar{R}$, the set $\{x \text{ in } A_0 : f(x) < y\}$ belongs to $\mathscr{A}$;*
(iii) *for each y in $\bar{R}$, the set $\{x \text{ in } A_0 : f(x) \geqq y\}$ belongs to $\mathscr{A}$;*
(iv) *for each y in $\bar{R}$, the set $\{x \text{ in } A_0 : f(x) > y\}$ belongs to $\mathscr{A}$.*

Suppose that (i) holds and that y is in $\bar{R}$. If $y = -\infty$, the set $\{x \text{ in } A_0 : f(x) < y\}$ is empty, and it belongs to $\mathscr{A}$ by Lemma 1-2. If $y > -\infty$, we can and do choose an ascending sequence of real numbers $y_1 < y_2 < y_3 < \cdots$ with limit y. If x is in A_0 and $f(x) < y$, then $y_n \geqq f(x)$ for all large n; if $y_n \geqq f(x)$ for some n, then $f(x) < y$. So

$$\{x \text{ in } A_0 : f(x) < y\} = \bigcup_{n=1}^{\infty} \{x \text{ in } A_0 : f(x) \leqq y_n\}.$$

The sets named in the right member are in $\mathscr{A}$, since we have assumed that (i) holds, so their union is in $\mathscr{A}$ because $\mathscr{A}$ is a σ-algebra. Therefore (ii) holds.

If (ii) holds and y is in $\bar{R}$, the set $\{x \text{ in } A_0 : f(x) < y\}$ belongs to $\mathscr{A}$ by hypothesis, and by Lemma 1-2 so does $A_0 \setminus \{x \text{ in } A_0 : f(x) < y\}$, which is $\{x \text{ in } A_0 : f(x) \geqq y\}$. Therefore (iii) holds.

If (iii) holds, let y belong to $\bar{R}$. If $y = \infty$, the set $\{x \text{ in } A_0 : f(x) > y\}$ is empty and therefore belongs to $\mathscr{A}$. If $y < \infty$, we can and do choose a sequence of real numbers $y_1 > y_2 > y_3 > \cdots$ with limit y. The sets $\{x \text{ in } A_0 : f(x) \geqq y_n\}$ $(n = 1, 2, 3, \ldots)$ belong to $\mathscr{A}$ by hypothesis, and their union is $\{x \text{ in } A_0 : f(x) > y\}$. Since $\mathscr{A}$ is a σ-algebra, this union belongs to $\mathscr{A}$, so (iv) holds.

If (iv) holds, and y is in $\bar{R}$, then

$$\{x \text{ in } A_0 : f(x) \leqq y\} = A_0 \setminus \{x \text{ in } A_0 : f(x) > y\}.$$

Both sets named in the right member belong to $\mathscr{A}$ by hypothesis, so the difference belongs to $\mathscr{A}$ by Lemma 1-2, and (i) holds. The proof is complete.

COROLLARY 1-6 *Let $(X, \mathscr{A})$ be a measurable space and A_0 a set that belongs to $\mathscr{A}$. If an extended-real-valued function f is measurable on A_0, so are $-f$, f^+, and f^-.*

For each y in $\bar{R}$,

$$\{x \text{ in } A_0 : -f(x) \leqq y\} = \{x \text{ in } A_0 : f(x) \geqq -y\},$$

and the last-named set is in $\mathscr{A}$ by Lemma 1-5, so $-f$ is measurable with respect to $\mathscr{A}$ on A_0. If $y \geqq 0$,

$$\{x \text{ in } A_0 : f^+(x) \leqq y\} = \{x \text{ in } A_0 : f(x) \leqq y\},$$

and the last-named set is in $\mathscr{A}$ by hypothesis. If $y < 0$,

$$\{x \text{ in } A_0 : f^+(x) \leqq y\} = \varnothing,$$

which is in $\mathscr{A}$ by Lemma 1-2. So in any case the set $\{x \text{ in } A_0 : f^+(x) \leqq y\}$ belongs to $\mathscr{A}$, and f^+ is $\mathscr{A}$-measurable. Also, $f^- = (-f)^+$, which is $\mathscr{A}$-measurable by the part of the proof already completed, so the proof of the corollary is complete.

The next theorem is a ready consequence of the definitions.

THEOREM 1-7 *Let $(X, \mathscr{A})$ be a measurable space, and let A_0 belong to $\mathscr{A}$. Then*

(i) *every function constant on A_0 is $\mathscr{A}$-measurable on A_0;*

(ii) *if $f_1, f_2, f_3, \ldots$ are extended-real-valued functions, each $\mathscr{A}$-measurable on A_0, the functions*

$$\sup\{f_n : n = 1,2,3,\ldots\}, \qquad \inf\{f_n : n = 1,2,3,\ldots\},$$
$$\limsup_{n\to\infty} f_n, \qquad \liminf_{n\to\infty} f_n$$

are all $\mathscr{A}$-measurable on A_0.

Let f have the constant value c on A_0. Then for each y in $\bar{R}$ the set

$$\{x \text{ in } A_0 : f(x) \leqq y\}$$

is empty if $y < c$ and is A_0 if $y \geqq c$. In either case it belongs to $\mathscr{A}$, so f is $\mathscr{A}$-measurable on A_0. This completes the proof of (i).

Suppose that $f_1, f_2, f_3, \ldots$ are all $\mathscr{A}$-measurable on A_0. Let y be in $\bar{R}$, and for each positive integer n define

$$A_n = \{x \text{ in } A_0 : f_n(x) > y\}.$$

Since each f_n is $\mathscr{A}$-measurable, each A_n belongs to $\mathscr{A}$. If a point x belongs to the union of the A_n, there is a positive integer j such that x is in A_j, and therefore

$$\sup\{f_n(x) : n = 1,2,3,\ldots\} \geqq f_j(x) > y.$$

Conversely, if $\sup\{f_n(x) : n = 1,2,3,\ldots\} > y$, there is some integer j for which $f_j(x) > y$, so x is in A_j and is therefore in the union of the A_n. Therefore,

$$\{x \text{ in } A_0 : \sup\{f_x(x) : n = 1,2,3,\ldots\} > y\} = \bigcup_{n=1}^{\infty} A_n.$$

This union belongs to $\mathscr{A}$, so by definition $\sup\{f_n : n = 1,2,3,\ldots\}$ is $\mathscr{A}$-measurable.

By Corollary 1-6, the functions $-f_n$ are $\mathscr{A}$-measurable. By the part of the proof already completed, so is the supremum of the $-f_n$, and again by Corollary 1-6 so is

$$-\sup\{-f_n : n = 1,2,3,\ldots\}.$$

But this last-named function is $\inf\{f_n : n = 1,2,3,\ldots\}$.

By the part of the proof already completed, for each positive integer k the function

$$g_k = \inf\{f_n : n = k+1, k+2, k+3, \ldots\}$$

is $\mathscr{A}$-measurable, and the supremum of the g_k for all positive integers k is also measurable. This is $\liminf_{n\to\infty} f_n$. Similarly, $\limsup_{n\to\infty} f_n$ is $\mathscr{A}$-measurable.

The functions most intimately related to a σ-algebra $\mathscr{A}$ are the indicators of sets belonging to $\mathscr{A}$. Next in closeness of relationship are the *simple* functions, which are the (finite) linear combinations of indicators of sets that belong to $\mathscr{A}$.

DEFINITION 1-8 *Let $(X, \mathscr{A})$ be a measurable space. A function g on X is a **simple** function if it is defined and real-valued on X and has finitely many different values, each assumed on an $\mathscr{A}$-set.*

It is easy to see that a function s on X is a simple function if and only if there are finitely many pairwise disjoint sets that belong to $\mathscr{A}$ and have union X, such that s has a constant real value on each of these sets.

LEMMA 1-9 *Let $(X, \mathscr{A})$ be a measurable space. If f and g are simple functions on X and a and b are real numbers, the functions*

$$f^+, f^-, |f|, f \vee g, f \wedge g, af + bg, fg$$

are simple functions on X.

Let f have the constant values $c_1, \ldots, c_h$ on the respective pairwise disjoint $\mathscr{A}$-sets $A_1, \ldots, A_h$, and let g have the constant values $d_1, \ldots, d_k$ on the respective pairwise disjoint $\mathscr{A}$-sets $B_1, \ldots, B_k$. Then the functions f^+, f^-, $|f|$ have the respective constant values $c_i \vee 0$, $(-c_i) \vee 0$, $|c_i|$ on A_i for $i = 1, \ldots, h$, and therefore they are simple functions. The functions $f \vee g, f \wedge g, af + bg, fg$ have the respective constant values

$$c_i \vee d_j,\ c_i \wedge d_j,\ ac_i + bd_j,\ c_i d_j$$

on the sets $A_i \cap B_j$ $(i = 1, \ldots, h; j = 1, \ldots, k)$, and these are pairwise disjoint $\mathscr{A}$-sets whose union is X. Therefore they too are simple functions.

It is obvious that simple functions are $\mathscr{A}$-measurable, and by Theorem 1-7, every function that is the limit of a sequence of simple functions is $\mathscr{A}$-measurable. The converse is also true; in fact, we can establish a stronger statement.

THEOREM 1-10 *Let $(X, \mathscr{A})$ be a measurable space, and let f be nonnegative and $\mathscr{A}$-measurable on a set A that belongs to $\mathscr{A}$. Then there exists an ascending sequence of nonnegative simple functions $s_1, s_2, s_3, \ldots$ that converges to f_A on X.*

As in previous chapters, f_A is the function that coincides with f on A and is 0 on $X \setminus A$.

The positive rational numbers form a denumerable set, so we can and do arrange them in a sequence $r_1, r_2, r_3, \ldots$. For each positive integer n we define

$$\text{(A)} \qquad s_n = \sup\{r_j 1_{\{f_A > r_j\}} : j = 1, \ldots, n\}.$$

Each set $\{f_A > r_j\}$ belongs to $\mathscr{A}$ by Lemma 1-5, so by Lemma 1-9 the s_n are simple functions. Obviously they are nonnegative and ascend as n increases. Let x be any point of X. If $f_A(x) = 0$, $s_n(x) = 0$; if $f_A(x) > 0$, $s_n(x)$ is the largest one of the numbers $0, r_1, \ldots, r_n$ that is less than $f_A(x)$. So in any case,

$$0 \leqq s_n(x) \leqq f_A(x).$$

In particular, if $f_A(x) = 0$,

(B) $$\lim_{n\to\infty} s_n(x) = f_A(x).$$

If $f_A(x) > 0$, let c be any number less than $f_A(x)$. We can and do choose a rational number greater than c; this rational number is r_k for some k. Then for $n > k$, (A) defines $s_n(x)$ as the supremum of a collection of numbers including r_k, so

$$s_n(x) \geqq r_k > c.$$

Therefore (B) holds in this case also, and the proof is complete.

COROLLARY 1-11 *Let $(X, \mathscr{A})$ be a measurable space, and let f be a function that is $\mathscr{A}$-measurable on an $\mathscr{A}$-set A. Then there exists a sequence $s_1, s_2, s_3, \ldots$ of simple functions such that for each x in X,*

$$|s_n(x)| \leqq |f_A(x)|$$

and

$$\lim_{n\to\infty} s_n(x) = f_A(x).$$

By Corollary 1-6 and Theorem 1-10, there are two ascending sequences of nonnegative simple functions $s_1^+, s_2^+, s_3^+, \ldots$ and $s_1^-, s_2^-, s_3^-, \ldots$ that tend to f_A^+ and to f_A^-, respectively. Then the functions

$$s_n = s_n^+ - s_n^-$$

are simple functions that tend to f_A at every point in X. If $f_A(x) \geqq 0$, we have

$$0 \leqq s_n^+(x) \leqq f_A^+(x), \qquad 0 = s_n^-(x) = f_A^-(x),$$

and if $f_A(x) \leqq 0$, we have

$$0 = s_n^+(x) = f_A^+(x), \qquad 0 \leqq s_n^-(x) \leqq f_A^-(x).$$

These inequalities imply that

$$|s_n(x)| \leqq |f_A(x)|$$

for all x in X. The proof is complete.

COROLLARY 1-12 *Let $(X, \mathscr{A})$ be a measurable space and A an $\mathscr{A}$-set. If f and g are extended-real-valued functions that are $\mathscr{A}$-measurable on A, and a and b are real numbers, then fg is $\mathscr{A}$-measurable on A, and so is $af + bg$ if it is defined at each point of A.*

By Corollary 1-11, there are sequences $s'_1, s'_2, s'_3, \ldots$ and $s''_1, s''_2, s''_3, \ldots$ of simple functions that satisfy

(C) $$\lim_{n\to\infty} s'_n(x) = f_A(x), \qquad \lim_{n\to\infty} s''_n(x) = g_A(x)$$

and

(D) $$|s'_n(x)| \leqq |f_A(x)|, \qquad |s''_n(x)| \leqq |g_A(x)|$$

for all x in X and all positive integers n. Then by (C) and (D), $as'_n(x) + bs''_n(x)$ tends to $af_A(x) + bg_A(x)$ for all x in X, and by Theorem 1-7, $af + bg$ is $\mathscr{A}$-measurable on A. If neither $f_A(x)$ nor $g_A(x)$ is 0, (C) implies that $s'_n(x)s''_n(x)$ tends to $f_A(x)g_A(x)$. If $f_A(x) = 0$, then $s'_n(x) = 0$ for all n, so again we have

(E) $$\lim_{n\to\infty} s'_n(x)s''_n(x) = f_A(x)g_A(x);$$

and likewise (E) holds if $g_A(x) = 0$. So $s'_ns''_n$ tends to f_Ag_A everywhere in X, and by Theorem 1-7, fg is $\mathscr{A}$-measurable on A.

When m is an additive nonnegative function of left-open intervals in R^r, a function f on R^r is m-measurable on R^r if it is the limit of a sequence of m-integrable functions, by Definition III-10-1. If $\mathscr{M}$ denotes the σ-algebra of m-measurable subsets of R^r, the meaning of the statement that f is $\mathscr{M}$-measurable is given in Definition 1-4. We shall now show that for such interval-functions m, the two concepts of measurability are equivalent. Suppose, first, that f is m-measurable and vanishes outside a bounded interval B. Then by Lemma III-10-3, for each real y the function $[f \vee (y1_B)] \wedge [(y+1)1_B]$ is integrable over R^r, and so is

$$g = (y+1)1_B - [f \vee (y1_B)] \wedge [(y+1)1_B].$$

From here on we follow the proof of Theorem III-10-12 with only trivial changes and deduce that the set $B \cap \{f \leqq y\}$ is m-measurable for each y in R and therefore for each y in $\bar{R}$. If we define

(F) $$W[n] = (-n, n] \times \cdots \times (-n, n] \qquad (r \text{ factors}),$$

this shows that the set

$$W[n] \cap \{f1_{W[n]} \leqq y\}$$

is m-measurable for all y in $\bar{R}$ and all n, and the union of these sets for $n = 1, 2, 3, \ldots$ is the set $\{f \leqq y\}$. So $\{f \leqq y\}$ is in $\mathscr{M}$ for all y in $\bar{R}$, and f is $\mathscr{M}$-measurable.

Conversely, let f be $\mathscr{M}$-measurable. It is then the limit of a sequence of simple functions $s_1, s_2, s_3, \ldots$. From the Definition II-12-1 of measurable set if follows easily that if A is an m-measurable set and B is a bounded left-open interval, $A \cap B$ has finite measure, and its indicator is integrable. If s is a simple function, it has the form

$$s = \sum_{j=1}^{k} c_j 1_{A(j)},$$

in which the c_j are nonzero real numbers and the $A(j)$ are m-measurable sets.

Then if $W[n]$ is defined by (F),

$$s1_{W[n]} = \sum_{j=1}^{k} c_j 1_{W[n] \cap A(j)},$$

which is integrable with respect to m over R^r. This holds for all simple functions s, so, in particular, it holds for the s_n, and so each product $s_n 1_{W[n]}$ is integrable with respect to m over R^r. The product $s_n 1_{W[n]}$ tends everywhere in R^r to f, so by Definition III-10-1, f is m-measurable.

EXERCISE 1-1 Let $(X, \mathscr{A})$ be a measurable space and $A_1, A_2, A_3, \ldots$ a sequence of $\mathscr{A}$-sets. Prove that

(i) the set B of all points of X that belong to infinitely many sets A_n is an $\mathscr{A}$-set;
(ii) the set C of all points of X that belong to all but finitely many of the A_n is an $\mathscr{A}$-set.

EXERCISE 1-2 Prove the statement in the last sentence of the paragraph following Definition 1-3 by verifying the following assertions.

(i) If together with each real number x we class all real numbers whose difference from x is rational, every real number belongs to exactly one such class.
(ii) Each such class has nonempty intersection with $[0, 1]$.

Now from each class select just one member of that class that is in $[0, 1]$. Let B be the set of numbers thus selected. For each r, let $r + B$ denote the set of numbers $r + x$ with x in B.

(iii) If r and s are two different rationals, $r + B$ and $s + B$ are disjoint.
(iv) If $r_1, r_2, r_3, \ldots$ are all the rationals in $[-1, 1]$, the union of the sets $r_n + B$ $(n = 1, 2, 3, \ldots)$ contains $(0, 1)$ and is contained in $[-1, 2]$.
(v) If m is a nonnegative countably additive measure on the σ-algebra of all subsets of R, and m is invariant under translation, and $m[-1, 2]$ is finite, by (iv) we must have $mB = 0$, and then we cannot have $m[0, 1) = 1$.

2. Definition of the Lebesgue Integral

We have defined measure in Definition 1-3. It is clearly reasonable to define the integral of the indicator of a measurable set to be the measure of that set, as we did in previous chapters:

$$\text{(A)} \qquad \int_X 1_A(x)\, m(dx) = \int_A 1\, m(dx) = mA \qquad (A \text{ in } \mathscr{A}).$$

Suppose that we return to the situation in Chapter III, in which m is a nonnegative additive function of left-open intervals in a space R^r and this m is then extended to be a measure m on a family of m-measurable sets, which family we shall call $\mathscr{M}$. For each positive integer n the collection of all those m-measurable sets that are contained in the cube

$$\text{(B)} \qquad W[n] = (-n, n] \times \cdots \times (-n, n]$$

in R^r is a σ-algebra of subsets of $W[n]$, and m is countably additive on it, as can be shown by a trifling modification of the proofs of Theorems III-10-8 and III-10-10. If B is a subset of R^r such that $B \cap W[n]$ has finite m-measure for each n, the integrable functions $1_{B \cap W[n]}$ $(n = 1, 2, 3, \ldots)$ ascend and converge to 1_B, so by Theorem II-12-7 and Definition II-12-1, the set B is m-measurable and (A) is satisfied. Conversely, if B is m-measurable, by Lemma III-10-3 and Theorem III-10-5, the set $B \cap W[n]$ has finite m-measure for each positive integer n. From this it follows that M is a σ-algebra of sets even if $mR^r = \infty$.

Suppose next that s is a simple function with the respective values $c_1, \ldots, c_k$ on the pairwise disjoint m-measurable sets $B(1), \ldots, B(k)$. Then

$$\text{(C)} \qquad s(x) = \sum_{j=1}^{k} c_j 1_{B(j)}.$$

If A has finite m-measure, by applying (A) to each set $A \cap B(j)$ we obtain

$$\text{(D)} \qquad \int_A s(x)\, m(dx) = \sum_{j=1}^{k} c_j m[A \cap B(j)].$$

If A is any m-measurable set, we apply (D) to each intersection $A \cap W[n]$ and then let n increase; this yields (D) for such sets A also, provided that the sum in the right member of (D) exists.

This suggests the following definition of the integral of a simple function for any measure on any measurable space.

DEFINITION 2-1 *Let m be a measure on a measurable space $(X, \mathscr{A})$; let A belong to $\mathscr{A}$, and let s be a simple function given by equation* (C). *Then*

$$\int_A s(x)\, m(dx) = \sum_{j=1}^{k} c_j m[A \cap B(j)],$$

provided that the sum in the right member exists.

The sum in the equation in Definition 2-1 exists unless there are two values j', j'' of j such that $A \cap B(j')$ and $A \cap B(j'')$ both have measure ∞, while one of $c_{j'}$, $c_{j''}$ is positive and the other negative. According to the discussion preceding Definition 2-1, if m is defined by (A) in terms of integration with respect to a nonnegative additive function of left-open intervals in R^r and $\mathscr{A}$ is the class of all m-measurable sets in R^r, the integral of a simple function s

as defined in Definition 2-1 is the same as the gauge-integral of s as defined in Definition III-7-1.

We still have to show that the integral of s defined in Definition 2-1 is not dependent on the particular representation that we choose for the function s. Let s be defined by (C), and let $c'_1, \ldots, c'_h$ be the different nonzero values assumed by s. If we define

$$A'(i) = \{x \text{ in } X : s(x) = c'_i\},$$

we can write s in the form

$$s = \sum_{i=1}^{h} c'_i 1_{A'(i)}.$$

This we call the **canonical representation** of s. Since $A'(i)$ is the union of the sets $B(j)$ for which $c_j = c'_i$,

$$\sum_{j=1}^{k} c_j m[A \cap B(j)] = \sum_{i=1}^{h} \sum_{c_j = c'_i} c'_i m[A \cap B(j)] = \sum_{i=1}^{h} c'_i m[A'(i) \cap A].$$

The last sum is independent of the representation (C) of s with which we started.

It is obvious that if s is nonnegative, so is its integral. The integral is a linear functional of s. For let s' and s'' be the two simple functions

$$s' = \sum_{i=1}^{h} c'_i 1_{A'(i)}, \qquad s'' = \sum_{j=1}^{k} c''_j 1_{A''(j)},$$

and let a and b be real numbers. Then

$$as' + bs'' = \sum_{i=1}^{h} \sum_{j=1}^{k} [ac'_i + bc''_j] 1_{A'(i) \cap A''(j)},$$

so

$$\begin{aligned}
&\int_A [as'(x) + bs''(x)]\, m(dx) \\
&\quad = \sum_{i=1}^{h} \sum_{j=1}^{k} [ac'_i + bc''_j] m[A'(i) \cap A''(j) \cap A] \\
&\quad = a \sum_{i=1}^{h} c'_i \left\{ \sum_{j=1}^{k} m[A'(i) \cap A''(j) \cap A \right\} \\
&\qquad + b \sum_{j=1}^{k} c''_j \left\{ \sum_{i=1}^{h} m[A'(i) \cap A''(j) \cap A] \right\} \\
&\quad = a \sum_{i=1}^{h} c'_i m[A'(i) \cap A] + b \sum_{j=1}^{k} c''_j m[A''(j) \cap A] \\
&\quad = a \int_A s'(x)\, m(dx) + b \int_A s''(x)\, m(dx).
\end{aligned}$$

Theorem 1-10 suggests that we define the integral of a nonnegative $\mathscr{A}$-measurable function f to be the limit of the integrals of the simple functions in an ascending sequence that converges to f. We prefer to use a different but equivalent wording.

DEFINITION 2-2 *Let m be a measure on a measurable space $(X, \mathscr{A})$; let A be a set that belongs to $\mathscr{A}$, and let f be a function that is nonnegative and $\mathscr{A}$-measurable on A. Then the integral*

$$\int_A f(x)\, m(dx)$$

is defined to be the supremum of the integrals over A of all simple functions s such that on A,

$$0 \leqq s \leqq f.$$

This integral is evidently defined, unique, and nonnegative (possibly ∞) for all nonnegative measurable f and all sets A in $\mathscr{A}$. To compute with it, it is convenient to relate it to the integrals of ascending sequences of simple functions.

The conclusions of Corollary III-10-11 hold for measures in general. We prove one of them.

LEMMA 2-3 *If $A_1, A_2, A_3, \ldots$ is an expanding sequence of members of $\mathscr{A}$,*

$$\text{(E)} \qquad m \bigcup_{j=1}^{\infty} A_j = \lim_{j \to \infty} mA_j.$$

Let A denote the union of the sets $A_1, A_2, A_3, \ldots$. If the right member of (E) is ∞, for all positive integers j we have

$$mA = mA_j + m(A \setminus A_j) \geqq mA_j,$$

and since mA_j increases without bound as j increases, we must have $mA = \infty$, so that (E) holds. If the right member of (E) is a finite number c, we define A_0 to be the empty set, and for each positive integer n we define

$$D_n = A_n \setminus A_{n-1}.$$

The D_n are pairwise disjoint members of $\mathscr{A}$, and their union is A, so

$$mA = \sum_{n=1}^{\infty} mD_n = \lim_{n \to \infty} \sum_{j=1}^{n} [mA_n - mA_{n-1}] = \lim_{n \to \infty} mA_n,$$

so equation (E) holds in this case also.

LEMMA 2-4 *Let m be a measure on a measurable space $(X, \mathscr{A})$. Let $s, s_1, s_2, s_3, \ldots$ be nonnegative simple functions such that on a set A that belongs to $\mathscr{A}$,*

$$s_1(x) \leqq s_2(x) \leqq s_3(x) \leqq \cdots \qquad \textit{and} \qquad \lim_{n\to\infty} s_n(x) \geqq s(x).$$

Then

$$\text{(F)} \qquad \lim_{n\to\infty} \int_A s_n(x)\, m(dx) \geqq \int_A s(x)\, m(dx).$$

Let

$$s = \sum_{i=1}^{k} c_i 1_{A(i)}$$

be the canonical representation of s, so that the c_i are all positive and different. Without loss of generality we may assume that the notation has been chosen so that $A(i) \cap A$ has positive measure for $i = 1, \ldots, h$ and measure 0 for $i = h + 1, \ldots, k$. Let r be any number less than the right member of (F). Then for each i in $\{1, \ldots, h\}$ we can and do choose positive numbers c_i', m_i' such that

$$c_i' < c_i, \qquad m_i' < m[A(i) \cap A]$$

and

$$\text{(G)} \qquad \sum_{i=1}^{h} c_i' m_i' > r.$$

For $i = 1, \ldots, h$ the sets

$$B_{i,n} = \{x \text{ in } A(i) \cap A : s_n(x) > c_i'\} \qquad (n = 1, 2, 3, \ldots)$$

expand as n increases. If x is any point in $A(i) \cap A$, the numbers $s_n(x)$ tend to a limit that is at least as great as the value c_i of $s(x)$ and is therefore greater than c_i'. Then x is in $B_{i,n}$ for all large n. So $A(i) \cap A$ is the union of the $B_{i,n}$ for $n = 1, 2, 3, \ldots$, and by Lemma 2-3 the measures $mB_{i,n}$ tend to $m[A(i) \cap A]$ as n increases. Therefore we can and do choose an integer $n(i)$ such that if $n > n(i)$,

$$mB_{i,n} > m_i'.$$

Let n' be the greatest of the numbers $n(1), \ldots, n(h)$. Then for all n greater than n'

$$\int_A s_n(x)\, m(dx) \geqq \sum_{i=1}^{h} c_i' m B_{i,n} \geqq \sum_{i=1}^{h} c_i' m_i' > r.$$

This holds for all r less than the right member of (F), so (F) is satisfied, and the proof is complete.

THEOREM 2-5 *Let m be a measure on a measurable space $(X, \mathscr{A})$; let A be an $\mathscr{A}$-set and f a function that is nonnegative and $\mathscr{A}$-measurable on A. Let*

$s_1, s_2, s_3, \ldots$ be a sequence of simple functions nonnegative and ascending on A such that $s_n(x)$ converges to $f(x)$ for all x in A. Then

$$\text{(H)} \qquad \lim_{n\to\infty} \int_A s_n(x)\, m(dx) = \int_A f(x)\, m(dx).$$

By Definition 2-2, the left member of (H) cannot be greater than the right member. Let r be any number less than the right member of (H). By Definition 2-2, there exists a simple function s such that $0 \leqq s \leqq f$ on A and

$$\int_A s(x)\, m(dx) > r.$$

For the functions $s_1, s_2, s_3, \ldots$ of the theorem, the limit as n increases is $f(x)$ for all x in A, and $f(x) \geqq s(x)$, so by Lemma 2-4,

$$\lim_{n\to\infty} \int_A s_n(x)\, m(dx) \geqq \int_A s(x)\, m(dx) > r.$$

Therefore the left member of (H) is not less than the right member, and (H) is satisfied.

COROLLARY 2-6 *Let m be a measure on a measurable space $(X, \mathscr{A})$. Let A be a set that belongs to $\mathscr{A}$; let f and g be nonnegative functions measurable on A, and let a and b be nonnegative numbers. Then*

$$\int_A [af(x) + bg(x)]\, m(dx) = a \int_A f(x)\, m(dx) + b \int_A g(x)\, m(dx).$$

Let $s'_1, s'_2, s'_3, \ldots$ be a sequence of simple functions that on A are ascending and nonnegative and tend to f, and let $s''_1, s''_2, s''_3, \ldots$ be a sequence of simple functions that on A are ascending and nonnegative and tend to g. As proved just before Definition 2-2,

$$\int_A [as'_n(x) + bs''_n(x)]\, m(dx) = a \int_A s'_n(x)\, m(dx) + b \int_A s''_n(x)\, m(dx).$$

As n increases, the simple functions $as'_n + bs''_n$ are nonnegative and ascending on A, and they converge to $af + bg$. By Theorem 2-5, the conclusion of this corollary is valid.

We are now ready to extend the definition of the integral to a larger class of functions.

DEFINITION 2-7 *Let m be a measure on a measurable space $(X, \mathscr{A})$; let A be an $\mathscr{A}$-set and f an extended-real-valued function that is $\mathscr{A}$-measurable on A. Unless* *the integrals of both f^+ and f^- over A are both ∞, we define*

$$\int_A f(x)\,m(dx) = \int_A f^+(x)\,m(dx) - \int_A f^-(x)\,m(dx).$$

If at least one of the integrals in the right member is finite, we say that f ***has a Lebesgue integral*** *with respect to m over A. If both f^+ and f^- have finite integrals over A, we say that f is (Lebesgue)* ***integrable*** *or is (Lebesgue)* ***summable*** *with respect to m over A.*

COROLLARY 2-8 *With $(X, \mathscr{A})$, A, and m as in Definition 2-7, let f be a function on A that is the difference $p - n$ of two functions both nonnegative and integrable over A and such that for no x in A are both $p(x)$ and $n(x)$ equal to ∞. Then*

$$\int_A f(x)\,m(dx) = \int_A p(x)\,m(dx) - \int_A n(x)\,m(dx).$$

Since p and n are $\mathscr{A}$-measurable on A, so is their difference f, and so therefore are f^+ and f^-. For all x in A

(I) $$p(x) - n(x) = f(x) = f^+(x) - f^-(x).$$

At each x in A at which $f(x)$ is finite, the other four numbers in (I) are also finite, and

(J) $$p(x) + f^-(x) = f^+(x) + n(x).$$

Where $f(x) = \infty$, both $p(x)$ and $f^+(x)$ are ∞ and neither $n(x)$ nor $f^-(x)$ is ∞, so (J) holds. Similarly (J) holds where $f(x) = -\infty$. Since

$$f^+(x) = \max[p(x) - n(x), 0] \leqq p(x),$$

and likewise

$$f^-(x) \leqq n(x),$$

the integrals of both f^+ and f^- are finite, and f is integrable. By integrating both members of (J) and recalling Corollary 2-6, we obtain

$$\int_A p(x)\,m(dx) + \int_A f^-(x)\,m(dx) = \int_A f^+(x)\,m(dx) + \int_A n(x)\,m(dx).$$

Hence,

$$\begin{aligned}\int_A p(x)\,m(dx) - \int_A n(x)\,m(dx) &= \int_A f^+(x)\,m(dx) - \int_A f^-(x)\,m(dx)\\ &= \int_A f(x)\,m(dx),\end{aligned}$$

which completes the proof.

COROLLARY 2-9 *Let $(X, \mathscr{A})$, A, and m be as in Definition 2-7. Let f and g be extended-real-valued functions that are $\mathscr{A}$-measurable over A, and let c be a real number. Then*

(i) *if f has an integral over A, so has cf, and*

$$\int_A cf(x)\, m(dx) = c \int_A f(x)\, m(dx);$$

(ii) *if f and g are integrable over A, and the sum $f(x) + g(x)$ exists for every x in A, then $f + g$ is integrable over A, and*

$$\int_A [f(x) + g(x)]\, m(dx) = \int_A f(x)\, m(dx) + \int_A g(x)\, m(dx).$$

If $c = 0$, (i) is trivial. If $c < 0$,

$$\text{(K)} \qquad (cf)^+ = (-c)f^-, \qquad (cf)^- = (-c)f^+.$$

Since f has an integral over A, the integrals of f^+ and f^- over A are not both ∞, so by Corollary 2-6 the integrals of $(-c)f^-$ and $(-c)f^+$ are not both ∞. By (K), $(cf)^+$ and $(cf)^-$ have integrals over A that are not both ∞. So cf has an integral, and

$$\begin{aligned} \int_A (cf(x))\, m(dx) &= \int_A (cf(x))^+\, m(dx) - \int_A (cf(x))^-\, m(dx) \\ &= (-c) \int_A f^-(x)\, m(dx) - (-c) \int_A f^+(x)\, m(dx) \\ &= c \int_A f(x)\, m(dx). \end{aligned}$$

Therefore (i) holds if $c < 0$.

If $c > 0$, we apply the proof just completed first to $(-c)f$ and then to $(-1)[(-c)f]$.

For (ii) we have at all x in A

$$\text{(L)} \qquad f(x) + g(x) = [f^+(x) + g^+(x)] - [f^-(x) + g^-(x)].$$

Each of the four functions named in the right member is integrable over A, so by Corollary 2-6 the nonnegative functions $f^+ + g^+$ and $f^- + g^-$ are integrable over A. By (L) and Corollary 2-8, $f + g$ is integrable over A; and by Corollaries 2-8 and 2-6,

$$\begin{aligned} \int_A [f(x) + g(x)]\, m(dx) &= \int_A [f^+(x) + g^+(x)]\, m(dx) \\ &\quad - \int_A [f^-(x) + g^-(x)]\, m(dx) \\ &= \int_A f(x)\, m(dx) + \int_A g(x)\, m(dx). \end{aligned}$$

This completes the proof.

We can now prove many of the well-known theorems about integrals; the proofs are easy, and we list them among the exercises. However, the monotone convergence theorem played a central role in preceding chapters, and we shall exhibit its proof for the Lebesgue integral also. The proof is quite short because when we assumed that the measure is countably additive, we had already come close to assuming that the monotone convergence theorem holds. The proof in Chapter II is longer, which is not surprising. There we started from a function of intervals that was assumed only to be finitely additive and nonnegative, and it was that proof of the monotone convergence theorem that brought countable additivity into the theory.

THEOREM 2-10 *Let m be a measure on a measurable space $(X, \mathscr{A})$, and let A be a set that belongs to $\mathscr{A}$. Let $f_1, f_2, f_3, \ldots$ be an ascending sequence of functions integrable over $\mathscr{A}$, and let f be their limit. Then f has an integral with respect to m over A, and*

$$\text{(M)} \qquad \int_A f(x)\, m(dx) = \lim_{n\to\infty} \int_A f_n(x)\, m(dx).$$

(Note that the integrals of the f_n are assumed to be finite, but the integral of f may be ∞.)

For simplicity we shall prove this only under the added hypothesis that $f_1(x)$ is never $-\infty$. This restriction can easily be removed with the help of Exercises 2-1 and 2-3. It is enough to prove the theorem for nonnegative f_n since if it holds for nonnegative f_n we can apply it to the sequence $f_1 - f_1, f_2 - f_1, f_3 - f_1, \ldots$ and obtain the conclusion for arbitrary ascending sequences with f_1 never $-\infty$.

Since each f_n is integrable, we can and do select an ascending sequence of nonnegative simple functions $s_{n,1}, s_{n,2}, s_{n,3}, \ldots$ that converge everywhere on A to f_n. Then by Theorem 2-5,

$$\text{(N)} \qquad \lim_{j\to\infty} \int_A s_{n,j}(x)\, m(dx) = \int_A f_n(x)\, m(dx).$$

For each positive integer j we define

$$s'_j = s_{1,j} \vee s_{2,j} \vee \cdots \vee s_{j,j}.$$

Then at each x in X, and for each positive integer j,

$$\begin{aligned} s'_{j+1}(x) &= \max\{s_{1,j+1}(x), \ldots, s_{j+1,j+1}(x)\} \\ &\geqq \max\{s_{1,j+1}(x), \ldots, s_{j,j+1}(x)\} \\ &\geqq \max\{s_{1,j}(x), \ldots, s_{j,j}(x)\} \\ &= s'_j(x). \end{aligned}$$

Thus, the s'_j form an ascending sequence of simple functions. For each x in A and each positive integer j, $s'_j(x)$ is one of the numbers $s_{1,j}(x), \ldots, s_{j,j}(x)$, so we can and do choose a k such that $1 \leqq k \leqq j$ and

$$s'_j(x) = s_{k,j}(x).$$

Since the numbers $s_{k,1}(x)$, $s_{k,2}(x)$, $s_{k,3}(x), \ldots$ ascend and converge to $f_k(x)$, they are all at most equal to $f_k(x)$. Therefore,

$$\text{(O)} \qquad s'_j(x) = s_{k,j}(x) \leqq f_k(x) \leqq f(x).$$

On the other hand, if r is any number less than $f(x)$, for some positive integer k we have

$$f_k(x) > r.$$

We choose such a k. Since $s_{k,j}(x)$ converges to $f_k(x)$ as j increases, we can and do choose a j such that

$$s_{k,j}(x) > r.$$

For each integer i greater than $\max\{k,j\}$,

$$s'_i(x) \geqq s_{k,i}(x) \geqq s_{k,j}(x) > r.$$

This and (O) show that $s'_j(x)$ tends to $f(x)$ as j increases, at each x in A. Therefore, by Theorem 2-5,

$$\text{(P)} \qquad \lim_{j \to \infty} \int_A s'_j(x)\, m(dx) = \int_A f(x)\, m(dx).$$

We have shown that for each positive integer j, (O) holds at each x in A, so with the k of (O),

$$s'_j(x) \leqq f_k(x) \leqq f_j(x).$$

Therefore,

$$\text{(Q)} \qquad \int_A s'_j(x)\, m(dx) \leqq \int_A f_j(x)\, m(d) \leqq \int_A f(x)\, m(dx).$$

Statements (P) and (Q) imply (M), and the proof is complete.

Suppose that m is a nonnegative additive function of left-open intervals in a space R^r and that $\mathscr{M}$ is the family of m-measurable subsets of R^r. We have seen that $(R^r, \mathscr{M})$ is a measurable space. We can now prove that an extended-real-valued function f on R^r has a gauge-integral with respect to m over R^r if and only if it has a Lebesgue integral with respect to the measure m generated by the interval-function m. Suppose first that f is nonnegative. If f has a gauge-integral over R^r, it is m-measurable (hence $\mathscr{M}$-measurable) over R^r, and therefore there exists an ascending sequence of simple functions $s_1, s_2, s_3, \ldots$ converging to f. We saw in the discussion at the beginning of this section that the gauge-integral of s_n and its integral as defined in Definition 2-1 are equal. These integrals converge to the gauge-integral of f by Theorem II-12-6 and to the Lebesgue integral of f by Theorem 2-5 or Theorem 2-10, so the Lebesgue integral exists and is equal to the gauge-integral of f. Conversely, if the Lebesgue integral of f exists, there is an ascending sequence of nonnegative simple functions $s_1, s_2, s_3, \ldots$ that converge to f. Their integrals converge to the Lebesgue integral of f by Theorem 2-5 and to the gauge-integral of f by Theorem II-12-6, so the gauge-integral exists and is equal to the Lebesgue integral.

If f is Lebesgue-integrable, so are f^+ and f^-. As just proved, these have gauge-integrals with the same value, so f has a gauge-integral, and its value is

$$\int_{R^r} f^+(x)\,m(dx) - \int_{R^r} f^-(x)\,m(dx),$$

which is also the value of the Lebesgue integral. A similar proof applies if f has a gauge-integral with respect to m. So, for integrals of the sort defined in Chapter III the two kinds of integral do not have to be distinguished. This can also be shown to be true, by almost exactly the same proof, when m is a probability measure on a space R^T with T countable, as in Section IV-14.

An example of a different type, in which the space X does not have to have the form R^T for any T, is provided by what is called *counting measure*. Let X be any nonempty set. We define $\mathscr{A}$ to be the family of all subsets of X; this is evidently a σ-algebra. For each set A in $\mathscr{A}$ we define $m_{\#}A$ to be the number of points in A if that number is finite and to be ∞ if there are infinitely many points in A. This is easily seen to be nonnegative and countably additive on $\mathscr{A}$, so it is a measure on $(X, \mathscr{A})$.

If s is a nonnegative simple function, its canonical representation has the form

$$s = \sum_{j=1}^{k} c_j 1_{B(j)},$$

in which all c_j are positive. The integral

$$\int_X s(x)\,m_{\#}(dx)$$

is finite if and only if each $B(j)$ is a finite set, in which case the union

$$C(S) = \{x \text{ in } X : s(x) > 0\}$$

is a finite set. If f is nonnegative and integrable with respect to $m_{\#}$ over X, it is the limit of an ascending sequence of nonnegative simple functions $s_1, s_2, s_3, \ldots$, each integrable over X. Then each of the sets

$$C(s_n) = \{x \text{ in } X : s_n(x) > 0\}$$

is finite, and their union

$$\bigcup_{n=1}^{\infty} C(s_n) = \{x \text{ in } X : f(x) > 0\}$$

is countable. So if f is nonnegative and integrable with respect to $m_\#$ over X, it is 0 except at countably many points of X. If f is any function integrable with respect to $m_\#$ over X, both f^+ and f^- are 0 except on a countable set, so $f(x) = 0$ except on a countable set.

Suppose that f is nonnegative and is 0 except on the set $\{x_1, x_2, x_3, \ldots\}$. For each positive integer n we define

$$s_n(x) = f(x) \qquad (x = x_1, \ldots, x_n),$$
$$s_n(x) = 0 \qquad (\text{all other } x \text{ in } X).$$

These form an ascending sequence of simple functions, and they converge everywhere to $f(x)$, so

$$\begin{aligned}\int_X f(x)\, m_\#(dx) &= \lim_{n\to\infty} \int_X s_n(x)\, m_\#(dx) \\ &= \lim_{n\to\infty} [f(x_1) + \cdots + f(x_n)] \\ &= \sum_{j=1}^{\infty} f(x_j).\end{aligned}$$

Therefore, an extended-real-valued function f on X is integrable over X with respect to $m_\#$ if and only if the series

$$f(x_1) + f(x_2) + f(x_3) + \cdots$$

is absolutely convergent, and in that case the integral is equal to the sum of the series. Thus, the theory of the integral with respect to counting measure is a generalization of the theory of absolutely convergent series.

EXERCISE 2-1 Prove that if f is integrable with respect to m over an $\mathscr{A}$-set A, $f(x)$ is finite almost everywhere in A. *Suggestion*: Let N be the set $\{x \text{ in } A : f(x) = \infty\}$. For each n, $n1_N$ is a simple function $\leqq |f|$ on A, so $|f| - n1_N$ has a nonnegative integral over A.

EXERCISE 2-2 Prove that if f is nonnegative and $\int_A f(x)m(dx) = 0$, then $f(x) = 0$ for almost all x in A.

EXERCISE 2-3 Prove that if f is integrable over A and $g = f$ almost everywhere in A, then g is integrable over A and has the same integral as f.

EXERCISE 2-4 Generalize the dominated converge theorem (Theorem II-12-8) to the Lebesgue integral over an arbitrary X.

EXERCISE 2-5 Prove Fatou's lemma: if $g, f_1, f_2, f_3, \ldots$ are functions integrable with respect to m over an $\mathscr{A}$-set A and

$$f_n(x) \geqq g(x) > -\infty \qquad (n = 1, 2, 3, \ldots;\ x \text{ in } A),$$

then

$$\int_A \left\{ \liminf_{n\to\infty} f_n(x) \right\} m(dx) \leqq \liminf_{n\to\infty} \int_A f_n(x)\, m(dx).$$

EXERCISE 2-6 Prove that if f is integrable over an $\mathscr{A}$-set A, it is integrable over every $\mathscr{A}$-set contained in A.

EXERCISE 2-7 Prove that if f is nonnegative and is integrable over X and F is the function on $\mathscr{A}$ defined by

$$F(A) = \int_A f(x) m(dx),$$

then F is a measure on $\mathscr{A}$.

EXERCISE 2-8 Let f be integrable over X, and for each $\mathscr{A}$-set A define $F(A)$ as in the preceding exercise. Prove that to each positive ε there corresponds a positive δ such that if A belongs to $\mathscr{A}$ and $mA < \delta$, then $F(A) < \varepsilon$. *Suggestion*: Choose a simple function s such that $0 \leqq s \leqq f$ and

$$\int_X s(x)\, m(dx) > \int_X f(x)\, m(dx) - \frac{\varepsilon}{2}.$$

This s is bounded.

EXERCISE 2-9 Let $m_{\#}$ be counting measure on an infinite set X, and let p be a nonnegative function on X whose integral with respect to $m_{\#}$ over X is 1. Define

$$P(A) = \int_A p(x)\, m_{\#}(dx)$$

for all subsets A of X. Let $\{x_1, x_2, x_3, \ldots\}$ be the set of points of X at which p is positive. Prove that a function f on X is integrable with respect to P over X if and only if the series

$$f(x_1)p(x_1) + f(x_2)p(x_2) + f(x_3)p(x_3) + \cdots$$

is absolutely convergent; and in that case

$$\int_X f(x)\, P(dx) = \int_X f(x)p(x)\, m_{\#}(dx) = \sum_{j=1}^{\infty} f(x_j)p(x_j).$$

EXERCISE 2-10 Let m be a measure on a measure space $(X, \mathscr{A})$ that is **complete**, meaning that if A is an $\mathscr{A}$-set and $mA = 0$, then every subset of A is an $\mathscr{A}$-set and has measure 0. Prove that with this m, if f is defined on X and $f(x) = 0$ except on a set of measure 0, its integral over X is 0.

EXERCISE 2-11 If a function f is integrable over an $\mathscr{A}$-set A, and $c \geqq 0$, then $f \wedge c$ is integrable over A. *Suggestion*: Prove this first for simple functions, then use the monotone convergence theorem.

3. Borel Sets and Borel-Measurable Functions

If $\mathscr{A}$ is a σ-algebra of subsets of a set X, and we know that a certain collection $\mathscr{K}$ of subsets of X is contained in the σ-algebra $\mathscr{A}$, automatically some other subsets of X must belong to $\mathscr{A}$. For example, every set that is the union of a sequence of members of $\mathscr{K}$ must belong to $\mathscr{A}$, and so must every set that is the intersection of a sequence of such unions of members of $\mathscr{K}$. Instead of pursuing this construction through an infinite sequence of operations of forming unions and intersections, we define the **σ-algebra generated by** K to be the smallest σ-algebra of sets that contains $\mathscr{K}$, in the sense that it is contained in every σ-algebra that contains K. We shall shortly show that such a smallest σ-algebra necessarily exists. Assuming that it does, we see readily that every set in it belongs to $\mathscr{A}$; for $\mathscr{A}$ is a σ-algebra that contains $\mathscr{K}$, and by definition the σ-algebra generated by $\mathscr{K}$ is contained in every σ-algebra that contains $\mathscr{K}$, and, in particular, it is contained in $\mathscr{A}$.

To show that there exists a smallest σ-algebra that contains $\mathscr{K}$, we define $\mathscr{B}[\mathscr{K}]$ to be the collection of subsets of X defined as follows.

(A) A subset B of X belongs to $\mathscr{B}[\mathscr{K}]$ if and only if it belongs to every σ-algebra of subsets of X that contains $\mathscr{K}$.

There are σ-algebras of subsets of X that contain $\mathscr{K}$; for example, the collection of all subsets of X is a σ-algebra of subsets of X that contains $\mathscr{K}$. If K is any member of the family $\mathscr{K}$, it belongs to every subalgebra that contains $\mathscr{K}$, and therefore it qualifies as a member of $\mathscr{B}[\mathscr{K}]$. The empty set is in every σ-algebra and, in particular, in every σ-algebra that contains $\mathscr{K}$, and therefore it is in $\mathscr{B}[\mathscr{K}]$. If $B_1, B_2, B_3, \ldots$ is any sequence of sets all belonging to $\mathscr{B}[\mathscr{K}]$, and $\mathscr{A}^*$ is any σ-algebra of subsets of X that contains $\mathscr{K}$, by definition (A) the B_n all belong to $\mathscr{A}^*$. Since $\mathscr{A}^*$ is a σ-algebra of subsets of X, $X \setminus B_1$ and the union of the B_n both belong to $\mathscr{A}^*$. This holds for every σ-algebra $\mathscr{A}^*$ of subsets of X that contains $\mathscr{K}$, so $X \setminus B_1$ and the union of the B_n both belong to $\mathscr{B}[\mathscr{K}]$. Therefore $\mathscr{B}[\mathscr{K}]$ is a σ-algebra of subsets of X that contains $\mathscr{K}$. By (A), every set in $\mathscr{B}[\mathscr{K}]$ is contained in every σ-algebra of subsets of X that contains $\mathscr{K}$, so $\mathscr{B}[\mathscr{K}]$ is the σ-algebra generated by $\mathscr{K}$.

The Definition 1-4 of measurability of a function f (Definition 1-4) can be worded thus: if $(X, \mathscr{A})$ is a measurable space, an extended-real-valued function f on an $\mathscr{A}$-set A is $\mathscr{A}$-measurable on A if and only if for each closed half-line $[-\infty, y]$ in $\bar{R}$ the inverse image $f^{-1}([-\infty, y])$ belongs to $\mathscr{A}$. We shall now show that this implies that $\mathscr{A}$ also contains the inverse images of a much larger family of subsets of $\bar{R}$, namely, of all subsets of $\bar{R}$ that are in the σ-algebra generated by the closed half-lines.

LEMMA 3-1 *Let $(X, \mathscr{A})$ be a measurable space, Y a set, f a function on X with values in Y, and $\mathscr{K}$ a family of subsets of Y such that $f^{-1}(K)$ is in $\mathscr{A}$ whenever K is in $\mathscr{K}$. Then $f^{-1}(B)$ is in $\mathscr{A}$ whenever B belongs to the σ-algebra of subsets of Y generated by $\mathscr{K}$.*

Let $\mathscr{M}$ be the family consisting of all subsets M of Y such that $f^{-1}(M)$ belongs to $\mathscr{A}$. By hypothesis, $\mathscr{K} \subset \mathscr{M}$. Clearly $\varnothing$ belongs to $\mathscr{M}$. If M is in $\mathscr{M}, f^{-1}(M)$ belongs to the σ-algebra $\mathscr{A}$ and therefore so does $X \setminus f^{-1}(M)$, which is $f^{-1}(Y \setminus M)$. Therefore, if M is in $\mathscr{M}$, so is $Y \setminus M$. If $M_1, M_2, M_3, \ldots$ belong to $\mathscr{M}$, each of the inverse images $f^{-1}(M_n)$ $(n = 1,2,3,\ldots)$ belongs to $\mathscr{A}$. So, therefore, does

$$\bigcup_{n=1}^{\infty} f^{-1}(M_n) = f^{-1}\left(\bigcup_{n=1}^{\infty} M_n\right),$$

and therefore the union of the M_n is in $\mathscr{M}$. We have now shown that $\mathscr{M}$ is a σ-algebra that contains $\mathscr{K}$. It therefore contains the σ-algebra generated by $\mathscr{K}$, and so every set B in that σ-algebra belongs to $\mathscr{M}$ and has an inverse image $f^{-1}(B)$ that belongs to $\mathscr{A}$.

When X is a topological space, the σ-algebra generated by the open sets is called the σ-algebra of **Borel sets**. In particular, suppose that X is a Cartesian product-space $\bar{R}^T$ with T countable (finite or denumerably infinite). By a **rational half-space** in $\bar{R}^T$ we shall mean a set of the form

(B) $$J = \mathop{\times}_{t \text{ in } T} [-\infty, b^t]$$

in which all the b^t but one are ∞ and that one is either a rational number or is also ∞. The set $\mathscr{J}$ of all rational half-spaces is denumerably infinite. By forming set-differences of two such half-spaces and then forming intersections of finitely many such differences, we arrive at a denumerably infinite collection $\mathscr{J}^+$ of left-open intervals in $\bar{R}^T$ that is obviously contained in the σ-algebra generated by $\mathscr{J}$. If G is an open subset of $\bar{R}^T$ and x is a point in G, there exists an open interval A in $\bar{R}^T$ that contains x and is contained in G. This A has the form

$$A = \mathop{\times}_{t \text{ in } T} A^t$$

in which each A^t is an open interval A in $\bar{R}^T$ that contains x^t, and for all t except those in a finite set $\{t_1, \ldots, t_k\}$, A^t is $\bar{R}^T$. Then there exists for each t in $\{t_1, \ldots, t_k\}$ a left-open interval J^t that is contained in A^t, contains x^t, and has the form $(a, b]$ if x^t is finite, the form $(a, \infty]$ if $x^t = \infty$, and the form $[-\infty, b]$ if $x^t = \infty$, a and b being rational numbers. For all other t we take $J^t = \bar{R}^T$; then the Cartesian product

$$\mathop{\times}_{t \text{ in } T} J^t$$

belongs to $\mathscr{J}^+$, contains x, and is contained in G. So G is the union of countably many members of the class $\mathscr{J}^+$ and belongs to the σ-algebra generated by $\mathscr{J}$. Therefore the σ-algebra generated by the open sets, which is the σ-algebra of Borel sets, is contained in the σ-algebra generated by $\mathscr{J}$. Conversely, each rational half-space is the intersection of a sequence of open half-spaces, which are open intervals, so rational half-spaces are Borel sets. Therefore the σ-algebra generated by J is contained in the σ-algebra of Borel sets. The two statements together prove that when T is countable, the σ-algebra generated by the rational half-spaces in $\bar{R}^T$ is the σ-algebra of Borel sets in $\bar{R}^T$.

By a similar proof, when T is countable the σ-algebra generated by the rational half-spaces in R^T, or by the left-open intervals in R^T with rational end-points, is the same as the σ-algebra of Borel sets in R^T.

A function f that is defined and extended-real-valued on a topological space X is **Borel-measurable** (on X) if it is measurable with respect to the σ-algebra of Borel sets in X.

THEOREM 3-2 *Let $(X, \mathscr{A})$ be a measurable space and T a nonempty set. For each t in T, let f^t be an extended-real-valued $\mathscr{A}$-measurable function on X. For each x in X, let $f(x)$ be the point of $\bar{R}^T$ with coordinates $(f^t(x) : t$ in $T)$. Then for every set B in the σ-algebra generated by the left-open intervals in $\bar{R}^T$, $f^{-1}(B)$ belongs to $\mathscr{A}$. In particular, if T is countable, for every Borel set B in $\bar{R}^T$, $f^{-1}(B)$ belongs to $\mathscr{A}$.*

Let J be a left-open interval in $\bar{R}^T$. Then J is the Cartesian product

$$J = \bigtimes_{t \text{ in } T} J^t,$$

where for every t in T the set J^t is a left-open interval in $\bar{R}$, and for all t except those in a finite set $\{t_1, \ldots, t_k\}$, J^t is $\bar{R}$. For $i = 1, \ldots, k$, the inverse image

$$\{x \text{ in } X : f^{t_i}(x) \text{ in } J^{t_i}\}$$

belongs to $\mathscr{A}$, since f^{t_i} is $\mathscr{A}$-measurable. The intersection of these sets is $f^{-1}(J)$, which therefore belongs to $\mathscr{A}$. By Lemma 3-1, for every set B in the σ-algebra generated by the left-open intervals in $\bar{R}^T$, $f^{-1}(B)$ belongs to $\mathscr{A}$. In particular, if T is countable, that σ-algebra is the σ-algebra of Borel sets in $\bar{R}^T$.

COROLLARY 3-3 *Let $(X, \mathscr{A})$ be a measurable space and T a countable set. Let g be a function on X with values in $\bar{R}^T$ such that each coordinate function g^t is $\mathscr{A}$-measurable on X, and let f be an extended-real-valued function that is Borel-measurable on $\bar{R}^T$. Then the composite function $x \to f(g(x))$ $(x$ in $X)$ is $\mathscr{A}$-measurable.*

For each number b in $\bar{R}$ the set

$$\{y \text{ in } \bar{R}^T : f(y) \leqq b\}$$

is a Borel set in $\bar{R}^T$, by hypothesis. By Theorem 3-2, the set

$$g^{-1}\{y \text{ in } \bar{R}^T : f(y) \leqq b\}$$

belongs to $\mathscr{A}$. But this is the same as the set $\{x \text{ in } X : f(g(x)) \leqq b\}$. This completes the proof.

EXERCISE 3-1 Let f be Borel-measurable on an interval A in R; let g be Borel-measurable on an interval B in $\bar{R}$; and let $f(x)$ be in B for all x in A. Prove that $g \circ f$ is Borel-measurable on A.

EXERCISE 3-2 Show that if f and g are real-valued and $\mathscr{A}$-measurable on a set X, so are f^+, f^-, $|f|$, $f \vee g$, $f \wedge g$, $f + g$, and fg.

EXERCISE 3-3 Prove that if X is a topological space, every function f continuous on X is Borel-measurable on X.

EXERCISE 3-4 Prove that if X is a topological space, every function f that is upper semicontinuous or lower semicontinuous on X is Borel-measurable on X.

EXERCISE 3-5 Let m be a nonnegative additive function of left-open intervals in R^r. Prove that every Borel set in R^r is m-measurable and every Borel-measurable function on R^r is m-measurable.

EXERCISE 3-6 Although the concept of "function" was not very clearly defined two centuries ago, in a rough way it meant the assignment of a value $f(x)$ to each x in an interval by means of some allowed computational process. The allowed processes were: forming polynomials, taking roots, and passing to the limits of sequences (with less freedom than we now allow). Show that if this assertion about the definition of "function" is correct, the functions in the literature of that time were all Borel-measurable.

4. Integration with Respect to Other Functions of Sets

Heretofore we have considered integrals with respect to measures that by definition are nonnegative. However, in applications we encounter functions of sets that are positive for some sets and negative for others, and we need to define integrals with respect to such functions of sets.

If $\mathscr{A}$ is a σ-algebra of sets and F is an additive extended-real-valued function on $\mathscr{A}$, $F(A)$ may be ∞ for some sets A in $\mathscr{A}$ or it may be $-\infty$ for some sets in $\mathscr{A}$, but not both. For suppose that there exist $\mathscr{A}$-sets A and B such that $F(A) = \infty$ and $F(B) = -\infty$. Then $F(A \cap B)$ cannot be finite, for if it is, then from the equation $F(A) = F(A \setminus B) + F(A \cap B)$ we obtain $F(A \setminus B) = \infty$, and then the equation

$$F(A \setminus B) + F(B) = F([A \setminus B] \cup B)$$

is meaningless; $F(A \cap B)$ cannot be ∞, for then the equation

$$F(B) = F(A \cap B) + F(B \setminus A)$$

could not be valid; and it cannot be $-\infty$, for then the equation

$$F(A) = F(A \cap B) + F(A \setminus B)$$

could not be valid. So whenever we assume that an extended-real-ralued function on a σ-algebra of sets is countably additive, we imply that either it never takes on the value ∞ or else it never takes on the value $-\infty$.

The principal device in extending the theory of integration to such functions of sets is to represent them as differences of nonnegative functions of sets, such as we have already used as measures. This representation as differences rests on the *Hahn decomposition* of the space, which we now describe.

THEOREM 4-1 *Let $(X, \mathscr{A})$ be a measurable space and F an extended-real-valued function that is countably additive on $\mathscr{A}$. Then*

(A) *if F never assumes the value ∞, it has a finite upper bound, and if it never assumes the value $-\infty$, it has a finite lower bound;*

(B) *there exists an $\mathscr{A}$-set P such that for every $\mathscr{A}$-set A contained in P, $F(A) \geqq 0$, and for every $\mathscr{A}$-set A contained in $X \setminus P$, $F(A) \leqq 0$.*

By the remarks before the theorem, there are only two possible cases; either $F(A)$ is never ∞ or it is never $-\infty$.

Case 1. For all sets A in $\mathscr{A}$, $F(A) < \infty$.

Let M be the supremum of the values of $F(A)$ for all A in $\mathscr{A}$. If $M = 0$, we define P to be the empty set, and then (A) and (B) are satisfied. So we suppose that $M > 0$, and begin by proving an auxiliary statement.

(C) To each number c such that $0 < c < M$ there corresponds an $\mathscr{A}$-set $P[c]$ such that (i) $F(P[c]) > c$ and (ii) for each $\mathscr{A}$-set A contained in $X \setminus P[c]$, $F(A) \leqq 0$.

We can and do choose an $\mathscr{A}$-set C_0 such that $F(C_0) > c$. Then we select inductively a sequence $c_1, c_2, c_3, \ldots$ of nonnegative numbers and a sequence $C_1, C_2, C_3, \ldots$ of $\mathscr{A}$-sets as follows. For each positive integer n, suppose that the sets $C_0, \ldots, C_{n-1}$ have been chosen. We define

$$c_n = \sup\{F(A) : A \text{ an } \mathscr{A}\text{-set, } A \text{ disjoint from } C_0 \cup C_1 \cup \cdots \cup C_{n-1}\}.$$

We then can and do select an $\mathscr{A}$-set C_n disjoint from $C_0, \ldots, C_{n-1}$ such that

(D) $$F(C_n) \geqq 1 \wedge (c_n/2).$$

(If $c_n = 0$, we can choose $C_n = \varnothing$.) We define $P[c]$ to be the set

$$P[c] = \bigcup_{n=0}^{\infty} C_n.$$

Since the C_n all belong to $\mathscr{A}$, so does $P[c]$; and since the C_n are pairwise disjoint,

(E) $$F(P[c]) = \sum_{n=0}^{\infty} F(C_n).$$

The terms in the right member are all nonnegative, so

$$F(P[c]) \geqq F(C_0) > c,$$

establishing the first statement in (C). If the series in the right member of (E) failed to converge, $F(P[c])$ would be ∞, which by hypothesis is impossible. So the series converges, and therefore its nth term tends to 0 as n increases. By (D), this implies that c_n also tends to 0. Now let A be any $\mathscr{A}$-set disjoint from $P[c]$. It is then disjoint from $C_0, \ldots, C_{n-1}$ for every positive integer n, and so by definition of c_n,

$$F(A) \leqq c_n.$$

Since c_n tends to 0, this implies $F(A) \leqq 0$, and the proof of statement (C) is complete.

Now let $b_1, b_2, b_3, \ldots$ be a sequence of numbers such that

$$0 < b_1 < b_2 < b_3 < \cdots$$

and

$$\lim_{n \to \infty} b_n = M.$$

For each n we choose an $\mathscr{A}$-set $P[b_n]$ such that $F(P[b_n]) > b_n$ and $F(A) \leqq 0$ for every $\mathscr{A}$-set A disjoint from $P[b_n]$; this is possible by (C). We next prove

(F) If $P'[b_n] = P[b_1] \cap P[b_2] \cap \cdots \cap P[b_n]$, then (i) $F(P'[b_n]) > b_n$ and (ii) for every $\mathscr{A}$-set A contained in $X \setminus P'[b_n]$, $F(A) \leqq 0$.

The proof is by induction. The statement is obvious for $n = 1$. If it is true for $n = k - 1$, the sets $P'[b_k]$ and $P[b_k] \setminus P'[b_{k-1}]$ are disjoint $\mathscr{A}$-sets, and the latter is disjoint from $P'[b_{k-1}]$, and the union of the two sets is $P[b_k]$, so by the induction hypothesis,

$$b_k < F(P[b_k]) = F(P'[b_k]) + F(P[b_k] \setminus P'[b_{k-1}]) \leqq F(P'[b_k]).$$

This establishes the first statement in (F). For the second, let A be an $\mathscr{A}$-set contained in $X \setminus P'[b_k]$. Then A is the union of the disjoint $\mathscr{A}$-sets $A \setminus P'[b_{k-1}]$ and $A \cap (P'[b_{k-1}] \setminus P'[b_k])$. By the induction hypothesis,

(G) $$F(A \setminus P'[b_{k-1}]) \leqq 0.$$

Every point in $P'[b_{k-1}] \setminus P'[b_k]$ is in $P[b_1] \cap \cdots \cap P[b_{k-1}]$ but not in $P[b_1] \cap \cdots \cap P[b_k]$, so it is absent from $P[b_k]$, and

$$A \cap (P'[b_{k-1}] \setminus P'[b_k]) \subset X \setminus P[b_k].$$

Therefore, by the choice of $P[b_k]$,

$$F(A \cap (P'[b_{k-1}] \setminus P'[b_k])) \leqq 0.$$

This and (G) prove that $F(A) \leqq 0$, and the proof of (F) is complete.

As n increases, the sets $P'[b_n]$ shrink. We define P to be their intersection. For each positive integer k and each x in $P'[b_k]$, x is either in the intersection P or else there is a single integer j such that x is in $P'(b_j]$ but not in $P'[b_{j+1}]$. So $P'[b_k]$ is the union of the pairwise disjoint $\mathscr{A}$-sets

$$P, P'[b_k] \setminus P'[b_{k+1}], P'[b_{k+1}] \setminus P'[b_{k+2}], \ldots,$$

and therefore

$$F(P'[b_k]) = F(P) + \sum_{j=k}^{\infty} F(P'[b_k] \setminus P'[b_{k+1}]).$$

Since $P'[b_k] \setminus P'[b_{k+1}]$ is disjoint from $P'[b_{k+1}]$, by (F) every term in the infinite sum in the last equation is nonpositive, so

$$F(P) \geqq F(P'[b_k]) > b_k.$$

This holds for all b_k, and the b_k converge to M, so

(H) $$F(P) \geqq M.$$

By hypothesis, $F(P)$ is not ∞, so the upper bound M of $F(A)$ is finite, and equality holds in (H). If A is an $\mathscr{A}$-set contained in $X \setminus P$, we have by (H),

$$M \geqq F(P \cup A) = F(P) + F(A) = M + F(A),$$

so $F(A)$ is nonpositive. If A is an $\mathscr{A}$-set contained in P,

$$F(A) = F(P) - F(P \setminus A) = M - F(P \setminus A) \geqq 0.$$

This completes the proof in Case 1.

Case 2 For all sets A in $\mathscr{A}$, $F(A) > -\infty$.

Define $F' = -F$. This satisfies the hypotheses of Case 1, so there is a finite number M' and an $\mathscr{A}$-set P' such that $F'(A) \leqq M'$ for all $\mathscr{A}$-sets A, and $F'(A) \geqq 0$ for all $\mathscr{A}$-sets A contained in P', and $F'(A) \leqq 0$ for all $\mathscr{A}$-sets A contained in $X \setminus P'$. We now define

$$M = -M', \qquad P = X \setminus P'.$$

Then the conclusions of the theorem are obviously satisfied. The proof is complete.

If F is countably additive on a σ-algebra $\mathscr{A}$, for each set A in $\mathscr{A}$ we define

(I) $$F^+(A) = \sup\{F(B) : B \text{ in } \mathscr{A},\ B \subset A\},$$

$$F^-(A) = \sup\{-F(B) : B \text{ in } \mathscr{A},\ B \subset A\}.$$

We also define the total variation $T_F(A)$ of F over A to be the supremum of the sums

$$|F(B_1)| + \cdots + |F(B_k)|$$

for all finite collections $\{B_1, \ldots, B_k\}$ of pairwise disjoint $\mathscr{A}$-sets contained in A. These set-functions are closely related to F through the Hahn decomposition.

COROLLARY 4-2 *Let F be extended-real-valued and countably additive on a σ-algebra $\mathscr{A}$ of subsets of X. Let P, $X \setminus P$ be a Hahn decomposition of X, as in* Theorem 4-1. *Then F^+, F^-, and T_F are countably additive on $\mathscr{A}$ and for every set B in $\mathscr{A}$*

$$\begin{aligned} F^+(B) &= F(P \cap B), \\ F^-(B) &= -F(B \setminus P), \\ T_F(B) &= F^+(B) + F^-(B), \\ F(B) &= F^+(B) - F^-(B). \end{aligned}$$

Let P be a set with the properties specified in Theorem 4-1. Then whenever A and B are $\mathscr{A}$-sets with $B \subset A$,

$$\begin{aligned} F(B) &= F(B \cap P) + F(B \setminus P) \leqq F(B \cap P) \\ &= F(A \cap P) - F([A \setminus B] \cap P) \leqq F(A \cap P). \end{aligned}$$

So $F(A \cap P)$ is an upper bound for $F(B)$ for all $\mathscr{A}$-sets B contained in A, and by definition,

$$F(A \cap P) \geqq F^+(A).$$

On the other hand, $A \cap P$ is an $\mathscr{A}$-set contained in A, so

$$F^+(A) \geqq F(A \cap P).$$

So the two are equal, and the first equation in the conclusion is established. The second is similarly proved, and the fourth follows from the first and the second because F is additive. We leave the proof of the third equation as an exercise.

If $A_1, A_2, A_3, \ldots$ are pairwise disjoint $\mathscr{A}$-sets and A^* is their union, by the first equation in the conclusion and the countable additivity of F,

$$\begin{aligned} F^+(A^*) &= F(A^* \cap P) \\ &= F([A_1 \cap P] \cup [A_2 \cap P] \cup [A_3 \cap P] \cup \cdots) \\ &= \sum_{j=1}^{\infty} F(A_j \cap P) \\ &= \sum_{j=1}^{\infty} F^+(A_j). \end{aligned}$$

So F^+ is countably additive. Similarly, we prove that F^- is countably additive, and by the third equation, so is T_F.

By Corollary 4-2, F^+ and F^- are measures on $(X, \mathscr{A})$, so integration with respect to F^+ and integration with respect to F^- have already been defined. The definition of integration with respect to F suggests itself.

DEFINITION 4-3 *Let $(X, \mathscr{A})$ be a measurable space and F an extended-real-valued countably additive function on $\mathscr{A}$. Let g be an extended-real-valued function on X. Then g is integrable with respect to F over X if and only if g is integrable over X with respect to both measures F^+ and F^-, and in that case*

$$\int_X g(x)\, F(dx) = \int_X g(x)\, F^+(dx) - \int_X g(x)\, F^-(dx).$$

COROLLARY 4-4 *Let $(X, \mathscr{A})$ be a measurable space, F a countably additive extended-real-valued function on $\mathscr{A}$, and g an extended-real-valued $\mathscr{A}$-measurable function on X. Then g is integrable with respect to F over X if and only if it is integrable with respect to T_F over X.*

Suppose first that g is nonnegative. There exists an ascending sequence of nonnegative simple functions $s_1, s_2, s_3, \ldots$ converging everywhere to g. By Corollary 4-2, for each n

$$\int_X s_n(x)\, T_F(dx) = \int_X s_n(x)\, F^+(dx) + \int_X s_n(x)\, F^-(dx).$$

All these integrals are nonnegative, so the integral in the left member is bounded for all n if and only if the two integrals in the right member are bounded for all n. Therefore, g is integrable with respect to T_F if and only if it is integrable with respect to F^+ and with respect to F^-, and in that case

$$\text{(J)} \qquad \int_X g(x)\, T_F(dx) = \int_X g(x)\, F^+(dx) + \int_X g(x)\, F^-(dx).$$

If g is any $\mathscr{A}$-measurable function, then if g is integrable with respect to T_F so are g^+ and g^-. As just proved, g^+ and g^- are integrable with respect to F^+ and to F^-, so g is integrable with respect to both. By Definition 4-3, it is then integrable with respect to F, and (J) is satisfied. Conversely, if g is integrable with respect to F, it is integrable with respect to F^+ and to F^-, so the same is true of g^+ and g^-. As proved above, g^+ and g^- are then integrable with respect to T_F, and so therefore is their difference g.

EXERCISE 4-1 Let $(X, \mathscr{A})$ be a measurable space, F_1 and F_2 countably additive functions on A, and g_1 and g_2 extended-real-valued functions on X that are $\mathscr{A}$-measurable. Prove that if g_2 is integrable with respect to F_2 over X, and $|g_1(x)| \leqq g_2(x)$ for all x in X, and $|F_1(A)| \leqq F_2(A)$ for all A in $\mathscr{A}$, then g_1 is integrable with respect to F_1 over X, and

$$\left| \int_X g_1(x)\, F_1(dx) \right| \leqq \int_X g_2(x)\, F_2(dx).$$

EXERCISE 4-2 Let $(X, \mathscr{A})$ be a measurable space and m a measure on it, and let f be integrable with respect to m over X. For each set A in $\mathscr{A}$, define

$$F(A) = \int_A f(x)\, m(dx).$$

Prove that for each A in $\mathscr{A}$,

$$F^+(A) = \int_A f^+(x)\, m(dx), \qquad F^-(A) = \int_A f^-(x)\, m(dx),$$

$$T_F(A) = \int_A |f(x)|\, m(dx).$$

EXERCISE 4-3 Let $(X, \mathscr{A})$, m, f, and F be as in Exercise 4-2, and let g be an extended-real-valued function that is $\mathscr{A}$-measurable on X. Prove that g is integrable with respect to F over X if and only if gf is integrable with respect to m over X, and in that case

$$\int_X g(x)\, F(dx) = \int_X g(x) f(x)\, m(dx).$$

Suggestion: Assume to start with that f and g are nonnegative. Prove the statement first for nonnegative simple functions g.

EXERCISE 4-4 Let F be countably additive on a σ-algebra $\mathscr{A}$ of subsets of X, and let P, $X \setminus P$ be the Hahn decomposition of X with respect to F, as in Theorem 4-1. Prove that an $\mathscr{A}$-set P' has the properties specified for P if and only if $F(A) = 0$ for every $\mathscr{A}$-set A contained in $(P \setminus P') \cup (P' \setminus P)$.

5. The Radon–Nikodým Theorem

In preceding chapters we gave the principal role in our theories to integrals. Newton put differentiation in first place; he regarded integration as the process inverse to differentiation. For us, integration is in first place, and it would be consistent to regard differentiation as the process inverse to integration. In many theorems in preceding chapters there is a hypothesis, not that some function F

has a derivative, but that F is the indefinite integral of some function that takes over the role of the derivative of F. This gives importance to the problem of distinguishing those functions that are indefinite integrals.

Suppose that $(X, \mathscr{A})$ is a measurable space and that f is a function integrable with respect to m over X. Then the indefinite integral of f, whose value for each set A in $\mathscr{A}$ is

$$F(A) = \int_A f(x)\, m(dx),$$

is first of all a function on a σ-algebra of sets. Second, it is countably additive; this follows at once from the dominated convergence theorem and the integrability of $|f|$. Third, it is *m-continuous*, which is defined next.

DEFINITION 5-1 *Let $(X, \mathscr{A})$ be a measurable space, m a measure on $(X, \mathscr{A})$, and F an extended-real-valued function on $\mathscr{A}$. Then F is* **m-continuous** *if to each positive ε there corresponds a positive δ such that for every $\mathscr{A}$-set A with $mA < \delta$ it is true that $|F(A)| < \varepsilon$.*

It is evident that if F is m-continuous and A is an $\mathscr{A}$-set with $mA = 0$, then $F(A) = 0$. For let ε be positive and let δ correspond to ε as in Definition 5-1. Then $mA < \delta$, so $|F(A)| < \varepsilon$. This holds for all positive ε, so $F(A) = 0$.

The principal result of this section is the important *Radon–Nikodým theorem*, which states in part that if F is m-continuous and countably additive and X has finite measure or is the union of countably many sets of finite measure, then F is the indefinite integral of some function on X. This has the corollary that when A is an interval in R and F is a function on A that is absolutely continuous on A in the sense of Definition II-9-1, F has a derivative almost everywhere in A and is the indefinite integral of its derivative.

We first prove a corollary of Theorem 4-1.

LEMMA 5-2 *Let $(X, \mathscr{A})$ be a measurable space, m a measure on $(X, \mathscr{A})$, and F a countably additive real-valued function on $\mathscr{A}$. Then to each real number c there corresponds an $\mathscr{A}$-set $P[c]$ such that*

(A) *for every $\mathscr{A}$-set A contained in $P[c]$, $F(A) \geqq cmA$;*

(B) *for every $\mathscr{A}$-set B contained in $X \setminus P[c]$, $F(B) \leqq cmA$;*

(C) *if $c_1 \leqq c_2$, then $P[c_1] \supset P[c_2]$.*

Since F is countably additive and finite-valued, we can easily verify that for each real number c the function $A \mapsto F(A) - cmA$ is also countably additive. So by Theorem 4-1, to each real c there corresponds an $\mathscr{A}$-set $Q[c]$ such that for every $\mathscr{A}$-set A contained in $Q[c]$,

(D) $$F(A) - cmA \geqq 0,$$

and for every $\mathscr{A}$-set B contained in $X \setminus Q[c]$,

(E) $$F(B) - cmB \leqq 0.$$

We define

(F) $$P[c] = \bigcap \{Q[r] : r \text{ rational}, r \leqq c\}.$$

This is the intersection of countably many $\mathscr{A}$-sets, so it is an $\mathscr{A}$-set. Conclusion (C) is evidently valid. Let A be an $\mathscr{A}$-set contained in $P[c]$. Then for every rational number r in $(-\infty, c]$, A is contained in $Q[r]$, so by (D),

$$F(A) \geqq rmA.$$

This holds for all rational numbers r not greater than c, so it holds with c in place of r, and conclusion (A) is established.

Let B be an $\mathscr{A}$-set contained in $X \setminus P[c]$. The rational numbers in $(-\infty, c]$ can be arranged in a sequence $r_1, r_2, r_3, \ldots$. If x is a point of B, it is not in $P[c]$, so by (F) it is absent from some $Q[r]$ with $r \leqq c$ and is hence absent from some $Q[r_n]$. There is then a first n such that x is in $B \setminus Q[r_n]$. Define

$$B_1 = B \setminus Q[r_1],$$
$$B_n = (B \setminus Q[r_n]) \cap Q[r_1] \cap \cdots \cap Q[r_{n-1}] \qquad (n = 2, 3, 4, \ldots).$$

These are pairwise disjoint $\mathscr{A}$-sets, and as we have just seen, their union is B. For each positive integer n, B_n is contained in $X \setminus Q[r_n]$, so by (E),

$$F(B_n) \leqq r_n mB_n \leqq cmB_n.$$

By adding for $n = 1, 2, 3, \ldots$, we obtain conclusion (B). The proof is complete.

The next lemma contains the essence of the Radon–Nikodým theorem.

Lemma 5-3 *Let $(X, \mathscr{A})$ be a measurable space, m a measure on $(X, \mathscr{A})$, and F a real-valued countably additive m-continuous function on $\mathscr{A}$. Then there exists an extended-real-valued function f on X such that for every $\mathscr{A}$-set A with $mA < \infty$, f is integrable with respect to m over A, and*

$$F^+(A) = \int_A f^+(x)\, m(dx), \qquad F^-(A) = \int_A f^-(x)\, m(dx),$$

$$F(A) = \int_A f(x)\, m(dx).$$

Let $P[c]$ (c real) be a family of $\mathscr{A}$-sets with the properties listed in the conclusion of Lemma 5-2. By Corollary 4-2, for every $\mathscr{A}$-set A

$$F^+(A) = F(A \cap P[0]), \qquad F^-(A) = -F(A \setminus P[0]).$$

By Theorem 4-1, these are bounded. We discuss them separately, starting with F^+.

For each nonnegative rational number r, $P[r]$ is an $\mathscr{A}$-set, so the function

(G) $$f' = \sup\{r1_{P[r]} : r \text{ rational}, r \geqq 0\}$$

is a nonnegative $\mathscr{A}$-measurable function that vanishes on $X \setminus P[0]$. So f' has an integral with respect to m over every $\mathscr{A}$-set.

Let B be an $\mathscr{A}$-set contained in $P[0]$ and having $mB < \infty$, and let n be a positive integer. We define

(H) $$B(j) = B \cap \{P[2^{-n}j] \setminus P[2^{-n}(j+1)]\} \qquad (j = 0, 1, 2, \ldots),$$

$$B^* = B \cap \left\{ \bigcap_{j=0}^{\infty} P[2^{-n}j] \right\}.$$

These are pairwise disjoint $\mathscr{A}$-sets whose union is B.

Since $B(j)$ is contained in $P[2^{-n}j]$ and in $X \setminus P[2^{-n}(j+1)]$, by (A) and (B),

(I) $$2^{-n}jmB(j) \leqq F(B(j)) \leqq 2^{-n}(j+1)mB(j).$$

For each nonnegative integer j and each x in $B(j)$ the rational numbers for which x is in $P[r]$ include $2^{-n}j$ but do not include $2^{-n}(j+1)$ or any larger rational. So, by (G),

(J) $$2^{-n}j \leqq f'(x) \leqq 2^{-n}(j+1).$$

Since f' is $\mathscr{A}$-measurable and (J) holds, we can integrate over $B(j)$ (which has finite m-measure) and obtain

(K) $$\begin{aligned} 2^{-n}jmB(j) &\leqq \int_{B(j)} f'(x)\, m(dx) \\ &= \int_X f'(x) 1_{B(j)}(x)\, m(dx) \\ &\leqq 2^{-n}(j+1)mB(j). \end{aligned}$$

From (I) and (K),

(L) $$-2^{-n}mB(j) \leqq \int_X f'(x) 1_{B(j)}(x)\, m(dx) - F(B(j)) \leqq 2^{-n}mB(j).$$

Let us define

(M) $$U(k) = B(0) \cup B(1) \cup \cdots \cup B(k), \qquad U(\infty) = \bigcup_{j=0}^{\infty} B(j).$$

By adding inequalities (L) for $j = 0, 1, \ldots, k$, we obtain

(N) $$-2^{-n}mU(k) = \int_X f'(x) 1_{U(k)}(x)\, m(dx) - \sum_{j=0}^{k} F(B(j)) \leqq 2^{-n}mU(k).$$

As k increases, the sum of the $F(B(j))$ converges to $F(U(\infty))$, and $mU(k)$ converges to $mU(\infty)$. The integrands in the integral in (N) ascend, so by the monotone convergence theorem, their limit $f'(x)1_{U(\infty)}(x)$ is integrable, and

$$\text{(O)} \qquad -2^{-n}mU(\infty) \leqq \int_X f'(x)1_{U(\infty)}(x)\,m(dx) - F(U(\infty)) \leqq 2^{-n}mU(\infty).$$

For each positive integer j, B^* is contained in $P[2^{-n}j]$, by (H), so by (A) we have

$$F(B^*) \geqq 2^{-n}jmB^*.$$

But $F(B^*)$ is finite, so this implies that $mB^* = 0$. By the m-continuity of F, $F(B^*)$ is also 0. The sets B^*, $U(\infty)$ are disjoint $\mathscr{A}$-sets whose union is B. Since $F(B^*) = 0$, the middle member of (O) is unchanged if we replace $F(U(\infty))$ by $F(B)$ $(= F(U(\infty)) + F(B^*))$, and the integral in that middle term is unchanged if we replace $1_{U(\infty)}$ by 1_B, which differs from it only on the set B^* of m-measure 0. Hence from (O) we conclude

$$\text{(P)} \qquad -2^{-n}mB \leqq \int_X f'(x)1_B(x)\,m(dx) - F(B) \leqq 2^{-n}mB.$$

Since this holds for all positive integers n, we have for all $\mathscr{A}$-sets B contained in $P[0]$ and having finite m-measure

$$\text{(Q)} \qquad F(B) = \int_X f'(x)1_B(x)\,m(dx).$$

If B is an $\mathscr{A}$-set with finite mB, by Corollary 4-2, $F^+(B) = F(B \cap P[0])$, and

$$f'(x)1_{B \cap P[0]}(x) = f'(x)1_B(x)$$

for all x in X because all points x at which $1_{B \cap P[0]}(x) \neq 1_B(x)$ are in $X \setminus P[0]$ and have $f'(x) = 0$. So from (Q) we obtain

$$\text{(R)} \qquad F^+(B) = \int_X f'(x)1_B(x)\,m(dx),$$

which is valid for all $\mathscr{A}$-sets B with finite mB.

We now apply the same reasoning to the function $-F$, with $X \setminus P[0]$ taking the place of $P[0]$, and we find that there exists an $\mathscr{A}$-measurable function f'' on X that vanishes on $P[0]$ and has the property that whenever B is an $\mathscr{A}$-set with finite m-measure,

$$\text{(S)} \qquad F^-(B) = \int_X f''(x)1_B(x)\,m(dx).$$

We now define

$$f = f' - f''.$$

The equation

$$f^+(x) = f'(x)$$

holds at each x in X; for if x is in $P[0]$, $f'(x) \geqq 0$ and $f''(x) = 0$, and if x is in $X \setminus P[0]$, $f'(x) = 0$ and $f(x) = f'(x) - f''(x) \leqq 0$. Similarly, at each x in X

$$f^-(x) = f''(x).$$

So equations (R) and (S) are the first two equations in the conclusion, and the third equation follows from them by subtraction. The proof is complete.

In stating the Radon–Nikodým theorem it is convenient to introduce a widely used expression.

DEFINITION 5-4 *Let $(X, \mathscr{A})$ be a measurable space and m a measure on $(X, \mathscr{A})$. A subset A of X* **has σ-finite measure** *if it is the union of countably many $\mathscr{A}$-sets $A_1, A_2, A_3, \ldots$ such that for each n, mA_n is finite.*

If A is the union of a sequence $A_1, A_2, A_3, \ldots$ of sets of finite m-measure, it is also the union of the expanding sequence $U(1), U(2), U(3), \ldots$ of sets of finite m-measure, where

$$\text{(T)} \qquad U(n) = A_1 \cup \cdots \cup A_n \qquad (n = 1, 2, 3, \ldots).$$

It is also the union of the sequence of pairwise disjoint $\mathscr{A}$-sets

$$U(1),\ U(2) \setminus U(1),\ U(3) \setminus U(2), \ldots,$$

each having finite m-measure.

THEOREM 5-5 (Radon–Nikodým) *Let $(X, \mathscr{A})$ be a measurable space, m a measure on $(X, \mathscr{A})$, and F a real-valued countably additive m-continuous function on $\mathscr{A}$. Then there exists an extended-real-valued function f on X with the following properties.*

(U) *If A is an $\mathscr{A}$-set with σ-finite m-measure and g is an $\mathscr{A}$-measurable extended-real-valued function on A, g is F-integrable over A if and only if gf is m-integrable over A, and in that case*

$$\int_A g(x)\, F(dx) = \int_A g(x) f(x)\, m(dx).$$

(V) *If A is in $\mathscr{A}$ and mA is σ-finite, f is m-integrable over A, and*

$$F^+(A) = \int_A f^+(x)\, m(dx),$$

$$F^-(A) = \int_A f^-(x)\, m(dx),$$

$$F(A) = \int_A f(x)\, m(dx).$$

We can and do choose a function f with the properties specified in the conclusion of Lemma 5-3. We shall show that it then has the properties (U) and (V).

Let A be an $\mathscr{A}$-set with σ-finite m-measure. Then there exists a sequence of pairwise disjoint sets $A_1, A_2, A_3, \ldots$ of finite m-measure whose union is A; and if we define

$$U(n) = A_1 \cup \cdots \cup A_n \qquad (n = 1, 2, 3, \ldots),$$

we have by Lemma 5-3,

$$\sum_{j=1}^{n} F^+(A_j) = F^+(U(n)) = \int_X f^+(x) 1_{U(n)}(x)\, m(dx).$$

As n increases, the first expression in this equation tends to $F^+(A)$, since F^+ is countably additive. By the monotone convergence theorem, the last expression in the equation tends to the integral of $f^+ 1_A$ over X, so the first equation in (V) is established. The second is similarly proved, and by subtraction we obtain the last equation in (V) from the first two.

Let s be a simple function on X that vanishes outside a set A of σ-finite m-measure. Then s has a canonical representation

$$s = \sum_{i=1}^{k} c_i 1_{B(i)},$$

in which the c_i are distinct nonzero numbers and the $B(i)$ are pairwise disjoint $\mathscr{A}$-sets that are contained in A and therefore have σ-finite m-measure. Then by (V),

$$\text{(W)} \qquad \int_X s(x)\, F(dx) = \sum_{i=1}^{k} c_i F(B(i))$$
$$= \sum_{i=1}^{k} c_i \int_X 1_{B(i)}(x) f(x)\, m(dx)$$
$$= \int_X s(x) f(x)\, m(dx).$$

Analogous equations hold with F^+ and f^+ in place of F and f, and also with F^- and f^- in place of F and f.

If g is a nonnegative $\mathscr{A}$-measurable function on X that vanishes outside an $\mathscr{A}$-set A with σ-finite m-measure, g is the limit of an ascending sequence $s_1, s_2, s_3, \ldots$ of nonnegative simple functions. These all have to vanish outside A, so by (W) we have for each positive integer n

$$\int_X s_n(x)\, F^+(dx) = \int_X s_n(x) f^+(x)\, m(dx).$$

By the monotone convergence theorem,

$$\text{(X)} \qquad \int_X g(x)\,F^+(dx) = \int_X g(x) f^+(x)\,m(dx).$$

In particular, if g is integrable with respect to F^+, the left member is finite, so the right member is finite, and gf^+ is integrable with respect to m; and conversely, if gf^+ is integrable with respect to m, the right member is finite, so the left member is finite, and g is integrable with respect to F^+.

Let g be a function that is $\mathscr{A}$-measurable on X, and let A be a set of σ-finite m-measure. As usual, we denote by g_A the function that coincides with g on A and is 0 on $X \setminus A$. Then both g_A^+ and g_A^- are $\mathscr{A}$-measurable and vanish outside A. By four applications of (X) in which F^+ can be replaced by F^-, we obtain

$$\text{(Y)} \qquad \begin{aligned} \int_X g_A^+(x)\,F^+(dx) &= \int_X g_A^+(x) f^+(x)\,m(dx), \\ -\int_X g_A^-(x)\,F^+(dx) &= -\int_X g_A^-(x) f^+(x)\,m(dx), \\ -\int_X g_A^+(x)\,F^-(dx) &= -\int_X g_A^+(x) f^-(x)\,m(dx), \\ \int_X g_A^-(x)\,F^-(dx) &= \int_X g_A^-(x) f^-(x)\,m(dx). \end{aligned}$$

If g is integrable with respect to F over A, by definition g_A is integrable with respect to F^+ and with respect to F^- over X. Since F^+ and F^- are measures, g_A^+ and g_A^- are integrable with respect to F^+ and to F^- over X, so the left members of all four equations (Y) are finite. Therefore, all four right members are finite. By adding the four equations member-by-member, we find that $g_A f$ is integrable with respect to m over X, and the equation in (U) is satisfied. Conversely, if $g_A f$ is m-integrable over X, so is its product with any bounded $\mathscr{A}$-measurable function. In particular, if C is the set of x on which $f^+(x) > 0$, $g_A f 1_C$ is integrable over X. This is the same as $g_A f^+$. Since $g_A f^+$ is integrable, so are $(g_A f^+)^+$ and $(g_A f^+)^-$. These are respectively equal to $g_A^+ f^+$ and to $g_A^- f^+$, so the right members of the first and second equations (Y) are finite. In the same way the right members of the third and fourth equations (Y) are finite. Therefore, all four left members are finite. Then g is integrable with respect to F over A, and by adding the equations (Y) member-by-member, we obtain the equation in (U). The proof is complete.

The process of finding an f that corresponds to a given F as in Theorem 5-5 is the operation inverse to integration, so the function f that corresponds to F as in Theorem 5-5 is often called the *Radon–Nikodým derivative* of F and is denoted by the symbol dF/dm. In general, this cannot be compared with any ordinary derivative because in general, X has no structure that would allow us to define a derivative in the ordinary sense. But when X is R^1 and m is Lebesgue measure

m_L, a function $x \mapsto F(x)$ on an interval A gives rise to a function of intervals ΔF, where $\Delta F(a, b] = F(b) - F(a)$, and now the Radon–Nikodým derivative of ΔF is a real-valued function on A, and so is DF, if it exists. In this case the two can be compared, and this we shall do. The argument extends with little difficulty to the case of functions F of intervals in R^r, m again being m_L.

Let F be real-valued on an interval $[a, b]$ in R. For each subinterval $A = (c, d]$ we define $T(A)$, $P(A)$, $N(A)$ to be the suprema of the respective sums

$$\text{(Z)} \qquad \sum_{i=1}^{k} |F(x_i) - F(x_{i-1})|, \qquad \sum_{i=1}^{k} [F(x_i) - F(x_{i-1})] \vee 0,$$

$$\sum_{i=1}^{k} [F(x_{i-1}) - F(x_i)] \vee 0$$

for all finite sets of numbers

$$x_0 = c < x_1 < x_2 < \cdots < x_k = d.$$

These are named, respectively, the **total variation**, the **positive variation**, and the **negative variation** of F over A. If in the first sum in (Z) we insert another point x^*, say between x_{j-1} and x_j, the single term $|F(x_j) - F(x_{j-1})|$ is replaced by the sum of two terms $|F(x^*) - F(x_{j-1})| + |F(x_j) - F(x^*)|$, and the sum is not decreased. Likewise, the two other sums in (Z) are not decreased if we insert another point among the x_i. If we subdivide A into two intervals $A' = (c, e]$, $A'' = (e, d]$ and choose three numbers c', c'', c''' that are less than $T(A')$, $T(A'')$, $T(A)$, respectively, we can choose three finite sets $x'_0, \ldots, x'_h$, $x''_0, \ldots, x''_k$, $x'''_0, \ldots, x'''_n$ (each in increasing order) such that $x'_0 = x'''_0 = c$, $x'_h = x''_0 = e$, $x''_k = x'''_n = d$, and

$$c' < \sum_{i=1}^{h} |F(x'_i) - F(x'_{i-1})| \leqq T(A'),$$

$$c'' < \sum_{i=1}^{k} |F(x''_i) - F(x''_{i-1})| \leqq T(A''),$$

$$c''' < \sum_{i=1}^{n} |F(x'''_j) - F(x'''_{j-1})| \leqq T(A).$$

If the numbers $\{x_0, \ldots, x_q\}$ are the numbers in the union of the three sets arranged in increasing order, e is among the x_i (it is x''_0), so $e = x_r$ for some integer r. Inserting the numbers $x_0, \ldots, x_r$ among the numbers $x'_0, \ldots, x'_h$ does not diminish the sum, and likewise for the other two sums. Hence,

$$c' < \sum_{i=1}^{r} |F(x_i) - F(x_{i-1})| \leqq T(A'),$$

$$c'' < \sum_{i=r+1}^{q} |F(x_i) - F(x_{i-1})| \leqq T(A''),$$

$$c''' < \sum_{i=1}^{q} |F(x_i) - F(x_{i-1})| \leqq T(A).$$

From these inequalities we deduce

$$c' + c'' < T(A), \qquad c''' < T(A') + T(A'').$$

Since these are valid whenever c''', c', and c'' are less than $T(A)$, $T(A')$, and $T(A'')$, respectively, it follows that

$$T(A) = T(A') + T(A''),$$

so T is a (finitely) additive function of left-open subintervals of A. In the same way we prove that P and N are also additive.

Let c', c'', c''' be any numbers less than $P(A)$, $N(A)$, $T(A)$, respectively. We can find three sets of numbers in A (with the same notation as before) such that

(AA)
$$c' < \sum_{i=1}^{h} [F(x_i') - F(x_{i-1}')] \vee 0 \leqq P(A),$$

$$c'' < \sum_{i=1}^{k} [F(x_{i-1}'') - F(x_i'')] \vee 0 \leqq N(A),$$

$$c''' < \sum_{i=1}^{n} |F(x_i''') - F(x_{i-1}''')| \leqq T(A).$$

Again we denote by $\{x_0, \ldots, x_q\}$ the union of the three sets of points x_i', x_i'', and x_i''', arranged in increasing order. If we insert the x_i among the x_i', the sum in the first of inequalities (AA) is not decreased, and similarly for the other two inequalities in (AA). Hence,

(BB)
$$c' < \sum_{i=1}^{q} [F(x_i) - F(x_{i-1})] \vee 0 \leqq P(A),$$

$$c'' < \sum_{i=1}^{q} [F(x_{i-1}) - F(x_i)] \vee 0 \leqq N(A),$$

$$c''' < \sum_{i=1}^{q} |F(x_i) - F(x_{i-1})| \leqq T(A).$$

If we subtract the second of inequalities (BB) from the first, we obtain

$$c' - N(A) < F(x_q) - F(x_0) < P(A) - c''.$$

Since $x_0 = c$ and $x_q = d$, and c' and c'' are any numbers less than $P(A)$ and $N(A)$, respectively, this implies

$$F(d) - F(c) = P(A) - N(A).$$

Likewise, by adding the first two of inequalities (BB), we obtain

$$c' + c'' < \sum_{i=1}^{q} |F(x_i) - F(x_{i-1})| \leqq P(A) + N(A).$$

Comparing this with the last of inequalities (BB) yields

$$c' + c'' < T(A), \qquad c''' < P(A) + N(A).$$

Since this holds for all c' and c'' less than $P(A)$ and $N(A)$, respectively, and all c''' less than $T(A)$, it must be true that

$$T(A) = P(A) + N(A).$$

In particular, if $T(A)$ is finite, so are $P(A)$ and $N(A)$.

Suppose next that A is bounded and that F is absolutely continuous on A in the sense of Definition II-9-1. Then to each positive ε there corresponds a positive δ such that if $A_1, \ldots, A_k$ are pairwise disjoint subintervals of A with $A_i = (c_i, d_i]$, such that

$$\sum_{i=1}^{k} [d_i - c_i] < \delta,$$

then

$$\sum_{i=1}^{k} |F(d_i) - F(c_i)| < \frac{\varepsilon}{2}.$$

With this same ε, δ and intervals $A_1, \ldots, A_k$ we choose in each A_i a set of points

$$x_{i,0} = c_i < x_{i,1} < x_{i,2} < \cdots < x_{i,q} = d_i$$

(there is no loss of generality in assuming q to be the same for all the intervals) such that

$$\sum_{j=1}^{q} |F(x_{i,j}) - F(x_{i,j-1})| > T(A_i) - \varepsilon/2k.$$

The sum of the left members of these inequalities is less than $\varepsilon/2$ because the sum of the lengths of all the subintervals $(x_{i,j-1}, x_{i,j}]$ is less than δ, so

$$\sum_{i=1}^{k} T(A_i) < \varepsilon.$$

The same inequality holds for P and for N, since these cannot be greater than T. In particular, this implies that $T(A)$, $P(A)$, and $N(A)$ are continuous functions of the two end-points of A, so they are regular additive functions of intervals.

Now, as in Chapter II, we can extend P and N to be countably additive functions on a σ-algebra of sets that includes all intervals contained in A. (Or, to stay closer to the notation used in Chapter II, we can first extend P to be a function of left-open intervals in R by setting $P(B) = P(A \cap B)$ for all left-open intervals B in R and then extend this to be countably additive on a σ-algebra of subsets of R including all left-open intervals.) If G is an open subset of A, it is the union of pairwise disjoint left-open subintervals $A_1, A_2, A_3, \ldots$ of A. Let ε be positive and δ a positive number such that $P(A_1 \cup \cdots \cup A_n) < \varepsilon/2$

whenever $\Sigma m_L A_j < \delta$. Then if $m_L G < \delta$, for the intervals $A_1, A_2, A_3, \ldots$ whose union is G we have $m_L A_1 + \cdots + m_L A_n < \delta$ for all positive integers n, so

$$\sum_{j=1}^{n} P(A_j) < \frac{\varepsilon}{2}.$$

Letting n increase yields

$$P(G) = \sum_{j=1}^{\infty} P(A_j) \leqq \frac{\varepsilon}{2}.$$

If B is any Borel set with measure $m_L B < \delta$, there exists an open set G that contains B and has $m_L G < \delta$. Then

$$P(B) \leqq P(G) \leqq \varepsilon/2 < \varepsilon,$$

and the set-function P (as extended to all Borel sets) is m_L-continuous on the σ-algebra of Borel subsets of A. The same is true of the extensions of N and T to the σ-algebra of Borel subsets of A.

In particular, there is a positive δ that corresponds to $\varepsilon = 1$. If n is an integer greater than $(b - a)/\delta$, each subinterval B of A can be subdivided into n subintervals $B_1, \ldots, B_n$ each of length less than δ. For each of these, $T(B_k) < 1$, so $T(B) < n$. So T is bounded on the class of all subintervals of A; and since P and N do not exceed T, they too are bounded on that class.

Now by the Radon–Nikodým theorem, both P and N have Radon–Nikodým derivatives with respect to m_L. That is, there exist Borel-measurable functions f_1, f_2 on A such that for every Borel subset B of A,

$$P(B) = \int_B f_1(x)\,dx, \qquad N(B) = \int_B f_2(x)\,dx.$$

In particular, when B happens to be a left-open subinterval of A, we already know that

$$F(B) = P(B) - N(B),$$

whence

$$F(B) = \int_B [f_1(x) - f_2(x)]\,dx.$$

When we combine this with Theorem II-14-1, we find that we have proved the following theorem.

THEOREM 5-6 *Let F be a real-valued function that is absolutely continuous (in the sense of Definition* II-9-1) *on a bounded interval A in R. Then F has a derivative at almost all points of A; and if $\dot{F}$ is any function on A that coincides with $DF(x)$ at almost all points x at which $DF(x)$ exists, $\dot{F}$ is integrable with respect to m_L over A,* *and for all x and a in A*

$$F(x) - F(a) = \int_a^x \dot{F}(u)\,du.$$

EXERCISE 5-1 Prove that if F is any countably additive nonnegative function on a σ-algebra of sets, the Radon–Nikodým derivative of F with respect to F exists and can be chosen to be identically 1.

EXERCISE 5-2 Let $(X, \mathscr{A})$ be any measurable space and F any function on $\mathscr{A}$. Show that F is absolutely continuous with respect to counting measure.

EXERCISE 5-3 Prove that if M is the class of Lebesgue-measurable subsets of R, m_L is absolutely continuous with respect to counting measure $m_\#$ but does not have a Radon–Nikodým derivative with respect to $m_\#$. Why does this not contradict Theorem 5-5?

EXERCISE 5-4 State and prove a "chain rule" for Radon–Nikodým derivatives.

EXERCISE 5-5 Let f be integrable with respect to a measure m over R^r, and for each left-open interval A in R^r define

$$F(A) = \int_A f(x)\,m(dx).$$

Show that the total variation of F over a left-open interval B is

$$\int_B |f(x)|\,m(dx).$$

6. Conditional Expectations

An important application of the Radon–Nikodým theorem is in the theory of conditional probabilities, which in its general form is central in probability theory. To avoid the appearance of abstruseness, we begin with an example. Suppose that a nonprofit life insurance company is offering a 10-year policy that in the event of the death of the insured n years after the purchase of the policy will pay to his estate an amount $a(n)$, which is 0 if $n > 10$. The population to which the policy is offered is the set Ω of all inhabitants of the United States who are over a certain minimum age and are capable of passing certain health tests. If for each subset B of Ω we denote by $m_\# B$ the number of individuals in B, we assign to B the probability measure $P(B) = m_\# B/m_\# \Omega$. This corresponds to regarding all individuals as having the same chance of being selected. For each individual ω, we shall know 10 years from now the present value $C(\omega)$ of the amount that his policy will have cost the company; if he dies n years after purchase of the policy, this will be the sum of the amount $a(n)$ discounted back n years to the time of purchase, plus the various expenses of administering the policy, each discounted back to the time of purchase. For simplicity we shall henceforth refer to $C(\omega)$ as the *cost* of the policy, asking the reader to keep in mind the discounting back to time of purchase. Of course, this cost is not known at the time of purchase. All that is known is a collection of data about the individual ω – for example, ω's age, sex, height, weight, and occupation. The nonnumerical data in the collection are converted into numbers by some coding scheme. If T is the set of labels for the data, consisting of "age," "weight," etc., then to each individual ω there corresponds a point or vector $Y(\omega)$ in R^T. This vector $Y(\omega)$ we call the *data vector* for ω. In insurance practice there will be only finitely many possible data vectors; for example, ages will be given only as integers between 10 and 95. But we can imagine that there are in fact infinitely many data vectors, and even that T is infinite; for example, the data could conceivably include ω's weight at each instant of his or her life. The amount that will be charged ω for the policy is determined by the data vector $Y(\omega)$. That is, there is a function Z on R^T such that the amount charged will be

$$\text{(A)} \qquad Z^*(\omega) = Z(Y(\omega)).$$

The function Z is the company's rate book. In practice there can be no question of measurability, since Z need only be defined on a finite set; but even if we allow the possibility of an infinite collection of values for Y, it is still reasonable to assume that Z is Borel-measurable. The function Z^* provides the answer to ω's question, "How much do I have to pay for this?" Given the function Y, Z^* is determined by Z through equation (A). Since Y is $\mathscr{A}$-measurable and Z is Borel-measurable, Z^* is $\mathscr{A}$-measurable; it is a random variable. But more than this is true. The set of individuals ω who have to pay an amount $Z^*(\omega)$ in an interval $(c, d]$ is the same as the set

$$\{\omega \text{ in } \Omega : Z(Y(\omega)) \text{ in } (c, d]\} = \{\omega \text{ in } \Omega : Y(\omega) \text{ in } Z^{-1}(c, d]\}$$
$$= Y^{-1}(Z^{-1}(c, d]).$$

Since Z is Borel-measurable, $Z^{-1}(c, d]$ is a Borel set, and $Y^{-1}(Z^{-1}(c, d])$ is in $\mathscr{A}$. Therefore the collection of sets $\{\omega \text{ in } \Omega : Z^*(\omega) \text{ in } (c, d]\}$ is contained in the collection $\mathscr{B}$ of all sets of the form $Y^{-1}(B')$ with B' a Borel set in R^T, and this collection $\mathscr{B}$ is easily shown to be a σ-subalgebra of $\mathscr{A}$. If B is any set that belongs to $\mathscr{B}$, a point ω belongs to B if and only if the data vector $Y(\omega)$ concerning ω is in a certain set B'. For this reason, $\mathscr{B}$ is called the *information algebra* corresponding to the data-function Y.

We have just shown that whenever Z^* is a composite function of the form (A) with Z a Borel-measurable function, Z^* is $\mathscr{B}$-measurable. The converse is also true, as we shall prove in Theorem 6-11. This gives us a complete interchangeability between Borel-measurable functions Z on R^T and $\mathscr{B}$-measurable functions Z^* on Ω. The choice between them is only a matter of convenience.

We said at the start that the company is a "nonprofit organization." There was no need to mention that the administrators also want it to be a nonloss organization. We shall assume that they also want it to be equitable in its pricing, in the sense that for each set B in the information algebra they wish the total amount paid by the individuals ω in B to be equal to the total cost $\Sigma\{C(\omega): \omega$ in $B\}$ of the policies issued to the individuals ω in B. The amount paid by the individuals in B is the sum of the prices $Z^*(\omega)$ for all individuals ω in B. This is the same as the sum of $Z^*(\omega)1_B(\omega)$ for all ω in Ω, and this in turn is the same as the integral over Ω of $Z^*(\omega)1_B(\omega)$ with respect to counting measure. By definition of P, the total amount paid by the individuals in B is

(B) $$[m_\#\Omega]\int_\Omega Z^*(\omega)1_B(\omega)\,P(d\omega).$$

The total cost to the company of the policies issued to individuals in set B is the sum of $C(\omega)$ for all ω in B, which is the same as the sum of $C(\omega)1_B(\omega)$ for all ω in Ω, and this in turn is the same as

(C) $$[m_\#\Omega]\int_\Omega C(\omega)1_B(\omega)\,P(d\omega).$$

The criterion of fairness is that (B) and (C) be equal; that is,

(D) For every set B in the information algebra $\mathscr{B}$,

$$\int_B Z^*(\omega)\,P(d\omega) = \int_B C(\omega)\,P(d\omega).$$

This is clearly a strong requirement. There are as many equations (D) to be satisfied as there are sets B in the information algebra $\mathscr{B}$, and it is not immediately clear that a single $\mathscr{B}$-measurable function Z^* can be chosen that will satisfy all of them. One of the principal results of this section will be the proof that such a choice of Z^* is always possible.

Suppose, then, that Z^* satisfies (D). Let ω be any member of Ω, and let B be the set of all ω' in Ω such that $Y(\omega') = Y(\omega)$. Evidently B is not empty. If we multiply both members of (D) by $m_\#\Omega/m_\#B$, we obtain

(E) $$\Big[\sum_{\omega' \text{ in } B} Z^*(\omega')\Big]\Big/ m_\#B = \Big[\sum_{\omega' \text{ in } B} C(\omega')\Big]\Big/ m_\#B.$$

For all ω' in B,

$$Z^*(\omega') = Z(Y(\omega')) = Z(Y(\omega)) = Z^*(\omega),$$

so the left member of (E) is $Z^*(\omega)$. The right member of (E) is the average cost to the company of the policies issued to all individuals in B. Hence, by (E), $Z^*(\omega)$ is equal to the average cost of the policies issued to all individuals that satisfy the condition $Y(\omega') = Y(\omega)$. Therefore $Z^*(\omega)$ is called the *conditional expectation* of the cost C when conditioned by (knowledge of the value of) $Y(\omega)$. Alternatively, it is called the *conditional expectation* of C when conditioned by the information algebra $\mathscr{B}$. We shall prefer the latter term, and we shall use either of the symbols

$$E^{\mathscr{B}}C \quad \text{or} \quad E(C \mid \mathscr{B}) \tag{F}$$

to denote it. Note that each of these is a typographically complicated symbol for a function on Ω, namely, the function we have been calling Z^*. For each ω in Ω, $E^{\mathscr{B}}C(\omega)$ is the value at ω of the function named $E^{\mathscr{B}}C$, and $E(C \mid \mathscr{B})(\omega)$ is the value at ω of the function named $E(C \mid \mathscr{B})$.

We extend all these ideas from the example to a general situation by simple imitation. In doing this, it is easier to give the principal role to the information algebras $\mathscr{B}$. Later we shall return and relate these to data vectors Y.

DEFINITION 6-1 *Let $(\Omega, \mathscr{A}, P)$ be a probability triple, and let $\mathscr{B}$ be a σ-algebra contained in $\mathscr{A}$. Let C be a random variable that has an expectation (possibly infinite). Then the name "conditional expectation of C given $\mathscr{B}$" is given to every $\mathscr{B}$-measurable function Z^* on Ω such that for every set B in $\mathscr{B}$,*

$$\int_B Z^*(\omega)\, P(d\omega) = \int_B C(\omega)\, P(d\omega).$$

Every function that satisfies this condition is designated by either of the symbols

$$E^{\mathscr{B}}C,\ E(C \mid \mathscr{B}).$$

This is a generalization of the idea of integration. Let C have finite expectation, and let the information algebra $\mathscr{B}$ correspond to total lack of information; it contains only the two sets $\varnothing$ and Ω. Then a function Z^* is $\mathscr{B}$-measurable if and only if it is constant, and the conditional expectation of C given $\mathscr{B}$ is the constant Z^* such that

$$\int_\Omega Z^*\, P(d\omega) = \int_\Omega C(\omega)\, P(d\omega).$$

That is, $Z^* = E(C)$, which is the integral of C over Ω. At the other extreme, the greatest possible amount of information would correspond to the greatest σ-subalgebra of $\mathscr{A}$, which is $\mathscr{B} = \mathscr{A}$. When $\mathscr{B} = \mathscr{A}$, for $E^{\mathscr{B}}C$ we can choose C itself; for this is $\mathscr{B}$-measurable, and its integral over every set B in $\mathscr{B}$ is the same as the integral of C over that set.

The equation in Definition 6-1 can be written in the alternate form

$$E[1_B E^{\mathscr{B}}C] = E[1_B C] \qquad (B \text{ in } \mathscr{B}). \tag{G}$$

This implies in particular, choosing $B = \Omega$, that

(H) $$E[E^{\mathscr{B}}C] = E[C].$$

We shall follow the custom in probability theory of using the phrase "almost surely" (abbreviated "a.s.") to mean "almost everywhere," that is "at all points except those in a set of P-measure 0."

Since the conditional expectation generalizes the idea of integral, it is natural that many of the theorems about integrals generalize to it.

LEMMA 6-2 *If X and Y are random variables with conditional expectations $E^{\mathscr{B}}X$ and $E^{\mathscr{B}}Y$, respectively, and $X \leqq Y$ a.s., then $E^{\mathscr{B}}X \leqq E^{\mathscr{B}}Y$ a.s.*

Let B be the set $\{\omega \text{ in } \Omega : E^{\mathscr{B}}X(\omega) > E^{\mathscr{B}}Y(\omega)\}$. Since $E^{\mathscr{B}}X$ and $E^{\mathscr{B}}Y$ are $\mathscr{B}$-measurable, B belongs to $\mathscr{B}$, so by Definition 6-1,

$$\int_B [E^{\mathscr{B}}X(\omega) - E^{\mathscr{B}}Y(\omega)]\, P(d\omega) = \int_B [X(\omega) - Y(\omega)]\, P(d\omega).$$

The integrand in the right member is a.s. nonpositive by hypothesis, so the right member is at most 0. The integrand in the left member is positive on B by definition of B, and we have just seen that its integral is nonpositive, so by Exercise 2-2, the integrand is 0 at all points of B except those in a set of P-measure 0. This can happen only if $P(B) = 0$, so $E^{\mathscr{B}}X(\omega) \leqq E^{\mathscr{B}}Y(\omega)$ a.s.

As a corollary,

LEMMA 6-3 *If Z^* is a function that satisfies the conditions in Definition* 6-1, *and Z^{**} is a $\mathscr{B}$-measurable function on Ω, then Z^{**} is also one of the functions $E^{\mathscr{B}}C$ if and only if $Z^{**} = Z^*$ a.s.*

If Z^{**} is $\mathscr{B}$-measurable and is a.s. equal to Z^*, it obviously satisfies the conditions for being a version of $E^{\mathscr{B}}C$. Conversely, if Z^{**} is also a version of $E^{\mathscr{B}}C$, by Lemma 6-2 with $X = Y = C$ we have both $Z^{**} \leqq Z^*$ a.s. and $Z^* \leqq Z^{**}$ a.s., so they are a.s. equal.

LEMMA 6-4 *Let X and Y be random variables with expectations and with the respective conditional expectations $E^{\mathscr{B}}X$, $E^{\mathscr{B}}Y$. Let a and b be real numbers. If both $aX + bY$ and $aE^{\mathscr{B}}X + bE^{\mathscr{B}}Y$ are defined at all points of Ω, then a.s.*

$$aE^{\mathscr{B}}X + bE^{\mathscr{B}}Y = E^{\mathscr{B}}[aX + bY].$$

The proof is trivial.

LEMMA 6-5 (Monotone Convergence) *Let $X_1, X_2, X_3, \ldots$ be an ascending sequence of random variables such that X_1 has finite expectation, and let $E^{\mathscr{B}}X_n$ be* *the conditional expectation of* X_n *given* $\mathscr{B}$, *where* $\mathscr{B}$ *is a* σ-*subalgebra of* $\mathscr{A}$. *Then except on a subset* N *of* Ω *with* $PN = 0$, *the limit of* $E^{\mathscr{B}}X_n(\omega)$ *exists (finite or infinite), and*

$$\lim_{n\to\infty} E^{\mathscr{B}}X_n(\omega) = E^{\mathscr{B}}\left[\lim_{n\to\infty} X_n\right](\omega).$$

For each pair of integers i, j such that $0 < i < j$, let $N_{i,j}$ be the set of all ω such that $E^{\mathscr{B}}X_i(\omega) > E^{\mathscr{B}}X_j(\omega)$. By Lemma 6-2, each set $N_{i,j}$ is a member of $\mathscr{B}$ with $P(N_{i,j}) = 0$. Let N be the union of all of them; then $P(N) = 0$. For all ω in $\Omega \setminus N$, the numbers $E^{\mathscr{B}}X_n(\omega)$ ascend and therefore approach a limit. If we define

$$Z^*(\omega) = \liminf_{n\to\infty} E^{\mathscr{B}}X_n(\omega)$$

for all ω in Ω, Z^* is a $\mathscr{B}$-measurable function that is the limit of $E^{\mathscr{B}}X_n(\omega)$ at all points ω of $\Omega \setminus N$. Let B be any set that belongs to $\mathscr{B}$. Then $B \setminus N$ also belongs to $\mathscr{B}$, and by Definition 6-1,

$$\int_{B\setminus N} E^{\mathscr{B}}X_n(\omega)\, P(d\omega) = \int_{B\setminus N} X_n(\omega)\, P(d\omega).$$

Both integrands ascend. By the monotone convergence theorem, the right member converges to

$$\int_{B\setminus N} \left[\lim_{n\to\infty} X_n(\omega)\right] P(d\omega).$$

Again by the monotone convergence theorem, the limit Z^* of the $E^{\mathscr{B}}X_n$ has the same integral over $B \setminus N$, so

$$\int_{B\setminus N} Z^*(\omega)\, P(d\omega) = \int_{B\setminus N} \left[\lim_{n\to\infty} X_n(\omega)\right] P(d\omega).$$

Since $PN = 0$, we can delete the N in the last two integrals and obtain the statement that Z^* satisfies the conditions in Definition 6-1 for being a version of $E^{\mathscr{B}}[\lim_{n\to\infty} X_n]$.

From this we can easily deduce a form of the dominated convergence theorem. We shall not even state it.

We shall now prove that every random variable that has an expectation also has a conditional expectation given $\mathscr{B}$ whenever $\mathscr{B}$ is a σ-subalgebra of $\mathscr{A}$.

THEOREM 6-6 *Let* $(\Omega, \mathscr{A}, P)$ *be a probability triple,* $\mathscr{B}$ *a* σ-*algebra contained in* $\mathscr{A}$, *and* X *a random variable that has expectation, finite or infinite. Then a function* $E^{\mathscr{B}}X$ *exists.*

We prove this first for random variables X with finite expectation. The function F defined on $\mathscr{B}$ by

$$F(B) = \int_B X(\omega)\, P(d\omega)$$

is countably additive and P-continuous. So by the Radon–Nikodým theorem, there exists a finite-valued function Z^* that is $\mathscr{B}$-measurable on Ω, such that

$$F(B) = \int_B Z^*(\omega)\, P(d\omega)$$

for all sets B in $\mathscr{B}$. This satisfies the conditions in Definition 6-1 and can be chosen for $E^{\mathscr{B}}X$.

If X is a nonnegative random variable, for each positive integer n the function $X_n - X \wedge n$ is a random variable with finite expectation. Therefore, by the preceding paragraph, it has a conditional expectation $E^{\mathscr{B}}X_n$. By Lemma 6-5, X has a conditional expectation, and we can choose

$$E^{\mathscr{B}}X = \liminf_{n\to\infty} E^{\mathscr{B}}X_n.$$

If X has an expectation, either X^+ or X^- has a finite expectation. To be specific, we assume that X^- has a finite expectation. Then, by the first paragraph of this proof, X^- has a finite-valued conditional expectation, and by the second paragraph, X^+ has a conditional expectation. By Lemma 6-4, $E^{\mathscr{B}}X^+ - E^{\mathscr{B}}X^-$ is the conditional expectation of $X = X^+ - X^-$. The proof is complete.

In connection with the operation $E^{\mathscr{B}}$, $\mathscr{B}$-measurable functions act in the way that constants act in ordinary integration, as the next lemma shows.

LEMMA 6-7 *Let X and Y be random variables such that Y has expectation, finite or infinite, and X is finite-valued and $\mathscr{B}$-measurable. Then a.s.*

$$E^{\mathscr{B}}[XY] = XE^{\mathscr{B}}Y. \tag{I}$$

We prove this first under the extra hypothesis that X and Y are nonnegative. If X is the indicator of a set B' that belongs to $\mathscr{B}$, for all sets B in $\mathscr{B}$ we have

$$\begin{aligned}
\int_B E^{\mathscr{B}}[XY](\omega)\, P(d\omega) &= \int_B 1_{B'}(\omega) Y(\omega)\, P(d\omega) \\
&= \int_{B\cap B'} Y(\omega)\, P(d\omega) \\
&= \int_{B\cap B'} E^{\mathscr{B}}Y(\omega)\, P(d\omega) \\
&= \int_B X(\omega) E^{\mathscr{B}}Y(\omega)\, P(d\omega).
\end{aligned}$$

By Definition 6-1, (I) is valid. This extends as usual to nonnegative simple functions X. If X is nonnegative and $\mathscr{B}$-measurable, it is the limit of an ascending sequence of nonnegative $\mathscr{B}$-measurable simple functions $X_1 = 0, X_2, X_3, \ldots$. For each of these we have just shown that

$$E^{\mathscr{B}}[X_n Y] = X_n E^{\mathscr{B}} Y.$$

By Lemma 6-5, this implies that (I) is valid for the nonnegative random variables X and Y.

If Y has expectation, finite or infinite, either Y^+ or Y^- has finite expectation. To be specific, we suppose the latter. We can then choose a finite-valued function for $E^{\mathscr{B}} Y^-$. By the preceding proof, the four equations

$$\begin{aligned} E^{\mathscr{B}}[X^+ Y^+] &= X^+ E^{\mathscr{B}} Y^+, \\ -E^{\mathscr{B}}[X^+ Y^-] &= -X^+ E^{\mathscr{B}} Y^-, \\ -E^{\mathscr{B}}[X^- Y^+] &= -X^- E^{\mathscr{B}} Y^+, \\ E^{\mathscr{B}}[X^- Y^-] &= X^- E^{\mathscr{B}} Y^- \end{aligned}$$

are valid. The right members of the second and fourth of these equations are finite. The right member of the first equation can be ∞ and that of the third can be $-\infty$, but not simultaneously for any ω, since at each ω at least one of $X^+(\omega)$, $X^-(\omega)$ is 0. So the right members of the four equations can be added. By Lemma 6-4 we obtain (I), and the proof is complete.

As a corollary, by taking $Y = 1$ we find that if X is finite-valued and $\mathscr{B}$-measurable,

$$E^{\mathscr{B}} X = X \qquad \text{a.s.}$$

But we proved this, even without the assumption that X is finite-valued, just after Definition 6-1.

LEMMA 6-8 *Let $\mathscr{B}$ and $\mathscr{B}^*$ be σ-algebras such that $\mathscr{B}^* \subset \mathscr{B} \subset A$, and let X be a random variable that has expectation. Then a.s.*

$$E^{\mathscr{B}^*}(E^{\mathscr{B}} X) = E^{\mathscr{B}^*} X = E^{\mathscr{B}}(E^{\mathscr{B}^*} X).$$

The second equation is trivial; since $E^{\mathscr{B}^*} X$ is $\mathscr{B}^*$-measurable and $B^* \subset B$, it is $\mathscr{B}$-measurable, and by the remark just before this lemma we can choose $E^{\mathscr{B}^*} X$ for $E^{\mathscr{B}}(E^{\mathscr{B}^*} X)$. For the first equation in the conclusion, let B^* be any set that belongs to $\mathscr{B}^*$. By Definition 6-1,

$$\text{(J)} \qquad \int_{B^*} E^{\mathscr{B}^*}(E^{\mathscr{B}} X)(\omega)\, P(d\omega) = \int_{B^*} E^{\mathscr{B}} X(\omega)\, P(d\omega).$$

But B^* belongs to $\mathscr{B}^*$ and therefore to $\mathscr{B}$, and again by Definition 6-1,

$$\text{(K)} \qquad \int_{B^*} E^{\mathscr{B}} X(\omega)\, P(d\omega) = \int_{B^*} X(\omega)\, P(d\omega).$$

So the left member of (J) is equal to the right member of (K) for every set B^* in $\mathscr{B}^*$. Therefore $E^{\mathscr{B}^*}(E^{\mathscr{B}}X)$ satisfies the requirements in Definition 6-1 for a version of $E^{\mathscr{B}^*}X$.

COROLLARY 6-9 *If $\mathscr{B}^*$ and $\mathscr{B}$ are σ-algebras such that $\mathscr{B}^* \subset \mathscr{B} \subset \mathscr{A}$, and X and Y are random variables such that X is finite-valued and $\mathscr{B}$-measurable and both Y and XY have expectations, then a.s.*

$$\text{(L)} \qquad E^{\mathscr{B}^*}(XY) = E^{\mathscr{B}^*}(XE^{\mathscr{B}}Y).$$

In particular,

$$\text{(M)} \qquad E(XY) = E(XE^{\mathscr{B}}Y).$$

By Lemma 6-8, the left member of (L) is a.s. equal to $E^{\mathscr{B}^*}(E^{\mathscr{B}}(XY))$, and by Lemma 6-7 this is a.s. equal to the right member of (L). In particular, when $\mathscr{B}^*$ contains only the two elements $\varnothing$ and Ω, $E^{\mathscr{B}^*}X = EX$ for every random variable X that has expectation, so (L) takes the form (M).

We are now in a position to describe an illuminating geometric interpretation of conditional expectation that applies to all random variables with finite second moments. These functions constitute the space of $\mathscr{A}$-measurable functions on Ω whose squares are integrable over Ω. On this space we define the inner product

$$\langle X, Y \rangle = \int_{\Omega} X(\omega) Y(\omega)\, P(d\omega)$$

and the pseudo-norm

$$\|X\| = \langle X, X \rangle^{1/2}.$$

If we lump the functions in this space into equivalence-classes, the equivalence-classes form a Hilbert space, which we shall call $\mathscr{H}$. This formation of equivalence-classes is not merely mathematically possible; it is probabilistically reasonable. If two random variables X, Y are equivalent, there is only a set of probability measure 0 on which they differ, and in any sequence of trials we can ignore the chance that a point ω of that set will be encountered. If $\mathscr{B}$ is a σ-algebra contained in $\mathscr{A}$, we shall denote by $\mathscr{H}(\mathscr{B})$ the subset of $\mathscr{H}$ consisting of those equivalence-classes $[X]$ that contain a $\mathscr{B}$-measurable function; this class $[X]$ may contain other functions that are not $\mathscr{B}$-measurable, but this is harmless. For each σ-algebra $\mathscr{B}$ contained in $\mathscr{A}$ we now show that $\mathscr{H}(\mathscr{B})$ is a closed linear subspace of $\mathscr{H}$. That it is linear is evident. Suppose that $[X]$ is a point of $\mathscr{H}$ that is the limit in the pseudo-norm $\|\cdot\|$ (which is a norm on $\mathscr{H}$) of a sequence of members $[X_1], [X_2], [X_3], \ldots$ of $\mathscr{H}(\mathscr{B})$. From each of these equivalence-classes $[X_n]$ we can and do select a $\mathscr{B}$-measurable function, which we denote by X_n. Theorem VI-4-3 extends to the integrals we are considering in this chapter, so we can select a subsequence of the X_n that converges both in norm and a.s. to a function X that is $\mathscr{A}$-measurable and has X^2 integrable with respect to P over Ω. To simplify notation we suppose that $X_1, X_2, X_3, \ldots$ is already that subsequence. Then X differs only on a set of P-measure 0 from $\liminf X_n$, which is $\mathscr{B}$-measurable because each X_n is $\mathscr{B}$-measurable. Therefore the equivalence-class $[X]$ belongs to $\mathscr{H}(\mathscr{B})$, and $\mathscr{H}(\mathscr{B})$ is closed.

The discussion at the beginning of this section indicated that the random variables that we can utilize when the information available to us is information algebra $\mathscr{B}$ are those random variables that are $\mathscr{B}$-measurable. This idea will be given a precise form in Theorem 6-11. If we wish to use one of these as an estimate for a random variable X, it is clearly reasonable to choose an estimate that in some useful sense is as close as possible to X. One way of choosing that comes at once to mind and is in fact appropriate in many applications is to choose a $\mathscr{B}$-measurable function Z^* for which the mean square $E([X - Z^*]^2)$ of the error is as small as possible. This is the same as choosing the point $[Z^*]$ of $\mathscr{H}(\mathscr{B})$ nearest to $[X]$. Conveniently, this nearest point of $\mathscr{H}(\mathscr{B})$ turns out to be $[E^{\mathscr{B}}X]$, as we now prove.

Among the equivalent $\mathscr{B}$-measurable functions named $E^{\mathscr{B}}X$ we can and do choose one that is finite-valued. Let $[Y]$ be any point of $\mathscr{H}(\mathscr{B})$. From the class $[Y]$ we can and do choose a finite-valued $\mathscr{B}$-measurable function, and we call it Y. Then

$$
\text{(N)} \qquad \begin{aligned} \|X - Y\|^2 &= E(\{(X - E^{\mathscr{B}}X) + (E^{\mathscr{B}}X - Y)\}^2) \\ &= \|X - E^{\mathscr{B}}X\|^2 + 2E(\{X - E^{\mathscr{B}}X\}[E^{\mathscr{B}}X - Y]) \\ &\quad + \|E^{\mathscr{B}}X - Y\|^2. \end{aligned}
$$

Since both $E^{\mathscr{B}}X$ and Y are finite-valued and $\mathscr{B}$-measurable, by Corollary 6-9,

$$
\begin{aligned} E(\{X - E^{\mathscr{B}}X\}[E^{\mathscr{B}}X - Y]) &= E(E^{\mathscr{B}}(\{X - E^{\mathscr{B}}X\}[E^{\mathscr{B}}X - Y])) \\ &= E([E^{\mathscr{B}}X - Y]E^{\mathscr{B}}\{X - E^{\mathscr{B}}X\}) \\ &= E([E^{\mathscr{B}}X - Y]\{E^{\mathscr{B}}X - E^{\mathscr{B}}E^{\mathscr{B}}X\}). \end{aligned}
$$

Since $E^{\mathscr{B}}X$ is $\mathscr{B}$-measurable, $E^{\mathscr{B}}E^{\mathscr{B}}X = E^{\mathscr{B}}X$ a.s., and the last expression is 0. So the second term in the right member of (N) is 0, and

$$
\|X - Y\|^2 \geqq \|X - E^{\mathscr{B}}X\|^2.
$$

This holds for all Y in $\mathscr{H}(\mathscr{B})$, and our assertion is proved.

This furnishes us with a geometric interpretation of Lemma 6-4. For the perpendicular projection on $\mathscr{H}(\mathscr{B})$ of aX, which is $E^{\mathscr{B}}(aX)$, is clearly a times the perpendicular projection of X, which is $aE^{\mathscr{B}}X$. A similar interpretation holds for sums of random variables. Also, it immediately suggests the theorem that if $X_1, X_2, X_3, \ldots$ converge in L_2-norm to X, their perpendicular projections $E^{\mathscr{B}}X_n$ $(n = 1, 2, 3, \ldots)$ will converge in L_2-norm to $E^{\mathscr{B}}X$. The conclusion of Lemma 6-8 is also obvious; for if $\mathscr{B}^* \subset \mathscr{B}$, $\mathscr{H}(\mathscr{B}^*)$ is a subspace of $\mathscr{H}(\mathscr{B})$, and we can obtain the projection $E^{\mathscr{B}^*}X$ of X on $\mathscr{H}(\mathscr{B}^*)$ by first projecting X onto $\mathscr{H}(\mathscr{B})$ and then projecting that projection onto $\mathscr{H}(\mathscr{B}^*)$.

Another important consequence of the geometric interpretation is that it becomes obvious that increasing information never increases the mean square of the error of the estimate $E^{\mathscr{B}}X$. For let $\mathscr{B}$ and $\mathscr{B}^*$ be two information algebras with $\mathscr{B}^* \subset \mathscr{B}$. Then $\mathscr{H}(\mathscr{B}^*) \subset \mathscr{H}(\mathscr{B})$, so the distance from X to the nearest point $E^{\mathscr{B}}X$ of $\mathscr{H}(\mathscr{B})$ is not greater than the distance from X to the nearest point $E^{\mathscr{B}^*}X$ of $\mathscr{H}(\mathscr{B}^*)$.

All our developments so far have been in terms of the information algebras $\mathscr{B}$; we have said nothing, after the introductory remarks, about the data vectors. We now return to them and show that when the information algebras are defined by means of data vectors, we can use either interchangeably.

Suppose, then, that T is a nonempty set and that to each t in T there corresponds a random variable Y^t on Ω. For each ω in Ω we denote by $Y(\omega)$ the point in R^T whose t-coordinate is the number $Y^t(\omega)$. Let $\mathscr{B}'$ be the σ-algebra in R^T generated by the rational half-spaces, which is the same as the σ-algebra generated by the left-open intervals in R^T; if T is countable, it is also the same as the σ-algebra of Borel sets in R^T. We define $\mathscr{B}$ to be the family of all sets B in Ω such that $B = Y^{-1}(B')$ for some B' in $\mathscr{B}'$.

By Lemma 3-1, each set B in $\mathscr{B}$ is $\mathscr{A}$-measurable. Also, $\varnothing$ belongs to $\mathscr{B}$, and if $B = Y^{-1}(B')$ belongs to $\mathscr{B}$, so does $\Omega \setminus B$, which is $Y^{-1}(R^T \setminus B')$. If $B_1, B_2, B_3, \ldots$ belong to $\mathscr{B}$, for each n there is a set B'_n in $\mathscr{B}'$ such that $B_n = Y^{-1}(B'_n)$. Then the union of the B'_n belongs to $\mathscr{B}'$, and

$$\bigcup_{n=1}^{\infty} B_n = Y^{-1}\left(\bigcup_{n=1}^{\infty} B'_n\right),$$

so $\mathscr{B}$ is a σ-algebra.

We can convert the measurable space $(R^T, \mathscr{B}')$ into a probability triple by defining, for each set B' in $\mathscr{B}'$,

$$\text{(O)} \qquad P'(B') = P(Y^{-1}(B')).$$

It is easy to show that P' is a countably additive nonnegative measure with $P'(R^T) = 1$, so $(R^T, \mathscr{B}', P')$ is a probability triple. This is a generalization of the idea of the joint distribution of several random variables, and for it we can prove a generalization of Theorem IV-10-5.

THEOREM 6-10 *If f is a $\mathscr{B}'$-measurable function on R^T, the composite $f \circ Y$ is a $\mathscr{B}$-measurable random variable; and if either of the integrals*

$$\int_{\Omega} f(Y(\omega))\, P(d\omega), \quad \int_{R^T} f(y)\, P'(dy)$$

exists so does the other, and then they are equal.

If c is any number in $\bar{R}$, the set $\{\omega \text{ in } \Omega : f(Y(\omega)) \leqq c\}$ is the same as $Y^{-1}\{f \leqq c\}$. The set $\{f \leqq c\}$ belongs to $\mathscr{B}'$ because f is $\mathscr{B}'$-measurable, so the set $Y^{-1}\{f \leqq c\}$ (which is $\{f \circ Y \leqq c\}$) belongs to $\mathscr{B}$ by definition, and $f \circ Y$ is $\mathscr{B}$-measurable. Since $B \subset A$, $f \circ Y$ is a $\mathscr{B}$-measurable random variable.

If f is the indicator of a set $B'(1)$ that belongs to $\mathscr{B}'$, $f(Y(\omega))$ is equal to 1 on $Y^{-1}(B'(1))$ and is 0 outside that set. So by (O),

$$\begin{aligned}\int_{R^T} f(y)\,P'(dy) &= P'(B'(1))\\ &= P(Y^{-1}(B'(1)))\\ &= \int_{R^T} 1_{Y^{-1}(B'(1))}(\omega)\,P(d\omega)\\ &= \int_{R^T} f(Y(\omega))\,P(d\omega).\end{aligned}$$

So the equation

$$\text{(P)}\qquad \int_{R^T} f(y)\,P'(dy) = \int_\Omega f(Y(\omega))\,P(d\omega)$$

holds when f is the indicator of a set belonging to $\mathscr{B}'$.

By the usual argument, (P) holds also when f is a $\mathscr{B}'$-measurable simple function. If f is a nonnegative $\mathscr{B}'$-measurable function, it is the limit of an ascending sequence of nonnegative $\mathscr{B}'$-measurable simple functions f_1, f_2, $f_3, \ldots$. For each n we have

$$\text{(Q)}\qquad \int_{R^T} f_n(y)\,P'(dy) = \int_\Omega f_n(Y(\omega))\,P(d\omega).$$

If either member of (P) is finite, both members of (Q) approach finite limits as n increases, and by the monotone convergence theorem, both members of (P) are finite and they are equal. We remove the restriction to nonnegative f by considering f^+ and f^- separately.

For mathematical study of the properties of conditional expectations it is most convenient to define conditional expectation in terms of the data algebras, but for numerical computations it is often preferable to express the conditional expectation as a function of the random data vectors. The next theorem allows us to pass back and forth between the two sets of ideas.

THEOREM 6-11 *Let Ω, $\mathscr{A}$, Y, $\mathscr{B}$, and $\mathscr{B}'$ be as in the preceding paragraphs. Then an extended-real-valued function Z^* on Ω is $\mathscr{B}$-measurable if and only if there exists a $\mathscr{B}'$-measurable function Z on R^T such that $Z^* = Z \circ Y$.*

Suppose that there exists a $\mathscr{B}'$-measurable function Z on R^T such that $Z^* = Z \circ Y$. For every number c in $\bar{R}$

$$\{\omega \text{ in } \Omega : Z^*(\omega) \leqq c\} = \{\omega \text{ in } \Omega : Z(Y(\omega)) \leqq c\} = \{Y^{-1}\{Z \leqq c\}\}.$$

Since Z is $\mathscr{B}'$-measurable, the set $\{Z \leqq c\}$ belongs to $\mathscr{B}'$, so by definition $Y^{-1}\{Z \leqq c\}$ belongs to $\mathscr{B}$. Therefore Z^* is $\mathscr{B}$-measurable.

To prove the converse, we observe that

(R) if B' is in $\mathscr{B}'$ and $Y^{-1}(B') = B$, then $1_{B'}$ is $\mathscr{B}'$-measurable, and $1_{B'} \circ Y = 1_B$.

If Z^* is a simple B-measurable function on Ω with values $c_1, \ldots, c_k$ on the respective $\mathscr{B}$-measurable sets $B(1), \ldots, B(k)$, it can be written as

$$Z^* = \sum_{j=1}^{k} c_j 1_{B(j)}.$$

We define Z to be the simple function

$$Z = \sum_{j=1}^{k} c_j 1_{B'}(j),$$

where $B'(j)$ is the set such that

$$B(j) = Y^{-1}(B'(j)).$$

By definition of $\mathscr{B}$ this is $\mathscr{B}'$-measurable, and

$$Z \circ Y = \left(\sum_{j=1}^{k} c_j 1_{B'(j)} \right) \circ Y = \sum_{j=1}^{k} c_j 1_{B(j)} = Z^*.$$

If Z^* is any $\mathscr{B}$-measurable function on Ω, it is the limit of a sequence of $\mathscr{B}$-measurable simple functions $Z_1^*, Z_2^*, Z_3^*, \ldots$. By the preceding sentences, for each n we can choose a $\mathscr{B}'$-measurable simple function Z_n on R^T such that

$$Z_n^* = Z_n \circ Y.$$

We choose such Z_n, and we define

$$Z = \liminf_{n \to \infty} Z_n.$$

This is a $\mathscr{B}'$-measurable function on R^T, and for every ω in Ω

$$Z(Y(\omega)) = \liminf_{n \to \infty} Z_n(Y(\omega)) = \liminf_{n \to \infty} Z_n^*(\omega) = Z^*(\omega).$$

This completes the proof.

When the information algebra $\mathscr{B}$ is defined by means of data vectors Y, we have the choice of two ways of regarding the conditional expectation. We can think of it as a $\mathscr{B}$-measurable function on Ω, as we did in most of this section, or we can transfer it to R^T and regard it as a $\mathscr{B}'$-measurable function on R^T. Each point of view has certain advantages. Transferring to R^T is much the same as treating several random variables by means of their joint distribution, since Theorem 6-10 generalizes Theorem IV-10-5. This is likely to be convenient in computations. But for studying the properties of conditional expectations in general, as, for example, when we wish to find the effect of adding new information, the approach by way of information algebras is often more convenient.

EXERCISE 6-1 In elementary probability, when A and C are events with $0 < P(A) < 1$, the conditional probability of C given A is defined to be $P(A \cap C)/P(A)$. Show that this is consistent with our definition of conditional expectation by verifying the following statements.

(i) The information algebra $\mathscr{B}(A)$ determined by 1_A consists of the four sets Ω, A, $\Omega \setminus A$, and $\varnothing$.

(ii) A function g on Ω is $\mathscr{B}(A)$-measurable if and only if it has a constant value g_1 on A and a constant value g_2 on $\Omega \setminus A$; and this is true if and only if there is a Borel-measurable function φ on R such that

$$g(\omega) = \varphi(1_A(\omega))$$

for all ω in Ω.

(iii) The $\mathscr{B}(A)$-measurable function s on Ω that minimizes $E([1_C - s]^2)$ has the constant value $P(A \cap C)/P(A)$ on A and the constant value $P(C \setminus A)/P(\Omega \setminus A)$ on $\Omega \setminus A$.

(iv) The minimizing function s in (iii) is $E^{\mathscr{B}(A)}(1_C)$.

(v) If, consistently with previous usage, we define the conditional probability of an event C given $\mathscr{B}(A)$ to be the conditional expectation of 1_C given $\mathscr{B}(A)$, then the conditional probability of C given A is $P(A \cap C)/P(A)$ on the set A.

EXERCISE 6-2 Let Y be a random variable that assumes finitely many distinct values $c_1, c_2, \ldots, c_k$. Show that

(i) the σ-algebra $\mathscr{B}$ generated by Y consists of all finite unions of sets

$$B(j) = \{Y = c_j\} \qquad (j = 1, 2, \ldots, k);$$

(ii) if f is any random variable, $E^{\mathscr{B}}f$ is the function whose value on each set $B(j)$ is constant and on each $B(j)$ with $P(B(j)) > 0$ satisfies

$$E^{\mathscr{B}}f(\omega) = E(f 1_{B(j)})/PB(j).$$

Verify that among all functions s that are constant on each $B(j)$, $E^{\mathscr{B}}f$ gives the least value to $E([f - s]^2)$.

EXERCISE 6-3 Let Ω be the set of points $\omega = (x, y)$ in R^2 such that

$$x^2/9 + y^2/4 \leqq 1,$$

and let $X(\omega) = x$. Let the events be the Lebesgue-measurable subsets of Ω, and for each such set A define

$$P(A) = m_L A / m_L \Omega.$$

For brevity, define

$$e(x) = 2[1 - x^2/9]^{1/2} \qquad (-3 \leqq x \leqq 3).$$

Show that the information algebra $\mathscr{B}$ determined by X is the family of all sets

$$\{(x, y) \text{ in } R^r : x \text{ in } B_1, \ -e(x) \leqq y \leqq e(x)\}$$

with B_1 a Borel subset of $[-3, 3]$. Show that a function f on Ω is integrable with respect to P if and only if it is Lebesgue-integrable over the ellipse Ω, and in that case $E^{\mathscr{B}}f$ is any function g on Ω that has the form

$$g(x, y) = b(x) \qquad ((x, y) \text{ in } \Omega)$$

with b a Borel-measurable function on $[-3, 3]$ that satisfies

$$b(x) = \left\{ \int_{-e(x)}^{e(x)} f(x, v)\, dv \right\} \Big/ 2e(x)$$

for almost all x in $[-3, 3]$. Show that if for each Borel set B_1 in $[-3, 3]$ we define

$$P_1(B_1) = \int_{B_1} 2e(x)\, dx / m_L \Omega,$$

then P_1 is a probability measure on $[-3, 3]$, and if f is Lebesgue-integrable over Ω and $\varphi(x) = E^{\mathscr{B}}f(x, 0)$,

$$\int_{-3}^{3} \varphi(x)\, P_1(dx) = Ef.$$

7. Brownian Motion

So far, the only examples we have had of measures that are not necessarily on a space R^T are counting measure and modifications of it, and these are closely related to the summation of infinite series and hence are not basically new. In this section we shall study a measure on a space that is neither of the form R^T nor is it invented just to show that the theory of integration has great generality; the measure is one that is used with great frequency in applications.

In the early nineteenth century it was observed by several biologists that a microscopic grain of pollen in a fluid underwent ceaseless motion. Robert Brown wrote about this phenomenon in 1827, and it is known as the *Brownian motion*. In 1905, when the hypothesis of the molecular structure of matter was still rejected by some scientists, Einstein, who then knew nothing of the Brownian motion, computed the motion of a visibly large molecule in a fluid of invisibly small molecules and subsequently found that his results agreed with the Brownian motion of the "large molecule" that was the pollen grain. This gave solid credibility to the molecular theory; it also stimulated mathematical study of the Brownian motion and caused physicists and mathematicians to concentrate on the displacements rather than on the velocities, as previously. Fundamentally, the explanation is that in any time-interval of the order of a second or so the microscopic particle is struck many times by molecules of the fluid, and the combined effects of the impacts produce the random motion. Although the original Brownian motion no longer interests physicists very much, it is one example of a large and important family of phenomena. As a help in constructing mathematical models for these phenomena we shall consider the Brownian motion with some simplifying assumptions.

The large molecule or body has at time t a position $(x(t), y(t), z(t))$ in three-space. We suppose it to be at rest at time 0. By moving the origin if necessary, we can make $x(0) = y(0) = z(0) = 0$. The body is in a region that contains many small particles that move independently of each other, the distribution of positions and momenta being the same at all times. We first choose a positive δ small enough so that the chance that the body is struck by more than, say, five particles in any time-interval of length δ is negligibly small. We subdivide $(0, \infty)$ into intervals $A_n = ([n-1]\delta, n\delta]$ $(n = 1, 2, 3, \ldots)$, and we define

$$\Delta_n x = x(n\delta) - x([n-1]\delta).$$

If a particle strikes the body at a time t', let $x'(t)$ denote the x-coordinate of the position the body would have had at time t' if that particle had not struck the body. Clearly $x'(t) = x(t)$ for $0 \leqq t \leqq t'$. The collision imparts momentum to the body. We assume that by a time $t' + \delta'$ this extra momentum has become imperceptibly small because of friction. Then at all times $t > t' + \delta'$ the displacement $x(t) - x'(t)$ caused by this particular collision will be the same as $x(t' + \delta') - x'(t' + \delta')$. In particular, if for some n the collision occurred at time t' in $([n-1]\delta, n\delta - \delta')$, we will have $\Delta_j x = \Delta_j x'$ for all integers j except n. We now add the assumption that δ' is so small that there is only a negligible chance that among the (few) collisions that occur during time-interval A_n, one or more will occur in the subinterval $(n\delta - \delta', n\delta]$. If we neglect these, we are left with the statement that for each positive integer n, the displacement $\Delta_n x$ is (very nearly) the sum of the changes in $\Delta_n x$ caused by those collisions that occurred during time-interval A_n.

If the experiment is repeated many times, each $\Delta_n x$ is a random variable whose distribution is determined by the distribution of positions and momenta of the particles, which does not change with time. So all the $\Delta_n x$ have the same distribution. The value of each $\Delta_n x$ is determined by the coordinates and momenta of the particles that strike the body during interval A_n, and the particles move independently of each other, so the $\Delta_n x$ $(n = 1, 2, 3, \ldots)$ are statistically independent. If we define

$$V = [\text{Var}\, \Delta_1 x]/\delta, \qquad m = E(\Delta_1 x)/\delta,$$

each $\Delta_n x$ will have expectation $m\delta$ and variance $V\delta$.

Now let $(s, t]$ be an interval whose ends are integral multiples $n\delta, p\delta$ of δ. Then

$$x(t) - x(s) = \sum_{j=n+1}^{p} \Delta_n x,$$

so the expectation and variance of $x(t) - x(s)$ are

$$E(x(t) - x(s)) = (p - n - 1)E(\Delta_1 x) = (t - s)m,$$

$$\text{Var}(x(t) - x(s)) = (p - n - 1)[\text{Var}\, \Delta_1 x] = (t - s)V.$$

If $(s_1, t_1], \ldots, (s_k, t_k]$ are pairwise disjoint intervals with all s_i and t_i integral multiples of δ, the increments $x(t_i) - x(s_i)$ are independent random variables. Moreover, by the central limit theorem, if $t - s$ is large, the distribution of $x(t) - x(s)$ is approximately normal.

The distributions of the (independent) $\Delta_n x$ generate a distribution on the set of all sequences $(\Delta_1 x, \Delta_2 x, \ldots)$, and each such sequence corresponds to a set of functions $t \mapsto x(t)$ continuous on $[0, \infty)$. So what we need is a probability measure on the space Ω of all functions continuous on $[0, \infty)$ and having $x(0) = 0$, with the properties that whenever s and t are integral multiples of δ, the increment $x(t) - x(s)$ is normally distributed with expectation $m(t - s)$ and variance $V(t - s)$, and if $(s_1, t_1], \ldots, (s_k, t_k)$ are pairwise disjoint intervals of that type, the increments $x(t_i) - x(s_i)$ are independent.

However, the restriction to intervals $(s, t]$ in which s and t are integral multiples of some fixed δ would be mathematically inconvenient and physically unreasonable. We therefore try to construct a probability measure on the space of all functions x on $[0, \infty)$ such that for all intervals $[s, t]$ in $[0, \infty]$ the movement $x(t) - x(s)$ is normally distributed with variance $V(r - s)$ and expectation $m(t - s)$, the movements of x over pairwise disjoint intervals being independent. It is harmless to restrict our attention to the case $m = 0$ and $V = 1$; if we replace each function $t \mapsto x(t)$ by $t \mapsto V^{-1/2}x(t) - mt$ each increment $x(t) - x(s)$ is replaced by $V^{-1/2}[x(t) - x(s)] - m(t - s)$, the expectation being changed to 0 and the variance multiplied by V^{-1}. It is far from obvious that there exists any such a probability measure on the space of continuous functions. In 1922 Norbert Wiener proved that it does exist. It is usually called *Wiener measure*. We are going to show that such a measure exists, assuming for notational simplicity that $V = 1$ and $m = 0$. The proof will make use both of gauge-integral ideas and of measure-space arguments.

Let Ω denote the family of real-valued functions x continuous on $[0, \infty)$ and having $x(0) = 0$. Each such function determines a point $t \mapsto x(t)$ $(t > 0)$ in $\bar{R}^{(0,\infty)}$. Let J be an interval in $\bar{R}^{(0,\infty)}$. Then

$$\text{(A)} \qquad J = \mathop{\times}_{t>0} J^t,$$

where for all but a finite set $t_1 < t_2 < \cdots < t_k$ of positive t the set J^t is $\bar{R}$, and for $i = 1, \ldots, k$, J^t is a left-open interval $(a, b]$ or $[-\infty, b]$ in $\bar{R}$. If we write t_0 for 0, the Wiener measure (if it exists) must have the property that the increments $x(t_i) - x(t_{i-1})$ $(i = 1, \ldots, k)$ are normally distributed independent random variables z_i, with means 0 and variances $t_i - t_{i-1}$. So the probability measure that we assign to J must be

(B) $$P(J) = P(z_1 + \cdots + z_j \text{ in } J^{t_j} : j = 1, \ldots, k),$$

where $z_i, \ldots, z_k$ are independent random variables with mean 0 and variances $t_1 - t_0, \ldots, t_k - t_{k-1}$.

Our first task is to show that this really is a function of the interval J, independent of the particular representation of J that we have chosen. If an interval J has a representation (A), it will also have other representations in which there are additional t_j, each with $\bar{R}$ for the corresponding J^{t_j}. Suppose that in representation (A) of J there is a particular j, say $j = h$, for which $J^{t_j} = \bar{R}$. Then J also has the representation

(C) $$J = \underset{t>0}{\times} J^t,$$

in which $J^t = \bar{R}$ for all t except those in

$$\{t_1, \ldots, t_{h-1}, t_{h+1}, \ldots, t_k\},$$

the J^{t_i} being as before for $i = 1, \ldots, h-1, h+1, \ldots, k$. We write

$$t'_0 = t_0, \qquad t'_1 = t_1, \ldots, t'_{h-1} = t_{h-1}, \qquad t'_h = t_{h+1}, \ldots, \qquad t'_{k-1} = t_k.$$

Our specification of the measures of intervals gives for this representation of J the probability measure

(D) $$P(J) = P(w_1 + \cdots + w_j \text{ in } J^{t'_j} : j + 1, \ldots, k-1),$$

where the w_j are normally distributed independent random variables with mean 0 and variances $t'_i - t'_{i-1}$. To compare this with (B), we observe that the hth of the conditions in the right member of (B) is

$$z_1 + \cdots + z_h \text{ in } \bar{R},$$

which can be omitted because it is always satisfied. If we define

$$w_1 = z_1, \ldots, w_{h-1} = z_{h-1}, \qquad w_h = z_h + z_{h+1},$$

$$w_{h+1} = z_{h+2}, \ldots, w_{k-1} = z_k,$$

the w_i are normally distributed independent random variables with mean 0 and variances $t'_i - t'_{i-1}$, so (B) implies (D). The probability assigned by (B) to J does not depend on the representation of J.

With the z_i as in (B), the density of the distribution of z_i is p_i, where

(E) $$p_i(y) = [2\pi(t_i - t_{i-1})]^{-1/2} \exp[-y^2/2(t_i - t_{i-1})].$$

Since $z_1, \ldots, z_k$ are independent, their joint distribution has the density p defined by

$$\text{(F)} \qquad p(y) = p_1(y_1) \cdots p_k(y_k) \qquad (y = (y_1, \ldots, y_k) \text{ in } R^k).$$

If J^{t_i} has end-points a_i and b_i, the right member of (B) is equal to the integral

$$\text{(G)} \qquad P(J) = \int_{G^*} p(y)\, dy,$$

where

$$\text{(H)} \qquad G^* = \{y \text{ in } R^k : a_j < y_1 + \cdots + y_j < b_j,\ j = 1, \ldots, k\}.$$

The omission of the end-points of the J^{t_i} does not affect the value of the integral. In the right member of (G) we make the change of variables $y = h(u)$, where

$$y_1 = u_1, \qquad y_i = u_i - u_{i-1} \qquad (i = 2, \ldots, k).$$

Then

$$u_i = y_1 + \cdots + y_i \qquad (i = 1, \ldots, k).$$

The image of G^* under h is

$$G = (a_1, b_1) \times \cdots \times (a_k, b_k).$$

Since the determinant of the transformation (H) is 1, by Theorem IV-5-14

$$\begin{aligned} \text{(I)} \qquad P(J) &= \int_G p(h(u))\, du \\ &= \left[(2\pi)^k \prod_{i=1}^{k} (t_i - t_{i-1})\right]^{-1/2} \\ &\quad \times \int_{a_1}^{b_1} \cdots \int_{a_k}^{b_k} \exp\left[-\frac{u_1^2}{2t_1}\right] \exp\left[-\frac{(u_2 - u_1)^2}{2(t_2 - t_1)}\right] \cdot \\ &\quad \times \exp\left[-\frac{(u_k - u_{k-1})^2}{2(t_k - t_{k-1})}\right] du_k \cdots du_1 . \end{aligned}$$

By (B) or by (I), we have defined the measures of all left-open intervals in $\bar{R}^T$. However, we still face two problems. We cannot use Section IV-14 to extend the measure from intervals to a σ-algebra of sets because T is not countable. Also, even if we could extend the theorem of that sectiono to uncountable T, we still would have a measure on $\bar{R}^T$ instead of on the space Ω of continuous functions on $[0, \infty)$ that vanish at 0. So, to proceed, we first replace T by the set D of positive dyadic rationals; that is, D is the set of numbers $j/2^n$, where j and n are positive integers. This is countable. We shall always understand that a function that belongs to $\bar{R}^D$ is extended to the nonnegative dyadic rationals by setting $x(0) = 0$. For all intervals J in $\bar{R}^D$ we already have a definition (I) for $P(J)$, which is easily seen to be nonnegative and finitely additive and to have $P(\bar{R}^D) = 1$, and also to be regular in the sense defined in Definition IV-14-2. So, as we showed in Section IV-14, the measure P can be extended to a complete countably additive measure on a σ-algebra of subsets of $\bar{R}^D$, coinciding with the measure P defined by (I) for all left-open intervals. By (I), for each t in D the set of x in $\bar{R}^t$ with $x(t) = \pm\infty$ has P-measure 0_1. Since P is countably additive and D is a countable set, the subset of $\bar{R}^D$ consisting of all x with one or more $x(t)$ infinite has P-measure 0. We discard these and are left with a probability measure P on a σ-algebra of subsets of R^D.

We shall now prove that all x in R^D except those in a set of P-measure 0 possess a local form of "Hölder continuity." (A more customary form is stated in (Z).)

(J) There exists a set U^* in R^D with $P(U^*) = 0$ and a function $(x, T, h) \mapsto \delta(x \mid T, h)$, defined for all x in R^D, all positive T, and all h in $(0, \frac{1}{2})$, such that for each T and h, $x \mapsto \delta(x \mid T, h)$ is a random variable that is positive for all x in $R^D \setminus U^*$, and if x is in $R^D \setminus U^*$ and s and t are dyadic rationals in $[0, T]$ and $|t - s| < \delta(x \mid T, h)$, then

$$|x(t) - x(s)| \leqq |t - s|^h.$$

Let q and n be positive integers. For each positive integer j the difference

$$\Delta_j(x) = x(2^{-n}j) - x(2^{-n}(j-1))$$

is a random variable (with respect to P) that is normally distributed and has mean 0 and variance 2^{-n}. By Theorem IV-10-1, in which we take $f(y) = y^{8q}$, we obtain

$$E(\Delta_j^{8q}) = \int_{R^D} \Delta_j(x)^{8q}\, P(dx)$$

$$= \int_R u^{8q}[2\pi \cdot 2^{-n}]^{-1/2} \exp\left[-\frac{u^2}{2 \cdot 2^{-n}}\right] du.$$

By this and Lemma II-11-3,

$$E(\Delta_j^{8q}) = (2^{-n})^{4q} C_q,$$

where

(K) $$C_q = 1 \cdot 3 \cdot 5 \cdots (8q - 1).$$

So, by Chebyshev's inequality,

$$P(|\Delta_j|^{8q} \geqq 2^{-4nq+2n-32q}) \leqq (2^{-n})^{4q} C_q / 2^{-4nq+2n-32q}.$$

To simplify, we define

(L) $$r = 2^{-1/2+1/4q}.$$

Then the preceding inequality implies

(M) $$P(|\Delta_j| \geqq r^n/16) \leqq C_q 2^{-2n+32q}.$$

The points $j/2^n$ $(j = 0, 1, \ldots, 2^{n+q})$ subdivide $[0, 2^q]$ into 2^{n+q} intervals of length 2^{-n}. Let us define

(N) $$A(q, n) = \{x \text{ in } R^D : \sup_{1 \leqq j \leqq 2^{n+q}} |x(2^{-n}j) - x(2^{-n}(j-1))| \geqq r^n/16\},$$

$$U(q, k) = \bigcup_{n=k}^{\infty} A(q, n) \qquad (k = 1, 2, 3, \ldots),$$

$$U(q, \infty) = \bigcap_{k=1}^{\infty} U(q, k).$$

As k increases, the sets $U(q, k)$ contract. By (M) and (N),

$$P(A(q, n)) \leqq \sum_{j=1}^{2^{n+q}} P\left(|\Delta_j| \geqq \frac{r^n}{16}\right)$$
$$\leqq 2^{n+q}(C_q \cdot 2^{-2n+32q}) = C_q 2^{33q-n},$$

$$P(U(q, k)) \leqq \sum_{n=k}^{\infty} C_q 2^{33q-n} = C_q 2^{33q+1-k}.$$

The last inequality implies

$$P(U(q, \infty)) = 0.$$

We now define a random variable N_q on R^D by

$$N_q(x) = 1 + \sum_{k=1}^{\infty} 1_{U(q,k)}(x).$$

Then

(O) $$N_q(x) \geqq 1 + h \text{ if and only if } x \text{ is in } U(q, h).$$

Except on the set $U(q, \infty)$ of P-measure 0, the random variable δ_q defined by

(P) $$\delta_q(x) = 2^{-N_q(x)}$$

is in the interval $(0, \frac{1}{2}]$. Let x be any member of $R^D \setminus U(q, \infty)$ and n an integer such that

$$2^{-n} < \delta_q(x).$$

Let t^* be a nonnegative dyadic rational $2^{-n}k$, and let t be a nonnegative dyadic rational in $(t^* - 2^{-n}, t^* + 2^{-n})$. Then

$$t = t^* + [\operatorname{sgn}(t - t^*)] \sum_{j=n+1}^{n+n'} \tau_j 2^{-j},$$

where n' is a positive integer, each τ_j is either 0 or 1, and $\operatorname{sgn}(t - t^*)$ is $+1$ if $t > t^*$ and -1 if $t < t^*$. We define $t_0 = t^*$, and successively

$$t_{i+1} = t_i + [\operatorname{sgn}(t - t^*)]\tau_{n+i} 2^{-n-i} \qquad (i = 0, 1, \ldots, n' - 1).$$

Then $t_{n'} = t$. Since

$$2^{-n} < \delta_q(x) = 2^{-N_q(x)},$$

we must have $N_q(x) < n$, and by (O), x cannot belong to $U(q, n-1)$. Therefore it cannot belong to any set $A(q,j)$ with $j \geqq n-1$. Since

$$t_{i+1} - t_i = \tau_{n+i}2^{-n-i} \qquad (i = 0, \ldots, n'-1),$$

if $\tau_{n+i} = 1$ there is an integer j such that, if $t > t^*$,

$$t_i = 2^{-n-i}j, \qquad t_{i+1} = 2^{-n-i}(j+1),$$

or, if $t < t^*$,

$$t_{i+1} = 2^{-n-i}j, \qquad t_i = 2^{-n-i}(j+1).$$

Since x is not in $A(q, n+1)$,

$$|x(t_{i+1}) - x(t_i)| < r^{n+i}/16.$$

Evidently this is also true in the case $\tau_{n+1} = 0$, so it holds for all i in $\{0, \ldots, n'-1\}$. By adding,

$$\text{(Q)} \qquad |x(t) - x(t^*)| \leqq \sum_{i=0}^{n'-1} |x(t_{i+1}) - x(t_i)|$$

$$\leqq \sum_{i=0}^{n'-1} \frac{r^{n+i}}{16} < \frac{r^n}{16(1-r)}.$$

Next let s and t be any two dyadic rationals such that $0 \leqq s \leqq t \leqq 2^q$ and $t - s < \delta_q(x)/2$. There exists a positive integer n such that

$$2^{-n-1} \leqq |t - s| < 2^{-n}.$$

One of the intervals $(2^{-n}j, 2^{-n}(j+1))$ has its center at distance not more than 2^{-n-1} from $(t+s)/2$; let the left end-point of this interval be $t^* = 2^{-n}j^*$. Then

$$|t^* - s| < 2^{-n}, \qquad |t^* - t| < 2^{-n}.$$

Since $2^{-n} \leqq 2|t - s| < \delta_q(x)$, we can apply (Q) and obtain

$$|x(t) - x(t^*)| < r^n/16(1-r), \qquad |x(s) - x(t^*)| < r^n/16(1-r),$$

hence

$$\text{(R)} \qquad |x(t) - x(s)| < r^n/8(1-r).$$

Since

$$r^n = (2^{-n})^{1/2-1/4q} \leqq [2|t-s|]^{1/2-1/4q} = |t-s|^{1/2-1/4q}/r,$$

(R) yields

$$|x(t) - x(s)| < [\tfrac{1}{8}(r - r^2)]\,|t-s|^{1/2-1/4q}$$

$$= [\tfrac{1}{8}\{\tfrac{1}{4} - (r - 1/2)^2\}]\,|t - s|^{1/2 - 1/4q}.$$

The quantity in square brackets is less than 1. Therefore,

(S) $$|x(t) - x(s)| < |t - s|^{1/2 - 1/4q}$$

whenever s and t are dyadic rational numbers in $[0, 2^q]$ and $|t - s| < \delta_q(x)/2$.

We now define

$$U^* = \bigcup_{q=1}^{\infty} U(q, \infty).$$

Then $P(U^*) = 0$. Let T be positive and let h be a number in $(0, \frac{1}{2})$. We choose for q the least positive integer such that

$$2^q > T \qquad \text{and} \qquad \tfrac{1}{2} - \tfrac{1}{4q} > h,$$

and we define

$$\delta(x \mid T, h) = \delta_q(x)/2.$$

This is positive for x in $R^D \setminus U^*$. By (S), if x is in $R^D \setminus U^*$, and s and t are dyadic rationals in $[0, T]$, and $|t - s| < \delta(x \mid T, h)$,

$$|x(t) - x(s)| \leqq |t - s|^h,$$

establishing (J).

To each function $t \mapsto x(t)$ (t in $[0, \infty)$) in Ω there corresponds a function in R^D obtained by restricting t to D. This we denote by $\Pi(x)$:

$$\Pi(x) = (t \mapsto x(t) : t \text{ in } D).$$

Obviously, not all members of R^D have this form; for example, the function $t \mapsto 1_{[0,\pi]}(t)$ (t in D) is not $\Pi(x)$ for any x in Ω. On the other hand, if $\tilde{x}$ is in R^D but not in the set U^* of statement (J), it is $\Pi(x)$ for some x in Ω, as we now show. Let T be positive and let t be in $[0, T)$. To T and $h = \frac{1}{4}$ corresponds a $\delta(\tilde{x} \mid T, \frac{1}{4})$ as in (J). If $t_1, t_2, t_3, \ldots$ is any sequence of points of D tending to t, for all j greater than a certain j^*, t_j is in $[0, T)$ and $|t_j - t| < \delta(\tilde{x} \mid T, \frac{1}{4})/2$. So by (J), if i and j are greater than j^*,

$$|\tilde{x}(t_i) - \tilde{x}(t_j)| \leqq |t_i - t_j|^{1/4}.$$

That is, the numbers $\tilde{x}(t_1), \tilde{x}(t_2), \tilde{x}(t_3), \ldots$ satisfy the Cauchy condition and must convrge to a limit, which we call $x(t)$. Since this holds for every sequence of t_j tending to t,

(T) $$\lim_{t' \to t} \tilde{x}(t') = x(t).$$

If t is in D we can take all t_j equal to t and find that $x(t) = \tilde{x}(t)$ on $D \cap [0, T)$. If s

and t are in $[0, T)$ and $|t - s| < \delta(\tilde{x} \mid T, \frac{1}{4})$, we can find sequences $s_1, s_2, s_3, \ldots$ and $t_1, t_2, t_3, \ldots$ of points of D tending to s and t, respectively. For all large j, $|t_j - s_j| < \delta(\tilde{x} \mid T, \frac{1}{4})$, so

$$|\tilde{x}(t_j) - \tilde{x}(s_j)| \leqq |t_j - s_j|^{1/4}.$$

This implies $|x(t) - x(s)| \leqq |t - s|^{1/4}$. Thus we have proved

(U) For each x in Ω and each t in $[0, T]$ such that $|t - s| \leqq \delta(\Pi(x) \mid T, h)$ it is true that

$$|x(t) - x(s)| \leqq |t - s|^{1/4}.$$

In particular, x is continuous on every interval $[0, T]$ and hence on $[0, \infty)$. It has $x(0) = 0$, so x is in Ω, and we have already shown that $x(t) = \tilde{x}(t)$ on D, so $\Pi(x) = \tilde{x}$.

We now define a subset A of Ω to be **Wiener measurable** if $\Pi(A)$ has P-measure, and in that case we define the Wiener measure of A to be

$$W(A) = P(\Pi(A)).$$

It is obvious that $W(\varnothing) = 0$. Since every $\tilde{x}$ in $R^D \setminus U^*$ is $\Pi(x)$ for some x in Ω,

$$R^D \setminus U^* \subset \Pi(\Omega) \subset R^D.$$

The first and last of these have P-measure 1, so $\Pi(\Omega)$ has P-measure 1, and Ω has Wiener measure 1. The countable additivity of W follows at once from the countable additivity of P, so W is a probability measure defined on the σ-algebra of Wiener-measurable subsets of Ω. It remains to show that for all intervals J in $R^{[0,\infty)}$ the measure $W(J)$ is given by the right member of equation (I). For convenience, we shall denote the right member of (I) by $I(t_1, a_1, b_1, \ldots, t_k, a_k, b_k)$.

A set $\tilde{J}$ in R^D is a left-open interval if $\tilde{J} = \times_{t \text{ in } D} (a(t), b(t)]$ where $a(t) = -\infty$ and $b(t) = \infty$ for all t except those in a finite set $t_1, \ldots, t_k$ with $0 < t_1 < \cdots < t_k$. This interval in R^D determines a subset of Ω consisting of all x in Ω such that $\Pi(x)$ is in $\tilde{J}$. We denote this subset of Ω by

$$\begin{aligned} &J(t_1, a(t_1), b(t_1), \ldots, t_k, a(t_k), b(t_k)) \\ &\quad = \Omega \cap \Pi^{-1}(\tilde{J}(t_1, a(t_1), b(t_1), \ldots, t_k, a(t_k), b(t_k)), \end{aligned}$$

and we abbreviate it to $J(t_1, \ldots, b(t_k))$ when convenient. We have defined

$$W(J(t_1, \ldots, b(t_k)) = P(\tilde{J}(t_1, \ldots, b(t_k));$$

and we have to prove that

(V)
$$\begin{aligned} &W(J(t_1, a(t_1), b(t_1), \ldots, t_k, a(t_k), b(t_k)) \\ &\quad = I(t_1, a(t_1), b(t_1), \ldots, t_k, a(t_k), b(t_k)). \end{aligned}$$

If the t_i are all dyadic rationals,

$$\begin{aligned} W(J(t_1,\ldots,b(t_k)) &= P(\Pi(J(t_1,\ldots,b(t_k)) \\ &= P(\tilde{J}(t_1,\ldots,b(t_k)) \\ &= I(t_1,a(t_1),b(t_1),\ldots,t_k,a(t_k),b(t_k)). \end{aligned}$$

So (V) holds in this case.

If the t_i are positive real numbers, let ε be positive and let T be some number greater than t_k. Since the random variable $\tilde{x} \mapsto \delta(\tilde{x} \mid T, \frac{1}{4})$ is positive on $R^D \setminus U^*$, and $P(U^*) = 0$, we can and do choose a positive δ' such that the set

$$\tilde{B}(\delta') = \{\tilde{x} \text{ in } R^D : \delta(\tilde{x} \mid T, \tfrac{1}{4}) \leqq \delta'\}$$

has measure

$$P(\tilde{B}(\delta')) < \varepsilon/2.$$

Then the set in Ω defined by

$$B(\delta') = \{x \text{ in } \Omega : \Pi(x) \text{ in } \tilde{B}(\delta')\}$$

has Wiener measure

$$W(B(\delta')) < \varepsilon/2,$$

and by (U), if x is in $\Omega \setminus B(\delta')$ and s and t are in $[0, T]$ with $|t - s| < \delta'$, then

$$|x(t) - x(s)| \leqq |t - s|^{1/4}.$$

The function $I(t_1,\ldots,b(t_k))$ depends continuously on all $3k$ variables, so there exists a positive number c such that if

$$\text{(W)} \qquad |t_i' - t_i| < c \qquad (i = 1,\ldots,k),$$

then

$$\begin{aligned} \text{(X)} \qquad & I(t_1', a(t_1) + c^{1/4}, b(t_1) - c^{1/4},\ldots,t_k', a(t_k) + c^{1/4}, b(t_k) - c^{1/4}) \\ & \quad > I(t_1,a(t_1),b(t_1),\ldots,t_k,a(t_k),b(t_k)) - \varepsilon/2, \\ & I(t_1', a(t_1) - c^{1/4}, b(t_1) + c^{1/4},\ldots,t_k', a(t_k) - c^{1/4}, b(t_k) + c^{1/4}) \\ & \quad < I(t_1,a(t_1),b(t_1),\ldots,t_k,a(t_k),b(t_k)) + \varepsilon/2. \end{aligned}$$

We can and do choose c less than δ', and we choose dyadic rationals $t_1',\ldots,t_k'$ that satisfy (W). Then for every x in $\Omega \setminus B(\delta')$ we have

$$|x(t_i') - x(t_i)| < c^{1/4} \qquad (i = 1,\ldots,k).$$

So if x is in $\Omega \setminus B(\delta')$, the inequalities

$$a(t_i) < x(t_i) \leqq b(t_i)$$

imply

$$a(t_i) - c^{1/4} < x(t_i') \leqq b(t_i) + c^{1/4},$$

and the inequalities

$$a(t_i) + c^{1/4} < x(t_i') \leqq b(t_i) - c^{1/4}$$

imply

$$a(t_i) < x(t_i) \leqq b(t_i).$$

That is,

(Y) $$J(t_1, a(t_1), b(t_1), \ldots, b(t_k))$$

$$\subset J[t_1', a(t_1) - c^{1/4}, b(t_1) + c^{1/4}, \ldots, b(t_k) + c^{1/4}] \cup (\Omega \setminus B(\delta')),$$

$$J(t_1', a(t_1) + c^{1/4}, b(t_1) - c^{1/4}, \ldots, b(t_k) - c^{1/4}) \setminus B(\delta')$$

$$\subset J(t_1, a(t_1), b(t_1), \ldots, b(t_k)).$$

The right member of the first of relations (Y) is the union of two sets, the first with Wiener measure less than $I(t_1, a(t_1), b(t_1), \ldots, b(t_k)) + \varepsilon/2$, and the second with Wiener measure less than $\varepsilon/2$. The left member of the second of relations (Y) is the set that results from removing the set $B(\delta')$ with $W(B(\delta')) < \varepsilon/2$ from a set of measure greater than $I(t_1, A(t_1), b(t_1), \ldots, b(t_k)) - \varepsilon/2$. Since ε is arbitrary, (V) must be valid. The proof is complete.

As an exercise, the reader should modify the proof, beginning between (T) and (U), to show that there is a set U^{**} in Ω with $W(U^{**}) = 0$ such that for each $T > 0$ and h in $(0, \frac{1}{2})$ there is a positive random variable $x \mapsto \delta'(x \mid T, h)$ on $\Omega \setminus U^{**}$ such that

(Z) if s and t are in $[0, T]$ and $|t - s| < \delta'(x \mid T, h)$, then

$$|x(t) - x(s)| \leqq |t - s|^h.$$

From this it is easy to deduce that for each x in $\Omega \setminus U^{**}$, each $T > 0$, and each h in $[0, \frac{1}{2})$ there is a K such that if s and t are in $[0, T]$ then

$$|x(t) - x(s)| \leqq K|t - s|^h.$$

It is true, but we shall not prove it, that this becomes false if we choose for h any number $\geqq \frac{1}{2}$. Moreover, the set of x in Ω for which $Dx(t)$ exists for even a single t has Wiener measure 0. Wiener measure has great importance in applications involving random functions, such as in finding the effects of random noise in mechanical or electrical systems. But caution is needed when it is combined with any procedure that involves arbitrarily small t-intervals, such as differentiation or forming integrals of the form $\int f(t)\, dx(t)$.

INDEX

D

E

F

G

H

Pure and Applied Mathematics

A Series of Monographs and Textbooks

Editors **Samuel Eilenberg and Hyman Bass**

Columbia University, New York

RECENT TITLES

CARL L. DEVITO. Functional Analysis

MICHIEL HAZEWINKEL. Formal Groups and Applications

SIGURDUR HELGASON. Differential Geometry, Lie Groups, and Symmetric Spaces

ROBERT B. BURCKEL. An Introduction to Classical Complex Analysis: Volume 1

JOSEPH J. ROTMAN. An Introduction to Homological Algebra

C. TRUESDELL AND R. G. MUNCASTER. Fundamentals of Maxwell's Kinetic Theory of a Simple Monatomic Gas: Treated as a Branch of Rational Mechanics

BARRY SIMON. Functional Integration and Quantum Physics

GRZEGORZ ROZENBERG AND ARTO SALOMAA. The Mathematical Theory of L Systems

DAVID KINDERLEHRER and GUIDO STAMPACCHIA. An Introduction to Variational Inequalities and Their Applications

H. SEIFERT AND W. THRELFALL. A Textbook of Topology; H. SEIFERT. Topology of 3-Dimensional Fibered Spaces

LOUIS HALLE ROWEN. Polynominal Identities in Ring Theory

DONALD W. KAHN. Introduction to Global Analysis

DRAGOS M. CVETKOVIC, MICHAEL DOOB, AND HORST SACHS. Spectra of Graphs

ROBERT M. YOUNG. An Introduction to Nonharmonic Fourier Series

MICHAEL C. IRWIN. Smooth Dynamical Systems

JOHN B. GARNETT. Bounded Analytic Functions

EDUARD PRUGOVEČKI. Quantum Mechanics in Hilbert Space, Second Edition

M. SCOTT OSBORNE AND GARTH WARNER. The Theory of Eisenstein Systems

K. A. ZHEVLAKOV, A. M. SLIN'KO, I. P. SHESTAKOV, AND A. I. SHIRSHOV. Translated by HARRY SMITH. Rings That Are Nearly Associative

JEAN DIEUDONNÉ. A Panorama of Pure Mathematics; Translated by I. Macdonald

JOSEPH G. ROSENSTEIN. Linear Orderings

AVRAHAM FEINTUCH AND RICHARD SAEKS. System Theory: A Hilbert Space Approach

ULF GRENANDER. Mathematical Experiments on the Computer

HOWARD OSBORN. Vector Bundles: Volume 1, Foundations and Stiefel-Whitney Classes

K. P. S. BHASKARA RAO AND M. BHASKARA RAO. Theory of Charges

RICHARD V. KADISON AND JOHN R. RINGROSE. Fundamentals of the Theory of Operator Algebras, Volume I

EDWARD B. MANOUKIAN. Renormalization

BARRETT O'NEILL. Semi-Riemannian Geometry: With Applications to Relativity

LARRY C. GROVE. Algebra

E. J. MCSHANE. Unified Integration

IN PREPARATION

ROBERT B. BURCKEL. An Introduction to Classical Complex Analysis: Volume 2

RICHARD V. KADISON AND JOHN R. RINGROSE. Fundamentals of the Theory of Operator Algebras, Volume II